# Applications of Mathematics

Peter E. Kloeden   Eckhard Platen

# Numerical Solution of Stochastic Differential Equations

With 85 Figures

Springer-Verlag
Berlin  Heidelberg  New York
London  Paris  Tokyo
Hong Kong  Barcelona
Budapest

Peter E. Kloeden
Department of Computing and Mathematics, Deakin University
Geelong 3217, Victoria, Australia

Eckhard Platen
Institut of Advanced Studies, Australian National University
GPO Box 4, Canberra, ACT 2601, Australia
*and* Institut für Angewandte Analysis und Stochastik
Mohrenstraße 39, 1086 Berlin, FRG

*Managing Editors*

A.V. Balakrishnan
Systems Science Department
University of California
Los Angeles, CA 90024, USA

I. Karatzas
Department of Statistics
Columbia University
New York, NY 10027, USA

M. Yor
Laboratoire de Probabilités
Université Pierre et Marie Curie
4 Place Jussieu, Tour 56
F-75230 Paris Cedex, France

---

Mathematics Subject Classification (1980): 60 H 10, 65 C 05

---

ISBN 3-540-54062-8 Springer-Verlag Berlin Heidelberg New York
ISBN 0-387-54062-8 Springer-Verlag New York Berlin Heidelberg

Library of Congress Cataloging-in-Publication Data
Kloeden, Peter E. Numerical solution of stochastic differential equations / Peter E. Kloeden,
Eckhard Platen. p. cm. - (Applications of mathematics; 23)
Includes bibliographical references and indexes.
ISBN 3-540-54062-8
1. Stochastic differential equations - Numerical solutions. I. Platen. Eckhard. II. Title.
III. Series. QA274.23.K56 1992 519.2 - dc20 92-15916 CIP

© Springer-Verlag Berlin Heidelberg 1992
Printed in the United States of America

41/3140 - 5 4 3 2 1 0 - Printed on acid-free paper

Dedicated to Our Parents

# Preface

The aim of this book is to provide an accessible introduction to stochastic differential equations and their applications together with a systematic presentation of methods available for their numerical solution.

During the past decade there has been an accelerating interest in the development of numerical methods for stochastic differential equations (SDEs). This activity has been as strong in the engineering and physical sciences as it has in mathematics, resulting inevitably in some duplication of effort due to an unfamiliarity with the developments in other disciplines. Much of the reported work has been motivated by the need to solve particular types of problems, for which, even more so than in the deterministic context, specific methods are required. The treatment has often been heuristic and ad hoc in character. Nevertheless, there are underlying principles present in many of the papers, an understanding of which will enable one to develop or apply appropriate numerical schemes for particular problems or classes of problems.

The present book does not claim to be a complete or an up to date account of the state of the art of the subject. Rather, it attempts to provide a systematic framework for an understanding of the basic concepts and of the basic tools needed for the development and implementation of numerical methods for SDEs, primarily time discretization methods for initial value problems of SDEs with Ito diffusions as their solutions. In doing so we have selected special topics and many recent results to illustrate these ideas, to help readers see potential developments and to stimulate their interest to contribute to the subject from the perspective of their own discipline and its particular requirements. The book is thus directed at readers from quite different fields and backgrounds. We envisage three broad groups of readers who may benefit from the book:

(i)  those just interested in modelling and applying standard methods, typically from the social and life sciences and often without a strong background in mathematics;

(ii)  those with a technical background in mathematical methods typical of engineers and physicists who are interested in developing new schemes as well as implementing them;

(iii)  those with a stronger, advanced mathematical background, such as stochasticians, who are more interested in theoretical developments and underlying mathematical issues.

The book is written at a level that is appropriate for a reader with an engineer's or physicist's undergraduate training in mathematical methods. Many chapters begin with a descriptive overview of their contents which may be accessible to those from the first group of readers mentioned above. There are also several more theoretical sections and chapters for the more mathematically inclined reader. In the "Suggestions for the Reader" we provide some hints for each of the three groups of readers on how to use the different parts of the book.

We have tried to make the exposition as accessible to as wide a readership as possible. The first third of the book introduces the reader to the theory of stochastic differential equations with minimal use of measure theoretic concepts. The reader will also find an extensive list of explicit solutions for SDEs. The application of SDEs in important fields such as physics, engineering, biology, communications, economics, finance, ecology, hydrology, filtering, control, genetics, etc, is emphasized and examples of models involving SDEs are presented. In addition, the use of the numerical methods introduced in the book is illustrated for typical problems in two separate chapters.

The book consists of 17 Chapters, which are grouped into 6 Parts. Part I on Preliminaries provides background material on probability, stochastic processes and statistics. Part II on Stochastic Differential Equations introduces stochastic calculus, stochastic differential equations and stochastic Taylor expansions. These stochastic Taylor expansions provide a universally applicable tool for SDEs which is analogous to the deterministic Taylor formula in ordinary calculus. Part III on Applications of Stochastic Differential Equations surveys the application of SDEs in a diversity of disciplines and indicates the essential ideas of control, filtering, stability and parametric estimation for SDEs. The investigation of numerical methods begins in Part IV on Time Discrete Approximations with a brief review of time discretization methods for ordinary differential equations and an introduction to such methods for SDEs. For the latter we use the simple Euler scheme to highlight the basic issues and types of problems and objectives that arise when SDEs are solved numerically. In particular, we distinguish between strong and weak approximations, depending on whether good pathwise or good probability distributional approximations are sought. In the remaining two parts of the book different classes of numerical schemes appropriate for these tasks are developed and investigated. Stochastic Taylor expansions play a central role in this development. Part V is on Strong Approximations and Part VI on Weak Approximations. It is in these two Parts that the schemes are derived, their convergence orders and stability established, and various applications of the schemes considered.

Exercises are provided in most sections to nurture the reader's understanding of the material under discussion. Solutions of the Exercises can be found at the end of the book.

Many PC-Exercises are included throughout the book to assist the reader to develop "hands on" numerical skills and an intuitive understanding of the basic concepts and of the properties and the issues concerning the implementation of the numerical schemes introduced. These PC-Exercises often build on earlier ones and reappear later in the text and applications, so the reader is encouraged to work through them systematically. The companion book

P. E. Kloeden, E. Platen and H. Schurz: *The Numerical Solution of Stochastic Differential Equations through Computer Experiments.* Springer (1992).

contains programs on a floppy disc for these PC–Exercises and a more detailed discussion on their implementation and results. Extensive simulation studies can also be found in this book.

To simplify the presentation we have concentrated on Ito diffusion processes and have intentionally not considered some important advanced concepts and results from stochastic analysis such as semimartingales with jumps or boundaries or SDEs on manifolds. For a more theoretical and complete treatment of stochastic differential equations than we give here we refer readers to the monograph

N. Ikeda and S. Watanabe: *Stochastic Differential Equations and Diffusion Processes.* North-Holland, Amsterdam (1981; 2nd Edition, 1989).

In the few instances that we shall require advanced results in a proof we shall state a reference explicitly in the text. In addition, in the case studies of different applications of SDEs and numerical methods in Chapters 7, 13 and 17 we shall indicate the names of the authors of the papers that we have consulted. Otherwise, and in general, further information and appropriate references for the section under consideration will be provided in the Bibliographical Remarks at the end of the book.

Two types of numbering system are used throughout the book. Equations are numbered by their section and number in the section, for example (2.1), and are referred to as such in this section and within the chapter which includes it; the chapter number appears as a prefix when the equation is referred to in other chapters. The resulting numbers, (2.1) or (3.2.1) say, will always appear in parentheses. Examples, Exercises, PC–Exercises, Remarks, Theorems and Corollaries are all numbered by their chapter, section and order of occurrence regardless of qualifier. They will always be prefixed by their qualifier and never appear in parentheses, for example Theorem 3.2.1. Figures and Tables are each, and separately, numbered by the same three number system, with the third number now referring only to the occurrence of the Figure or the Table, respectively. The only exception to these numbering systems is in the "Brief Survey of Stochastic Numerical Methods" at the beginning of the book, where just a single number is used for each equation.

During the writing of this book we have received much encouragement, support and constructive criticism from a large number of sources. In particular, we mention with gratitude L. Arnold, H. Föllmer, J. Gärtner, C. Heyde, G. Kallianpur, A. Pakes, M. Sørenson and D. Talay, as well as each others' institutions, the Institute for Dynamical Systems at the University of Bremen, the Institute of Advanced Studies at the Australian National University and the Institute for Applied Mathematics of the University of Hamburg. Special thanks also go to H. Schurz and N. Hofmann who programed and tested the PC-Exercises in the book and produced the figures.

Berlin, May 1991

*Peter E. Kloeden*
*Eckhard Platen*

# Contents

# Suggestions for the Reader

We mentioned in the Preface that we have tried to arrange the material of this book in a way that would make it accessible to as wide a readership as possible. Since prospective readers will undoubtedly have different backgrounds and objectives, the following hints may facilitate their use of the book.

(i)    We begin with those readers who require only sufficient understanding of stochastic differential equations to be able to apply them and appropriate numerical methods in different fields of application. The deeper mathematical issues are avoided in the following flowchart which provides a reading guide to the book for those without a strong background in mathematics.

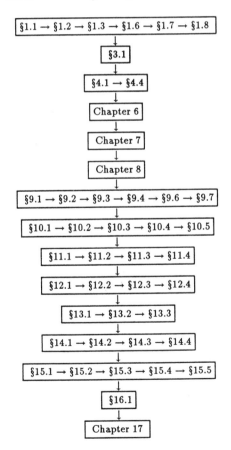

(ii)    Engineers, physicists and others with a more technical background in mathematical methods who are interested in applying stochastic differential equations and in implementing efficient numerical schemes or developing new schemes for specific classes of applications, could use the book according to the following flowchart. This now includes more material on the underlying mathematical techniques without too much emphasis on proofs.

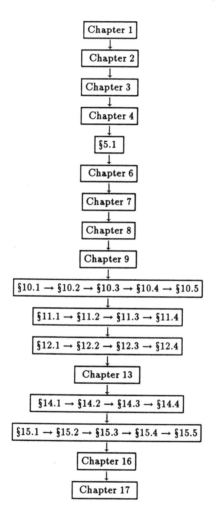

(iii)    Mathematicians and other readers with a stronger mathematical background may omit the introductory parts of the book. The following flowchart emphasizes the deeper, more theoretical aspects of the numerical approximation of Ito diffusion processes while avoiding well known or standard topics.

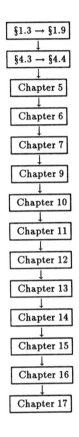

# Basic Notation

| | |
|---|---|
| $\emptyset$ | the empty set |
| $a \in A$ | $a$ is an element of the set $A$ |
| $a \notin A$ | $a$ is not an element of the set $A$ |
| $A^c$ | the complement of the set $A$ |
| $A \cup B$ | the intersection of sets $A$ and $B$ |
| $A \cap B$ | the intersection of sets $A$ and $B$ |
| $A \setminus B$ | the set of elements of set $A$ that are not in set $B$ |
| $:=$ | defined as or denoted by |
| $\equiv$ | identically equal to |
| $\approx$ | approximately equal to |
| $\sim$ | with distribution |
| $\Re$ | the set of real numbers |
| $\Re^+$ | the set of non-negative real numbers |
| $(a,b)$ | the open interval $a < x < b$ in $\Re$ |
| $[a,b]$ | the closed interval $a \le x \le b$ in $\Re$ |
| $a \vee b$ | the maximum of $a$ and $b$ |
| $a \wedge b$ | the minimum of $a$ and $b$ |
| $n!$ | the factorial of the positive integer $n$ |
| $[a]$ | the largest integer not exceeding $a$ |
| $\Re^d$ | the $d$-dimensional Euclidean space |
| $x = (x^1, \ldots, x^d)$ | a vector $x \in \Re^d$ with $i$th component $x^i$ for $i = 1, \ldots, d$ |
| $(x,y)$ | the scalar product of vectors $x, y \in \Re^d$ |
| $|x|$ | the Euclidean norm of a vector $x \in \Re^d$ |
| $x^\top$ | transpose of the vector $x$ |
| $A = [a^{i,j}]$ | a matrix $A$ with $ij$th component $a^{i,j}$ |
| $\imath$ | the square root of $-1$ |
| $\text{Re}(z)$ | the real part of a complex number $z$ |
| $\text{Im}(z)$ | the imaginary part of a complex number $z$ |

| | |
|---|---|
| $f : Q_1 \rightarrow Q_2$ | a function $f$ from $Q_1$ into $Q_2$ |
| $\text{Im}(z)$ | the imaginary part of a complex number $z$ |
| $f : Q_1 \rightarrow Q_2$ | a function $f$ from $Q_1$ into $Q_2$ |
| $1_A$ | the indicator function of the set $A$ |
| $f'$ | the first derivative of a function $f : \Re^1 \rightarrow \Re^1$ |
| $f''$ | the second derivative of a function $f : \Re^1 \rightarrow \Re^1$ |
| $f^{(k)}$ | the $k$th derivative of a function $f : \Re^1 \rightarrow \Re^1$ |
| $\partial_{x^i} u,\ \frac{\partial u}{\partial x^i}$ | the $i$th partial derivative of a function $u : \Re^d \rightarrow \Re^1$ |
| $\partial_{x^i}^k u,\ \left(\frac{\partial}{\partial x^i}\right)^k u$ | the $k$th order partial derivative of $u$ with respect to $x^i$ |
| $C\left(\Re^m, \Re^n\right)$ | the space of continuous functions $f : \Re^m \rightarrow \Re^n$ |
| $C^k\left(\Re^m, \Re^n\right)$ | the space of $k$ times continuously differentiable functions $f : \Re^n \rightarrow \Re^n$ |
| $\mathcal{B}$ | the $\sigma$-algebra of Borel subsets of $\Re^1$ |
| $\mathcal{L}$ | the $\sigma$-algebra of Lebesgue subsets of $\Re^1$ |
| $E(X)$ | the expectation of the random variable $X$ |
| $\delta_{i,j}$ | the Kronecker delta symbol |
| $O\left(r^p\right)$ | expression divided by $r^p$ remains bounded as $r \rightarrow 0$ |
| $o\left(r^p\right)$ | expression divided by $r^p$ converges to zero as $r \rightarrow 0$ |
| a.s. | almost surely |
| w.p.1 | with probability 1 |

Other notation will be defined where it is first used and can be located with the help of the Index of Symbols at the end of the book. Note that vectors and matrices will usually be indexed with superscipts. Parentheses will then be used when taking powers of their components, for example with $\left(x^i\right)^3$ denoting the cube of $x^i$. Square brackets [·] will often be used to visually simplify nested expressions, with the few instances where it denotes the integer part of a real number being indicated in the text. Function space norms will always be written with double bars $\|\cdot\|$, often with a distinguishing subscript.

# A Brief Survey
# of Stochastic Numerical Methods

An Ito process $X = \{X_t, t \geq 0\}$ has the form

$$(1) \qquad X_t = X_0 + \int_0^t a(X_s)\,ds + \int_0^t b(X_s)\,dW_s$$

for $t \geq 0$. It consists of an initial value $X_0 = x_0$, which may be random, a slowly varying continuous component called the drift and a rapidly varying continuous random component called the diffusion. The second integral in (1) is an Ito stochastic integral with respect to the Wiener process $W = \{W_t, t \geq 0\}$. The integral equation (1) is often written in the differential form

$$(2) \qquad dX_t = a(X_t)\,dt + b(X_t)\,dW_t$$

and is then called an Ito stochastic differential equation (SDE). For simplicity, in this survey we shall restrict attention to a 1-dimensional Ito process $X$ with a 1-dimensional driving Wiener process $W$.

Unfortunately explicitly solvable SDEs such as those listed in Section 4 of Chapter 4 are rare in practical applications. There are, however, now a number of papers which deal with numerical methods for SDEs, yet the gap between the well developed theory of stochastic differential equations and its application is still wide. A crucial task in bridging this gap is the development of efficient numerical methods for SDEs, a task to which this book is addressed. Obviously such methods should be implementable on digital computers. They typically involve the simulation of a large number of different sample paths in order to estimate various statistical features of the desired solution. Modern supercomputers with their parallel architecture are well suited to such calculations; see Petersen (1987).

Here we shall survey various time discrete numerical methods which are appropriate for the simulation of sample paths or functionals of Ito processes on digital computers.

## Numerical Approaches to Stochastic Differential Equations

To begin we shall briefly mention several different approaches that have been suggested for the numerical solution of SDEs. On the very general level there is a method due to Boyce (1978) by means of which one can investigate, in principle at least, general random systems by Monte Carlo methods. For SDEs this method is somewhat inefficient because it does not use the special structure of these equations, specifically their characterization by their drift and diffusion coefficients.

Kushner (1977) proposed the discretization of both time and space variables, so the approximating processes are then finite state Markov chains. These can be handled on digital computers through their transition matrices. In comparison with the information encompassed succinctly in the drift and diffusion coefficients of an SDE, transition matrices contain a considerable amount of superfluous information which must be repeatedly reprocessed during computations. Consequently such a Markov chain approach seems applicable only for low dimensional problems on bounded domains. Similar disadvantages also arise, in higher dimensions at least, when standard numerical methods are used to solve parabolic partial differential equations, such as the Fokker-Planck equation and its adjoint, associated with functionals of the solutions of SDEs. These are, of course, also methods for computing the probability densities of Ito diffusions.

The most efficient and widely applicable approach to solving SDEs seems to be the simulation of sample paths of time discrete approximations on digital computers. This is based on a finite discretization of the time interval $[0, T]$ under consideration and generates approximate values of the sample paths step by step at the discretization times. The simulated sample paths can then be analysed by usual statistical methods to determine how good the approximation is and in what sense it is close to the exact solution. The state variables here are not discretized as in Kushner's Markov chain approach and the structure of the SDE as provided by the drift and diffusion coefficients is used in a natural way. An advantage of considerable practical importance of this approach is that the computational costs such as time and memory required increase only polynomially with the dimension of the problem. Variance reduction methods allow a considerable decrease in the required sample size.

## Time Discrete Approximations

Early simulation studies and theoretical investigations by Clements & Anderson (1973), Wright (1974), Clark & Cameron (1980) and others showed that not all heuristic time discrete approximations of an SDE (2) converge in a useful sense to the solution process as the maximum step size $\delta$ tends to zero. In particular, it was found that one cannot simply use a deterministic numerical method for ordinary differential equations, such as a higher order Runge-Kutta method. Consequently a careful and systematic investigation of different methods is needed in order to select a sufficiently efficient method for the task at hand.

We shall consider a time discretization $(\tau)_\delta$ with

$$(3) \qquad 0 = \tau_0 < \tau_1 < \cdots < \tau_n < \cdots < \tau_N = T$$

of a time interval $[0, T]$, which in the simplest equidistant case has step size

$$(4) \qquad \delta = \frac{T}{N}.$$

We shall see in Chapter 9 that general time discretizations, even with random times, are possible, but usually a maximum step size $\delta$ must be specified.

The simplest heuristic time discrete approximation is the stochastic generalization of the Euler approximation which is sometimes called the *Euler-Maruyama approximation* (see Maruyama (1955)), but often just the *Euler approximation*. For the SDE (2) it has the form

$$(5) \qquad Y_{n+1} = Y_n + a\left(Y_n\right)\Delta_n + b\left(Y_n\right)\Delta W_n$$

for $n = 0, 1, \ldots, N-1$ with initial value

$$(6) \qquad Y_0 = x_0,$$

where

$$(7) \qquad \Delta_n = \tau_{n+1} - \tau_n = \delta$$

and

$$(8) \qquad \Delta W_n = W_{\tau_{n+1}} - W_{\tau_n}$$

for $n = 0, 1, \ldots, N-1$. Essentially, it is formed by fixing the integrands in (1) to their values at the beginning of each discretization time subinterval. The recursive scheme (5) obviously gives values of the approximation only at the discretization times. If values are required at intermediate instants, then either piecewise constant values from the preceding discretization point or some interpolation, especially a linear interpolation, of the values of the two immediate enclosing discretization points could be used.

The random variables $\Delta W_n$ defined in (8) are independent $N(0; \Delta_n)$ normally distributed random variables, that is with means and variances

$$E\left(\Delta W_n\right) = 0 \qquad \text{and} \qquad E\left(\left(\Delta W_n\right)^2\right) = \Delta_n,$$

respectively, for $n = 0, 1, \ldots, N-1$. In simulations we can generate such random variables from independent, uniformly distributed random variables on $[0, 1]$, which are usually provided by a pseudo-random number generator on a digital computer. We shall discuss and test random number generators in Sections 3 and 9 of Chapter 1.

## The Strong Convergence Criterion

In problems such as those that we shall consider in Chapter 13 involving direct simulations, filtering or testing estimators of Ito processes it is important that the trajectories, that is the sample paths, of the approximation be close to those of the Ito process. This suggests that a criterion involving some form of strong convergence should be used. Mathematically it is advantageous to consider the absolute error at the final time instant $T$, that is

$$(9) \qquad \epsilon\left(\delta\right) = E\left(\left|X_T - Y_N\right|\right),$$

which can be estimated from the root mean square error via the Lyapunov inequality

$$(10) \qquad \epsilon(\delta) = E(|X_T - Y_N|) \leq \sqrt{E\left(|X_T - Y_N|^2\right)}.$$

The absolute error (9) is certainly a criterion for the closeness of the sample paths of the Ito process $X$ and the approximation $Y$ at time $T$.

We shall say that an approximating process $Y$ *converges in the strong sense with order* $\gamma \in (0, \infty]$ if there exists a finite constant $K$ and a positive constant $\delta_0$ such that

$$(11) \qquad E(|X_T - Y_N|) \leq K \delta^\gamma$$

for any time discretization with maximum step size $\delta \in (0, \delta_0)$. In the deterministic case with zero diffusion coefficient $b \equiv 0$ this strong convergence criterion reduces to the usual deterministic criterion for the approximation of ordinary differential equations. The order of a scheme is sometimes less in the stochastic case than in the corresponding deterministic one, essentially because the increments $\Delta W_n$ of the Wiener process are of root mean square order $\delta^{1/2}$ and not $\delta$. In fact, the Euler approximation (5) for SDEs has strong order $\gamma = 0.5$ in contrast with the order 1.0 of the Euler approximation for ordinary differential equations.

Important contributions which also influenced the future development of higher order strong approximations include Milstein (1974), McShane (1974), Rao, Borwankar & Ramakrishna (1974), Kloeden & Pearson (1977), Wagner & Platen (1978), Clark (1978), Platen (1981), Rümelin (1982), Talay (1982a), Newton (1986a) and Chang (1987).

## The Weak Convergence Criterion

In many practical situations, some of which will be described in Chapter 17, it is not necessary to have a close pathwise approximation of an Ito process. Often one may only be interested in some function of the value of the Ito process at a given final time $T$ such as one of the first two moments $E(X_T)$ and $E((X_T)^2)$ or, more generally, the expectation $E(g(X_T))$ for some function $g$. In simulating such a functional it suffices to have a good approximation of the probability distribution of the random variable $X_T$ rather than a close approximation of sample paths. Thus the type of approximation required here is much weaker than that provided by the strong convergence criterion.

We shall say that a time discrete approximation $Y$ *converges in the weak sense with order* $\beta \in (0, \infty]$ if for any polynomial $g$ there exists a finite constant $K$ and a positive constant $\delta_0$ such that

$$(12) \qquad |E(g(X_T)) - E(g(Y_N))| \leq K \delta^\beta$$

for any time discretization with maximum step size $\delta \in (0, \delta_0)$. In Section 7 of Chapter 9 we shall generalize slightly the class of test fuctions $g$ used here. When the diffusion coefficient in (1) vanishes identically this weak convergence criterion with $g(x) \equiv x$ also reduces to the usual deterministic convergence criterion for ordinary differential equations.

Under assumptions of sufficient smoothness Milstein (1978) and Talay (1984) showed that an Euler approximation of an Ito process converges with weak order $\beta = 1.0$, which is greater than its strong order of convergence $\gamma = 0.5$. On the other hand, Mikulevicius & Platen (1986) showed that the Euler scheme converges, but with weak order less than 1.0, when the coefficients of (1) are only Hölder continuous, that is Lipschitz-like with a fractional power.

Higher order weak approximations have been extensively investigated by Milstein (1978), Talay (1984), Platen (1984) and Mikulevicius & Platen (1988). In particular, weak approximations of the Runge-Kutta type have been proposed and studied by Greenside & Helfand (1981), Talay (1984), Platen (1984), Klauder & Petersen (1985a), Milstein (1985), Haworth & Pope (1986) and Averina & Artemev (1986). In addition, Wagner (1987) has investigated the use of unbiased weak approximations, that is with $\beta = \infty$, for estimating functionals of Ito diffusions.

## Stochastic Taylor Formulae

Another natural way of classifying numerical methods for SDEs is to compare them with strong and weak Taylor approximations. The increments of such approximations are obtained by truncating the stochastic Taylor formula, which was derived in Wagner & Platen (1978) by an iterated application of the Ito formula. It was then generalized and investigated in Azencott (1982), Platen (1982) and Platen & Wagner (1982). A Stratonovich version of the stochastic Taylor formula was presented in Kloeden & Platen (1990) and can be found together with results on multiple stochastic integrals in Chapter 5.

The stochastic Taylor formula allows a function of an Ito process, that is $f(X_t)$, to be expanded about $f(X_{t_0})$ in terms of multiple stochastic integrals weighted by coefficients which are evaluated at $X_{t_0}$. These coefficients are formed from the drift and diffusion coefficients of the Ito process and their derivatives up to some specified order. The remainder term in the formula contains a finite number of multiple stochastic integrals of the next higher multiplicity, but now with nonconstant integrands. For example, a stochastic Taylor formula for the Ito process (1) for $t \in [t_0, T]$ may have the form

$$(13) \qquad f(X_t) \;=\; f(X_{t_0}) + c_1(X_{t_0}) \int_{t_0}^{t} ds + c_2(X_{t_0}) \int_{t_0}^{t} dW_s$$

$$+ c_3(X_{t_0}) \int_{t_0}^{t}\int_{t_0}^{s_2} dW_{s_1}\, dW_{s_2} + R$$

with coefficients

$$c_1(x) \;=\; a(x)f'(x) + \frac{1}{2}(b(x))^2 f''(x),$$

$$c_2(x) \;=\; b(x)f'(x),$$

$$c_3(x) \;=\; b(x)\{b(x)f''(x) + b'(x)f'(x)\}.$$

Here the remainder $R$ consists of higher order multiple stochastic integrals with nonconstant integrands involving the function $f$, the drift and diffusion coefficients and their derivatives. A stochastic Taylor formula can be thought of as a generalization of both the Ito formula and the deterministic Taylor formula.

If we use the function $f(x) = x$ in the stochastic Taylor formula (13) we obtain the following representation for the Ito process (1):

$$
(14) \qquad X_t = X_{t_0} + a(X_{t_0}) \int_{t_0}^{t} ds + b(X_{t_0}) \int_{t_0}^{t} dW_s
$$

$$
+ b(X_{t_0}) b'(X_{t_0}) \int_{t_0}^{t} \int_{t_0}^{s_2} dW_{s_1} dW_{s_2} + R.
$$

By truncating stochastic Taylor expansions such as (14) about successive discretization points we can form time discrete Taylor approximations which we may interpret as basic numerical schemes for an SDE. In addition we can compare other schemes, such as those of the Runge-Kutta type, with the time discrete Taylor approximations to determine their order of strong or weak convergence. We shall see that we must include the appropriate terms from the corresponding stochastic Taylor expansion, that is the necessary higher multiple stochastic integrals, to obtain a numerical scheme with a higher order of strong or weak convergence.

## Strong Taylor Approximations

The simplest strong Taylor approximation of an Ito diffusion is the *Euler approximation*

$$
(15) \qquad Y_{n+1} = Y_n + a\,\Delta_n + b\,\Delta W_n
$$

for $n = 0, 1, \ldots, N-1$ with initial condition (6), where $\Delta_n$ and $\Delta W_n$ are defined by (7) and (8), respectively, with the $\Delta W_n$ independent $N(0; \Delta_n)$ normally distributed random variables. Here we have written $a$ for $a(Y_n)$ and $b$ for $b(Y_n)$, a convention which we shall henceforth use for any function. In addition, as here, we shall not repeat the standard initial condition (6) in what follows. It was shown in Gikhman & Skorokhod (1972a) that the Euler scheme converges with strong order $\gamma = 0.5$ under Lipschitz and bounded growth conditions on the coefficients $a$ and $b$.

If we include the next term from the stochastic Taylor formula (14) in the scheme (15) we obtain the *Milstein scheme*

$$
(16) \qquad Y_{n+1} = Y_n + a\,\Delta_n + b\,\Delta W_n + \frac{1}{2}bb' \left\{ (\Delta W_n)^2 - \Delta_n \right\}
$$

for $n = 0, 1, \ldots, N-1$; see Milstein (1974). The additional term here is from the double Wiener integral in (14), which can be easily computed from the Wiener increment $\Delta W_n$ since

$$
(17) \qquad \int_{\tau_n}^{\tau_{n+1}} \int_{\tau_n}^{s_2} dW_{s_1} dW_{s_2} = \frac{1}{2}\left\{ (\Delta W_n)^2 - \Delta_n \right\}.
$$

We shall see that the Milstein scheme (16) converges with strong order $\gamma = 1.0$ under the assumption that $E((X_0)^2) < \infty$, that $a$ and $b$ are twice continuously differentiable, and that $a$, $a'$, $b$, $b'$ and $b''$ satisfy a uniform Lipschitz condition.

For a multi-dimensional driving Wiener process $W = (W^1, \ldots, W^m)$ the generalization of the Milstein scheme (16) involves the multiple Wiener integrals

$$(18) \qquad I_{(j_1,j_2)} = \int_{\tau_n}^{\tau_{n+1}} \int_{\tau_n}^{s_2} dW_{s_1}^{j_1} \, dW_{s_2}^{j_2},$$

for $j_1, j_2 \in \{1, \ldots, m\}$ with $j_1 \neq j_2$, which cannot be expressed simply as in (16) in terms of the increments $\Delta W_n^{j_1}$ and $\Delta W_n^{j_2}$ of the corresponding Wiener processes. In Section 8 of Chapter 5 we shall suggest one possible way of approximating higher order multiple stochastic integrals like (18); see also Kloeden, Platen & Wright (1991).

Generally speaking we obtain more accurate strong Taylor approximations by including additional multiple stochastic integral terms from the stochastic Taylor expansion. Such integrals contain additional information about the sample paths of the Wiener process over the discretization subintervals. Their presence is a fundamental difference between the numerical analysis of stochastic and ordinary differential equations.

For example, the *strong Taylor approximation of order* $\gamma = 1.5$ is given by

$$(19) \quad Y_{n+1} = Y_n + a\,\Delta_n + b\,\Delta W_n + \frac{1}{2}bb' \left\{ (\Delta W_n)^2 - \Delta_n \right\}$$

$$+ ba'\,\Delta Z_n + \frac{1}{2}\left\{ aa' + \frac{1}{2}b^2 a'' \right\} \Delta_n^2$$

$$+ \left\{ ab' + \frac{1}{2}b^2 b'' \right\} \{\Delta W_n \Delta_n - \Delta Z_n\}$$

$$+ \frac{1}{2}b \left\{ bb'' + (b')^2 \right\} \left\{ \frac{1}{3}(\Delta W_n)^2 - \Delta_n \right\} \Delta W_n$$

for $n = 0, 1, \ldots, N - 1$. Here the additional random variable $\Delta Z_n$ is required to represent the double integral

$$(20) \qquad \Delta Z_n = \int_{\tau_n}^{\tau_{n+1}} \int_{\tau_n}^{s_2} dW_{s_1} \, ds_2,$$

which is normally distributed with mean, variance and correlation

$$E\left(\Delta Z_n\right) = 0, \qquad E\left((\Delta Z_n)^2\right) = \frac{1}{3}(\Delta_n)^3 \quad \text{and} \quad E\left(\Delta W_n \Delta Z_n\right) = \frac{1}{2}(\Delta_n)^2,$$

respectively. All other multiple stochastic integrals appearing in the truncated Taylor expansion used to derive (19) can be expressed in terms of $\Delta_n$, $\Delta W_n$ and $\Delta Z_n$, thus resulting in (19). It was shown in Wagner & Platen (1978) and Platen (1981a) that the scheme (19) converges with strong order $\gamma = 1.5$ when the coefficients $a$ and $b$ are sufficiently smooth and satisfy Lipschitz and bounded growth conditions. We note that there is no difficulty in generating the pair of correlated normally distributed random variables $\Delta W_n$, $\Delta Z_n$ using the transformation

$$(21) \quad \Delta W_n = \zeta_{n,1} \Delta_n^{1/2} \quad \text{and} \quad \Delta Z_n = \frac{1}{2} \left( \zeta_{n,1} + \frac{1}{\sqrt{3}} \zeta_{n,2} \right) \Delta_n^{3/2},$$

where $\zeta_{n,1}$ and $\zeta_{n,2}$ are independent normally $N(0;1)$ distributed random variables.

Following Platen (1981a), we shall describe in Chapter 10 how schemes of any desired order of strong convergence can be constructed from the corresponding strong Taylor approximations. The implementation of such schemes requires the generation of multiple stochastic integrals such as $I_{(j_1,j_2)}$ and of higher multiplicity, which can be done by means of an approximation method which we shall describe in Chapter 5. Those readers who do not wish to use such multiple stochastic integrals could follow Clark (1978) and Newton (1986), in which schemes only involving the increments of the Wiener process are proposed. These schemes, which we shall describe in Section 4 of Chapter 13, are similar to the strong Taylor approximations above, but with the random variables modified. Moreover, they are optimal within the classes of strong orders $\gamma = 0.5$ or $1.0$, respectively.

## Strong Runge-Kutta, Two-Step and Implicit Approximations

A practical disadvantage of the above strong Taylor approximations is that the derivatives of various orders of the drift and diffusion coefficients must be determined and then evaluated at each step in addition to the coefficients themselves. There are time discrete approximations which avoid the use of derivatives, which we shall call Runge-Kutta schemes in analogy with similar schemes for ordinary differential equations. However, we emphasize that it is not always possible to use heuristic adaptations of deterministic Runge-Kutta schemes for SDEs because of the difference between ordinary and stochastic calculi.

A *strong order* $1.0$ *Runge-Kutta scheme* is given by

$$(22) \quad Y_{n+1} = Y_n + a\,\Delta_n + b\,\Delta_n W_n + \frac{1}{2}\left\{ b\left(\hat{\Upsilon}_n\right) - b \right\} \left\{ (\Delta W_n)^2 - \Delta_n \right\} \Delta_n^{-1/2}$$

with supporting value

$$\hat{\Upsilon}_n = Y_n + b\,\Delta_n^{1/2}$$

for $n = 0, 1, \ldots, N-1$. This scheme can be obtained heuristically from the Milstein scheme (16) simply by replacing the derivative there by the corresponding finite difference; see Platen (1984). Clark & Cameron (1980) and Rümelin (1982) have shown that Runge-Kutta schemes like (22) converge strongly with at most order $\gamma = 1.0$. More general Runge-Kutta schemes can be found in Chapter 11, but they have usually only the strong order of convergence $\gamma =$

1.0 if just the increments $\Delta W_n$ of the Wiener process are used; higher multiplicity stochastic integrals must be used to obtain a higher order of strong convergence.

For the case of additive noise, $b \equiv const.$, a *two-step order* 1.5 *strong scheme* takes the form

$$(23) \quad Y_{n+1} = Y_n + 2a \, \Delta_n - a' \, (Y_{n-1}) \, b \, (Y_{n-1}) \, \Delta W_{n-1} \, \Delta_n + V_n + V_{n-1}$$

with

$$V_n = b \, \Delta W_n + a' b \, \Delta Z_n,$$

where $\Delta W_n$ and $\Delta Z_n$ are the same as in (21); see Kloeden & Platen (1991c). Additional multi-step schemes will be described in Chapter 12.

A typical *implicit order* 1.5 *strong scheme* for additive noise is

$$(24) \quad Y_{n+1} \;=\; Y_n + \frac{1}{2} \left\{ a \, (Y_{n+1}) + a \right\} \Delta_n + b \, \Delta W_n$$

$$+ \frac{1}{2} \left\{ a \left( \hat{\Upsilon}_n^+ \right) - a \left( \hat{\Upsilon}_n^- \right) \right\} \left\{ \Delta Z_n - \frac{1}{2} \Delta W_n \, \Delta_n \right\} \Delta_n^{-1/2}$$

with supporting values

$$\hat{\Upsilon}_n^\pm = Y_n + a \, \Delta_n \pm b \, \Delta_n^{1/2},$$

where $\Delta W_n$ and $\Delta Z_n$ are the same as in (21); see Kloeden & Platen (1991c). Implicit or fully implicit schemes are needed to handle stiff SDEs, which will be discussed in Section 8 of Chapter 9 and in Chapter 12; see Petersen (1987), Drummond & Mortimer (1990) and Hernandez & Spigler (1990).

Another type of strong approximation was investigated in Gorostiza (1980) and Newton (1990b). In the 1-dimensional case the time variable is discretized in such a way that a random walk takes place on a prescribed set of threshholds in the state space, with the approximating process remaining on a fixed level for a random duration of time and then switching with given intensity to the next level above or below it.

Finally, the reader is referred to Doss (1978), Sussmann (1978) and Talay (1982a) for other investigations of strong approximations of Ito diffusions.

# Weak Taylor Approximations

When we are interested only in weak approximations of an Ito process, that is a process with approximately the same probability distribution, then we have many more degrees of freedom than with strong approximations. For example, it suffices to use an initial value $Y_0 = \hat{X}_0$ with a convenient probability distribution which approximates that of $X_0$ in an appropriate way. In addition the random increments $\Delta W_n$ of the Wiener process can be replaced by other more convenient approximations $\Delta \hat{W}_n$ which have similar moment properties to the $\Delta W_n$. In a weak approximation of order $\beta = 1.0$ we could, for instance, choose independent $\Delta \hat{W}_n$ for $n = 0, 1, \ldots, N - 1$ with moments

(25)
$$E\left(\left(\Delta \hat{W}_n\right)^r\right) = \begin{cases} 0 & : \quad r = 1 \text{ and } 3 \\ \Delta_n & : \quad r = 2 \\ Z_r(\Delta_n) & : \quad r = 4, 5, \ldots \end{cases}$$

where
(26)
$$|Z_r(\Delta_n)| \le K \Delta_n^2$$

for $r = 4, 5, \ldots$ and some finite constant $K > 0$. This means we could use an easily generated two-point distributed random variable taking values $\pm\sqrt{\Delta_n}$ with equal probabilities, that is with

(27)
$$P\left(\Delta \hat{W}_n = \pm\sqrt{\Delta_n}\right) = \frac{1}{2}.$$

The simplest useful weak Taylor approximation is the *weak Euler scheme*

(28)
$$Y_{n+1} = Y_n + a\,\Delta_n + b\,\Delta \hat{W}_n$$

for $n = 0, 1, \ldots, N - 1$. It follows from results in Talay (1984) that (28) has weak order $\beta = 1.0$ if the coefficients $a$ and $b$ are four times continuously differentiable with these derivatives satisfying a growth bound. This contrasts with the order $\gamma = 0.5$ of the strong Euler scheme (15).

We can construct weak Taylor approximations of higher order $\beta = 2.0, 3.0,$ $\ldots$ by truncating appropriate stochastic Taylor expansions. For example, the *weak Taylor approximation of order $\beta = 2.0$* has, following Milstein (1978) and Talay (1984), the form

(29)
$$\begin{aligned}
Y_{n+1} &= Y_n + a\,\Delta_n + b\,\Delta \hat{W}_n + \frac{1}{2}bb'\left\{\left(\Delta \hat{W}_n\right)^2 - \Delta_n\right\} \\
&\quad + ba'\,\Delta \hat{Z}_n + \frac{1}{2}\left\{aa' + \frac{1}{2}b^2 a'\right\}\Delta_n^2 \\
&\quad + \left\{ab' + \frac{1}{2}b^2 b''\right\}\left\{\Delta \hat{W}_n \Delta_n - \Delta \hat{Z}_n\right\}
\end{aligned}$$

for $n = 0, 1, \ldots, N - 1$. Here $\Delta \hat{W}_n$ approximates $\Delta W_n$ and $\Delta \hat{Z}_n$ the multiple stochastic integral (20). As with the weak Euler scheme (28) we can choose random variables $\Delta \hat{W}_n$ and $\Delta \hat{Z}_n$ which have approximately the same moment properties as $\Delta W_n$ and $\Delta Z_n$, For example, we could use

(30)
$$\Delta \hat{W}_n = \Delta W_n \qquad \text{and} \qquad \Delta \hat{Z}_n = \frac{1}{2}\Delta W_n \Delta_n$$

with the $\Delta W_n$ independent $N(0; \Delta_n)$ normally distributed, or we could use

(31)
$$\Delta \hat{W}_n = \Delta_n^{1/2} T_n \qquad \text{and} \qquad \Delta \hat{Z}_n = \frac{1}{2}\Delta_n^{3/2} T_n$$

where the $T_n$ are independent three-point distributed random variables with

$$(32) \qquad P\left(T_n = \pm\sqrt{3}\right) = \frac{1}{6} \quad \text{and} \quad P\left(T_n = 0\right) = \frac{2}{3}.$$

Multi-dimensional and higher order weak Taylor approximations also involve additional random variables, but these are much simpler than those in strong approximations as will be seen in Chapter 14.

It was shown under appropriate assumptions in Platen (1984) that a Taylor approximation converges with any desired weak order $\beta = 1.0, 2.0, \ldots$ when the multiple stochastic integrals up to multiplicity $\beta$ are included in the truncated stochastic Taylor expansion used to construct the scheme.

## Weak Runge-Kutta and Extrapolation Approximations

It is often convenient computationally to have weak approximations of the Runge-Kutta type which avoid the use of derivatives of the drift and diffusion coefficients, particularly the higher order derivatives. An *order* 2.0 *weak Runge-Kutta scheme* proposed by Talay (1984) is of the form

$$(33) \quad Y_{n+1} = Y_n + \left\{ a\left(\hat{\Upsilon}_n\right) - \frac{1}{2} b\left(\hat{\Upsilon}_n\right) b'\left(\hat{\Upsilon}_n\right) \right\} \Delta_n$$

$$+ \left\{ \frac{1}{\sqrt{2}} b\left(A_n - B_n\right) + \sqrt{2} b\left(\hat{\Upsilon}_n\right) B_n \right\} \Delta_n^{1/2}$$

$$+ \left\{ \frac{1}{2} \left( b\left(\hat{\Upsilon}_n\right) b'\left(\hat{\Upsilon}_n\right) - bb' \right) B_n^2 - bb' A_n B_n \right\} \Delta_n$$

with supporting value

$$\hat{\Upsilon}_n = Y_n + \frac{1}{2}\left(a - \frac{1}{2} bb'\right) \Delta_n + \frac{1}{\sqrt{2}} bA_n \Delta_n^{1/2} + \frac{1}{4} bb' A_n^2 \Delta_n$$

for $n = 0, 1, \ldots, N-1$, where the $A_n$ and $B_n$ are independent random variables which are, for example, standard normally distributed or as in (32).

The scheme (33) still uses the derivative $b'$ of the diffusion coefficient $b$. It is also possible to avoid using this derivative, as in the following *order* 2.0 *weak Runge-Kutta scheme* from Platen (1984):

$$(34) \quad Y_{n+1} = Y_n + \frac{1}{2}\left\{ a\left(\hat{\Upsilon}_n\right) + a \right\} \Delta_n + \frac{1}{4}\left\{ b\left(\Upsilon_n^+\right) + b\left(\Upsilon_n^-\right) + 2b \right\} \Delta\hat{W}_n$$

$$+ \frac{1}{4}\left\{ b\left(\Upsilon_n^+\right) - b\left(\Upsilon_n^-\right) \right\} \left\{ \left(\Delta\hat{W}_n\right)^2 - \Delta_n \right\} \Delta_n^{-1/2}$$

with supporting values

$$\hat{\Upsilon}_n = Y_n + a\Delta_n + b\Delta\hat{W}_n \quad \text{and} \quad \Upsilon_n^{\pm} = Y_n + a\Delta_n \pm b\Delta_n^{1/2}$$

for $n = 0, 1, \ldots, N - 1$, where the $\Delta \hat{W}_n$ can be chosen as in (30) or (31).

Higher order approximations of functionals can also be obtained with lower order weak schemes by extrapolation methods. Talay & Tubaro (1990) proposed an *order* 2.0 *weak extrapolation method*

$$(35) \qquad V^{\delta}_{g,2}(T) = 2E\left(g\left(Y^{\delta}(T)\right)\right) - E\left(g\left(Y^{2\delta}(T)\right)\right)$$

where $Y^{\delta}(T)$ and $Y^{2\delta}(T)$ are the Euler approximations at time $T$ for the step sizes $\delta$ and $2\delta$, respectively. Higher order extrapolation methods from Kloeden & Platen (1991b) will also be presented in Section 3 of Chapter 15. Essentially, many order $\beta$ weak scheme can be extrapolated with formulae similar to (35) to provide order $2\beta$ accuracy for $\beta = 1.0, 2.0, \ldots$.

An *order* 2.0 *weak predictor-corrector scheme* for SDEs proposed in Platen (1991), which has the corrector

$$(36) \qquad Y_{n+1} = Y_n + \frac{1}{2}\left\{a\left(\hat{\Upsilon}_{n+1}\right) + a\right\}\Delta_n + \Psi_n$$

with

$$\Psi_n = b\,\Delta \hat{W}_n + \frac{1}{2}\,bb'\left\{\left(\Delta \hat{W}_n\right)^2 - \Delta_n\right\} + \frac{1}{2}\left\{ab' + \frac{1}{2}b^2b''\right\}\Delta \hat{W}_n\,\Delta_n$$

and the predictor

$$(37) \quad \hat{\Upsilon}_{n+1} = Y_n + a\,\Delta_n + \Psi_n + \frac{1}{2}\,a'b\,\Delta \hat{W}_n\,\Delta_n + \frac{1}{2}\left\{aa' + \frac{1}{2}b^2a''\right\}\Delta_n^2,$$

where $\Delta \hat{W}_n$ can be as in (30) or (31)–(32), is an example of a simplified weak Taylor scheme, where $\Delta_n$ can be as in (30) or (31)–(32). The corrector (36) resembles the *implicit order* 2.0 *weak scheme*

$$(38) \qquad Y_{n+1} = Y_n + \frac{1}{2}\left\{a\left(Y_{n+1}\right) + a\right\}\Delta_n + \Psi_n.$$

Higher order Runge-Kutta, predictor-corrector and implicit weak schemes as well as extrapolation methods will be examined in Chapter 15.

Runge-Kutta schemes with convergence only in the first two moments have been considered in Greenside & Helfand (1984), Haworth & Pope (1986), Helfand (1979), Klauder & Petersen (1985a) and Petersen (1987). This convergence criterion is weaker than the weak convergence criterion (10) considered here. Obviously, a scheme which converges with some weak order $\beta$ will not only converge in the first two moments, but also in all higher moments with this same order $\beta$ when they exist.

Finally, Wagner (1987) has proposed another way of approximating weak approximations of diffusion processes which is based on the Monte Carlo simulation of functional integrals and uses unbiased, variance reduced approximations to estimate functionals of Ito diffusion processes. This will be described in

Chapter 16. Milstein (1988) and Chang (1987) have also investigated variance reduction techniques.

We conclude this brief survey with the remark that the theoretical understanding and practical application of numerical methods for stochastic differential equations are still in their infancy. An aim of this book is to stimulate an interest and further work on such methods. For this the Bibliographical Notes at the end of the book may be of assistance.

# Chapter 1
# Probability and Statistics

The basic concepts and results of probability and stochastic processes needed later in the book are reviewed here. The emphasis is descriptive and PC-Exercises (PC= Personal Computer), based on pseudo-random number generators introduced in Section 3, are used extensively to help the reader to develop an intuitive understanding of the material. Statistical tests are discussed briefly in the final section.

## 1.1   Probabilities and Events

If we toss a die, then, excluding absurd situations, we always observe one of six basic outcomes; it lands with its upper face indicating one of the numbers 1, 2, 3, 4, 5 or 6. We shall denote these outcomes by $\omega_1$, $\omega_2$, $\omega_3$, $\omega_4$, $\omega_5$ and $\omega_6$, respectively, and call the set of outcomes $\Omega = \{\omega_1, \omega_2, \omega_3, \omega_4, \omega_5, \omega_6\}$ the *sample space*. If we toss the die $N$ times and count the number of times $N_i$ that outcome $\omega_i$ occurs, we obtain a *relative frequency* $f_i(N) = N_i/N$. This number usually varies considerably with $N$, but experience tells us that as $N$ becomes larger it approaches a limit $\lim_{N \to \infty} f_i(N) = p_i$, which we call the *probability* of outcome $\omega_i$. Clearly $0 \leq p_i \leq 1$ for each $i = 1, 2, \ldots, 6$ and $\sum_{i=1}^{6} p_i = 1$; for a fair die each $p_i = 1/6$, giving a uniform distribution of probabilities over the outcomes.

Often we are interested in combinations of outcomes, that is subsets of the sample space $\Omega$ such as the subset $\{\omega_1, \omega_3, \omega_5\}$ of odd indexed outcomes. If we can distinguish such a combination by either its occurence or its nonoccurence we call it an *event*. Clearly if a subset $A$ is an event, then its complement $A^c = \{\omega_i \in \Omega : \omega_i \notin A\}$ must also be an event. In particular, the whole sample space $\Omega$ is an event, which we call the *sure event* since one of its outcomes must always occur; its complement, the empty set $\emptyset$, is also an event but can never occur. We might think that every subset $A$ of $\Omega$ should be an event, in which case we could determine its probability $P(A)$ from those of its constituent outcomes, that is as $P(A) = \sum_{\omega_i \in A} p_i$. However this corresponds to a situation of complete information about each of the outcomes, information which we may not always possess. For example, we may have only kept records of the occurences of odd or even indexed outcomes, but not of the actual outcomes themselves. Then we only distinguish and determine probabilities for the four subsets $\emptyset$, $O = \{\omega_1, \omega_3, \omega_5\}$, $E = \{\omega_2, \omega_4, \omega_6\}$ and $\Omega$, which are thus the only events in this case. Actually, we could introduce new basic outcomes $O$ (odd) and $E$ (even) here;

then the sample space is $\{O, E\}$ and all of its subsets are events. This shows that we have some flexibility in the choice of the sample space.

Whatever the sample space $\Omega$ and the collection $\mathcal{A}$ of events, we would always expect that $\emptyset$ and $\Omega$ are events and that

(1.1)     $A^c, A \cap B$ and $A \cup B$ are events if $A$ and $B$ are events.

Here the event $A \cup B$ occurs if either the event $A$ or the event $B$ occurs, whereas $A \cap B$ occurs if both $A$ and $B$ occur. Then, supposing that we have determined the probability of each event from frequency records either of the event itself or of its constituent components, we would always expect these probabilities to satisfy

(1.2)     $0 \leq P(A) \leq 1,\quad P(A^c) = 1 - P(A),\quad P(\emptyset) = 0,\quad P(\Omega) = 1$

and

(1.3)     $P(A \cup B) = P(A) + P(B)$  if $A \cap B = \emptyset$

for any events $A$ and $B$. From these we could then deduce for any positive integer $n$ that

(1.4)     $\displaystyle\bigcup_{i=1}^{n} A_i$ and $\displaystyle\bigcap_{i=1}^{n} A_i$ are events if $A_1, A_2, \ldots, A_n$ are events

and that

(1.5)     $\displaystyle P\left(\bigcup_{i=1}^{n} A_i\right) = \sum_{i=1}^{n} P(A_i)$ if $A_1, A_2, \ldots, A_n$ are mutually exclusive,

that is if $A_i \cap A_j = \emptyset$ for all $i, j = 1, 2, \ldots, n$ with $i \neq j$.

A similar situation applies when we have a countably infinite number of outcomes. Suppose we count the number of telephone calls arriving at an exchange during a specified time period. This will be a finite nonnegative integer $i = 1, 2, \ldots$, but, in principle, it can be arbitrarily large. An appropriate choice of sample space here is the countably infinite set $\Omega = \{\omega_0, \omega_1, \omega_2, \ldots\}$, where the outcome $\omega_i$ corresponds to the arrival of $i$ calls. If we repeat these counts over sufficiently many time periods, we could then use the limits of relative frequencies of occurence to determine the probability $p_i$ for each outcome $\omega_i$, obtaining $0 \leq p_i \leq 1$ for each $i = 0, 1, 2, \ldots$ with $\sum_{i=0}^{\infty} p_i = 1$. As in the die example we call a subset of $\Omega$ an event if we can distinguish it by either its occurence or nonoccurence; not every outcome $\omega_i$ need be an event. (Strictly speaking, no outcome $\omega_i$ can be an event since it is an element and not a subset of the sample space $\Omega$. However the singleton set $\{\omega_i\}$ may be an event, in which case we call it an elementary event, although it need not be an event). There may now be an infinite number of events, such as $E_n = \{\omega_0, \omega_2, \ldots, \omega_{2n}\}$ corresponding to an even number of calls not exceeding $2n$, for each $n = 0, 1, 2, \ldots$. Their countable union $E = \bigcup_{n=1}^{\infty} E_n = \{\omega_0, \omega_2, \omega_4, \ldots\}$

occurs if there is an even number of calls, no matter how large, and so should also be an event if all of the $E_n$ are events. In general, we would expect that

$$(1.6) \qquad \bigcup_{i=1}^{\infty} A_i \text{ and } \bigcap_{i=1}^{\infty} A_i \text{ are events if } A_1, A_2, A_3, \ldots \text{ are events}$$

and that

$$(1.7) \qquad P\left(\bigcup_{i=1}^{\infty} A_i\right) = \sum_{i=1}^{\infty} P(A_i) \text{ if } A_1, A_2, A_3, \ldots \text{ are mutually exclusive}$$

for any countable collection $A_1$, $A_2$, $A_3$, ... of events. Apart from this modification, things are then much the same as for a finite number of basic outcomes. An obvious difference is that a uniform distribution of probabilities over all outcomes is now impossible.

For uncountably many basic outcomes matters are not so straightforward. Suppose that we wish to determine the speed of a car from its speedometer which ranges from 0 to 250 km/h. The interval $0 \leq \omega \leq 250$ seems to be an appropriate choice for the sample space $\Omega$, but here the intrinsic limitations in the calibration of the speedometer prevent us from ever reading with complete accuracy any specific nonzero speed $\bar{\omega}$; the best we can do is to assertain that the speed lies within some small interval $(\bar{\omega} - \epsilon, \bar{\omega} + \epsilon)$ around $\bar{\omega}$. Such subintervals are the natural events here. The singleton set $\{0\}$ might also be an event since we can easily detect zero speed, though possible not solely from a speedometer. As in the previous examples, we would expect to obtain events when we took unions or intersections of events, and to be able to combine their probabilities accordingly. However, there may now be uncountably many events and this can cause problems. For example, if each singleton set $A_\omega = \{\omega\}$ is an event and has zero probability $P(A_\omega) = 0$, then $\Omega = \cup\{A_\omega; \omega \in \Omega\}$ is an uncountable union if $\Omega$ is uncountable. Obviously, $P(\Omega) = 1$, yet for any countable collection of the $A_\omega$ we would have $\sum_{i=1}^{\infty} P(A_{\omega_i}) = 0$. Worse still, the uncountable union or intersection of events need not itself be an event. We can fortunately avoid these difficulties by restricting attention to countable combinations of events.

In each of the preceding examples the essential probabilistic information can be succinctly summarized in the corresponding triplet $(\Omega, \mathcal{A}, P)$ consisting of the sample space $\Omega$, the collection of events $\mathcal{A}$ and the probability measure $P$, where $\mathcal{A}$ and $P$ satisfy the properties (1.1)–(1.7). We call such a triplet $(\Omega, \mathcal{A}, P)$ a *probability space*. In our discussion above we have glossed over the conceptual subtleties associated with the use of limits of relative frequencies to define and to determine probabilities. These can lead to serious problems if we try to develop a logically consistent theory of probability from this direction. To circumvent these difficulties it is now usual to develop probability theory axiomatically, with a probability space as its starting point. The probabilities are now just numbers assigned to each event and the relative frequencies only suggestive of how such numbers might be obtained. This axiomatic approach to probability theory was first propounded by Kolmogorov in the 1930s. We shall return to it in Chapter 2 and present definitions there. Some terminology will however be useful before then. In particular, the collection of events $\mathcal{A}$ is known technically as a *σ-algebra* or *σ-field*. While $P(\emptyset) = 0$ always holds, there may also be nonempty events $A$ with $P(A) = 0$; we call these *null events*. The sample space $\Omega$ is the *sure event*, and we say that any other event $A$ with $P(A) = 1$ occurs *almost surely* (a.s.) or *with probability one* (w.p.1).

Regardless of how we actually evaluate it, the probability $P(A)$ of an event $A$ is an indicator of the likelihood that $A$ will occur. Our estimate of this likelihood may change if we possess some additional information, such as that another event has occured. For example, if we toss a fair die the probability of obtaining a 6 is $P(\{\omega_6\}) = p_6 = 1/6$ and the probability of obtaining an even number, that is the probability of the event $E = \{\omega_2, \omega_4, \omega_6\}$, is $P(E) = p_2 + p_4 + p_6 = 1/2$. If we know that an even number has been thrown, then, since this occurs in one of three equally likely ways, we might now expect that the probability of its being the outcome $\omega_6$ is $1/3$. We call this the conditional probability of the event $\{\omega_6\}$ given that the event $E$ has occured and denote it by $P(\{\omega_6\}|E)$, noting that

$$P(\{\omega_6\}|E) = \frac{P(\{\omega_6\} \cap E)}{P(E)}$$

where $P(E) > 0$. In general, we define the *conditional probability* $P(A|B)$ of $A$ given that an event $B$ has occured by

$$(1.8) \qquad\qquad P(A|B) = \frac{P(A \cap B)}{P(B)}$$

provided $P(B) > 0$ and define it to be equal to 0 (or we leave it undefined) in the vacuous case that $P(B) = 0$. This definition is readily suggested from the relative frequencies

$$\frac{N_{A \cap B}}{N_B} = \frac{N_{A \cap B}}{N} \Big/ \frac{N_B}{N},$$

where $N_{A \cap B}$ and $N_B$ are the numbers of times that the events $A \cap B$ and $B$, respectively, occur out of $N$ repetitions of what we usually call an *experiment*.

It is possible that the occurence or not of an event $A$ is unaffected by whether or not another event $B$ has occured. Then its conditional probability $P(A|B)$ should be the same as $P(A)$, which with (1.8) implies that

$$(1.9) \qquad\qquad P(A \cap B) = P(A)P(B)$$

In this case we say that the events $A$ and $B$ are *independent*. For example, events $A$ and $B$ are independent if $P(A) = P(B) = 1/2$ and $P(A \cap B) = 1/4$. This particular situation occurs if we toss a fair coin twice, with $A$ the event that we obtain a head on the first toss and $B$ a head on the second toss, provided that the way we toss the coin the second time is not biased by the outcome of the first toss. This may seem tautological; in fact, we shall use the independence of outcomes of a repeated experiment to define *independent repetitions* of the experiment. Finally, we say that $n$ events $A_1$, $A_2$, ..., $A_n$ are *independent* if

$$(1.10) \qquad P(A_{i_1} \cap A_{i_2} \cap \cdots \cap A_{i_k}) = P(A_{i_1})P(A_{i_2}) \cdots P(A_{i_k})$$

for all nonempty subsets $\{i_1, i_2, \ldots, i_k\}$ of the set of indices $\{1, 2, \ldots, n\}$.

## 1.2 Random Variables and Distributions

We are often interested in numerical quantities associated with the outcome of a probabilistic experiment, such as our winnings in a gambling game based on tossing a die or the revenue made by a telephone company based on the number of calls made. These numbers, $X(\omega)$ say, provide us with information about the experiment, which, of course, can never exceed that already summarized in its probability space $(\Omega, \mathcal{A}, P)$. They correspond to the values taken by some function $X : \Omega \to \Re$, which we call a random variable if its information content is appropriately restricted.

Consider the *indicator function* $I_A$ of a subset $A$ of $\Omega$ defined by

$$(2.1) \qquad I_A(\omega) = \begin{cases} 0 & : \quad \omega \notin A \\ 1 & : \quad \omega \in A, \end{cases}$$

which is thus a function from $\Omega$ into $\Re$. For $I_A$ to be a random variable we require $A$ to be an event, or equivalently the subset

$$\{\omega \in \Omega : I_A(\omega) \leq a\} = \begin{cases} \emptyset & : \quad a < 0 \\ A^c & : \quad 0 \leq a < 1 \\ \Omega & : \quad 1 \leq a \end{cases}$$

to be an event for each $a \in \Re$.

In general, for a probability space $(\Omega, \mathcal{A}, P)$ we say that a function $X : \Omega \to \Re$ is a *random variable* if

$$(2.2) \qquad \{\omega \in \Omega : X(\omega) \leq a\} \in \mathcal{A} \quad \text{for each} \quad a \in \Re,$$

that is if $\{\omega \in \Omega : X(\omega) \leq a\}$ is an event for each $a \in \Re$. This is not quite so restrictive as it may seem because it implies that $\{\omega \in \Omega : X(\omega) \in B\}$ is an event for any *Borel subset* $B$ of $\Re$, that is any subset of $\Re$ in the $\sigma$-algebra $\mathcal{B}$ is generated from countable unions, intersections or complements of the semi-infinite intervals $\{x \in \Re; -\infty < x \leq a\}$.

We call the function $P_X$ defined for all $B \in \mathcal{B}$ by

$$(2.3) \qquad P_X(B) = P(\{\omega \in \Omega : X(\omega) \in B\})$$

the *distribution* of the random variable $X$, and note that the ordered triple $(\Re, \mathcal{B}, P_X)$ is a probability space which contains all of the essential information associated with the random variable $X$. Since point functions are simpler than set functions we often restrict attention to the function $F_X : \Re \to \Re$ defined for each $x \in \Re$ by

(2.4)        $F_X(x) = P_X((-\infty, x)) = P(\{\omega \in \Omega : X(\omega) < x\})$,

which we call the *distribution function* of $X$. For example, if $A$ is an event the distribution function for its indicator function $I_A$ is

(2.5)        $F_{I_A}(x) = \begin{cases} 0 & : \ x < 0 \\ 1 - P(A) & : \ 0 \le x < 1 \\ 1 & : \ 1 \le x \end{cases}$

Since for any $x < y$ we have $\{\omega \in \Omega : X(\omega) \le x\} \subseteq \{\omega \in \Omega : X(\omega) \le y\}$ and hence $P(\{\omega \in \Omega : X(\omega) \le x\}) \le P(\{\omega \in \Omega : X(\omega) \le y\})$, we can see that any distribution function $F_X$ satisfies:

(2.6)        $\lim_{x \to -\infty} F_X(x) = 0$   and   $\lim_{x \to +\infty} F_X(x) = 1$

with

(2.7)        $F_X(x)$   nondecreasing in   $x$.

The example (2.5) shows that $F_X$ need not be a continuous function, but from properties (1.1)–(1.7) and (2.4) we can show that $F_X$ is always *continuous from the right*, that is

(2.8)        $\lim_{h \to 0+} F_X(x + h) = F_X(x)$   for all   $x \in \Re$.

Conversely, for any function $F : \Re \to \Re$ satisfying properties (2.6)–(2.8), we can define a random variable $X$ which has $F$ as its distribution function. We can use $(\Re, \mathcal{B}, P_X)$ as the underlying probability space for this random variable, with $P_X$ defined for subintervals $(-\infty, x]$ in terms of $F$ using (2.4) and then extended appropriately to the more general Borel subsets. In this setting the

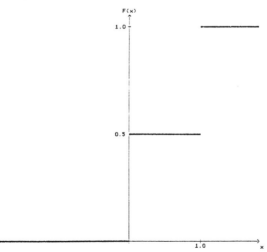

**Figure 1.2.1**    Distribution (2.5) for $P(A) = \frac{1}{2}$.

random variable is the identity function on $\Re$. Often we omit mention of this probability space and concentrate on the distribution function.

In applications the following examples are frequently encountered.

**Example 1.2.1** *The simplest nontrivial random variable $X$ takes just two distinct real values $x_1$ and $x_2$, where $x_1 < x_2$, with probabilities $p_1$ and $p_2$ $= 1 - p_1$, respectively. It is often called a* two-point random variable *and its distribution function is given by*

$$F_X(x) = \begin{cases} 0 & : \quad x < x_1 \\ p_1 & : \quad x_1 \leq x < x_2 \\ 1 & : \quad x_2 \leq x \end{cases}$$

For instance, the indicator function $I_A$ of an event $A$ is such a random variable with $x_1 = 0$ and $x_2 = 1$ (see $(2.5)$). Another two-point random variable arises in the gambling game where we win a dollar when a tossed coin shows a head and lose a dollar when it shows a tail; here $x_1 = -1$ and $x_2 = +1$.

**Example 1.2.2** *In radioactive decay the number of atoms decaying per unit time is a random variable $X$ taking values 0, 1, 2, ... without any upper bound. The probabilities $p_n = P(X = n)$ often satisfy the* Poisson distribution *with*

$$p_n = \frac{\lambda^n}{n!} \exp(-\lambda)$$

*for $n = 0$, 1, 2, ..., where $\lambda > 0$ is a given parameter.*

The above two examples are typical of a discrete random variable $X$ taking a finite or countably infinite number of distinct values $x_0 < x_1 < \cdots < x_n < \cdots$ with probabilities $p_n = P(X = x_n)$ for $n = 0$, 1, 2, .... The distribution function $F_X$ here satisfies

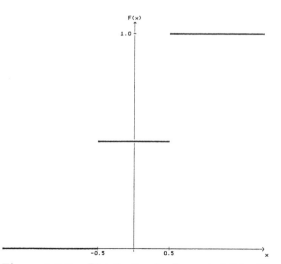

**Figure 1.2.2** Distribution for Example 1.2.1 with $-x_1 = x_2$ and $p_1 = 0.5$.

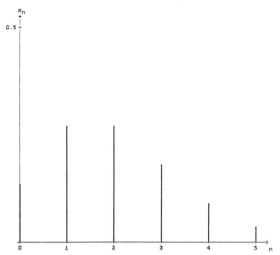

**Figure 1.2.3**    Poisson probabilities for $\lambda = 2$.

$$(2.9) \qquad F_X(x) = \begin{cases} 0 & : \quad x < x_0 \\ \sum_{i=0}^{n} p_i & : \quad x_n \leq x < x_{n+1},\ n = 0, 1, \ldots \end{cases}$$

$F_X$ is a step-function with steps of height $p_n$ at $x = x_n$. For such a random variable the set $\{x_0, x_1, x_2, \ldots\}$ could be used as the sample space $\Omega$, with all of its subsets being events.

In sharp contrast are the random variables taking all possible values in $\Re$. We call such a random variable $X$ a *continuous random variable* if the probability $P(\{\omega \in \Omega : X(\omega) = x\})$ is zero for all $x \in \Re$. In this case the distribution function $F_X$ is often differentiable, that is there exists a nonnegative function $p$, called the *density function*, such that $F'_X(x) = p(x)$ for each $x \in \Re$; when $F_X$ is only piecewise differentiable this holds everywhere except at certain isolated points. Then

$$(2.10) \qquad F_X(x) = \int_{-\infty}^{x} p(s)\,ds$$

for all $x \in \Re$, including the above mentioned exceptional points in the piecewise differentiable case. Such a distribution function is usually said to be *absolutely continuous*. The following are commonly occuring examples.

**Example 1.2.3**    *Consider a random variable $X$ which only takes values in a finite interval $a \leq x \leq b$, such that the probability of its being in a given subinterval is proportional to the length of the subinterval. Then the distribution function is given by*

$$F_X(x) = \begin{cases} 0 & : \quad x < a \\ \frac{x-a}{b-a} & : \quad a \leq x \leq b \\ 1 & : \quad b < x \end{cases}$$

*which is differentiable everywhere except at $x = a$ and $x = b$. The corresponding density function is given by*

$$p(x) = \begin{cases} 0 & : \quad x \notin [a, b] \\ \dfrac{1}{b - a} & : \quad x \in [a, b] \end{cases}$$

We say that the random variable $X$ in Example 1.2.3 is *uniformly distributed* on $[a, b]$ and denote this by $X \sim U(a, b)$. Alternatively, we say that it has a *rectangular* density function.

The $U(0, 1)$ random variables are a special case of the beta- distributed random variable with parameters $\alpha = \beta = 1$. In general, the *beta-distribution* with positive parameters $\alpha$ and $\beta$ has the density function

(2.11)
$$p(x) = \begin{cases} 0 & : \quad x \notin [0, 1] \\ \dfrac{x^{\alpha - 1}(1 - x)^{\beta - 1}}{B(\alpha, \beta)} & : \quad x \in [0, 1] \end{cases}$$

where

$$B(\alpha, \beta) = \int_0^1 x^{\alpha - 1}(1 - x)^{\beta - 1}\, dx.$$

**Example 1.2.4**  *The life-span of a light bulb is a random variable $X$ which is often modelled by the* exponential distribution

$$F_X(x) = \begin{cases} 0 & : \quad x < 0 \\ 1 - \exp(-\lambda x) & : \quad x \geq 0 \end{cases}$$

*for some intensity parameter $\lambda > 0$. $F_X$ is differentiable everywhere except for $x = 0$ and has the density function*

$$p(x) = \begin{cases} 0 & : \quad x < 0 \\ \lambda \exp(-\lambda x) & : \quad x \geq 0. \end{cases}$$

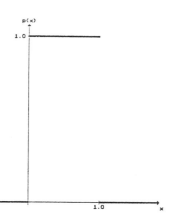

**Figure 1.2.4**  The rectangular density with $a = 0$ and $b = 1$.

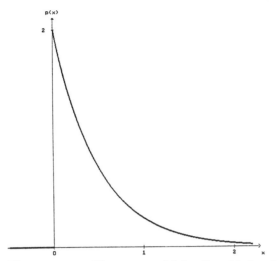

**Figure 1.2.5**    The exponential density with $\lambda = 2$.

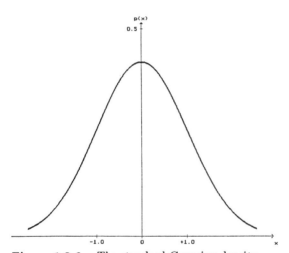

**Figure 1.2.6**    The standard Gaussian density.

**Example 1.2.5**    *The density function*

$$(2.12) \qquad\qquad p(x) = \frac{1}{\sqrt{2\pi}} \exp\left(-\frac{1}{2}x^2\right)$$

*has a bell-shaped graph which is symmetric about $x = 0$. The corresponding distribution function $F_X(x)$ is differentiable everywhere and has a sigmoidal-shaped graph, but must be evaluated numerically or taken from tables since no anti-derivative is known in analytical form for (2.12). A random variable with this density function is called a* standard Gaussian random variable.

Gaussian random variables occur so commonly in applications, for example as measurement errors in laboratory experiments, that they are often said to be *no mally distributed.* Their ubiquity is explained by the fundamental theorem of probability and statistics, the Central Limit Theorem, which will be discussed in Section 5.

In anticipation of the next section, we say that two random variables $X$ and $Y$ are *independent* if the events $\{\omega : X(\omega) \in A\}$ and $\{\omega : Y(\omega) \in B\}$ are independent for all Borel sets $A$ and $B$. Essentially, the values taken by either of the random variables are uninfluenced by those taken by the other. More will be said about this concept in Section 4.

## 1.3   Random Number Generators

The numerical simulation of a mathematical model of a complicated probabilistic system often provides information about the behaviour of the model, and hopefully of the original system itself, which cannot be obtained directly or easily by other means. Numerical values of each of the random variables must be provided for a test run of the model, and then the outputs of many test runs are analysed statistically. This procedure requires the generation of large quantities of random numbers with the specified statistical properties. Originally such numbers were taken directly from actual random variables, generated, for example, mechanically by tossing a die or electronically by the noisy output of a valve, and often listed in random number tables. This proved impractical for large scale simulations and the numbers were not always statistically reliable. In addition, a particular sequence of random numbers could not always be reproduced, an important feature for comparative studies, and so had to be stored. The advent of electronic computers lead to the development of simple deterministic algorithms to generate sequences of random variables, quickly and reproducably. Such numbers are consequently not truly random, but with sufficient care they can be made to resemble random numbers in most properties, in which case they are called *pseudo-random numbers.*

These days most digital computers include a *linear congruential pseudorandom number generator.* These have the recursive form

$$(3.1) \qquad\qquad X_{n+1} = aX_n + b \pmod{c}$$

where $a$ and $c$ are positive integers and $b$ a nonnegative integer. For an integer initial value or *seed* $X_0$, the algorithm (3.1) generates a sequence taking integer values from 0 to $c - 1$, the remainders when the $aX_n + b$ are divided by $c$. When the coefficients $a, b$ and $c$ are chosen appropriately the numbers

$$(3.2) \qquad\qquad U_n = X_n/c$$

seem to be uniformly distributed on the unit interval $[0, 1]$. Since only finitely many different numbers occur, the modulus $c$ should be chosen as large as possible, and perhaps also as a power of 2 to take advantage of the binary arithmetic used in computers. To prevent cycling with a period less than $c$ the multiplier $a$ should also be taken relatively prime to $c$. Typically $b$ is chosen equal to zero, the resulting generator then being called a *multiplicative generator*. A much used example was the *RANDU* generator of the older *IBM* Scientific Subroutine Package with multiplier $a = 65,539 = 2^{16} + 3$ and modulus $c = 2^{31}$; the *IBM* System 360 Uniform Random Number Generator uses the multiplier $a = 16,807 = 7^5$ and modulus $c = 2^{31} - 1$, which is a prime number.

The reader is referred to specialist textbooks for an extensive discussion on pseudo-random number generators and their properties. In Section 9 we shall mention some basic tests for checking their statistical properties and reliability. For the remainder of this section we shall assume that we have a subroutine *RDN* which provides us with $U(0,1)$ uniformly distributed pseudo-random numbers by means of (3.2) and a generator (3.1). We shall show how we can then use this subroutine to generate pseudo-random numbers with other commonly encountered distributions, in particular those described in Section 2.

A two-point random variable $X$ (see Example 1.2.1) taking values $x_1 < x_2$ with probabilities $p_1$ and $p_2 = 1 - p_1$ can be generated easily from a $U(0,1)$ random variable $U$, namely with

$$(3.3) \qquad X = \begin{cases} x_1 & : \quad 0 \le U \le p_1 \\ x_2 & : \quad p_1 < U \le 1. \end{cases}$$

This idea extends readily to an $N$-state random variable $X$ taking values $x_1 < x_2 < \cdots < x_N$ with nonzero probabilities $p_1, p_2, \ldots, p_N$ where $\sum_{i=1}^{N} p_i = 1$. With $s_0 = 0$ and $s_j = \sum_{i=1}^{j} p_i$ for $j = 1, 2, \ldots, N$ we set $X = x_{j+1}$ if $s_j < U \le s_{j+1}$ for $j = 0, 1, 2, \ldots, N - 1$. We could approximate an infinite-state Poisson distributed random variable (see Example 1.2.2) by an $N$-state random variable for some large $N$ by coalescing all of the remaining, less probable, states into the final state. However, computationally more efficient algorithms for Poisson random variables have been developed, but will not be given here.

The corresponding method for a continuous random variable $X$ requires the probability distribution function $F_X$ to be inverted when this is possible. For a number $0 < U < 1$ we define $x(U)$ by $U = F_X(x(U))$, so $x(U) = F_X^{-1}(U)$ if $F_X^{-1}$ exists, or in general

$$(3.4) \qquad x(U) = \inf\{x : U \le F_X(x)\}.$$

This is called the *inverse transform method* and is best used when (3.4) is easy to evaluate. For example, the exponential random variable with parameter $\lambda > 0$ (see Example 1.2.4) has an invertible distribution function with

$$x(U) = F_X^{-1}(U) = -\ln(1 - U)/\lambda \quad \text{for} \quad 0 < U < 1.$$

When $U$ is $U(0,1)$ distributed, then so too is $1 - U$ and a little computational effort can be spared by using *RDN* directly to generate $1 - U$.

In principle the inverse transform method could be used for any continuous random variable, but may require too much computational effort to evaluate

(3.4). This is the situation with the standard Gaussian random variable, since the integrals for its distribution function must be evaluated numerically. The *Box-Muller method* for generating standard Gaussian random variables avoids this problem. It is based on the observation that if $U_1$ and $U_2$ are two independent $U(0,1)$ uniformly distributed random variables, then $N_1$ and $N_2$ defined by

$$(3.5) \qquad \begin{aligned} N_1 &= \sqrt{-2\ln(U_1)}\cos(2\pi U_2) \\ N_2 &= \sqrt{-2\ln(U_1)}\sin(2\pi U_2) \end{aligned}$$

are two independent standard Gaussian random variables. This can be verified with a change of coordinates from cartesian coordinates $(N_1, N_2)$ to polar coordinates $(r, \theta)$ and then to $U_1 = \exp(-\frac{1}{2}r^2)$ and $U_2 = \theta/2\pi$ (see Exercise 1.4.11).

A variation of the Box-Muller method which avoids the time consuming calculation of trigonometric functions is the *Polar Marsaglia method*. It is based on the facts that $V = 2U - 1$ is $U(-1,1)$ uniformly distributed if $U$ is $U(0,1)$ distributed and that for two such random variables $V_1$ and $V_2$ with $W = V_1^2 + V_2^2 \leq 1$, $W$ is $U(0,1)$ distributed and $\theta = \arctan(V_1/V_2)$ is $U(0,2\pi)$ distributed. Since the inscribed unit circle has $\pi/4$ of the area of the square $[-1,1]^2$, the point $(V_1, V_2)$ will take values inside this circle with probability $\pi/4 \approx 0.7864816\cdots$. We only consider these points, discarding the others. Using

$$\cos\theta = \frac{V_1}{\sqrt{W}}, \quad \sin\theta = \frac{V_2}{\sqrt{W}}$$

when $W = V_1^2 + V_2^2 \leq 1$ we can rewrite (3.5) (see Exercise 1.4.11) as

$$(3.6) \qquad \begin{aligned} N_1 &= V_1\sqrt{-2\ln(W)/W} \\ N_2 &= V_2\sqrt{-2\ln(W)/W} \end{aligned}$$

Although a proportion of the generated uniformly distributed numbers are discarded, this method is often computationally more efficient than the Box-Muller method when a large quantity of numbers is to be generated.

We shall use the above methods in exercises in the following sections, assuming that they do indeed generate random numbers with the asserted properties. In Section 9 we shall examine the validity of this assumption.

**PC-Exercise 1.3.1**     *Write a program for two-point, exponential and Gaussian random variables based on the above methods to generate a list of pseudorandom numbers.*

## 1.4   Moments

The values taken by a random variable $X$ can vary considerably. It is useful if we can isolate some salient features of this variability, such as the average value and the way in which the values spread out about this average value. The first of these is an average weighted by the likelihood of occurence and is usually called the *mean value* or *expected value* and denoted by $E(X)$. For a discretely distributed random variable it is defined as

$$(4.1) \qquad\qquad E(X) = \sum_{i \in I} x_i p_i,$$

where $I = \{0, \pm 1, \pm 2, \ldots\}$, which is readily suggested by the relative frequency interpretation of the probabilities; the summation is over all possible index values taken by the random variable.

When the random variable has an absolutely continuous distribution the corresponding definition for its mean value is

$$(4.2) \qquad\qquad E(X) = \int_{-\infty}^{\infty} x\, p(x)\, dx,$$

since, roughly speaking, $p(x)\, dx$ is the probability that $X$ takes its value in the interval $(x, x + dx)$. This assumes that the improper integral (4.2), and the infinite series (4.1) if countably infinite indices are involved, actually converge. This does not happen, for example, with the Cauchy probability density function

$$(4.3) \qquad\qquad p(x) = \frac{1}{\pi(1 + x^2)}$$

since it gives too much weighting to large values of $x$.

**Exercise 1.4.1**    *Show that the improper integral (4.2) diverges for the Cauchy density function (4.3).*

For a particular random variable $X$ we often use the notation $\mu = E(X)$ for the mean value. A measure of the spread about $\mu$ of the values taken by $X$ is given by its *variance* which is defined as

$$(4.4) \qquad\qquad \mathrm{Var}(X) = E\left((X - \mu)^2\right),$$

provided that the infinite series or improper integral involved converges. The variance is consequently always nonnegative and is often denoted by $\sigma^2 = \mathrm{Var}(X)$, where $\sigma$ is called the standard deviation of $X$. For the commonly encountered probability distributions considered in Section 2 the mean values and variances all exist and are listed in Table 1.4.1.

**Exercise 1.4.2**    *Expand $E\left((X - \mu)^2\right)$ to show that $\mathrm{Var}(X) = E(X^2) - \mu^2$.*

**Exercise 1.4.3**    *Verify the means and variances in Table 1.4.1.*

| Distribution | Example | Mean $\mu$ | Variance $\sigma^2$ |
|:---:|:---:|:---:|:---:|
| Poisson | 1.2.2 | $\lambda$ | $\lambda$ |
| Uniform $U(a,b)$ | 1.2.3 | $(a+b)/2$ | $(b-a)^2/12$ |
| Exponential | 1.2.4 | $\lambda^{-1}$ | $\lambda^{-2}$ |
| Standard Gaussian | 1.2.5 | 0 | 1 |

**Table 1.4.1**   Means and variances of some common probability distributions.

**PC-Exercise 1.4.4**    *Use the RDN generator on your PC to generate $N = 10^4$ $U(0,1)$ uniformly distributed pseudo-random numbers. Partition the interval $[0, 1]$ into subintervals of equal length $5 \times 10^{-2}$ and count the number of generated numbers falling into each subinterval. Plot a histogram of the relative frequencies divided by the subinterval length. Does this histogram resemble the graph of the density function of a $U(0,1)$ random variable? In addition, evaluate the sample average $\hat{\mu}_N$ and sample variance $\hat{\sigma}_N^2$ for the generated numbers $x_1, x_2, \ldots, x_N$, where*

$$(4.5) \qquad \hat{\mu}_N = \frac{1}{N} \sum_{i=1}^{N} x_i \quad and \quad \hat{\sigma}_N^2 = \frac{1}{N-1} \sum_{i=1}^{N} (x_i - \hat{\mu}_N)^2 .$$

*How do these compare with the mean $1/2$ and variance $1/12$ of a truly $U(0,1)$ distributed random variable? What can you conclude about the statistical reliability of your RDN generator from this simple test? (Note that we divide by $N - 1$ here rather than by $N$ in the sample variance in (4.5 ) as this provides an unbiased estimator of the true variance.)*

**PC-Exercise 1.4.5**    *Repeat PC-Exercise 1.4.4 for exponentially distributed random numbers with the parameter value $\lambda = 2.0$. Since the density function $p(x) > 0$ for all $x \geq 0$, you could partition the finite interval $[0, 2]$, say, into subintervals of equal length, with all of the values larger than 2 being discarded.*

**PC-Exercise 1.4.6**    *Repeat PC-Exercise 1.4.4 for standard Gaussian random numbers generated by the Box-Muller and Polar Marsaglia methods. For this, partition the interval $[-2.5, 2.5]$, say, into $10^2$ subintervals of equal length, with fixed semi-infinite intervals $(-\infty, -2.5)$ and $(2.5, \infty)$ for the other values.*

The standard Gaussian distribution considered in Example 1.2.5 is a special case of the general Gaussian distribution with mean $\mu$ and variance $\sigma^2$. Its density function is given by

$$(4.6) \qquad p(x) = \frac{1}{\sqrt{2\pi}\,\sigma} \exp\left(\frac{-(x-\mu)^2}{2\sigma^2}\right)$$

and still has the bell-shaped graph, which is now centered on the mean value $x = \mu$ and is stretched or compressed according to the magnitude of $\sigma^2$, its maximum value being $1/(\sqrt{2\pi}\,\sigma)$ and its points of inflection at $\mu \pm \sigma$. We often write $X \sim N(\mu; \sigma^2)$ for a random variable with this distribution. It is not hard to see then that the transformed random variable $Z = (X - \mu)/\sigma$ satisfies the standard Gaussian distribution. The inverse transformation $X = \sigma Z + \mu$ converts a standard Gaussian random variable into an $N(\mu; \sigma^2)$ distributed random variable $X$. We can use this to obtain an $N(\mu; \sigma^2)$ distributed pseudo-random variable $X$ from the output of the Box-Muller or Polar Marsaglia methods.

We remark that in general we obtain another random variable when we combine random variables by the usual arithmetic operations. For a general transformation of a random variable, however, we need some restriction on the transforming function to ensure that the resulting variable satisfies the condition (2.2). Let us define $Y(\omega) = g(X(\omega))$ for all $\omega \in \Omega$, where $X$ is a random variable and $g : \Re \to \Re$ a function. Then in order that $Y$ be a random variable we require $\{\omega \in \Omega : Y(\omega) \in B\}$, or equivalently $\{\omega \in \Omega : X(\omega) \in g^{-1}(B)\}$, to be an event for every Borel subset $B$ of $\Re$. This is certainly the case when $g$ is continuous or piecewise continuous; in fact, $g$ could be even more irregular. When $Y$ is a random variable its expected value, if it exists, is just

$$(4.7) \qquad \sum_{i \in I} g(x_i)p_i \quad \text{or} \quad \int_{-\infty}^{\infty} g(x)p(x)\,dx$$

when the distribution of $X$ is discrete or absolutely continuous. Typical functions often considered are the polynomials $g(x) = x^p$ or $g(x) = (x - \mu)^p$ for integers $p \geq 1$. The resulting expected values of $Y = g(X)$ are then called the *pth-moment* or the *pth-central moment*; the variance $\text{Var}(X)$ is thus the 2nd *order* or *squared central moment* of $X$. Such moments convey information about the random variable, but the higher order moments need not always provide additional information; for example, the Gaussian distribution is completely characterized by its first two moments, its mean $\mu$ and variance $\sigma^2$, and the Poisson distribution by its mean $\lambda$.

**Exercise 1.4.7**   *For $X \sim N(\mu; \sigma^2)$ and $k = 0, 1, 2, \ldots$ show that $E((X - \mu)^{2k+1}) = 0$ and $E((X - \mu)^{2k}) = 1 \cdot 3 \cdot 5 \cdots (2k - 1)\sigma^{2k}$, but $E(|X|^{-1}) = \infty$.*

From the basic properties of integrals, or of infinite series in the discrete case, we always have

$$(4.8) \qquad E(\alpha X_1 + \beta X_2) = \alpha E(X_1) + \beta E(X_2)$$

for any two random variables $X_1$, $X_2$ and any two real numbers $\alpha$, $\beta$, provided the mean values here all exist. Also when $X_1(\omega) \leq X_2(\omega)$, almost surely, we have

$$(4.9) \qquad E(X_1) \leq E(X_2)$$

Moreover, *Jensen's inequality*

$$(4.10) \qquad g(E(X)) \leq E(g(X))$$

holds for any convex function $g : \Re \to \Re$, that is satisfying

$$g(\lambda x + (1 - \lambda)y) \leq \lambda g(x) + (1 - \lambda)g(y)$$

for all $x$, $y \in \Re$ and $0 \leq \lambda \leq 1$. In particular, for $g(x) = |x|$ and $g(x) = x^2$ it gives, respectively,

(4.11) $$|E(X)| \leq E(|X|) \quad \text{and} \quad |E(X)| \leq \sqrt{E(X^2)}$$

A generalization of this is the *Lyapunov inequality*: if $X$ is not concentrated on a single point and if $E(|X|^s)$ exists for some $s > 0$, then for all $0 < r < s$ and $a \in \Re$

(4.12) $$\left(E(|X - a|^r)\right)^{1/r} \leq \left(E(|X - a|^s)\right)^{1/s} .$$

For a nonnegative random variable $X$, that is with $X(\omega) \geq 0$ almost surely, we have the *Markov inequality*

(4.13) $$P\left(\{\omega : X(\omega) \geq a\}\right) \leq \frac{1}{a} E(X) \quad \text{for all} \quad a > 0.$$

From this we can deduce the *Chebyshev inequality*

(4.14) $$P\left(\{\omega; |X(\omega)|^2 \geq a\}\right) \leq \frac{1}{a} E(X^2) \text{ for all} \quad a > 0.$$

**Exercise 1.4.8**   *Prove the Markov and Chebyshev inequalities.*

The mean value or expectation $E(X)$ is the coarsest estimate that we have for a random variable $X$. If we know that some event has occured we may be able to improve on this estimate. For instance, suppose that the event $A = \{\omega \in \Omega : a \leq X(\omega) \leq b\}$ has occured. Then in evaluating our estimate of the value of $X$ we need only consider these values of $X$ and weight them according to their likelihood of occurence, which is now the conditional probability given this event. The resulting estimate is called the *conditional expectation* of $X$ given event $A$ and is denoted by $E(X|A)$. For a discretely distributed random variable the conditional probabilities for the event $A = \{\omega \in \Omega : a \leq X(\omega) \leq b\}$ satisfy

$$P(X = x_i|A) = \begin{cases} 0 & : \quad x_i < a \text{ or } b < x_i \\ p_i \Big/ \sum_{a \leq x_j \leq b} p_j & : \quad a \leq x_i \leq b \end{cases}$$

and so the conditional expectation

(4.15) $$E(X|A) = \sum_{i \in I} x_i P(X = x_i|A) = \sum_{a \leq x_i \leq b} x_i \, p_i \Big/ \sum_{a \leq x_i \leq b} p_i .$$

For a continuously distributed random variable with a density function $p$ the corresponding *conditional density* is

$$p(x|A) = \begin{cases} 0 & : \quad x < a \text{ or } b < x \\ p(x) \Big/ \int_a^b p(s) \, ds & : \quad a \leq x \leq b \end{cases}$$

with the conditional expectation

(4.16)     $E(X|A) = \int_{-\infty}^{\infty} x p(x|A)\, dx = \int_a^b x\, p(x)\, dx \Big/ \int_a^b p(x)\, dx$ .

Similar results hold for other kinds of events provided that they occur with nonzero probability. The conditional expectations so obtained satisfy properties analogous to (4.8)–(4.11) for the usual expectations. Additional properties of this important concept can be found in Section 2 of Chapter 2.

**Exercise 1.4.9**   *Show that a random variable $X$ distributed exponentially with parameter $\lambda$ (see Exercise 1.2.4) has conditional expectations*

$$E(X|X \ge a) = a + E(X) = a + \lambda^{-1}$$

*for any $a > 0$. Here the event $A = \{\omega \in \Omega : X(\omega) \ge a\}$ is abbreviated to $\{X \ge a\}$. (This identity characterizes the exponential distribution; it says, for example, that the expected remaining life of a light bulb is independent of how old the bulb is.)*

**PC-Exercise 1.4.10**   *Generate $10^3$ exponentially distributed random numbers with parameter $\lambda = 0.5$. Calculate the average of these numbers and the averages of those numbers $\ge a$ where $a = 1$, 2, 3 and 4. Compare these with the identity in Exercise 1.4.9.*

Often we need to consider several random variables $X_1$, $X_2$, ..., $X_n$ at the same time. These may, for example, be the components of a vector-valued random variable $X : \Omega \to \Re^n$ or of a matrix-valued random variable, which are then called a *random vector* or a *random matrix*, respectively. As with a single random variable, we can form a distribution function for $n$ random variables $X_1$, $X_2$, ..., $X_n$ which are defined on the same probability space; if they are not, we can always modify them to have a common probability space. The distribution function $F_{X_1 X_2 \cdots X_n} : \Re^n \to [0, 1]$, called the *joint distribution function*, is defined by

(4.17)        $F_{X_1 X_2 \cdots X_n}(x_1, x_2, \ldots, x_n)$

$$= P(\{\omega \in \Omega : X_i(\omega) \le x_i,\ i = 1, 2, \ldots, n\})$$

Its properties are most apparent in the case of two random variables $X_1$ and $X_2$. Then $F_{X_1 X_2}$ satisfies

(4.18)                    $\lim_{x_i \to -\infty} F_{X_1 X_2}(x_1, x_2) = 0$

for $i = 1$ or 2,

(4.19)                    $\lim_{x_1, x_2 \to \infty} F_{X_1 X_2}(x_1, x_2) = 1$

and

(4.20)  $F_{X_1 X_2}$ is nondecreasing and continuous from the right in $x_1$ and $x_2$.

Moreover

(4.21) $$F_{X_1 X_2}(x_1, x_2) = F_{X_2 X_1}(x_2, x_1)$$

and for $i_1 = 1$ or $2$ and $i_2 = 2$ or $1$ the *marginal distribution* $F_{X_{i_1}}$ satisfies

(4.22) $$F_{X_{i_1}}(x_{i_1}) = \lim_{x_{i_2} \to \infty} F_{X_{i_1} X_{i_2}}(x_{i_1}, x_{i_2}).$$

Unlike the case of a distribution function of a single variable, properties (4.18)–(4.22) for a collection of functions must be strengthened to ensure that there exists a pair of random variables $X_1$, $X_2$ which has these functions as its joint and marginal distributions. For continuous random variables the joint distribution function is often differentiable, except possibly at some isolated or boundary points, and there is a density function $p : \Re^2 \to \Re^+$ given by

$$p(x_1, x_2) = \frac{\partial^2 F_{X_1 X_2}}{\partial x_1 \partial x_2}(x_1, x_2),$$

in which case

$$F_{X_1 X_2} = \int_{-\infty}^{x_1} \int_{-\infty}^{x_2} p(s_1, s_2) \, ds_1 ds_2.$$

When the density function is

$$p(x_1, x_2) = \frac{\sqrt{\det C}}{2\pi} \exp\left( -\frac{1}{2} \sum_{i,j=1}^{2} c^{i,j}(x_i - \mu_i)(x_j - \mu_j) \right)$$

for some vector $\mu = (\mu_1, \mu_2) \in \Re^2$ and some 2×2 positive definite and symmetric matrix (thus with two real, positive eigenvalues) $C = [c^{i,j}]$ with determinant $\det C$, we say that the random variables are *jointly Gaussian* with *mean vector* $\mu$ and *covariance matrix* $C^{-1}$, the *ijth* component of $C^{-1}$ being $E((X_i - \mu_i)(X_j - \mu_j))$ for $i, j = 1, 2$. Since there exists an invertible $2 \times 2$ matrix $S$ such that $C = S^\top S$, the vector $X = S^\top Z + \mu$ is jointly Gaussian with mean vector $\mu$ and covariance matrix $C^{-1} = S^{-1}(S^{-1})^\top$ whenever $Z = (Z_1, Z_2)$ has independent standard Gaussian components $Z_1$ and $Z_2$. The Box-Muller or Polar Marsaglia methods can thus be used to generate pairs of Gaussian pseudo-random numbers with any mean vector and covariance matrix. This reflects the important property that linear and orthogonal transformations of Gaussian random variables yield Gaussian random variables.

**Exercise 1.4.11** *Show that the random variables $N_1$ and $N_2$ generated by the Box-Muller and Polar Marsaglia methods (see (3.5) and (3.6)) are Gaussian with zero mean vector and identity covariance matrix when $U_1$ and $U_2$ are independent $U(0, 1)$ uniformly distributed random variables.*

**PC-Exercise 1.4.12** *Write a program to generate a pair of Gaussian pseudo-random numbers $X_1$, $X_2$ with means $\mu_1 = E(X_1) = 0$, $\mu_2 = E(X_2) = 0$ and covariances $E(X_1^2) = h$, $E(X_2^2) = h^3/3$ and $E(X_1 X_2) = h^2/2$ for any $h > 0$. Generate $10^3$ pairs of such numbers for $h = 0.1$, $1$ and $10$, and evaluate the sample averages and sample covariances, comparing these with the above values.*

We say that two random variables $X_1$ and $X_2$ are *independent* if their joint and marginal distribution functions satisfy

$$(4.23) \qquad F_{X_1 X_2}(x_1, x_2) = F_{X_1}(x_1) F_{X_2}(x_2)$$

for all $x_1$, $x_2 \in \Re$. This is equivalent to the independent- event definition given at the end of Section 2. If both $F_{X_1}$ and $F_{X_2}$ have density functions $p_1$ and $p_2$, respectively, and if $X_1$ and $X_2$ are independent, then their joint distribution function $F_{X_1 X_2}$ has a density function $p$ which satisfies

$$(4.24) \qquad p(x_1, x_2) = p_1(x_1) p_2(x_2).$$

Moreover, the product $X_1 X_2$ of two independent random variables $X_1$ and $X_2$ has expectation given by

$$(4.25) \qquad E(X_1 X_2) = E(X_1) E(X_2)$$

and the sum $X_1 + X_2$ has variance

$$(4.26) \qquad \text{Var}(X_1 + X_2) = \text{Var}(X_1) + \text{Var}(X_2),$$

if these moments exist. Hence for two independent random variables both the means and variances are additive, and the means are also multiplicative.

**PC-Exercise 1.4.13**    *Check numerically whether or not the jointly Gaussian random variables $X_1$, $X_2$ with means and covariances given in PC-Exercise 1.4.12 might be independent.*

**Exercise 1.4.14**    *Verify analytically that the random variables $X_1$ and $X_2$ in PC-Exercise 1.4.12 are dependent.*

The properties (4.18)–(4.22) generalize readily to any number $n \geq 2$ of random variables $X_1, X_2, \ldots, X_n$, but require somewhat more complicated terminology if they are to be expressed succinctly. We let $\{i_1, i_2, \ldots, i_k\}$ be any subset of $\{1, 2, \ldots, n\}$ for $2 \leq k \leq n$. Then the joint distributions $F_{X_{i_1} X_{i_2} \cdots X_{i_n}}$ and the marginal distributions $F_{X_{i_1} X_{i_2} \cdots X_{i_k}}$ defined by (4.17) satisfy

$$(4.27) \qquad \lim_{\substack{x_i \to -\infty \\ \exists i = 1, 2, \ldots, n}} F_{X_1 X_2 \cdots X_n}(x_1, x_2, \ldots, x_n) = 0;$$

$$(4.28) \qquad \lim_{\substack{x_i \to +\infty \\ \forall i = 1, 2, \ldots, n}} F_{X_1 X_2 \cdots X_n}(x_1, x_2, \ldots, x_n) = 1;$$

(4.29) $F_{X_1 X_2 \cdots X_n}$ is nonincreasing and continuous from the right in $x_i$

for $i = 1, 2, \ldots, n$;

$$(4.30) \qquad F_{X_{i_1} X_{i_2} \cdots X_{i_n}}(x_{i_1}, x_{i_2}, \cdots, x_{i_n}) = F_{X_1 X_2 \cdots X_n}(x_1, x_2, \ldots, x_n);$$

with the marginal distributions $F_{X_{i_1} X_{i_2} \cdots X_{i_k}}$ for $1 \leq k \leq n$ satisfying

(4.31)
$$F_{X_{i_1} X_{i_2} \cdots X_{i_k}}(x_{i_1}, x_{i_2}, \cdots, x_{i_k})$$
$$= \lim_{\substack{x_i \to +\infty \\ \forall i \neq i_1, i_2, \ldots, i_k}} F_{X_1 X_2 \cdots X_n}(x_1, x_2, \ldots, x_n).$$

Similarly the definition and properties (4.23)–(4.26) generalize to $n \geq 2$ independent random variables $X_1$, $X_2$, ..., $X_n$, with independence occuring if all of the joint and marginal distribution functions satisfy

(4.32)  $F_{X_{i_1} X_{i_2} \cdots X_{i_k}}(x_{i_1}, x_{i_2}, \cdots, x_{i_k}) = F_{X_{i_1}}(x_{i_1}) F_{X_{i_2}}(x_{i_2}) \cdots F_{X_{i_k}}(x_{i_k})$

for all $\{i_1, i_2, \ldots, i_k\} \subseteq \{1, 2, \ldots, n\}$ and $1 \leq k \leq n$. If each $F_{X_i}$ has density function $p_i$, then $F_{X_1 X_2 \cdots X_n}$ has density function $p$ given by

(4.33)
$$p(x_1, x_2, \ldots, x_n) = p(x_1) p(x_2) \cdots p(x_n).$$

In addition, for $n$ independent random variables the product $X_1 X_2 \cdots X_n$ has expectation

(4.34)
$$E(X_1 X_2 \cdots X_n) = E(X_1) E(X_2) \cdots E(X_n),$$

whereas the sum $\sum_{i=1}^{n} X_i$ has variance

(4.35)
$$\operatorname{Var}\left(\sum_{i=1}^{n} X_i\right) = \sum_{i=1}^{n} \operatorname{Var}(X_i),$$

provided these quantities exist.

When we are considering $n$ different random variables $X_1$, $X_2$, ..., $X_n$ at the same time it is often convenient to use vector notation. For vectors $x = (x_1, \ldots, x_n)$ and $y = (y_1, \ldots, y_n)$ in $\Re^n$ we recall that the *Euclidean norm* $|x|$ and *inner product* $(x, y)$ are defined by

(4.36)
$$|x| = \sqrt{\sum_{i=1}^{n} x_i^2} \quad \text{and} \quad (x, y) = \sum_{i=1}^{n} x_i y_i,$$

respectively. We note that for $n = 1$ the Euclidean norm coincides with the absolute value. It can be shown algebraically that these satisfy the *triangle inequality*

(4.37)
$$|x + y| \leq |x| + |y|$$

and the *Cauchy-Schwarz inequality*

(4.38)
$$|(x, y)| \leq |x| \, |y|$$

for all $x$ and $y$ in $\Re^n$. The former also holds for other norms on $\Re^n$ and indeed for many vector spaces, such as a function space, on which a norm can be defined; the latter holds whenever the vector space has an inner product compatible with a norm, that is with $|x| = \sqrt{(x, x)}$ for all vectors $x$ in the space. An integral version of (4.38) is

$$(4.39) \qquad \left| \int_a^b f(x)g(x)\,dx \right|^2 \leq \int_a^b f^2(x)\,dx \int_a^b g^2(x)\,dx,$$

provided these integrals are meaningful. A generalization of this is the *Hölder inequality*

$$(4.40) \qquad \left| \int_a^b f(x)g(x)\,dx \right| \leq \left( \int_a^b f^p(x)\,dx \right)^{1/p} \left( \int_a^b g^q(x)\,dx \right)^{1/q}$$

where $1/p + 1/q = 1$ with $p, q > 1$, provided the integrals exist.

The following moment inequalities for random vectors $X = (X_1, \ldots, X_n)$ and $Y = (Y_1, \ldots, Y_n)$ are often useful:

$$(4.41) \qquad E(|X + Y|^r) \leq c_r \left( E(|X|^r) + E(|Y|^r) \right)$$

with $c_r = 1$ for $r \leq 1$ and $c_r = 2^{r-1}$ for $r \geq 1$;

$$(4.42) \qquad (E(|X + Y|^r))^{1/r} \leq (E(|X|^r))^{1/r} + (E(|Y|^r))^{1/r}$$

for $r \geq 1$; and

$$(4.43) \qquad E(|(X, Y)|) \leq (E(|X|^p))^{1/p} (E(|Y|^q))^{1/q}$$

for $p, q > 1$ with $1/p + 1/q = 1$.

**Exercise 1.4.15**    *Prove inequalities (4.41)–(4.43).*

## 1.5    Convergence of Random Sequences

In many situations we have an infinite sequence $X_1, X_2, \ldots, X_n, \ldots$ of random variables and are interested in their asymptotic behaviour, that is in the existence of a random variable $X$ which is the limit of the $X_n$ in some sense. There are several different ways in which such convergence can be defined. Broadly speaking these fall into two classes, a stronger one in which the realizations of $X_n$ are required to be close in some way to those of $X$ and a weaker one in which only their probability distributions need be close. We can assume without loss of generality that the random variables are all defined on a common probability space.

The following three convergences are the most commonly used ones in the strong class.

**I.**  *Convergence with probability one* (w.p.1):

$$(5.1) \qquad P\left( \left\{ \omega \in \Omega : \lim_{n \to \infty} |X_n(\omega) - X(\omega)| = 0 \right\} \right) = 1.$$

This is also known as *almost sure convergence*.

**II.** *Mean-square convergence*: $E(X_n^2) < \infty$ for $n = 1, 2, \ldots$, $E(X^2) < \infty$ and

(5.2) $$\lim_{n \to \infty} E\left(|X_n - X|^2\right) = 0.$$

**III.** *Convergence in probability*:

(5.3) $$\lim_{n \to \infty} P\left(\{\omega \in \Omega; |X_n(\omega) - X(\omega)| \geq \epsilon\}\right) = 0 \text{ for all } \epsilon > 0.$$

Convergence in probability is also called *stochastic convergence*. It is equivalent to

(5.4) $$\lim_{n \to \infty} E\left(\frac{|X_n - X|}{1 + |X_n - X|}\right) = 0,$$

which is sometimes more convenient since the expectation in (5.4) defines a distance function, known as a metric, between the random variables.

Both **(I)** convergence w.p.1 and **(II)** mean-square convergence imply **(III)** convergence in probability, the latter following directly from the Chebyshev inequality (4.14):

$$P\left(|X_n - X| \geq \epsilon\right) = P\left(|X_n - X|^2 \geq \epsilon^2\right) \leq \frac{1}{\epsilon^2} E\left(|X_n - X|^2\right).$$

Generally **(III)** does not imply either **(I)** or **(II)**, nor does either of **(I)** or **(II)** imply the other. This can be demonstrated by counterexamples, such as in the following exercise.

**Exercise 1.5.1**  Let $\Omega = [0, 1]$ and let $P([a, b]) = |b - a|$ for any subinterval $[a, b] \subseteq [0, 1]$. For $n = 1, 2, 3, \ldots$ let $A_n = \{\omega \in \Omega : 0 \leq \omega \leq 1/n\}$ and define $X_n = \sqrt{n} I_{A_n}$. Evaluate $E(X_n^2)$ and $P(X_n \geq \epsilon)$ for all $\epsilon > 0$ and $n = 1, 2, 3, \ldots$. Hence deduce that $X_n$ converges to 0 in probability, but not in the mean-square sense.

If, however, $|X_n(\omega)| \leq |Y(\omega)|$, w.p.1, for some random variable $Y$ with $E(Y^2) < \infty$, then **(I)** implies **(II)** and **(II)** is equivalent to **(III)**. This is a consequence of the Dominated Convergence Theorem, which will be stated in Section 2 of Chapter 2 (Theorem 2.2.3).

For the class of weaker convergences we do not need to know the actual random variables or the underlying probability space(s), but rather we only need to know the distribution functions. We mention here the following convergences from this class.

**IV.** *Convergence in distribution*:

(5.5) $$\lim_{n \to \infty} F_{X_n}(x) = F_X(x) \quad \text{at all continuity points of } F_X.$$

This is also known as *convergence in law* and refers to the distributions of the random variables $X_n$ and $X$.

**V.** *Weak convergence*:

(5.6) $$\lim_{n \to \infty} \int_{-\infty}^{\infty} f(x) \, dF_{X_n}(x) = \int_{-\infty}^{\infty} f(x) \, dF(x)$$

for all *test functions* $f : \Re \to \Re$, usually continuous functions vanishing outside
of a bounded interval which may depend on the particular function. Sometimes
it is useful to consider weak convergence with respect to other sets of test
functions. The integrals in (5.6) are (improper) Riemann-Stieltjes integrals,
about which we shall say more in Chapter 2. If the distribution functions have
densities, then (5.6) reduces to the more familiar (improper) Riemann integrals:

$$(5.7) \qquad \lim_{n \to \infty} \int_{-\infty}^{\infty} f(x) p_n(x)\, dx = \int_{-\infty}^{\infty} f(x) p(x)\, dx.$$

While they may appear more complicated, the integral limits (5.6) and (5.7)
are often easier to verify in theoretical contexts than are the pointwise limits
(5.5).

Convergence in probability (**III**) implies convergence in distribution (**IV**),
but the converse implication does not hold as the following counterexample
shows.

**Example 1.5.2**    *Consider two independent two-point distributed random
variables $X$ and $Y$, both taking values 0 and 1 with equal probabilities 1/2,
and let $X_n = Y$ for $n = 1, 2, 3, \ldots$. Then $X_n$ and $X$ have the same distributions, so they obviously converge in distribution, However, they do not converge
in probability because $P(|X_n - X| = 1) = 1/2$ for $n = 1, 2, 3, \ldots$.*

**PC-Exercise 1.5.3**    *Define $Y_n = X + Z_n$ where $X$ is a $U(0,1)$ uniformly
distributed random variable and $Z_n$ an independent $N(0; 1/n)$ Gaussian random
variable for $n = 1, 2, 3, \ldots$. For fixed but large $n$, say $n = 10^3$, use the
programs from PC-Exercise 1.3.1 to generate sufficiently many realizations of
$Y_n$. Calculate relative frequencies and plot a histogram to obtain an estimation
of the graph of the density function of $Y_n$. Then estimate the mean-square error
$E(|Y_n - X|^2)$ for large $n$. Does this suggest that the $Y_n$ converge to $X$ in the
mean-square sense?*

The intuitive idea of defining probabilities as the limits of relative frequencies determined from many repetitions of a given probabilistic experiment re-
ceives some theoretical justification from an asymptotic analysis of sequences
of *independent identically distributed* (i.i.d.) random variables $X_1, X_2, X_3, \ldots$.
Let $\mu = E(X_n)$ and $\sigma^2 = \text{Var}(X_n)$. From this independence it follows (see
(4.35)) that the averaged random variables

$$A_n = \frac{1}{n} S_n = \frac{1}{n} (X_1 + X_2 + \cdots + X_n)$$

also have mean $E(A_n) = \mu$ and variance $\text{Var}(A_n) = \sigma^2/n$. The *Law of Large
Numbers*, one of the fundamental theorems of probability and mathematical
statistics, says that

$$(5.8) \qquad\qquad A_n \to \mu \quad \text{as} \quad n \to \infty$$

with convergence in probability (**III**) in its weak version and convergence with
probability one (**I**) or in mean-square (**II**) in its strong versions.

**PC-Exercise 1.5.4**   *Use the random number generators in Section 3 for uniform, two-point, exponential and Gaussian random numbers to verify the limit (5.8) in the mean-square sense by averaging $(A_n - \mu)^2$ for sufficiently many different runs, say $10^2$, and for $n = 10$, $10^2$ and $10^3$.*

Another fundamental result, the *Central Limit Theorem*, says that the normalized random variables

$$(5.9) \qquad Z_n = \frac{(S_n - n\mu)}{\sigma\sqrt{n}},$$

for which $E(Z_n) = 0$ and $\text{Var}(Z_n) = 1$, converge in distribution to a standard Gaussian random variable $Z$. This is also true under weaker assumptions than the i.i.d. assumption on the original random variables $X_1$, $X_2$, $X_3$, ... and provides an explanation for the dominant role of the Gaussian distribution in probability and statistics.

**Exercise 1.5.5**   *Show that the Central Limit Theorem provides more information than the Law of Large Numbers about the limiting behaviour of the sequence $A_1$, $A_2$, $A_3$, ....*

**PC-Exercise 1.5.6**   *As in PC-Exercise 1.5.4 generate sequences of random numbers and calculate their $Z_n$ values from (5.9). Using histograms of relative frequency plots obtain simulated approximations to the graphs of the density functions for the $Z_n$ and show that these approach the graph of the density function of the standard Gaussian distribution as n is taken larger and larger, specifically for $n = 100$, $500$ and $1000$.*

The *Bernoulli trials* are an illustration of the Law of Large Numbers and the Central Limit Theorem. They are independent repetitions of an experiment with two basic outcomes, such as heads and tails when a coin is tossed, with probabilities $p$ and $1 - p$, respectively. If we let $X_n = 1$ for a head and $X_n = 0$ for a tail, then we have an i.i.d. sequence $X_1$, $X_2$, $X_3$, ... with mean $E(X_n) = p$ and variance $\text{Var}(X_n) = p(1 - p)$. The sum $S_n = X_1 + X_2 + \cdots + X_n$ is then just the number $n(H)$ of heads occuring out of $n$ tosses and $A_n = S_n/n = n(H)/n$ is the relative frequency of tossing a head in $n$ tosses. Of course, these relative frequencies are themselves random variables, but the Law of Large Numbers says that they converge to the common mean $p$, which is the probability of throwing a head in a single trial.

As another illustration consider a fair coin and let $X_n = +1$ for a head and $X_n = -1$ for a tail. Then $E(X_n) = 0$ and $\text{Var}(X_n) = 1$, so by the Central Limit Theorem the random variables $Z_n = S_n/\sqrt{n}$ defined by (5.9) converge in distribution to a standard Gaussian random variable $Z$. This is more than we get from the Law of Large Numbers for the averages $A_n = S_n/n$, namely convergence to 0. Alternatively, the random variables $S_n$ are approximately $N(0; n)$ distributed for large $n$. Such a sequence of sums of i.i.d. random variables is called a *random walk* on account of the appearance of its graph when plotted against $n$.

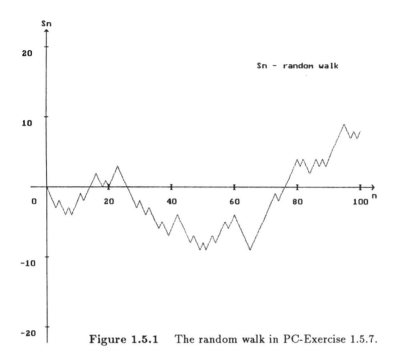

**Figure 1.5.1**    The random walk in PC-Exercise 1.5.7.

**PC-Exercise 1.5.7**    *Form a random walk $S_n$ using the two-point random number generator (3.3) and plot $S_n$ against $n$, joining the successive points in the $(n, S_n)$ plane by straight line segments. Repeat this for other sequences corresponding to different initial seeds and compare the plotted paths.*

## 1.6    Basic Ideas About Stochastic Processes

A sequence of random variables $X_1$, $X_2$, ..., $X_n$, ... often describes, or may be conveniently imagined to describe, the evolution of a probabilistic system over discrete instants of time $t_1 < t_2 < \cdots < t_n < \cdots$. We then say that it is a *stochastic process* and call the totality of its joint distribution functions $\{F_{X_{i_1} X_{i_2} \cdots X_{i_j}} : i_j = 1, 2, \ldots \text{ and } j = 1, 2, \ldots\}$ its *probability law*. Henceforth we shall write these distributions as $F_{t_{i_1} t_{i_2} \cdots t_{i_j}}$ to emphasize the role of the time instants. A sequence of i.i.d. random variables is a trivial example of a stochastic process in the sense that what happens at one instant is completely unaffected by what happens at any other, past or future, instant. Moreover its probability law is quite simple, since any joint distribution is given by

$$(6.1) \qquad F_{t_{i_1} t_{i_2} \cdots t_{i_j}}(x_{i_1}, x_{i_2}, \ldots, x_{i_j}) = F(x_{i_1}) F(x_{i_2}) \cdots F(x_{i_j}),$$

where $F$ is the common distribution function of the random variables. Another simple example are the iterates of a first order difference equation such as

$$(6.2) \qquad\qquad X_{n+1} = 4X_n(1 - X_n)$$

for $n = 1$, 2, 3, ...; here each random variable is determined exactly from its predecessor, and ultimately from the initial random variable $X_1$. For most interesting stochastic processes encountered in applications the relationship between the random variables at different time instants lies somewhere between these two extremes. For instance, the joint distributions may all be Gaussian in which case we call the process a *Gaussian process;* they may satisfy (6.1), but generally need not.

Stochastic processes may also be defined for all time instants in a bounded interval such as $[0, 1]$ or in an unbounded interval such as $[0, \infty)$, in which case we call them *continuous time* stochastic processes. As an example, let $\lambda > 0$ be a fixed parameter and for each $t > 0$ define $X(t)$ to be a Poisson distributed random variable with parameter $\lambda t$. Continuing with the radioactive decay interpretation of Example 1.2.2, $X(t)$ then represents the number of atoms decaying in the time interval $(0, t]$.

In general we shall denote the time set under consideration by $T$ and assume that there is a common underlying probability space $(\Omega, \mathcal{A}, P)$. A stochastic process $X = \{X(t), t \in T\}$ is thus a function of two variables $X : T \times \Omega \to \Re$ where $X(t) = X(t, \cdot)$ is a random variable for each $t \in T$. For each $\omega \in \Omega$ we call $X(\cdot, \omega) : T \to \Re$ a *realization*, a *sample path* or a *trajectory* of the stochastic process. The curves plotted in Figure 1.5.1 can be interpreted as the realizations of a stochastic process. The functions $X$ cannot be completely arbitrary, but must satisfy information restrictions, both at specific time instants, to ensure that random variables result, and also between different time instants. The latter are particularly crucial, mathematically speaking, for continuous time processes and will be discussed in some detail in Chapter 2.

For both continuous and discrete time sets $T$ it is useful to distinguish various classes of stochastic processes according to their specific temporal relationships. Assuming that the expressions exist, the expectations and variances

$$(6.3) \qquad \mu(t) = E(X(t)), \quad \sigma^2(t) = \text{Var}(X(t))$$

at each instant $t \in T$ and the covariances

$$(6.4) \qquad C(s, t) = E\left((X(s) - \mu(s))(X(t) - \mu(t))\right)$$

at distinct instants $s$, $t \in T$ provide some information about the time variability of a stochastic process.

**Example 1.6.1** *A* Poisson process *with intensity parameter* $\lambda > 0$ *is a stochastic process* $X = \{X(t), t \geq 0\}$ *such that (i)* $X(0) = 0$; *(ii)* $X(t) - X(s)$ *is a Poisson distributed random variable with parameter* $\lambda(t - s)$ *for all* $0 \leq s < t$; *and (iii) the increments* $X(t_2) - X(t_1)$ *and* $X(t_4) - X(t_3)$ *are independent for all* $0 < t_1 < t_2 \leq t_3 < t_4$. *Its means, variances and covariances are* $\mu(t) = \lambda t$, $\sigma^2(t) = \lambda t$ *and* $C(s, t) = \lambda \min\{s, t\}$, *respectively, for all* $s$, $t > 0$.

An important class of stochastic processes are those with *independent increments*, that is for which the random variables $X(t_{j+1}) - X(t_j)$, $j = 0$, 1, 2, ..., $n - 1$ are independent for any finite combination of time instants $t_0 <$

$t_1 < \cdots < t_n$ in $T$. If $t_0$ is the smallest time instant in $T$ then the random variables $X(t_0)$ and $X(t_j) - X(t_0)$ for any other $t_j$ in $T$ are also required to be independent. A Poisson process is an example of a continuous time stochastic process with independent increments. Another is the *standard Wiener process* $W = \{W(t), t \geq 0\}$, that is a Gaussian process with independent increments for which

$$(6.5) \qquad W(0) = 0 \quad \text{w.p.1}, \quad E(W(t)) = 0, \quad \text{Var}(W(t) - W(s)) = t - s$$

for all $0 \leq s \leq t$. In fact, it can be shown that any continuous time stochastic process with independent increments and finite second moments $E(X^2(t))$ for all $t$ is a Gaussian process provided $X(t_0)$ is Gaussian for some $t_0 \in T$.

The *stationary processes* are another interesting class of stochastic processes since they represent a form of probabilistic equilibrium in the sense that the particular instants at which they are examined are not important. We say that a process is *strictly stationary* if its joint distributions are all invariant under time displacements, that is if

$$(6.6) \qquad\qquad F_{t_1+h, t_2+h, \cdots, t_n+h} = F_{t_1, t_2, \cdots, t_n}$$

for all $t_i$, $t_i + h \in T$ where $i = 1, 2, \ldots, n$ and $n = 1, 2, 3, \ldots$; in particular, the $X(t)$ have the same distribution for all $t \in T$. For example, a sequence of i.i.d. random variables is strictly stationary. On the other hand, if there is a constant $\mu$ and a function $c : \Re \to \Re$ such that the means, variances and covariances of a stochastic process satisfy

$$(6.7) \qquad\qquad \mu(t) = \mu, \quad \sigma^2(t) = c(0) \quad \text{and} \quad C(s,t) = c(t-s)$$

for all $s$, $t \in T$, then we call the process *wide-sense stationary*. This means the process is only stationary with respect to its first and second moments. It is straightforward to show that a strictly stationary process is wide-sense stationary if its means, variances and covariances are all finite, but a wide-sense stationary process need not be strictly stationary.

**Example 1.6.2**    *The Ornstein-Uhlenbeck process $X = \{X(t), t \in \Re^+\}$ with parameter $\gamma > 0$ and initial value $X_0 \in N(0;1)$ is a Gaussian process with means $E(X(t)) = 0$ and covariances $E(X_s X_t) = \exp(-\gamma|t - s|)$ for all $s$, $t \in \Re^+$. Hence it is a wide-sense stationary process. It is also strictly stationary, as will be seen from equation (7.4).*

**Exercise 1.6.3**    *Show that a standard Wiener process $W$ has covariances $C(s,t) = \min\{s,t\}$ and hence is not wide-sense stationary.*

Conditional probabilities provide a more refined tool than mean values and covariances for analysing the relationships between the random variables of a stochastic process at different time instants. To illustrate this let us consider a sequence of random variables $X_1, X_2, X_3, \ldots$ taking values in a discrete set $\mathcal{X} = \{x_1, x_2, x_3, \ldots\}$, which may be finite or infinite. We shall suppose that we are at a time instant $n$ and know the present outcome $x_{i_n}$, that is the value of $X_n$, as well as the past outcomes $X_1 = x_{i_1}, X_2 = x_{i_2}, \ldots, X_{n-1} = x_{i_{n-1}}$. Using this information we wish to predict something about the future outcomes, the values of $X_{n+1}, X_{n+2}, \ldots$, in particular the value of the immediate future random variable $X_{n+1}$. For this the conditional probabilities

$$P\left(X_{n+1} = x_j | X_1 = x_{i_1}, X_2 = x_{i_2}, \ldots, X_n = x_{i_n}\right)$$

$$= \frac{P\left(X_1 = x_{i_1}, X_2 = x_{i_2}, \ldots, X_n = x_{i_n}, X_{n+1} = x_j\right)}{P\left(X_1 = x_{i_1}, X_2 = x_{i_2}, \ldots, X_n = x_{i_n}\right)}$$

for each $x_j \in \mathcal{X}$ are useful. Here we have abbreviated our notation for events using, for example, $\{X_1 = x_{i_1}\}$ for $\{\omega \in \Omega : X_1(\omega) = x_{i_1}\}$. In principle, we could determine such conditional probabilities for all $x_{i_1}, x_{i_2}, \ldots, x_{i_n}$ in $\mathcal{X}$ and all $n = 1, 2, 3, \ldots$. In a deterministic system governed, for instance, by a first order difference equation such as (6.2) only the present value $x_{i_n}$ of $X_n$ is needed to determine the future value of $X_{n+1}$; the past values of $X_1, X_2, \ldots, X_{n-1}$ are involved only indirectly in that they determine the value of $X_n$. This is just the common law of causality and there is a stochastic analogue called the *Markov property*, that is

(6.8) $\quad P\left(X_{n+1} = x_j | X_n = x_{i_n}\right)$

$$= \quad P\left(X_{n+1} = x_j | X_1 = x_{i_1}, X_2 = x_{i_2}, \ldots, X_n = x_{i_n}\right)$$

for all possible $x_j, x_{i_1}, x_{i_2}, \ldots, x_{i_n}$ in $\mathcal{X}$ with $n = 1, 2, 3, \ldots$. A sequence of discrete valued random variables with this property is an example of a *Markov process*; in particular, we call it a *discrete time Markov chain*.

While they are peripheral to the theme of this book, we shall look a little more closely at discrete time Markov chains with a finite number of states $\mathcal{X} = \{x_1, x_2, \ldots, x_N\}$ because they allow some important concepts to be introduced and understood in a relatively simple setting. For each $n = 1, 2, 3, \ldots$ we define an $N \times N$ matrix $P(n) = [p^{i,j}(n)]$ componentwise by

(6.9) $\qquad\qquad p^{i,j}(n) = P(X_{n+1} = x_j | X_n = x_i)$

for $i, j = 1, 2, \ldots, N$. Obviously $0 \le p^{i,j}(n) \le 1$ and, since $X_{n+1}$ can only attain states in $\mathcal{X}$,

(6.10) $\qquad\qquad \sum_{j=1}^{n} p^{i,j}(n) = 1$

for each $i = 1, 2, \ldots, N$ and $n = 1, 2, 3, \ldots$. We call the matrix $P(n)$ the *transition matrix* of the Markov chain and its components (6.9) the *transition probabilities* at time $n$. A *probability vector* on $\mathcal{X}$ is a vector $p = (p_1, p_2, \ldots, p_N) \in \Re^N$ with $0 \le p_i \le 1$ for $i = 1, 2, \ldots, N$ and $\sum_{i=1}^{N} p_i = 1$. Thus if $p(n)$ is the probability vector corresponding to the distribution of the random variable $X_n$, that is if $p_i(n) = P(X_n = x_i)$ for $i = 1, 2, \ldots, N$, then the probability vector $p(n + 1)$ corresponding to $X_{n+1}$ is related to it through the vector equation

(6.11) $$p(n + 1) = p(n)P(n).$$

Hence, if we know the initial distribution probability vector $p(1)$, then we have

(6.12) $$p(n) = p(1)P(1)P(2) \cdots P(n - 1)$$

for $n = 2, 3, \ldots$ by applying (6.11) recursively. In the case that the transition matrices are all the same , that is $P(n) \equiv P$ for $n = 1, 2, 3, \ldots$, we say that the Markov chain is *homogeneous*. For a homogeneous Markov chain (6.12) takes the form $p(n) = p(1)P^{n-1}$ or, more generally,

(6.13) $$p(n + k - 1) = p(k)P^{n-1}$$

for any $k = 1, 2, 3, \ldots$ and $n = 2, 3, 4, \ldots$.

**Example 1.6.4**    *Suppose that a city with constant population has two districts, E and W, and that in any given year $100a\%$ of the inhabitants of E move to W and $100b\%$ of those of W move to E, with the others remaining where they are. Taking E as the first state and W as the second, the matrix*

$$\begin{bmatrix} 1-a & a \\ b & 1-b \end{bmatrix}$$

*is the transition matrix of a homogeneous Markov chain for the (fractional) population distribution $p = (p_E, p_W)$ of the two districts. If the population distribution of the city is initially $p(1)$, then by (6.13) after n years it is $p(n+1) = p(1)P^n$.*

**PC-Exercise 1.6.5**    *Let $a = 0.1$ and $b = 0.01$ in the Markov chain described in Example 1.6.4 and consider a person originally living in district E of the city. Use the two-point random number generator to simulate this person's yearly district of residence. Assuming a lifespan of 100 years, estimate from 100 simulations the probabilities of this person's residing in districts E and W. Repeat the calculations for a person originally living in district W.*

An interpretation of (6.13) is that the probability distribution depends only on the time that has elapsed rather on the actual time itself. This does not however say that a homogeneous Markov chain is a strictly stationary stochastic process. That requires the probability distributions for all instants of time to be identical, or equivalently the existence of a probability vector $\bar{p}$ such that $p(n) = \bar{p}$ for each $n = 1, 2, 3, \ldots$. The transition relationship (6.11) implies that such a probability vector $\bar{p}$ must satisfy the system of simultaneous equations, written here in vector form,

(6.14) $$\bar{p} = \bar{p}\, P.$$

Property (6.11) for a constant transition matrix $P$ ensures that (6.14) has at least one solution which is a probability vector. Such a solution is called a *stationary probability vector* or distribution for the Markov chain. For example

$$\bar{p} = \left( \frac{b}{a+b}, \frac{a}{a+b} \right)$$

is a stationary probability vector for the Markov chain described in Example 1.6.4, a fact which should have been discovered by those readers who did PC-Exercise 1.6.5. This particular $\bar{p}$ is the only stationary probability vector for this transition matrix. Other transition matrices may have more than one stationary probability vector.

**Exercise 1.6.6** *Show that all of the probability vectors $\bar{p}_\lambda = (\lambda/2, \lambda/2, 1-\lambda)$ for each $0 \le \lambda \le 1$ satisfies (6.14) with the transition matrix*

$$P = \begin{bmatrix} 0.5 & 0.5 & 0 \\ 0.5 & 0.5 & 0 \\ 0 & 0 & 1 \end{bmatrix}.$$

A homogeneous Markov chain thus describes a strictly stationary stochastic process only if the initial random variable $X_1$ is distributed according to one of its stationary probability vectors.

Certain homogeneous Markov chains with a unique stationary probability vector $\bar{p}$ enjoy a powerful property called *ergodicity* relating the long-term time averages of its realizations to the spatial averaging with respect to the stationary distribution. For any function $f : \mathcal{X} \to \Re$ we can form the time average

$$(6.15) \qquad \frac{1}{T} \sum_{t=1}^{T} f(X_t)$$

of the values $f(X_t)$ taken by a sequence of random variables $X_1, X_2, \ldots, X_t,$ $\ldots$ generated by the Markov chain. An obvious question is: what happens with these averages as $T \to \infty$? In fact, in the special case that $f(x_k) = 1$ for some $k$ and $f(x_j) = 0$ for all other $j \ne k$ the time average (6.15) is just the relative frequency of the chain's being in state $x_k$. We recall that for i.i.d. random variables $X_1, X_2, X_3, \ldots$ the Law of Large Numbers would tell us that the limit exists and is equal to the probability $p_k$. In general, we say that a Markov chain is *ergodic* if for every initial $X_1$ the limits of the time averages (6.15) exist and are equal to the average of $f$ over $\mathcal{X}$ with respect to the stationary probability vector $\bar{p}$, that is if

$$(6.16) \qquad \lim_{T \to \infty} \frac{1}{T} \sum_{t=1}^{T} f(X_t) = \sum_{i=1}^{N} f(x_i) \bar{p}_i.$$

A condition guaranteeing the ergodicity of a Markov chain is that all of the components of its unique stationary probability vector $\bar{p}$ are nonzero, that is

$\bar{p}_i > 0$ for $i = 1, 2, \ldots, N$. A weaker condition is that all of the components of some $k$th power $P^k$ of the transition matrix are nonzero. Ergodicity is thus an extension of the Law of Large Numbers to stochastic processes, but says much more about the stochastic process, for example, implying its eventual mixing over the states in $\mathcal{X}$ according to the stationary distribution.

**PC-Exercise 1.6.7**    *Verify (6.16) numerically for the Markov chain in Example 1.6.4, taking into account PC-Exercise 1.6.5. Consider functions $f : \{E, W\} \rightarrow \Re$ with: (i) $f(E) = 0$, $f(W) = 1$; (ii) $f(E) = 1$, $f(W) = 0$; (iii) $f(E) = 1$, $f(W) = -1$; (iv) $f(E) = -1$, $f(W) = 1$. In each case try a variety of initial distributions $p(1) = (p_E(1), p_W(1)) = (p, 1 - p)$ for $X_1$, say with $p = 0$, 0.1, 0.2,..., 0.9, 1. Also try $p = 1/11 = 0.09090909\ldots$.*

We can handle *continuous time Markov chains* in a similar way. Let $X(t)$ be distributed over a finite state space $\mathcal{X} = \{x_1, x_2, \ldots, x_N\}$ according to an $N$-dimensional probability vector $p(t)$ for each $t \geq 0$. In this context the Markov property (6.8) takes the form

$$(6.17) \qquad P\left(X(t_1) = x_j | X(s_1) = x_{i_1}, \ldots, X(s_n) = x_{i_n}, X(t_0) = x_i\right)$$

$$= P\left(X(t_1) = x_j | X(t_0) = x_i\right)$$

for all $0 \leq s_1 \leq s_2 \leq \cdots \leq s_n < t_0 \leq t_1$ and all $x_i, x_j, x_{i_1}, x_{i_2}, \ldots, x_{i_n} \in \mathcal{X}$ where $n = 1, 2, 3, \ldots$. For each $0 \leq t_0 \leq t_1$ we can define an $N \times N$ transition matrix $P(t_0; t_1) = [p^{i,j}(t_0; t_1)]$ componentwise by

$$p^{i,j}(t_0; t_1) = P\left(X(t_1) = x_j | X(t_0) = x_i\right)$$

for $i, j = 1, 2, \ldots, N$. Clearly then $P(t_0; t_0) = I$ and the probability vectors $p(t_0)$ and $p(t_1)$ are related by

$$p(t_1) = p(t_0)P(t_0; t_1).$$

Applying this for $t_0 \leq t_1 \leq t_2$ we have $p(t_2) = p(t_0)P(t_0; t_2)$ and

$$p(t_2) = p(t_1)P(t_1; t_2) = p(t_0)P(t_0; t_1)P(t_1; t_2)$$

for any probability vector $p(t_0)$. From this we can conclude that the transition matrices satisfy the relationship

$$(6.18) \qquad\qquad P(t_0; t_2) = P(t_0; t_1)P(t_1; t_2)$$

for all $t_0 \leq t_1 \leq t_2$. In the special case that the transition matrices $P(t_0; t_1)$ depend only on the time difference $t_1 - t_0$, that is $P(t_0; t_1) = P(0; t_1 - t_0)$ for all $0 \leq t_0 \leq t_1$, we say that the continuous time Markov chain is *homogeneous* and write $P(t)$ for $P(0; t)$. Then (6.18) reduces to

$$(6.19) \qquad\qquad P(s + t) = P(s)P(t) = P(t)P(s)$$

for all $s, t \geq 0$.

There exists an $N \times N$ *intensity matrix* $A = (a^{i,j})$ with

$$a^{i,j} = \begin{cases} \lim_{t \to 0} \dfrac{p^{i,j}(t)}{t} & : \quad i \neq j \\[2mm] \lim_{t \to 0} \dfrac{p^{i,i}(t) - 1}{t} & : \quad i = j \end{cases}$$

which, together with the initial probability vector $p(0)$, completely character-izes the homogeneous continuous time Markov chain. If the diagonal compo-nents $a^{i,i}$ are finite for each $i = 1, \ldots, N$, then the transition probabilities satisfy the *Kolmogorov forward equation*

(6.20) $$\frac{dp^{i,j}}{dt}(t) - \sum_{k=1}^{N} p^{i,k}(t)\, a^{k,j} = 0$$

and the *Kolmogorov backward equation*

(6.21) $$\frac{dp^{i,j}}{dt}(t) - \sum_{k=1}^{N} a^{k,i}\, p^{k,j}(t) = 0$$

for all $i, j = 1, 2, \ldots, N$. The *waiting time* of a homogeneous continuous time Markov chain, that is the time between transitions from a state $x_i$ to any other state, is thus exponentially distributed with intensity parameter $\lambda_i = \sum_{j \neq i} a^{i,j}$.

**Example 1.6.8** *Consider a stochastic process $X(t)$ taking only the two val-ues $+1$ and $-1$ with probabilities $(p^{+}(t), p^{-}(t)) = p(t)$ and switching according to the homogeneous transition matrix*

$$P(t) = \begin{bmatrix} (1 + e^{-t})/2 & (1 - e^{-t})/2 \\ (1 - e^{-t})/2 & (1 + e^{-t})/2 \end{bmatrix}$$

*for $t \geq 0$. The intensity matrix here is*

$$A = \begin{bmatrix} -0.5 & 0.5 \\ 0.5 & -0.5 \end{bmatrix}.$$

*For the initial probability vector $p(0) = \bar{p} = (0.5, 0.5)$ we find that $\bar{p}P(t) = p(t) \equiv \bar{p}$ for all $t \geq 0$, so $\bar{p}$ is a stationary probability vector for this Markov chain; the corresponding strictly stationary stochastic process is known as* random telegraphic noise.

**PC-Exercise 1.6.9** *Use exponentially distributed waiting times to simulate the telegraphic noise of Example 1.6.8 on the time interval $[0, T]$ with $T = 10$. From a sample of 100 simulations calculate the frequency of being in the state 1 at time $T$.*

We remark that a continuous time Markov chain is called *ergodic* if

$$(6.22) \qquad \lim_{T \to \infty} \frac{1}{T} \int_0^T f(X(t)) \, dt = \sum_{i=1}^N f(x_i) \bar{p}_i,$$

that is with the time averages (6.15) in (6.16) now written in integral form. The random telegraphic noise process is an ergodic continuous time Markov chain.

**Exercise 1.6.10**   *For Example 1.6.8 show that the probability vectors $p(t) = p(0)P(t) \to \bar{p}$ as $t \to \infty$ for any initial probability vector $p(0)$.*

As a final exercise we return to the Poisson process defined in Example 1.6.1, which is a continuous time Markov chain on a countably infinite state space.

**Exercise 1.6.11**   *Show that the Poisson process is a continuous time homogeneous Markov chain on the countably infinite state space $\{0, 1, 2, \ldots\}$. Determine its transition matrix $P(t)$.*

## 1.7   Diffusion Processes

Markov processes taking continuous values in $\Re$ require a somewhat more complicated mathematical framework than their discrete state counterparts, especially when they also involve continuous time values. A rich and useful class of such Markov processes are the diffusion processes, which we shall define below after first considering some more general background material. In what follows we shall always suppose that for $k = 1, 2, \ldots$ every joint distribution $F_{t_1 t_2 \ldots t_k}(x_1, x_2, \ldots, x_k)$ of the process $X = \{X(t), t \geq 0\}$ under consideration has a density $p(t_1, x_1; t_2, x_2; \ldots; t_k, x_k)$. Then we define the conditional probabilities

$$(7.1) \qquad P\left(X(t_{n+1}) \in B \,|\, X(t_1) = x_1, X(t_2) = x_2, \ldots, X(t_n) = x_n\right)$$

$$= \frac{\displaystyle\int_B p(t_1, x_1; t_2, x_2; \ldots; t_n, x_n; t_{n+1}, y) \, dy}{\displaystyle\int_{-\infty}^{\infty} p(t_1, x_1; t_2, x_2; \ldots; t_n, x_n; t_{n+1}, y) \, dy}$$

for all Borel subsets $B$ of $\Re$, time instants $0 < t_1 < t_2 < \ldots < t_n < t_{n+1}$ and all states $x_1, x_2, \ldots, x_n \in \Re$, provided the denominators are nonzero. (When the denominator in (7.1) is zero we either define the conditional probability to be zero or we leave it undefined). In this context the *Markov property* takes the form

$$(7.2) \qquad P\left(X(t_{n+1}) \in B \,|\, X(t_1) = x_1, X(t_2) = x_2, \ldots, X(t_n) = x_n\right)$$

$$= P\left(X(t_{n+1}) \in B \,|\, X(t_n) = x_n\right)$$

for all Borel subsets $B$ of $\Re$, time instants $0 < t_1 < t_2 < \ldots < t_n < t_{n+1}$ and all states $x_1, x_2, \ldots, x_n \in \Re$ for which the conditional probabilities are defined. If (7.2) is satisfied we call the stochastic process $X(t)$ a *Markov process* and write its transition probabilities as

$$P(s, x; t, B) = P(X(t) \in B | X(s) = x),$$

where $s < t$. For fixed $s, x$ and $t$, $P(s, x; t, \cdot)$ is a probability function (measure) on the $\sigma$-algebra $\mathcal{B}$ of Borel subsets of $\Re$. Under our assumption above it has a density $p(s, x; t, \cdot)$, called a *transition density*, so

$$P(s, x; t, B) = \int_B p(s, x; t, y) \, dy$$

for all $B \in \mathcal{B}$. For technical convenience we usually also define $P(s, x; s, B) = I_B(x)$ for $t = s$, where $I_B$ is the indicator function of the set $B$. Analogously with Markov chains, we say that a continuous time Markov process is *homogeneous* if all of its transition densities $p(s, x; t, y)$ depend only on the time difference $t - s$ rather than on the specific values of $s$ and $t$. Examples of homogeneous Markov processes are the standard Wiener process (see equation 6.5) with transition density

$$(7.3) \qquad p(s, x; t, y) = \frac{1}{\sqrt{2\pi(t-s)}} \exp\left(-\frac{(y-x)^2}{2(t-s)}\right)$$

and the Ornstein-Uhlenbeck process with $\gamma = 1$ (see Example 1.6.2) with transition density

$$(7.4) \qquad p(s, x; t, y) = \frac{1}{\sqrt{2\pi\left(1 - e^{-2(t-s)}\right)}} \exp\left(-\frac{\left(y - xe^{-(t-s)}\right)^2}{2\left(1 - e^{-2(t-s)}\right)}\right).$$

Since $p(s, x; t, y) = p(0, x; t-s, y)$ we usually omit the superfluous first variable and simply write $p(x; t - s, y)$, with $P(x; t - s, B) = P(0, x; t - s, B)$ for the transition probabilities.

The transition matrix equation (6.19) for continuous time Markov chains has a counterpart for the transition densities of a continuous state Markov process. This also follows from the Markov property and is called the *Chapman-Kolmogorov equation*:

$$(7.5) \qquad p(s, x; t, y) = \int_{-\infty}^{\infty} p(s, x; \tau, z) p(\tau, z; t, y) \, dz$$

for all $s \leq \tau \leq t$ and $x, y \in \Re$. For transition probabilities it takes the form

$$(7.6) \qquad P(s, x; t, B) = \int_{-\infty}^{\infty} P(\tau, z; t, B) P(s, x; \tau, dz),$$

where $B \in \mathcal{B}$ and the integral is an improper Riemann- Stieltjes integral.

**Exercise 1.7.1**    *Verify that the transition density (7.3) of a standard Wiener process satisfies the Chapman-Kolmogorov equation (7.5). What happens to the $\tau$ variable in the integrand here?*

The transition density of the standard Wiener process is obviously a smooth function of its variables for $t > s$. Evaluating the partial derivatives of (7.3) explicitly, we find that they satisfy the partial differential equations

$$(7.7) \qquad \frac{\partial p}{\partial t} - \frac{1}{2}\frac{\partial^2 p}{\partial y^2} = 0, \qquad (s, x) \text{ fixed,}$$

and

$$(7.8) \qquad \frac{\partial p}{\partial s} + \frac{1}{2}\frac{\partial^2 p}{\partial x^2} = 0, \qquad (t, y) \text{ fixed.}$$

The transition density (7.4) of the Ornstein-Uhlenbeck process satisfies related, but more complicated partial differential equations (see Exercise 1.7.2 below). Equation (7.7) is an example of a heat equation which describes the variation in temperature as heat diffuses through a physical medium. The same equation describes the diffusion of a chemical substance such as an inkspot. It should thus not be surprising that a standard Wiener process serves as a prototypical example of a (stochastic) diffusion process. To be specific, a Markov process with transition densities $p(s, x; t, y)$ is called a *diffusion process* if the following three limits exist for all $\epsilon > 0$, $s \geq 0$ and $x \in \Re$ :

$$(7.9) \qquad \lim_{t \downarrow s} \frac{1}{t-s} \int_{|y-x|>\epsilon} p(s, x; t, y)\, dy = 0,$$

$$(7.10) \qquad \lim_{t \downarrow s} \frac{1}{t-s} \int_{|y-x|<\epsilon} (y-x)p(s, x; t, y)\, dy = a(s, x)$$

and

$$(7.11) \qquad \lim_{t \downarrow s} \frac{1}{t-s} \int_{|y-x|<\epsilon} (y-x)^2 p(s, x; t, y)\, dy = b^2(s, x),$$

where $a$ and $b$ are well-defined functions. Condition (7.9) prevents a diffusion process from having instantaneous jumps. The quantity $a(s, x)$ is called the *drift* of the diffusion process and $b(s, x)$ its *diffusion coefficient* at time $s$ and position $x$. (7.10) implies that

$$(7.12) \qquad a(s, x) = \lim_{t \downarrow s} \frac{1}{t-s} E\left(X(t) - X(s) | X(s) = x\right),$$

so the drift $a(s, x)$ is the instantaneous rate of change in the mean of the process given that $X(s) = x$. Similarly, it follows from (7.11) that the squared diffusion coefficient

$$(7.13) \qquad b^2(s, x) = \lim_{t \downarrow s} \frac{1}{t-s} E\left((X(t) - X(s))^2 | X(s) = x\right)$$

denotes the instantaneous rate of change of the squared fluctuations of the process given that $X(s) = x$.

**Exercise 1.7.2** *Show that the standard Wiener process is a diffusion process with drift $a(s, x) \equiv 0$ and diffusion coefficient $b(s, x) \equiv 1$, and that the Ornstein-Uhlenbeck process with $\gamma = 1$ is a diffusion process with drift $a(s, x) = -x$ and diffusion coefficient $b(s, x) \equiv \sqrt{2}$.*

When the drift $a$ and diffusion coefficient $b$ of a diffusion process are moderately smooth functions, then its transition density $p(s, x; t, y)$ also satisfies partial differential equations, which reduce to (7.7) and (7.8) for a standard Wiener process. These are the *Kolmogorov forward equation*

$$(7.14) \quad \frac{\partial p}{\partial t} + \frac{\partial}{\partial y} \{a(t, y)p\} - \frac{1}{2} \frac{\partial^2}{\partial y^2} \{b^2(t, y)p\} = 0, \quad (s, x) \text{ fixed,}$$

and the *Kolmogorov backward equation*

$$(7.15) \quad \frac{\partial p}{\partial s} + a(s, x)\frac{\partial p}{\partial x} + \frac{1}{2}b^2(s, x)\frac{\partial^2 p}{\partial x^2} = 0, \quad (t, y) \text{ fixed,}$$

with the former giving the forward evolution with respect to the final state $(t, y)$ and the latter giving the backward evolution with respect to the initial $(s, x)$. The forward equation (7.14) is the formal adjoint of the backward equation (7.15) and is commonly called the *Fokker-Planck equation*. Both follow from the Chapman-Kolmogorov equation (7.5) and (7.9)–(7.11). Instead of presenting a rigorous proof, which is intricate, we shall derive the backward equation (7.15) in a rough, yet illustrative way. For this we approximate the diffusion process by a discrete-time continuous state process with two equally probable jumps from $(s, x)$ to $(s + \Delta s, x + \bar{a}\Delta s \pm \sqrt{\bar{b}^2\Delta s})$ where $\bar{a} = a(s, x)$ and $\bar{b}^2 = b^2(s, x)$, which is consistent with the interpretations (7.11) and (7.12) of the drift and diffusion coefficients. For this approximate process we have

$$\tilde{p}(s, x; t, y) = \frac{1}{2} \{\tilde{p}(s + \Delta s, x^+; t, y) + \tilde{p}(s + \Delta s, x^-; t, y)\}$$

where $x^{\pm} = x + \bar{a}\Delta s \pm \sqrt{\bar{b}^2\Delta s}$. Taking Taylor expansions up to the first order in $\Delta s$ about $(s, x; t, y)$ we find that

$$0 = \frac{\partial \tilde{p}}{\partial s} \Delta s + \bar{a}\frac{\partial \tilde{p}}{\partial x} \Delta s + \frac{1}{2} \bar{b}^2 \frac{\partial^2 \tilde{p}}{\partial x^2} \Delta s + O((\Delta s)^{3/2})$$

where the partial derivatives are evaluated at $(s, x; t, y)$. Since this discrete-time process converges in distribution to the diffusion process, we obtain (7.15) when we take the limit $\Delta s \to 0$.

**Exercise 1.7.3** *Use the results of Exercise 1.7.2 to write down the Kolmogorov forward and backward equations for the Ornstein-Uhlenbeck process with parameter $\gamma = 1$. Then verify that the transition density (7.4) satisfies the backward equation.*

For a stationary diffusion process there usually exists a *stationary probability density* $\bar{p}(y)$ such that

$$\bar{p}(y) = \int_{-\infty}^{\infty} p(s, x; t, y)\bar{p}(x)\,dx$$

for all $0 \leq s \leq t$ and $y \in \Re$. This density $\bar{p}$ then satifies the corresponding stationary or time-independent Fokker-Planck equation, which, in this 1-dimensional case, is the ordinary differential equation

(7.16)
$$\frac{d}{dy}\{a(y)\bar{p}(y)\} - \frac{1}{2}\frac{d^2}{dy^2}\{b^2(y)\bar{p}(y)\} = 0$$

with drift $a$ and diffusion coefficient $b$ independent of time $t$. Naturally $\bar{p}(y) \geq 0$ for all $y \in \Re$ and $\int_{-\infty}^{\infty} \bar{p}(y)\,dy = 1$. Such a diffusion process $X = \{X(t), t \geq 0\}$ is said to be *ergodic* if the following time average limit exists and equals the spatial average with respect to $\bar{p}$, w.p.1, that is if

(7.17)
$$\lim_{T \to \infty} \frac{1}{T}\int_0^T f(X(t))\,dt = \int_{-\infty}^{\infty} f(x)\bar{p}(x)\,dx,$$

for all bounded measurable functions $f : \Re \to \Re$. However, unlike its counterparts (6.16) and (6.20) for Markov chains, (7.17) is usually quite difficult to verify directly for a diffusion process.

**Exercise 1.7.4**    *Solve (7.16) for the stationary probability density of the diffusion process with drift $a(y) = \nu y - y^2$ and diffusion coefficient $b(y) = \sqrt{2y}$ for $y \geq 0$. Assume that $\bar{p}$ and its derivative $\bar{p}'$ converge to 0 exponentially fast as $y \to \infty$. Distinguish the cases where $\nu < 1$, $1 \leq \nu < 2$ and $\nu \geq 2$.*

We would expect from (7.9) that a diffusion process should have well - behaved sample paths. In fact these are, almost surely, continuous functions of time, although they need not be differentiable, as we shall show later for the Wiener process. We can define the continuity of a stochastic process $X(t)$ in $t$ in several different ways corresponding to the different convergences of sequences of random variables introduced in Section 5. In particular we have at a fixed instant $t$:

**I.**  *Continuity with probability one:*

(7.18)
$$P\left(\{\omega \in \Omega : \lim_{s \to t} |X(s, \omega) - X(t, \omega)| = 0\}\right) = 1;$$

**II.**  *Mean-square continuity:* $E(X(t)^2) < \infty$ and

(7.19)
$$\lim_{s \to t} E\left(|X(s) - X(t)|^2\right) = 0;$$

**III.**  *Continuity in probability:*

(7.20)
$$\lim_{s \to t} P(\{\omega \in \Omega : |X(s, \omega) - X(t, \omega)| \geq \epsilon\}) = 0 \text{ for all } \epsilon > 0; \quad \text{and}$$

**IV.** *Continuity in distribution:*

(7.21) $\qquad \lim_{s \to t} F_s(x) = F_t(x) \quad$ for all continuity points of $F_t$.

These are related to each other in the same way as the convergences of random sequences. In particular, **(I)** and **(II)** both imply **(III)**, but not each other, and **(III)** implies **(IV)**. Moreover **(III)** can be written in terms of the metric (5.4) as

(7.22) $\qquad \lim_{s \to t} E \left( \dfrac{|X(s) - X(t)|}{1 + |X(s) - X(t)|} \right) = 0.$

The random telegraphic noise process (Example 1.6.8) is a continuous time Markov chain with two states $-1$ and $+1$, so it is obviously not a diffusion process. Its sample paths are piecewise continuous functions jumping between the two values $\pm 1$ with equal probability. It follows from its transition matrices that the covariances are

$$E\left(|X(s) - X(t)|^2\right) = 2\left(1 - \exp(-|s - t|)\right)$$

for all $s$, $t \geq 0$, so this process is mean-square continuous at any instant $t$ and hence also continuous in probability and in distribution. It is also known that the telegraphic noise process is continuous with probability one at any instant $t$. This may seem surprising in view of the fact that its sample paths are actually discontinuous at the times when a jump occurs. There is, however, a simple explanation. Continuity with probability one at the time instant $t$ means that the probability $P(A_t) = 0$ where $A_t = \{\omega \in \Omega : \lim_{s \to t} |X(s, \omega) - X(t, \omega)| \neq 0\}$. Continuity of almost all of the sample paths, which we call *sample-path continuity*, requires $P(A) = 0$ where $A = \cup_{t \geq 0} A_t$. Since $A$ is the uncountable union of events $A_t$ it need not be an event, in which case $P(A)$ is not defined, or when it is an event it could have $P(A) > 0$ even though $P(A_t) = 0$ for every $t \geq 0$. The second of these is what happens for the telegraphic noise process.

For a diffusion process the appropriate continuity concept is sample-path continuity. There is a criterion due to Kolmogorov which implies the sample-path continuity of a continuous time stochastic process $X = \{X(t), t \in T\}$, namely the existence of positive constants $\alpha, \beta, C$ and $h$ such that

(7.23) $\qquad E\left(|X(t) - X(s)|^\alpha\right) \leq C|t - s|^{1+\beta}$

for all $s$, $t \in T$ with $|t - s| \leq h$. For example, this is satisfied by the standard Wiener process $W = \{W(t), t \geq 0\}$, for which it can be shown that

(7.24) $\qquad E\left(|W(t) - W(s)|^4\right) = 3|t - s|^2$

for all $s$, $t \geq 0$. Hence the Wiener process has, almost surely, continuous sample paths.

**Exercise 1.7.5** *Use properties (6.5) of a standard Wiener process to prove (7.24).*

We conclude this section with the remark that the uncountability of the time set is responsible for many subtle problems concerning continuous time stochastic processes. These require the more rigorous framework of measure and integration theory for their clarification and resolution. We shall consider some of these matters in Chapter 2.

## 1.8   Wiener Processes and White Noise

We have already introduced the standard Wiener process in Section 6 and have considered some of its basic properties. Recalling (6.5), we define a *standard Wiener process* $W = \{W(t), t \geq 0\}$ to be a Gaussian process with independent increments such that

$$(8.1) \quad W(0) = 0 \text{ w.p.1}, \quad E(W(t)) = 0, \quad \text{Var}(W(t) - W(s)) = t - s$$

for all $0 \leq s \leq t$. This process was proposed by Wiener as a mathematical description of Brownian motion, the erratic motion of a grain of pollen on a water surface due to its being continually bombarded by water molecules. The Wiener process is sometimes called *Brownian motion,* but we will use separate terminology to distinguish between the mathematical and physical processes.

We can approximate a standard Wiener process in distribution on any finite time interval by a scaled random walk. For example, we can subdivide the unit interval $[0, 1]$ into $N$ subintervals

$$0 = t_0^{(N)} < t_1^{(N)} < \cdots < t_k^{(N)} < \cdots < t_N^{(N)} = 1$$

of equal length $\Delta t = 1/N$ and construct a stepwise continuous random walk $S_N(t)$ by taking independent, equally probable steps of length $\pm\sqrt{\Delta t}$ at the end of each subinterval. As already mentioned in the paragraph preceding PC- Exercise 1.5.7, we start with independent two-point random variables $X_n$ taking values $\pm 1$ with equal probability. Then we define

$$(8.2) \qquad S_N(t_n^{(N)}) = (X_1 + X_2 + \cdots + X_n)\sqrt{\Delta t}$$

with

$$(8.3) \qquad\qquad S_N(t) = S_N(t_n^{(N)})$$

on $t_n^{(N)} \leq t < t_{n+1}^{(N)}$ for $n = 0, 1, \ldots, N - 1$, where $S_N(0) = 0$. This random walk has independent increments $X_1\sqrt{\Delta t}$, $X_2\sqrt{\Delta t}$, $X_3\sqrt{\Delta t}$, ... for the given subintervals, but is not a process with independent increments. It follows that

$$E\left(S_N(t)\right) = 0, \quad \text{Var}\left(S_N(t)\right) = \Delta t \left[\frac{t}{\Delta t}\right]$$

for $0 \leq t \leq 1$, where $[\tau]$ denotes the integer part of $\tau \in \Re$, that is $[\tau] = k$ if $k \leq \tau < k + 1$ for some integer $k$. Now $\text{Var}\left(S_N(t)\right) \to t$ as $N = 1/\Delta t \to \infty$ for any

$0 \leq t \leq 1$. Similarly, it can be shown for any $0 \leq s < t \leq 1$ that $\text{Var}(S_N(t) - S_N(s)) \to t - s$ as $N \to \infty$. Hence it follows by the Central Limit Theorem (see Section 5) that $S_N(t)$ converges in distribution as $N \to \infty$ to a process with independent increments satisfying conditions (8.1), that is a standard Wiener process.

**Exercise 1.8.1**    *Verify that* $\text{Var}(S_N(t) - S_N(s)) \to t - s$ *as* $N \to \infty$ *where* $0 \leq s < t \leq 1$ *for the process* $S_N(t)$ *defined by (8.2) and (8.3).*

**PC-Exercise 1.8.2**    *Generate and plot sample paths of the process* $S_N(t)$, *defined on* $0 \leq t \leq 1$ *by (8.2) and (8.3), for increasing values of* $N = 10, 20,$ *..., 100. To compare approximations of the same sample path with, say, $N = 50$ and $N = 100$, generate 100 random numbers $X_1, X_2, ..., X_{100}$ and use them to determine a sample path of $S_{100}(t)$. Then add successive pairs to form $\tilde{X}_1 = X_1 + X_2, \tilde{X}_2 = X_3 + X_4, ..., \tilde{X}_{50} = X_{99} + X_{100}$ and use $\tilde{X}_1, \tilde{X}_2, ..., \tilde{X}_{50}$ to determine the corresponding sample path of $S_{50}(t)$.*

**Exercise 1.8.3**    *Construct an interpolated random walk on $0 \leq t \leq 1$ by using linear interpolation within each subinterval instead of (8.3). Determine its means and variances and their limits. Show that this process is not a process with independent increments, but still converges in distribution to a standard Wiener process.*

We saw at the end of Section 7 that a Wiener process is sample-path continuous, that is its sample paths are almost surely continuous functions of time. From (8.1) we see that the variance grows without bound as time increases while the mean always remains zero. This says that many sample paths must attain larger and larger values, both positive and negative, as time increases. In analogy with the strong version of the Law of Large Numbers one finds that with probability one

$$(8.4) \qquad \lim_{t \to \infty} \frac{W(t)}{t} = 0$$

**Exercise 1.8.4**    *Use the properties (8.1) of a Wiener process to prove (8.4).*

A much sharper growth rate is given by the *Law of the Iterated Logarithm*, which says that

$$(8.5) \qquad \limsup_{t \to \infty} \frac{W(t)}{\sqrt{2t \ln \ln t}} = +1, \qquad \liminf_{t \to \infty} \frac{W(t)}{\sqrt{2t \ln \ln t}} = -1$$

with probabilty one. This can be interpreted as saying that there is a sequence $t_n \to \infty$ for which the values of the ratios converge to $+1$ and another sequence $t'_n \to \infty$ for which they converge to $-1$. Since it can be shown that $\{tW(1/t), t \geq 0\}$ is also a Wiener process, we can rewrite (8.5) in the form

$$(8.6) \qquad \limsup_{t \downarrow 0} \frac{W(t)}{\sqrt{2t \ln \ln(1/t)}} = +1, \qquad \liminf_{t \downarrow 0} \frac{W(t)}{\sqrt{2t \ln \ln(1/t)}} = -1.$$

**Exercise 1.8.5**    *Show that* $t\,W(1/t)$ *and* $W(t+s) - W(s)$*, for any fixed* $s$ $\geq 0$*, are Wiener processes when* $W(t)$ *is a Wiener process.*

**PC-Exercise 1.8.6**    *As in PC-Exercise 1.8.2 generate random walks* $S_N(t)$ *on* $0 \leq t \leq 1$ *for increasing* $N$ *to approximate the same sample path in the limiting process. For fixed* $N$*, say 20, evaluate the ratios*

$$\frac{S_N(h+0.5) - S_N(0.5)}{h}$$

*for successively smaller values of* $h$ *and plot the linearly interpolated ratios against* $h$*. Repeat this for larger values of* $N$*, say 50 and 100. What do these results suggest about the differentiability of the limiting sample path at* $t = 0.5$*?*

Using Exercise 1.8.5, let us now apply (8.6) to show that the process $\{W(t+s)$ $- W(s),\ t \geq 0\}$ is not differentiable at $t = 0$. From (8.6) for any $\epsilon > 0$ there exist sequences $t_n, t_n' \downarrow 0$ (depending on the sample path) such that

$$\frac{W(t_n + s) - W(s)}{t_n} \geq (1 - \epsilon)\sqrt{\frac{2\ln\ln(1/t_n)}{t_n}}$$

and

$$\frac{W(t_n' + s) - W(s)}{t_n'} \leq (-1 + \epsilon)\sqrt{\frac{2\ln\ln(1/t_n')}{t_n'}}$$

when $n$ is sufficiently large. Now we have

$$2\ln\ln(1/t))/t \to \infty \quad \text{as} \quad t \downarrow 0,$$

so

$$\left(W(t_n + s) - W(s)\right)/t_n \to \infty$$

and

$$\left(W(t_n' + s) - W(s)\right)/t_n' \to -\infty$$

as $t_n, t_n' \downarrow 0$. Hence the sample path of $W(t)$ under consideration cannot be differentiable at $t = s$. Since the limits hold almost surely and $s \geq 0$ was arbitrary, we have thus shown that the sample paths of a Wiener process are, almost surely, nowhere differentiable functions of time. We shall give another, more direct and rigorous proof of this unusual property in Section 4 of Chapter 2.

A Wiener process $W(t)$ has by definition the initial value $W(0) = 0$, w.p.1. By sample-path continuity and the Law of the Iterated Logarithm (8.5) it follows that, almost surely, the sample paths sweep in time repeatedly through the real numbers, both positive and negative. In some applications it is useful to have a modification of a Wiener process for which the sample paths all pass through the same initial point $x$, not necessarily 0, and a given point $y$ at a later time $t = T$. Such a process $B_{0,x}^{T,y}(t)$ is defined sample pathwise for $0 \leq t \leq T$ by

$$(8.7) \qquad B_{0,x}^{T,y}(t,\omega) = x + W(t,\omega) - \frac{t}{T}\{W(T,\omega) - y + x\}$$

and is called a *Brownian bridge* or a *tied-down Wiener process*. It is a Gaussian process satisfying the constraints $B_{0,x}^{T,y}(0,\omega) = x$ and $B_{0,x}^{T,y}(T,\omega) = y$, so in a manner of speaking can be considered as to be a kind of conditional Wiener process. Since it is Gaussian it is determined uniquely by its means and covariances, which are

$$(8.8) \qquad \mu(t) = x - \frac{t}{T}(x-y) \quad \text{and} \quad C(s,t) = \min\{s,t\} - \frac{st}{T}$$

for $0 \le s, t \le T$ (see (6.3) and (6.4) ).

**Exercise 1.8.7** *Show that a Brownian bridge $B_{0,x}^{T,y}(t)$ for $0 \le t \le T$ is Gaussian with means and covariances (8.8).*

**PC-Exercise 1.8.8** *Modify the random walks $S_N(t)$ on $0 \le t \le 1$ in PC-Exercise 1.8.2 to obtain approximations of the Brownian bridge $B_{0,0}^{1,0}$. For $N = 10, 20, \ldots, 100$ plot approximations to the same limiting sample path against time $t$ for $0 \le t \le 1$.*

In many time-invariant engineering systems the (time-independent) variance of a stochastic process $X(t)$ can be interpreted as an average power (or energy) and is written as

$$(8.9) \qquad \text{Var}(X(t)) = c(0) = \int_{-\infty}^{\infty} S(\nu)\,d\nu$$

where $c(0)$ is the value of the covariance $c(t-s)$ at $s = t$ and $S(\nu)$ denotes the *spectral density* measuring the average power per unit frequency at frequency $\nu$. $S(\nu)$ is real-valued and nonnegative with $S(-\nu) = S(\nu)$ for all $\nu$, and can be extracted from (8.9) by an inverse Fourier transform giving

$$(8.10) \qquad S(\nu) = \int_{-\infty}^{\infty} c(s)\exp(-2\pi i \nu s)\,ds = \int_{-\infty}^{\infty} c(s)\cos(2\pi \nu s)\,ds.$$

This brings us to *Gaussian white noise*, which can be thought of as a zero-mean wide-sense stationary process with constant nonzero spectral density $S(\nu) = S_0$. The name white noise comes from the fact that its average power is uniformly distributed in frequency, which is a characteristic of white light. Hence its covariances $c(s)$ satisfy formally

$$(8.11) \qquad c(s) = S_0\,\delta(s)$$

for all $s$, where $\delta(s)$ is the *Dirac delta function*, a generalized function with $\delta(s) = 0$ for all $s \ne 0$ such that

$$\int_{-\infty}^{\infty} f(s)\delta(s)\,ds = f(0)$$

for all functions $f$ continuous at $s = 0$. This suggests that Gaussian white noise $\dot{W}$ is an unusual stochastic process. To elaborate on this, let $W = \{W(t), t \ge$

0} be a standard Wiener process and for fixed $h > 0$ define a new process $X^h$ = $\{X^h(t), t \geq 0\}$ by

(8.12) $$X^h(t) = \frac{W(t+h) - W(t)}{h}$$

for all $t \geq 0$. This is a wide-sense stationary Gaussian process with zero means and with covariances

(8.13) $$c(t - s) = \frac{1}{h} \max\left\{0, 1 - \frac{1}{h}|t - s|\right\};$$

it thus has spectral density

(8.14) $$S_h(\nu) = \frac{1}{h} \int_{-h}^{h} \left(1 - \frac{|s|}{h}\right) \cos(2\pi\nu s)\,ds = \left(\frac{\sin(2\pi\nu h)}{\pi\nu h}\right)^2.$$

This density is very broad for small $h$ and, indeed, converges to 1 for all $\nu \neq 0$ as $h$ converges to 0, which suggests that the processes $X^h$ converge in some sense to a Gaussian white noise process $\dot{W}$ as $h$ converges to 0 and hence that a Gaussian white noise process is the derivative of a Wiener process. We have however already noticed above that, almost surely, the sample paths of a Wiener process are not differentiable anywhere. Thus a Gaussian white noise process cannot be a stochastic process in the usual sense, but must be interpreted in the sense of generalized functions like the Dirac delta function. It cannot be realized physically, but it can be approximated to any desired degree of accuracy by conventional stochastic processes with broad banded spectra, such as (8.12), which are commonly called *coloured noise processes*.

**Exercise 1.8.9**  *Verify (8.13) and (8.14).*

**Exercise 1.8.10**    *Show that the Ornstein-Uhlenbeck process with parameter $\gamma > 0$, that is with covariance function $c(s) = \exp(-\gamma |s|)$, has spectral density*

$$S(\nu) = \frac{2\gamma}{\gamma^2 + 4\pi^2\nu^2}.$$

*What happens as $\gamma \to \infty$? What does this suggest about the Ornstein-Uhlenbeck process?*

## 1.9   Statistical Tests and Estimation

So far we have glossed over some very basic issues concerning the use of pseudo-random number generators. In the preceding PC-Exercises we have assumed that the pseudo-random number generators actually do generate independent random numbers with the desired distribution and have never really specified, for example, just how many terms of a random sequence are required to provide a good approximation or estimation of their limit. These issues are closely interconnected and an extensive theory and array of tests have been proposed for their resolution. In principle we should only interpret a given sequence of numbers as the realization of a sequence of random variables with respect to a set of specified tests that it can pass. Unfortunately none of the tests is completely definitive and an element of subjectivity is often involved in their use. Moreover, where answers can be provided they are in the form of confidence intervals and levels of significance rather than certainties. We shall give a brief sketch of this important topic here and refer the reader to the extensive literature for more details.

We can calculate an estimate of an unknown mean value and determine the number of terms of an approximating sequence needed for a good estimate with the aid of the Law of Large Numbers (5.8), or the Central Limit Theorem (5.9) and the Chebyshev inequality (4.14). To see how, consider the Bernoulli trials with a sequence of i.i.d. random variables $X_1, X_2, \ldots, X_n, \ldots$ taking the value 1 with probability $p$ and the value 0 with probabilty $1 - p$, where $p$ is unknown. Then $E(X_n) = p$ and $\text{Var}(X_n) = p(1 - p)$ and so, by independence, $E(S_n) = np$ and $\text{Var}(S_n) = np(1 - p)$ for the sum $S_n = X_1 + X_2 + \ldots + X_n$. Alternatively, $E(A_n) = p$ and $\text{Var}(A_n) = p(1 - p)/n$ for the sample averages $A_n = S_n/n$. The weak version of the Law of Large Numbers tells us that $A_n$ converges to $p$ in distribution, but gives no more information about the rate of convergence. For this we can apply the Chebyshev inequality (4.14) to $A_n - p$ and the approximation $p(1 - p) \leq 1/4$ for $0 \leq p \leq 1$ to obtain

$$(9.1) \qquad P\left(\{\omega \in \Omega : |A_n - p| \geq a\}\right) \leq \frac{p(1 - p)}{na^2} \leq \frac{1}{4na^2}$$

for all $a > 0$ and so (omitting the $\omega$)

$$P\left(|A_n - p| < a\right) = 1 - P\left(|A_n - p| \geq a\right) \geq 1 - \frac{1}{4na^2}.$$

Thus for any $0 < \alpha < 1$ and $a > 0$ we can conclude that the unknown mean $p$ lies in the interval $(A_n(\omega) - a, A_n(\omega) + a)$ with at least probability $1 - \alpha$ when $n \geq n(a, \alpha) = 1/(4\alpha a^2)$. In statistical terminology we say that the hypothesis that $p$ lies in the interval $(A_n(\omega) - a, A_n(\omega) + a)$ for $n \geq n(a, \alpha)$ is acceptable at a $100\alpha\%$ *level of significance*, and call $(A_n - a, A_n + a)$ a $100(1 - \alpha)\%$ *confidence interval*. For example, $(A_n - 0.1, A_n + 0.1)$ is a 95% confidence interval when $n \geq n(0.1, 0.05) = 500$.

The number $n(a, \alpha)$ above is usually larger than necessary because of the coarseness of the inequalities in (9.1). We can sometimes determine a lower value by using the Central Limit Theorem instead of the Law of Large Numbers. Since the inequality $|A_n - p| \geq a$ is equivalent to $|Z_n| \geq b$ where

$$Z_n = (S_n - np)/\sqrt{np(1 - p)}$$

and $b = a\sqrt{n/p(1 - p)}$ and since, by the Central Limit Theorem, $Z_n$ is approximately standard Gaussian for large $n$ we have

$$(9.2) \qquad P\left(|A_n - p| < a\right) = P\left(|Z_n| < b\right) \approx 2\Phi(b)$$

for sufficiently large $n$, where

$$\Phi(b) = \frac{1}{\sqrt{2\pi}} \int_0^b \exp\left(-\frac{1}{2}x^2\right) dx.$$

For a given $100\alpha\%$ significance level we can read the standard Gaussian statistical tables backwards to determine a value $b = b(\alpha) > 0$ satisfying $\Phi(b) = (1-\alpha)/2$. If we know that, say, $p \in (\frac{1}{4}, \frac{3}{4})$, then for a given $a$ we solve the inequality

$$b(\alpha) = a\sqrt{n/p(1-p)} \leq 4a\sqrt{n}$$

for

$$n \geq \bar{n}(a, \alpha) = b^2(\alpha)/16a^2.$$

This will give us a $100(1-\alpha)\%$ confidence interval $(A_n(\omega)-a,\ A_n(\omega)+a)$ when $n \geq \bar{n}(a, \alpha)$. For example, when $\alpha = 0.05$ we solve $\Phi(b) = 0.475$ for $b \approx 1.96$, so for $a = 0.1$ we calculate $\bar{n}(0.1, 0.05) \approx 25$, which is considerably smaller than the $n(0.1, 0.05) = 500$ determined above.

**Exercise 1.9.1**  *Repeat the above analysis for a sequence of independent $U(0, 1)$ uniformly distributed random variables to obtain the corresponding estimates for $n(a, \alpha)$ and $\bar{n}(a, \alpha)$.*

If we do not know the variance $\sigma^2$ or do not have an estimate for it as with the Bernoulli trials, we can sometimes use the sample variance $\hat{\sigma}^2$ instead. Let $X_1, X_2, \ldots, X_n$ be $n$ independent Gaussian random variables with known mean $\mu$ and unknown variance $\sigma^2$. As before we define the *sample mean* $\hat{\mu}_n = A_n = \sum_{j=1}^n X_j/n$ and the *sample variance* as

$$(9.3) \qquad \hat{\sigma}_n^2 = \frac{1}{n-1} \sum_{j=1}^n (X_j - \hat{\mu}_n)^2,$$

where we divide by $n - 1$ instead of by $n$ since this has been found to give an unbiased estimate of the true variance. Henceforth we shall use a hat " $\hat{\ }$ " on sample statistics to distinguish them from the true statistics. Then for $n > 3$ the random variable

$$T_n = \frac{\hat{\mu}_n - \mu}{\sqrt{\hat{\sigma}_n^2/n}}$$

satisfies the *Student t-distribution* with $n - 1$ degrees of freedom, mean zero and variance $(n-1)/(n-3)$. Similarly to (9.2) we have

| n | 10 | 20 | 30 | 40 | 60 | 100 | 200 |
|---|---|---|---|---|---|---|---|
| $t_{0.9, n-1}$ | 1.83 | 1.73 | 1.70 | 1.68 | 1.67 | 1.66 | 1.65 |
| $t_{0.99, n-1}$ | 3.25 | 2.86 | 2.76 | 2.70 | 2.66 | 2.62 | 2.58 |

**Table 1.9.1**  Values of $t_{1-\alpha, n-1}$ for the Student t-distribution with $n - 1$ degrees of freedom for $\alpha = 0.1$ and $0.01$.

$$P\left(|\hat{\mu}_n - \mu| < a\right) = P\left(|T_n| < t\right)$$

where $t = a\sqrt{n/\hat{\sigma}_n^2}$. For a given $100\alpha\%$ significance level, we check whether or not the test variable

$$T_n^0 = \frac{\hat{\mu}_n - \mu_0}{\sqrt{\hat{\sigma}_n^2/n}}$$

with hypothesized mean $\mu_0$ satisfies the inequality

$$|T_n^0| < t_{1-\alpha, n-1}$$

where $t_{1-\alpha, n-1}$ can be found in statistical tables; some typical values of $t_{1-\alpha, n-1}$ are given in Table 9.1.1. If this is not the case, then we reject the *null hypothesis* $H_0$ that $\mu = \mu_0$. Otherwise, we accept it on the basis of this test. In addition, we form the corresponding $100(1 - \alpha)\%$ confidence interval $(\hat{\mu}_n - a, \hat{\mu}_n + a)$ with

$$a = t_{1-\alpha, n-1}\sqrt{\hat{\sigma}_n^2/n}.$$

This contains all of the values of $\mu_0$ for which the null hypothesis would be accepted by this test.

The t-test requires the original random variables to be Gaussian. When they are not, we can resort to the Central Limit Theorem and use the test asymptotically. We take $n$ batches of $m$ random variables $X_1^{(j)}$, $X_2^{(j)}$, ..., $X_m^{(j)}$ for $j = 1, 2, \ldots, n$, which are independent and identically distributed (i.i.d.) with mean $\mu$ and variance $\sigma^2$. Then we form the sample means $\hat{A}_m^{(j)}$ and the sample variances $(\hat{\sigma}_m^{(j)})^2$ for each batch $j = 1, 2, \ldots, n$ and use the Central Limit Theorem to conclude that the $\hat{A}_m^{(j)}$ are approximately Gaussian. The preceding t-test is then approximately valid for these batch averages rather than for the original $X_i^{(j)}$; in practice it has been found that each batch should have at least $m \geq 15$ terms. For pseudo-random number generators each batch could be determined from a different seed or starting value, thus allowing different sequences to be tested. If a parallel computer is available, the batches and their sample statistics can be calculated simultaneously.

**PC-Exercise 1.9.2** *Simulate $M = 20$ batches of length $N = 100$ of $U(0,1)$ distributed random numbers and evaluate the 90% confidence interval for their mean value.*

PC-Exercises 1.4.4, 1.4.5 and 1.4.6 required frequency histograms for the outputs of various random generators to be plotted and compared visually with the graphs of the density functions that they were supposed to simulate. There are various statistical tests which allow less subjective comparisons, the $\chi^2$-*goodness- of- fit test* being one of the most commonly used. To apply it we need a large number $N$ of values of i.i.d. random variables. From these we form a cumulative frequency histogram $F_N(x)$ which we wish to compare with the supposed distribution $\bar{F}(x)$. We subdivide our data values into $k + 1$ mutually exclusive categories and count the numbers $N_1, N_2, \ldots, N_{k+1}$ terms falling into these categories; obviously $N_1 + N_2 + \cdots + N_{k+1} = N$. We compare these with the expected numbers $N\bar{p}_1$, $N\bar{p}_2$, ..., $N\bar{p}_{k+1}$ for each category for the distribution $\bar{F}$. To do this we calculate the Pearson statistic

$$(9.4) \qquad \chi^2 = \sum_{j=1}^{k+1} \frac{(N_j - N\bar{p}_j)^2}{N\bar{p}_j}$$

which should be small if our null hypothesis $H_0$ that the data generating mechanism has $\bar{F}(x)$ as its distribution function is to be acceptable at a reasonable significance level. It is known that the Pearson statistic is distributed asymptotically in $N$ according to the $\chi^2$-*distribution* with $k$-degrees of freedom with $E(\chi^2) = k$ and $\text{Var}(\chi^2) = 2k$. To complete the test we pick a $100\alpha\%$ significance level and determine from statistical tables a value $\chi^2_{1-\alpha,k}$ such that $P(\chi^2 < \chi^2_{1-\alpha,k}) = 1 - \alpha$. If our $\chi^2$ value from (9.4) satisfies $\chi^2 < \chi^2_{1-\alpha,k}$, then we accept our null hypothesis at the significance level $100\alpha\%$; otherwise we reject this hypothesis. For instance, $\chi^2_{0.95,30} \approx 43.8$ and $\chi^2_{0.99,30} \approx 50.9$.

**PC-Exercise 1.9.3**    *Use the $\chi^2$-goodness-of-fit test with $k = 30$ degrees of freedom, $N = 10^3$ generated numbers and significance levels 1% and 5% to test the goodness-of-fit of the $U(0,1)$ uniformly distributed pseudo-random number generator on your PC. Repeat these tests for the exponentially distributed random number generator with parameter $\lambda = 1.0$ and for the Box-Muller and Polar Marsaglia $N(0;1)$ generators. In each case repeat the calculations using different seeds for the random number generator.*

For continuously distributed random variables the discrete categories of the $\chi^2$-goodness-of-fit test are artificial, subjective and do not take fully into account the variability in the data. These disadvantages are avoided in the *Kolmogorov-Smirnov test*. This is based on the Glivenko-Cantelli theorem which says that $D_N \rightarrow 0$ almost surely as $N \rightarrow \infty$, where

$$D_N = \sup_{-\infty < x < \infty} |F_N(x) - \bar{F}(x)|.$$

We recall that the sample frequency histograms $F_N$ here are themselves random variables. Kolmogorov and Smirnov proved for any continuous distribution $\bar{F}$ that $\sqrt{N} D_N$ converges in distribution as $N \rightarrow \infty$ to a random variable with the *Kolmogorov distribution*

$$H(x) = 1 - 2 \sum_{j=1}^{\infty} (-1)^{j-1} \exp\left(-2j^2 x^2\right)$$

for $x > 0$ with $H(0) = 0$, the values of which can be found in statistical tables. To apply the test at a $100\alpha\%$ significance level we compare the value of $\sqrt{N} D_N$ calculated from our data with the value $x_{1-\alpha}$ satisfying $H(x_{1-\alpha}) = 1 - \alpha$. If $\sqrt{N} D_N \leq x_{1-\alpha}$ we accept at the $100\alpha\%$ significance level the null hypothesis $H_0$ that the data generating mechanism has $\bar{F}(x)$ as its distribution function; otherwise we reject it. In general $N > 35$ suffices for this test, but for pseudo-random number generators a larger value should be taken in order to test a more representative sample of generated numbers. For instance, $x_{0.95} \approx 1.36$ and $x_{0.99} \approx 1.63$.

**PC-Exercise 1.9.4** *Repeat PC-Exercise 1.9.3 using the Kolmogorov-Smirnov test at 1% and 5% significance levels and $N = 10^3$ $U(0,1)$ random numbers. Compare the results with those for the $\chi^2$-test in PC-Exercise 1.9.3.*

Over recent decades it has been found that most commonly used pseudo-random number generators fit their supposed distributions reasonably well. The results of PC-Exercise 1.9.3 should concur with this perception. In contrast, the generated numbers often seem not to be independent as they are supposed to be, which is not surprising since, for congruential generators at least, each number is determined exactly by its predecessor. However, in practice statistical independence is an elusive property to confirm definitively for pseudo-random numbers generated by digital computers, and tests for it are nowhere near as satisfactory as those above for the goodness-of-fit of distributions. We shall restrict our remarks here to $U(0,1)$ uniformly distributed linear congruential generators as described by (3.1) which have the form $U_n = X_n/c$ where

$$X_{n+1} = aX_n + b \pmod{c}.$$

A simple test for independence involves plotting the successive pairs $(U_{2n-1}, U_{2n})$ for $n = 1, 2, \ldots$ as points on the unit square with the $U_{2n-1}$ as the $x$-coordinate and the $U_{2n}$ as the $y$-coordinate. These points lie on one of $c$ different straight lines of slope $a/c$ and a large number of them should fairly evenly fill the unit square. The presence of patches without any of these points is an indication of bias in the generator. One way to avoid it is to introduce a *shuffling procedure*. For this we generate a string of 20 or more numbers and choose one number with equal probability from the string. We take this number as the output of our shuffling procedure and then generate a new number to replace it in the string. Repeating this step as often as required, we obtain a shuffled sequence of pseudo-random numbers. We note that this requires more numbers to be generated than for an unshuffled sequence of the same length since a random number must also be generated at each step in order to choose the number to be taken from the string. Shuffling procedures have been found to be quite effective in reducing patchiness in poor generators.

**PC-Exercise 1.9.5** *Plot $10^3$ points $(U_{2n-1}, U_{2n})$ using the linear congruential generator (3.1) with parameters $a = 1229$, $b = 1$ and $c = 2048$, using the seed $X_0 = 0$. Add a shuffling procedure to the generator and repeat the above plots. Does the shuffling subroutine make a noticeable difference?*

**PC-Exercise 1.9.6** *Repeat PC-Exercise 1.9.5 using the $U(0,1)$ random number generator on your PC.*

The preceding test is useful in eliminating glaringly biased generators, but is no guarantee of an unbiased generator. For example, the *RANDU* generator

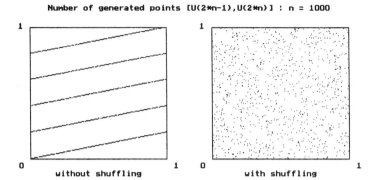

Figure 1.9.1    Result of PC-Exercise 1.9.5.

$X_{n+1} = 65,539X_n$    (mod $2^{31}$) appears relatively unbiased, but successive triples $(X_n, X_{n+1}, X_{n+2})$ satisfy $X_{n+2} = (6X_n - 9X_{n+1})$    (mod $2^{31}$). This relationship was, however, discovered long after the $RANDU$ generator was introduced and lead to its demise. There are many other tests for independence, including the *runs test* in which the number and lengths of runs of successively increasing numbers are analysed statistically, but their validity is still unclear.

To conclude this section we remark that there are generators which are fairly successful at mimicking the salient properties of truly independent random sequences on digital computers and that in applications the negative properties of such sequences often play no role. Thus, while they are far from perfect, pseudo-random number generators are often adequate for the task at hand. We shall assume in our simulations that they do have the asserted independence and distributions. An advantage of pseudo-random generators over random number generators based on natural physical noise sources is that they are reproducible.

# Chapter 2

# Probability Theory
# and Stochastic Processes

Like Chapter 1, the present chapter also reviews the basic concepts and results
on probability and stochastic processes for later use in the book, but now the
emphasis is more mathematical. Integration and measure theory are sketched
and an axiomatic approach to probability is presented. Apart form briefly
perusing the chapter, the general reader could omit this chapter on the first
reading.


## 2.1   Aspects of Measure
##        and Probability Theory

The axiomatic development of probability theory was initiated by Kolmogorov
in the early 1930's. It is based on measure theory, but it has developed characteristics and methods of its own. The fundamental concept in this approach to
probability theory is the *probability space,* an ordered triple $(\Omega, \mathcal{A}, P)$ consisting of an arbitrary nonempty set $\Omega$ called the *sample space,* a $\sigma$-algebra $\mathcal{A}$ of
subsets of $\Omega$ called *events* and a *probability measure* $P$ on $\mathcal{A}$ assigning to each
event $A \in \mathcal{A}$ a number $P(A) \in [0,1]$ called its *probability.*

**Definition 2.1.1**   *A collection $\mathcal{A}$ of subsets of $\Omega$ is a $\sigma$-algebra if*

$$(1.1) \qquad \Omega \ \in \ \mathcal{A}$$

$$(1.2) \qquad A^c \ \in \ \mathcal{A} \quad if \ A \in \mathcal{A}$$

$$(1.3) \qquad \bigcup_{n=1}^{\infty} A_n \ \in \ \mathcal{A} \quad if \ A_1, A_2, \ldots, A_n, \ldots \in \mathcal{A}.$$

A $\sigma$-algebra $\mathcal{A}$ is thus a collection of subsets of $\Omega$ which contains $\Omega$ and
is closed under the set operations of complementation and countable unions.
These defining properties are minimal and imply that $\emptyset \in \mathcal{A}$ and that $\bigcap_{n=1}^{\infty} A_n$
$\in \mathcal{A}$ whenever $A_1, A_2, \ldots, A_n, \ldots \in \mathcal{A}$, that is $\mathcal{A}$ is also closed under countable
intersections. The possible $\sigma$-algebras for a given set $\Omega$ range from the coarsest
$\{\emptyset, \Omega\}$ to the finest $\mathcal{P}(\Omega)$ consisting of all of the subsets of $\Omega$. Often we have
a collection $\mathcal{C}$ of subsets of $\Omega$ which is not a $\sigma$-algebra and require a $\sigma$-algebra
containing $\mathcal{C}$. Obviously $\mathcal{C}$ is contained in $\mathcal{P}(\Omega)$, but there is a smallest $\sigma$-
algebra containing $\mathcal{C}$ which we denote by $\mathcal{A}(\mathcal{C})$ and call the *$\sigma$-algebra generated
by $\mathcal{C}$.* For example, if $\mathcal{C} = \{C\}$ then $\mathcal{A}(\mathcal{C}) = \{\emptyset, C, C^c, \Omega\}$. Another example

is the $\sigma$-algebra $\mathcal{B}$ of Borel subsets of $\Re$, generated by the collection $\mathcal{C}$ of semi-infinite intervals $\{x \in \Re : -\infty < x \le c\}$ for all $c \in \Re$.

A *measurable space* is an ordered pair $(\Omega, \mathcal{A})$ consisting of an arbitrary nonempty set $\Omega$ and a $\sigma$-algebra $\mathcal{A}$ of subsets of $\Omega$. A measure $\mu$ on $(\Omega, \mathcal{A})$ is a countably additive set function assigning to each set $A \in \mathcal{A}$ a nonnegative, possibly infinite, number $\mu(A)$. To be specific

**Definition 2.1.2**   *A measure $\mu$ on a measurable space $(\Omega, \mathcal{A})$ is a nonnegative valued set function on $\mathcal{A}$ satisfying*

$$(1.4) \qquad\qquad \mu(\emptyset) = 0$$

*and*

$$(1.5) \qquad\qquad \mu\left(\bigcup_{n=1}^{\infty} A_n\right) = \sum_{n=1}^{\infty} \mu(A_n)$$

*for any sequence $A_1, A_2, \ldots, A_n, \ldots \in \mathcal{A}$ with $A_i \cap A_j = \emptyset$ for $i \ne j$ where $i$, $j = 1, 2, 3, \ldots$.*

For any measure $\mu$ on a measurable space $(\Omega, \mathcal{A})$ we have a *measure space* $(\Omega, \mathcal{A}, P)$. From (1.5) it follows that $\mu(A) \le \mu(B)$ for all $A \subseteq B$ in $\mathcal{A}$. We say that $\mu$ is *finite* if $\mu(\Omega) < \infty$, in which case $\mu(A) \le \mu(\Omega) < \infty$ for all $A \in \mathcal{A}$. Definition 2.1.2 does not exclude the trivial case that $\mu(A) = 0$ for all $A \in \mathcal{A}$, but we shall not consider such trivial measures here. For $0 < \mu(\Omega) < \infty$ we can normalize the measure $\mu$ and obtain a probability measure $P$ with $P(A) = \mu(A)/\mu(\Omega) \in [0, 1]$ for all $A \in \mathcal{A}$.

**Definition 2.1.3** *A probability measure $P$ is a measure on a measurable space $(\Omega, \mathcal{A})$ for which $P(\Omega) = 1$.*

A probability space $(\Omega, \mathcal{A}, P)$ is thus a measurable space with a probability measure. From Definitions 2.1.1, 2.1.2 and 2.1.3 we clearly have

$$(1.6) \qquad\qquad P(\emptyset) = 0, \qquad P(\Omega) = 1;$$

$$(1.7) \qquad\qquad 0 \le P(A) \le P(B) \le 1$$

for all $A, B \in \mathcal{A}$ with $A \subseteq B$;

$$(1.8) \qquad\qquad P(A^c) = 1 - P(A)$$

for all $A \in \mathcal{A}$; and

$$(1.9) \qquad\qquad P\left(\bigcup_{n=1}^{\infty} A_n\right) = \sum_{n=1}^{\infty} P(A_n)$$

for any mutually exclusive $A_1, A_2, \ldots, A_n, \ldots \in \mathcal{A}$, that is with $A_i \cap A_j = \emptyset$ for all $i \ne j$ and $i, j = 1, 2, 3, \ldots$. A little harder to prove is the continuity result

$$(1.10) \qquad\qquad P\left(\bigcap_{n=1}^{\infty} A_n\right) = \lim_{n \to \infty} P(A_n)$$

for any $A_1 \supset A_2 \supset \cdots \supset A_n \supset \cdots$ with $A_1, A_2, \ldots, A_n, \ldots \in \mathcal{A}$.

Another elementary consequence is the deceptively powerful *Borel- Cantelli Lemma*:

**Lemma 2.1.4** *For any sequence of events* $A_1, A_2, \ldots, A_n \ldots \in \mathcal{A}$

$$(1.11) \qquad P\left(\bigcap_{n=1}^{\infty} \bigcup_{k=n}^{\infty} A_k\right) = 0 \quad \text{if} \quad \sum_{n=1}^{\infty} P(A_n) < \infty.$$

The event $\bigcap_{n=1}^{\infty} \bigcup_{k=n}^{\infty} A_k$ consists of all those outcomes occuring in infinitely many of the events $A_1, A_2, \ldots, A_n, \ldots$ and is thus a null event if the series $\sum_{n=1}^{\infty} P(A_n)$ converges.

An important measure is the *Borel measure* $\mu_B$ on the $\sigma$-algebra $\mathcal{B}$ of Borel subsets of $\Re$, which assigns to each finite interval $[a, b] = \{x \in \Re : a \leq x \leq b\} =$ its length, that is $\mu_B([a, b]) = b - a$. Consequently $\mu_B$ is not a finite measure, but it is *$\sigma$-finite* on $\Re$ in that we can write $\Re = \bigcup_{n=1}^{\infty} [-n, n]$ with $\mu_B([-n, n]) = 2n < \infty$ for each $n = 1, 2, 3, \ldots$.

The measure space $(\Re, \mathcal{B}, \mu_B)$ is not complete in the sense that there exist subsets $B^*$ of $\Re$ with $B^* \notin \mathcal{B}$ but $B^* \subset B$ for some $B \in \mathcal{B}$ with $\mu(B) = 0$. Intuitively we would expect such subsets to have zero measure too. A procedure of measure theory allows us to enlarge the $\sigma$-algebra $\mathcal{B}$ to a $\sigma$-algebra $\mathcal{L}$ and to extend the measure $\mu_B$ uniquely to a measure $\mu_L$ on $\mathcal{L}$ so that $(\Re, \mathcal{L}, \mu_L)$ is *complete*, that is $L^* \in \mathcal{L}$ with $\mu_L(L^*) = 0$ whenever $L^* \subset L$ for some $L \in \mathcal{L}$ with $\mu_L(L) = 0$. In particular, we have $\mu_L(B) = \mu_B(B)$ for each $B \in \mathcal{B}$, whereas for each $L \in \mathcal{L} \setminus \mathcal{B}$ there exists a $B \in \mathcal{B}$ and an $L^* \in \mathcal{L}$ so that $L = B \cup L^*$ with $\mu_L(L) = \mu_B(B)$ and $\mu_L(L^*) = 0$. We call $\mathcal{L}$ the $\sigma$-algebra of *Lebesgue subsets* of $\Re$ and $\mu_L$ the *Lebesgue measure* on $\Re$.

Many other measures can also be defined on $(\Re, \mathcal{L})$. We say that a measure $\mu$ is *singular* with respect to the Lebesgue measure $\mu_L$, written $\mu \perp \mu_L$, if there exists a subset $S \in \mathcal{L}$ with $\mu(S) > 0$, but $\mu_L(S) = 0$. On the other hand, we say that $\mu$ is *absolutely continuous* with respect to $\mu_L$, written $\mu \ll \mu_L$, if $\mu(L) = 0$ whenever $\mu_L(L) = 0$ for any $L \in \mathcal{L}$. In general a measure $\mu$ on $(\Re, \mathcal{L})$ will be neither singular nor absolutely continuous with respect to $\mu_L$, but in these cases the *Lebesgue Decomposition Theorem* says that it can always be written as the sum of a singular and an absolutely continuous measure.

If $(\Omega_1, \mathcal{A}_1)$ and $(\Omega_2, \mathcal{A}_2)$ are two measurable spaces we call a function $f : \Omega_1 \to \Omega_2$ measurable, to be precise $\mathcal{A}_1 : \mathcal{A}_2$-measurable, if $f^{-1}(A_2) = \{\omega_1 \in \Omega_1 : f(\omega_1) \in A_2\} \subseteq \mathcal{A}_1$ for all $A_2 \in \mathcal{A}_2$. In particular, we say that a function is *Borel* (or *Lebesgue*) *measurable* when $\Omega_2 = \Re$ and $\mathcal{A}_2 = \mathcal{B}$ (or $\mathcal{L}$). When $(\Omega_1, \mathcal{A}_1, P)$ is a probability space and $(\Omega_2, \mathcal{A}_2)$ is either $(\Re, \mathcal{B})$ or $(\Re, \mathcal{L})$ we call a measurable function $f : \Omega_1 \to \Re$ a *random variable*, and usually denote it by $X$. For a Borel measurable random variable $X$ we can express measurability equivalently by requiring $X^{-1}((-\infty, x]) \in \mathcal{A}_1$, that is to be an event, for every $x \in \Re$. As we saw in Section 2 of Chapter 1 a random variable $X$ induces a probability distribution $P_X$ on $(\Re, \mathcal{B})$ defined by $P_X(B) = P(X^{-1}(B))$ for all $B \in \mathcal{B}$. $P_X$ is a probability measure, but the resulting probability space $(\Re, \mathcal{B}, P_X)$ is not complete. However we can complete it to $(\Re, \mathcal{L}, P_X)$ by extending $P_X$ to the Lebesgue subsets of $\Re$, where for convenience we use the same symbol $P_X$ for the extended measure. One advantage of working with this completed probability space is that every subset of a null event is itself a null event. Thus if $X$ is a (Lebesgue measurable) random variable and if $Y(\omega) = X(\omega)$, w.p.1, then $Y$ is also a random variable; $Y$ is essentially indistinguishable from $X$ and in this sense can be considered to be equivalent to $X$. Another advantage is that $P_X$ and, of greater practical significance, the corresponding distribution function $F_X$ defined by $F_X(x) = P_X((-\infty, x])$ for all $x \in \Re$, can be related to the Lebesgue measure $\mu_L$ on $(\Re, \mathcal{L})$. For example, when we have an absolutely continuous probability measure $P_X \ll \mu_L$, then there is a density function and the integral relationship (1.2.10) holds. On the other hand, when $X$ is a discrete random variable then we have a singular probability measure $P_X \perp \mu_L$.

From measure theory we know that a measurable function results when we perform countably many basic arithmetic operations on measurable functions defined on a common space, where the operations are defined pointwise with respect to the common domain of the functions. The same thus also holds for random variables. In particular, if $X$ and $Y$ are random variables on a probability space $(\Omega, \mathcal{A}, P)$ and if $c \in \Re$, then $X + c$, $cX$, $X + Y$, $X - Y$, $XY$ and $X/Y$ (provided for the last one that $Y = 0$ only on null events) are random variables. Moreover, for a sequence of random variables $X_1, X_2, \ldots, X_n, \ldots$ the functions defined by

$$\max_{1 \le k \le n} X_k, \quad \min_{1 \le k \le n} X_k, \quad \sup_{k \ge 1} X_k \quad \text{and} \quad \inf_{k \ge 1} X_k$$

are random variables, as are

(1.12)    $\displaystyle \limsup X_n = \inf_{n \ge 1} \sup_{k \ge n} X_k \quad \text{and} \quad \liminf X_n = \sup_{n \ge 1} \inf_{k \ge n} X_k.$

If the limit supremum and limit infimum in (1.12) are equal, except possibly on a null event, then their common value $\lim X_n$ is just the limit for convergence with probability one.

Often the $\sigma$-algebra $\mathcal{A}$ of events in a probability space $(\Omega, \mathcal{A}, P)$ contains much more information than is detectable by looking at the values taken by a particular random variable $X$. For this reason it is often useful to consider the $\sigma$-algebra $\mathcal{A}(X)$ generated by subsets of the form $X^{-1}(L)$ for all $L \in \mathcal{L}$. This is the coarsest $\sigma$-algebra consistent with the measurability of $X$, or equivalently minimal with respect to the events detectable by $X$. For example, $\mathcal{A}(I_A) = \{\emptyset, A, A^c, \Omega\}$ for the indicator $I_A$ of an event $A \in \mathcal{A}$ (see 1.2.1). The $\sigma$-algebra $\mathcal{A}(X_1, X_2, \ldots, X_n)$ for a collection of random variables $X_1, X_2, \ldots, X_n$ can be defined in an analogous way.

Finally, we consider product $\sigma$-algebras and product measures. Let $(\Omega_1, \mathcal{A}_1, \mu_1)$ and $(\Omega_2, \mathcal{A}_2, \mu_2)$ be two measure spaces. Then we define the *product $\sigma$-algebra* $\mathcal{A}_1 \otimes \mathcal{A}_2$ on $\Omega_1 \times \Omega_2$ to be the $\sigma$-algebra generated by the subsets $A_1 \times A_2$ for all $A_1 \in \mathcal{A}_1$ and $A_2 \in \mathcal{A}_2$. The *product measure* $\mu_1 \times \mu_2$ is the extension to a measure on $\mathcal{A}_1 \otimes \mathcal{A}_2$ of a set function satisfying $\mu_1 \times \mu_2 (A_1 \times A_2) = \mu_1(A_1)\mu_2(A_2)$ for any $A_1 \in \mathcal{A}_1$ and $A_2 \in \mathcal{A}_2$. In the case that the measurable spaces coincide, we write $\Omega^2$ for $\Omega \times \Omega$ and $\mathcal{A}^2$ for $\mathcal{A} \otimes \mathcal{A}$; note that the product

$\sigma$-algebra $\mathcal{A}^2$ also contains sets which are not product sets, for example the diagonal set $D = \{(\omega, \omega) : \omega \in \Omega\}$. When $\Omega = \Re$ and $\mathcal{A} = \mathcal{L}$ we obtain the $\sigma$-algebra $\mathcal{L}^2$ of two-dimensional Lebesgue subsets and call the product measure $\mu_L \times \mu_L$ the *2-dimensional Lebesgue measure*. Analogous definitions apply for the $n$-dimensional products $\Re^n$ and $\mathcal{L}^n$.

## 2.2 Integration and Expectations

Mathematical expectations and stochastic integration play a central role in this book. Hence in this section we shall summarize the basic concepts and results for deterministic integration, expectations and conditional expectations.

The standard integral of deterministic calculus is the *Riemann integral*, which is defined as the limit of the sums of areas of approximating rectangles. Let $f : [a, b] \to \Re$ be a continuous function on a bounded interval $[a, b]$ and let

$$a = x_1^{(n)} < x_2^{(n)} < \cdots < x_i^{(n)} < \cdots < x_{n+1}^{(n)} = b$$

be an arbitrary partition of $[a, b]$ into $n$ subintervals $\left[x_i^n, x_{i+1}^n\right]$ of (not necessarily equal) length $\delta_i^{(n)} = x_{i+1}^{(n)} - x_i^{(n)}$ where $i = 1, 2, \ldots, n$. On each subinterval we determine the minimum and maximum values of function $f$,

$$(2.1) \qquad m_i^{(n)} = \min_{x_i^{(n)} \leq x \leq x_{i+1}^{(n)}} f(x), \qquad M_i^{(n)} = \max_{x_i^{(n)} \leq x \leq x_{i+1}^{(n)}} f(x),$$

and form the sums of the areas of the lower and the upper rectangular approximations to the graph of $f$ on the subintervals, namely

$$L_n = \sum_{i=1}^{n} m_i^{(n)} \delta_i^{(n)}, \qquad U_n = \sum_{i=1}^{n} M_i^{(n)} \delta_i^{(n)}.$$

If we choose the partitions so that $\delta^{(n)} = \max_{1 \leq i \leq n} \delta_i^{(n)} \to 0$ as $n \to \infty$, then both of the limits

$$L = \lim_{n \to \infty} L_n, \qquad U = \lim_{n \to \infty} U_n$$

exist and are equal. We call their common value the Riemann integral of $f$ on $[a, b]$ and denote it by

$$(R) \int_a^b f(x) \, dx.$$

Since $m_i^{(n)} \leq f\left(\xi_i^{(n)}\right) \leq M_i^{(n)}$ for any $\xi_i^{(n)} \in \left[x_i^{(n)}, x_{i+1}^{(n)}\right]$ for $i = 1, 2, \ldots, n$, we have

$$L_n \leq \sum_{i=1}^{n} f\left(\xi_i^{(n)}\right) \leq U_n$$

and hence

$$\lim_{n \to \infty} \sum_{i=1}^{n} f\left(\xi_i^{(n)}\right) \delta_i^{(n)} = (R) \int_a^b f(x) \, dx;$$

that is, when a Riemann integral exists, the approximating rectangles can be evaluated at any point within a partition subinterval.

**Exercise 2.2.1**      *Determine the lower and upper rectangular sums of the function $f(x) = 2x$ on the interval $[0, 1]$ and use them to evaluate its Riemann integral on this interval.*

Every continuous function is Riemann integrable on a closed and bounded interval, but not all discontinuous bounded functions are, even if we replace the minimum and maximum in (2.1) by the infimum and supremum. For example, the indicator function $I_Q$ for the set $Q$ of rational numbers has $L_n = 0$ and $U_n = b - a$ for $n = 1, 2, 3, \ldots$ on any bounded interval $[a, b]$, so is not Riemann integrable there. It is obviously a Lebesgue measurable function, and so in a probabilistic context could be a random variable with an expectation and moments, properties which are usually expressed in terms of integrals. To handle such functions, we must use the definition of an integral due to Lebesgue. The Lebesgue integral has the same value as the Riemann integral when the latter exists, but has the added advantage of being the appropriate definition for the development of a unified, abstract theory of integration.

Let $(\Omega, \mathcal{A}, \mu)$ be a measure space and suppose that $\mu$ is finite, for otherwise we can restrict attention to the subsets in $\mathcal{A}$ with finite $\mu$ measure. A simple function on $(\Omega, \mathcal{A}, \mu)$ is a real valued function of the form $\phi = \sum_{i=1}^{N} a_i I_{A_i}$, where $a_1, a_2, \ldots, a_N \in \Re$ and the sets $A_1, A_2, \ldots, A_N \in \mathcal{A}$ are pairwise disjoint. The Lebesgue integral of such a simple function $\phi$ over a measurable subset $A \in \mathcal{A}$ is defined as

$$(2.2) \qquad \int_A \phi \, d\mu = \sum_{i=1}^{N} a_i \, \mu (A \cap A_i) .$$

The *Lebesgue integral* of a real-valued $\mathcal{A} : \mathcal{L}$- measurable function $f : \Omega \to \Re$ is then built up systematically from this definition. We shall henceforth omit the qualifier Lebesgue unless it is necessary to distinguish the integral from another kind of integral. Suppose that the function $f$ is nonnegative and bounded, that is with $0 \le f(\omega) < N$ for all $\omega \in \Omega$ and some finite $N$. Then we define

$$\int_A f \, d\mu = \sup_{\phi \le f} \int_A \phi \, d\mu,$$

where the supremum is taken over all simple functions $\phi$ with $\phi(\omega) \le f(\omega)$ for all $\omega \in A$; it is finite since the integrals of these simple functions are bounded from above by $N\mu(A)$. As an example of such simple functions consider $\phi_n = \sum_{i=0}^{n-1} (iN/n) I_{E_i(A;n)}$, where the subsets

$$(2.3) \qquad E_i(A; n) = A \cap \left\{ \omega \in \Omega : \frac{i}{n} N \le f(\omega) < \frac{i+1}{n} N \right\}$$

belong to $\mathcal{A}$ since $A$ does and $f$ is measurable; these $\phi_n$ satisfy $\phi_n(\omega) \le f(\omega)$ with $\phi_n(\omega) \to f(\omega)$ as $n \to \infty$ for all $\omega \in A$.

In general, when $f$ is nonnegative but unbounded we define

$$\int_A f \, d\mu = \sup_N \int_A f_N \, d\mu$$

where $f_N(\omega) = \min\{f(\omega), N\}$, but here the supremum may be infinite. Finally, for a measurable function $f$ of unrestricted sign we write $f = f^+ - f^-$, where $f^+(\omega) = \max\{f(\omega), 0\}$ and $f^-(\omega) = \max\{-f(\omega), 0\}$, and define

$$(2.4) \qquad \int_A f \, d\mu = \int_A f^+ \, d\mu - \int_A f^- \, d\mu,$$

provided at least one of the integrals on the right side of (2.4) is finite. If both are finite we say that $f$ is (Lebesgue) integrable, which is equivalent to $|f|$ being integrable since $|f| = f^+ + f^-$. We denote the set of all $\mathcal{A}:\mathcal{L}$-measurable functions which are integrable on $\Omega$ (and hence on any subset $A \in \mathcal{A}$) with respect to $\mu$ by $L^1(\Omega, \mathcal{A}, \mu)$.

**Exercise 2.2.2** *Use simple functions approximating the function $f(x) = 2x$ on the interval $[0, 1]$ to evaluate its Lebesgue integral.*

From the linearity of the summation in (2.2) we can easily show for any $f, g \in L^1(\Omega, \mathcal{A}, \mu)$ and real constants $\alpha, \beta$ that $\alpha f + \beta g \in L^1(\Omega, \mathcal{A}, \mu)$ with

$$(2.5) \qquad \int_A (\alpha f + \beta g) \, d\mu = \alpha \int_A f \, d\mu + \beta \int_A g \, d\mu,$$

$$(2.6) \qquad \int_A f \, d\mu \le \int_A g \, d\mu \quad \text{if} \quad f(\omega) \le g(\omega) \quad \text{for all} \quad \omega \in A$$

and

$$(2.7) \qquad \int_{A \cup B} f \, d\mu = \int_A f \, d\mu + \int_B f \, d\mu$$

for any disjoint subsets $A, B \in \mathcal{A}$ with finite $\mu$-measure. The pointwise inequality in (2.6) can be weakened to all $\omega$ in $A$ except those in a subset $A_0 \subset A$ of zero $\mu$-measure, that is for almost all $\omega \in A$. In addition, we obviously have

$$(2.8) \qquad \left| \int_A f \, d\mu \right| \le \int_A |f| \, d\mu.$$

A deeper result is the *Dominated Convergence Theorem of Lebesgue:*

**Theorem 2.2.3** *Suppose that $f, g \in L^1(\Omega, \mathcal{A}, \mu)$ where $\mu(\Omega) < \infty$ and that $f_1, f_2, f_3, \ldots$ is a sequence of $\mathcal{A}:\mathcal{L}$-measurable functions with $|f_n(\omega)| \le |g(\omega)|$ for almost all $\omega \in \Omega$ and $n = 1, 2, 3, \ldots$. Then*

$$\lim_{n \to \infty} \int_\Omega f_n \, d\mu = \int_\Omega f \, d\mu$$

*if*

$$\lim_{n \to \infty} f_n(\omega) = f(\omega)$$

*for almost all $\omega \in \Omega$ and $n = 1, 2, 3, \ldots$.*

If a Lebesgue integrable function $f : \Re \to \Re$ is Riemann integrable on a bounded interval $[a, b]$, then it is Lebesgue integrable on $[a, b]$ (with respect to the Lebesgue measure $\mu_L$) and the two integrals are equal, that is

$$(2.9) \qquad (R) \int_a^b f(x) \, dx = \int_{[a,b]} f(x) \, d\mu_L.$$

For a continuous function $f$ with $0 \leq f(x) \leq N$ for $x \in [a, b]$ this follows from the fact that the subset

$$E_i(n) = \left\{ x \in [a, b] : \frac{i}{n} N \leq f(x) \leq \frac{i+1}{n} N \right\}$$

is the union of a finite number of subintervals $[x_{i_j}, x_{i_j+1}]$ for $j = 1, 2, \ldots,$ $K(i, n)$. In each of these we can find an $\xi_{i_j}$ such that $f(\xi_{i_j}) = iN/n$ and then calculate the sum of the areas of the corresponding rectangles for all of the subintervals, obtaining

$$\sum_{j=1}^{K(i,n)} f\left(\xi_{i_j}\right) \left(x_{i_j+1} - x_{i_j}\right) = \frac{i}{n} N \, \mu_L \left(E_i(n)\right).$$

The general result is proved in much the same way. As a consequence we shall often write the Lebesgue integral (2.9) in the more convenient form $\int_a^b f(x) \, dx$. There are however quite elementary functions which are Lebesgue integrable but not Riemann integrable; for example, the indicator function of the rational numbers $I_{\mathbb{Q}}$ has Lebesgue integral $\int_a^b I_{\mathbb{Q}} \, dx = b - a$, but is not Riemann integrable.

As mentioned in Section 1 there are many measures on $(\Re, \mathcal{L})$ other than the Lebesgue measure $\mu_L$. For such measures an integral $\int_A f \, d\mu$ is sometimes called a *Lebesgue- Stieltjes integral*. Of particular interest are those measures $\mu$ which are absolutely continuous with respect to $\mu_L$, that is $\mu \ll \mu_L$. This certainly holds if there is a measurable function $p : \Re \to \Re$ which is nonnegative, except possibly on a set of $\mu_L$-measure zero, such that $\mu(A) = \int_A p \, d\mu_L$ for all $A \in \mathcal{L}$. The converse of this statement also holds and is a major result of integration theory. It is known as the *Radon-Nikodym Theorem* and, in fact, asserts that the function $p$ is unique in the sense that any other function with the same property differs from $p$ on at most a set of $\mu_L$-measure zero. We often write this relationship between measures symbolically as $d\mu/d\mu_L = p$. If $f \in L^1(\Omega, \mathcal{L}, \mu)$, we have $fp \in L^1(\Omega, \mathcal{L}, \mu_L)$ and

$$\int_A f \, d\mu = \int_A fp \, d\mu_L$$

for each $A \in \mathcal{L}$. In the case that $P_X \ll \mu_L$ for a probability measure $P_X$ induced on $(\Re, \mathcal{L})$ by a random variable $X$ the function $p = dP_X/d\mu_L$ is the density of the distribution function $F_X$ and is sometimes called a Radon-Nikodym derivative.

The above discussion also applies for measures on a measurable product space such as $(\Re^n, \mathcal{L}^n)$. This raises the question, which is of some practical significance, as to whether or not an integral with respect to a product measure $\mu_1 \times \mu_2 \times \cdots \times \mu_n$ on such a space can be evaluated as a succession of integrals with respect to each of the constituent measures $\mu_1, \mu_2, \ldots, \mu_n$. To be specific, let $(\Omega_1, \mathcal{A}_1, \mu_1)$ and $(\Omega_2, \mathcal{A}_2, \mu_2)$ be two complete measure spaces and let $f \in L^1(\Omega_1 \times \Omega_2, \mathcal{A}_1 \times \mathcal{A}_2, \mu_1 \times \mu_2)$. Can the integral $\int_{A_1 \times A_2} f \, d\mu_1 \times \mu_2$, where $A_1 \in \mathcal{A}_1$ and $A_2 \in \mathcal{A}_2$, be evaluated in terms of one or both of the iterated integrals

$$\int_{A_1} \left( \int_{A_2} f \, d\mu_2 \right) d\mu_1 \quad \text{and} \quad \int_{A_2} \left( \int_{A_1} f \, d\mu_1 \right) d\mu_2?$$

The answer is affirmative and is provided by *Fubini's Theorem*, which says that the three integrals are equal. It also says, necessarily, that the intermediate functions and integrals of a single variable are integrable on the appropriate subset $A_1$ or $A_2$. There is a catch here: the function $f(\omega_1, \cdot) \in L^1(\Omega^2, \mathcal{A}_2, \mu_2)$ only for almost all ($\mu_1$-measure) $\omega_1 \in \Omega_1$ rather than for all $\omega_1 \in \Omega_1$, and analogously for the functions $f(\cdot, \omega_2)$. Apart from technicalities in proofs this is not a problem because the exceptional points have no effect on the value of the final integral.

The preceding concepts and results specialize to probability spaces and random variables. Let $X$ be a random variable on a probability space $(\Omega, \mathcal{A}, P)$ and let $P_X$ be its probability distribution on $(\Re, \mathcal{L})$. If $X$ is integrable, that is if $X$ or equivalently $|X|$ belongs to $L^1(\Omega, \mathcal{A}, P)$, then we define the expectation of $X$ as

$$(2.10) \qquad E(X) = \int_\Omega X \, dP = \int_\Re x \, dP_X$$

where these integrals are of the Lebesgue or Lebesgue-Stieltjes types. This definition holds for both continuous and discrete valued random variables, for if $X$ takes only the discrete values $x_1, x_2, x_3, \ldots$ then the integrals in (2.10) reduce to a finite or an infinite sum and

$$(2.11) \qquad E(X) = \sum_{i \geq 1} x_i P(A_i)$$

where $A_i = \{\omega \in \Omega : X(\omega) = x_i\}$ for $i = 1, 2, 3, \ldots$. The absolute convergence of the sum (2.11), that is the "integrability" of $|X|$, is still necessary here in the infinite case to ensure that we can rearrange terms without causing difficulties. Obviously here $P_X \perp \mu_L$. In the case that $P_X \ll \mu_L$ we obtain integral expressions for the expectations involving density functions as in Section 4 of Chapter 1. In both the continuous and discrete valued cases the properties (2.5) and (2.6) of integrals give us

$$(2.12) \qquad E(\alpha X + \beta Y) = \alpha E(X) + \beta E(Y)$$

and

$$(2.13) \qquad E(X) \leq E(Y) \quad \text{if} \quad X \leq Y \quad \text{w.p.1}$$

for any $X, Y \in L^1(\Omega, \mathcal{A}, P)$ and $\alpha, \beta \in \Re$, and also Jensen's inequality (1.4.10), provided the functions involved are integrable. In addition, since a probability space is finite, we have a generalization of the Markov inequality (1.4.13)

$$(2.14) \qquad P\left(\{\omega \in \Omega : |X(\omega)| \geq a\}\right) \leq \frac{1}{a^p} E\left(|X|^p\right)$$

for any $a, p > 0$; this requires $|X|^p \in L^1(\Omega, \mathcal{A}, P)$, which we write as $X \in L^p(\Omega, \mathcal{A}, P)$. Using Jensen's inequality we obtain

$$(2.15) \qquad E\left(|X|^q\right) \leq \left(E\left(|X|^p\right)\right)^{q/p}$$

for all $0 < q \leq p$, so $X \in L^q(\Omega, \mathcal{A}, P)$ for all $0 < q \leq p$ whenever $L^p(\Omega, \mathcal{A}, P)$. As mentioned in Section 4 of Chapter 1, the number $E\left(|X|^p\right)$ is called the *pth-moment* of the random variable $X$. One consequence of (2.14) is that mean-square convergence of a sequence of random variables implies its convergence in probability. These results also apply for vector and matrix valued random variables, with the expectation being defined componentwise.

The Radon-Nikodym Theorem plays a central, and informative, role in the definition of conditional expectations of a random variable. Let $X$ be an integrable random variable on a complete probability space $(\Omega, \mathcal{A}, P)$ and let $\mathcal{S}$ be a sub-$\sigma$-algebra of $\mathcal{A}$, thus representing a coarser profile of information than is available in $\mathcal{A}$. We define the *conditional expectation* of $X$ with respect to $\mathcal{S}$ or the expectation of $X$ conditioned on $\mathcal{S}$, which we denote by $E(X|\mathcal{S})$, as any $\mathcal{S}$- measurable random variable $Y$ satisfying

$$(2.16) \qquad \int_S Y \, dP = \int_S X \, dP,$$

w.p.1, for all $S \in \mathcal{S}$. The existence of $Y = E(X|\mathcal{S})$ and its uniqueness, w.p.1, are guaranteed by the Radon-Nikodym Theorem. In the special case that $X \geq 0$, w.p.1, we can define a measure $\mu_X$ on $(\Omega, \mathcal{S})$ by

$$\mu_X(S) = \int_S X \, dP$$

for each $S \in \mathcal{S}$. Clearly $\mu_X \ll P$ and the function $Y = d\mu_X/dP$ has the desired properties. In the general case, we write $X = X^+ - X^-$ where $X^\pm \geq 0$ and take $Y = Y^+ - Y^-$ where the $Y^\pm$ correspond to the $X^\pm$ as above. As examples we have $E\left(X|\{\emptyset, \Omega\}\right) = E(X)$;

$$(2.17) \qquad E(X|\mathcal{S})(\omega) = \begin{cases} \frac{1}{P(A)} \int_A X \, dP & : \quad \omega \in A \\ \frac{1}{P(A^c)} \int_{A^c} X \, dP & : \quad \omega \in A^c \end{cases}$$

where $\mathcal{S} = \{\emptyset, A, A^c, \Omega\}$ for some $A \in \mathcal{A}$ with $0 < P(A) < 1$; and $E(X|\mathcal{A}) = X$, w.p.1. The conditional probability $E(X|\mathcal{S})$ is a random variable on the coarser probability space $(\Omega, \mathcal{S}, P)$, and hence on $(\Omega, \mathcal{A}, P)$ too. In contrast, $X$ need not be a random variable with respect to $(\Omega, \mathcal{S}, P)$ since it need not be $\mathcal{S} : \mathcal{L}$-

measurable, but when it is we have $E(X|\mathcal{S}) = X$, w.p.1. Roughly speaking, the conditional expectation $E(X|\mathcal{S})$ is the "best" approximation of $X$ detectable by the events in $\mathcal{S}$ and is obtained by smoothening $X$ over these events, as in (2.17). Thus the finer the $\sigma$- algebra $\mathcal{S}$, the more $E(X|\mathcal{S})$ resembles the random variable $X$. In addition, for nested $\sigma$-algebras $\mathcal{S} \subset \mathcal{T} \subset \mathcal{A}$ we have

$$(2.18) \qquad E\left(E\left(X|\mathcal{T}\right)|\mathcal{S}\right) = E\left(X|\mathcal{S}\right), \quad \text{w.p.1,}$$

and when $X$ is independent of the events in $\mathcal{S}$ we have $E(X|\mathcal{S}) = E(X)$, w.p.1.

**Exercise 2.2.4**    *Verify the identity (2.18) for the $\sigma$-algebras $\mathcal{S} = \mathcal{A}(\{A\})$ and $\mathcal{T} = \mathcal{A}(\{A, B\})$, where $A$ and $B$ are two arbitrary nonempty sets.*

In general, conditional expectations have similar properties to those of ordinary integrals, for example counterparts of the linearity (2.12) and order preserving (2.13) properties, Jensen's inequality (1.4.10) and the Dominated Convergence Theorem (Theorem 2.2.3).

If $Y$ is a second random variable on $(\Omega, \mathcal{A}, P)$ we define the *expectation of $X$ conditioned on $Y$*, written $E(X|Y)$, as the conditional expectation $E(X|\mathcal{A}(Y))$ where $\mathcal{A}(Y)$ is the (sub-) $\sigma$-algebra generated by $Y$. Then we have

$$(2.19) \qquad E\left(|X - E(X|Y)|^2\right) \le E\left(|X - f(Y)|^2\right)$$

for all $\mathcal{A}: \mathcal{L}$-measurable functions $f : \Omega \to \Re$, so we can think of $E(X|Y)$ as the best mean-square approximation for $X$ amongst the random variables $f(Y)$ for all such functions $f$. Moreover, when $X = I_A$ for some event $A \in \mathcal{A}$ and $\mathcal{S}$ is a sub-$\sigma$-algebra of $\mathcal{A}$, we write $P(A|\mathcal{S})$ for $E(I_A|\mathcal{S})$ and call it the *probability of $A$ conditioned on $\mathcal{S}$*. While $P(A|\mathcal{S})(\omega)$ appears to satisfy the properties of a probability measure on $\mathcal{A}$ for each $\omega \in \Omega$, it is not a probability measure because it is only defined with probability one on $\Omega$, with the exceptional set depending on the particular event $A$.

**Exercise 2.2.5**    *Prove inequality (2.19).*

Let us now consider a continuous valued random variable $X$ on a complete probabilty space $(\Omega, \mathcal{A}, P)$. Then, we define the distribution function $F_X$ of $X$ as a Lebesgue-Stieltjes integral

$$(2.20) \qquad F_X(x) = \int_{(-\infty, x]} 1 \, dP_X$$

for each $x \in \Re$, where $P_X$ is the probability distribution of $X$ on $(\Re, \mathcal{L})$. (In Chapter 1 we found it more convenient to write $P_X(dx)$ instead of $dP_X$ in integrals similar to (2.20), particularly on those involving transition properties of Markov processes). When $P_X \ll \mu_L$ there is an integrable density function $p = dP_X/d\mu_L$ and we can write (2.20) as

$$(2.21) \qquad F_X(x) = \int_{(-\infty, x]} p \, d\mu_L$$

for each $x \in \Re$. Moreover, if $p$ is sufficiently regular, say piecewise continuous, we can evaluate the integral (2.21) as an improper Riemann integral. We could use a proper Riemann integral if instead we evaluated $F_X(x) - F_X(a)$ for a bounded interval $[a, x]$. In other cases we can often use a generalization of the Riemann integral called the Riemann-Stieltjes integral, which is defined on a bounded interval $[a, b]$ with respect to a function $F$ of bounded variation on $[a, b]$. Let $a = x_1^{(n)} < x_2^{(n)} < \ldots < x_{n+1}^{(n)} = b$ be an arbitrary partition of $[a, b]$ with $\delta^{(n)} = \max_{1 \leq i \leq n} \delta_i^{(n)} \to 0$ as $n \to \infty$ where $\delta_i^{(n)} = x_{i+1}^{(n)} - x_i^{(n)}$ for $i = 1, 2, \ldots, n$. We define the *total variation* $V_a^b(F)$ of $F$ on $[a, b]$ as the supremum over all such partitions of the sums

$$\sum_{i=1}^{n} \left| F\left(x_{i+1}^{(n)}\right) - F\left(x_i^{(n)}\right) \right|$$

and if $V_a^b(F) < \infty$ we say that $F$ is of *bounded variation on* $[a, b]$, which we denote by $F \in BV(a, b)$. Now it can be shown that a function $F \in BV(a, b)$ if and only if its derivative $F'(x)$ exists for almost all $(\mu_L$-measure$)$ $x \in [a, b]$. Consequently, an absolutely continuous distribution function $F_X$, that is one with a density function, is of bounded variation. There are however distribution functions of bounded variation which are not absolutely continuous, for instance if $F_X$ has jumps in its graph. We define the *Riemann-Stieltjes integral* of a function $f$ with respect to an $F \in BV(a, b)$ on a bounded interval $[a, b]$ as the limit

$$(2.22) \qquad \lim_{n \to \infty} \sum_{i=1}^{n} f\left(\xi_i^{(n)}\right) \left(F\left(x_{i+1}^{(n)}\right) - F\left(x_i^{(n)}\right)\right),$$

if these exist and are equal, for arbitrary $\xi_i^{(n)} \in \left[x_i^{(n)}, x_{i+1}^{(n)}\right]$ where $\delta^{(n)} \to 0$ as $n \to \infty$, and denote it by

$$(2.23) \qquad \int_a^b f(x) \, dF(x).$$

This integral exists, for example, when $f$ is continuous on $[a, b]$, in which case it equals the Lebesgue integral $\int_{[a,b]} fF' \, d\mu_L$ or the Lebesgue-Stieltjes integral $\int_{[a,b]} f \, d\tilde{F}$ when $\tilde{F}$ is defined by $\tilde{F}(L) = \int_L F' \, d\mu_L$ for any $L \in \mathcal{L}$ is a measure on $(\Re, \mathcal{L})$ (this happens if $F'(x) \geq 0$ for almost all $x \in [a, b]$ ). When the derivative $F'$ is continuous on $[a, b]$ the Riemann-Stieltjes integral (2.23) reduces to the usual Riemann integral $\int_a^b f(x) F'(x) \, dx$.

**Exercise 2.2.6**   *Show that a bounded function $F$ is of bounded variation on an interval $[a, b]$ if and only if $F$ is the difference of two monotonic real valued functions on $[a, b]$.*

## 2.3 Stochastic Processes

We shall now look more closely at stochastic processes from the perspective of measure theory. Recall that a stochastic process $X = \{X(t), t \in T\}$ is a collection of random variables on a common probability space $(\Omega, \mathcal{A}, P)$ indexed by a parameter $t \in T \subset \Re$, which we usually interpret as time. It can thus be formulated as a function $X : T \times \Omega \to \Re$ such that $X(t, \cdot)$ is $\mathcal{A} : \mathcal{L}$-measurable in $\omega \in \Omega$ for each $t \in T$; henceforth we shall often follow established convention and write $X_t$ for $X(t)$. When $T$ is a countable set, the stochastic process is really just a sequence of random variables $X_{t_1}, X_{t_2}, \ldots, X_{t_n}, \ldots$, the analysis of which is relatively straightforward. In contrast, when $T$ is an interval, bounded or unbounded, the relationship between the $X_t$ random variables at different instants $t$ can lead to some delicate mathematical problems. We shall address these in this section, assuming from now that $T$ is an interval.

In many practical situations involving stochastic processes we are not given the probability space $(\Omega, \mathcal{A}, P)$, but rather the finite dimensional distributions, that is the totality of distribution functions $F_{t_1 t_2 \cdots t_n}$ for all finite combinations of time instants $t_1 < t_2 < \ldots < t_n$ in $T$ with $n = 1, 2, 3, \ldots$. Kolmogorov showed how we can then construct a probability space $(\Omega, \mathcal{A}, P)$ and a function $X$ such that $X$ is a stochastic process with the given finite dimensional distributions, provided these distribution functions satisfy consistency and compatibility relationships like (1.4.27)–(1.4.31). Essentially he took $\Omega = \Re^T$, the set of all functions $\omega : T \to \Re$, and defined $X(t, \omega) = \omega(t)$, so that $\omega$ not only labels the sample path but is the sample path. Then he defined $\mathcal{A}$ to be the $\sigma$-algebra generated by cylinder sets which typically have the form

$$A = \{\omega \in \Omega : X(t_i, \omega) \in L_i \quad \text{for} \quad i = 1, 2, \ldots, n\}$$

where $t_i \in T$ and $L_i \in \mathcal{L}$, to each of which he assigned the probability

$$(3.1) \qquad P(A) = \int_{L_1 \times L_2 \times \cdots \times L_n} I_A \, dF_{t_1 t_2 \ldots t_n}(x_1, x_2, \ldots, x_n).$$

Finally, he extended this set function (3.1) to the whole $\sigma$-algebra to obtain a probability measure. Other probability spaces are possible, and for a given space there may exist distinct stochastic processes $X = \{X_t, t \in T\}$ and $Y = \{Y_t, t \in T\}$ with the same distribution functions. We say that such processes are *equivalent* and that one is a *version* of the other. For example, $X_t(\omega) \equiv 0$ and $Y_t(\omega) = I_{\{\omega\}}(t)$, the indicator function of the singleton set $\{\omega\}$, are equivalent stochastic processes for $t \in [0, 1]$ on the probability space $([0, 1], \mathcal{L}, \mu_L)$. For equivalent processes we have $P(A_t) = 0$ for all $t \in T$ where $A_t = \{\omega \in \Omega : X_t(\omega) \neq Y_t(\omega)\}$. As the example shows such processes need not have the same sample paths, even with probability one, for that requires the set $A = \{\omega \in \Omega : X_t(\omega) \neq Y_t(\omega) \text{ for some } t \in T\}$ to be an event. However, as an uncountable union of events $A = \cup_{t \in T} A_t$ may not even be an event, let alone a null event.

While at the most general level no restrictions are imposed on the relationship between the random variables $X_t$ at different time instants $t \in T$, in practice it is useful to do so. A very broad class of stochastic processes, called separable processes, was introduced by Doob in order to circumvent difficulties arising from the uncountability of the time interval $T$. He called a stochastic process $X = \{X_t, t \in T\}$ defined on a complete probability space $(\Omega, \mathcal{A}, P)$ a *separable process* if there is a countably dense subset $S = \{s_1, s_2, s_3, \ldots\}$ of $T$, called a *separant set,* such that for any open interval $I_o$ and any closed interval $I_c$ the subset

$$\tilde{A} = \bigcup_{t \in T \cap I_o} \{\omega \in \Omega : X_t(\omega) \in I_c\}$$

of $\Omega$ differs from the event

$$A = \bigcup_{s_j \in S \cap I_o} \{\omega \in \Omega : X_{s_j}(\omega) \in I_c\}$$

by a subset of a null event. By the completeness of the probability space the set $\tilde{A}$ is thus itself an event and $P(\tilde{A}) = P(A)$. Consequently, for a separable process $X = \{X_t, t \in T\}$ sets like $\{\omega \in \Omega : X_t(\omega) \geq 0 \text{ for all } t \in T\}$ are events and the functions defined, w.p.1, by

$$\inf_{s \leq t} X_s(\omega), \qquad \sup_{s \leq t} X_s(\omega), \qquad \lim_{s \to t} X_s(\omega)$$

are random variables, if they exist. In our previous example the process defined by $X_t(\omega) \equiv 0$ is a separable process, whereas $Y_t(\omega) = I_{\{\omega\}}(t)$ is not.

Loève showed that for any stochastic process $X = \{X_t, t \in T\}$ there is always an equivalent process $\tilde{X}_t = \{\tilde{X}_t, t \in T\}$ defined on the same probability space which is separable, although his proof required the $\tilde{X}_t$ to be extended random variables, that is possibly taking values $\pm\infty$. When the process $X = \{X_t, t \in T\}$ is continuous in probability for each $t \in T$ matters are much nicer: there is an equivalent separable process $\tilde{X}_t = \{\tilde{X}_t, t \in T\}$ which is *jointly measurable,* that is the function $\tilde{X} : T \times \Omega \to \Re$ is measurable with respect to the product measure $(T \times \Omega, \mathcal{L} \times \mathcal{A}, \mu_L \times P)$, and for this process any countably dense subset of $T$ is admissible as a separant set. In this situation, the typical one encountered in this book, we can without any loss of generality replace the original process by such a separable version; indeed, we shall always assume that this has been done.

If a jointly measurably process $X = \{X_t, t \in T\}$ is integrable on $T \times \Omega$, that is if $X \in L^1(T \times \Omega, \mathcal{L} \times \mathcal{A}, \mu_L \times P)$, then we can apply Fubini's Theorem to any $[t_0, t] \times \Omega \subseteq T \times \Omega$ to conclude that

$$(3.2) \qquad \int_{t_0}^t E(X_s)\, ds = E\left( \int_{t_0}^t X_s\, ds \right).$$

In addition we have that $Z_t(\omega) = \int_{t_0}^t X_s(\omega)\, ds$ is well-defined, w.p.1, for almost all ($\mu_L$-measure) $t \in T$ and is integrable in $\omega$. Thus $Z_t$ is itself a random variable on $(\Omega, \mathcal{A}, P)$.

In Section 7 of Chapter 1 we mentioned the Kolmogorov criterion (1.7.23) for a stochastic process to be sample-path continuous, that is with almost surely continuous functions of $t$ as its sample paths. We saw that this criterion was satisfied by the standard Wiener process and thus concluded that this process is sample-path continuous. Strictly speaking, the criterion assures that there is a jointly measurable and separable version of the process which is sample-path continuous, the process itself need not be. Compare $X_t(\omega) \equiv 0$ and $Y_t(\omega) = I_{\{w\}}(t)$. For convenience, we usually talk about this separable, sample-path continuous version as if it were the given process. In fact, the standard Wiener process is now commonly defined on a canonical probability space $\Omega = C_0([0, \infty), \Re)$, the space of continuous functions $\omega : [0, \infty) \to \Re$ with $\omega(0) = 0$.

Conditional expectations with respect to a family of sub-$\sigma$- algebras offer a succinct way of expressing the temporal relationship of a stochastic process. Let $X = \{X_t, t \in T\}$ be a jointly measurable and separable stochastic process on a probability space $(\Omega, \mathcal{A}, P)$. Then $X_t$ is $\mathcal{A} : \mathcal{L}$-measurable for each $t \in T$, but in general the $\sigma$-algebra $\mathcal{A}$ contains many more events in addition to those detectable by $X_t$ at a particular instant $t \in T$, events of the form $X_t^{-1}(L)$ for $L \in \mathcal{L}$. For the sake of discussion we shall suppose that $T = [0, \infty)$ and denote by $\mathcal{A}_t$ the sub-$\sigma$-algebra of $\mathcal{A}$ generated by the totality of subsets

(3.3)        $A = \{\omega \in \Omega : X_{s_i}(\omega) \in L_i \quad \text{for} \quad i = 1, 2, \ldots, n\}$

of $\Omega$ for any $0 \le s_1 < s_2 < \ldots < s_n \le t$ and $L_1, L_2, \ldots, L_n \in \mathcal{L}$ where $n = 1, 2, 3, \ldots$. Thus $\mathcal{A}_s \subseteq \mathcal{A}_t$ for any $0 \le s \le t$ and we have an increasing family $\{\mathcal{A}_t, t \ge 0\}$ of sub-$\sigma$-algebras of $\mathcal{A}$, this corresponding to the fact that more information about the stochastic process becomes available with increasing time. Obviously $X_t$ is $\mathcal{A}_t : \mathcal{L}$-measurable for each $t \ge 0$. Conversely, we may be given an increasing family $\{\mathcal{A}_t, t \ge 0\}$ of sub-$\sigma$-algebras of $\mathcal{A}$ and require that a stochastic process $\{X_t, t \ge 0\}$ defined on a probability space $(\Omega, \mathcal{A}, P)$ have $X_t \mathcal{A}_t : \mathcal{L}$-measurable for each $t \ge 0$. In that case we say that the process $\{X_t, t \ge 0\}$ is *adapted* to the family $\{\mathcal{A}_t, t \ge 0\}$ of sub-$\sigma$-algebras. Generally, by enlargening the $\sigma$-algebras, we can assume that such an increasing family $\{\mathcal{A}_t, t \ge 0\}$ of sub-$\sigma$-algebras is right-continuous, that is it satisfies $\mathcal{A}_t = \cap_{\epsilon > 0} \mathcal{A}_{t+\epsilon}$ for all $t \ge 0$. Further, we can always assume that $\mathcal{A}_t$ is complete for each $t \ge 0$. These assumptions allow many technical simplifications in proofs. Such a family of sub-$\sigma$-algebras is then called a *filtration*. Usually we are given a filtration in addition to the underlying probability space. For a given stochastic process $X$ the simplest filtration is formed from sets defined by (3.3) and is often written $\{\mathcal{A}_t^X, t \ge 0\}$.

Let $X = \{X_t, t \ge 0\}$ be a stochastic process with

(3.4)        $E(X_t - X_s | \mathcal{A}_s) = 0, \quad \text{w.p.1},$

for all $0 \le s < t$, where $\{\mathcal{A}_t, t \ge 0\}$ is a filtration to which the process is adapted. Since $E(X_s | \mathcal{A}_s) = X_s$, w.p.1, we can write (3.4) as

(3.5)        $E(X_t | \mathcal{A}_s) = X_s, \quad \text{w.p.1},$

for all $0 \le s < t$. We call such a process a *martingale with respect to* $\{\mathcal{A}_t, t \ge 0\}$ or simply a martingale. If the equality sign in (3.5) is replaced by a $\le$ then we call $X$ a *supermartingale*, or a *submartingale* if it is replaced by a $\ge$. Thus a

Wiener process is a martingale, but there are also many other processes which are martingales.

**Example 2.3.1**    *For a standard Wiener process $W = \{W_t, t \geq 0\}$ the processes $W_t^2 - t$ and $\exp\left(W_t - \frac{1}{2}t\right)$ are martingales with respect to any family $\{A_t, t \geq 0\}$ of sub-$\sigma$- algebras to which $W$ is adapted.*

A separable martingale $X = \{X_t, t \geq 0\}$ with finite $p$th- moment satisfies the *maximal martingale inequality*

$$(3.6) \qquad P\left(\left\{\omega \in \Omega : \sup_{0 \leq s \leq t} |X_s(\omega)| \geq a\right\}\right) \leq \frac{1}{a^p} E\left(|X_t|^p\right)$$

for any $a > 0$ and, for $p > 1$, the *Doob inequality*

$$(3.7) \qquad E\left(\sup_{0 \leq s \leq t} |X_t|^p\right) \leq \left(\frac{p}{p-1}\right)^p E\left(|X_t|^p\right).$$

These important inequalities find many appplications since they give powerful uniform estimates of the sample paths of a process over a finite time interval rather than at a specific time instant, such as provided by the generalized Markov inequality (2.14), namely

$$P\left(\{\omega \in \Omega : |X_t(\omega)| \geq a\}\right) \leq \frac{1}{a^p} E\left(|X_t|^p\right).$$

**Exercise 2.3.2**    *Apply the maximal martingale and Doob inequalities (3.6) and (3.7) with $p = 2$ to $W_t^2 - t$ on the interval $[0, 1]$. In both cases simplify the bound on the right hand side of the inequalities.*

Variations of inequalities (3.6) and (3.7) also hold for supermartingales and submartingales. Taking limits as $t \to \infty$, we can then obtain estimates for the sample paths over all time $0 \leq s < \infty$ provided the limit $\lim_{t \to \infty} E(|X_t|^2)$ exists. In fact, martingale convergence theorems guarantee the existence of such limits. For example, for a positive supermartingale $\{X_t, t \geq 0\}$ there exists a nonnegative random variable $X_\infty$ with finite mean such that $X_t \to X_\infty$, w.p.1, and $E(X_t) \to E(X_\infty)$ as $t \to \infty$. For a discrete-time martingale $X_0, X_1, X_2, \ldots$ the identity (3.5) simplifies to

$$E\left(X_n | X_{n-1}\right) = X_{n-1}, \quad \text{w.p.1},$$

for $n = 1, 2, 3, \ldots$. Interpreted in a gambling context, it says that the expected winnings of the next game conditioned by knowledge of what was won in games up to the present game is exactly the winnings of the present game. Inequalities (3.6) and (3.7) also have analogues for discrete-time martingales.

**Example 2.3.3**    *Let $X$ be a random variable on a probability space $(\Omega, A, P)$ and let $\{A_n, n = 0, 1, 2, \ldots\}$ be an increasing family of sub-$\sigma$-algebras of $A$. Then $X_0, X_1, X_2, \ldots$ defined by*

$$X_n = E\left(X | A_n\right)$$

*for $n = 0, 1, 2, \ldots$ is a discrete-time martingale.*

The Markov property (see (1.6.5), (1.6.14), (1.7.2)) implies a useful expression in terms of expectations conditioned on time-dependent $\sigma$-algebras of events. This is possible for both discrete-time and continuous-time processes, but here we shall concentrate on the technically more complicated continuous-time case. We consider a separable stochastic process $X = \{X_t, t \geq 0\}$ defined on a probability space $(\Omega, \mathcal{A}, P)$, which is at least continuous in probability and is adapted to a filtration $\{\mathcal{A}_t, t \geq 0\}$. Then for each $t \geq 0$ we define $\mathcal{A}_t^+$ to be the sub-$\sigma$-algebra of $\mathcal{A}$ generated by the totality of subsets of the form (3.3) for any $t \leq s_1 \leq s_2 \leq \cdots \leq s_n$ and $L_1, L_2, \ldots, L_n$ where $n = 1, 2, 3, \ldots$. Consequently $X_s$ is $\mathcal{A}_t^+ : \mathcal{L}$-measurable for each $s \geq t$ and, in fact, $\mathcal{A}_t^+$ is the smallest $\sigma$-algebra with this property. It is the collection of all events detectable by the stochastic process at some future time if $t$ is considered to be the present. The Markov property implies that

$$(3.8) \qquad E(Y|\mathcal{A}_t) = E(Y|X_t), \quad \text{w.p.1},$$

for all $t \geq 0$ and all $\mathcal{A}_t^+ : \mathcal{L}$-measurable $Y$, for example $X_s$ for any $s \geq t$ and $I_B$ for any $B \in \mathcal{A}_t^+$. Essentially (3.8) says that the expectation of some future event given the past and the present is always the same as if given only the present. It follows from this that

$$(3.9) \qquad E(ZY|X_t) = E(Z|X_t) E(Y|X_t), \quad \text{w.p.1},$$

for all $\mathcal{A}_t : \mathcal{L}$-measurable $Z$, all $\mathcal{A}_t^+ : \mathcal{L}$-measurable $Y$ and all $t \geq 0$. This says that the future and the past are conditionally independent given the present. For a Markov process $X = \{X_t, t \geq 0\}$ it also follows from (3.8) and (2.18) that

$$E(Y|X_{t_0}) = E(E(Y|X_t)|X_{t_0}), \quad \text{w.p.1},$$

for any $0 \leq t_0 \leq t$ and any $\mathcal{A}_t^+ : \mathcal{L}$- measurable $Y$. This corresponds to the Chapman-Kolmogorov equation (1.7.6).

Let $\{\mathcal{A}_t, t \geq 0\}$ be a filtration on a probability space $(\Omega, \mathcal{A}, P)$. We call a nonnegative random variable $\tau$ on $(\Omega, \mathcal{A}, P)$ a *Markov time* if the event $\{\omega \in \Omega : \tau(\omega) \leq t\} \in \mathcal{A}_t$ for each $t \geq 0$. This means that by observing a sample path $X_s(\omega_0)$ over an interval $0 \leq s \leq t$ for a stochastic process adapted to $\{\mathcal{A}_t, t \geq 0\}$ we can always determine whether $\tau(\omega_0) \leq t$ or $\tau(\omega_0) > t$. In defining Markov times it is convenient to allow $\tau$ to take the value $+\infty$ in addition to finite values. Then for a sample-path continuous process $\{X_t, t \geq 0\}$ and a closed subset $F \subset \Re$

$$(3.10) \qquad \tau(\omega) = \inf\{t \geq 0; X_t(\omega) \in F\}$$

defines a Markov time with respect to a family of sub-$\sigma$- agebras to which the process is adapted, which may take infinite values. We call it the *hitting time* of the set $F$ or the *first exit time* of its complement $F^c$. Often we say that a Markov time is a *stopping time*, particularly when we are only interested in observing a process until a certain time which depends on the sample path under consideration. We remark that $\tau + t$, $\tau \wedge t = \min\{\tau, t\}$, $\tau \wedge \sigma$, $\tau \vee t = \max\{\tau, t\}$ and $\tau \vee \sigma$ are Markov times when $\tau$ and $\sigma$ are Markov times.

## 2.4   Diffusion and Wiener Processes

Let $X = \{X_t, t \geq 0\}$ be a vector-valued stochastic process on a probability space $(\Omega, \mathcal{A}, P)$ and taking values in $\Re^d$ with $X_t$ being $\mathcal{A} : \mathcal{L}^d$-measurable for each $t \geq 0$. Similarly to Section 7 of Chapter 1 for a 1-dimensional process, we define the transition probabilities for such an $d$-dimensional process by

$$(4.1) \qquad P(s, x; t, L) = P(X_t \in L | X_s = x)$$

for all $0 \leq s < t$, $x \in \Re^d$ and $L \in \mathcal{L}^d$. Apart from the obvious change in dimension, we are now using the slightly more general Lebesgue subsets of $\Re^d$ instead of the Borel subsets. The Markov property (1.7.2) can be restated with the obvious changes. We note that the individual components of a vector Markov process need not themselves be Markov processes. A similar generalization applies for the definition of a vector diffusion process. Since the transition densities need not always exist, we shall reformulate the defining properties (1.7.9)–(1.7.11) in terms of the transition probabilities (4.1) using the Lebesgue-Stieltjes integrals over subsets of $\Re^d$. We require the following limits to exist for any $\epsilon > 0$, $s \geq 0$ and $x \in \Re^d$:

$$(4.2) \qquad \lim_{t \downarrow s} \frac{1}{t - s} \int_{|y - x| > \epsilon} P(s, x; t, dy) = 0,$$

$$(4.3) \qquad \lim_{t \downarrow s} \frac{1}{t - s} \int_{|y - x| \leq \epsilon} (y - x) \, P(s, x; t, dy) = a(s, x)$$

and

$$(4.4) \qquad \lim_{t \downarrow s} \frac{1}{t - s} \int_{|y - x| \leq \epsilon} (y - x)(y - x)^\top P(s, x; t, dy) = B(s, x) B(s, x)^\top$$

where $a$ is an $d$-dimensional vector valued function and $D = BB^\top$ is a symmetric, positive definite $d \times d$-matrix valued function. We are using the Euclidean norm (1.4.36) here and interpret the vectors as column vectors, so $YY^\top$ is an $d \times d$-matrix with $ij$th component $y_i y_j$. If its transition probabilities satisfy (4.2)–(4.4) we call the process a *vector diffusion process*. The *drift vector* $a$ and the *diffusion matrix* $D$ have similar interpretations to their 1-dimensional counterparts, except here the off-diagonal components of $D$ are the instantaneous rates of change in the conditioned covariances between the corresponding components of the vector process, namely

$$d^{i,j}(s, x) = \lim_{t \downarrow s} \frac{1}{t - s} E\left( (X_t^i - X_s^i)(X_t^j - X_s^j) | X_s = x \right).$$

Compare this with (1.7.13) and note that we are using superscripts to index the components of vector-valued stochastic processes here.

When the drift vector and the diffusion matrix are moderately regular functions, the transition probabilities (4.1) have densities $p(s, x; t, y)$ which satisfy the *Kolmogorov forward equation*, perhaps better known as the *Fokker-Planck equation*,

$$(4.5) \qquad \frac{\partial p}{\partial t} + \sum_{i=1}^{d} \frac{\partial}{\partial y_i} \{a^i(t,y)p\} - \frac{1}{2} \sum_{i,j=1}^{d} \frac{\partial^2}{\partial y_i \partial y_j} \{d^{i,j}(t,y)p\} = 0$$

$(s,x$ in $p$ fixed) with the initial condition

$$\lim_{t \downarrow s} p(s,x;t,y) = \delta(x-y),$$

where $\delta$ is the Dirac delta function on $\Re^d$. The density $p$ is thus a fundamental solution of the parabolic partial differential equation (4.5), which we can write more compactly in operator form as

$$\frac{\partial p}{\partial t} - \mathcal{L}^* p = 0$$

where $\mathcal{L}^*$ is the formal adjoint of the elliptic operator $\mathcal{L}$ defined as

$$(4.6) \qquad \mathcal{L}u(s,x) = \sum_{i=1}^{d} a^i(s,x) \frac{\partial u}{\partial x_i}(s,x) + \frac{1}{2} \sum_{i,j=1}^{d} d^{i,j}(s,x) \frac{\partial^2 u}{\partial x_i \partial x_j}(s,x).$$

The *Kolmogorov backward equation* is then

$$(4.7) \qquad \frac{\partial u}{\partial s} + \mathcal{L}u = 0$$

and is satisfied by $u(s,x) = p(s,x;t,y)$ for fixed $t$ and $y$. It also has the solution

$$(4.8) \qquad u(s,x) = E\left(f(X_t) \,|\, X_s = x\right) = \int_{\Re^d} f(y)p(s,x;t,y)\,dy$$

corresponding to the final time condition

$$\lim_{s \uparrow t} u(s,x) = f(x)$$

for any sufficiently smooth function $f : \Re^d \to \Re$. Equation (4.8) is often called the *Kolmogorov formula*.

For a given (bounded) domain $\mathcal{D} \subset \Re^d$ the inhomogeneous equation

$$(4.9) \qquad \frac{\partial u}{\partial s} + \mathcal{L}u = -1 \quad \text{for} \quad x \in \mathcal{D}$$

with boundary condition $u(s,x) = 0$ for $x \in \mathcal{D}$ is satisfied by

$$u(s,x) = E\left(\tau_{s,x}\right) - s$$

where

$$\tau_{s,x}(\omega) = \inf\{t \geq s : X_t(\omega) \in \partial\mathcal{D} \text{ where } X_s(\omega) = x\}$$

is the *first exit time* of the process from $\mathcal{D}$ given that it starts at $x$ at time $s$.

**Exercise 2.4.1**    *Use integration by parts to show that*

$$\int_{\Re^d} f\, \mathcal{L}g\, dx - \int_{\Re^d} g\, \mathcal{L}^* f\, dx = 0,$$

*where $f, g$ and their derivatives vanish as $x \to \infty$. Hence derive $\mathcal{L}^*$ from $\mathcal{L}$ when $d = 1$.*

The simplest nontrivial $d$-dimensional vector diffusion process corresponds to the zero drift vector $a(s, x) \equiv 0$ and the identity diffusion matrix $D(s, x) \equiv I$. This is the *d- dimensional vector Wiener process* $W_t = (W_t^1,\, W_t^2,\, \ldots,\, W_t^d)$, the components of which are pairwise independent standard Wiener processes satisfying conditions (1.8.1). In fact, similar conditions for a vector process with independent increments completely characterize a vector Wiener process. A theorem of Doob says that such a process $\{X_t = (X_t^1, \ldots, X_t^d),\ t \geq 0\}$ adapted to an increasing family $\{\mathcal{A}_t, t \geq 0\}$ of $\sigma$-algebras is a vector standard Wiener process if and only if

(4.10)               $$X_0^i = 0, \qquad E\left(X_t^i - X_s^i | \mathcal{A}_s\right) = 0,$$

$$E\left((X_t^i - X_s^i)(X_t^j - X_s^j)|\mathcal{A}_s\right) = \delta_{i,j}\, (t - s),$$

w.p.1, for all $i, j = 1, \ldots, d$ and $0 \leq s \leq t$, where $\delta_{i,j}$ is the *Kronecker delta symbol* defined by

$$\delta_{i,j} = \begin{cases} 1 & : \quad i = j \\ 0 & : \quad i \neq j \end{cases}$$

for $i, j = 1, 2, \ldots, d$. These conditions are weaker than (1.8.1) in that they involve conditional expectations with respect to the coarser $\sigma$-algebras $\mathcal{A}_t$ rather than $\mathcal{A}$. The identity (1.7.24) for a 1-dimensional standard Wiener process now takes the form

$$E\left(|W_t - W_s|^4\right) = \left(d^2 + 2d\right) |t - s|^2$$

for a $d$-dimensional Wiener process. Hence we can use Kolmogorov's criterion (1.7.23) directly to conclude that there is a jointly measurable and separable version which is sample-path continuous. As for the one-dimensional case we shall always assume that we are using such a version.

A standard Wiener process $W = \{W_t, t \geq 0\}$ consists of uncountably many random variables. However, it is possible to represent it on any bounded interval $0 \leq t \leq T$ in terms of only countably many independent Gaussian random variables. This representation is very similar to a Fourier series with the random variables as its coefficients and is called the *Karhunen-Loève expansion* of the process. In the sense of mean-square convergence, we have

(4.11)               $$W_t(\omega) = \sum_{n=0}^{\infty} Z_n(\omega)\phi_n(t) \quad \text{for} \quad 0 \leq t \leq T$$

where the $Z_0, Z_1, \ldots, Z_n, \ldots$ are independent standard Gaussian random variables and the $\phi_0, \phi_1, \ldots, \phi_n, \ldots$ are the nonrandom functions

$$(4.12) \qquad \phi_n(t) = \frac{2\sqrt{2T}}{(2n+1)\pi} \sin\left(\frac{(2n+1)\pi t}{2T}\right)$$

for $n = 0, 1, 2, \ldots$. These time functions are themselves orthogonal with respect to an integral inner product, namely

$$(\phi_i, \phi_j) = \int_0^T \phi_i(t)\phi_j(t)\, dt = 0$$

if $i \neq j$ for any $i, j = 0, 1, 2, \ldots$, and satisfy

$$\sum_{n=0}^{\infty} \phi_n(s)\phi_n(t) = \min\{s, t\},$$

which is the covariance function of the process. In our case the random variables $Z_0, Z_1, Z_2, \ldots$ are determined in an almost identical way to the coefficients of a Fourier series, with

$$(4.13) \qquad Z_n(\omega) = \frac{2}{T}\left(\frac{(2n+1)\pi}{2\sqrt{2T}}\right)^2 \int_0^T W_t(\omega)\phi_n(t)\, dt$$

for $n = 0, 1, 2, \ldots$ and are independent standard Gaussian random variables. Obviously, both the random variables (4.13) and the time functions (4.12) depend on the particular time interval $0 \leq t \leq T$.

The expansion (4.11) has both theoretical and practical uses. For example, if we differentiate the series term by term the resulting series is divergent. This again suggests the nondifferentiability of the Wiener process. In addition, we can use a truncation of the series to generate an approximation of a Wiener process.

**PC-Exercise 2.4.2**   *Generate 50 independent standard Gaussian pseudo-random numbers for use as realizations of the first 50 random coefficients $Z_0$, $Z_1, \ldots, Z_{49}$ in the series expansion (4.11). Then plot the graphs of the partial sums*

$$\sum_{n=0}^{49} Z_n(\omega)\phi_n(t)$$

*against $t$ on the interval $0 \leq t \leq 1$.*

We can also form Karhunen-Loève expansions using other nonrandom functions than those in (4.12), for example the Haar functions defined in the next exercise.

**Exercise 2.4.3**   *The Haar functions $H_n : [0, 1] \rightarrow \Re$ defined by $H_1(x) \equiv 1$ and*

$$H_{2^m+1}(x) = \begin{cases} 2^{m/2} & : \ 0 \leq x < 2^{-m-1} \\ -2^{m/2} & : \ 2^{-m-1} \leq x < 2^{-m} \\ 0 & : \ \text{otherwise} \end{cases}.$$

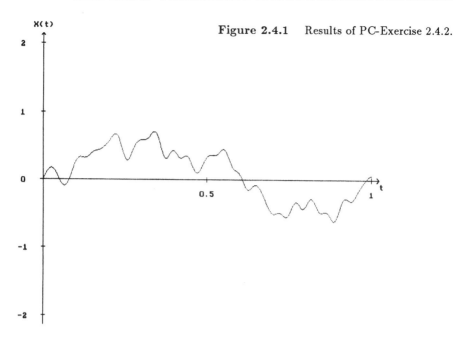

**Figure 2.4.1**   Results of PC-Exercise 2.4.2.

*with $H_{2^m+j}(x) = H_{2^m+1}(x - (j-1)/2^m)$ for $j = 1, 2, \ldots, 2^m$ and $m = 0, 1, 2,$*
*... are a complete and orthonormal system with*

$$\int_0^1 H_i(x)H_j(x)\,dx = \left\{ \begin{array}{lll} 1 & : & i = j \\ 0 & : & i \neq j \end{array} \right.$$

*Sketch the first eight Haar functions and verify directly that they satisfy the*
*preceding orthonormality condition. Then determine the Karhunen-Loève ex-*
*pansion for a standard Wiener process on the interval $0 \leq t \leq 1$ with respect*
*to the system of Haar functions.*

In Section 8 of Chapter 1 we used the Law of the Iterated Logarithm property (1.8.5) of a 1-dimensional standard Wiener process to show that its sample
paths are nowhere differentiable functions of time. We shall now show directly
that they do not have bounded variation on any finite time interval. We take
a bounded interval $a \leq t \leq b$ which we partition into subintervals $[t_k^{(n)}, t_{k+1}^{(n)}]$
of equal length $2^{-n}(b - a)$, where $t_k^{(n)} = a + k2^{-n}(b - a)$ for $k = 1, 2, \ldots, 2^n$.
Then we define

$$S_n(\omega) = \sum_{k=0}^{2^n-1} \left( W_{t_{k+1}^{(n)}}(\omega) - W_{t_k^{(n)}}(\omega) \right)^2 - (b - a)$$

$$= \sum_{k=0}^{2^n-1} \left( \Delta_{n,k}(\omega) - 2^{-n}(b - a) \right)$$

where we have written

$$\Delta_{n,k}(\omega) = \left( W_{t_{k+1}^{(n)}}(\omega) - W_{t_k^{(n)}}(\omega) \right)^2.$$

Since $S_n$ is the sum of independent random variables, each of which has zero mean, we have $E(S_n) = 0$ and

$$(4.14) \quad E\left(S_n^2\right) = E\left( \sum_{k=0}^{2^n-1} \left( \Delta_{n,k}(\omega) - 2^{-n}(b-a) \right)^2 \right)$$

$$= \sum_{k=0}^{2^n-1} \left( E\left(\Delta_{n,k}^2\right) - 2^{-n+1}(b-a)E\left(\Delta_{n,k}\right) + 2^{-2n}(b-a)^2 \right)$$

$$= \sum_{k=0}^{2^n-1} (b-a)^2 2^{-2n+1} \quad \text{by (1.7.24) and (1.8.1)}$$

$$= (b-a)^2 2^{-n+1}.$$

Obviously, then $S_n \to 0$ as $n \to \infty$ in mean-square convergence. More importantly, by the Markov inequality (2.14) for any $\epsilon > 0$ we have

$$P\left(\{\omega \in \Omega : |S_n(\omega)| \geq \epsilon\}\right) \leq \frac{1}{\epsilon^2} E\left(S_n^2\right) \leq (b-a)^2 \frac{1}{\epsilon^2} 2^{-n+1}.$$

Hence the positive series $\sum_{n=1}^{\infty} P(\{\omega \in \Omega : |S_n(\omega)| \geq \epsilon\})$ is bounded above by $(b-a)^2/\epsilon^2$ and so is convergent. By the Borel-Cantelli Lemma 2.1.4 we can then conclude that the events $\{\omega \in \Omega : |S_n(\omega)| \geq \epsilon\}$ occur for at most finitely many $n$, w.p.1. Consequently we have with probability one

$$(4.15) \quad \lim_{n \to \infty} \sum_{k=0}^{2^n-1} \left( W_{t_{k+1}^{(n)}}(\omega) - W_{t_k^{(n)}}(\omega) \right)^2 = b - a.$$

In fact, this holds for any partitioning of the interval $[a, b]$ with

$$\delta^{(n)} = \max_{k \geq 0} \left| t_{k+1}^{(n)} - t_k^{(n)} \right| \to 0 \quad \text{as} \quad n \to \infty$$

provided $\sum_{n=1}^{\infty} \delta^{(n)} < \infty$. From it we can deduce that $W_t(\omega)$ is, with probability one, not of bounded variation on $[a, b]$. Indeed from (4.15) we have

$$b - a \leq \limsup_{n \to \infty} \max_{0 \leq k \leq 2^n-1} \left| W_{t_{k+1}^{(n)}}(\omega) - W_{t_k^{(n)}}(\omega) \right| \sum_{k=0}^{2^n-1} \left| W_{t_{k+1}^{(n)}}(\omega) - W_{t_k^{(n)}}(\omega) \right|$$

and since, from sample-path continuity,

$$\max_{0 \leq k \leq 2^n-1} \left| W_{t_{k+1}^{(n)}}(\omega) - W_{t_k^{(n)}}(\omega) \right| \to 0$$

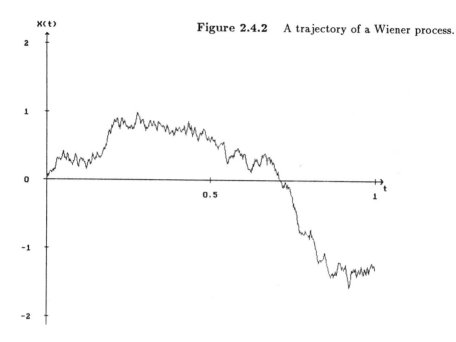

**Figure 2.4.2**    A trajectory of a Wiener process.

w.p.1 as $n \to \infty$, we have

$$\sum_{k=0}^{2^n-1} \left| W_{t_{k+1}^{(n)}}(\omega) - W_{t_k^{(n)}}(\omega) \right| \to \infty$$

w.p.1 as $n \to \infty$. Thus, almost surely, the sample paths do not have bounded variation on $a \le t \le b$. They cannot then be differentiable, except possibly on a time subset of Lebesgue measure zero. Roughly speaking, (4.15) says that $W_{t+\delta}(\omega) - W_t(\omega)$ is of order $\sqrt{\delta}$ and so

$$\frac{W_{t+\delta}(\omega) - W_t(\omega)}{\delta}$$

is of order $1/\sqrt{\delta}$, which suggests that its limit as $\delta \to 0^+$ cannot exist.

**PC-Exercise 2.4.4**    *Generate and plot the linearly interpolated trajectory of a Wiener process on $[0,1]$ at the time instants $t_k = k2^{-9}$ for $k = 0, 1, \ldots, 2^9$ using independent Gaussian increments.*

# Chapter 3

# Ito Stochastic Calculus

This chapter provides an introduction to stochastic calculus, in particular to stochastic integration. A fundamental result, the Ito formula, is also derived. This is a stochastic counterpart of the chain rule of deterministic calculus and will be used repeatedly throughout the book. Section 1 summarizes the key concepts and results and should be read by nonspecialists. Mathematical proofs are presented in the subsequent sections.

## 3.1   Introduction

The stochastic calculus of Ito originated with his investigation of conditions under which the local properties of a Markov process could be used to characterize this process. By local properties here we mean quantities such as the drift and the diffusion coefficient of a diffusion process. These had been used some time earlier by Kolmogorov to derive the partial differential equations (1.7.14)–(1.7.15), which now bear his name, for the transition probabilities of a diffusion process. In contrast, Ito's approach focused on the functional form of the processes themselves and resulted in a mathematically meaningful formulation of stochastic differential equations, which until then had been heuristic and inadequate. A similar theory was developed independently at about the same time by Gikhman.

An ordinary differential equation

$$(1.1) \qquad \dot{x} = \frac{dx}{dt} = a(t, x)$$

may be thought of as a degenerate form of a stochastic differential equation, as yet undefined, in the absence of randomness. It is therefore useful to review some of its basic properties. We could write (1.1) in the symbolic differential form

$$(1.2) \qquad dx = a(t, x)\, dt,$$

or more accurately as an integral equation

$$(1.3) \qquad x(t) = x_0 + \int_{t_0}^{t} a(s, x(s))\, ds$$

where $x(t) = x(t; x_0, t_0)$ is a solution satisfying the initial condition $x(t_0) = x_0$. Some regularity assumption is usually made on $a$, such as Lipschitz continuity, to the ensure existence of a unique solution $x(t; x_0, t_0)$ for each initial condition. These solutions are then related by the evolutionary property

(1.4)                          $$x(t; x_0, t_0) = x(t; x(s; x_0, t_0), s)$$

for all $t_0 \leq s \leq t$. Together these say that the future is determined completely by the present, with the past's being involved only in that it determines the present. This is a deterministic version of the Markov property (1.7.2); indeed, (1.4) is just the Chapman-Kolmogorov equation (1.7.5) for degenerate transition probabilities with densities $p(t_0, x_0; t, x) = \delta(x - x(t; x_0, t_0))$, where $\delta$ is the Dirac delta function (see (1.8.11)).

Following Einstein's explanation of observed Brownian motion during the first decade of this century, attempts were made by Langevin and others to formulate the dynamics of such motions in terms of differential equations. The resulting equations were written in the form as

(1.5)                          $$dX_t = a(t, X_t)\, dt + b(t, X_t)\, \xi_t\, dt$$

with a deterministic or averaged drift term (1.1) perturbed by a noisy, diffusive term $b(t, X_t)\xi_t$, where the $\xi_t$ were standard Gaussian random variables for each $t$ and $b(t, x)$ a (generally) space-time dependent intensity factor. This symbolic differential was interpreted as an integral equation

(1.6)     $$X_t(\omega) = X_{t_0}(\omega) + \int_{t_0}^{t} a(s, X_s(\omega))\, ds + \int_{t_0}^{t} b(s, X_s(\omega))\xi_s(\omega)\, ds$$

for each sample path. When extrapolated to a limit, the observations of Brownian motion seemed to suggest that the covariance $c(t) = E(\xi_s \xi_{s+t})$ of the process $\xi_t$ had a constant spectral density, that is with all time frequencies equally weighted in any Fourier transform of $c(t)$. Such a process became known as Gaussian white noise, particular in the engineering literature. For the special case of (1.6) with $a \equiv 0$, $b \equiv 1$ we see that $\xi_t$ should be the derivative of pure Brownian motion, that is the derivative of a Wiener process $W_t$, thus suggesting that we could write (1.6) alternatively as

(1.7)     $$X_t(\omega) = X_{t_0}(\omega) + \int_{t_0}^{t} a(s, X_s(\omega))\, ds + \int_{t_0}^{t} b(s, X_s(\omega))dW_s(\omega).$$

The problem with this is, as we saw in Section 4 of Chapter 2, that a Wiener process $W_t$ is nowhere differentiable, so strictly speaking the white noise process $\xi_t$ does not exist as a conventional function of $t$; indeed, a flat spectral density implies that its covariance function $c(t)$ is a constant multiple of the Dirac delta function $\delta(t)$. Thus the second integral in (1.7) cannot be an ordinary Riemann or Lebesgue integral. Worse still, the continuous sample paths of a Wiener process are not of bounded variation on any bounded time interval, so the second integral in (1.7) cannot even be interpreted as a Riemann-Stieltjes integral for each sample path.

For constant $b(t, x) \equiv b$ we would expect the second integral in (1.7), however it is to be defined, to equal $b\{W_t(\omega) - W_{t_0}(\omega)\}$. This is the starting point for Ito's definition of a stochastic integral. To fix ideas, we shall consider such an integral of a random function $f$ over the unit time interval $0 \leq t \leq 1$, denoting it by $I(f)$, where

$$(1.8) \qquad I(f)(\omega) = \int_0^1 f(s, \omega)\, dW_s(\omega).$$

For a nonrandom step function $f(t, \omega) = f_j$ on $t_j \leq t < t_{j+1}$ for $j = 1, 2, \ldots, n$ where $0 = t_1 < t_2 < \cdots < t_{n+1} = 1$ we should obviously take, at least w.p.1,

$$(1.9) \qquad I(f)(\omega) = \sum_{j=1}^n f_j \left\{ W_{t_{j+1}}(\omega) - W_{t_j}(\omega) \right\}.$$

This is a random variable with zero mean since it is the sum of random variables with zero mean. For random step functions appropriate measurability conditions must be imposed to ensure the nonanticipativeness of the integrand. To be specific, suppose that $\{\mathcal{A}_t, t \geq 0\}$ is an increasing family of $\sigma$-algebras such that $W_t$ is $\mathcal{A}_t$-measurable for each $t \geq 0$. Then we consider a random step function $f(t, \omega) = f_j(\omega)$ on $t_j \leq t < t_{j+1}$ for $j = 1, 2, \ldots, n$ where $0 = t_1 < t_2 < \cdots < t_{n+1} = 1$ and $f_j$ is $\mathcal{A}_{t_j}$-measurable, that is observable by events that can be detected at or before time $t_j$. We shall also assume that each $f_j$ is mean-square integrable over $\Omega$, hence $E(f_j^2) < \infty$ for $j = 1, 2, \ldots, n$. Since $E(W_{t_{j+1}} - W_{t_j} | \mathcal{A}_{t_j}) = 0$, w.p.1, it follows that the product $f_j\{W_{t_{j+1}} - W_{t_j}\}$, which is $\mathcal{A}_{t_{j+1}}$-measurable and integrable, has expectation

$$E\left( f_j \left\{ W_{t_{j+1}} - W_{t_j} \right\} \right) = E\left( f_j\, E\left( W_{t_{j+1}} - W_{t_j} | \mathcal{A}_{t_j} \right) \right) = 0$$

for each $j = 1, 2, \ldots, n$. Analogously to (1.9) we define the integral $I(f)$ by

$$(1.10) \qquad I(f)(\omega) = \sum_{j=1}^n f_j(\omega) \left\{ W_{t_{j+1}}(\omega) - W_{t_j}(\omega) \right\},$$

w.p.1. Since the $j$th term in this sum is $\mathcal{A}_{t_{j+1}}$-measurable and hence $\mathcal{A}_1$-measurable, it follows that $I(f)$ is $\mathcal{A}_1$-measurable. In addition, $I(f)$ is integrable over $\Omega$ and has zero mean; in fact, it is mean-square integrable with

$$(1.11) \qquad E\left( I(f)^2 \right) = \sum_{j=1}^n E\left( f_j^2\, E\left( \left| W_{t_{j+1}} - W_{t_j} \right|^2 | \mathcal{A}_{t_j} \right) \right)$$

$$= \sum_{j=1}^n E\left( f_j^2 \right) (t_{j+1} - t_j)$$

on account of the mean-square property of the increments $W_{t_{j+1}} - W_{t_j}$ for $j = 1, 2, \ldots, n$. Finally, from (1.10) we obviously have

$$(1.12) \qquad I(\alpha f + \beta g) = \alpha I(f) + \beta I(g),$$

w.p.1, for any $\alpha, \beta \in \Re$ and any random step functions $f, g$ satisfying the above properties, that is, the integration operator $I$ is linear in the integrand.

For a general integrand $f : [0,1] \times \Omega \to \Re$ we shall define $I(f)$ as the limit of integrals $I(f^{(n)})$ of random step functions $f^{(n)}$ converging to $f$. To do this we need to specify conditions on $f$ and determine an appropriate mode of convergence for which such an approximating sequence of step functions and limit exist. For the moment we shall assume that $f$ is continuous in $t$ for all $\omega \in \Omega$ and that for each $0 \le t \le 1$ the random variable $f(t, \cdot)$ is $\mathcal{A}_t$-measurable and mean-square integrable with $E(f(t, \cdot))^2$ continuous in $t$. Then we form a partition $0 = t_1^{(n)} < t_2^{(n)} < \cdots < t_{n+1}^{(n)} = 1$ with

$$\delta^{(n)} = \max_{1 \le j \le n} \left\{ t_{j+1}^{(n)} - t_j^{(n)} \right\} \to 0 \quad \text{as} \quad n \to \infty$$

and define a step function $f^{(n)}$ by $f^{(n)}(t, \omega) = f(\tau_j^{(n)}, \omega)$ on $t_j^{(n)} \le t < t_{j+1}^{(n)}$ for some choice of $\tau_j^{(n)}$ satisfying $t_j^{(n)} \le \tau_j^{(n)} < t_{j+1}^{(n)}$, where $j = 1, 2, \ldots, n$. For a choice $t_j^{(n)} < \tau_j^{(n)} < t_{j+1}^{(n)}$, the random variable $f^{(n)}(t, \cdot)$ need not be $\mathcal{A}_t$-measurable for each $t_j^{(n)} \le t < t_{j+1}^{(n)}$ nor independent of the increment $W_{t_{j+1}} - W_{t_j}$, that is the step function $f^{(n)}$ may depend on future events. To avoid this we must take $\tau_j^{(n)} = t_j^{(n)}$ for all $j = 1, 2, \ldots, n$ and $n = 1, 2, 3, \ldots$. We shall assume that we have done this and, for now, that the step functions $f^{(n)}$ converge to the integrand $f$ in an appropriate mode of convergence. The problem is to characterize the limit of the finite sums

$$(1.13) \qquad I\left(f^{(n)}\right)(\omega) = \sum_{j=1}^{n} f^{(n)}\left(t_j^{(n)}, \omega\right) \left\{ W_{t_{j+1}^{(n)}}(\omega) - W_{t_j^{(n)}}(\omega) \right\}$$

with respect to an appropriate mode of convergence. The Wiener process has well-behaved mean-square properties, in particular $E((W_t - W_s)^2) = t - s$. Moreover for the step functions $f^{(n)}$ the equality (1.11) gives

$$(1.14) \qquad E\left( I\left(f^{(n)}\right)^2 \right) = \sum_{j=1}^{n} E\left( f\left(t_j^{(n)}, \cdot\right)^2 \right) (t_{j+1} - t_j)$$

and this converges to the Riemann integral $\int_0^1 E(f(s, \cdot)^2)\, ds$ for $n \to \infty$ since $E(f(t, \cdot)^2)$ has been assumed to be continuous in $t$. Together these suggest that we should use mean-square convergence, and this is exactly what Ito did. If we assume that $E(|f^{(n)}(t, \cdot) - f(t, \cdot)|^2) \to 0$ as $n \to \infty$ for all $t \in [0, 1]$, then it is not hard to show that the mean-square limit of the $I(f^{(n)})$ exists and is unique, w.p.1. We shall denote it by $I(f)$ and call it the *Ito stochastic integral*, or just the *Ito integral*, of $f$ on $0 \le t \le 1$. $I(f)$ turns out to be an $\mathcal{A}_1$-measurable random variable which is mean-square integrable on $\Omega$ with $E(I(f)) = 0$ and

$$(1.15) \qquad E\left( I(f)^2 \right) = \int_0^1 E\left( f(s, \cdot)^2 \right) ds.$$

In addition, $I(f)$ inherits the linearity property (1.12) from the sums (1.13). Later in this chapter we shall drop the requirement that $f$ has continuous sample paths, where the Ito integral is defined as above and has the same properties.

The Ito integral is defined similarly on any bounded interval $[t_0, t]$, resulting in a random variable

(1.16)
$$X_t(\omega) = \int_{t_0}^{t} f(s, \omega) \, dW_s(\omega)$$

which is $\mathcal{A}_t$-measurable and mean-square integrable with zero mean and

$$E\left(X_t^2\right) = \int_{t_0}^{t} E\left(f(s, \cdot)^2\right) \, ds.$$

From the independence of nonoverlapping increments of a Wiener process in the step function sum (1.13), and in their mean-square limit, we have

$$E\left(X_{t_2} - X_{t_1} | \mathcal{A}_{t_1}\right) = 0,$$

w.p.1, for any $t_0 \leq t_1 \leq t_2$; hence $\{X_t, t \geq 0\}$ is a martingale (2.3.5). Then, for example, the maximal martingale inequality (2.3.6) gives

$$P\left(\sup_{t_0 \leq s \leq t} |X_s| \geq a\right) \leq \frac{1}{a^2} E\left(|X_t|^2\right) \leq \frac{1}{a^2} \int_{t_0}^{t} E\left(f(s, \cdot)^2\right) \, ds$$

for any $a > 0$.

The time-dependent Ito integrals (1.16) also satisfy the linearity property (1.12) and the additivity property

$$\int_{t_0}^{t_2} f(s, \cdot) \, ds = \int_{t_0}^{t_1} f(s, \cdot) \, ds + \int_{t_1}^{t_2} f(s, \cdot) \, ds,$$

w.p.1, for any $t_0 \leq t_1 \leq t_2$, properties which it shares with the conventional Riemann and Riemann-Stieltjes integrals. However the Ito integral also has the peculiar property, amongst others, that

(1.17)
$$\int_{0}^{t} W_s(\omega) \, dW_s(\omega) = \frac{1}{2} W_t^2(\omega) - \frac{1}{2} t,$$

w.p.1, in contrast to

$$\int_{0}^{t} w(s) \, dw(s) = \int_{0}^{w(t)} d\left(\frac{1}{2} w^2\right) = \frac{1}{2} w^2(t)$$

from classical calculus for a differentiable function $w(t)$ with $w(0) = 0$. The equality (1.17) follows from the algebraic rearrangement

$$\sum_{j=1}^{n} W_{t_j} \left\{W_{t_{j+1}} - W_{t_j}\right\} = \frac{1}{2} W_t^2 - \frac{1}{2} \sum_{j=1}^{n} \left\{W_{t_{j+1}} - W_{t_j}\right\}^2$$

for any $0 = t_1 < t_2 < \ldots < t_{n+1} = 1$ and the fact that the mean-square limit of the sum of squares on the right is equal to $t$.

We could write the equality (1.17) symbolically in terms of differentials as $E((dW_t)^2) = dt$, which has an interesting consequence in the following stochastic counterpart of the chain rule. For each $t \geq t_0$ we define a stochastic process $Y_t$ by

$$Y_t(\omega) = U(t, X_t(\omega))$$

where $U(t, x)$ has continuous second order partial derivatives and $X_t$ is given by (1.16) or, equivalently, by the stochastic differential $dX_t = f dW_t$. If $X_t$ were continuously differentiable, the *chain rule* of classical calculus would give as the differential for $Y_t$

(1.18)                    $$dY_t = \frac{\partial U}{\partial t}(t, X_t)\, dt + \frac{\partial U}{\partial x}(t, X_t)\, dX_t.$$

This follows from the first few terms in the Taylor expansion for $U$ in $\Delta Y_t = U(t + \Delta t, X_t + \Delta X_t) - U(t, X_t)$, with the second and higher order terms in $\Delta t$ and $\Delta X_t$ being discarded. In this case the only first order partial derivatives of $U$ remain. In contrast, when $X_t$ is given by (1.16) we need to take into account that $(dX_t)^2 = f^2(dW_t)^2$ and hence $E((dX_t)^2) = E(f^2)dt$, giving us a first order "$dt$" term coming from the second order part of the Taylor expansion for $U$. To be specific we have

$$
\begin{aligned}
\Delta Y_t &= U(t + \Delta t, X_t + \Delta X_t) - U(t, X_t) \\
&= \left\{ \frac{\partial U}{\partial t} \Delta t + \frac{\partial U}{\partial x} \Delta x \right\} \\
&\quad + \frac{1}{2} \left\{ \frac{\partial^2 U}{\partial t^2}(\Delta t)^2 + 2\frac{\partial^2 U}{\partial t \partial x}\Delta t\, \Delta x + \frac{\partial^2 U}{\partial x^2}(\Delta x)^2 \right\} \\
&\quad + \cdots
\end{aligned}
$$

where the partial derivatives are evaluated at $(t, X_t)$. Thus, as we shall see in Section 3, we obtain

(1.19)      $$dY_t = \left\{ \frac{\partial U}{\partial t}(t, X_t) + \frac{1}{2}f^2\frac{\partial^2 U}{\partial x^2}(t, X_t) \right\} dt + \frac{\partial U}{\partial x}(t, X_t)\, dX_t$$

with equality interpreted in the mean-square sense. This is a stochastic chain rule and is known as the *Ito formula*.   It contains an additional term not present in the usual chain rule (1.18), and this gives rise to the extra term in integrals like (1.17). For example, with $X_t = W_t$ and $Y_t = X_t^2$, so $f \equiv 1$ and $U(t, x) = x^2$, we have

$$dY_t = d\left(X_t^2\right) = 1\, dt + 2X_t\, dX_t$$

or

$$W_t\, dW_t = \frac{1}{2}d\left(W_t^2\right) - \frac{1}{2}\, dt.$$

Hence in integral form

$$\int_0^t W_s \, dW_s = \frac{1}{2} \int_0^t d\left(W_s^2\right) - \frac{1}{2} \int_0^t 1 \, ds$$
$$= \frac{1}{2} W_t^2 - \frac{1}{2} t,$$

since $W_0 = 0$, w.p.1.

It is usual to express the Ito formula in terms of the differentials $dt$ and $dW_t$. This is easily done in (1.19) since $dX_t = f \, dW_t$. In the general case a stochastic differential can also include a "$dt$" term, that is, it has the form

$$dX_t(\omega) = e(t, \omega) \, dt + f(t, \omega) \, dW_t(\omega).$$

Then the Ito formula has the form

$$(1.20) \qquad dY_t = \left\{ \frac{\partial U}{\partial t} + e \frac{\partial U}{\partial x} + \frac{1}{2} f^2 \frac{\partial^2 U}{\partial x^2} \right\} dt + f \frac{\partial U}{\partial x} \, dW_t$$

where the partial derivatives of $U$ are evaluated at $(t, X_t)$.

When $U$ is linear in $x$ we have

$$\frac{\partial^2 U}{\partial x^2} = 0$$

and the Ito formula (1.20) reduces to the usual chain rule (1.18).

## 3.2    The Ito Stochastic Integral

In this section we shall consider the Ito stochastic integral and its properties from a more thorough mathematical perspective, at the same time extending the definition to a wider class of integrands than that mentioned in Section 1. For this we suppose that we have a probability space $(\Omega, \mathcal{A}, P)$, a Wiener process $W = \{W_t, t \geq 0\}$ and an increasing family $\{\mathcal{A}_t, t \geq 0\}$ of sub -$\sigma$-algebras of $\mathcal{A}$ such that $W_t$ is $\mathcal{A}_t$- measurable with

$$E\left(W_t | \mathcal{A}_0\right) = 0 \quad \text{and} \quad E\left(W_t - W_s | \mathcal{A}_s\right) = 0,$$

w.p.1, for all $0 \leq s \leq t$. Here the $\sigma$-algebra $\mathcal{A}_t$ may be thought of as a collection of events that are detectable prior to or at time $t$, so that the $\mathcal{A}_t$-measurability of $Z_t$ for a stochastic process $\{Z_t, t \geq 0\}$ indicates its nonanticipativeness with respect to the Wiener process $W$.

For $0 < T < \infty$ we define a class $\mathcal{L}_T^2$ of functions $f : [0, T] \times \Omega \to \Re$ satisfying

$$(2.1) \qquad f \quad \text{is jointly } \mathcal{L} \times \mathcal{A}\text{-measurable};$$

$$(2.2) \qquad \int_0^T E\left(f(t, \cdot)^2\right) dt < \infty;$$

(2.3)                    $E\left(f(t,\cdot)^2\right) < \infty$   for each   $0 \le t \le T$;

and

(2.4)                    $f(t,\cdot)$ is $\mathcal{A}_t$-measurable for each   $0 \le t \le T$.

In addition we consider two functions in $\mathcal{L}_T^2$ to be identical if they are equal for all $(t,\omega)$ except possibly on a subset of $\mu_L \times P$-measure zero. Then with the norm

(2.5)                    $$\|f\|_{2,T} = \sqrt{\int_0^T E\left(f(t,\cdot)^2\right)\,dt}$$

$\mathcal{L}_T^2$ is a complete normed linear space, that is a Banach space, provided we identify functions which differ only on sets of measure zero (in fact, it is a Hilbert space).

We remark that conditions (2.1)–(2.4) are stronger than $f \in L^2([0,T]\times\Omega,$ $\mathcal{L}\times\mathcal{A},\ \mu_L \times P)$ which, by Fubini's Theorem guarantees (2.3) only for almost all ($\mu_L$-measure) $t \in [0,T]$, rather than for all $t \in [0,T]$, as will be required below.

For any partition $0 = t_1 < t_2 < \cdots < t_{n+1} = T$ and any mean-square integrable $\mathcal{A}_{t_j}$-measurable random variables $f_j,\ j = 1,2,\ldots,n$, we define a step function $f \in \mathcal{L}_T^2$ by $f(t,\omega) = f_j(\omega)$, w.p.1, for $t_j \le t < t_{j+1}$ and $j = 1,2,\ldots,n$. Here the integral in (2.2) takes the form

(2.6)                    $$\int_0^T E\left(f(t,\cdot)^2\right)\,dt = \sum_{j=1}^n E\left(f_j^2\right)(t_{j+1} - t_j).$$

We denote by $\mathcal{S}_T^2$ the subset of all step functions in $\mathcal{L}_T^2$. Then we can approximate any function in $\mathcal{L}_T^2$ by step functions in $\mathcal{S}_T^2$ to any desired degree of accuracy in the norm (2.5). To be specific we have

**Lemma 3.2.1**   $\mathcal{S}_T^2$ *is dense in* $(\mathcal{L}_T^2, \|\cdot\|_{2,T})$.

**Proof**   We shall consider partitions of $[0,T]$ of the form $0 = t_1^{(n)} < t_2^{(n)}$ $< \cdots < t_{n+1}^{(n)} = T$ with $t_{j+1}^{(n)} - t_j^{(n)} \to 0$ for $j = 1,2,\ldots,n$ as $n \to \infty$. When $f$ is mean-square continuous in $t$, that is when $E(f(t,\cdot)^2)$ is continuous, we define a sequence of step functions $f^{(n)}$ by $f^{(n)}(t,\omega) = f(t_j^{(n)},\omega)$, w.p.1, in $t_j^{(n)} \le t$ $< t_{j+1}^{(n)}$ for $j = 1,2,\ldots,n$ and $n = 1,2,3,\ldots$. Clearly then $f^{(n)} \in \mathcal{S}_T^2$ for each $n = 1,2,3,\ldots$ and

$$E\left(\left|f^{(n)}(t,\cdot) - f(t,\cdot)\right|^2\right) \to 0 \quad \text{as} \quad n \to \infty$$

for each $t \in [0,T]$. Hence by the Lebesgue Dominated Convergence Theorem 2.2.3 applied to the space $L^1([0,T],\mathcal{L},\mu_L)$ we have

$$\int_0^T E\left(\left|f^{(n)}(t,\cdot) - f(t,\cdot)\right|^2\right)\,dt \to 0 \quad \text{as} \quad n \to \infty.$$

Now a function $f \in \mathcal{L}_T^2$ is generally not mean-square continuous, but we can approximate it arbitrarily closely in the norm (2.5) by one that is. To begin, we approximate $f$ by a bounded function $f_N \in \mathcal{L}_T^2$ defined by

$$f_N(t, \omega) = \max\{-N, \min\{f(t, \omega), N\}\}$$

for some $N > 0$. Obviously $|f_N(t, \omega)| \leq N$, with $f_N(t, \omega) = f(t, \omega)$ for those $(t, \omega)$ for which $|f(t, \omega)| \leq N$. Moreover

$$\int_0^T E\left(|f_N(t, \cdot) - f(t, \cdot)|^2\right) dt \leq 4 \int_0^T E\left(|f(t, \cdot)|^2\right) dt < \infty,$$

so by the Dominated Convergence Theorem 2.2.3 applied to the functions $E(|f_N(t, \cdot) - f(t, \cdot)|^2)$ in $L^1([0, T], \mathcal{L}, \mu_L)$ it follows that

$$\int_0^T E\left(|f_N(t, \cdot) - f(t, \cdot)|^2\right) dt \to 0 \quad \text{as} \quad N \to \infty.$$

Then for such an $f_N$ we define a function $g_k$ for $k > 0$ by

$$g_k(t, \omega) = k e^{-kt} \int_0^t e^{ks} f_N(s, \omega) \, ds.$$

From the properties of $f_N$ and the fact that the above integrand does not involve values of $f_N$ for times exceeding $t$, it follows that $g_k$ is jointly $\mathcal{L} \times \mathcal{A}$-measurable and that $g_k(t, \cdot)$ is $\mathcal{A}_t$-measurable for each $t \in [0, T]$. Also from the bound on $|f_N|$ we have

$$|g_k(t, \omega)| \leq N\left(1 - e^{-kt}\right),$$

so $E(g_k(t, \cdot)^2)$ is finite and integrable over $0 \leq t \leq T$; hence $g_k \in \mathcal{L}_T^2$. Finally, the sample paths of $g_k$ satisfy

$$|g_k(t, \omega) - g_k(s, \omega)| \leq 2kN|t - s|$$

and are thus continuous. In fact this bound also implies that $E(g_k(t, \cdot)^2)$ is continuous, that is $g_k$ is mean-square continuous. Consequently we can approximate it by a step function $f^{(n)} \in \mathcal{S}_T^2$ as in the first part of the proof. For any given $\epsilon > 0$ we can choose $f_N$, $g_k$ and $f^{(n)}$ successively so that

$$\|f - f_N\|_{2,T} < \frac{1}{3}\epsilon, \quad \|f_N - g_k\|_{2,T} < \frac{1}{3}\epsilon,$$

$$\|g_k - f^{(n)}\|_{2,T} < \frac{1}{3}\epsilon.$$

Then by the triangle inequality (1.4.37) we have

$$\|f - f^{(n)}\|_{2,T} < \epsilon,$$

which is what we were required to prove. $\square$

Now let $f$ be a step function in $\mathcal{S}_T^2$ corresponding to a partition $0 = t_1 < t_2 < \cdots < t_{n+1} = T$ and random variables $f_1, f_2, \ldots, f_n$. We define the Ito stochastic integral for this $f$ over the interval $[0, T]$ by

$$(2.7) \qquad I(f)(\omega) = \sum_{j=1}^{n} f_j(\omega) \left\{ W_{t_{j+1}}(\omega) - W_{t_j}(\omega) \right\},$$

w.p.1. Since $f_j$ is $\mathcal{A}_{t_j}$-measurable and $W_{t_{j+1}} - W_{t_j}$ is $\mathcal{A}_{t_{j+1}}$-measurable, where $\mathcal{A}_{t_j} \subseteq \mathcal{A}_{t_{j+1}}$, their product $f_j\{W_{t_{j+1}} - W_{t_j}\}$ is $\mathcal{A}_{t_{j+1}}$-measurable for $j = 1, 2, \ldots, n$; hence $I(f)$ is $\mathcal{A}_T$-measurable. In addition, each product is integrable over $\Omega$, which follows from the Cauchy-Schwarz inequality (1.4.38) and the fact that each term is mean-square integrable; hence $I(f)$ is integrable. In fact

$$E(I(f)) = \sum_{j=1}^{n} E\left( f_j \left\{ W_{t_{j+1}} - W_{t_j} \right\} \right)$$

$$= \sum_{j=1}^{n} E(f_j) E\left( W_{t_{j+1}} - W_{t_j} \big| \mathcal{A}_{t_j} \right) = 0$$

since $E(W_{t_{j+1}} - W_{t_j} | \mathcal{A}_{t_j}) = 0$. Also $f_j$ and $f_j f_i \{ W_{t_{i+1}} - W_{t_i} \}$ are $\mathcal{A}_{t_j}$-measurable for any $i < j$. Thus

$$E\left( I(f)^2 \right) = \sum_{j=1}^{n} E\left( f_j^2 \left\{ W_{t_{j+1}} - W_{t_j} \right\}^2 \right)$$

$$+ 2 \sum_{j=1}^{n} \sum_{i=j+1}^{n} E\left( f_j f_i \left\{ W_{t_{j+1}} - W_{t_j} \right\} \left\{ W_{t_{i+1}} - W_{t_i} \right\} \right)$$

$$= \sum_{j=1}^{n} E(f_j^2) E\left( \left\{ W_{t_{j+1}} - W_{t_j} \right\}^2 \big| \mathcal{A}_{t_j} \right)$$

$$+ 2 \sum_{j=1}^{n} \sum_{i=j+1}^{n} E\left( f_j f_i \left\{ W_{t_{j+1}} - W_{t_j} \right\} \right) E\left( W_{t_{i+1}} - W_{t_i} \big| \mathcal{A}_{t_j} \right)$$

$$= \sum_{j=1}^{n} E(f_j^2)(t_{j+1} - t_j)$$

$$= \int_0^T E\left( f(t, \cdot)^2 \right) dt,$$

where we have used $E(W_{t_{j+1}} - W_{t_j} | \mathcal{A}_{t_j}) = 0$, $E(\{W_{t_{j+1}} - W_{t_j}\}^2 | \mathcal{A}_{t_j}) = t_{j+1} - t_j$ and the definition of the Lebesgue (or Riemann) integral for the nonrandom step function $E(f(t, \cdot)^2)$. Finally we note that for $f, g \in \mathcal{S}_T^2$ and $\alpha, \beta \in \Re$ we have $\alpha f + \beta g \in \mathcal{S}_T^2$, with the combined step points of $f$ and $g$, so by algebraic rearrangement we obtain, w.p.1,

$$I\left(\alpha f + \beta g\right) = \alpha I\left(f\right) + \beta I\left(g\right).$$

Collecting these results we thus proved

**Lemma 3.2.2** *For any $f, g \in \mathcal{S}_T^2$ and $\alpha, \beta \in \Re$ the Ito stochastic integral (2.7) satisfies*

$$(2.8) \qquad I\left(f\right) \quad \textit{is } \mathcal{A}_T\textit{-measurable},$$

$$(2.9) \qquad E\left(I\left(f\right)\right) = 0,$$

$$(2.10) \qquad E\left(I\left(f\right)^2\right) = \int_0^T E\left(f(t, \cdot)^2\right) dt,$$

*and*

$$(2.11) \qquad I\left(\alpha f + \beta g\right) = \alpha I\left(f\right) + \beta I\left(g\right), \quad w.p.1.$$

For an arbitrary function $f \in \mathcal{L}_T^2$ Lemma 3.2.1 provides us with a sequence of step functions $f^{(n)} \in \mathcal{S}_T^2$ for which

$$\int_0^T E\left(\left|f^{(n)}(t, \cdot) - f(t, \cdot)\right|^2\right) dt \to 0 \quad \text{as} \quad n \to \infty.$$

The Ito integrals $I(f^{(n)})$ are well-defined by (2.7) and, since $I(f^{(n)}) - I(f^{(n+m)}) = I(f^{(n)} - f^{(n+m)})$, they satisfy

$$E\left(\left|I\left(f^{(n)}\right) - I\left(f^{(n+m)}\right)\right|^2\right)$$

$$= E\left(\left|I\left(f^{(n)} - f^{(n+m)}\right)\right|^2\right)$$

$$= \int_0^T E\left(\left|f^{(n)}(t, \cdot) - f^{(n+m)}(t, \cdot)\right|^2\right) dt,$$

which, by the inequality $(a + b)^2 \leq 2(a^2 + b^2)$ gives

$$(2.12) \qquad E\left(\left|I\left(f^{(n)}\right) - I\left(f^{(n+m)}\right)\right|^2\right)$$

$$\leq 2\int_0^T E\left(\left|f^{(n)}(t, \cdot) - f(t, \cdot)\right|^2\right) dt$$

$$+ 2\int_0^T E\left(\left|f(t, \cdot) - f^{(n+m)}(t, \cdot)\right|^2\right) dt.$$

This says that $I(f^{(n)})$ is a Cauchy sequence in the Banach space $L^2(\Omega, \mathcal{A}, P)$, and so there exists a unique, w.p.1, random variable $I$ in $L^2(\Omega, \mathcal{A}, P)$ such that $E(|I(f^{(n)}) - I|^2) \to 0$ as $n \to \infty$. This $I$ is $\mathcal{A}_T$-measurable since it is the limit of $\mathcal{A}_T$-measurable random variables. Moreover we obtain the same limit $I$ for any choice of step functions converging to $f$ in $\mathcal{L}_T^2$. To see this let $\tilde{f}^{(n)}$ be another sequence of step functions converging to $f$ and suppose that $I(\tilde{f}^{(n)})$ converges to $\tilde{I}$. Then

$$E\left(\left|I - \tilde{I}\right|^2\right) \leq 2E\left(\left|I - I\left(f^{(n)}\right)\right|^2\right) + 2E\left(\left|\tilde{I} - I\left(\tilde{f}^{(n)}\right)\right|^2\right)$$

where we estimate the second term as in (2.12) with $\tilde{f}^{(n)}$ replacing $f^{(n+m)}$. Taking limits as $n \to \infty$ we obtain $E(|I - \tilde{I}|^2) = 0$, and hence $I = \tilde{I}$, w.p.1.

We define the *Ito stochastic integral* $I(f)$ of a function $f \in \mathcal{L}_T^2$ to be the common mean-square limit of sequences of the sums (2.7) for any sequence of step functions in $\mathcal{S}_T^2$ converging to $f$ in the norm (2.5). It obviously inherits the properties listed in Lemma 3.2.2 for the step functions, so we have

**Theorem 3.2.3**    *The Ito stochastic integral $I(f)$ satisfies properties (2.8)–(2.11) for functions $f \in \mathcal{L}_T^2$.*

From the identity $4ab = (a + b)^2 - (a - b)^2$, the linearity of Lebesgue and Ito integrals and property (2.10) of the Ito integral we can show the following relationship.

**Corollary 3.2.4**    *For any $f, g \in \mathcal{L}_T^2$*

$$E\left(I\left(f\right)I\left(g\right)\right) = \int_0^T E\left(f(t, \cdot)g(t, \cdot)\right)\,dt.$$

So far we have only considered the Ito integral $I(f)$ of a function $f \in \mathcal{L}_T^2$ over a fixed time interval $[0, T]$. We shall continue to assume that $f \in \mathcal{L}_T^2$ and take any Borel subset $B$ of $[0, T]$. Then the Ito integral of $f$ over the subset $B$ is just the Ito integral $I(fI_B)$ of $fI_B$ over $[0, T]$, where $I_B$ is the indicator function of $B$; clearly $fI_B \in \mathcal{L}_T^2$. Usually we consider subintervals $[t_0, t_1]$ of $[0, T]$ and denote the resulting Ito integral by $\int_{t_0}^{t_1} f\,dW_s$. We could alternatively define this directly in terms of step functions defined only on $[t_0, t_1]$. For $0 \leq t_0 < t_1 < t_2 \leq T$ we have

$$fI_{[t_0, t_2]} = fI_{[t_0, t_1]} + fI_{[t_1, t_2]},$$

except at the instant $t = t_1$, and so from the linearity property (2.11) we obtain, w.p.1,

$$(2.13) \qquad \int_{t_0}^{t_2} f(s, \omega)\,dW_s(\omega) = \int_{t_0}^{t_1} f(s, \omega)\,dW_s(\omega) + \int_{t_1}^{t_2} f(s, \omega)\,dW_s(\omega).$$

For a variable subinterval $[t_0, t] \subseteq [0, T]$ we form a stochastic process $Z = \{Z_t, t_0 \leq t \leq T\}$, defined by

$$(2.14) \qquad\qquad Z_t(\omega) = \int_{t_0}^t f(s, \omega)\,dW_s(\omega),$$

w.p.1, for $t_0 \leq t \leq T$. Replacing 0 by $t_0$ and $T$ by $t$ in Theorem 3.2.3, we see that $Z_t$ is $\mathcal{A}_t$-measurable with $E(Z_t) = 0$ and

$$(2.15) \qquad E\left(Z_t^2\right) = \int_{t_0}^{t} E\left(f(s, \cdot)^2\right) ds.$$

From (2.13) and (2.14) we then obtain for any $0 \leq t' \leq t \leq T$

$$E\left(|Z_t - Z_{t'}|^2\right) = \int_{t'}^{t} E\left(f(s, \cdot)^2\right) ds,$$

from which it follows that $Z_t$ is mean-square continuous. Thus it has a separable and jointly $\mathcal{L} \times \mathcal{A}$-measurable version, which we shall use from now on. In fact this version has, almost surely, continuous sample paths, which we shall prove in Theorem 3.2.6 using the following martingale property of $Z$.

**Theorem 3.2.5** *For $t_0 \leq s \leq t \leq T$ we have*

$$E\left(Z_t - Z_s | \mathcal{A}_s\right) = 0, \qquad w.p.1.$$

**Proof** We shall write $Z_t^{(n)}$ for the integral (2.14) when the integrand is a step function $f^{(n)}$. Then we have, w.p.1,

$$Z_t^{(n)}(\omega) - Z_s^{(n)}(\omega) = \int_{s}^{t} f^{(n)}(u, \omega) \, dW_u(\omega).$$

Restricting attention to the subinterval $[s, t]$ with a partition $s = t_1^{(n)} < t_2^{(n)} < \cdots < t_{n+1}^{(n)} = t$ and $f^{(n)}(t_j^{(n)}, \omega) = f_j^{(n)}(\omega)$, we see that $f_j^{(n)}$ is $\mathcal{A}_{t_j^{(n)}}$-measurable, whereas $E(W_{t_{j+1}^{(n)}} - W_{t_j^{(n)}} | \mathcal{A}_{t_j^{(n)}}) = 0$, w.p.1, for $j = 1, 2, \ldots, n$. Hence

$$E\left(f_j^{(n)}\left\{W_{t_{j+1}^{(n)}} - W_{t_j^{(n)}}\right\} | \mathcal{A}_{t_j^{(n)}}\right) = f_j^{(n)} E\left(W_{t_{j+1}^{(n)}} - W_{t_j^{(n)}} | \mathcal{A}_{t_j^{(n)}}\right) = 0,$$

w.p.1, for $j = 1, 2, \ldots, n$. Taking successive conditional expectations with respect to the coarser $\sigma$-algebras $\mathcal{A}_{t_n^{(n)}}, \mathcal{A}_{t_{n-1}^{(n)}}, \ldots, \mathcal{A}_{t_2^{(n)}}, \mathcal{A}_s$ of the corresponding sum (2.7) for the Ito integral of the step function $f^{(n)}$ over $[s, t]$, we obtain

$$(2.16) \qquad E\left(Z_t^{(n)} - Z_s^{(n)} | \mathcal{A}_s\right) = 0, \qquad w.p.1,$$

on account of the properties of nested conditional expectations. The stated result then follows because (2.16) is preserved in the mean-square limit. $\square$

Since $Z_t$ defined by (2.14) is a mean-square continuous martingale, the maximal martingale inequality (2.3.6) with (2.15) implies that

$$(2.17) \qquad P\left(\sup_{t_0 \leq s \leq t} |Z_s| \geq a\right) \leq \frac{1}{a^2} \int_{t_0}^{t} E\left(f(s, \cdot)^2\right) ds$$

for any $a > 0$, and the Doob inequality (2.3.7) with (2.15) yields

$$(2.18) \qquad E\left(\sup_{t_0 \leq s \leq t} |Z_s|^2\right) \leq 4 \int_{t_0}^{t} E\left(f(s, \cdot)^2\right) ds.$$

The difference of two martingales with respect to the same increasing family of $\sigma$-algebras is itself a martingale. Hence for $f \in \mathcal{L}_T^2$ and $\{f^{(n)}\}$ a sequence of step functions converging to $f$ in $\mathcal{L}_T^2$, the difference $Z_t - Z_t^{(n)}$ is a martingale, where $Z_t$ and $Z_t^{(n)}$ are defined as in (2.14) for $f$ and $f^{(n)}$, respectively. Then by the inequality (2.17) we have

$$P\left(\sup_{t_0 \leq s \leq T} \left|Z_t - Z_t^{(n)}\right| \geq \frac{1}{n}\right) \leq n^2 \int_{t_0}^{T} E\left(\left|f(s,\cdot) - f^{(n)}(s,\cdot)\right|^2\right) ds \leq \frac{1}{n^2}$$

if we choose the step functions $f^{(n)}$ such that

$$\int_{t_0}^{T} E\left(\left|f(s,\cdot) - f^{(n)}(s,\cdot)\right|^2\right) ds \leq \frac{1}{n^4}$$

for $n = 1, 2, 3, \ldots$. The infinite series

$$\sum_{n=1}^{\infty} P\left(\sup_{t_0 \leq s \leq T} \left|Z_t - Z_t^{(n)}\right| \geq \frac{1}{n}\right)$$

is thus convergent, and so by the Borel-Cantelli Lemma 2.1.4

$$N = \bigcap_{n \geq 1} \bigcup_{k \geq n} \left\{\omega \in \Omega : \sup_{t_0 \leq s \leq T} \left|Z_t(\omega) - Z_t^{(n)}(\omega)\right| \geq \frac{1}{n}\right\}$$

is a null event, that is for $\omega \in N$ the inequality

$$\sup_{t_0 \leq s \leq T} \left|Z_t(\omega) - Z_t^{(n)}(\omega)\right| \geq \frac{1}{n}$$

can occur for at most a finite number of $n$. Thus for $\omega \notin N$

$$\lim_{n \to \infty} \sup_{t_0 \leq s \leq T} \left|Z_t(\omega) - Z_t^{(n)}(\omega)\right| = 0.$$

Now the sample paths of $Z_t^{(n)}(\omega)$ are obviously continuous on $t_0 \leq t \leq T$. Since $Z_t(\omega)$ is the uniform limit of continuous functions, it is itself continuous. Thus we have proven

**Theorem 3.2.6**    *A separable, jointly measurable version of $Z_t$ defined by*

$$Z_t(\omega) = \int_{t_0}^{t} f(s,\omega) \, dW_s(\omega)$$

*for $t \in [t_0, T]$ has, almost surely, continuous sample paths.*

We recall that $\mathcal{A}(Z_t)$ denotes the $\sigma$-algebra generated by the random variable $Z_t$. When $Z_t$ is defined by (2.14) it is $\mathcal{A}(Z_t)$-measurable and $\mathcal{A}(Z_t) \subseteq \mathcal{A}_t$. Thus from the properties of conditional expectations we have for any $t_0 \leq t_1 \leq t_2 \leq T$

$$E\left(Z_{t_2} - Z_{t_1}|\mathcal{A}(Z_{t_1})\right)$$
$$= E\left(E\left(Z_{t_2} - Z_{t_1}|\mathcal{A}_{t_1}\right)|\mathcal{A}(Z_{t_1})\right)$$
$$= E\left(0|\mathcal{A}(Z_{t_1})\right) = 0,$$

w.p.1, since $Z_t$ is an $\mathcal{A}_t$-martingale. Consequently

(2.19) $$E\left(Z_{t_2}|\mathcal{A}(Z_{t_1})\right) = E\left(Z_{t_1}|\mathcal{A}(Z_{t_1})\right) = Z_{t_1},$$

w.p.1, so $Z_t$ is also an $\mathcal{A}(Z_t)$-martingale.

**Exercise 3.2.7** *Show that*

$$\int_0^T f(t)\,dW_t = f(T)W_T - \int_0^T f'(t)W_t\,dt$$

*for any continuously differentiable function $f : [0,T] \to \Re$. Find the random functions, if there are any, for which this formula is valid.*

In many ways a sample-path continuous version $Z_t$ of an Ito integral (2.14) resembles a Wiener process. It is $\mathcal{A}_t$- measurable with $Z_{t_0} = 0$ and $E(Z_t - Z_s|\mathcal{A}_s) = 0$, w.p.1, for $t_0 \le s \le t \le T$. The main difference is that we have

$$E\left(|Z_t - Z_s|^2\,|\mathcal{A}_s\right) = \int_s^t E\left(f(u,\cdot)^2|\mathcal{A}_s\right)\,du,$$

w.p.1, instead of equalling $t - s$ as it would for a Wiener process. When the integrand $f$ is nonrandom, that is $f(t,\omega) \equiv f(t)$, we can transform the time variable to convert $Z$ into a Wiener process. To show this we shall suppose that $t_0 = 0$ and define

$$\tilde{t}(t) = \int_0^t f(u)^2\,du$$

This is nondecreasing, but for simplicity we shall assume that it is strictly increasing and hence invertible. We define $\tilde{Z}_{\tilde{t}} = Z_{t(\tilde{t})}$ and $\tilde{\mathcal{A}}_{\tilde{t}} = \mathcal{A}_{t(\tilde{t})}$, where $t(\tilde{t})$ is the inverse of $\tilde{t}(t)$. Then $\tilde{Z}_{\tilde{t}}$ is $\tilde{\mathcal{A}}_{\tilde{t}}$-measurable with

$$\tilde{Z}_0 = 0, \qquad E\left(\tilde{Z}_{\tilde{t}} - \tilde{Z}_{\tilde{s}}|\tilde{\mathcal{A}}_{\tilde{s}}\right) = 0$$

and

$$E\left(\left|\tilde{Z}_{\tilde{t}} - \tilde{Z}_{\tilde{s}}\right|^2|\tilde{\mathcal{A}}_{\tilde{s}}\right) = \int_s^t f(u)^2\,du = \tilde{t} - \tilde{s},$$

w.p.1. Hence by a theorem of Doob (see 2.4.10) the process $\{\tilde{Z}_{\tilde{t}}, \tilde{t} \ge 0\}$ is a Wiener process with respect to the family of $\sigma$-algebras $\{\tilde{\mathcal{A}}_{\tilde{t}}, \tilde{t} \ge 0\}$, at least for $0 \le \tilde{t} \le \tilde{t}(T)$.

**Exercise 3.2.8** *Is $\tilde{Z}_{\tilde{t}} = \int_0^t 2s\,dW_s$ with $\tilde{t} = \frac{4}{3}t^3$ a standard Wiener process with respect to the $\tilde{\mathcal{A}}_{\tilde{t}}$?*

To conclude this section we shall extend the Ito integral to a wider class of integrands than those in the space $\mathcal{L}_T^2$. We shall say that $f$ belongs to $\mathcal{L}_T^w$ if $f$ is jointly $\mathcal{L} \times \mathcal{A}$-measurable, $f(t, \cdot)$ is $\mathcal{A}_t$-measurable for each $t \in [0, T]$ and

$$(2.20) \qquad \int_0^T f(s, \omega)^2 \, ds < \infty,$$

w.p.1; hence $\mathcal{L}_T^2 \subset \mathcal{L}_T^w$. We then define $f_n \in \mathcal{L}_T^w$ by

$$f_n(t, \omega) = \begin{cases} f(t, \omega) & : \quad \int_0^t f(s, \omega)^2 \, ds \le n \\ 0 & : \quad \text{otherwise} \end{cases}$$

The Ito stochastic integrals $I(f_n)$ of the $f_n$ over $0 \le t \le T$ are thus well-defined. It can then be shown that they converge in probability to a unique, w.p.1, random variable, which we shall denote by $I(f)$ and call the Ito stochastic integral of $f \in \mathcal{L}_T^w$ over the interval $0 \le t \le T$. Apart from those properties explicitly involving expectations, which may now not exist, the Ito integrals of integrands $f \in \mathcal{L}_T^w$ satisfy analogous properties to those of integrands $f \in \mathcal{L}_T^2$. It is no longer a mean-square integrable martingale, but it is the convergence in probability limit of such martingales. Useful information can be deduced from this, such as the following counterpart to the estimate (2.17)

$$(2.21) \qquad P\left( \sup_{t_0 \le s \le t} |Z_s| > N \right) \le P\left( \int_{t_0}^t f(s, \omega)^2 \, ds > M \right) + \frac{M}{N^2}$$

for any $f \in \mathcal{L}_T^w$ where $Z_t$ is defined as in (2.14). This inequality can be used to show that $Z_t$ has, almost surely, continuous sample paths.

**Exercise 3.2.9**    *Prove inequality (2.21).*

## 3.3    The Ito Formula

The martingale property $E(Z_t | \mathcal{A}_s) = Z_s$, which follows from Theorem 3.2.5, and its useful technical consequences, is one of the most advantageous features of the Ito stochastic integral. There is however a price to be paid for this, namely that stochastic differentials, which are interpreted in terms of stochastic integrals, do not transform according to the chain rule of classical calculus. Instead an additional term appears and the resulting expression is called the Ito formula. Roughly speaking, the difference is due to the fact that the stochastic differential $(dW_t)^2$ is equal to $dt$ in the mean-square sense, which we used in our formal derivation of the Ito formula in Section 1. We shall now present a rigorous justification for the Ito formula.

Let $e$ and $f$ be two functions with $\sqrt{|e|}$ and $f \in \mathcal{L}_T^w$, so that $e$ and $f$ satisfy the properties required of a function in $\mathcal{L}_T^w$ except that the integral of $e(t, \omega)^2$ is replaced by

$$\int_0^T |e(s, \omega)| \, ds < \infty, \quad \text{w.p.1.}$$

Then, by a *stochastic differential* we mean an expression

$$dX_t(\omega) = e(t,\omega)\, dt + f(t,\omega)\, dW_t(\omega),$$

which is just a symbolical way of writing

$$(3.1) \qquad X_t(\omega) - X_s(\omega) = \int_s^t e(u,\omega)\, du + \int_s^t f(u,\omega)\, dW_u(\omega),$$

w.p.1, for any $0 \le s \le t \le T$. The first integral in (3.1) is an ordinary Riemann or Lebesgue integral for each $\omega \in \Omega$ and the second is an Ito integral.

Since the Ito integral for an integrand $f \in \mathcal{L}_T^w$ is defined as the limit of Ito integrals for integrands in $\mathcal{L}_T^2$, we shall also consider the special case that $\sqrt{|e|}$ and $f \in \mathcal{L}_T^2$. In addition we shall always suppose that $X_t$ is a separable, jointly measurable version of (3.1) with, almost surely, continuous sample paths. When $e$ and $f$ do not depend on $t$ they are $\mathcal{A}_0$-measurable random variables. The increments $X_t - X_s$ are then Gaussian and independent on nonoverlapping intervals with

$$(3.2) \quad E\left(X_t - X_s\right) = E(e)(t - s), \quad \mathrm{Var}\left(X_t - X_s\right) = E\left(f^2\right)(t - s).$$

Other properties, or their analogues, considered in the previous section for the special case $e \equiv 0$ also hold. For example, $X_t$ is $\mathcal{A}_t$-measurable provided $X_0$ is assumed to be $\mathcal{A}_0$-measurable. In addition, from the linearity of the Lebesgue and Ito integrals, we have

$$(3.3) \quad d\left(\alpha X_t^{(1)} + \beta X_t^{(2)}\right) = \left(\alpha e^{(1)} + \beta e^{(2)}\right) dt + \left(\alpha f^{(1)} + \beta f^{(2)}\right) dW_t$$

for any $\alpha, \beta \in \Re$, where

$$(3.4) \qquad\qquad dX_t^{(i)} = e^{(i)}\, dt + f^{(i)}\, dW_t$$

for $i = 1$ and 2.

For nonlinear combinations or transformations of stochastic differentials we must use the Ito formula. We shall prove it now for a scalar transformation of a single stochastic differential, and then consider its multi-component (vector) form in the next section. For our proof we shall use the following lemma, which is a simple consequence of the Taylor and Mean Value Theorems of classical calculus.

**Lemma 3.3.1** *Let $U : [0,T] \times \Re \to \Re$ have continuous partial derivatives $\frac{\partial U}{\partial t}$, $\frac{\partial U}{\partial x}$ and $\frac{\partial^2 U}{\partial x^2}$. Then for any $t, t + \Delta t \in [0,T]$ and $x, x + \Delta x \in \Re$ there exist constants $0 \le \alpha \le 1$, $0 \le \beta \le 1$ such that*

$$(3.5) \qquad U\left(t + \Delta t, x + \Delta x\right) - U\left(t, x\right)$$

$$= \frac{\partial U}{\partial t}\left(t + \alpha\Delta t, x\right)\Delta t + \frac{\partial U}{\partial x}\left(t, x\right)\Delta x$$

$$+ \frac{1}{2}\frac{\partial^2 U}{\partial x^2}\left(t, x + \beta\Delta x\right)\left(\Delta x\right)^2.$$

We shall write $e_t$ and $f_t$ for $e(t, \omega)$ and $f(t, \omega)$, respectively in the following theorem and proof. In it we would gain little from the stronger assumption that $\sqrt{|e|}, f \in \mathcal{L}_T^2$ because that does not guarantee that the integrand $f_t \frac{\partial U}{\partial x}(t, X_t)$ of the Ito integral there belongs to $\mathcal{L}_T^2$ too; we can only conclude that it is in $\mathcal{L}_T^w$, but that also holds when $\sqrt{|e|}, f \in \mathcal{L}_T^w$. A multi-dimensional version of the Ito formula will be given in the next section; see (3.4.6) and (3.4.7).

**Theorem 3.3.2 (The Ito Formula)**      *Let $Y_t = U(t, X_t)$ for $0 \le t \le T$ where $U$ is as in Lemma 3.3.1 and $X_t$ satisfies (3.1) with $\sqrt{|e|}, f \in \mathcal{L}_T^w$. Then*

$$(3.6) \qquad Y_t - Y_s = \int_s^t \left\{ \frac{\partial U}{\partial t}(u, X_u) + e_u \frac{\partial U}{\partial x}(u, X_u) \right.$$
$$\left. + \frac{1}{2} f_u^2 \frac{\partial^2 U}{\partial x^2}(u, X_u) \right\} du$$
$$+ \int_s^t f_u \frac{\partial U}{\partial x}(u, X_u) \, dW_u,$$

*w.p.1, for any $0 \le s \le t \le T$.*

**Proof**   To begin we suppose that $e$ and $f$ do not depend on $t$, so they are $\mathcal{A}_0$-measurable random variables. We choose a sample-path continuous version of $X_t$ and fix a subinterval $[s, t] \subseteq [0, T]$, of which we consider partitions of the form $s = t_1^{(n)} < t_2^{(n)} < \ldots < t_{n+1}^{(n)} = t$ with $\Delta t_j^{(n)} = t_{j+1}^{(n)} - t_j^{(n)}$. Then

$$Y_t - Y_s = U(t, X_t) - U(s, X_s) = \sum_{j=1}^n \Delta U_j^{(n)}$$

where
$$\Delta U_j^{(n)} = U\left(t_{j+1}^{(n)}, X_{t_{j+1}^{(n)}}\right) - U\left(t_j^{(n)}, X_{t_j^{(n)}}\right)$$

for $j = 1, 2, \ldots, n$. Applying Lemma 3.3.1 on each subinterval $[t_j^{(n)}, t_{j+1}^{(n)}]$ for each $\omega \in \Omega$, we have $\alpha_j^{(n)}(\omega), \beta_j^{(n)}(\omega) \in [0, 1]$ such that

$$(3.7) \qquad \Delta U_j^{(n)} = \frac{\partial U}{\partial t}\left(t_j^{(n)} + \alpha_j^{(n)} \Delta t_j^{(n)}, X_{t_j^{(n)}}\right) \Delta t_j^{(n)}$$
$$+ \frac{\partial U}{\partial x}\left(t_j^{(n)}, X_{t_j^{(n)}}\right) \Delta X_j^{(n)}$$
$$+ \frac{1}{2} \frac{\partial^2 U}{\partial x^2}\left(t_j^{(n)}, X_{t_j^{(n)}} + \beta_j^{(n)} \Delta X_j^{(n)}\right) \left(\Delta X_j^{(n)}\right)^2,$$

w.p.1, where $\Delta X_j^{(n)} = X_{t_{j+1}^{(n)}} - X_{t_j^{(n)}}$ for $j = 1, 2, \ldots, n$. By the continuity of $\frac{\partial U}{\partial t}$ and $\frac{\partial^2 U}{\partial x^2}$, and the sample-path continuity of $X_t$, we have for each $j = 1, 2, \ldots, n$

(3.8) $\quad \dfrac{\partial U}{\partial t}\left(t_j^{(n)} + \alpha_j^{(n)}\Delta t_j^{(n)}, X_{t_j^{(n)}}\right) - \dfrac{\partial U}{\partial t}\left(t_j^{(n)}, X_{t_j^{(n)}}\right) \to 0,\quad$ w.p.1,

and

(3.9) $\quad \dfrac{\partial^2 U}{\partial x^2}\left(t_j^{(n)}, X_{t_j^{(n)}} + \beta_j^{(n)}\Delta X_j^{(n)}\right) - \dfrac{\partial^2 U}{\partial x^2}\left(t_j^{(n)}, X_{t_j^{(n)}}\right) \to 0,\quad$ w.p.1,

where $\delta^{(n)} = \max_{1 \le j \le n}\Delta t_j^{(n)} \to 0$ as $n \to \infty$.

For $e$ and $f$ independent of $t$, the increments are of the form $\Delta X_j^{(n)} = e\Delta t_j^{(n)} + f\Delta W_j^{(n)}$ where $\Delta W_j^{(n)} = W_{t_{j+1}}^{(n)} - W_{t_j}^{(n)}$ for $j = 1, 2, \ldots, n$. Hence

(3.10)
$$\sum_{j=1}^{n}\left\{\left(\Delta X_j^{(n)}\right)^2 - \left(f\Delta W_j^{(n)}\right)^2\right\}$$
$$= e^2 \sum_{j=1}^{n}\left(\Delta t_j^{(n)}\right)^2 + 2ef \sum_{j=1}^{n}\Delta W_j^{(n)}\Delta t_j^{(n)},$$

which tends to 0 in probability for $\delta^{(n)} \to 0$ as $n \to \infty$. Combining (3.7)–(3.10) we find for convergence in probability that

(3.11) $\quad Y_t - Y_s = \displaystyle\lim_{\substack{\delta^{(n)}\to 0 \\ n\to\infty}} \sum_{j=1}^{n}\Delta U_j^{(n)}$

$$= \lim_{\substack{\delta^{(n)}\to 0 \\ n\to\infty}} \sum_{j=1}^{n}\left\{\frac{\partial U}{\partial t}\left(t_j^{(n)}, X_{t_j^{(n)}}\right) + e\frac{\partial U}{\partial x}\left(t_j^{(n)}, X_{t_j^{(n)}}\right)\right.$$
$$\left. + \frac{1}{2}f^2\frac{\partial^2 U}{\partial x^2}\left(t_j^{(n)}, X_{t_j^{(n)}}\right)\right\}\Delta t_j^{(n)}$$
$$+ \lim_{\substack{\delta^{(n)}\to 0 \\ n\to\infty}} \sum_{j=1}^{n} f\frac{\partial U}{\partial x}\left(t_j^{(n)}, X_{t_j^{(n)}}\right)\Delta W_j^{(n)}$$
$$+ \lim_{\substack{\delta^{(n)}\to 0 \\ n\to\infty}} \sum_{j=1}^{n}\frac{1}{2}f^2\frac{\partial^2 U}{\partial x^2}\left(t_j^{(n)}, X_{t_j^{(n)}}\right)\left(\left(\Delta W_j^{(n)}\right)^2 - \Delta t_j^{(n)}\right).$$

The first two limits on the right side of (3.11) are the terms on the right side of (3.6), so we need to show that the last limit vanishes. For this we shall write $\Gamma_j^{(n)} = (\Delta W_j^{(n)})^2 - \Delta t_j^{(n)}$ and denote by $I_{n,j}^{(N)}$ the indicator function of the set

$$A_{n,j}^{(N)} = \left\{\omega \in \Omega : |X_{t_i^{(n)}}| \le N \quad \text{for}\quad i = 1, 2, \ldots, j\right\}$$

for $j = 1, 2, \ldots, n$. For fixed $n$ the random variables $\Gamma_j^{(n)}$ are independent with

$E\left(\Gamma_j^{(n)}\right) = 0$ and $E\left(\left(\Gamma_j^{(n)}\right)^2\right) = 2\left(\Delta t_j^{(n)}\right)^2$. From this it follows that

$$E\left(\left|\sum_{j=1}^{n}\frac{\partial^2 U}{\partial x^2}\left(t_j^{(n)},X_{t_j^{(n)}}\right)I_{n,j}^{(N)}\,\Gamma_j^{(n)}\right|^2\right)$$

$$=\quad\sum_{j=1}^{n}E\left(\left|\frac{\partial^2 U}{\partial x^2}\left(t_j^{(n)},X_{t_j^{(n)}}\right)I_{n,j}^{(N)}\,\Gamma_j^{(n)}\right|^2\right)$$

$$\leq\quad C\sum_{j=1}^{n}2\left(\Delta t_j^{(n)}\right)^2$$

$$\leq\quad 2C|t-s|\,\delta^{(n)}\to 0$$

since $\delta^{(n)}\to 0$ as $n\to\infty$; here

$$C=\max_{\substack{s\leq u\leq t\\|x|\leq N}}\left|\frac{\partial^2 U}{\partial x^2}(u,x)\right|^2<\infty.$$

This implies that the last limit in (3.11) vanishes provided that $P\left(A_{n,j}^{(N)}\right)\to 1$ as $N\to\infty$, but this follows from the fact that

$$\bigcup_{j=1}^{n}\left(A_{n,j}^{(N)}\right)^c\subseteq B^{(N)}=\left\{\omega\in\Omega:\sup_{s\leq u\leq t}|X_u(\omega)|>N\right\},$$

where $P\left(B^{(N)}\right)\to 0$ as $N\to\infty$ by inequality similar to (2.21). The proof is thus complete for functions $e$ and $f$ which do not depend on $t$. It is similar for step functions since these do not vary within common partition subintervals.

For general intergrands $e$ and $f$ with $\sqrt{|e|}$, $f\in\mathcal{L}_T^w$ we can find sequences of step functions $\{e^{(n)}\}$, $\{f^{(n)}\}$ in $\mathcal{L}_T^2$ such that the integrals

$$\int_s^t\left|e^{(n)}(u,\omega)-e(u,\omega)\right|du,\qquad\int_s^t\left|f^{(n)}(u,\omega)-f(u,\omega)\right|^2 du$$

converge in probability to zero. Then the sequence defined by

$$X_r^{(n)}=X_s+\int_s^r e_u^{(n)}du+\int_s^r f_u^{(n)}dW_u$$

converges in probability to $X_r$ as $n\to\infty$ for each $s\leq r\leq t$. By taking sub-sequences if necessary, but retaining the original index for simplicity, we can replace each of these convergences in probability by convergence with probability one; moreover we can do this uniformly on the interval $[s,t]$. As Ito's formula holds for step functions we have

$$(3.12)\quad Y_t^{(n)}-Y_s^{(n)}\quad=\quad U\left(t,X_t^{(n)}\right)-U\left(s,X_s^{(n)}\right)$$

$$+\int_s^t\left\{\frac{\partial U}{\partial t}\left(u,X_u^{(n)}\right)+e_u^{(n)}\frac{\partial U}{\partial x}\left(u,X_u^{(n)}\right)\right.$$

$$+ \frac{1}{2} \left( f_u^{(n)} \right)^2 \frac{\partial^2 U}{\partial x^2} \left( u, X_u^{(n)} \right) \Big\} \, du$$

$$+ \int_s^t f_u^{(n)} \frac{\partial U}{\partial x} (u, X_u^{(n)}) \, dW_u,$$

w.p.1, for each $n$. Now $X_u^{(n)} \to X_u$ in probability as $n \to \infty$. By the triangle inequality it thus follows for convergence in probability that

$$\int_s^t \left( \frac{\partial U}{\partial t} \left( u, X_u^{(n)} \right) + e_u^{(n)} \frac{\partial U}{\partial x} \left( u, X_u^{(n)} \right) + \frac{1}{2} \left( f_u^{(n)} \right)^2 \frac{\partial^2 U}{\partial x^2} \left( u, X_u^{(n)} \right) \right) \, du$$

$$\longrightarrow \int_s^t \left( \frac{\partial U}{\partial t} (u, X_u) + e_u \frac{\partial U}{\partial x} (u, X_u) + \frac{1}{2} f_u^2 \frac{\partial^2 U}{\partial x^2} (u, X_u) \right) \, du$$

and

$$\int_s^t f_u^{(n)} \frac{\partial U}{\partial x} \left( u, X_u^{(n)} \right) \, dW_u \longrightarrow \int_s^t f_u \frac{\partial U}{\partial x} (u, X_u) \, dW_u.$$

In fact, taking subsequences if necessary, these can be considered to hold with probability one. Now each path of the process $X_t$ is continuous, w.p.1, and thus bounded, w.p.1. This means that for each path all of the terms appearing in (3.11) are bounded, so we can apply the Lebesgue Dominated Convergence Theorem 2.2.3 to each continuous sample path to conclude that the first integral in (3.12) converges, w.p.1, to the first integral in (3.6). Similarly we have

$$\int_s^t \left| f_u^{(n)} \frac{\partial U}{\partial x} \left( u, X_u^{(n)} \right) \right|^2 \, du \longrightarrow \int_s^t \left| f_u \frac{\partial U}{\partial x} (u, X_u) \right|^2 \, du,$$

w.p.1. Thus the second integral in (3.12) converges in probability and hence, using a subsequence if necessary, also with probability one to the second integral in (3.6). Combining these results we have thus shown that the right side of (3.12) converges with probability one to the right side of (3.6). This completes the proof of Theorem 3.3.2. □

**Example 3.3.3** Let $dX_t = f_t \, dW_t$ and $Y_t = U(t, X_t)$. With $U(t, x) = e^x$ the Ito formula gives

$$(3.13) \qquad\qquad dY_t = \frac{1}{2} f_t^2 Y_t \, dt + f_t Y_t \, dW_t,$$

whereas with $U(t, x) = \exp \left( x - \frac{1}{2} \int_0^t f_u^2 \, du \right)$ it gives

$$(3.14) \qquad\qquad dY_t = f_t Y_t \, dW_t.$$

Equations (3.13) and (3.14) are examples of stochastic differential equations. The latter shows that the counterpart of the exponential in the Ito calculus is

$$\exp \left( X_t - \frac{1}{2} \int_0^t f_u^2 \, du \right) = \exp \left( \int_0^t f_u \, dW_u - \frac{1}{2} \int_0^t f_u^2 \, du \right)$$

rather than the $\exp(X_t)$ of conventional calculus. This indicates that solving Ito stochastic differential equations directly by quadrature will be more complicated than for ordinary differential equations.

**Exercise 3.3.4**   *Use the Ito formula to show that*

$$d\left(X_t^{2n}\right) = n(2n-1) f_t^2 X_t^{2n-2} dt + 2n f_t X_t^{2n-1} dW_t$$

*for $n \geq 1$, where $dX_t = f_t dW_t$. Hence determine $d\left(W_t^{2n}\right)$ for $n \geq 1$.*


## 3.4   Vector Valued Ito Integrals

Let $W = \{W_t, t \geq 0\}$ be an $m$-dimensional Wiener process with independent components associated with an increasing family of $\sigma$-algebras $\{\mathcal{A}_t, t \geq 0\}$. That is, $W_t = (W_t^1, W_t^2, \ldots, W_t^m)$ where the $W^j$ for $j = 1, 2, \ldots, m$ are scalar Wiener processes with respect to $\{\mathcal{A}_t, t \geq 0\}$, which are pairwise independent. Thus each $W_t^j$ is $\mathcal{A}_t$-measurable with

$$E\left(W_t^j | \mathcal{A}_0\right) = 0, \qquad E\left(W_t^j - W_s^j | \mathcal{A}_s\right) = 0,$$

w.p.1, for $0 \leq s \leq t$ and $j = 1, 2, \ldots, m$. In addition,

$$(4.1) \qquad E\left((W_t^i - W_s^i)(W_t^j - W_s^j) | \mathcal{A}_s\right) = (t-s)\delta_{i,j},$$

w.p.1, for $0 \leq s \leq t$ and $i, j = 1, 2, \ldots, m$, where $\delta_{i,j}$ is the Kronecker delta symbol (see (2.4.10)).

We shall consider $d$-dimensional vector functions $e : [0, T] \times \Omega \to \Re^d$ with components $e^k$ satisfying $\sqrt{|e^k|} \in \mathcal{L}_T^w$ (or $\mathcal{L}_T^2$) for $k = 1, 2, \ldots, d$ and $d \times m$-matrix functions $F : [0, T] \times \Omega \to \Re^{d \times m}$ with components $F^{i,j} \in \mathcal{L}_T^w$ (or $\mathcal{L}_T^2$) for $k = 1, 2, \ldots, d$ and $j = 1, 2, \ldots, m$. In analogy with the scalar case we denote by $e_t$ and $F_t$ the vector and matrix valued random variables taken by $e$ and $F$ at an instant $t$. Then we write symbolically as a *d-dimensional vector stochastic differential*

$$(4.2) \qquad\qquad dX_t = e_t dt + F_t dW_t$$

the vector stochastic integral expression

$$(4.3) \qquad X_t - X_s = \int_s^t e_u du + \int_s^t F_u dW_u$$

for any $0 \leq s \leq t \leq T$, which we interpret componentwise as

$$(4.4) \qquad X_t^k) - X_s^k = \int_s^t e_u^k du + \sum_{j=1}^m \int_s^t F_u^{k,j} dW_u^j,$$

w.p.1, for $k = 1, 2, \ldots, d$. When $d = 1$ this covers the scalar case with several independent noise processes. For a preassigned $\mathcal{A}_0$-measurable $X_0$ the resulting

$d$-dimensional stochastic process $X = \{X_t = (X_t^1, X_t^2, \ldots, X_t^d), t \geq 0\}$ enjoys similar properties componentwise to those listed in the previous sections for scalar differentials involving a single Wiener process, with additional properties relating the different components. The actual properties depend on whether the $\sqrt{|e^k|}$ and $F^{k,j}$ belong to $\mathcal{L}_T^2$ for all components or just to $\mathcal{L}_T^w$. In the former case with $e \equiv 0$, for example, we have

$$E\left(X_t^k - X_s^k \mid \mathcal{A}_s\right) = 0$$

and

$$(4.5) \qquad E\left((X_t^k - X_s^k)(X_t^i - X_s^i) \mid \mathcal{A}_s\right) = \sum_{j=1}^m \int_s^t E\left(F_u^{k,j} F_u^{i,j}\right) du,$$

w.p.1, for $0 \leq s \leq t \leq T$ and $k, i = 1, 2, \ldots, d$. Here (4.5) follows from the independence of the components of $W$ and the identity (4.1), which we could write symbolically as $E(dW_t^i dW_t^j) = \delta_{i,j}\, dt$. As in the scalar case this leads to additional terms in the chain rule formula for the transformation of the vector stochastic differential (4.2).

Let $U : [0, T] \times \Re^d \to \Re$ have continuous partial derivatives $\frac{\partial U}{\partial t}, \frac{\partial U}{\partial x_k}, \frac{\partial^2 U}{\partial x_k \partial x_i}$ for $k, i = 1, 2, \ldots, d$, and define a scalar process $\{Y_t, 0 \leq t \leq T\}$ by

$$Y_t = U(t, X_t) = U\left(t, X_t^1, X_t^2, \ldots, X_t^d\right),$$

w.p.1, where $X_t$ satisfies the differential (4.2). Then the stochastic differential for $Y_t$ is given by

$$(4.6) \quad dY_t = \left\{ \frac{\partial U}{\partial t} + \sum_{k=1}^d e_t^k \frac{\partial U}{\partial x_k} + \frac{1}{2} \sum_{j=1}^m \sum_{i,k=1}^d F_t^{i,j} F_t^{k,j} \frac{\partial^2 U}{\partial x_i \partial x_k} \right\} dt$$

$$+ \sum_{j=1}^m \sum_{i=1}^d F_t^{i,j} \frac{\partial U}{\partial x_i} dW_t^j,$$

where the partial derivatives are evaluated at $(t, X_t)$. This is the multi-component analogue of the *Ito formula* (3.6), by which name it is also known. In vector-matrix notation it has the condensed form

$$(4.7) \quad dY_t = \left\{ \frac{\partial U}{\partial t} + e_t^\top \nabla U + \frac{1}{2} \text{tr} \left( F_t F_t^\top \nabla[\nabla U] \right) \right\} dt + \nabla U^\top F_t\, dW_t,$$

where $\nabla$ is the gradient operator, $^\top$ the vector or matrix transpose operation and "tr" the trace of the inscribed matrix, that is the sum of its diagonal components. Thus $\nabla U$ is the vector of the first order spatial partial derivatives of $U$ and $\nabla[\nabla U]$ the matrix of the second order spatial partial derivatives of $U$. The proof of this vector version of the Ito formula is a straightforward modification of the proof in the scalar case (Theorem 3.3.2). A formal derivation similar to that in Section 1 is quicker and insightful; it uses the equality in mean of the differentials $dW_t^i dW_t^j$ and $\delta_{ij}\, dt$ in the Taylor expansion for $U$.

**Example 3.4.1**    *Let $X_t^1$ and $X_t^2$ satisfy the scalar stochastic differentials*

$$(4.8) \qquad\qquad dX_t^i = e_t^i \, dt + f_t^i \, dW_t^i$$

*for $i = 1, 2$ and let $U(t, x_1, x_2) = x_1 x_2$. Then the stochastic differential for the product process*

$$Y_t = X_t^1 \, X_t^2$$

*depends on whether the Wiener processes $W_t^1$ and $W_t^2$ are independent or dependent. In the former case the differentials (4.8) can be written as the vector differential*

$$d \begin{pmatrix} X_t^1 \\ X_t^2 \end{pmatrix} = \begin{pmatrix} e_t^1 \\ e_t^2 \end{pmatrix} dt + \begin{bmatrix} f_t^1 & 0 \\ 0 & f_t^2 \end{bmatrix} d \begin{pmatrix} W_t^1 \\ W_t^2 \end{pmatrix}$$

*and the transformed differential is*

$$(4.9) \qquad dY_t = \left( e_t^1 X_t^2 + e_t^2 X_t^1 \right) dt + f_t^1 X_t^2 \, dW_t^1 + f_t^2 X_t^1 \, dW_t^2.$$

*In contrast, when $W_t^1 = W_t^2 = W_t$ the vector differential for (4.8) is*

$$d \begin{pmatrix} X_t^1 \\ X_t^2 \end{pmatrix} = \begin{pmatrix} e_t^1 \\ e_t^2 \end{pmatrix} dt + \begin{pmatrix} f_t^1 \\ f_t^2 \end{pmatrix} dW_t$$

*and there is an extra term $f_t^1 f_t^2 \, dt$ in the differential of $Y_t$, which is now*

$$(4.10) \qquad dY_t = \left( e_t^1 X_t^2 + e_t^2 X_t^1 + f_t^1 f_t^2 \right) dt + \left( f_t^1 X_t^2 + f_t^2 X_t^1 \right) dW_t.$$

**Exercise 3.4.2**    *Show that*

$$d \left( W_t^1 W_t^2 \right) = W_t^2 \, dW_t^1 + W_t^1 \, dW_t^2$$

*for independent Wiener processes $W_t^1$ and $W_t^2$, whereas*

$$d \left( (W_t)^2 \right) = 1 \, dt + 2 W_t \, dW_t$$

*when $W_t^1 = W_t^2 = W_t$.*

The Ito formula also holds for a vector valued transformation $U : [0, T] \times \Re^d \to \Re^k$ resulting in a vector valued process $Y_t = U(t, X_t)$. For such processes the Ito formula (4.6) is applied separately to each component $Y_t^l = U^l(t, X_t)$ for $l = 1, 2, \ldots, k$.

**Exercise 3.4.3**    *Show that*

$$d \left( \cos W_t \right) = -\frac{1}{2} \cos W_t \, dt - \sin W_t \, dW_t$$

*and*

$$d \left( \sin W_t \right) = -\frac{1}{2} \sin W_t \, dt + \cos W_t \, dW_t.$$

The process $Y_t = (\cos W_t, \sin W_t)$ is a Wiener process on the unit circle. Rewriting the differentials in Exercise 3.4.3 in terms of the components $(Y_t^1, Y_t^2)$ of $Y_t$, we see that $Y_t$ is a solution of the *system of stochastic differential equations*

$$dY_t^1 = -\frac{1}{2} Y_t^1 \, dt - Y_t^2 \, dW_t$$

$$dY_t^2 = -\frac{1}{2} Y_t^2 \, dt + Y_t^1 \, dW_t$$

or, equivalently, of the *vector stochastic differential equation*

$$(4.11) \qquad dY_t = -\frac{1}{2} Y_t \, dt + \begin{bmatrix} 0 & -1 \\ 1 & 0 \end{bmatrix} Y_t \, dW_t$$

with the constraint that

$$|Y_t|^2 = \left(Y_t^1\right)^2 + \left(Y_t^2\right)^2 = 1.$$

**Exercise 3.4.4**   *Derive the vector stochastic differential equation satisfied by the process*

$$Y_t = \left(Y_t^1, Y_t^2\right) = \left(\exp\left(W_t\right), W_t \exp\left(W_t\right)\right).$$

## 3.5   Other Stochastic Integrals

The Ito integral $\int_0^T f(t, \omega) \, dW_t(\omega)$ for an integrand $f \in \mathcal{L}_T^2$ is equal to the mean-square limit of the sums

$$(5.1) \qquad S_n(\omega) = \sum_{j=1}^n f\left(\xi_j^{(n)}, \omega\right) \left\{ W_{t_{j+1}^{(n)}}(\omega) - W_{t_j^{(n)}}(\omega) \right\}$$

with evaluation points $\xi_j^{(n)} = t_j^{(n)}$ for partitions $0 = t_1^{(n)} < t_2^{(n)} < \cdots < t_{n+1}^{(n)} = T$ for which

$$\delta^{(n)} = \max_{1 \leq j \leq n} \left( t_{j+1}^{(n)} - t_j^{(n)} \right) \to 0 \quad \text{as} \quad n \to \infty.$$

Other choices of evaluation points $t_j^{(n)} \leq \xi_j^{(n)} \leq t_{j+1}^{(n)}$ are possible, but generally lead to different random variables in the limit. While arbitrarily chosen evaluation points have little practical or theoretical use, those chosen systematically by

$$(5.2) \qquad \xi_j^{(n)} = (1 - \lambda) t_j^{(n)} + \lambda t_{j+1}^{(n)}$$

for the same fixed $0 \leq \lambda \leq 1$ lead to limits, which we shall denote here by

$$(\lambda) \int_0^T f(t, \omega) \, dW_t(\omega).$$

These are related in a simple, but interesting manner. We note that the case $\lambda$ = 0 is just the Ito integral. The other cases $0 < \lambda \leq 1$ differ in that the process they define with repect to a variable upper integration endpoint is in general no longer a martingale.

When the integrand $f$ has continuously differentiable sample paths we obtain from Taylor's Theorem

$$f\left((1 - \lambda)t_j^{(n)} + \lambda t_{j+1}^{(n)}, \omega\right) = (1 - \lambda)f\left(t_j^{(n)}, \omega\right) + \lambda f\left(t_{j+1}^{(n)}, \omega\right)$$

$$+ O\left(\left|t_{j+1}^{(n)} - t_j^{(n)}\right|\right).$$

Since the higher order terms do not contribute to the limit as $\delta^{(n)} \to 0$, we see that the $(\lambda)$-integrals could then be evaluated alternatively as the mean-square limit of the sums

(5.3)        $$\tilde{S}_n(\omega) = \sum_{j=1}^n \left\{(1 - \lambda)f\left(t_j^{(n)}, \omega\right) + \lambda f\left(t_{j+1}^{(n)}, \omega\right)\right\}$$

$$\times \left\{W_{t_{j+1}^{(n)}}(\omega) - W_{t_j^{(n)}}(\omega)\right\}.$$

In the general case the $(\lambda)$-integrals are usually defined in terms of the sums (5.3) rather than (5.1) with evaluation points (5.2), and we shall follow this practice here. As an indication of how the value of these integrals vary with $\lambda$ we observe that for $f(t, \omega) = W_t(\omega)$ we have

(5.4)        $$(\lambda) \int_0^T W_t(\omega)\, dW_t(\omega) = \frac{1}{2}W_T(\omega)^2 + \left(\lambda - \frac{1}{2}\right)T.$$

This follows using (2.4.14) in the following mean-square limits

$$\sum_{j=1}^n W_{t_j^{(n)}}\left\{W_{t_{j+1}^{(n)}} - W_{t_j^{(n)}}\right\} \longrightarrow \frac{1}{2}W_T^2 - \frac{1}{2}T$$

and

$$\sum_{j=1}^n W_{t_{j+1}^{(n)}}\left\{W_{t_{j+1}^{(n)}} - W_{t_j^{(n)}}\right\}$$

$$= \sum_{j=1}^n \left(W_{t_{j+1}^{(n)}} - W_{t_j^{(n)}}\right)^2 + \sum_{j=1}^n W_{t_j^{(n)}}\left\{W_{t_{j+1}^{(n)}} - W_{t_j^{(n)}}\right\}$$

$$\longrightarrow T + \left(\frac{1}{2}W_T^2 - \frac{1}{2}T\right) = \frac{1}{2}W_T^2 + \frac{1}{2}T,$$

which are multiplied by $(1 - \lambda)$ and $\lambda$, respectively, to give (5.4). Unlike any of the others, the symmetric case $\lambda = \frac{1}{2}$ of the integral (5.4), which was introduced by Stratonovich, does not contain a term in addition to that given by classical calculus. It is now known as the *Stratonovich integral* and denoted by

$$\int_0^T f_t \circ dW_t$$

for an integrand $f \in \mathcal{L}_T^2$; it can be extended to integrands in $\mathcal{L}_T^w$ in the same way as for Ito integrals.

Usually only the Ito and Stratonovich integrals are widely used. As suggested by (5.4) the Stratonovich integral obeys the transformation rules of clasical calculus, and this is a major reason for its use. To see this, let $h : \Re \to \Re$ be continuously differentiable and consider the Stratonovich integral of $h(W_t)$. By the Taylor formula we have

$$h\left(W_{t_{j+1}^{(n)}}\right) = h\left(W_{t_j^{(n)}}\right) + h'\left(W_{t_j^{(n)}}\right)\left\{W_{t_{j+1}^{(n)}} - W_{t_j^{(n)}}\right\} + \text{ higher order terms},$$

so the sum (5.3) with $\lambda = \frac{1}{2}$ is

$$\tilde{S}_n = \sum_{j=1}^n h\left(W_{t_j^{(n)}}\right)\left\{W_{t_{j+1}^{(n)}} - W_{t_j^{(n)}}\right\} + \frac{1}{2}\sum_{j=1}^n h'\left(W_{t_j^{(n)}}\right)\left\{W_{t_{j+1}^{(n)}} - W_{t_j^{(n)}}\right\}^2$$

$$+ \text{ higher order terms}$$

$$\to \int_0^T h\left(W_t\right) dW_t + \frac{1}{2}\int_0^T h'\left(W_t\right) dt$$

in the mean-square sense. Hence

$$(5.5) \qquad \int_0^T h\left(W_t\right) \circ dW_t = \int_0^T h\left(W_t\right) dW_t + \frac{1}{2}\int_0^T h'\left(W_t\right) dt.$$

Now let $U$ be an anti-derivative of $h$, so $U'(x) = h(x)$ and hence $U''(x) = h'(x)$. Applying Ito's formula to the transformation $Y_t = U(W_t)$, we obtain

$$U\left(W_T\right) - U\left(W_0\right) = \frac{1}{2}\int_0^T h'\left(W_t\right) dt + \int_0^T h\left(W_t\right) dW_t.$$

Thus from (5.5) the Stratonovich integral

$$\int_0^T h\left(W_t\right) \circ dW_t = U\left(W_T\right) - U\left(W_0\right),$$

as in classical calculus.

**Example 3.5.1** *For $h(x) = e^x$ an anti-derivative is $U(x) = e^x$, so*

$$\int_0^T \exp\left(W_t\right) \circ dW_t = \exp\left(W_T\right) - 1,$$

*since $W_0 = 0$. Thus $Y_t = \exp(W_t)$ is a solution of the Stratonovich stochastic differential equation*

(5.6)                          $$dY_t = Y_t \circ dW_t.$$

*In contrast the Ito stochastic differential equation*

$$dY_t = Y_t \, dW_t$$

*has the solution $Y_t = \exp\left(W_t - \frac{1}{2}t\right)$ for the same initial value $Y_0 = 1$.*

The major advantage of the Stratonovich stochastic integral is that it obeys the usual transformation rules of calculus. For this reason it is often used to formulate stochastic differential equations on manifolds such as a circle or sphere, as is required, for example, for a stability analysis of stochastic dynamical systems. It is also useful via the identity (5.5) for explicitly evaluating certain Ito integrals. Stochastic processes defined by Stratonovich integrals over a varying time interval do not, however, satisfy the powerful martingale properties of their Ito integral counterparts as we have already mentioned.

**Exercise 3.5.2**    *Use deterministic calculus to solve the Stratonovich stochastic differential equation*

$$dY_t = \exp(-Y_t) \circ dW_t$$

*for $Y_0 = 0$. Hence determine the Ito stochastic differential equation that is satisfied by this solution.*

# Chapter 4

# Stochastic Differential Equations

The theory of stochastic differential equations is introduced in this chapter. The emphasis is on Ito stochastic differential equations, for which an existence and uniqueness theorem is proved and the properties of their solutions investigated. Techniques for solving linear and certain classes of nonlinear stochastic differential equations are presented, along with an extensive list of explicitly solvable equations. Finally, Stratonovich stochastic differential equations and their relationship to Ito equations are examined.

## 4.1 Introduction

The inclusion of random effects in differential equations leads to two distinct classes of equations, for which the solution processes have differentiable and nondifferentiable sample paths, respectively. They require fundamentally different methods of analysis. The first, and simpler, class arises when an ordinary differential equation has random coefficients, a random initial value or is forced by a fairly regular stochastic process, or when some combination of these holds. The equations are called *random differential equations* and are solved sample path by sample path as ordinary differential equations. The sample paths of the solution processes are then at least differentiable functions. As an example consider the linear random differential equation

$$(1.1) \qquad \dot{x} = \frac{dx}{dt} = a(\omega)x + b(t,\omega)$$

where the forcing process $b$ is continuous in $t$ for each $\omega$. For an initial value $x_0(\omega)$ at $t = 0$, the solution is given by

$$(1.2) \qquad x(t,\omega) = e^{a(\omega)t}\left( x_0(\omega) + \int_0^t e^{-a(\omega)s}\, b(s,\omega)\, ds \right).$$

Its sample paths are obviously differentiable functions of $t$.

The second class occurs when the forcing is an irregular stochastic process such as Gaussian white noise. The equations are then written symbolically as stochastic differentials, but are interpreted as integral equations with Ito or Stratonovich stochastic integrals. They are called *stochastic differential equations*, which we shall abbreviate by SDEs, and in general their solutions inherit the nondifferentiability of sample paths from the Wiener processes in the stochastic integrals. In many applications such equations result from the incorporation of either internally or externally originating random fluctuations in the dynamical description of a system. An example of the former is the molecular bombardment of a speck of dust on a water surface, which results in Brownian motion. The intensity of this bombardment does not depend on the state variables, for instance the position and velocity of the speck. Taking $X_t$ as one of the components of the velocity of the particle, Langevin wrote the equation

(1.3) $$\frac{dX_t}{dt} = -aX_t + b\,\xi_t$$

for the acceleration of the particle. This is the sum of a retarding frictional force depending on the velocity and the molecular forces represented by a white noise process $\xi_t$, with intensity $b$ which is independent of the velocity. Here $a$ and $b$ are positive constants. We now interpret the *Langevin equation* (1.3) symbolically as a stochastic differential

(1.4) $$dX_t = -aX_t\,dt + b\,dW_t,$$

that is as a stochastic integral equation

(1.5) $$X_t = X_{t_0} - \int_{t_0}^t aX_s\,ds + \int_{t_0}^t b\,dW_s$$

where the second integral is an Ito stochastic integral. Similar equations arise from electrical systems where $X_t$ is the current and $\xi_t$ represents thermal noise. Additional examples will be given in Chapter 7.

With external fluctuations the intensity of the noise usually depends on the state of the system. For example, the growth coefficient in an exponential growth equation $\dot{x} = \alpha x$ may fluctuate on account of environmental effects, taking the form $\alpha = a + b\,\xi_t$ where $a$ and $b$ are positive constants and $\xi_t$ is a white noise process. This results in the heuristically written equation

(1.6) $$\frac{dX_t}{dt} = aX_t + bX_t\,\xi_t.$$

To be precise, it is a stochastic differential

(1.7) $$dX_t = aX_t\,dt + bX_t\,dW_t$$

or, equivalently, a stochastic integral equation

(1.8) $$X_t = X_{t_0} + \int_{t_0}^t aX_s\,ds + \int_{t_0}^t bX_s\,dW_s.$$

The second integral is again an Ito integral, but now involves the unknown solution.

In the physical sciences the random forcing in (1.3)–(1.5) is called *additive noise,* whereas in (1.6)–(1.8) it is called *multiplicative noise.* Both cases are included in the general differential formulation

(1.9) $$dX_t = a(X_t)\,dt + b(X_t)\,dW_t$$

or equivalent integral formulation

$$(1.10) \qquad X_t = X_{t_0} + \int_{t_0}^t a(X_s)\, ds + \int_{t_0}^t b(X_s)\, dW_s,$$

for appropriate coefficients $a(x)$ and $b(x)$, which may be constants. Using Ito calculus we can verify that

$$(1.11) \qquad X_t = e^{-at} X_0 + e^{-at} \int_0^t e^{as} b\, dW_s$$

is a solution of (1.3)–(1.5) and that

$$(1.12) \qquad X_t = X_0 \exp\left(\left(a - \frac{1}{2}b^2\right)t + bW_t\right)$$

is a solution of (1.6)–(1.8). The former is similar to (1.2), but involves an Ito integral rather than a Riemann integral. We must however impose some restriction on the initial value $X_0$ here so the solution process $X_t$ is nonanticipative with respect to the Wiener process $W_t$, and hence so the Ito integrals in (1.5) and (1.8), the latter in particular, are meaningful. For this we need $X_0$ to be independent of $W_t$ for all $t > 0$, which follows if $X_0$ is $\mathcal{A}_0$-measurable, where $\{\mathcal{A}_t, t \geq 0\}$ is the family of increasing $\sigma$-algebras associated with the Wiener process $\{W_t, t \geq 0\}$, because $X_t$ is then $\mathcal{A}_t$-measurable for each $t \geq 0$.

The explicit solutions (1.11) and (1.12) are the only solutions for their respective equations and given initial values in the sense that any other solution is an equivalent stochastic process, that is with the same finite dimensional probability distributions. In fact, a stronger form of *uniqueness* holds here: any equivalent version $\tilde{X}_t$ with continuous sample paths has, almost surely, the same sample paths as $X_t$, that is

$$P\left(\sup_{0 \leq t \leq T} \left|\tilde{X}_t - X_t\right| > 0\right) = 0$$

for any $T > 0$. We then say that the solutions are *pathwise unique*.

In writing (1.11) and (1.12) we have assumed implicitly that we have a prescribed Wiener process $\{W_t, t \geq 0\}$. Were we to change the Wiener process we would again obtain a unique solution, given by the same formula with the new Wiener process in it. We call such a solution a *strong solution* of the stochastic differential equation and use the term *weak solution* for when we are free to select a Wiener process and then find a solution corresponding to this particular Wiener process. Some stochastic differential equations may only have weak solutions and no strong solutions.

As with most ordinary differential equations we cannot generally find explicit formulae like (1.11) and (1.12) for the solutions of stochastic differential equations and thus need to use a numerical method to determine the solutions approximately. For this we need to know that the equation actually does have a solution, a unique solution preferably, for a given initial value. For an ordinary differential equation

(1.13) $$\dot{x} = \frac{dx}{dt} = a(x)$$

this kind of information is provided by an *existence and uniqueness theorem*. A sufficient condition for the existence and uniqueness of a solution $x(t; x_0)$ with initial value $x(0; x_0) = x_0$ is that $a = a(x)$ satisfies the *Lipschitz condition*

(1.14) $$|a(x) - a(y)| \leq K\,|x - y|$$

for all $x, y \in \Re$, where $K$ is a positive constant. The usual proof involves the Picard-Lindelöf *method of successive approximations* $\{x^{(n)}\}$ with

$$x^{(n+1)}(t) = x_0 + \int_0^t a\left(x^{(n)}(s)\right)\,ds$$

for $n = 0, 1, 2, \ldots$, where $x^{(0)}(t) \equiv x_0$. The Lipschitz condition (1.14) provides a crucial inequality

$$\left|x^{(n+1)}(t) - x^{(n)}(t)\right| \leq K \int_0^t \left|x^{(n)}(s) - x^{(n-1)}(s)\right|\,ds.$$

This is used to establish the uniform convergence of the successive approximations to a continuous solution of the integral equation

$$x(t) = x_0 + \int_0^t a(x(s))\,ds,$$

which is thus a solution of the original differential equation (1.13). The uniqueness of the solution then follows by a similar application of the Lipschitz condition. A serious deficiency is that this solution may become unbounded after a small elapse of time. For example, the solution $x(t) = x_0/(1 - x_0 t)$ of the differential equation $\dot{x} = x^2$ blows up at time $t = T(x_0) = 1/x_0$. To ensure the *global existence* of a solution, that is existence for all time $t > 0$, we need a *growth bound* on $a = a(x)$ such as

(1.15) $$x\,a(x) \quad \text{or} \quad |a(x)|^2 \leq L\left(1 + |x|^2\right)$$

for all $x \in \Re$, where $L$ is a positive constant.

An analogous existence and uniqueness result holds for strong solutions of an SDE (1.8)–(1.9) provided both coefficients $a = a(x)$ and $b = b(x)$ satisfy a Lipschitz condition (1.14) and a growth bound (1.15). Here the Wiener process $\{W_t, t \geq 0\}$ with associated family of $\sigma$-algebras $\{\mathcal{A}_t, t \geq 0\}$ is preassigned and the initial value $X_0$ must be $\mathcal{A}_0$- measurable. The proof also uses successive approximations

$$X_t^{(n+1)} = X_{t_0} + \int_0^t a\left(X_s^{(n)}\right)\,ds + \int_0^t b\left(X_s^{(n)}\right)\,dW_s$$

for $n = 0, 1, 2, \ldots$ where $X_t^{(0)} \equiv X_0$. A simple case assumes that $E\left(X_0^2\right) < \infty$ and then the growth bound is used to show that

$$E\left(\sup_{0 \le t \le T} X_t^2\right) < \infty$$

for any fixed $0 < T < \infty$. The Lipschitz condition is used in a similar way to the deterministic case to show the mean-square convergence of the successive approximations and the mean-square uniqueness of the limiting solution. The Borel-Cantelli Lemma 2.1.4 is then applied to establish the stronger pathwise uniqueness of the solutions. We shall present the details of this proof in Section 5. Variations are possible, though technically more complicated. For example, the requirement that $E\left(X_0^2\right) < \infty$ can be dropped and the Lipschitz condition (1.14) weakened to a *local Lipschitz condition*

$$|a(x) - a(y)| \le K_N \, |x - y|$$

for all $|x|, |y| \le N$, where $K_N$ is a positive constant for each $N > 0$. (From the Mean-Value Theorem of Calculus the latter holds for any continuously differentiable function $a = a(x)$). Growth bounds such as (1.15) are not required for existence or uniqueness, but their absence may result in the sample paths blowing up in a finite time, that is

$$|X_t(\omega)| \to \infty \quad \text{as} \quad t \to T(X_0(\omega)).$$

Here $T(X_0(\omega))$ is called the *explosion time*. The coefficients of the SDE

$$(1.16) \qquad dX_t = -\frac{1}{2}\exp\left(-2X_t\right)dt + \exp\left(-X_t\right)dW_t$$

do not satisfy a growth bound for $x < 0$. The unique solution

$$X_t = \ln\left(W_t + \exp(X_0)\right)$$

of (1.16) exists only for $0 \le t < T(X_0(\omega))$ where

$$T(X_0(\omega)) = \min\left\{t \ge 0 : W_t(\omega) = -\exp(X_0(\omega))\right\}.$$

From their construction, the successive approximations $X_t^{(n)}$ above are obviously $\mathcal{A}_t$-measurable and have, almost surely, continuous sample paths. These properties are inherited by a limiting strong solution $X_t$ of the SDE (1.9)–(1.10). Such a solution, or more precisely the family of solutions $X_t^{0,x}$ with initial values $X_0^{0,x} = x$, w.p.1, for all $x \in \Re$, is a homogeneous Markov process and is often called an *Ito diffusion*.

When the coefficients $a = a(x)$ and $b = b(x)$ of the SDE are sufficiently smooth the transition probabilities of this Markov process have a density $p = p(s, x; t, y)$ satisfying the Fokker-Planck equation

$$\frac{\partial p}{\partial t} + \frac{\partial}{\partial y}(ap) - \frac{1}{2}\frac{\partial^2}{\partial y^2}(\sigma p) = 0$$

with $\sigma = b^2$ (see $(2.4.5)$).

The question of whether or not a process with a density satisfying the Fokker-Planck equation is necessarily an Ito diffusion underlies the interest in weak solutions of stochastic differential equations. While it is not true in general, an affirmative answer can be obtained when the coefficients satisfy certain basic smoothness and boundedness properties which we shall mention in Section 7. There is, however, some ambiguity here for usually the diffusion term $\sigma(x) = b(x)^2$, but not $b(x)$ itself, is specified. Instead of the scalar coefficient $\sqrt{\sigma(x)}$, we could also have a vector function $b(x) = \left(b^1(x), b^2(x), \ldots, b^k(x)\right)$ with $\sum_{i=1}^{k}(b^i(x))^2 = \sigma(x)$ and $X_t$ may also be the solution of the SDE

$$(1.17) \qquad dX_t = a(X_t)\, dt + \sum_{i=1}^{k} b^i(X_t)\, dW_t^i$$

for some $k$-dimensional Wiener process $W_t = (W_t^1, W_t^2, \ldots, W_t^k)$ where $k \geq 1$. In this way the density may be solve the Fokker-Planck equation of several different stochastic differential equations. Equations such as $(1.17)$ and more general vector stochastic differential equations will be considered in Section 8.

Stochastic differential equations can also be formed with more general coefficients than those in $(1.9)$. The most apparent generalization is to allow the coefficients to be nonautonomous, that is, to depend explicitly on $t$ so now $a = a(t, x)$ and $b = b(t, x)$. In this case the resulting solutions are now inhomogeneous Markov processes. The coefficients must be at least measurable in $t$, and the Lipschitz condition $(1.14)$ and the growth condition $(1.15)$ must hold uniformly on $0 \leq t \leq T$ to ensure the existence and uniqueness of strong solutions on $0 \leq t_0 \leq t \leq T$.

Another common extension is for the coefficients to be random, that is $a = a(t, x, \omega)$ and $b = b(t, x, \omega)$. Appropriate measurability restrictions, such as $\mathcal{A}_t$-measurability, must be imposed to ensure that the integrands are nonanticipative. This situation occurs in stability analysis when we linearize about a solution $\bar{X}_t$ of $(1.9)$. With $Z_t = X_t - \bar{X}_t$ we then obtain the linear SDE

$$(1.18) \qquad dZ_t = a'\left(\bar{X}_t\right) Z_t\, dt + b'\left(\bar{X}_t\right) Z_t\, dW_t$$

with coefficients $a(t, z, \omega) = a'(\bar{X}_t(\omega))\, z$ and $b(t, z, \omega) = b'(\bar{X}_t(\omega))\, z$.

We shall restrict our attention now to stochastic differential equations

$$(1.19) \qquad dX_t = a(t, X_t)\, dt + b(t, X_t)\, dW_t$$

with nonrandom coefficients. A simple modification of the Ito formula $(3.3.6)$ shows that a function $Y_t = U(t, X_t)$ of a strong solution of $(1.19)$ satisfies

$$dY_t = \left\{ \frac{\partial U}{\partial t} + a\frac{\partial U}{\partial x} + \frac{1}{2}b^2\frac{\partial^2 U}{\partial x^2} \right\} dt + b\frac{\partial U}{\partial x}\, dW_t,$$

for $U = U(t, x)$ sufficiently smooth, where the coefficients are evaluated at $(t, X_t)$. We can use this to solve some elementary stochastic differential equations explicitly. For example, we know from Chapter 3 that

$$X_t = X_0 \exp\left(W_t - \frac{1}{2}t\right)$$

is a solution of the SDE $dX_t = X_t \, dW_t$. Then $Y_t = U(X_t) = (X_t)^2$ satisfies the SDE

$$dY_t = Y_t \, dt + 2Y_t \, dW_t,$$

so

$$Y_t = Y_0 \exp\left(2W_t - t\right)$$

is a solution of this SDE. In fact, this is a special case of (1.12).

In some applications it is more appropriate to formulate stochastic differential equations in terms of Stratonovich rather than Ito stochastic integrals. We call such an equation a *Stratonovich stochastic differential equation*, writing it in differential form as

$$(1.20) \qquad dX_t = \underline{a}(t, X_t) \, dt + b(t, X_t) \circ dW_t$$

or in the equivalent integral equation form

$$(1.21) \qquad X_t = X_{t_0} + \int_{t_0}^{t} \underline{a}(t, X_t) \, dt + \int_{t_0}^{t} b(t, X_t) \circ dW_t$$

The "∘" notation here denotes the use of Stratonovich calculus. It turns out that the solutions of the Stratonovich SDE (1.20)–(1.21) also satisfy an Ito SDE with the same diffusion coefficient $b(t, x)$, but with the modified drift coefficient

$$a(t, x) = \underline{a}(t, x) + \frac{1}{2} b(t, x) \frac{\partial b}{\partial x}(t, x).$$

For example, the Stratonovich SDE

$$dX_t = 2X_t \circ dW_t$$

and the Ito SDE

$$dX_t = 2X_t \, dt + 2X_t \, dW_t$$

have the same solutions

$$X_t = X_{t_0} \exp\left(2(W_t - W_{t_0})\right).$$

In the case of additive noise the corresponding Ito and Stratonovich SDEs have the same drift coefficients $a \equiv \underline{a}$.

Various considerations, to be discussed in Chapters 6 and 7, determine whether the Ito or the Stratonovich interpretation of an SDE is appropriate in a particular context. However, we can always switch to the corresponding SDE in the other interpretation to take advantage of the special properties of that stochastic calculus. For instance, simple Stratonovich SDEs can be solved directly by methods of classical calculus, whereas the appropriate drift and diffusion coefficients for the Fokker-Planck equation are those of the Ito SDE.

## 4.2   Linear Stochastic Differential Equations

As with linear ordinary differential equations, the general solution of a linear stochastic differential equation can be found explicitly. The method of solution also involves an integrating factor or, equivalently, a fundamental solution of an associated homogeneous differential equation. We shall describe it here for scalar equations and consider their vector counterparts in Section 8.

The general form of a scalar *linear stochastic differential equation* is

$$(2.1) \qquad dX_t = (a_1(t)X_t + a_2(t))\, dt + (b_1(t)X_t + b_2(t))\, dW_t$$

where the coefficients $a_1$, $a_2$, $b_1$, $b_2$ are specified functions of time $t$ or constants. Provided they are Lebesgue measurable and bounded on an interval $0 \le t \le T$, the existence and uniqueness theorem applies, ensuring the existence of a strong solution $X_t$ on $t_0 \le t \le T$ for each $0 \le t_0 < T$ and each $\mathcal{A}_{t_0}$-measurable initial value $X_{t_0}$ corresponding to a given Wiener process $\{W_t, t \ge 0\}$ and associated family of $\sigma$-algebras $\{\mathcal{A}_t, t \ge 0\}$. When the coefficients are all constants the SDE is *autonomous* and its solutions, which exist for all $t - t_0 \ge 0$, are homogeneous Markov processes. In this case it suffices to consider $t_0 = 0$. When $a_2(t) \equiv 0$ and $b_2(t) \equiv 0$, (2.1) reduces to the *homogeneous* linear SDE

$$(2.2) \qquad dX_t = a_1(t)X_t\, dt + b_1(t)X_t\, dW_t,$$

Obviously $X_t \equiv 0$ is a solution of (2.2). Of far greater significance is the *fundamental solution* $\Phi_{t,t_0}$ which satisfies the initial condition $\Phi_{t_0,t_0} = 1$ since all other solutions can be expressed in terms of it. The problem is to find such a fundamental solution.

When $b_1(t) \equiv 0$ in (2.1) the SDE has the form

$$(2.3) \qquad dX_t = (a_1(t)X_t + a_2(t))\, dt + b_2(t)\, dW_t,$$

that is the noise appears additively. In this case we say that the SDE is *linear in the narrow-sense*. The homogeneous equation obtained from (2.3) is then an ordinary differential equation

$$(2.4) \qquad \frac{dX_t}{dt} = a_1(t)X_t$$

and its fundamental solution is

$$\Phi_{t,t_0} = \exp\left(\int_{t_0}^t a_1(s)\, ds\right).$$

Applying the Ito formula (3.3.6) to the transformation $U(t, x) = \Phi_{t,t_0}^{-1}\, x$ and the solution $X_t$ of (2.3), we obtain

$$d\left(\Phi_{t,t_0}^{-1}X_t\right) = \left(\frac{d\Phi_{t,t_0}^{-1}}{dt}X_t + (a_1(t)X_t + a_2(t))\,\Phi_{t,t_0}^{-1}\right)dt + b_2(t)\Phi_{t,t_0}^{-1}\,dW_t$$

$$= a_2(t)\Phi_{t,t_0}^{-1}\,dt + b_2(t)\Phi_{t,t_0}^{-1}\,dW_t,$$

since

$$\frac{d\Phi_{t,t_0}^{-1}}{dt} = -\Phi_{t,t_0}^{-1}a_1(t).$$

The right hand side of

$$d\left(\Phi_{t,t_0}^{-1}X_t\right) = a_2(t)\Phi_{t,t_0}^{-1}\,dt + b_2(t)\Phi_{t,t_0}^{-1}\,dW_t$$

only involves known functions of $t$ and $\omega$, so can be integrated to give

$$\Phi_{t,t_0}^{-1}X_t = \Phi_{t_0,t_0}^{-1}X_{t_0} + \int_{t_0}^{t}a_2(s)\Phi_{s,t_0}^{-1}\,ds + \int_{t_0}^{t}b_2(s)\Phi_{s,t_0}^{-1}\,dW_s.$$

Since $\Phi_{t_0,t_0} = 1$ this leads to the solution

$$(2.5) \qquad X_t = \Phi_{t,t_0}\left(X_{t_0} + \int_{t_0}^{t}a_2(s)\Phi_{s,t_0}^{-1}\,ds + \int_{t_0}^{t}b_2(s)\Phi_{s,t_0}^{-1}\,dW_s\right)$$

of the narrow-sense linear SDE (2.3) where

$$(2.6) \qquad\qquad \Phi_{t,t_0} = \exp\left(\int_{t_0}^{t}a_1(s)\,ds\right).$$

**Example 4.2.1** *The Langevin equation (1.4) is linear in the narrow-sense with coefficients $a_1(t) \equiv -a$, $a_2(t) \equiv 0$, $b_1(t) \equiv 0$ and $b_2(t) \equiv b$. Thus the fundamental solution $\Phi_{t,t_0} = \exp(-a(t-t_0))$ and the solution (2.5) reduces to (1.11) for $t_0 = 0$. This solution is an Ornstein-Uhlenbeck process.*

The solution (2.5) is a Gaussian process whenever the initial value $X_{t_0}$ is either a constant or a Gaussian random variable. Its mean and second order moment then both satisfy ordinary differential equations. These are stated below in (2.10) and (2.11) for the general linear SDE (2.1), but the solution is now generally not Gaussian.

**Exercise 4.2.2** *Verify that the solution (2.5) is a Gaussian process when $X_{t_0}$ is a constant or a Gaussian random variable.*

The general linear case is more complicated because the associated homogeneous equation (2.2) is a genuine stochastic differential equation. The fundamental solution (2.6) of the narrow-sense linear case satisfies the (ordinary) differential equation $d(\ln \Phi_{t,t_0}) = a_1(t)\,dt$. Using this as a clue, it follows by the Ito formula (3.3.6) that the transformed process $\ln \Phi_{t,t_0}$ for the fundamental solution $\Phi_{t,t_0}$ of (2.2) satisfies

$$d\left(\ln \Phi_{t,t_0}\right) = \left(a_1(t)\Phi_{t,t_0}\Phi_{t,t_0}^{-1} - \frac{1}{2}b_1^2(t)\Phi_{t,t_0}^2\Phi_{t,t_0}^{-2}\right) dt + b_1(t)\Phi_{t,t_0}\Phi_{t,t_0}^{-1} dW_t$$

$$= \left(a_1(t) - \frac{1}{2}b_1^2(t)\right) dt + b_1(t) dW_t,$$

which consists only of known functions of $t$ and $\omega$. Hence

$$\ln \Phi_{t,t_0} = \int_{t_0}^t \left(a_1(s) - \frac{1}{2}b_1^2(s)\right) ds + \int_{t_0}^t b_1(s) dW_s,$$

since $\Phi_{t_0,t_0} = 1$, or

$$(2.7) \qquad \Phi_{t,t_0} = \exp\left(\int_{t_0}^t \left(a_1(s) - \frac{1}{2}b_1^2(s)\right) ds + \int_{t_0}^t b_1(s) dW_s\right),$$

which in fact reduces to (2.6) when $b_1(t) \equiv 0$. Similarly, applying the Ito formula to $\Phi_{t,t_0}^{-1}$ we obtain

$$(2.8) \qquad d\left(\Phi_{t,t_0}^{-1}\right) = \left(-a_1(t) + b_1^2(t)\right)\Phi_{t,t_0}^{-1} dt - b_1(t)\Phi_{t,t_0}^{-1} dW_t.$$

Then, as with the narrow-sense case considered above, the process $\Phi_{t,t_0}^{-1} X_t$ for a solution $X_t$ of the general linear equation (2.1) has an explicitly integrable stochastic differential. However, here both of the terms $\Phi_{t,t_0}$ and $X_t$ have stochastic differentials involving the same Wiener process $W_t$, so the Ito formula must be used with the two component transformation $U(X_t^{(1)}, X_t^{(2)}) = X_t^{(1)} X_t^{(2)}$, as in Example 3.4.1, with $X_t^{(1)} = \Phi_{t,t_0}^{-1}$ and $X_t^{(2)} = X_t$. The result is equation (3.4.10) with the coefficients of (2.1) and (2.8), that is

$$d(\Phi_{t,t_0}^{-1} X_t) = \left[\left(-a_1(t) + b_1^2(t)\right) X_t + (a_1(t)X_t + a_2(t))\right]\Phi_{t,t_0}^{-1} dt$$

$$-b_1(t)\left[b_1(t)X_t + b_2(t)\right]\Phi_{t,t_0}^{-1} dt$$

$$+\left[-b_1(t)\Phi_{t,t_0}^{-1} X_t + (b_1(t)X_t + b_2(t))\Phi_{t,t_0}^{-1}\right] dW_t$$

$$= \left(a_2(t) - b_1(t)b_2(t)\right)\Phi_{t,t_0}^{-1} dt + b_2(t)\Phi_{t,t_0}^{-1} dW_t.$$

Integrating and using $\Phi_{t_0,t_0} = 1$ we obtain

$$\Phi_{t,t_0}^{-1} X_t = X_{t_0} + \int_{t_0}^t \left(a_2(s) - b_1(s)b_2(s)\right)\Phi_{s,t_0}^{-1} ds + \int_{t_0}^t b_2(s)\Phi_{s,t_0}^{-1} dW_s$$

and hence

$$(2.9) X_t = \Phi_{t,t_0}\left(X_{t_0} + \int_{t_0}^t \left(a_2(s) - b_1(s)b_2(s)\right)\Phi_{s,t_0}^{-1} ds + \int_{t_0}^t b_2(s)\Phi_{s,t_0}^{-1} dW_s\right)$$

where $\Phi_{t,t_0}$ is given by (2.8). Applying this to the linear SDE (1.7), where $a_1(t) \equiv a$, $b_1(t) \equiv b$ and $a_2(t) \equiv b_2(t) \equiv 0$, yields the solution (1.11). We observe that (2.9) reduces to the narrow-sense solution (2.5) when $b_1(t) \equiv 0$.

If we take the expectation of the integral form of equation (2.1) and use the zero expectation property (3.2.9) of an Ito integral, we obtain an ordinary differential equation for the mean $m(t) = E(X_t)$ of its solution, namely

$$(2.10) \qquad \frac{dm(t)}{dt} = a_1(t)m(t) + a_2(t).$$

We also find that the second order moment $P(t) = E(X_t^2)$ satisfies the ordinary differential equation

$$(2.11) \qquad \frac{dP(t)}{dt} = \left(2a_1(t) + b_1^2(t)\right) P(t) + 2m(t)\left(a_2(t) + b_1(t)b_2(t)\right)$$
$$+ b_2^2(t).$$

To derive (2.11) we use the Ito formula to obtain an SDE for $X_t^2$ and then take the expectation of the integral form of this equation. Both (2.10) and (2.11) are linear and can be solved using integrating factors. In the special case of a narrow-sense linear SDE (2.3) equation (2.10) remains the same, whereas equation (2.11) simplifies to

$$(2.12) \qquad \frac{dP(t)}{dt} = 2a_1(t)P(t) + 2m(t)a_2(t) + b_2^2(t).$$

**Exercise 4.2.3**  *Derive the ordinary differential equation (2.11) for the second order moment $P(t)$.*

**Exercise 4.2.4**  *Solve (2.10) and (2.11) for the Langevin equation (1.4) and for the SDE (1.7).*

In Section 8 the corresponding results are stated for vector linear SDEs. These also apply to scalar SDEs driven by a vector Wiener process such as (1.17).

## 4.3   Reducible Stochastic Differential Equations

With an appropriate substitution $X_t = U(t, Y_t)$ certain nonlinear stochastic differential equations

$$(3.1) \qquad dY_t = a(t, Y_t)\, dt + b(t, Y_t)\, dW_t$$

can be reduced to a linear SDE in $X_t$

$$(3.2) \qquad dX_t = (a_1(t)X_t + a_2(t))\, dt + (b_1(t)X_t + b_2(t))\, dW_t.$$

If $\frac{\partial U}{\partial y}(t, y) \neq 0$, the Inverse Function Theorem ensures the existence of a local inverse $y = V(t, x)$ of $x = U(t, y)$, that is with $x = U(t, V(t, x))$ and $y = V(t, U(t, y))$. A solution $Y_t$ of (3.1) then has the form $Y_t = V(t, X_t)$ where $X_t$ is given by (3.2) for appropriate coefficients $a_1$, $a_2$, $b_1$ and $b_2$. From the Ito formula (3.3.6)

$$dU(t, Y_t) = \left( \frac{\partial U}{\partial t} + a \frac{\partial U}{\partial y} + \frac{1}{2} b^2 \frac{\partial^2 U}{\partial y^2} \right) dt + b \frac{\partial U}{\partial y} dW_t,$$

where the coefficients and the partial derivatives are evaluated at $(t, Y_t)$. This coincides with a linear SDE of the form (3.2) if

$$(3.3) \qquad \frac{\partial U}{\partial t}(t, y) + a(t, y) \frac{\partial U}{\partial y}(t, y) + \frac{1}{2} b^2(t, y) \frac{\partial^2 U}{\partial y^2}(t, y)$$

$$= a_1(t) U(t, y) + a_2(t)$$

and

$$(3.4) \qquad b(t, y) \frac{\partial U}{\partial y}(t, y) = b_1(t) U(t, y) + b_2(t).$$

At this level of generality there is little more that can be said that is straightforward or specific. Specializing to the case where $a_1(t) \equiv b_1(t) \equiv 0$ and writing $a_2(t) = \alpha(t)$ and $b_2(t) = \beta(t)$, we obtain from (3.3) the identity

$$\frac{\partial^2 U}{\partial t \partial y}(t, y) = -\frac{\partial}{\partial y} \left( a(t, y) \frac{\partial U}{\partial y}(t, y) + \frac{1}{2} b^2(t, y) \frac{\partial^2 U}{\partial y^2}(t, y) \right)$$

and, from (3.4), the identities

$$\frac{\partial}{\partial y} \left( b(t, y) \frac{\partial U}{\partial y}(t, y) \right) = 0$$

and

$$b(t, y) \frac{\partial^2 U}{\partial t \partial y}(t, y) + \frac{\partial b}{\partial t}(t, y) \frac{\partial U}{\partial y}(t, y) = \beta'(t).$$

Assume for now that $b(t, y) \neq 0$. Then, eliminating $U$ and its derivatives we obtain

$$\beta'(t) = \beta(t) b(t, y) \left( \frac{1}{b^2(t, y)} \frac{\partial b}{\partial t}(t, y) - \frac{\partial}{\partial y} \left( \frac{a(t, y)}{b(t, y)} \right) + \frac{1}{2} \frac{\partial^2 b}{\partial y^2}(t, y) \right).$$

Since the left hand side is independent of $y$ this means

$$\frac{\partial \gamma}{\partial y}(t, y) = 0$$

where

$$(3.5) \qquad \gamma(t,y) = \frac{1}{b(t,y)} \frac{\partial b}{\partial t}(t,y) - b(t,y) \frac{\partial}{\partial y} \left( \frac{a(t,y)}{b(t,y)} - \frac{1}{2} \frac{\partial b}{\partial y}(t,y) \right).$$

This is a sufficient condition for the reducibility of the nonlinear SDE (3.1) to the explicitly integrable SDE

$$(3.6) \qquad dX_t = \alpha(t)\, dt + \beta(t)\, dW_t$$

by means of a transformation $x = U(t,y)$. It can be determined from (3.3) and (3.4) which, in this special case, reduce to

$$\frac{\partial U}{\partial t}(t,y) + a(t,y)\frac{\partial U}{\partial y}(t,y) + \frac{1}{2}b^2(t,y)\frac{\partial^2 U}{\partial y^2}(t,y) = \alpha(t)$$

and

$$b(t,y)\frac{\partial U}{\partial y}(t,y) = \beta(t),$$

resulting in

$$U(t,y) = C\,\exp\left( \int_0^t \gamma(s,y)\, ds \right) \int_0^y \frac{1}{b(t,z)}\, dz$$

where $C$ is an arbitrary constant. We remark that this method can also be used to reduce certain linear SDEs to stochastic differentials of the form (3.6).

A variation of the procedure is applicable to reduce a nonlinear autonomous SDE

$$(3.7) \qquad dY_t = a(Y_t)\, dt + b(Y_t)\, dW_t$$

to the autonomous linear SDE

$$(3.8) \qquad dX_t = (a_1 X_t + a_2)\, dt + (b_1 X_t + b_2)\, dW_t.$$

by means of a time-independent transformation $X_t = U(Y_t)$. In this case the identities (3.3) and (3.4) take the form

$$(3.9) \qquad a(y)\frac{dU}{dy}(y) + \frac{1}{2}b^2(y)\frac{d^2 U}{dy^2}(y) = a_1 U(y) + a_2$$

and

$$(3.10) \qquad b(y)\frac{dU}{dy}(y) = b_1 U(y) + b_2.$$

Assuming that $b(y) \not\equiv 0$ and $b_1 \neq 0$, it follows from (3.10) that

$$(3.11) \qquad U(y) = C\exp\left( b_1 B(y) \right) - \frac{b_2}{b_1}$$

where

$$B(y) = \int_{y_0}^y \frac{ds}{b(s)}$$

and $C$ is an arbitrary constant. Substituting this expression for $U(y)$ into (3.9) gives

$$(3.12) \qquad \left(b_1 A(y) + \frac{1}{2} b_1^2 - a_1\right) C \exp\left(b_1 B(y)\right) = a_2 - a_1 \frac{b_2}{b_1}$$

where

$$A(y) = \frac{a(y)}{b(y)} - \frac{1}{2} \frac{db}{dy}(y).$$

Differentiating (3.12), multiplying the result by $b(y) \exp\left(-b_1 B(y)\right)/b_1$ and then differentiating again, we obtain the relation

$$(3.13) \qquad b_1 \frac{dA}{dy} + \frac{d}{dy}\left(b \frac{dA}{dy}\right) = 0.$$

This is certainly satisfied if $\frac{dA}{dy} = 0$ or if

$$(3.14) \qquad \frac{d}{dy}\left(\frac{\dfrac{d}{dy}\left(b \dfrac{dA}{dy}\right)}{\dfrac{dA}{dy}}\right) = 0,$$

provided $b_1$ is chosen so that

$$b_1 = -\frac{\dfrac{d}{dy}\left(b \dfrac{dA}{dy}\right)}{\dfrac{dA}{dy}}.$$

If $b_1 \neq 0$ the appropriate transformation is

$$(3.15) \qquad U(y) = C \exp\left(b_1 B(y)\right),$$

whereas if $b_1 = 0$ it takes the form

$$(3.16) \qquad U(y) = b_2 B(y) + C$$

where $b_2$ is chosen so that (3.10) is satisfied.

**Example 4.3.1**     *For the nonlinear SDE*

$$(3.17) \qquad dY_t = -\frac{1}{2} \exp\left(-2Y_t\right) dt + \exp\left(-Y_t\right) dW_t$$

*$a(y) = -\frac{1}{2} \exp(-2y)$ and $b(y) = \exp(-y)$, so $A(y) \equiv 0$. Thus (3.13) is satisfied with any $b_1$. For $b_1 = 0$ and $b_2 = 1$ a solution of (3.10) is $U = \exp(y)$ by (3.16). Substituting this into (3.9) results in $a_1 = a_2 = 0$. Hence $X_t = \exp(Y_t)$ and (3.8) reduces to the stochastic differential $dX_t = dW_t$, which has solution*

$$X_t = W_t + X_0 = W_t + \exp(Y_0).$$

*The original nonlinear SDE (3.7) thus has solution*

$$Y_t = \ln \left( W_t + \exp(Y_0) \right),$$

*which is valid until the sample-path dependent explosion time*

$$T(Y_0(\omega)) = \min \left\{ t \geq 0 : W_t(\omega) + \exp(Y_0(\omega)) = 0 \right\}.$$

**Exercise 4.3.2**    *Show that the SDE*

$$dX_t = \frac{1}{2} g(X_t) g'(X_t) \, dt + g(X_t) \, dW_t,$$

*where $g$ is a given differentiable function, is reducible and has the general solution*

$$X_t = h^{-1}(W_t + h(X_0))$$

*with*

(3.18)
$$y = h(x) = \int^x \frac{ds}{g(s)}.$$

**Exercise 4.3.3**    *Repeat Exercise 4.3.2 for the SDE*

$$dX_t = \left( \alpha g(X_t) + \frac{1}{2} g(X_t) g'(X_t) \right) dt + g(X_t) \, dW_t$$

*to show that its general solution is*

$$X_t = h^{-1}(\alpha t + W_t + h(X_0)).$$

**Exercise 4.3.4**    *Determine the general solution of the SDE*

$$dX_t = \left( \beta g(X_t) h(X_t) + \frac{1}{2} g(X_t) g'(X_t) \right) dt + g(X_t) \, dW_t,$$

*where $h(x)$ is given by (3.18), by first reducing it to a Langevin equation.*

**Exercise 4.3.5**    *Write the SDE in Exercise 4.3.2 as a Stratonovich SDE and solve it directly using the rules of classical calculus.*

## 4.4    Some Explicitly Solvable SDEs

In this section we shall list some explicitly solvable stochastic differential equations and their general solutions, which we have found in various books and papers. These are presented here primarily for use in case studies to check the accuracy of numerical methods, but can also be used for practicing solution techniques, and for verifying the theoretical estimates that arise, for instance, in existence and uniqueness considerations. For convenience we include the linear SDEs already treated in Section 2. All of the nonlinear SDEs in our list are reducible to these linear SDEs. We classify them according to the relationship between their drift and diffusion coefficients, and remark that in many cases where the solution is a standard function of the Wiener process the drift coefficient is simply the Stratonovich correction term and the corresponding Stratonovich SDE is drift free. Most of the other examples are variations of this relationship (see Exercises 4.3.2–4.3.5). We also include some complex valued SDEs. For brevity we shall not indicate the interval of existence of the general solutions or the restrictions on the solutions required for them to be meaningful.

## Linear SDEs: Additive Noise

*Constant coefficients: homogeneous*

(4.1) $$dX_t = -\alpha X_t \, dt + \sigma \, dW_t$$

$$X_t = e^{-\alpha t} \left( X_0 + \sigma \int_0^t e^{\alpha s} \, dW_s \right) ;$$

*Constant coefficients: inhomogeneous*

(4.2) $$dX_t = (aX_t + b) \, dt + c \, dW_t$$

$$X_t = e^{at} \left( X_0 + \frac{b}{a}(1 - e^{-at}) + c \int_0^t e^{-as} \, dW_s \right) ;$$

*Variable coefficients:*

(4.3) $$dX_t = (a(t)X_t + b(t)) \, dt + c(t) \, dW_t$$

$$X_t = \Phi_{t,t_0} \left( X_{t_0} + \int_{t_0}^t \Phi_{s,t_0}^{-1} b(s) \, ds + \int_{t_0}^t \Phi_{s,t_0}^{-1} c(s) \, dW_s \right)$$

with fundamental solution

$$\Phi_{t,t_0} = \exp\left( \int_{t_0}^t a(s) \, ds \right).$$

For example,

(4.4) $$dX_t = \left( \frac{2}{1+t} X_t + b(1+t)^2 \right) dt + b(1+t)^2 \, dW_t$$

has fundamental solution $\Phi_{t,t_0} = \left( \frac{1+t}{1+t_0} \right)^2$ and general solution

$$X_t = \left( \frac{1+t}{1+t_0} \right)^2 X_0 + b(1+t)^2 (W_t - W_{t_0} + t - t_0).$$

Usually the Wiener process will appear in an integral, as for

(4.5)
$$dX_t = \left(\frac{b - X_t}{T - t}\right) dt + dW_t$$

which is satisfied by the process

$$X_t = X_0 \left(1 - \frac{t}{T}\right) + b\frac{t}{T} + (T - t) \int_0^t \frac{1}{T - s} dW_s$$

on the interval $0 \leq t \leq T$.

## Linear SDEs: Multiplicative Noise

*Constant coefficients: homogeneous*

(4.6)
$$dX_t = aX_t \, dt + bX_t \, dW_t$$

$$X_t = X_0 \exp\left(\left(a - \frac{1}{2}b^2\right)t + b\, W_t\right).$$

The two most important examples are the Ito exponential SDE

(4.7)
$$dX_t = \frac{1}{2}X_t \, dt + X_t \, dW_t$$

$$X_t = X_0 \exp(W_t)$$

and the corresponding drift-free SDE

(4.8)
$$dX_t = X_t \, dW_t$$

$$X_t = X_0 \exp\left(W_t - \frac{1}{2}t\right).$$

*Constant coefficients: inhomogeneous*

(4.9)
$$dX_t = (aX_t + c) \, dt + (bX_t + d) \, dW_t$$

$$X_t = \Phi_t \left(X_0 + (c - bd) \int_0^t \Phi_s^{-1} \, ds + d \int_0^t \Phi_s^{-1} \, dW_s\right)$$

with fundamental solution

$$\Phi_t = \exp\left(\left(a - \frac{1}{2}b^2\right)t + bW_t\right).$$

*Variable coefficients: homogeneous*

(4.10)
$$dX_t = a(t)X_t \, dt + b(t)X_t \, dW_t$$

$$X_t = X_0 \exp\left( \int_0^t \left( a(s) - \frac{1}{2} b^2(s) \right) ds + \int_0^t b(s)\, dW_s \right).$$

*Variable coefficients: inhomogeneous*

(4.11)          $$dX_t = (a(t)X_t + c(t))\, dt + (b(t)X_t + d(t))\, dW_t$$

$$X_t = \Phi_{t,t_0}\left( X_{t_0} + \int_{t_0}^t \Phi_{s,t_0}^{-1}(c(s) - b(s)d(s))\, ds + \int_{t_0}^t \Phi_{s,t_0}^{-1}\, d(s)\, dW_s \right)$$

with fundamental solution

$$\Phi_{t,t_0} = \exp\left( \int_{t_0}^t \left( a(s) - \frac{1}{2} b^2(s) \right) ds + \int_{t_0}^t b(s)\, dW_s \right).$$

## Reducible SDEs:  Case 1

The Ito SDE

(4.12)          $$dX_t = \frac{1}{2} b(X_t) b'(X_t)\, dt + b(X_t)\, dW_t$$

for a given differentiable function $b$ is equivalent to the Stratonovich SDE

(4.13)          $$dX_t = b(X_t) \circ dW_t.$$

We can either reduce the Ito SDE (4.12) to a linear SDE (see Exercise 4.3.2) or integrate the Stratonovich SDE (4.13) directly (see Exercise 4.3.5) to obtain the general solution

$$X_t = h^{-1}(W_t + h(X_0))$$

where

(4.14)          $$y = h(x) = \int^x \frac{ds}{b(s)}.$$

Most standard functions of a Wiener process satisfy SDEs of the form (4.12). Examples include the linear SDE (4.7) and the following SDEs.

(4.15)          $$dX_t = \frac{1}{2} a(a-1) X_t^{1-2/a}\, dt + a X_t^{1-1/a}\, dW_t$$

$$X_t = \left( W_t + X_0^{1/a} \right)^a ;$$

(4.16)          $$dX_t = \frac{1}{2} a^2 X_t\, dt + a X_t\, dW_t$$

$$X_t = X_0 \exp\left( a W_t \right);$$

(4.17)          $$dX_t = \frac{1}{2} (\ln a)^2 X_t\, dt + (\ln a) X_t\, dW_t$$

$$X_t = X_0 a^{W_t} = X_0 \exp\left( W_t \ln a \right);$$

(4.18)
$$dX_t = (2 \ln b)^{-1} b^{-2X_t} dt + (\ln b)^{-1} b^{-X_t} dW_t$$
$$X_t = \log_b \left( aW_t + b^{X_0} \right);$$

(4.19)
$$dX_t = -\frac{1}{2} a^2 X_t dt + a \sqrt{1 - X_t^2} \, dW_t$$
$$X_t = \sin \left( aW_t + \arcsin X_0 \right);$$

(4.20)
$$dX_t = -\frac{1}{2} a^2 X_t dt - a \sqrt{1 - X_t^2} \, dW_t$$
$$X_t = \cos \left( aW_t + \arccos X_0 \right);$$

(4.21)
$$dX_t = a^2 X_t \left( 1 + X_t^2 \right) dt + a \left( 1 + X_t^2 \right) dW_t$$
$$X_t = \tan \left( aW_t + \arctan X_0 \right);$$

(4.22)
$$dX_t = a^2 X_t \left( 1 + X_t^2 \right) dt - a \left( 1 + X_t^2 \right) dW_t$$
$$X_t = \cot \left( aW_t + \text{arccot } X_0 \right);$$

(4.23)
$$dX_t = \frac{1}{2} a^2 X_t \left( 2X_t^2 - 1 \right) dt + a X_t \sqrt{X_t^2 - 1} \, dW_t$$
$$X_t = \sec \left( aW_t + \text{arcsec } X_0 \right);$$

(4.24)
$$dX_t = \frac{1}{2} a^2 X_t \left( 2X_t^2 - 1 \right) dt - a X_t \sqrt{X_t^2 - 1} \, dW_t$$
$$X_t = \csc \left( aW_t + \text{arccsc } X_0 \right);$$

(4.25)
$$dX_t = \frac{1}{2} a^2 \tan X_t \sec^2 X_t \, dt + a \sec X_t \, dW_t$$
$$X_t = \arcsin \left( aW_t + \sin X_0 \right);$$

(4.26)
$$dX_t = -\frac{1}{2} a^2 \cot X_t \csc^2 X_t \, dt - a \csc X_t \, dW_t$$
$$X_t = \arccos \left( aW_t + \cos X_0 \right);$$

(4.27)
$$dX_t = -a^2 \sin X_t \cos^3 X_t \, dt + a \cos^2 X_t \, dW_t$$
$$X_t = \arctan \left( aW_t + \tan X_0 \right);$$

(4.28)
$$dX_t = a^2 \cos X_t \sin^3 X_t \, dt - a \sin^2 X_t \, dW_t$$
$$X_t = \text{arccot} \, (aW_t + \cot X_0) ;$$

(4.29)
$$dX_t = \frac{1}{2} a^2 X_t \, dt + a \sqrt{X_t^2 + 1} \, dW_t$$
$$X_t = \sinh \, (aW_t + \text{arcsinh} \, X_0) ;$$

(4.30)
$$dX_t = \frac{1}{2} a^2 X_t \, dt + a \sqrt{X_t^2 - 1} \, dW_t$$
$$X_t = \cosh \, (aW_t + \text{arccosh} \, X_0) ;$$

(4.31)
$$dX_t = -a^2 X_t \left(1 - X_t^2\right) \, dt + a \left(1 - X_t^2\right) \, dW_t$$
$$X_t = \tanh \, (aW_t + \text{arctanh} \, X_0) , \qquad (\text{see } (4.40));$$

(4.32)
$$dX_t = a^2 X_t \left(1 - X_t^2\right) \, dt - a \left(1 - X_t^2\right) \, dW_t$$
$$X_t = \coth \, (a W_t + \text{arccoth} \, X_0) ;$$

(4.33)
$$dX_t = -\frac{1}{2} a^2 \tanh X_t \, \text{sech}^2 X_t \, dt + a \, \text{sech} \, X_t \, dW_t$$
$$X_t = \text{arcsinh} \, (a W_t + \sinh X_0) ;$$

(4.34)
$$dX_t = -\frac{1}{2} a^2 \coth X_t \, \text{csch}^2 X_t \, dt + a \, \text{csch} \, X_t \, dW_t$$
$$X_t = \text{arccosh} \, (a W_t + \cosh X_0) ;$$

(4.35)
$$dX_t = a^2 \sinh X_t \, \cosh^3 X_t \, dt + a \, \cosh^2 X_t \, dW_t$$
$$X_t = \text{arctanh} \, (a W_t + \tanh X_0) ;$$

(4.36)
$$dX_t = a^2 \cosh X_t \, \sinh^3 X_t \, dt - a \, \sinh^2 X_t \, dW_t$$
$$X_t = \text{arccoth} \, (a W_t + \coth X_0) ;$$

(4.37)
$$dX_t = 1 \, dt + 2 \sqrt{X_t} \, dW_t$$
$$X_t = \left(W_t + \sqrt{X_0}\right)^2 ;$$

$$(4.38) \qquad dX_t = -X_t \left(2\ln X_t + 1\right) dt - 2X_t\sqrt{-\ln X_t}\, dW_t$$

$$X_t = \exp\left(-\left(W_t + \sqrt{-\ln X_0}\right)^2\right);$$

$$(4.39) \qquad dX_t = \frac{1}{2}a^2 m X_t^{2m-1}\, dt + aX_t^m\, dW_t \qquad m \neq 1$$

$$X_t = \left(X_0^{1-m} - a(m-1)\,W_t\right)^{1/(1-m)};$$

$$(4.40) \qquad dX_t = -\beta^2 X_t\left(1 - X_t^2\right) dt + \beta\left(1 - X_t^2\right) dW_t$$

$$X_t = \frac{(1+X_0)\exp(2\beta W_t) + X_0 - 1}{(1+X_0)\exp(\beta W_t) + 1 - X_0}, \qquad (\text{see } (4.31));$$

$$(4.41) \qquad dX_t = \frac{1}{3}X_t^{1/3}\, dt + X_t^{2/3}\, dW_t$$

$$X_t = \left(X_0^{1/3} + \frac{1}{3}W_t\right)^3.$$

This last SDE has nonunique solutions for $X_0 = 0$; for example, $X_t \equiv 0$ is also a solution.

## Reducible SDEs: Case 2

The Ito SDE

$$(4.42) \qquad dX_t = \left(\alpha\, b(X_t) + \frac{1}{2}b(X_t)b'(X_t)\right) dt + b(X_t)\, dW_t$$

is equivalent to the Stratonovich SDE

$$(4.43) \qquad dX_t = \alpha\, b(X_t)\, dt + b(X_t) \circ dW_t$$

and is reducible to the stochastic differential

$$dY_t = \alpha\, dt + dW_t$$

for $Y_t = h(X_t)$, where $h$ is given by (4.14). Its general solution is thus

$$X_t = h^{-1}(\alpha t + W_t + h(X_0)).$$

All of the examples in Case 1 can be modified to provide examples for this case. In particular we consider

(4.44)          $dX_t = (1 + X_t)\left(1 + X_t^2\right)\,dt + \left(1 + X_t^2\right)\,dW_t$

$$X_t = \tan\left(t + W_t + \arctan X_0\right);$$

(4.45)          $dX_t = \left(\frac{1}{2}X_t + \sqrt{X_t^2 + 1}\right)\,dt + \sqrt{X_t^2 + 1}\,dW_t$

$$X_t = \sinh\left(t + W_t + \operatorname{arcsinh} X_0\right);$$

(4.46)          $dX_t = -(\alpha + \beta^2 X_t)(1 - X_t^2)\,dt + \beta(1 - X_t^2)\,dW_t$

$$X_t = \frac{(1 + X_0)\exp(-2\alpha t + 2\beta W_t) + X_0 - 1}{(1 + X_0)\exp(-2\alpha t + 2\beta W_t) + 1 - X_0}.$$

## Reducible SDEs: Case 3

The Ito SDE

(4.47)     $dX_t = \left(\alpha\, b(X_t)h(X_t) + \frac{1}{2}b(X_t)b'(X_t)\right)\,dt + b(X_t)\,dW_t$

where $h$ is given by (4.14) is reducible to the Langevin SDE (1.4) with $b = 1$ in the variable for $Y_t = h(X_t)$. Its general solution is thus

$$X_t = h^{-1}\left(e^{\alpha t}h(X_0) + e^{\alpha t}\int_0^t e^{-\alpha s}\,dW_s\right).$$

All of the examples in Case 1 can be modified to provide examples for this case. In particular we consider

(4.48)     $dX_t = -\left(\sin 2X_t + \frac{1}{4}\sin 4X_t\right)\,dt + \sqrt{2}\,\cos^2 X_t\,dW_t$

$$X_t = \arctan\left(e^{-t}\tan X_0 + \sqrt{2}\,e^{-t}\int_0^t e^s\,dW_s\right);$$

(4.49)     $dX_t = -\tanh X_t\left(a + \frac{1}{2}b^2\operatorname{sech}^2 X_t\right)\,dt + b\operatorname{sech} X_t\,dW_t$

$$X_t = \operatorname{arcsinh}\left(e^{-at}\sinh X_0 + {}_-e^{-at}\int_0^t e^{as}\,dW_s\right).$$

## Reducible SDEs: Miscellaneous

We shall give some examples of nonlinear reducible SDEs not included in the preceding three cases. The first is the most general form of a reducible SDE with polynomial drift of degree $n$.

$$(4.50) \qquad dX_t = (aX_t^n + bX_t)\, dt + cX_t\, dW_t$$

$$X_t = \Theta_t \left( X_0^{1-n} + a(1-n) \int_0^t \Theta_s^{n-1}\, ds \right)^{1/(1-n)}$$

with

$$\Theta_t = \exp\left( \left(b - \frac{1}{2}c^2\right) t + cW_t \right).$$

The substitution $y = h(x) = x^{1-n}$ reduces the SDE (4.50) to a linear SDE with multiplicative noise. A special case for $n = 2$ is the *stochastic Verhulst equation*

$$(4.51) \qquad dX_t = \left(\lambda X_t - X_t^2\right) dt + \sigma X_t\, dW_t$$

$$X_t = \frac{X_0 \exp\left( \left(\lambda - \frac{1}{2}\sigma^2\right) t + \sigma W_t \right)}{1 + X_0 \int_0^t \exp\left( \left(\lambda - \frac{1}{2}\sigma^2\right) s + \sigma W_s \right) ds};$$

For $n = 3$ we have the *stochastic Ginzburg-Landau equation*

$$(4.52) \qquad dX_t = \left( -X_t^3 + \left(\alpha + \frac{1}{2}\sigma^2\right) X_t \right) dt + \sigma X_t\, dW_t$$

$$X_t = \frac{X_0 \exp\left(\alpha t + \sigma W_t\right)}{\sqrt{1 + 2X_0^2 \int_0^t \exp\left(2\alpha s + 2\sigma W_s\right) ds}};$$

Another example, which uses the exponential substitution $y = h(x) = \exp(-cx)$, is

$$(4.53) \qquad dX_t = (a \exp(cX_t) + b)\, dt + \sigma\, dW_t$$

$$X_t = X_0 + bt + \sigma W_t - \frac{1}{c} \ln\left( 1 - ac \int_0^t \exp(cX_0 + bcs + \sigma c W_s)\, ds \right).$$

## Complex-Valued SDEs

Here $\imath = \sqrt{-1}$, $Z_t$ is a complex-valued Ito process and $W_t$ is a real-valued Wiener process.

$$(4.54) \qquad dZ_t = -\frac{1}{2} Z_t\, dt + \imath\, Z_t\, dW_t$$

$$Z_t = Z_0 \exp\left(\imath\, W_t\right);$$

(4.55) $$dZ_t = \imath Z_t \, dW_t$$

$$Z_t = Z_0 \exp\left(\imath W_t + \frac{1}{2}t\right);$$

(4.56) $$dZ_t = \frac{1}{2}\imath Z_t \, dt + \sqrt{\imath}\, Z_t \, dW_t$$

$$Z_t = Z_0 \exp\left(\sqrt{\imath}\, W_t\right);$$

(4.57) $$dZ_t = \lambda Z_t \, dt + \gamma Z_t \, dW_t$$

$$Z_t = Z_0 \exp\left(\gamma\alpha W_t + \left(\lambda - \frac{1}{2}\gamma^2\right)t\right);$$

(4.58) $$dZ_t = Z_t^3 \, dt - Z_t^2 \, dW_t$$

$$Z_t = \frac{Z_0}{1 + Z_0 W_t}.$$

These complex-valued SDEs can be written alternatively as 2-dimensional vector SDEs in the real and imaginary parts.

## Linear SDEs with 2-dimensional Noise

As a generalization of (4.6) we have

(4.59) $$dX_t = aX_t \, dt + b^1 X_t \, dW_t^1 + b^2 X_t \, dW_t^2$$

$$X_t = X_0 \exp\left(\left(a - \frac{1}{2}\left(\left(b^1\right)^2 + \left(b^2\right)^2\right)\right)t + b^1 W_t^1 + b^2 W_t^2\right).$$

To compare the output of numerical schemes applied to the above SDEs with their explicit solutions we must evaluate these solutions numerically. This is easy to do when the explicit solutions do not contain stochastic integrals, as in the following PC-Exercise.

**PC-Exercise 4.4.1**    *Use a Gaussian random number generator to simulate the increments* $\Delta W_i = W_{(i+1)h} - W_{ih}$ *for* $i = 0, 1, \ldots, n-1$ *with* $n = 2^9$ *of a standard Wiener process* $\{W_t, t \geq 0\}$ *on the unit interval* $[0,1]$ *for time steps* $h$ *appropriate for the resolution of your PC screen. Hence evaluate the explicit solution* $X_t$ *of the SDE (4.37) with the initial value* $X_0 = 100$ *and plot it against* $t \in [0,1]$.

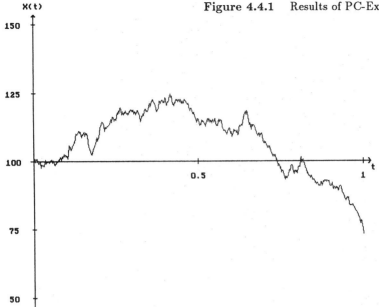

**Figure 4.4.1**    Results of PC-Exercise 4.4.1.

# 4.5    The Existence and Uniqueness of Strong Solutions

A statement and proof of an existence and uniqueness theorem for strong solutions of Ito stochastic differential equations, foreshadowed in Section 1, will be presented in this section. We shall do this in the slightly more general context of a nonautonomous scalar stochastic differential equation

$$(5.1) \qquad dX_t = a(t, X_t)\, dt + b(t, X_t)\, dW_t.$$

As mentioned earlier, (5.1) is interpreted as a stochastic integral equation

$$(5.2) \qquad X_t = X_{t_0} + \int_{t_0}^t a(s, X_s)\, ds + \int_{t_0}^t b(s, X_s)\, dW_s,$$

where the first integral is a Lebesgue (or Riemann) integral for each sample path and the second integral is an Ito integral. A solution $\{X_t,\ t \in [t_0, T]\}$ of (5.2) must thus have properties which ensure that these integrals are meaningful. This holds if, for $e$ and $f$ defined by

$$e(t, \omega) = f(t, \omega) = 0 \quad \text{for} \quad 0 \le t < t_0$$

and

$$e(t, \omega) = a(t, X_t(\omega)), \quad f(t, \omega) = b(t, X_t(\omega)) \quad \text{for} \quad t_0 \le t \le T,$$

the functions $\sqrt{|e|}$ and $f$ belong to the spaces $\mathcal{L}_T^2$ or $\mathcal{L}_T^w$, which were defined in Section 2 of Chapter 3. In turn, this follows if the coefficient functions $a$ and $b$ are sufficiently regular and if the process $X_t$ is regular and nonanticipative with respect to the given Wiener process $W = \{W_t, t \geq 0\}$, that is $\mathcal{A}^*$-adapted where $\mathcal{A}^* = \{\mathcal{A}_t, t \geq 0\}$ is the family of $\sigma$-algebras associated with the Wiener process. In addition, the integrals in (5.2) should exist, at least w.p.1, for each $t \in [t_0, T]$. We then call the process $X = \{X_t, t_0 \leq t \leq T\}$ a *solution* of (5.2) on $[t_0, T]$. For fixed coefficients $a$ and $b$, any solution $X$ will depend on the particular initial value $X_{t_0}$ and Wiener process $W$ under consideration. If there is a solution for each given Wiener process we say that the stochastic differential equation has a *strong solution*. Such a solution can be roughly thought of as a functional of the initial value $X_{t_0}$ and of the values $W_s$ of the Wiener process over the subinterval $t_0 \leq s \leq t$. For a specified initial value $X_{t_0}$ the *uniqueness* of solutions of (5.1) refers to the equivalence (P almost surely) of the solution processes that satisfy the stochastic integral equation (5.2). If there is a solution, then there will be a separable version which has, almost surely, continuous sample paths. We shall subsequently consider only this kind of solution. If any two such solutions $X_t$ and $\tilde{X}_t$ have, almost surely, the same sample paths on $[t_0, T]$, that is if

$$(5.3) \qquad P\left(\sup_{t_0 \leq t \leq T} \left|X_t - \tilde{X}_t\right| > 0\right) = 0,$$

we say that the solutions of (5.1) are *pathwise unique*.

The hypotheses of an existence and uniqueness theorem are usually sufficient, but not necessary, conditions to ensure the conclusion of the theorem. Some of those that we shall use here are quite strong, but can be weakened in several ways. In what follows the initial instant $0 \leq t_0 < T$ is arbitrary, but fixed, and the coefficients functions $a$, $b : [t_0, T] \times \Re \to \Re$ are given. Most of the assumptions concern these coefficients.

**A1** (Measurability):   $a = a(t, x)$ and $b = b(t, x)$ are jointly ($\mathcal{L}^2$-) measurable in $(t, x) \in [t_0, T] \times \Re$;

**A2** (Lipschitz condition):   There exists a constant $K > 0$ such that

$$|a(t, x) - a(t, y)| \leq K |x - y|$$

and

$$|b(t, x) - b(t, y)| \leq K |x - y|$$

for all $t \in [t_0, T]$ and $x, y \in \Re$;

**A3** (Linear growth bound):   There exists a constant $K > 0$ such that

$$|a(t, x)|^2 \leq K^2 (1 + |x|^2)$$

and

$$|b(t, x)|^2 \leq K^2 (1 + |x|^2)$$

for all $t \in [t_0, T]$ and $x, y \in \Re$.

We shall henceforth hold fixed a Wiener process $W = \{W_t, t \geq 0\}$ and an associated family of $\sigma$-algebras $\{\mathcal{A}_t, t \geq 0\}$ for which the properties listed in the first paragraph of Section 4 of Chapter 3 are satisfied. The remaining assumption concerns the initial value.

**A4** (Initial value): $X_{t_0}$ is $\mathcal{A}_{t_0}$-measurable with $E(|X_{t_0}|^2) < \infty$.

The Lipschitz condition A2 provides the key estimates in both the proofs of uniqueness and of existence by the method of successive approximations. To these estimates we shall then apply the following inequality of Gronwall, which is easily proved.

**Lemma 4.5.1 (The Gronwall Inequality)** *Let $\alpha$, $\beta : [t_0, T] \to \Re$ be integrable with*

$$0 \leq \alpha(t) \leq \beta(t) + L \int_{t_0}^{t} \alpha(s)\, ds$$

*for $t \in [t_0, T]$ where $L > 0$. Then*

$$\alpha(t) \leq \beta(t) + L \int_{t_0}^{t} e^{L(t-s)} \beta(s)\, ds$$

*for $t \in [t_0, T]$.*

Assuming that strong solutions of (5.1) exist, we can show their pathwise uniqueness using just the measurability assumption A1 and the Lipschitz condition A2. The proof would be simpler if we could also assume that the initial value had finite second moment, as in assumption A4, because, by Theorem 4.5.3 below, the solutions would then have finite second moments. Without this property we are forced to use a truncation procedure.

**Lemma 4.5.2** *If A1 and A2 hold, then the solutions of (5.2) corresponding to the same initial value and the same Wiener process are pathwise unique.*

**Proof** Let $X_t$ and $\tilde{X}_t$ be two such solutions of (5.2) on $[t_0, T]$ with, almost surely, continuous sample paths. Since they may not have finite second moments, we shall use the following *truncation procedure*: for $N > 0$ and $t \in [t_0, T]$ we define

$$I_t^{(N)}(\omega) = \begin{cases} 1 & : \quad |X_u(\omega)|, |\tilde{X}_u(\omega)| \leq N \quad \text{for} \quad t_0 \leq u \leq t \\ 0 & : \quad \text{otherwise.} \end{cases}$$

Obviously $I_t^{(N)}$ is $\mathcal{A}_t$-measurable and $I_t^{(N)} = I_t^{(N)} I_s^{(N)}$ for $t_0 \leq s \leq t$. Consequently the integrals in the following expression are meaningful:

$$(5.4) \qquad Z_t^{(N)} = I_t^{(N)} \int_{t_0}^{t} I_s^{(N)} \left( a(s, X_s) - a(s, \tilde{X}_s) \right) ds$$

$$+ I_t^{(N)} \int_{t_0}^{t} I_s^{(N)} \left( b(s, X_s) - b(s, \tilde{X}_s) \right) dW_s$$

where $Z_t^{(N)} = I_t^{(N)}(X_t - \tilde{X}_t)$. From the Lipschitz condition A2 we then have

(5.5)   $\max \left\{ \left| I_s^{(N)} \left( a(s, X_s) - a(s, \tilde{X}_s) \right) \right|, \left| I_s^{(N)} \left( b(s, X_s) - b(s, \tilde{X}_s) \right) \right| \right\}$

$$\leq K I_s^{(N)} \left| X_s - \tilde{X}_s \right| \leq 2KN$$

for $t_0 \leq s \leq t$. Thus the second order moments exist for $Z_t^{(N)}$ and the two integrals in (5.4). Using the inequality $(a+b)^2 \leq 2(a^2+b^2)$, the Cauchy-Schwarz inequality (1.4.38) and the property (3.2.10) of an Ito integral we obtain from

(5.4)   $E\left( \left| Z_t^{(N)} \right|^2 \right)$

$$\leq 2E\left( \left| \int_{t_0}^t I_s^{(N)} \left( a(s, X_s) - a(s, \tilde{X}_s) \right) ds \right|^2 \right)$$

$$+ 2E\left( \left| \int_{t_0}^t I_s^{(N)} (b(s, X_s) - b(s, \tilde{X}_s) \, dW_s \right|^2 \right)$$

$$\leq 2(T - t_0) \int_{t_0}^t E\left( \left| I_s^{(N)} \left( a(s, X_s) - a(s, \tilde{X}_s) \right) \right|^2 \right) ds$$

$$+ 2 \int_{t_0}^t E\left( \left| I_s^{(N)} \left( b(s, X_s) - b(s, \tilde{X}_s) \right) \right|^2 \right) ds,$$

which we combine with (5.5) to get

(5.6)   $$E\left( \left| Z_t^{(N)} \right|^2 \right) \leq L \int_{t_0}^t E\left( \left| Z_s^{(N)} \right|^2 \right) ds$$

for $t \in [t_0, T]$ where $L = 2(T - t_0 + 1)K^2$. We then apply the Gronwall inequality (Lemma 4.5.1) with $\alpha(t) = E(|Z_t^{(N)}|^2)$ and $\beta(t) \equiv 0$ to conclude that

$$E\left( \left| Z_t^{(N)} \right|^2 \right) = E\left( \left| I_t^{(N)} \left( X_t - \tilde{X}_t \right) \right|^2 \right) = 0,$$

and hence that $I_t^{(N)} X_t = I_t^{(N)} \tilde{X}_t$, w.p.1, for each $t \in [t_0, T]$. Since the sample paths are continuous surely they are bounded almost surely. Thus we can make the probability

$$P\left( I_t^{(N)} \not\equiv 1 \, \forall t \in [t_0, T] \right) \leq P\left( \sup_{t_0 \leq t \leq T} |X_t| > N \right) + P\left( \sup_{t_0 \leq t \leq T} \left| \tilde{X}_t \right| > N \right)$$

arbitrarily small by taking $N$ sufficiently large. This means that $P\left( X_t \neq \tilde{X}_t \right) = 0$ for each $t \in [t_0, T]$, and hence that $P\left( X_t \neq \tilde{X}_t : t \in D \right) = 0$ for any countably dense subset $D$ of $[t_0, T]$. As the solutions are continuous and coincide on a countably dense subset of $[t_0, T]$, they must coincide, almost surely, on the entire interval $[t_0, T]$. Thus (5.3) holds, that is the solutions of (5.2) are pathwise unique. $\square$

**Theorem 4.5.3** *Under assumptions A1–A4 the stochastic differential equation (5.1) has a pathwise unique strong solution $X_t$ on $[t_0, T]$ with*

$$\sup_{t_0 \leq t \leq T} E\left(|X_t|^2\right) < \infty.$$

**Proof** In view of Lemma 4.5.2 we have only to establish the existence of a continuous solution $X_t$ on $[t_0, T]$ for the given Wiener process $W = \{W_t, t \geq 0\}$. We shall do this by the method of successive approximations, first defining $X_t^{(0)} \equiv X_{t_0}$ and then

$$(5.7) \qquad X_t^{(n+1)} = X_{t_0} + \int_{t_0}^t a\left(s, X_s^{(n)}\right) ds + \int_{t_0}^t b\left(s, X_s^{(n)}\right) dW_s$$

for $n = 0, 1, 2, \ldots$. If for a fixed $n \geq 0$ the approximation $X_t^{(n)}$ is $\mathcal{A}_t$-measurable and continuous on $[t_0, T]$, then it follows from assumptions A1, A2 and A3 that the integrals in (5.7) are meaningful and that the resulting process $X_t^{(n+1)}$ is $\mathcal{A}_t$-measurable and continuous on $[t_0, T]$. As $X_t^{(0)}$ is obviously $\mathcal{A}_t$-measurable and continuous on $[t_0, T]$, it follows by induction that so too is each $X_t^{(n)}$ for $n = 1, 2, 3, \ldots$.

From assumption A4 and the definition of $X_t^{(0)}$ it is clear that

$$\sup_{t_0 \leq t \leq T} E\left(\left|X_t^{(0)}\right|^2\right) < \infty.$$

Applying the inequality $(a + b + c)^2 \leq 3(a^2 + b^2 + c^2)$, the Cauchy-Schwarz inequality (1.4.38), the identity (3.2.10) and the linear growth bound A3 to (5.7) we obtain

$$E\left(\left|X_t^{(n+1)}\right|^2\right)$$

$$\leq 3E\left(|X_{t_0}|^2\right) + 3E\left(\left|\int_{t_0}^t a\left(s, X_s^{(n)}\right) ds\right|^2\right)$$

$$+ 3E\left(\left|\int_{t_0}^t b\left(s, X_s^{(n)}\right) dW_s\right|^2\right)$$

$$\leq 3E\left(|X_{t_0}|^2\right) + 3(T - t_0)E\left(\int_{t_0}^t \left|a\left(s, X_s^{(n)}\right)\right|^2 ds\right)$$

$$+ 3E\left(\int_{t_0}^t \left|b\left(s, X_s^{(n)}\right)\right|^2 ds\right)$$

$$\leq \ 3E\left(|X_{t_0}|^2\right) + 3(T - t_0 + 1)K^2 E\left(\int_{t_0}^t \left(1 + \left|X_s^{(n)}\right|^2\right) ds\right)$$

for $n = 0, 1, 2, \ldots$. By induction we thus have

(5.8) 
$$\sup_{t_0 \leq t \leq T} E\left(\left|X_t^{(n)}\right|^2\right) \leq C_0 < \infty$$

for $n = 1, 2, 3, \ldots$.

Similarly to the derivation of the inequality (5.4), except now factors like $I_t^{(N)}$ are not required because of (5.8), we can show that

(5.9) 
$$E\left(\left|X_t^{(n+1)} - X_t^{(n)}\right|^2\right) \leq L \int_{t_0}^t E\left(\left|X_s^{(n+1)} - X_s^{(n)}\right|^2\right) ds$$

for $t \in [t_0, T]$ and $n = 1, 2, 3, \ldots$ where $L = 2(T - t_0 + 1)K^2$. Then using the Cauchy formula

$$\int_{t_0}^t \int_{t_0}^{t_{n-1}} \cdots \int_{t_0}^{t_1} f(s) \, ds \, dt_1 \ldots dt_{n-1} = \frac{1}{(n-1)!} \int_{t_0}^t (t - s)^{n-1} f(s) \, ds$$

in repeated iterations of (5.9), we obtain

(5.10) 
$$E\left(\left|X_t^{(n+1)} - X_t^{(n)}\right|^2\right)$$
$$\leq \frac{L^n}{(n-1)!} \int_{t_0}^t (t - s)^{n-1} E\left(\left|X_s^{(1)} - X_s^{(0)}\right|^2\right) ds$$

for $t \in [t_0, T]$ and $n = 1, 2, 3, \ldots$. Also, using the growth bound A3 instead of the Lipschitz condition A2 in the derivation of (5.9) for $n = 0$, we find that

$$E\left(\left|X_t^{(n+1)} - X_t^{(n)}\right|^2\right)$$
$$\leq L \int_{t_0}^t \left(1 + E\left(\left|X_s^{(0)}\right|^2\right)\right) ds$$
$$\leq L(T - t_0)\left(1 + E\left(|X_{t_0}|^2\right)\right) = C_1.$$

On inserting this into (5.10) we get

$$E\left(\left|X_t^{(n+1)} - X_t^{(n)}\right|^2\right) \leq \frac{C_1 L^n (t - t_0)^n}{n!}$$

for $t \in [t_0, T]$ and $n = 0, 1, 2, \ldots$, and hence

(5.11) 
$$\sup_{t_0 \leq t \leq T} E\left(\left|X_t^{(n+1)} - X_t^{(n)}\right|^2\right) \leq \frac{C_1 L^n (T - t_0)^n}{n!}$$

for $n = 0, 1, 2, \ldots$. This implies the mean-square convergence of the successive approximations uniformly on $[t_0, T]$, but we need the almost sure convergence of their sample paths uniformly on $[t_0, T]$. To show this we define

$$Z_n = \sup_{t_0 \leq t \leq T} \left| X_t^{(n+1)} - X_t^{(n)} \right|$$

for $n = 0, 1, 2, \ldots$ and so from (5.5) obtain

$$Z_n \leq \int_{t_0}^t \left| a(s, X_s^{(n)}) - a(s, X_s^{(n-1)}) \right| \, ds$$

$$+ \sup_{t_0 \leq t \leq T} \left| \int_{t_0}^t \left( b(s, X_s^{(n)}) - b(s, X_s^{(n-1)}) \right) \, dW s \right|.$$

Using the Doob inequality (2.3.7), the Cauchy-Schwarz inequality (1.4.38) and the Lipschitz condition A2 we determine

$$E\left(|Z_n|^2\right) \leq 2(T - t_0)K^2 \int_{t_0}^T E\left(\left| X_s^{(n)} - X_s^{(n-1)} \right|^2\right) \, ds$$

$$+ 8K^2 \int_{t_0}^T E\left(\left| X_s^{(n)} - X_s^{(n-1)} \right|^2\right) \, ds$$

$$\leq 2(T - t_0 + 4)K^2 \int_{t_0}^T E\left(\left| X_s^{(n)} - X_s^{(n-1)} \right|^2\right) \, ds,$$

which we combine with (5.11) to conclude that

$$E\left(|Z_n|^2\right) \leq \frac{C_2 L^{n-1}(T - t_0)^{n-1}}{(n-1)!}$$

for $n = 1, 2, 3, \ldots$ where $C_2 = 2C_1 K^2(T - t_0 + 4)(T - t_0)$. Then, after applying the Markov inequality (2.2.14) to each term and summing, we have

$$\sum_{n=1}^{\infty} P\left(Z_n > \frac{1}{n^2}\right) \leq C_2 \sum_{n=1}^{\infty} \frac{n^4}{(n-1)!} L^{n-1}(T - t_0)^{n-1},$$

where the series on the right side converges by the ratio test. Hence the series on the left side also converges, so by the Borel-Cantelli Lemma 2.1.4 we conclude that the $Z_n$ converge to 0, almost surely, that is the successive approximations $X_t^{(n)}$ converge, almost surely, uniformly on $[t_0, T]$ to a limit $\tilde{X}_t$ defined by

$$\tilde{X}_t = X_{t_0} + \sum_{n=0}^{\infty} \left\{ X_t^{(n+1)} - X_t^{(n)} \right\}.$$

It follows from (5.8) that $\tilde{X}$ is mean square bounded on $[0, T]$. As the limit of $\mathcal{A}^*$-adapted processes, $\tilde{X}$ is $\mathcal{A}^*$-adapted and as the uniform limit of continuous processes it is continuous. In view of this and the growth bound A3, the right side of the integral equation (5.2) is well defined for this process $\tilde{X}_t$. It remains to show that it then equals the left side. Taking the limit as $n \to \infty$ in (5.7) we see that $\tilde{X}$ is a solution of (5.2). The left side of (5.7) converges to $\tilde{X}_t$ uniformly on $[t_0, T]$. Comparing the right sides, we have by the Lipschitz condition A2

$$\left| \int_{t_0}^t a(s, X_s^{(n)}) \, ds - \int_{t_0}^t a(s, \tilde{X}_s) \, ds \right| \leq K \int_{t_0}^t \left| X_s^{(n)} - \tilde{X}_s \right| \, ds \longrightarrow 0$$

and

$$\int_{t_0}^t \left| b(s, X_s^{(n)}) - b(s, \tilde{X}_s) \right|^2 \, ds \leq K^2 \int_{t_0}^t \left| X_s^{(n)} - \tilde{X}_s \right|^2 \, ds \longrightarrow 0,$$

w.p.1, which imply that

$$\int_{t_0}^t a(s, X_s^{(n)}) \, ds \longrightarrow \int_{t_0}^t a(s, \tilde{X}_s) \, ds,$$

w.p.1, and

$$\int_{t_0}^t b(s, X_s^{(n)}) \, dW_s \longrightarrow \int_{t_0}^t b(s, \tilde{X}_s) \, dW_s,$$

in probability, as $n \to \infty$ for each $t \in [t_0, T]$. Hence the right side of (5.7) converges to the right side of (5.2), and so the limit process $\tilde{X}$ satisfies the stochastic integral equation (5.2).

This completes the proof of the existence and uniqueness of a strong solution of the stochastic differential equation (5.1) for an initial value $X_{t_0}$ with $E(|X_{t_0}|^2) < \infty$. □

Variations of Theorem 4.5.3 are possible with weakened assumptions. The most obvious is to drop the requirement that the initial value $X_{t_0}$ satisfies $E(|X_{t_0}|^2) < \infty$. The proof must then be modified with a truncation procedure similar to that used in the proof of Lemma 4.5.2, because the successive approximations and their limit may now not be mean-square bounded. Another obvious generalization is to replace the global Lipschitz condition A2 by a local one, that is with the condition holding with possibly different constants $K_N$ for $|x|, |y| \leq N$ and each $N > 0$. This significantly enlarges the class of admissible coefficients since by the Mean Value Theorem every continuously differentiable function satisfies a local Lipschitz condition. In this case a truncation argument must also be used in the proof. A key step is the observation that if the coefficients of two stochastic differential equations coincide on some interval $-N \leq x \leq N$, then the first exit times $\tau_N$ and $\tilde{\tau}_N \geq t_0$ of their solutions (with the same initial value) from this interval are equal, w.p.1, and the solutions coincide until this common exit time. In the proof of existence here, for each $N > 0$ the successive approximations are stopped on reaching $\pm N$ and a limiting solution $X_t^{(N)}$ is obtained. As $N$ increases the exit times $\tau_N$ of the solutions $X_t^{(N)}$ increase, and for any two $N$, $N'$ the corresponding solutions $X_t^{(N)} = X_t^{(N')}$ for $0 \leq t \leq \min\{\tau_N, \tau_{N'}\}$, w.p.1. Taking the limit as $N \to \infty$

gives the desired global solution $X_t$, that is a solution existing on the entire interval $[t_0, T]$, provided a growth bound such as A3 is satisfied. As for ordinary differential equations a Lipschitz condition, global or local, is not necessary in order to prove the existence of solutions of stochastic differential equations, but without such a condition the method of successive approximations cannot be used and the uniqueness of solutions is not assured. For example, the SDE $dX_t = |X_t|^\alpha \, dW_t$ has a unique solution $X_t \equiv 0$ for $X_0 = 0$ when $\alpha \geq 1/2$, but has infinitely many solutions for this initial value when $0 < \alpha < 1/2$. This compares with the ordinary differential equation $\dot{x} = x^\alpha$ which has infinitely many solutions for $x_0 = 0$ when $0 < \alpha < 1$. The functions $x^\alpha$ and $|x|^\alpha$ are not differentiable at $x = 0$ for $0 < \alpha < 1$ and do not satisfy a local Lipschitz condition in a neighbourhood of $x = 0$.

In some cases it is possible to replace the Lipschitz condition on $b$ by the weaker *Yamada condition:* there exists an increasing function $\rho : [0, \infty) \to \Re$ with $\rho(0) = 0$ and $\int_{0+} \rho^{-2}(u) \, du = +\infty$ such that

$$|b(t, x) - b(t, y)| \leq \rho\left(|x - y|\right)$$

for all $x$, $y \in \Re$ and $t \in [t_0, T]$. This also implies an existence and uniqueness theorem of strong solutions. For example, $\rho$ may be $\rho(u) = u^\alpha$ for $1/2 \leq \alpha \leq 1$ or $\rho(u) = \sqrt{|u \ln u|}$. With the first of these it can be concluded that the SDE $dX_t = |X_t|^\alpha \, dW_t$ has the unique strong solution $X_t \equiv 0$ for $X_0 = 0$ when $1/2 \leq \alpha \leq 1$.

A useful extension to the existence and uniqueness theorem is to weaken the linear growth bound A3, which essentially says that the coefficients must not grow faster than linearly in magnitude for large $|x|$. This excludes drift terms such as $a(x) = x - x^3$. For the corresponding ordinary differential equation $\dot{x} = a(x)$ the solutions satisfy

$$\frac{1}{2} \frac{d}{dt} x^2 = xa(x) = x^2 - x^4 \leq x^2$$

for all $x \in \Re$, from which it follows that $x^2(t) \leq x^2(0)e^{2t}$ for all $t \geq 0$; hence its solutions cannot explode in finite time. In fact, the growth bound on $a = a(t, x)$ in A3 can be replaced by

$$x \, a(t, x) \leq K^2(1 + |x|^2),$$

with $b = b(t, x)$ still satisfying the original growth bound, and the existence of a global solution of the stochastic differential equation (5.1) remains true. The absence of some kind of growth bound on either of the coefficients may lead to explosions in the solutions.

In some instances stronger results can be obtained by strengthening the assumptions of the existence and uniqueness theorem. The next theorem provides useful upper bounds on the higher even order moments of the solution. When the initial value is a constant it implies the existence of all such moments.

**Theorem 4.5.4**     *Suppose that A1–A4 hold and that*

$$E\left(|X_{t_0}|^{2n}\right) < \infty$$

*for some integer $n \geq 1$. Then the solution $X_t$ of (5.1) satisfies*

(5.12) $$E\left(|X_t|^{2n}\right) \leq \left(1 + E\left(|X_{t_0}|^{2n}\right)\right) e^{C(t-t_0)}$$

*and*

(5.13) $$E\left(|X_t - X_{t_0}|^{2n}\right) \leq D\left(1 + E\left(|X_{t_0}|^{2n}\right)\right)(t - t_0)^n e^{C(t-t_0)}$$

*for $t \in [t_0, T]$ where $T < \infty$, $C = 2n(2n+1)K^2$ and $D$ is a positive constant depending only on $n$, $K$ and $T - t_0$.*

**Proof**  It follows from the Ito formula (3.3.6) that $|X_t|^{2n}$ satisfies the stochastic differential equation

(5.14) $$\begin{aligned} |X_t|^{2n} &= |X_{t_0}|^{2n} + \int_{t_0}^t 2n\, |X_s|^{2n-2}\, X_s\, a(s, X_s)\, ds \\ &\quad + \int_{t_0}^t n(2n-1)\, |X_s|^{2n-2}\, b^2(s, X_s)\, ds \\ &\quad + \int_{t_0}^t 2n\, |X_s|^{2n-2}\, X_s\, b(s, X_s)\, dW_s. \end{aligned}$$

Since $E(|X_{t_0}|^{2n}) < \infty$ we can use an analogous argument to that in the proof of Theorem 4.5.3 to show that $E(|X_t|^{2n}) < \infty$ for $t \in [t_0, T]$. Hence the function $f$ defined by $f(t, \omega) = 0$ for $0 \leq t < t_0$ and $f(t, \omega) = 2n\,|X_t(\omega)|^{2n-2}\, X_t(\omega)\, b(t, X_t(\omega))$ for $t_0 \leq t \leq T$ belongs to $\mathcal{L}_T^2$, so by (3.2.9)

$$E\left(\int_{t_0}^t 2n\, |X_s|^{2n-2}\, X_s\, b(s, X_s)\, dW_s\right) = 0$$

for each $t \in [t_0, T]$. Taking expectations on both sides of (5.14) we obtain

$$E\left(|X_t|^{2n}\right)$$

$$\leq E\left(|X_{t_0}|^{2n}\right) + \int_{t_0}^t E\left(2n\, |X_s|^{2n-2}\, X_s\, a(s, X_s)\right) ds$$

$$\quad + \int_{t_0}^t E\left(n(2n-2)\, |X_s|^{2n-2}\, b^2(s, X_s)\right) ds$$

$$\leq E\left(|X_{t_0}|^{2n}\right) + n(2n+1)K^2 \int_{t_0}^t E\left(|X_s|^{2n-2}\left(1 + |X_s|^2\right)\right) ds,$$

where we have used the growth bound A3. Then, since $(1+a^2)a^{2n-2} \leq 1 + 2a^{2n}$, we have

$$E\left(|X_t|^{2n}\right) \leq E\left(|X_{t_0}|^{2n}\right) + n(2n+1)K^2(t-t_0)$$

$$+2n(2n+1)K^2 \int_{t_0}^t E\left(|X_s|^{2n-2}\right) ds$$

and hence from the Gronwall inequality (Lemma 4.5.1)

$$E\left(|X_t|^{2n}\right) \leq \beta(t) + 2n(2n+1)K^2 \int_{t_0}^t \exp\left(2n(2n+1)K^2(t-s)\right)\beta(s)\,ds$$

where $\beta(t) = E(|X_{t_0}|^{2n}) + n(2n+1)K^2(t-t_0)$. We obtain (5.12) when we evaluate the integral here.

To establish (5.13) we use the inequality $(a+b)^{2n} \leq 2^{2n-1}(a^{2n}+b^{2n})$ and

$$E\left(|X_t - X_{t_0}|^{2n}\right)$$

$$= E\left(\left|\int_{t_0}^t a(s, X_s)\,ds + \int_{t_0}^t b(s, X_s)\,dW_s\right|^{2n}\right)$$

$$\leq 2^{2n-1}\left\{E\left(\left|\int_{t_0}^t a(s, X_s)\,ds\right|^{2n}\right) + E\left(\left|\int_{t_0}^t b(s, X_s)\,dW_s\right|^{2n}\right)\right\}$$

$$\leq 2^{2n-1}(t-t_0)^{2n-1}\int_{t_0}^t E\left(|a(s, X_s)|^{2n}\right)\,ds$$

$$+2^{2n-1}(t-t_0)^{n-1}[n(2n-1)]^n \int_{t_0}^t E\left(|b(s, X_s)|^{2n}\right)\,ds.$$

The last line follows from the Hölder inequality (1.4.40) and the fact that

(5.15)
$$E\left(\left|\int_{t_0}^t f(s,\omega)\,dW_s\right|^{2n}\right)$$

$$\leq (t-t_0)^{n-1}[n(2n-1)]^n \int_{t_0}^t E\left(|f(s,\omega)|^{2n}\right)\,ds,$$

for any $f$ with $f^n \in \mathcal{L}_T^2$, which can be verified inductively with the help of the Ito formula. Using the growth bound A3 in the above integrals and the inequality $(1+a^2)^n \leq 2^{2n-1}(1+a^{2n})$, we obtain

$$E\left(|X_t - X_{t_0}|^{2n}\right) \leq 2^{2n-1}K^{2n}\{(T-t_0)^n + [n(2n-1)]^n\}$$

$$\times (t-t_0)^{n-1}\int_{t_0}^t E\left(\left(1+|X_s|^2\right)^n\right)\,ds$$

$$\leq \frac{1}{2}D(t-t_0)^{n-1}\int_{t_0}^t E\left(1+|X_s|^{2n}\right)\,ds$$

$$\leq \ \frac{1}{2} D(t - t_0)^{n-1} \int_{t_0}^{t} \left(1 + E\left(1 + |X_{t_0}|^{2n}\right) e^{C(s-t_0)}\right) ds$$

where we have used (5.12), which has already been proved, and

$$D = 2^{2(2n-1)} K^{2n} \left\{ (T - t_0)^n + [n(2n - 1)]^n \right\}.$$

Finally, integrating and then using the inequality $(e^x - 1)/x \leq e^x$, we obtain

$$E\left(|X_t - X_{t_0}|^{2n}\right)$$

$$\leq \ \frac{1}{2} D(t - t_0)^n + \frac{1}{2} D(t - t_0)^n \left(1 + E\left(|X_{t_0}|^{2n}\right)\right) \frac{e^{C(t-t_0)} - 1}{C(t - t_0)}$$

$$\leq \ \frac{1}{2} D(t - t_0)^n + \frac{1}{2} D(t - t_0)^n \left(1 + E\left(|X_{t_0}|^{2n}\right)\right) e^{C(t-t_0)}$$

$$\leq \ D\left(1 + E\left(|X_{t_0}|^{2n}\right)\right)(t - t_0)^n e^{C(t-t_0)},$$

which is the desired result. $\square$

**Exercise 4.5.5**    *By modifying the proof of Theorem 4.5.4 and using the Doob inequality (2.3.7) derive the following uniform moment estimates:*

$$(5.16) \quad E\left(\sup_{t_0 \leq s \leq T} |X_s|^{2n}\right)$$

$$\leq \ D\left\{ E\left(|X_{t_0}|^{2n}\right) + \left(1 + E\left(|X_{t_0}|^{2n}\right)\right)(T - t_0)^n e^{C(T-t_0)}\right\}$$

*and*

$$(5.17) \quad E\left(\sup_{t_0 \leq s \leq T} |X_s - X_{t_0}|^{2n}\right)$$

$$\leq \ D\left(1 + E\left(|X_{t_0}|^{2n}\right)\right)(T - t_0)^n e^{C(T-t_0)},$$

*for positive constants $C$ and $D$ depending only on $n$, $K$ and $T - t_0$.*

With similar inequalities to those in the proof of the existence and uniqueness theorem (Theorem 4.5.3) we can show that the solution of a stochastic differential equation depends continuously on a parameter when the coefficients do. For a parameter $\nu$ we consider a family of stochastic differential equations in integral form

$$(5.18) \quad X_t^{(\nu)} = X_{t_0}^{(\nu)} + \int_{t_0}^{t} a^{(\nu)}\left(s, X_s^{(\nu)}\right) ds + \int_{t_0}^{t} b^{(\nu)}\left(s, X_s^{(\nu)}\right) dW_s,$$

where the initial value $X_{t_0}^{(\nu)}$ may also depend on the parameter. In the case that only the initial value depends on the parameter we obtain the continuous dependence on initial conditions. Analogous results to those in the following theorem also hold for sequences if we replace $\nu$ by $1/n$ and $\nu_0$ by $0$.

**Theorem 4.5.6**   *Suppose that A1–A4 are satisfied by (5.18) for each $\nu \in \Re$ and that*

(5.19)
$$\lim_{\nu \to \nu_0} E\left(\left|X_{t_0}^{(\nu)} - X_{t_0}^{(\nu_0)}\right|^2\right) = 0$$

*and*

(5.20)
$$\lim_{\nu \to \nu_0} \sup_{|x| \le N} \left|a^{(\nu)}(t, x) - a^{(\nu_0)}(t, x)\right| = 0,$$

$$\lim_{\nu \to \nu_0} \sup_{|x| \le N} \left|b^{(\nu)}(t, x) - b^{(\nu_0)}(t, x)\right| = 0$$

*for each $t \in [t_0, T]$ and $N > 0$. Then the solutions $X^{(\nu)}$ of (5.18) satisfy*

$$\lim_{\nu \to \nu_0} \sup_{t \in [t_0, T]} E\left(\left|X_t^{(\nu)} - X_t^{(\nu_0)}\right|^2\right) = 0.$$

**Proof**   The solutions $X_t^{(\nu)}$ exist and are unique and mean-square bounded by Theorem 4.5.3. Subtracting (5.18) for $\nu_0$, squaring and then taking expectations we obtain

$$E\left(\left|X_t^{(\nu)} - X_t^{(\nu_0)}\right|^2\right)$$

$$\le \ 3E\left(\left|X_{t_0}^{(\nu)} - X_{t_0}^{(\nu_0)}\right|^2\right)$$

$$+3E\left(\left|\int_{t_0}^t \left(a^{(\nu)}(s, X_s^{(\nu)}) - a^{(\nu_0)}(s, X_s^{(\nu_0)})\right) ds\right|^2\right)$$

$$+3E\left(\left|\int_{t_0}^t \left(b^{(\nu)}(s, X_s^{(\nu)}) - b^{(\nu_0)}(s, X_s^{(\nu_0)})\right) dW_s\right|^2\right)$$

$$\le \ 3E\left(\left|X_{t_0}^{(\nu)} - X_{t_0}^{(\nu_0)}\right|^2\right)$$

$$+3(T - t_0)\int_{t_0}^t E\left(\left|a^{(\nu)}(s, X_s^{(\nu)}) - a^{(\nu_0)}(s, X_s^{\nu_0})\right|^2\right) ds$$

$$+3\int_{t_0}^t E\left(\left|b^{(\nu)}(s, X_s^{(\nu)}) - b^{(\nu_0)}(s, X_s^{(\nu_0)})\right|^2\right) ds,$$

by the Cauchy-Schwarz inequality (1.4.38) and the linearity of Ito integrals in their integrands. Now

$$\left| a^{(\nu)}(t, X_t^{(\nu)}) - a^{(\nu_0)}(t, X_t^{(\nu_0)}) \right|^2$$

$$\leq 2 \left| a^{(\nu)}(t, X_t^{(\nu)}) - a^{(\nu)}(t, X_t^{(\nu_0)}) \right|^2 + 2 \left| a^{(\nu)}(t, X_t^{(\nu_0)}) - a^{(\nu_0)}(t, X_t^{(\nu_0)}) \right|^2$$

$$\leq 2K^2 \left| X_t^{(\nu)} - X_t^{(\nu_0)} \right|^2 + 2 \left| a^{(\nu)}(t, X_t^{(\nu_0)}) - a^{(\nu_0)}(t, X_t^{(\nu_0)}) \right|^2,$$

and similarly for the $b^{(\nu)}$ coefficients. Hence

$$E \left( \left| X_t^{(\nu)} - X_t^{(\nu_0)} \right|^2 \right)$$

$$\leq 3E \left( \left| X_{t_0}^{(\nu)} - X_{t_0}^{(\nu_0)} \right|^2 \right)$$

$$+ 6(T - t_0) \int_{t_0}^t E \left( \left| a^{(\nu)}(s, X_s^{(\nu_0)}) - a^{(\nu_0)}(s, X_s^{(\nu_0)}) \right|^2 \right) ds$$

$$+ 6 \int_{t_0}^t E \left( \left| b^{(\nu)}(s, X_s^{(\nu_0)}) - b^{(\nu_0)}(s, X_s^{(\nu_0)}) \right|^2 \right) ds$$

$$+ 6K^2(T - t_0 + 1) \int_{t_0}^t E \left( \left| X_s^{(\nu)} - X_s^{(\nu_0)} \right|^2 \right) ds$$

$$= \beta^{(\nu)}(t) + L \int_{t_0}^t E \left( \left| X_s^{(\nu)} - X_s^{(\nu_0)} \right|^2 \right) ds$$

where $\beta^{(\nu)}(t)$ and $L$ denote the corresponding expressions in the preceding line. Then by Gronwall's inequality (Lemma 4.5.1) we get

$$E \left( \left| X_t^{(\nu)} - X_t^{(\nu_0)} \right|^2 \right) \leq \beta^{(\nu)}(t) + L \int_{t_0}^t e^{L(t-s)} \beta^{(\nu)}(s) \, ds$$

for each $t \in [t_0, T]$. By the convergence assumptions (5.19)–(5.20) the $\beta^{(\nu)}(t) \to 0$ as $\nu \to \nu_0$, and thus the stated result follows. $\square$

We remark that as in Exercise 4.5.5 we can strengthen the convergence of the preceding theorem to the mean-square uniform convergence

$$\lim_{\nu \to \nu_0} E \left( \sup_{t_0 \leq t \leq T} \left| X_t^{(\nu)} - X_t^{(\nu_0)} \right|^2 \right) = 0.$$

In addition, we can obtain analogous convergence results for the derivatives $\frac{\partial}{\partial \nu} X_t^{(\nu)}$, which are defined in the mean-square sense, and also satisfy stochastic differential equations.

**Example 4.5.7** *Consider the solutions of the stochastic differential equations*

$$dX_t^{(\nu)} = \nu a \left( t, X_t^{(\nu)} \right) dt + dW_t$$

*with the same initial value* $X_0^{(\nu)} = X_0$ *for each* $\nu \geq 0$. *For* $\nu = 0$ *the solution is obviously* $X_t^{(0)} = X_0 + W_t$. *Thus by Theorem 4.5.6 we have*

$$\lim_{\nu \to 0} E\left(\left|X_t^{(\nu)} - X_0 - W_t\right|^2\right) = 0$$

*for each* $t \in [0, T]$.

**Exercise 4.5.8**    *Determine the mean-square limit as* $\nu \to 0$ *of the solutions of the stochastic differential equations*

$$dX_t^{(\nu)} = a\left(t, X_t^{(\nu)}\right) dt + \nu \, dW_t$$

*with the same deterministic initial value* $X_0^{(\nu)} = X_0$ *for each* $\nu \geq 0$.

# 4.6   Strong Solutions as Diffusion Processes

Possibly the most attractive and important property of the solutions of stochastic differential equations is that they are usually Markov processes, in fact in many cases diffusion processes. As a consequence of this we can apply the powerful analytical tools that have been developed for Markov and diffusion processes to the solutions of stochastic differential equations. In discussing such solutions below we shall always assume that we have chosen a separable version with, almost surely, continuous sample paths. In addition, we shall often denote by $X_t(t_0, x_0)$ the solution with fixed initial value $X_{t_0} = x_0$.

Under assumptions A1–A4, which were stated at the beginning of Section 5, the solution $X$ of the stochastic differential equation (5.1) is a Markov process on the interval $[t_0, T]$ with transition probabilities

$$P(s, x; t, B) = P\left(X_t \in B | X_s = x\right) = P\left(X_t(s, x) \in B\right)$$

for all $t_0 \leq s \leq t \leq T$, $x \in \Re$ and Borel subsets $B$ of $\Re$.

For an autonomous stochastic differential equation

(6.1)  $$dX_t = a(X_t) \, dt + b(X_t) \, dW_t$$

the solutions $X$ are homogeneous Markov processes, that is their transition probabilities $P(s, x; t, B)$ depend only on the elapsed period of time $t - s \geq 0$ rather than on the specific values of $s$ and $t$. In this case we usually write $P(x; t - s, B)$ for $P(s, x; t, B)$ and can, without loss of generality, take $s = 0$. The solution of an autonomous SDE will be a stationary Markov process with stationary distribution $\tilde{P}$ if there exists a probability measure $\tilde{P}$ on $(\Re, \mathcal{B})$ satisfying

(6.2)  $$\tilde{P}(B) = \int_{\Re} P(x; t, B) \tilde{P}(dx)$$

for all $t \geq 0$, $x \in \Re$ and $B \in \mathcal{B}$. Various analytical conditions on the coefficients $a$ and $b$ are known which ensure the existence of a density function $\tilde{p}$ for such a stationary distribution as the solution of the stationary Fokker-Planck equation.

In general, the solutions of a stochastic differential equation

$$(6.3) \qquad dX_t = a(t, X_t)\, dt + b(t, X_t)\, dW_t$$

are diffusion processes with their transition probabilities satisfying the three limits (1.7.9)–(1.7.11) for the drift $a(t, x)$ and diffusion coefficient $b(t, x)$. We shall prove this below under assumptions A2–A4 and the additional assumption that the coefficients of (6.3) are continuous (which implies A1, and is implied by the Lipschitz condition A2 when the differential equation is autonomous). A weakening of assumptions A2–A4 is possible provided the existence of unique solutions still holds.

**Theorem 4.6.1**    *Assume that $a$ and $b$ are continuous and that A2-A4 hold. Then the solution $X_t$ of (6.3) for any fixed initial value $X_{t_0}$ is a diffusion process on $[t_0, T]$ with drift $a(t, x)$ and diffusion coefficient $b(t, x)$.*

**Proof**    From inequality (5.13) with $n = 2$ we have

$$E\left(|X_t(s, x) - x|^4\right) \leq C\, |t - s|^2$$

for $t_0 \leq s \leq t \leq T$ and some constant $C$ which depends on $t_0$, $T$ and $x$. Hence for any $\epsilon > 0$

$$\int_{|y-x|>\epsilon} P(s, x; t, dy) \;\leq\; \epsilon^{-4} \int_{\Re} |y - x|^4\, P(s, x; t, dy)$$

$$\leq\; C\epsilon^{-4}\, |t - s|^2,$$

so

$$\lim_{t \downarrow s} \frac{1}{t - s} \int_{|y-x|>\epsilon} P(s, x; t, dy) = 0,$$

which is the first of the three required limits (1.7.9)-(1.7.11). To verify the other two limits it suffices for us to show that

$$(6.4) \qquad \lim_{t \downarrow s} \frac{1}{t - s} E\left(X_t(s, x) - x\right) = a(s, x)$$

and

$$(6.5) \qquad \lim_{t \downarrow s} \frac{1}{t - s} E\left(|X_t(s, x) - x|^2\right) = b^2(s, x).$$

Taking the expectation of the integral equation form of (6.3) for the solution $X_t(s, x)$ and using property (3.2.9) of Ito integrals, we obtain

$$E\left(X_t(s, x) - x\right) \;=\; \int_s^t E\left(a(u, X_u(s, x))\right) du$$

$$= (t-s) \int_0^1 E\left(a(s+v(t-s), X_{s+v(t-s)}(s,x))\right) \, dv.$$

From the continuity of the sample paths of $X_t(s,x)$, almost surely, and the continuity of $a$ we have

$$\lim_{t\downarrow s} a\left(s+v(t-s), X_{s+v(t-s)}(s,x;\omega)\right) = a(s,x),$$

w.p.1. In addition, from the growth bound A3

$$\left|a(s+v(t-s), X_{s+v(t-s)}(s,x))\right|^2 \le K^2 \left(1 + \left|X_{s+v(t-s)}(s,x)\right|^2\right)$$

and the mean-square boundedness of the solution, we have

$$E \int_0^1 \left|a\left(s+v(t-s), X_{s+v(t-s)}(s,x)\right)\right|^2 \, dv < \infty.$$

Hence by a theorem of Lebesgue on the interchange of limit and integration operations we conclude that

$$\int_0^1 E\left(a\left(s+v(t-s), X_{s+v(t-s)}(s,x)\right)\right) \, dv \to a(s,x)$$

as $t \downarrow s$, from which (6.4) then follows.

The remaining limit (6.5) is established in a similar way after first using the Ito formula to obtain a stochastic integral equation for $(X_t(s,x) - x)^2$. The details are left to the interested reader as an exercise. $\square$

In simple cases we can verify directly that the solution of a stochastic differential equation is a diffusion process. For example, when the coefficients of (6.3) are both constants the solution is

$$X_t = X_{t_0} + a(t-s) + b(W_t - W_s), \qquad \text{so}$$

$$E(X_t - x | X_s = x) = a(t-s) \qquad \text{and}$$

$$E\left((X_t - x)^2 | X_s = x\right) = b^2(t-s) + a^2(t-s)^2.$$

Consequently the transition probabilities $P(s,x;t,\cdot)$ are Gaussian with means $x + a(t-s)$ and variances $b^2(t-s)$. If the initial value $X_{t_0}$ is Gaussian, then $X_t$ is Gaussian and the solution process is a Gaussian process, which is often called a *Gauss-Markov process*.

**Exercise 4.6.2** *Show directly that the solution of the Langevin equation*

$$dX_t = -X_t \, dt + dW_t$$

*is a diffusion process when its initial value $X_0$ is deterministic.*

**Exercise 4.6.3** *Prove (6.5).*

## 4.7   Diffusion Processes as Weak Solutions

We shall now turn to the converse problem of whether or not a given diffusion process with probability density satisfying the Fokker-Planck equation is the solution of some stochastic differential equation. Closely related to this is the question of the existence of weak solutions of a stochastic differential equation. From Section 1 we recall that a *weak solution* is a solution of a stochastic differential equation for which the coefficients of the equation, but not the Wiener process, are specified. For a diffusion process these coefficients are obtained from the drift $a(t, x)$ and diffusion term $\sigma(t, x) = b^2(t, x)$ of the process, although, as indicated in Section 1, not without ambiguity. On a superficial level, the question has a simple affirmative answer: if $Y_t$ is the diffusion process with drift $a$ and diffusion term $\sigma = b^2$, or even if just these coefficients are specified, then we can take the stochastic differential equation

$$(7.1) \qquad dX_t = a(t, X_t)\, dt + b(t, X_t)\, dW_t$$

with the initial value $X_{t_0} = Y_{t_0}$, for any Wiener process $W = \{W_t, t \geq 0\}$. Provided the coefficients satisfy appropriate properties, such as assumptions A1–A3 of Theorem 4.5.3, there will be a solution $X$ for each Wiener process. Such a solution is then an equivalent stochastic process to the given diffusion process $Y$, that is it has the same probability law. In general, however, it will not be sample path equivalent to $Y_t$. To guarantee this we must choose the Wiener process with more care.

Let $Y$ be a given diffusion process on $[0, T]$ with drift $a(t, y)$ and strictly positive diffusion coefficient $b(t, y)$. Under assumptions which we shall specify later we define functions $g$ and $\bar{a}$ by

$$(7.2) \qquad g(t, y) = \int_0^y \frac{dx}{b(t, x)}$$

and

$$(7.3) \qquad \bar{a}(t, z) = \left( \frac{\partial g}{\partial t} + a \frac{\partial g}{\partial y} + \frac{1}{2} b^2 \frac{\partial^2 g}{\partial y^2} \right) \left( t, g^{-1}(t, z) \right)$$

with $a$ and $b$ evaluated at $(t, y)$, where $y = g^{-1}(t, z)$ is the inverse of $z = g(t, y)$. Then we define a process $Z_t = g(t, Y_t)$, which is a diffusion process with drift $\bar{a}(t, z)$ and diffusion coefficient 1, and a process

$$(7.4) \qquad \tilde{W}_t = Z_t - Z_0 - \int_0^t \bar{a}(s, Z_s)\, ds,$$

which will turn out to be a Wiener process. Consequently (7.4) will be equivalent to the stochastic differential equation

$$dZ_t = \bar{a}(t, Z_t)\, dt + 1\, d\tilde{W}_t,$$

which, by (7.2), (7.3) and Ito's formula, will imply that $Y_t$ is a solution of the stochastic differential equation

$$(7.5) \qquad dY_t = a(t, Y_t)\, dt + b(t, Y_t)\, d\tilde{W}_t,$$

that is of (7.1) with the Wiener process $\tilde{W}_t$. In order to justify these steps we need to impose some regularity conditions on the diffusion process $Y_t$ and its coefficients $a$ and $b$.

**Theorem 4.7.1**    *Let $Y$ be a diffusion process on $[0, T]$ with coefficients $a(t, y)$ and $b(t, y) > 0$ satisfying for all $y \in \Re$ and $t \in [0, T]$ the following conditions:*

*(1)   $a(t, y)$ is continuous in both variables and*

$$|a(t, y)| \leq K (1 + |y|)$$

*for some positive constant $K$;*

*(2)    $b(t, y)$ is continuous in both variables, $b(t, y)^{-1}$ is bounded, and the partial derivatives $\frac{\partial b}{\partial t}$ and $\frac{\partial b}{\partial y}$ are continuous and bounded;*

*(3)    There exists a function $\psi(y) > 1 + |y|$ such that*

*(i)*                    $\sup_{0 \leq t \leq T} E\left(\psi(Y_t)\right) < \infty$,

*(ii)*      $E\left(|Y_t - Y_s| \mid Y_s = y\right) + E\left(|Y_t - Y_s|^2 \mid Y_s = y\right) \leq (t - s) \psi(y)$,

*(iii)*          $E\left(|Y_t| \mid Y_s = y\right) + E\left(|Y_t|^2 \mid Y_s = y\right) \leq \psi(y)$.

*Then $\tilde{W}$ defined by (7.4) is a Wiener process and $Y$ is a solution of the stochastic differential equation (7.5).*

**Proof**   We shall first show that $Z_t = g(t, Y_t)$ is a diffusion process with drift $\bar{a}$ and diffusion coefficient 1. For this we use a change of variables argument. Since $g$ is monotone in $y$, $Z_t$ is a Markov process and its transition probabilities are

$$\tilde{P}(s, z; t, A) = P\left(s, g^{-1}(s, z); t, g^{-1}(t, A)\right),$$

where the $P(s, y; t, B)$ are the transition probabilities of the given diffusion process $Y_t$ and $g^{-1}(t, A) = \{y : g(t, y) \in A\}$. From the assumptions the function $z = g(t, y)$ and its inverse $y = g^{-1}(t, z)$ are continuous, so with $y = g^{-1}(t, z)$, $\eta = g^{-1}(t, \xi)$ and $\delta = \delta(\epsilon) > 0$ we have

$$\lim_{t \downarrow s} \frac{1}{t - s} \int_{|z - \xi| > \delta} P\left(s, g^{-1}(t, z); t, g^{-1}(t, d\xi)\right)$$

$$= \lim_{t \downarrow s} \frac{1}{t - s} \int_{|g(s, y) - g(t, \eta)| > \delta} P(s, y; t, d\eta) = 0.$$

This is the first requirement (1.7.9) on the transition probabilities for the process $Z_t$ to be a diffusion process. The other two, (1.7.10) and (1.7.11), are also inherited by $Z_t$ from the original diffusion process $Y_t$. By the assumptions of the theorem, $g(t, y)$ is continuously differentiable in $t$ and twice continuously differentiable in $y$. Also by the Taylor Theorem there exist numbers $\theta_1, \theta_2 \in (0, 1)$ such that

(7.6) $$\frac{1}{t-s}\int_{|z-\xi|\leq\delta(\epsilon)}(\xi-z)P\left(s,g^{-1}(t,z);t,g^{-1}(t,d\xi)\right)$$

$$=\frac{1}{t-s}\int_{|g(s,y)-g(t,\eta)|\leq\delta(\epsilon)}(g(t,\eta)-g(s,y))\,P(s,y;t,d\eta)$$

$$=\frac{\partial g}{\partial s}(s+\theta_1(t-s),y)\int_{|g(s,y)-g(t,\eta)|\leq\delta(\epsilon)}P(s,y;t,d\eta)$$

$$+\frac{1}{t-s}\frac{\partial g}{\partial y}(s,y)\int_{|g(s,y)-g(t,\eta)|\leq\delta(\epsilon)}(\eta-y)P(s,y;t,d\eta)$$

$$+\frac{1}{t-s}\int_{|g(s,y)-g(t,\eta)|\leq\delta(\epsilon)}\frac{1}{2}\frac{\partial^2 g}{\partial y^2}(s+\theta_2(t-s),y)\,(\eta-y)^2P(s,y;t,d\eta).$$

The first integral in the last part of (7.6) can be written as

$$\int_{|g(s,y)-g(t,\eta)|\leq\delta(\epsilon)}P(s,y;t,d\eta)=1-\int_{|g(s,y)-g(t,\eta)|>\delta(\epsilon)}P(s,y;t,d\eta),$$

and thus converges to 1 as $t \downarrow s$ by the limit (1.7.9) for the transition probabilities of the given diffusion process $Y_t$ and the continuity of the function $g$. Similarly, by choosing $\epsilon$ sufficiently small, we can make $\frac{\partial^2 g}{\partial y^2}(s+\theta_2(t-s),y)$ arbitrarily close to $\frac{\partial^2 g}{\partial y^2}(s,y)$. Finally, from the limits (1.7.10) and (1.7.11) for the transition probabilities of $Y_t$ we see that the limit as $t \downarrow s$ of the right side of (7.6) equals

$$\left(\frac{\partial g}{\partial t}+a\frac{\partial g}{\partial y}+\frac{1}{2}b^2\frac{\partial^2 g}{\partial y^2}\right)\left(s,g^{-1}(t,z)\right),$$

which is just $\bar{a}(s,y)$ defined in (7.3). Thus the requirement (1.7.10) of a diffusion process is satisfied by the transition probabilities of $Z_t$. The final requirement (1.7.11) is satisfied with the diffusion coefficient 1 since, similarly to above, we can show that

$$\frac{1}{t-s}\int_{|g(s,y)-g(t,\eta)|\leq\delta(\epsilon)}(\eta-y)^2P(s,y;t,d\eta)$$

$$=\left(\frac{\partial g}{\partial y}b\right)^2\left(s,g^{-1}(s,z)\right)\equiv 1,$$

since $\frac{\partial g}{\partial y}=1/b$ from (7.2). This shows that $Z_t$ is a diffusion process and that the process $\tilde{W}_t$ is well defined by (7.4).

We need now to show that $\tilde{W}_t$ is a Wiener process. Let $\mathcal{A}_t$ denote the $\sigma$-algebra generated by $Z_s$ for $0\leq s\leq t$. By (7.4) $\tilde{W}_t$ is $\mathcal{A}_t$-measurable for each $t\in[0,T]$. Obviously $\tilde{W}_0=0$, w.p.1, and $\tilde{W}_t$ has, almost surely, continuous sample paths. Defining $\tilde{\psi}(t,z)=\psi(g^{-1}(t,z))$, we have by assumption (3i) that

$$\sup_{0\leq t\leq T}E\left(\tilde{\psi}(t,Z_t)\right)=\sup_{0\leq t\leq T}E\left(\psi(Y_t)\right)<\infty.$$

Then by the Taylor Theorem there are constants $\theta_1, \theta_2 \in (0,1)$ such that

$$E\left(Z_{t+h} - Z_t | Z_t\right) = E\left(g(t+h, Y_{t+h}) - g(t, Y_t) | Y_t\right)$$

$$= E\left(\frac{\partial g}{\partial t}(t + \theta_1 h, Y_t) | Y_t\right) h + E\left(\frac{\partial g}{\partial y}(t, Y_t)(Y_{t+h} - Y_t) | Y_t\right)$$

$$+ \frac{1}{2} E\left(\frac{\partial^2 g}{\partial y^2}(t, Y_t + \theta_2(Y_{t+h} - Y_t))\,(Y_{t+h} - Y_t)^2 | Y_t\right).$$

From the assumptions on the coefficient $b$ and the definition of function $g$, we have $|g(t, y)| \leq C|y|$ for some constant $C$ and, in addition, that the partial derivatives $\frac{\partial g}{\partial y}$ and $\frac{\partial^2 g}{\partial y^2}$ are bounded. Then from assumption (3) we obtain

$$E\left(Z_{t+h} - Z_t | Z_t\right) \leq C_1 \tilde{\psi}(t, Z_t) h$$

for some constant $C_1$. Using this, the definition (7.4) of $\tilde{W}_t$ and a growth bound on the coefficient $\bar{a}$, we have

$$\left| E\left(\tilde{W}_{t+h} - \tilde{W}_t | \mathcal{A}_t\right) \right|$$

$$= \left| E\left(Z_{t+h} - Z_t | Z_t\right) - \int_t^{t+h} E\left(\bar{a}(s, Z_s) | \mathcal{A}_s\right) ds \right|$$

$$\leq C_1 \tilde{\psi}(t, Z_t) h + C_2 \int_t^{t+h} E\left(1 + |Y_s| \,|\, \mathcal{A}_t\right) ds$$

$$\leq C_3 \left(\tilde{\psi}(t, Z_t) + 1\right) h$$

for appropriate constants $C_2$ and $C_3$. Hence

(7.7) $$\lim_{h \downarrow 0} E\left(\tilde{W}_{t+h} - \tilde{W}_t | \mathcal{A}_t\right) = 0.$$

Similarly, we can show that

$$E\left(\left(\tilde{W}_{t+h} - \tilde{W}_t\right)^2 | \mathcal{A}_t\right) \leq C_4 \tilde{\psi}(t, Z_t) h$$

and also that

(7.8) $$\lim_{h \downarrow 0} \frac{1}{h} E\left(\left(\tilde{W}_{t+h} - \tilde{W}_t\right)^2 | \mathcal{A}_t\right) = 1.$$

Finally, it can be deduced from (7.7) and (7.8) that

$$E\left(\tilde{W}_{t+h} - \tilde{W}_t | \mathcal{A}_t\right) = 0 \quad \text{and} \quad E\left(\left(\tilde{W}_{t+h} - \tilde{W}_t\right)^2 | \mathcal{A}_t\right) = h,$$

w.p.1. Hence, by the theorem of Doob mentioned in Section 4 of Chapter 2, $\tilde{W}_t$ is a Wiener process with respect to the family of $\sigma$-algebras $\{\mathcal{A}_t,\, 0 \le t \le T\}$. This completes the proof of the theorem. $\square$

The rather strong conditions in the preceding theorem are sufficient, but not necessary for the existence of weak solutions. We remark that in some cases the SDE may have only weak solutions and no strong solutions.

**Example 4.7.2**    *The stochastic differential equation*

$$dX_t = \operatorname{sgn} X_t\, dt + dW_t,$$

*where* $\operatorname{sgn} x = +1$ *if* $x \ge 0$ *and* $-1$ *if* $x < 0$, *only has weak solutions, but no strong solution for the initial value* $X_0 = 0$. *In fact, if* $X_t$ *is such a weak solution for the Wiener process* $W$, *then* $-X_t$ *is a weak solution for the Wiener process* $-W$. *These solutions have the same probability law, but not the same sample paths.*

Theorem 4.7.1 tells us that there is a Wiener process such that the given diffusion process satisfies a stochastic differential equation with its drift and diffusion coefficient and with this particular Wiener process. In the proof we used not only the drift and diffusion coefficient, but also the diffusion process itself. If we know only the coefficients we can determine the diffusion process by a method proposed by Stroock and Varadhan connected with the *martingale problem*. This requires us to solve the equation

$$(7.9) \qquad f(X_t) - f(X_0) + \int_0^t \mathcal{L}f(X_s)\, ds + M_t = 0$$

where $f$ is any twice continuously differentiable function vanishing outside of a bounded interval and $M_t$ is some continuous (locally-) square integrable martingale and $\mathcal{L}$ is the elliptic operator (2.4.6) in the Kolmogorov backward equation with the given drift and diffusion coefficients. It can be shown under suitable assumptions that (7.9) has a solution which is unique in law for all such functions $f$.

**Exercise 4.7.3**    *Use the Ito formula to derive equation (7.9) for the Ornstein-Uhlenbeck process with drift* $a(x) = -x$ *and diffusion coefficient* $b(x) \equiv \sqrt{2}$. *What is the martingale* $M_t$ *here?*

# 4.8    Vector Stochastic Differential Equations

The relationship between vector and scalar stochastic differential equations is analogous to that between vector and scalar stochastic differentials. In what follows we interpret a vector as a column vector and its transpose as a row vector. We consider an $m$-dimensional Wiener process $W = \{W_t, t \ge 0\}$ with components $W_t^1, W_t^2, \ldots, W_t^m$, which are independent scalar Wiener processes with respect to a common family of $\sigma$-algebras $\{\mathcal{A}_t, t \geq 0\}$. Then we take a $d$-dimensional vector function $a : [0,T] \times \Re^d \to \Re^d$ and a $d \times m$-matrix function $b : [0,T] \times \Re^d \to \Re^{d \times m}$ to form a $d$-dimensional *vector stochastic differential equation*

$$(8.1) \qquad\qquad dX_t = a(t, X_t)\, dt + b(t, X_t)\, dW_t.$$

We interpret this as a stochastic integral equation

$$(8.2) \qquad\qquad X_t = X_{t_0} + \int_{t_0}^{t} a(s, X_s)\, ds + \int_{t_0}^{t} b(s, X_s)\, dW_s$$

where the Lebesgue and Ito integrals are determined component by component, with the $i$th component of (8.2) being

$$X_t^i = X_{t_0}^i + \int_{t_0}^{t} a^i(s, X_s)\, ds + \sum_{j=1}^{m} \int_{t_0}^{t} b^{i,j}(s, X_s)\, dW_s^j.$$

Analogous definitions of strong and weak solutions apply here, with the resulting process $X_t$ required to be $\mathcal{A}_t$-measurable. The existence and uniqueness theorem for strong solutions, Theorem 4.5.3, carries over verbatim to the vector case provided the absolute values in the assumptions and proof are replaced by vector and matrix norms, such as the Euclidean norms (1.4.36). Theorems 4.5.4 and 4.5.5 on estimates for the $2n$th order moments and on the continuity in a parameter also carry over to the vector context.

Vector stochastic differential equations arise naturally in systems described by vector valued states, some examples of which will be presented in Chapter 7. They also occur when certain scalar equations not in the form (8.1) are reformulated to have this form. For example, when the coefficients of a scalar equation also depend explicitly on $W_t$ as in

$$dY_t = a(t, Y_t, W_t)\, dt + b(t, Y_t, W_t)\, dW_t,$$

we can rewrite the equation as the 2-dimensional vector SDE

$$dX_t = \begin{pmatrix} a(t, X_t) \\ 0 \end{pmatrix} dt + \begin{pmatrix} b(t, X_t) \\ 1 \end{pmatrix} dW_t$$

with state components $X_t^1 = Y_t$ and $X_t^2 = W_t$. Similarly, we can reformulate a second order differential equation disturbed by white noise $\xi_t$ such as

$$\ddot{Y}_t = a(t, Y_t, \dot{Y}_t) + b(t, Y_t, \dot{Y}_t)\, \xi_t$$

as the 2-dimensional vector stochastic differential equation

$$dX_t = \begin{pmatrix} X_t^2 \\ a(t, X_t) \end{pmatrix} dt + \begin{pmatrix} 0 \\ b(t, X_t) \end{pmatrix} dW_t$$

with state components $X_t^1 = Y_t$ and $X_t^2 = \dot{Y}_t$. We observe in this case that the scalar process $Y_t$ is continuously differentiable whereas its derivative $\dot{Y}_t$ is only continuous, so the vector process $X_t = (X_t^1, X_t^2)$ is continuous, but not differentiable.

Another situation in which vector stochastic differential equations arise is when processes are restricted to lie on certain manifolds, such as the unit circle $S^1$ which is a 1-dimensional compact manifold. We saw in Section 4 of Chapter 3 that a Wiener process on $S^1$ satisfies the vector differential equation

$$dX_t = -\frac{1}{2} X_t \, dt + \begin{bmatrix} 0 & -1 \\ 1 & 0 \end{bmatrix} X_t \, dW_t,$$

with the constraint $|X_t| = 1$, where $W_t$ is a scalar Wiener process.

For a sufficiently smooth transformation $U : [0, T] \times \Re^d \to \Re^k$ of the solution $X_t$ of (8.1) we obtain a $k$-dimensional process $Y_t = U(t, X_t)$. This process will have a vector stochastic differential which can be determined by applying the Ito formula (3.4.6) to each component. The resulting expression is more transparent in component form

$$(8.3) \quad dY_t^p = \left( \frac{\partial U^p}{\partial t} + \sum_{i=1}^d a^i \frac{\partial U^p}{\partial x_i} + \frac{1}{2} \sum_{i,j=1}^d \sum_{l=1}^m b^{i,l} b^{j,l} \frac{\partial^2 U^p}{\partial x_i \partial x_j} \right) dt$$

$$+ \sum_{l=1}^m \sum_{i=1}^d b^{i,l} \frac{\partial U^p}{\partial x_i} \, dW_t^l$$

for $p = 1, 2, \ldots, k$ where the terms are all evaluated at $(t, X_t)$. As in the scalar case, we can use this formula to determine the solutions of certain vector stochastic differential equations in terms of known solutions of other equations, for example linear equations.

The general form of a $d$-dimensional *linear stochastic differential equation* is

$$(8.4) \quad dX_t = (A(t)X_t + a(t)) \, dt + \sum_{l=1}^m (B^l(t)X_t + b^l(t)) \, dW_t^l$$

where $A(t)$, $B^1(t)$, $B^2(t)$, ..., $B^m(t)$ are $d \times d$-matrix functions and $a(t)$, $b^1(t)$, $b^2(t)$, ..., $b^m(t)$ are $d$-dimensional vector functions. When the $B^l$ are all identically zero, we say that (8.4) is *linear in the narrow-sense* and when $a$ and the $b^l$ are all zero we call it a *homogeneous equation*. Duplicating the argument used for the scalar case in Section 2, we find that the solution of (8.4) is

$$(8.5) \quad X_t = \Phi_{t,t_0} \left( X_{t_0} + \int_{t_0}^t \Phi_{s,t_0}^{-1} \left( a(s) - \sum_{l=1}^m B^l(s) b^l(s) \right) ds \right.$$

$$\left. + \sum_{l=1}^m \int_{t_0}^t \Phi_{s,t_0}^{-1} b^l(s) \, dW_s^l \right),$$

where $\Phi_{t,t_0}$ is the $d \times d$ fundamental matrix satisfying $\Phi_{t_0,t_0} = I$ and the homogeneous *matrix stochastic differential equation*

(8.6)
$$d\Phi_{t,t_0} = A(t)\Phi_{t,t_0}\,dt + \sum_{l=1}^{m} B^l(t)\Phi_{t,t_0}\,dW_t^l,$$

which we interpret column vector by column vector as vector stochastic differential equations. Unlike the scalar homogeneous linear equations, we cannot generally solve (8.5) explicitly for its fundamental solution, even when all of the matrices are constant matrices. If, however, the matrices $A$, $B^1$, $B^2$, ..., $B^m$ are constants and commute, that is if

(8.7)
$$AB^l = B^l A \quad \text{and} \quad B^l B^k = B^k B^l$$

for all $k$, $l = 1, 2, \ldots, m$, then we obtain the following explicit expression for the fundamental matrix solution

$$\Phi_{t,t_0} = \exp\left(\left(A - \frac{1}{2}\sum_{l=1}^{m}\left(B^l\right)^2\right)(t - t_0) + \sum_{l=1}^{m} B^l\left(W_t^l - W_{t_0}^l\right)\right).$$

In the special case that (8.4) is linear in the narrow-sense this reduces to

(8.8)
$$\Phi_{t,t_0} = \exp\left(A(t - t_0)\right),$$

which is the fundamental matrix of the deterministic linear system

$$\dot{x} = Ax.$$

In such autonomous cases $\Phi_{t,t_0} = \Phi_{t-t_0,0}$, so we need only to consider $t_0 = 0$ and can write $\Phi_t$ for $\Phi_{t,0}$.

**Exercise 4.8.1** *Solve the random oscillator equation*

$$dX_t = \begin{bmatrix} 0 & -1 \\ 1 & 0 \end{bmatrix} X_t\,dt + \begin{pmatrix} 0 \\ \sigma^2 \end{pmatrix} dW_t,$$

*where $X_t = (X_t^1, X_t^2)$ and $W_t$ is a scalar Wiener process, by first showing that the fundamental matrix is*

$$\Phi_t = \begin{bmatrix} \cos t & -\sin t \\ \sin t & \cos t \end{bmatrix}.$$

**Exercise 4.8.2**   *Verify (8.8).*

In the same way as for scalar linear SDEs in Section 2, we can derive vector and matrix ordinary differential equations for the vector mean $m(t) = E(X_t)$ and the $d \times d$ matrix second moment $P(t) = E\left(X_t X_t^\mathsf{T}\right)$ of a general vector linear SDE (8.4). Recall that for $d$-dimensional vectors $x$ and $y$, the product $xy^\mathsf{T}$ is a $d \times d$ matrix with $ij$th component $x^i y^j$. Thus $P(t)$ is a symmetric matrix. We then obtain

$$(8.9) \qquad \frac{dm}{dt} = A(t)m + a(t)$$

and

$$(8.10) \qquad \frac{dP}{dt} = A(t)P + PA(t)^{\mathsf{T}} + \sum_{l=1}^{m} B^l(t)PB^l(t)^{\mathsf{T}}$$

$$+ a(t)m(t)^{\mathsf{T}} + m(t)a(t)^{\mathsf{T}}$$

$$+ \sum_{l=1}^{m} \left( B^l(t)m(t)b^l(t)^{\mathsf{T}} + b^l(t)m(t)^{\mathsf{T}} B^l(t) + b^l(t)b^l(t)^{\mathsf{T}} \right),$$

with initial conditions $m(t_0) = E\left(X_{t_0}\right)$ and $P(t_0) = E\left(X_{t_0}X_{t_0}^{\mathsf{T}}\right)$.

**Exercise 4.8.3**   *Derive equations (8.9) and (8.10).*

**Exercise 4.8.4**   *Solve the ordinary differential equations (8.9) and (8.10) corresponding to the 2-dimensional linear SDE*

$$dX_t = -2X_t\, dt + \begin{bmatrix} 0 & -1 \\ 1 & 0 \end{bmatrix} X_t\, dW_t,$$

*where $X_t = (X_t^1, X_t^2)^{\mathsf{T}}$ and $W_t$ is a scalar Wiener process.*

**Exercise 4.8.5**   *Solve the scalar SDE*

$$dX_t = -\frac{1}{2}X_t\, dt + X_t\, dW_t^1 + X_t\, dW_t^2,$$

*where $W_t^1$ and $W_t^2$ are independent scalar Wiener processes.*

In the remaining part of this section we shall list several important results which we shall use later. The reader may wish to omit it now and refer back to it when the results are first used.

The solution $X_t$ of a vector stochastic differential equation (8.1) is a Markov process. Under smoothness conditions on its coefficients, like those in the Existence and Uniqueness Theorem 4.5.3, $X_t$ is also a diffusion process with drift vector $a(t, x)$ and $d \times d$ diffusion matrix $D(t, x) = b(t, x)b(t, x)^{\mathsf{T}}$, or in component form $d^{i,j}(t, x) = \sum_{l=1}^{m} b^{i,l}(t, x)b^{j,l}(t, x)$ for $i, j = 1, 2, \ldots, d$. The transition probabilities then satisfy the Kolmogorov backward equation with these coefficients. Generally, the diffusion matrices $D$ are only positive semi-definite rather than positive definite, and this may lead to singularities in the transition densities. Many difficulties can thus arise in functional analytical investigations of these partial differential equations, but, as we shall see, these can often be circumvented by the use of probabilistic methods. An important situation where this happens is in the verification of that the *Kolmogorov formula*

$$(8.11) \qquad u(s, x) = E\left(f(X_T)|X_s = x\right)$$

gives a solution of the Kolmogorov backward equation (2.4.7)

(8.12)
$$\frac{\partial u}{\partial s} + \mathcal{L}u = 0$$

for $0 \leq s \leq T$, where $\mathcal{L}$ is the elliptic operator (2.4.6), with the *final condition*

(8.13)
$$u(T, x) = f(x)$$

for a sufficiently smooth function $f : \Re^d \to \Re$.

To be more specific, let $C^l(\Re^d, \Re)$ denote the space of $l$ times continuously differentiable functions $w : \Re^d \to \Re$ and $C_P^l(\Re^d, \Re)$ the subspace of functions $w \in C^l(\Re^d, \Re)$ for which all partial derivatives up to order $l$ have *polynomial growth*, that is for which there exist constants $K > 0$ and $r \in \{1, 2, 3, \ldots\}$, depending on $w$, such that

$$\left| \partial_y^j w(y) \right| \leq K \left( 1 + |y|^{2r} \right)$$

for all $y \in \Re^d$ and any partial derivative $\partial_y^j w$ of order $j \leq l$. Then we have the following theorem due to Mikulevicius, which we shall use in the chapters on weak approximations.

**Theorem 4.8.6**    *Suppose that $f \in C_P^{2(\gamma+1)}(\Re^d, \Re)$ for some $\gamma = 1, 2, 3, \ldots$ and that $X_t$ is a homogeneous diffusion process for which the drift vector and diffusion matrix components $a^i$, $b^{ij} \in C_P^{2(\gamma+1)}(\Re^d, \Re)$ with uniformly bounded derivatives. Then $u : [0, T] \times \Re^d \to \Re$ defined by (8.11) satisfies the final value problem (8.12)–(8.13) with $\frac{\partial u}{\partial s}$ continuous and $u(s, \cdot) \in C_P^{2(\gamma+1)}(\Re^d, \Re)$ for each $0 \leq s \leq T$.*

We remark that the diffusion process may be degenerate here, that is with the diffusion matrix vanishing at various points in $\Re^d$. The Ito formula is used to show that the function $u$ given by (8.11) is a solution of the Kolmogorov backward equation (8.12). It can be used in a similar way under analogous smoothness assumptions to show that the *Feynman-Kac formula*

(8.14)
$$u(s, x) = E\left( f(X_T) \exp\left( \int_s^T g(X_u)\, du \right) \middle| X_s = x \right),$$

where $g$ is a bounded function, is a solution of the partial differential equation

(8.15)
$$\frac{\partial u}{\partial s} + \mathcal{L}u + gu = 0$$

with the final condition (8.13) and elliptic operator (2.4.6).

**Exercise 4.8.7**    *Use the Ito formula to verify that (8.11) satisfies (8.12).*

In nonlinear filtering and other applications we often encounter a "drifted" Wiener process $X$ on a probability space $(\Omega, \mathcal{A}, P)$ with a filtration $\mathcal{A}^* = \{\mathcal{A}_t, t \in [0, T]\}$. This is defined by

(8.16)
$$X_t = X_0 + \int_0^t A(s)\, ds + W_t$$

for $t \in [0, T]$, where $A$ is an $\mathcal{A}^*$-adapted, right-continuous process and $W$ is an $\mathcal{A}^*$-adapted Wiener process with respect to the probability measure $P$. In general $X$ is not a Wiener process with respect to the given probability measure, but it is sometimes useful to interpret it as one with respect to another probability measure. We can do this by using the *Girsanov transformation* to transform the underlying probability measure $P$ on the canonical sample space $\Omega = \mathcal{C}_0([0, T], \Re)$ to an absolutely continuous probability measure $P_X$ with the Radon-Nikodym derivative

$$(8.17) \qquad \frac{dP_X}{dP} = \exp\left( \int_0^T A(s) \, dX_s - \frac{1}{2} \int_0^T |A(s)|^2 \, ds \right).$$

The $\mathcal{A}^*$-adapted process $X$ turns out then to be a Wiener process on the canonical probability space $(\Omega, \mathcal{A}, P_X)$. A heuristic justification of the formula (8.17) will be given in Section 4 of Chapter 6.

We have already mentioned an ergodic theorem for Markov chains in Section 6 of Chapter 1 (see (1.6.22)). An analogous result also holds for Ito diffusions. We say that a $d$-dimensional Ito process $X = \{X_t, t \geq 0\}$ is *ergodic* if it has a unique invariant probability law $\mu$ such that

$$(8.18) \qquad \lim_{t \to \infty} \frac{1}{t} \int_0^t f\left( X_s^{0,x} \right) \, ds = \int_{\Re^d} f(y) \, d\mu(y),$$

w.p.1, for any $\mu$-integrable function $f : \Re^d \to \Re$ and any deterministic initial condition $X_0 = x$. The following sufficient condition for a process to be ergodic, which is quite strong, is due to Hasminski.

**Theorem 4.8.8**    *Suppose that the drift $a$ and diffusion coefficient $b$ of an autonomous Ito process $X$ are smooth with bounded derivatives of any order, $b$ is bounded and there exists a constant $\beta > 0$ and a compact subset $K \subset \Re^d$ such that*

$$(8.19) \qquad x^\top a(x) \leq -\beta |x|^2$$

*for all $x \in \Re^d \setminus K$. Then $X$ is ergodic.*

**Exercise 4.8.9**    *Show that the 1-dimensional Ito process with drift and diffusion coefficients $a(x) = 1 - x$ and $b(x) \equiv 1$ is ergodic.*

# 4.9    Stratonovich Stochastic Differential Equations

In Chapter 3 we saw that a stochastic differential equation

$$(9.1) \qquad dX_t = a(t, X_t) \, dt + b(t, X_t) \, dW_t$$

must be interpreted mathematically as a stochastic integral equation

$$(9.2) \qquad X_t = X_{t_0} + \int_{t_0}^t a(s, X_s) \, ds + \int_{t_0}^t b(s, X_s) \, dW_s$$

where the second integral is a stochastic integral. Apart from a brief discussion in Section 5 of Chapter 3, we have so far taken a stochastic integral to be an Ito stochastic integral, in which case we could call (9.1) an *Ito stochastic differential equation*.

With different choices of stochastic integrals we would generally obtain different solutions for the above integral equation, which would justify our saying that we have different stochastic differential equations, even though they have the same coefficients. Of all these integrals, the one proposed by Stratonovich, in which the integrand is evaluated, essentially, at the midpoint $\frac{1}{2}(t_j^{(n)} + t_{j+1}^{(n)})$ of each partition subinterval $[t_j^{(n)}, t_{j+1}^{(n)}]$, is the most appealing because it alone satisfies the usual transformation rules of classical calculus. If we use the Stratonovich stochastic integral, it is appropriate to say that we have a *Stratonovich stochastic differential equation*. We shall denote it by

$$(9.3) \qquad dX_t = \underline{a}(t, X_t) \, dt + b(t, X_t) \circ dW_t$$

or

$$(9.4) \qquad X_t = X_{t_0} + \int_{t_0}^t \underline{a}(s, X_s) \, ds + \int_{t_0}^t b(s, X_s) \circ dW_s,$$

and reserve (9.1) and (9.2) for the Ito case.

In Section 5 of Chapter 3 we defined the Stratonovich integral only for integrands of the form $f(t, \omega)$ and $h(W_t(\omega))$. Since there is a slight ambiguity as to how this should be extended to integrands of the form $h(t, X_t(\omega))$ for a function $h = h(t, x)$ and a diffusion process $X_t$, we shall define the Stratonovich integral

$$(9.5) \qquad \int_0^T h(s, X_s(\omega)) \circ dW_s$$

to be the mean-square limit of the sums

$$(9.6) \qquad S_n(\omega) = \sum_{j=1}^N h\left(t_j^{(n)}, \frac{1}{2}\left(X_{t_j^{(n)}} + X_{t_{j+1}^{(n)}}\right)\right) \left\{W_{t_{j+1}^{(n)}} - W_{t_j^{(n)}}\right\}$$

for partitions $0 = t_1^{(n)} < t_2^{(n)} < \cdots < t_{n+1}^{(n)} = T$ with

$$\delta^{(n)} = \max_{1 \le j \le n} \left|t_{j+1}^{(n)} - t_j^{(n)}\right| \to 0 \quad \text{as} \quad n \to \infty.$$

Alternatively, averaging $h(t, X_t)$ over its values at the endpoints $t_j^{(n)}$ and $t_{j+1}^{(n)}$ will give the same value if $h$ is sufficiently smooth. To ensure that the limit exists, we need to impose some restrictions on $h$ and $X_t$. Let $W = \{W_t, t \ge 0\}$ be a Wiener process with an associated family $\{\mathcal{A}_t, t \ge 0\}$ of increasing $\sigma$-algebras. We shall suppose that $X_t$ is a diffusion process in $\Re$ for $0 \le t \le T$ with continuous drift $a = a(t, x)$ and diffusion coefficient $b = b(t, x)$, and that

$h : [0, T] \times \Re \to \Re$ is continuous in $t$ with the partial derivative $\frac{\partial h}{\partial x}$ continuous in both $t$ and $x$. Finally, we shall suppose that the function $f$ defined by $f(t, \omega) = h(t, X_t(\omega))$ belongs to the function space $\mathcal{L}_T^2$, defined in Section 2 of Chapter 3, which in particular requires $X_t$ to be $\mathcal{A}_t$-measurable for each $0 \le t \le T$ and

$$\int_0^T E\left(|h(t, X_t)|^2\right) dt < \infty.$$

Under these conditions the above limits exist and are unique, w.p.1, so the integral (9.5) is meaningful. Moreover the identity (3.5.5) with $\lambda = 1/2$ generalizes to

$$(9.7) \quad \int_0^T h(t, X_t) \circ dW_t = \int_0^T h(t, X_t)\, dW_t + \frac{1}{2} \int_0^T b(t, X_t)\frac{\partial h}{\partial x}(t, X_t)\, dt$$

or, equivalently, in differential form

$$(9.8) \quad h(t, X_t) \circ dW_t = h(t, X_t)\, dW_t + \frac{1}{2} b(t, X_t)\frac{\partial h}{\partial x}(t, X_t)\, dt.$$

The proof is similar. Using the Taylor Theorem and the Mean Value Theorem, we have

$$h\left(t_j^{(n)}, \frac{1}{2}\left(X_{t_j^{(n)}} + X_{t_{j+1}^{(n)}}\right)\right) - h\left(t_j^{(n)}, X_{t_j^{(n)}}\right)$$

$$= \frac{1}{2}\frac{\partial h}{\partial x}\left(t_j^{(n)}, \frac{1}{2}\left((2 - \theta_j)X_{t_j^{(n)}} + \theta_j X_{t_{j+1}^{(n)}}\right)\right)\left\{X_{t_{j+1}^{(n)}} - X_{t_j^{(n)}}\right\}$$

for some (random) $\theta_j \in [0, 1]$ for $j = 1, 2, \ldots, n$. Since $X_t$ is a diffusion process

$$\Delta X_{t_j^{(n)}} = X_{t_{j+1}^{(n)}} - X_{t_j^{(n)}}$$

$$= a\left(t_j^{(n)}, X_{t_j^{(n)}}\right) \Delta t_j^{(n)} + b\left(t_j^{(n)}, X_{t_j^{(n)}}\right) \Delta W_{t_j^{(n)}}$$

$$+ \text{ higher order terms},$$

where $\Delta t_j^{(n)} = t_{j+1}^{(n)} - t_j^{(n)}$ and $\Delta W_{t_j^{(n)}} = W_{t_{j+1}^{(n)}} - W_{t_j^{(n)}}$ for $j = 1, 2, \ldots, n$.

Thus each term in the sum (9.6) equals

$$h\left(t_j^{(n)}, X_{t_j^{(n)}}\right) \Delta W_{t_j^{(n)}} + \frac{1}{2}\frac{\partial h}{\partial x}\bigg|_{\theta_j} \Delta X_{t_j^{(n)}} \Delta W_{t_j^{(n)}}$$

$$= h\left(t_j^{(n)}, X_{t_j^{(n)}}\right) \Delta W_{t_j^{(n)}} + \frac{1}{2}\frac{\partial h}{\partial x}\bigg|_{\theta_j} b\left(t_j^{(n)}, X_{t_j^{(n)}}\right) \left(\Delta W_{t_j^{(n)}}\right)^2$$

$$+ \frac{1}{2} \frac{\partial h}{\partial x} \bigg|_{\theta_j} a \left( t_j^{(n)}, X_{t_j^{(n)}} \right) \Delta t_j^{(n)} \Delta W_{t_j^{(n)}}$$

$$+ \text{ higher order terms,}$$

where

$$\frac{\partial h}{\partial x} \bigg|_{\theta_j} = \frac{\partial h}{\partial x} \left( t_j^{(n)}, \frac{1}{2} \left( (2 - \theta_j) X_{t_j^{(n)}} + \theta_j X_{t_{j+1}^{(n)}} \right) \right).$$

Taking into account the continuity of the coefficients and the facts that $E(\Delta W_{t_j^{(n)}})^2 = \Delta t_j^{(n)}$ and $E(\Delta t_j^{(n)} \Delta W_{t_j^{(n)}}) = 0$, we obtain (9.7) in the mean-square limit.

From property (3.2.9) of Ito integrals we have

$$(9.9) \qquad E \left( \int_0^T h(t, X_t) \, dW_t \right) = 0.$$

Such a property is not, generally, enjoyed by the Stratonovich integral. Taking expectations of both sides of (9.7) and using (9.9), we obtain

$$(9.10) \qquad E \left( \int_0^T h(t, X_t) \circ dW_t \right) = \frac{1}{2} \int_0^T E \left( b(t, X_t) \frac{\partial h}{\partial x}(t, X_t) \right) \, dt,$$

which need not be zero.

When the diffusion process $X_t$ satisfies an Ito SDE (9.1), we see from (9.8) with $h \equiv b$ that $X_t$ is also a solution of the Stratonovich SDE

$$(9.11) \qquad dX_t = \underline{a}(t, X_t) \, dt + b(t, X_t) \circ dW_t$$

with modified drift $\underline{a}$ defined by

$$\underline{a}(t, x) = a(t, x) - \frac{1}{2} b(t, x) \frac{\partial b}{\partial x}(t, x).$$

**Exercise 4.9.1**     *Let $Y_t = U(t, X_t)$, where $X_t$ is a solution of the Stratonovich SDE (9.3) and $U(t, x)$ is sufficiently smooth. Apply the Ito formula to the corresponding Ito SDE (9.1) and show that*

$$dY_t = \left( \frac{\partial U}{\partial t} + \underline{a} \frac{\partial U}{\partial x} \right) dt + b \frac{\partial U}{\partial x} \circ dW_t$$

$$= \frac{\partial U}{\partial t} \, dt + \underline{a} \frac{\partial U}{\partial x} \circ dX_t,$$

*where all of the functions are evaluated at $(t, X_t)$.*

The two stochastic differential equations (9.1) and (9.11) have the same coefficients if the diffusion coefficient $b$ is independent of $x$, that is when the noise appears additively. In general the drift terms will differ. For example, the diffusion process

$$X_t = X_{t_0} \exp\left(\left(a - \frac{1}{2}b^2\right)(t - t_0) + b\left(W_t - W_{t_0}\right)\right)$$

is a solution of the Ito SDE

(9.12)  $$dX_t = aX_t\, dt + bX_t\, dW_t$$

and of the Stratonovich SDE

$$dX_t = \left(a - \frac{1}{2}b^2\right)X_t\, dt + bX_t \circ dW_t.$$

If we start with the Stratonovich SDE (9.3), then the corresponding Ito SDE is

(9.13)  $$dX_t = a(t, X_t)\, dt + b(t, X_t)\, dW_t$$

with the drift modified to $a$ defined by

$$a(t, x) = \underline{a}(t, x) + \frac{1}{2}b(t, x)\frac{\partial b}{\partial x}(t, x).$$

For example, the diffusion process

$$X_t = X_{t_0} \exp\left(a(t - t_0) + b(W_t - W_{t_0})\right)$$

is a solution of the Stratonovich SDE

(9.14)  $$dX_t = aX_t\, dt + bX_t \circ dW_t$$

and of the Ito SDE

$$dX_t = \left(a + \frac{1}{2}b^2\right)X_t\, dt + bX_t\, dW_t.$$

The Ito and Stratonovich linear SDEs with the same drift and diffusion coefficients, (9.12) and (9.14) respectively, thus have distinct solutions. Moreover, these solutions may behave quite differently; for example, convergence to 0, w.p.1, as $t \to \infty$ holds for the Ito solution if $a < \frac{1}{2}b^2$, but only if $a < 0$ for the Stratonovich solution. How we interpret a stochastic differential equation may thus have critical consequences, which must be borne in mind when constructing models involving stochastic differential equations.

Whichever interpretation of an SDE is deemed appropriate in a particular situation, we can always switch to the corresponding SDE in the other interpretation when this is advantageous. For instance, we can use the existence and uniqueness results of Section 5 for an Ito SDE (9.13) to obtain analogous results for the corresponding Stratonovich SDE (9.11). Similarly we must use the corresponding Ito SDE to determine the appropriate coefficients of the Fokker-Planck equation for a diffusion process arising as the solution of a Stratonovich equation. On the other hand the Stratonovich stochastic calculus obeys the same transformation rules of classical calculus, so methods that have been developed to solve ordinary differential equations can sometimes be used successfully to solve Stratonovich SDEs.

**Exercise 4.9.2**   *Determine and solve the Stratonovich SDE correponding to the Ito SDE*

$$dX_t = 1\,dt + 2\sqrt{X_t}\,dW_t.$$

To conclude this section we state without proof the vector analogues of the above relationships between Ito and Stratonovich stochastic integrals and differential equations. In particular we take $X_t$ to be a solution of the vector Ito stochastic differential equation

$$(9.15) \qquad dX_t = a(t, X_t)\,dt + b(t, X_t)\,dW_t$$

where $a, X \in \Re^d$, $b \in \Re^{d \times m}$ and $W \in \Re^m$. For a $d \times m$ matrix valued function $h$ we define a $d$-dimensional vector valued function $c = c(t, X)$ componentwise by

$$c^i(t, X) = \frac{1}{2} \sum_{j=1}^{d} \sum_{k=1}^{m} b^{j,k}(t, X) \frac{\partial h^{i,k}}{\partial x_j}(t, X),$$

for $i = 1, 2, \ldots, d$. Then the Ito and Stratonovich integrals of $h(t, X_t)$ are related by

$$\int_0^T h(t, X_t) \circ dW_t = \int_0^T h(t, X_t)\,dW_t + \int_0^T c(t, X_t)\,dt$$

or in equivalent differential form

$$h(t, X_t) \circ dW_t = h(t, X_t)\,dW_t + c(t, X_t)\,dt.$$

With $h \equiv b$ it follows that the Stratonovich SDE corresponding to the Ito SDE (9.15) is

$$(9.16) \qquad dX_t = \underline{a}(t, X_t)\,dt + b(t, X_t) \circ dW_t$$

where the modified drift is defined componentwise by

$$
\begin{aligned}
\underline{a}^i(t, X) &= a^i(t, X) - c^i(t, X) \\
&= a^i(t, X) - \frac{1}{2} \sum_{j=1}^{d} \sum_{k=1}^{m} b^{j,k}(t, X) \frac{\partial b^{i,k}}{\partial x_j}(t, X)
\end{aligned}
$$

To obtain the Ito SDE corresponding to a given Stratonovich SDE we must modify the drift to

$$a(t, X) = \underline{a}(t, X) + c(t, X),$$

now adding rather than subtracting the correction term.

**Exercise 4.9.3**    *Determine the vector Stratonovich SDE describing a Wiener process on the unit circle, that is corresponding to the vector Ito SDE*

$$dX_t = -\frac{1}{2} X_t \, dt + \begin{bmatrix} 0 & -1 \\ 1 & 0 \end{bmatrix} X_t \, dW_t,$$

*with the constraint $|X_t| = 1$, where $W_t$ is a scalar Wiener process.*

Unlike in the scalar case, the vector Ito and Stratonovich SDEs may now have the same drift term when the diffusion coefficients depend on $X$.

**Example 4.9.4**    *The drift correction term $c(t, X) \equiv 0$ in 2-dimensional SDEs involving a scalar Wiener process occurs, for instance, when the diffusion coefficient $b = (b^1, b^2)$ satisfies*

$$b^1(x_1, x_2) = -b^2(x_1, x_2) = x_1 + x_2.$$

# Chapter 5

# Stochastic Taylor Expansions

In this chapter stochastic Taylor expansions are derived and investigated. They generalize the deterministic Taylor formula as well as the Ito formula and allow various kinds of higher order approximations of functionals of diffusion processes to be made. These expansions are the key to the stochastic numerical analysis which we shall develop in the second half of this book. Apart from Section 1, which provides an introductory overview, this chapter could be omitted at the first reading of the book.

## 5.1  Introduction

Deterministic Taylor expansions are well known. We shall review them here using terminology which will facilitate our presentation of their stochastic counterparts. To begin we shall consider the solution $X_t$ of a 1-dimensional ordinary differential equation

$$(1.1) \qquad \frac{d}{dt} X_t = a(X_t),$$

with initial value $X_{t_0}$, for $t \in [t_0, T]$ where $0 \le t_0 < T$. We can write this in the equivalent integral equation form

$$(1.2) \qquad X_t = X_{t_0} + \int_{t_0}^t a(X_s)\, ds.$$

To justify the following constructions we shall require the function $a$ to satisfy appropriate properties, for example to be sufficiently smooth and to have a linear growth bound. Let $f : \Re \to \Re$ be a continuously differentiable function. Then by the chain rule we have

$$(1.3) \qquad \frac{d}{dt} f(X_t) = a(X_t) \frac{\partial}{\partial x} f(X_t),$$

which, using the operator

$$(1.4) \qquad L = a \frac{\partial}{\partial x},$$

we can express as the integral relation

$$(1.5) \qquad f(X_t) = f(X_{t_0}) + \int_{t_0}^t Lf(X_s)\, ds$$

for all $t \in [t_0, T]$. When $f(x) \equiv x$ we have $Lf = a$, $L^2 f = La$, ... and (1.5) reduces to

$$(1.6) \qquad\qquad X_t = X_{t_0} + \int_{t_0}^t a(X_s) \, ds,$$

that is to equation (1.2).

If we now apply the relation (1.5) to the function $f = a$ in the integral in (1.6), we obtain

$$(1.7) \qquad X_t = X_{t_0} + \int_{t_0}^t \left( a(X_{t_0}) + \int_{t_0}^s La(X_z) \, dz \right) ds$$

$$= X_{t_0} + a(X_{t_0}) \int_{t_0}^t ds + \int_{t_0}^t \int_{t_0}^s La(X_z) \, dz \, ds,$$

which is the simplest nontrivial Taylor expansion for $X_t$. We can apply (1.5) again to the function $f = La$ in the double integral to derive

$$(1.8) \qquad X_t = X_{t_0} + a(X_{t_0}) \int_{t_0}^t ds + La(X_{t_0}) \int_{t_0}^t \int_{t_0}^s dz \, ds + R_3$$

with remainder

$$(1.9) \qquad R_3 = \int_{t_0}^t \int_{t_0}^s \int_{t_0}^z L^2 a(X_u) \, du \, dz \, ds,$$

for $t \in [t_0, T]$. For a general $r+1$ times continuously differentiable function $f : \Re \to \Re$ this method gives the classical *Taylor formula in integral form*:

$$(1.10) \qquad f(X_t) = f(X_{t_0}) + \sum_{l=1}^r \frac{(t - t_0)^l}{l!} L^l f(X_{t_0})$$

$$+ \int_{t_0}^t \cdots \int_{t_0}^{s_2} L^{r+1} f(X_{s_1}) \, ds_1 \ldots ds_{r+1}$$

for $t \in [t_0, T]$ and $r = 1, 2, 3, \ldots$.

The Taylor formula (1.10) has proven to be a very useful tool in both theoretical and practical investigations, particularly in numerical analysis. It allows the approximation of a sufficiently smooth function in a neighbourhood of a given point to any desired order of accuracy. This expansion depends on the values of the function and some of its higher derivatives at the expansion point, weighted by corresponding multiple time integrals. In addition, there is a remainder term which contains the next multiple time integral, but now with a time dependent integrand.

To expand the increments of smooth functions of Ito processes, for instance in the construction of numerical methods, it is advantageous to have stochastic expansion formulae with analogous properties to the deterministic Taylor formula. There are several possibilities for such a stochastic Taylor formula. One is based on the iterated application of the Ito formula (3.3.6), which we shall call the *Ito-Taylor expansion*. We shall indicate it here for the solution $X_t$ of the 1-dimensional Ito stochastic differential equation in integral form

$$(1.11) \qquad X_t = X_{t_0} + \int_{t_0}^t a(X_s)\,ds + \int_{t_0}^t b(X_s)\,dW_s$$

for $t \in [t_0, T]$, where the second integral in (1.11) is an Ito stochastic integral and the coefficients $a$ and $b$ are sufficiently smooth real-valued functions satisfying a linear growth bound. Then, for any twice continuously differentiable function $f : \Re \to \Re$ the Ito formula (3.3.6) gives

$$
\begin{aligned}
(1.12) \quad f(X_t) \;=\;& f(X_{t_0}) \\
&+ \int_{t_0}^t \left( a(X_s)\frac{\partial}{\partial x} f(X_s) + \frac{1}{2}b^2(X_s)\frac{\partial^2}{\partial x^2} f(X_s) \right) ds \\
&+ \int_{t_0}^t b(X_s)\frac{\partial}{\partial x} f(X_s)\,dW_s \\
=\;& f(X_{t_0}) + \int_{t_0}^t L^0 f(X_s)\,ds + \int_{t_0}^t L^1 f(X_s)\,dW_s,
\end{aligned}
$$

for $t \in [t_0, T]$. Here we have introduced the operators

$$(1.13) \qquad L^0 = a\frac{\partial}{\partial x} + \frac{1}{2}b^2\frac{\partial^2}{\partial x^2}$$

and

$$(1.14) \qquad L^1 = b\frac{\partial}{\partial x}.$$

Obviously, for $f(x) \equiv x$ we have $L^0 f = a$ and $L^1 f = b$, in which case (1.12) reduces to the original Ito equation for $X_t$, that is to

$$(1.15) \qquad X_t = X_{t_0} + \int_{t_0}^t a(X_s)\,ds + \int_{t_0}^t b(X_s)\,dW_s.$$

In analogy with the above deterministic expansions, if we apply the Ito formula (1.12) to the functions $f = a$ and $f = b$ in (1.15) we obtain

$$
\begin{aligned}
(1.16) \quad X_t \;=\;& X_{t_0} \\
&+ \int_{t_0}^t \left( a(X_{t_0}) + \int_{t_0}^s L^0 a(X_z)\,dz + \int_{t_0}^s L^1 a(X_z)\,dW_z \right) ds \\
&+ \int_{t_0}^t \left( b(X_{t_0}) + \int_{t_0}^s L^0 b(X_z)\,dz + \int_{t_0}^s L^1 b(X_z)\,dW_z \right) dW_s \\
=\;& X_{t_0} + a(X_{t_0}) \int_{t_0}^t ds + b(X_{t_0}) \int_{t_0}^t dW_s + R
\end{aligned}
$$

with remainder

$$R = \int_{t_0}^{t} \int_{t_0}^{s} L^0 a(X_z)\, dz\, ds + \int_{t_0}^{t} \int_{t_0}^{s} L^1 a(X_z)\, dW_z\, ds$$

$$+ \int_{t_0}^{t} \int_{t_0}^{s} L^0 b(X_z)\, dz\, dW_s + \int_{t_0}^{t} \int_{t_0}^{s} L^1 b(X_z)\, dW_z\, dW_s.$$

This is the simplest nontrivial Ito-Taylor expansion. We can continue it, for instance, by applying the Ito formula (1.12) to $f = L^1 b$ in (1.16), in which case we get

$$(1.17) \qquad X_t = X_{t_0} + a(X_{t_0}) \int_{t_0}^{t} ds + b(X_{t_0}) \int_{t_0}^{t} dW_s$$

$$+ L^1 b(X_{t_0}) \int_{t_0}^{t} \int_{t_0}^{s} dW_z\, dW_s + \bar{R}$$

with remainder

$$\bar{R} = \int_{t_0}^{t} \int_{t_0}^{s} L^0 a(X_z)\, dz\, ds + \int_{t_0}^{t} \int_{t_0}^{s} L^1 a(X_z)\, dW_z\, ds$$

$$+ \int_{t_0}^{t} \int_{t_0}^{s} L^0 b(X_z)\, dz\, dW_s + \int_{t_0}^{t} \int_{t_0}^{s} \int_{t_0}^{z} L^0 L^1 b(X_u)\, du\, dW_z\, dW_s$$

$$+ \int_{t_0}^{t} \int_{t_0}^{s} \int_{t_0}^{z} L^1 L^1 b(X_u)\, dW_u\, dW_z\, dW_s.$$

In Section 5 we shall formulate the Ito-Taylor expansion for a general function $f$ and arbitrarily high order. Nevertheless, its main properties are already apparent in the preceding example. In particular, we have an expansion with the multiple Ito integrals

$$\int_{t_0}^{t} ds, \qquad \int_{t_0}^{t} dW_s, \qquad \int_{t_0}^{t} \int_{t_0}^{s} dW_z\, dW_s$$

and a remainder term involving the next multiple Ito integrals, but now with nonconstant integrands. The Ito-Taylor expansion can, in this sense, be interpreted as a generalization of both the Ito formula and the deterministic Taylor formula.

We shall now consider another representation for the increments of a function of an Ito process $X$, which we shall call the *Stratonovich-Taylor expansion*. For this we start with the 1-dimensional Stratonovich stochastic differential equation in integral form

$$(1.18) \qquad X_t = X_{t_0} + \int_{t_0}^{t} \underline{a}(X_s)\, ds + \int_{t_0}^{t} b(X_s) \circ dW_s$$

for $t \in [t_0, T]$, where the second integral in (1.18) is a Stratonovich stochastic integral and the coefficients $\underline{a}$ and $b$ are sufficiently smooth real valued functions satisfying a linear growth bound. From Chapter 4 we recall that when

$$\underline{a} = a - \frac{1}{2} bb'$$

the Ito equation (1.11) and the Stratonovich equation (1.18) are equivalent and have the same solutions. We know from Exercise 4.9.1 that the solution of a Stratonovich SDE transforms according to the deterministic chain rule, so for any twice continuously differentiable function $f : \Re \rightarrow \Re$ we have

$$(1.19) \quad f(X_t) \; = \; f(X_{t_0}) + \int_{t_0}^t \underline{a}(X_s)\frac{\partial}{\partial x}f(X_s)\,ds$$

$$+ \int_{t_0}^t b(X_s)\frac{\partial}{\partial x}f(X_s) \circ dW_s$$

$$= \; f(X_{t_0}) + \int_{t_0}^t \underline{L}^0 f(X_s)\,ds + \int_{t_0}^t \underline{L}^1 f(X_s) \circ dW_s,$$

for $t \in [t_0, T]$, with the operators

$$(1.20) \qquad\qquad \underline{L}^0 = \underline{a}\frac{\partial}{\partial x}$$

and

$$(1.21) \qquad\qquad \underline{L}^1 = b\frac{\partial}{\partial x}.$$

Obviously, for $f(x) \equiv x$ we have $\underline{L}^0 f = \underline{a}$ and $\underline{L}^1 f = b$, in which case (1.19) reduces to the original Stratonovich equation

$$(1.22) \qquad X_t = X_{t_0} + \int_{t_0}^t \underline{a}(X_s)\,ds + \int_{t_0}^t b(X_s) \circ dW_s.$$

Analogously with the Ito case just considered, we can apply (1.19) to the integrand functions $f = \underline{a}$ and $f = b$ in (1.20). This gives

$$(1.23) \quad X_t \; = \; X_{t_0}$$

$$+ \int_{t_0}^t \left( \underline{a}(X_{t_0}) + \int_{t_0}^s \underline{L}^0\underline{a}(X_z)\,dz + \int_{t_0}^s \underline{L}^1\underline{a}(X_z) \circ dW_z \right)\,ds$$

$$+ \int_{t_0}^t \left( b(X_{t_0}) + \int_{t_0}^s \underline{L}^0 b(X_z)\,dz \right.$$

$$\left. + \int_{t_0}^s \underline{L}^1 b(X_z) \circ dW_z \right) \circ dW_s$$

$$= \; X_{t_0} + \underline{a}(X_{t_0})\int_{t_0}^t ds + b(X_{t_0})\int_{t_0}^t \circ dW_s + R$$

with remainder

$$R = \int_{t_0}^{t} \int_{t_0}^{s} \underline{L}^0 \underline{a}(X_z) \, dz \, ds + \int_{t_0}^{t} \int_{t_0}^{s} \underline{L}^1 \underline{a}(X_z) \circ dW_z \, ds$$
$$+ \int_{t_0}^{t} \int_{t_0}^{s} \underline{L}^0 b(X_z) \, dz \circ dW_s + \int_{t_0}^{t} \int_{t_0}^{s} \underline{L}^1 b(X_z) \circ dW_z \circ dW_s.$$

This is the simplest nontrivial Stratonovich-Taylor expansion of $f(X_t)$. We can continue expanding, for instance, by applying (1.19) to the integrand $f = \underline{L}^1 b$ in (1.23) to obtain

$$X_t = X_{t_0} + \underline{a}(X_{t_0}) \int_{t_0}^{t} ds + b(X_{t_0}) \int_{t_0}^{t} \circ dW_s$$
$$+ \underline{L}^1 b(X_{t_0}) \int_{t_0}^{t} \int_{t_0}^{s} \circ dW_z \circ dW_s + \bar{R}$$

with remainder

$$\bar{R} = \int_{t_0}^{t} \int_{t_0}^{s} \underline{L}^0 \underline{a}(X_z) \, dz \, ds + \int_{t_0}^{t} \int_{t_0}^{s} \underline{L}^1 \underline{a}(X_z) \circ dW_z \, ds$$
$$+ \int_{t_0}^{t} \int_{t_0}^{s} \underline{L}^0 b(X_z) \, dz \circ dW_s + \int_{t_0}^{t} \int_{t_0}^{s} \int_{t_0}^{z} \underline{L}^0 \underline{L}^1 b(X_u) \, du \circ dW_z \circ dW_s$$
$$+ \int_{t_0}^{t} \int_{t_0}^{s} \int_{t_0}^{z} \underline{L}^1 \underline{L}^1 b(X_u) \circ dW_u \circ dW_z \circ dW_s.$$

In Section 6 we shall formulate the Stratonovich-Taylor expansion for a general function $f$ and arbitrarily high order. It will be similar to the Ito-Taylor expansion, but instead of multiple Ito stochastic integrals it will involve multiple Stratonovich stochastic integrals. While it appears formally similar to the Ito-Taylor expansion, the Stratonovich-Taylor expansion will be seen to have a simpler structure which makes it a more natural generalization of the deterministic Taylor formula and more convenient to use in stochastic numerical analysis.

Similar expansions hold for multi-dimensional Ito processes satisfying nonautonomous stochastic differential equations. In the following sections we shall refer to the nonautonomous $d$-dimensional Ito equation

$$(1.24) \qquad X_t = X_{t_0} + \int_{t_0}^{t} a(s, X_s) \, ds + \sum_{j=1}^{m} \int_{t_0}^{t} b^j(s, X_s) \, dW_s^j$$

for $t \in [t_0, T]$ and the equivalent Stratonovich equation

$$(1.25) \qquad X_t = X_{t_0} + \int_{t_0}^{t} \underline{a}(s, X_s) \, ds + \sum_{j=1}^{m} \int_{t_0}^{t} b^j(s, X_s) \circ dW_s^j$$

for $t \in [t_0, T]$, where

$$(1.26) \qquad \underline{a}^i = a^i - \frac{1}{2} \sum_{j=1}^{m} \sum_{k=1}^{d} b^{k,j} \frac{\partial b^{i,j}}{\partial x^k}$$

for $i = 1, 2, \ldots, d$. In both of these SDEs $W = \{W_t, t \in [0, T]\}$ is a standard $m$-dimensional Wiener process adapted to an increasing family of sub-$\sigma$-algebras $\{A_t, t \in [0, T]\}$.

## 5.2 Multiple Stochastic Integrals

In this section we shall introduce some notation to allow us to formulate Ito-Taylor and Stratonovich-Taylor expansions in a way that will considerable simplify the presentation and proofs.

### Multi-indices

We shall call a row vector

$$(2.1) \qquad \alpha = (j_1, j_2, \ldots, j_l),$$

where

$$(2.2) \qquad j_i \in \{0, 1, \ldots, m\}$$

for $i \in \{1, 2, \ldots, l\}$ and $m = 1, 2, 3, \ldots$, a *multi-index* of length

$$(2.3) \qquad l := l(\alpha) \in \{1, 2, \ldots\}.$$

Here $m$ will denote the number of components of the Wiener process under consideration. For completeness we denote by $v$ the multi-index of length zero, that is with

$$(2.4) \qquad l(v) := 0.$$

Thus, for example,

$$l((1,0)) = 2 \quad \text{and} \quad l((1,0,1)) = 3.$$

In addition, we shall write $n(\alpha)$ for the number of components of a multi-index $\alpha$ which are equal to 0. For example,

$$n((1,0,1)) = 1, \quad n((0,1,0)) = 2, \quad n((0,0)) = 2.$$

We denote the set of all multi-indices by $\mathcal{M}$, so

$$(2.5) \qquad \mathcal{M} = \Big\{ (j_1, j_2, \ldots, j_l) : j_i \in \{0, 1, \ldots, m\}, \, i \in \{1, \ldots, l\},$$

$$\text{for } l = 1, 2, 3, \ldots \Big\} \cup \{v\}.$$

Given $\alpha \in \mathcal{M}$ with $l(\alpha) \geq 1$, we write $-\alpha$ and $\alpha-$ for the multi-index in $\mathcal{M}$ obtained by deleting the first and the last component, respectively, of $\alpha$. Thus

$$-(1,0) = (0), \qquad (1,0)- = (1),$$

$$-(0,1,1) = (1,1), \qquad (0,1,1)- = (0,1).$$

Finally, for any two multi-indices $\alpha = (j_1, j_2, \ldots, j_k)$ and $\bar{\alpha} = (\bar{j}_1, \bar{j}_2, \ldots, \bar{j}_l)$ we introduce an operation $*$ on $\mathcal{M}$ by

$$(2.6) \qquad \alpha * \bar{\alpha} = (j_1, j_2, \ldots, j_k, \bar{j}_1, \bar{j}_2, \ldots, \bar{j}_l),$$

the multi-index formed by adjoining the two given multi-indices. We shall call this the *concatenation* operation. For example, for $\alpha = (0,1,2)$ and $\bar{\alpha} = (1,3)$ we have

$$\alpha * \bar{\alpha} = (0,1,2,1,3) \quad \text{and} \quad \bar{\alpha} * \alpha = (1,3,0,1,2).$$

**Exercise 5.2.1**     *Determine $l(\alpha)$, $n(\alpha)$, $-\alpha$ and $\alpha-$ for $\alpha = (0,0,0)$, $(2,0,1)$ and $(0,1,0,0,2)$.*

## Multiple Ito Integrals

To begin we shall define three sets of adapted right continuous stochastic processes $f = \{f(t), t \geq 0\}$ with left hand limits. The first set, $\mathcal{H}_v$, is the totality of all such processes with

$$(2.7) \qquad |f(t,\omega)| < \infty,$$

w.p.1, for each $t \geq 0$; the second, $\mathcal{H}_{(0)}$, contains all those with

$$(2.8) \qquad \int_0^t |f(s,\omega)|\, ds < \infty,$$

w.p.1, for each $t \geq 0$; and the third, $\mathcal{H}_{(1)}$, all those with

$$(2.9) \qquad \int_0^t |f(s,\omega)|^2\, ds < \infty,$$

w.p.1, for each $t \geq 0$. In addition we write

$$(2.10) \qquad \mathcal{H}_{(j)} = \mathcal{H}_{(1)}$$

for each $j \in \{2, \ldots, m\}$ if $m \geq 2$. Below we shall define the sets $\mathcal{H}_\alpha$ for multi-indices $\alpha \in \mathcal{M}$ with length $l(\alpha) > 1$.

Let $\rho$ and $\tau$ be two stopping times with

$$(2.11) \qquad 0 \leq \rho(\omega) \leq \tau(\omega) \leq T,$$

w.p.1. Then, for a multi-index $\alpha = (j_1, j_2, \ldots, j_l) \in \mathcal{M}$ and a process $f \in \mathcal{H}_\alpha$ we define the *multiple Ito integral* $I_\alpha[f(\cdot)]_{\rho,\tau}$ recursively by

$$(2.12) \quad I_\alpha[f(\cdot)]_{\rho,\tau} := \begin{cases} f(\tau) & : \ l = 0 \\ \int_\rho^\tau I_{\alpha-}[f(\cdot)]_{\rho,s} \, ds & : \ l \geq 1 \ \text{and} \ j_l = 0 \\ \int_\rho^\tau I_{\alpha-}[f(\cdot)]_{\rho,s} \, dW_s^{j_l} & : \ l \geq 1 \ \text{and} \ j_l \geq 1. \end{cases}$$

For $\alpha = (j_1, j_2, \ldots, j_l)$ with $l \geq 2$ we define recursively the set $\mathcal{H}_\alpha$ to be the totality of adapted right continuous processes $f = \{f(t), \ t \geq 0\}$ with left hand limits such that the integral process $\{I_{\alpha-}[f(\cdot)]_{\rho,t}, \ t \geq 0\}$ considered as a function of $t$ satisfies

$$(2.13) \qquad\qquad I_{\alpha-}[f(\cdot)]_{\rho,\cdot} \in \mathcal{H}_{(j_l)}.$$

As an illustration of this terminolgy we consider the following examples:

$$I_v[f(\cdot)]_{0,t} = f(t),$$

$$I_{(0)}[f(\cdot)]_{\tau_i,\tau_{i+1}} = \int_{\tau_i}^{\tau_i+1} f(s) \, ds,$$

$$I_{(1)}[f(\cdot)]_{\rho,\tau} = \int_\rho^\tau f(s) \, dW_s^1,$$

$$I_{(0,1)}[f(\cdot)]_{0,t} = \int_0^t \int_0^{s_2} f(s_1) \, ds_1 dW_{s_2}^1$$

$$I_{(0,2,1)}[f(\cdot)]_{0,t} = \int_0^t \int_0^{s_3} \int_0^{s_2} f(s_1) \, ds_1 dW_{s_2}^2 dW_{s_3}^1$$

for an appropriate process $f$. For simpler notation in the case that $f(t) \equiv 1$, or when the stopping times $\rho$ and $\tau$ are obvious from the context, we shall often abbreviate $I_\alpha[f(\cdot)]_{\rho,\tau}$ to $I_{\alpha,\tau}$ or just $I_\alpha$.

**Exercise 5.2.2**   *Write out in full the multiple Ito stochastic integrals* $I_\alpha[f(\cdot)]_{\rho,\tau}$ *for* $f(t) \equiv 1$, $\rho = 0$, $\tau = T$ *and* $\alpha = (0,0,0)$, $(2,0,1)$ *and* $(1,2)$.

## Relations Between Multiple Ito Integrals

There is a recursive relationship for multiple Ito integrals, which we shall now derive. For convenience we write

$$(2.14) \qquad\qquad I_{\alpha,t} = I_\alpha[1]_{0,t}$$

and

$$(2.15) \qquad\qquad W_t^0 = t$$

for $\alpha \in \mathcal{M}$ and $t \geq 0$, and recall that $I_A$ denotes the indicator function of the set $A$.

**Propostion 5.2.3**    *Let $j_1, \ldots, j_l \in \{0, 1, \ldots, m\}$ and $\alpha = (j_1, \ldots, j_l) \in \mathcal{M}$ where $l = 1, 2, 3, \ldots$ Then*

$$(2.16) \qquad W_t^j \, I_{\alpha,t} \;=\; \sum_{i=0}^{l} I_{(j_1,\ldots,j_i,j,j_{i+1},\ldots,j_l),t}$$

$$+ \sum_{i=1}^{l} I_{\{j_i=j\neq 0\}} \, I_{(j_1,\ldots,j_{i-1},0,j_{i+1},\ldots,j_l),t}$$

*for all $t \geq 0$.*

**Proof**    We consider the multi-dimensional linear Ito process $X = \{X_t,\, t \geq 0\}$ defined by

$$X_t = \begin{pmatrix} X_t^{(0)} \\ \vdots \\ X_t^{(m)} \\ X_t^{(j_1,j_2)} \\ \vdots \\ X_t^{(j_1,j_2,j_3)} \\ \vdots \\ X_t^{(j_1,\ldots,j_l)} \end{pmatrix} = \begin{pmatrix} I_{(0),t} \\ \vdots \\ I_{(m),t} \\ I_{(j_1,j_2),t} \\ \vdots \\ I_{(j_1,j_2,j_3),t} \\ \vdots \\ I_{(j_1,\ldots,j_l),t} \end{pmatrix},$$

where each component represents a multiple Ito integral. Here for the $\beta$th component with $\beta = (j_1', \ldots, j_r')$ the drift coefficient

$$a^\beta = \begin{cases} x^{\beta-} & : \quad j_r' = 0 \\ 0 & : \quad \text{otherwise} \end{cases}$$

and the diffusion coefficient

$$b^{\beta,j} = \begin{cases} x^{\beta-} & : \quad j = j_r' \in \{1, \ldots, m\} \\ 0 & : \quad \text{otherwise,} \end{cases}$$

where $\beta- = (j_1, \ldots, j_{r-1})$ and $x^{\beta-}$ is the corresponding component of $x$. Obviously $X_0$ is the zero vector. Then, from (2.12) and the Ito formula (3.4.6) it can be shown that

$$(2.17) \qquad W_t^j \, I_{\alpha,t} \;=\; I_{(j),t} I_{\alpha,t}$$

$$= \int_0^t I_{\alpha,s} \, dI_{(j),s} + \int_0^t I_{(j),s} \, I_{\alpha-,s} \, dW_s^{j_l}$$

$$+ I_{\{j_l=j\neq 0\}} \int_0^t I_{\alpha-,s} \, ds.$$

For $l = 1$ the relation (2.16) follows directly from (2.17). When $l \geq 2$ we obtain

$$W_t^j I_{\alpha,t} = I_{(j_1,\ldots,j_l,j),t} + \int_0^t I_{(j),s} I_{\alpha-,s} \, dW_s^{j_l}$$

$$+ I_{\{j_l = j \neq 0\}} I_{(j_1,\ldots,j_{l-1},0),t}$$

from (2.17) and (2.12). The assertion (2.16) then follows by induction with respect to $l$ if we express $I_{(j),s} I_{\alpha-,s}$ in terms of (2.16) for lower indices. $\square$

Formula (2.16) describes a relation between different multiple Ito integrals. In the special case that the multiple integrals are all integrated with respect to the same component of the Wiener process it reduces to a simpler expression provided by the following corollary.

**Corollary 5.2.4** *Suppose that* $\alpha = (j_1, \ldots, j_l)$ *with* $j_1 = \cdots = j_l = j \in \{0, \ldots, m\}$ *where* $l \geq 2$. *Then for* $t \geq 0$

$$(2.18) \qquad I_{\alpha,t} = \begin{cases} \frac{1}{l!} t^l & : \ j = 0 \\ \frac{1}{l} \left( W_t^j I_{\alpha-,t} - t I_{(\alpha-)-,t} \right) & : \ j \geq 1 \end{cases}$$

**Proof** The case $j = 0$ follows from the usual deterministic integration rule. For $j \in \{1, \ldots, m\}$ the relation (2.16) gives

$$(2.19) \qquad t I_{(\alpha-)-,t} = \sum_{i=0}^{l-2} I_{(j_1,\ldots,j_i,0,j_{i+1},\ldots,j_{l-2}),t} \qquad \text{and}$$

$$(2.20) \qquad W_t^j I_{\alpha-,t} = l I_{\alpha,t} + \sum_{i=1}^{l-1} I_{(j_1,\ldots,j_{i-1},0,j_{i+1},\ldots,j_{l-1}),t}.$$

On renumbering and inserting (2.19) into (2.20) we get (2.18) for $j \in \{1, \ldots, m\}$ too. $\square$

The recursion formula (2.18) for $j \in \{1, \ldots, m\}$ has the same form as that for the Hermite polynomials. Using this we can express the multiple integrals $I_{(j,j,\ldots,j),t}$ as a Hermite polynomial in $I_{(j),t}$ for any $j \in \{1, \ldots, m\}$. In particular we have

$$(2.21) \qquad I_{(j,j),t} = \frac{1}{2!} \left( I_{(j),t}^2 - t \right),$$

$$I_{(j,j,j),t} = \frac{1}{3!} \left( I_{(j),t}^3 - 3t \, I_{(j),t} \right),$$

$$I_{(j,j,j,j),t} = \frac{1}{4!} \left( I_{(j),t}^4 - 6t \, I_{(j),t}^2 + 3t^2 \right),$$

$$I_{(j,j,j,j,j),t} = \frac{1}{5!} \left( I_{(j),t}^5 - 10t \, I_{(j),t}^3 + 15t^2 \, I_{(j),t} \right),$$

$$I_{(j,j,j,j,j,j),t} = \frac{1}{6!} \left( I_{(j),t}^6 - 15t \, I_{(j),t}^4 + 45t^2 \, I_{(j),t}^2 - 15t^3 \right),$$

$$I_{(j,j,j,j,j,j,j),t} = \frac{1}{7!} \left( I_{(j),t}^7 - 21t \, I_{(j),t}^5 + 105t^2 \, I_{(j),t}^3 - 105t^3 \, I_{(j),t} \right).$$

For any $j \in \{1, \ldots, m\}$ we also have the following special cases of the relation (2.16):

$$(2.22) \qquad \begin{aligned} t\, I_{(j),t} &= I_{(j,0),t} + I_{(0,j),t}, \\ t\, I_{(j,j),t} &= I_{(j,j,0),t} + I_{(j,0,j),t} + I_{(0,j,j),t}, \\ I_{(j),t} I_{(0,j),t} &= 2 I_{(0,j,j),t} + I_{(j,0,j),t} + I_{(0,0),t}, \\ I_{(j),t} I_{(j,0),t} &= I_{(j,0,j),t} + 2 I_{(j,j,0),t} + I_{(0,0),t}. \end{aligned}$$

We remark that there are many other useful relations between multiple Ito integrals, which can be derived as required. These mainly involve multiple Ito integrals with multi-indices of the same or shorter length.

**Exercise 5.2.5**   *Show that the multiple Ito integrals $I_\alpha$ with multi-indices $\alpha = (0,1)$, $(1,1)$ and $(1,1,1)$ can be expressed in terms of those with multi-indices $(0)$, $(1)$ and $(1,0)$.*

**Exercise 5.2.6**   *Find expressions for the multiple Ito integrals $I_\alpha$ with multi-indices $\alpha = (0,1)$, $(1,1)$, $(1,1,1)$ and $(1,1,1,1)$ in terms of those with multi-indices $(0)$, $(1)$ and $(1,0)$.*

**Exercise 5.2.7**   *Show that $I_{(1,0)}$ is normally distributed with mean, variance and correlation*

$$E\left(I_{(1,0)}\right) = 0, \quad E\left(I_{(1,0)}^2\right) = \frac{1}{3}\Delta^3, \quad E\left(I_{(1,0)} I_{(1)}\right) = \frac{1}{2}\Delta^2.$$

## Multiple Stratonovich Integrals

We shall denote by $\mathcal{H}_v$ and $\mathcal{H}_{(0)}$ the sets of functions $g : \Re^+ \times \Re^d \to \Re$ for which $g(\cdot, X.) \in \mathcal{H}_v$ and $g(\cdot, X.) \in \mathcal{H}_{(0)}$, respectively, where $X = \{X_t, t \geq 0\}$ is a $d$-dimensional Ito process which satisfies the Stratonovich stochastic differential equation (1.25). In addition for $j \in \{1, \ldots, m\}$ we define $\mathcal{H}_{(j)}$ to be the set of differentiable functions $g : \Re^+ \times \Re^d \to \Re$ such that $g(\cdot, X.) \in \mathcal{H}_{(j)}$ and

$$(2.23) \qquad \int_0^t \left| \underline{L}^j g(s, X_s) \right| ds < \infty,$$

w.p.1, for each $t \geq 0$, where

$$(2.24) \qquad \underline{L}^j = \sum_{k=1}^d b^{k,j} \frac{\partial}{\partial x^k}.$$

Below we shall define recursively sets $\mathcal{H}_\alpha$ for multi-indices $\alpha$ of length $l(\alpha) \geq 2$. Let $\rho$ and $\tau$ be two stopping times with

$$(2.25) \qquad 0 \leq \rho(\omega) \leq \tau(\omega) \leq T,$$

w.p.1. Then for any $j \in \{1, \ldots, m\}$ and $g \in \underline{\mathcal{H}}_{(j)}$ the relationship between Ito and Stratonovich integrals can be written as

$$(2.26) \qquad \int_\rho^\tau g(s, X_s) \circ dW_s^j$$

$$= \int_\rho^\tau g(s, X_s)\, dW_s^j + I_{\{j \neq 0\}} \int_\rho^\tau \frac{1}{2} \underline{L}^j g(s, X_s)\, ds$$

$$= I_{(j)}[g(\cdot, X.)]_{\rho,\tau} + I_{\{j \neq 0\}} I_{(0)} \left[\frac{1}{2} \underline{L}^j g(\cdot, X.)\right]_{\rho,\tau}.$$

To formulate a recursive definition of multiple Stratonovich integrals we shall adjoin to the given $d$-dimensional Ito process $X$, which satisfies the Stratonovich equation

$$(2.27) \qquad X_t = X_\rho + \int_\rho^t \underline{a}(s, X_s)\, ds + \sum_{j=1}^m \int_\rho^t \underline{b}^j(s, X_s) \circ dW_s^j,$$

a $(d+1)$th and $(d+2)$th component

$$(2.28) \qquad X_t^{d+1} = \int_\rho^t g(s, X_s) \circ dW_s^{j_1}, \qquad X_t^{d+2} = \int_\rho^t X_s^{d+1} \circ dW_s^{j_2}$$

for some $j_1, j_2 \in \{0, 1, \ldots, m\}$. Here $\circ dW_s^0 = ds$, $t \in [\rho, \tau]$ and the function $g$ is chosen so that the integrals exist. Interpreting $X$ now as a $(d+2)$-dimensional Ito process, it then follows from (2.26) that

$$(2.29) \quad X_t^{d+2} = \int_\rho^t \int_\rho^{s_2} g(s, X_{s_1}) \circ dW_s^{j_1} \circ dW_{s_2}^{j_2}$$

$$= \int_\rho^t X_{s_2}^{d+1} dW_{s_2}^{j_2} + \frac{1}{2} I_{\{j_2 \neq 0\}} \int_\rho^t I_{\{j_1 = j_2\}} g(s, X_s)\, ds$$

$$= \int_\rho^t \int_\rho^{s_2} g(s_1, X_{s_1}) \circ dW_{s_1}^{j_1} dW_{s_2}^{j_2} + \frac{1}{2} \int_\rho^t I_{\{j_1 = j_2 \neq 0\}} g(s_1, X_{s_1})\, ds_1$$

$$= \int_\rho^t \int_\rho^{s_2} g(s_1, X_{s_1})\, dW_{s_1}^{j_1} dW_{s_2}^{j_2} + \frac{1}{2} \int_\rho^t \int_\rho^{s_2} I_{\{j_1 \neq 0\}} \underline{L}^{j_1} g(s_1, X_{s_1}) ds_1\, dW_{s_2}^{j_2}$$

$$+ \frac{1}{2} \int_\rho^t I_{\{j_1 = j_2 \neq 0\}} g(s_1, X_{s_1})\, ds_1$$

for each $t \in [\rho, \tau]$. From this we have

$$(2.30) \qquad J_{(j_1, j_2)}[g(\cdot, X.)]_{\rho,t} = I_{(j_1, j_2)}[g(\cdot, X.)]_{\rho,t}$$

$$+ \frac{1}{2} I_{\{j_1 \neq 0\}} I_{(0, j_2)} \left[\underline{L}^{j_1} g(\cdot, X.)\right]_{\rho,t}$$

$$+ \frac{1}{2} I_{\{j_1 = j_2 \neq 0\}} I_{(0)} \left[g(\cdot, X.)\right]_{\rho,t}.$$

Then for two such stopping times, a multi-index $\alpha = (j_1, \ldots, j_l) \in \mathcal{M}$ and a function $g \in \underline{\mathcal{H}}_\alpha$ we define the *multiple Stratonovich integral* $J_\alpha[g(\cdot, X.)]_{\rho,\tau}$ recursively by

$$(2.31) \quad J_\alpha[g(\cdot, X.)]_{\rho,\tau} = \begin{cases} g(\tau, X_\tau) & : \quad l = 0 \\ \int_\rho^\tau J_{\alpha-}[g(\cdot, X.)]_{\rho,s} \, ds & : \quad l \geq 1, j_l = 0 \\ \int_\rho^\tau J_{\alpha-}[g(\cdot, X.)]_{\rho,s} \circ dW_s^{j_l} & : \quad l \geq 1, j_l \geq 1. \end{cases}$$

Here we also define recursively the set $\underline{\mathcal{H}}_\alpha$ for $\alpha = (j_1, \ldots, j_l)$ with $l \geq 2$ as the set of all functions $g : \Re^+ \times \Re^d \to \Re$ for which

$$(2.32) \qquad J_{\alpha-}[g(\cdot, X.)]_{0,\cdot} \in \mathcal{H}_{(j_l)}$$

and

$$(2.33) \qquad J_{(\alpha-)-}[I_{\{j_l=j_{l-1}\neq 0\}} g(\cdot, X.)]_{0,\cdot} \in \mathcal{H}_{(0)}.$$

Using (2.31), (2.26) and (2.29) we obtain a relationship between the multiple Ito and Stratonovich integrals $I_\alpha$ and $J_\alpha$ for $\alpha = v$ or $\alpha = (j_1, \ldots, j_l) \in \mathcal{M}$. We distinguish the three cases: $l = 0$, $l = 1$ and $l \geq 2$.

$$(2.34) \qquad J_v[g(\cdot, X.)]_{\rho,\tau} = I_v[g(\cdot, X.)]_{\rho,\tau} = g(\tau, X_\tau),$$

$$J_{(j_1)}[g(\cdot, X.)]_{\rho,\tau} = I_{(j_1)}[g(\cdot, X.)]_{\rho,\tau} + I_{\{j_1 \neq 0\}} I_{(0)} \left[ \frac{1}{2} \underline{L}^{j_1} g(\cdot, X.) \right]_{\rho,\tau},$$

and

$$J_\alpha[g(\cdot, X.)]_{\rho,\tau} = I_{(j_l)} \left[ J_{\alpha-}[g(\cdot, X.)]_{\rho,\cdot} \right]_{\rho,\tau}$$
$$+ I_{\{j_l=j_{l-1}\neq 0\}} I_{(0)} \left[ \frac{1}{2} J_{(\alpha-)-}[g(\cdot, X.)]_{\rho,\cdot} \right]_{\rho,\tau}$$

for $l \geq 2$.

**Remark 5.2.8**   It follows from (2.34) that any multiple Stratonovich integral $J_\alpha$ can be written as a finite sum of multiple Ito integrals $I_\beta$ with

$$l(\alpha) + n(\alpha) \begin{cases} = l(\beta) + n(\beta) & : \quad l(\alpha) \text{ even} \\ \leq l(\beta) + n(\beta) & : \quad l(\alpha) \text{ odd.} \end{cases}$$

To simplify the notation when $g(t, x) \equiv 1$, or when the stopping times $\rho$ and $\tau$ are obvious, we shall often write $J_\alpha[g(\cdot, X.)]_{\rho,\tau}$ as $J_{\alpha,\tau}$ or just $J_\alpha$. Then from (2.34) we have

$$(2.35) \qquad J_\alpha = I_\alpha$$

for $l(\alpha) \in \{0, 1\}$,

$$J_\alpha = I_\alpha + \frac{1}{2} I_{\{j_1=j_2\neq 0\}} I_{(0)}$$

for $l(\alpha) = 2$,

$$J_\alpha = I_\alpha + \frac{1}{2}\left(I_{\{j_1=j_2\neq 0\}}I_{(0,j_3)} + I_{\{j_2=j_3\neq 0\}}I_{(j_1,0)}\right)$$

for $l(\alpha) = 3$, and

$$
\begin{aligned}
J_\alpha &= I_\alpha + \frac{1}{4}I_{\{j_1=j_2\neq 0\}}I_{\{j_3=j_4\neq 0\}}I_{(0,0)} \\
&\quad + \frac{1}{2}\left(I_{\{j_1=j_2\neq 0\}}I_{(0,j_3,j_4)} + I_{\{j_2=j_3\neq 0\}}I_{(j_1,0,j_4)} + I_{\{j_3=j_4\neq 0\}}I_{(j_1,j_2,0)}\right)
\end{aligned}
$$

for $l(\alpha) = 4$.

**Exercise 5.2.9**  *Verify (2.35).*

Similarly, we have

(2.36) $$I_\alpha = J_\alpha$$

for $l(\alpha) \in \{0,1\}$,

$$I_\alpha = J_\alpha - \frac{1}{2}I_{\{j_1=j_2\neq 0\}}J_{(0)}$$

for $l(\alpha) = 2$,

$$I_\alpha = J_\alpha - \frac{1}{2}\left(I_{\{j_1=j_2\neq 0\}}J_{(0,j_3)} + I_{\{j_2=j_3\neq 0\}}J_{(j_1,0)}\right)$$

for $l(\alpha) = 3$, and

$$
\begin{aligned}
I_\alpha &= J_\alpha + \frac{1}{4}I_{\{j_1=j_2\neq 0\}}I_{\{j_3=j_4\neq 0\}}J_{(0,0)} \\
&\quad - \frac{1}{2}\left(I_{\{j_1=j_2\neq 0\}}J_{(0,j_3,j_4)} + I_{\{j_2=j_3\neq 0\}}J_{(j_1,0,j_4)} + I_{\{j_3=j_4\neq 0\}}J_{(j_1,j_2,0)}\right)
\end{aligned}
$$

for $l(\alpha) = 4$.

## Relations Between Multiple Stratonovich Integrals

There is a recursive relationship for multiple Stratonovich integrals analogous to that for multiple Ito integrals (2.16) when the integrand is identically equal to 1. In order to state it succinctly we shall use the abbreviation

(2.37) $$J_{\alpha,t} = J_\alpha[1]_{0,t}$$

and, as before, write

(2.38) $$W_t^0 = t.$$

Then, for $\alpha \in \mathcal{M}$ and $t \geq 0$ we have

**Propostion 5.2.10**  *Let $j_1, \ldots, j_l \in \{0, 1, \ldots, m\}$ and $\alpha = (j_1, \ldots, j_l) \in \mathcal{M}$ where $l = 1, 2, 3, \ldots$. Then*

(2.39)
$$W_t^j \, J_{\alpha,t} = \sum_{i=0}^{l} J_{(j_1,\ldots,j_i,j,j_{i+1},\ldots,j_l),t}$$

*for all $t \geq 0$.*

We note that (2.38) has a simpler structure than the corresponding relation (2.16) for multiple Ito integrals.

**Proof**  Since transformations of solutions of Stratonovich equations satisfy the deterministic chain rule, we find that

(2.40)
$$X_t^1 X_t^2 = \int_0^t X_s^2 \, b^{1j}(X_s) \circ dW_s^j + \int_0^t X_s^1 \, b^{2j_l}(X_s) \circ dW_s^{j_l}$$

for $X_t = (X_t^1, X_t^2)$, where

(2.41)
$$\begin{pmatrix} X_t^1 \\ X_t^2 \end{pmatrix} = \int_0^t \begin{pmatrix} b^{1j}(X_s) \\ 0 \end{pmatrix} \circ dW_s^j + \int_0^t \begin{pmatrix} 0 \\ b^{2j_l}(X_s) \end{pmatrix} \circ dW_s^{j_l}$$

for $b^{1j} \equiv 1$ and $b^{2j_l}(X_s) = J_{\alpha-,s}$ with $\alpha = (j_1, \ldots, j_l) \in \mathcal{M}$, $l \geq 1$. From this we obtain

(2.42)
$$W_t^j \, J_{\alpha,t} = J_{\alpha*(j),t} + \int_0^t J_{\alpha-,s} \, W_s^j \circ dW_s^{j_l},$$

where $*$ is the concatenation operation on multi-indices (2.6). For $l = 1$ this reduces to

(2.43)
$$W_t^j \, J_{(j_1),t} = J_{(j_1,j),t} + J_{(j,j_1),t},$$

which is just (2.39). The proof for $l \geq 2$ then follows by induction from (2.42).
□

The next corollary, which is a direct consequence of (2.39), gives a clear indication of the simpler structure offered by multiple Stratonovich integrals when compared with its counterpart for multiple Ito integrals, Corollary 5.2.4.

**Corollary 5.2.11**  *Suppose that $\alpha = (j_1, \ldots, j_l)$ with $j_1 = \cdots = j_l = j \in \{0, \ldots, m\}$ where $l \geq 0$. Then for $t \geq 0$*

(2.44)
$$J_{\alpha,t} = \frac{1}{l!} \left( J_{(j),t} \right)^l.$$

For later use we now state some special cases of (2.39):

(2.45)
$$\begin{aligned}
t \, J_{(j),t} &= J_{(j,0),t} + J_{(0,j),t}, \\
t \, J_{(j,j),t} &= J_{(j,j,0),t} + J_{(j,0,j),t} + J_{(0,j,j),t}, \\
J_{(j),t} J_{(0,j),t} &= 2 J_{(0,j,j),t} + J_{(j,0,j),t}, \\
J_{(j),t} J_{(j,0),t} &= J_{(j,0,j),t} + 2 J_{(j,j,0),t},
\end{aligned}$$

where $j \in \{0, 1, \ldots, m\}$. These are simpler than the relations (2.22) for the corresponding multiple Ito integrals.

As with multiple Ito integrals it is also possible to express multiple Stratonovich integrals in terms of others with multi-indices of the same or shorter length.

**Exercise 5.2.12** *Show that the multiple Stratonovich integrals $J_\alpha$ with multi-indices $\alpha = (0, 1)$, $(1, 1)$ and $(1, 1, 1)$ can be expressed in terms of those with multi-indices $(0)$, $(1)$ and $(1, 0)$.*

## 5.3 Coefficient Functions

In this section we shall introduce functions which will be needed later in defining the coefficients of stochastic Taylor expansions. We shall consider the Ito and Stratonovich cases separately.

### Ito Coefficient Functions

We shall write the diffusion operator for the Ito equation (1.24) as

$$(3.1) \qquad L^0 = \frac{\partial}{\partial t} + \sum_{k=1}^{d} a^k \frac{\partial}{\partial x^k} + \frac{1}{2} \sum_{k,l=1}^{d} \sum_{j=1}^{m} b^{k,j} b^{l,j} \frac{\partial^2}{\partial x^k \partial x^l}$$

and for $j \in \{1, \ldots, m\}$ introduce the operator

$$(3.2) \qquad L^j = \sum_{k=1}^{d} b^{k,j} \frac{\partial}{\partial x^k}.$$

For each $\alpha = (j_1, \ldots, j_l)$ and function $f \in C^h(\Re^+ \times \Re^d, \Re)$ with $h = l(\alpha) + n(\alpha)$ we define recursively the *Ito coefficient function*

$$(3.3) \qquad f_\alpha = \begin{cases} f & : \quad l = 0 \\ L^{j_1} f_{-\alpha} & : \quad l \geq 1. \end{cases}$$

If the function $f$ is not explicitly stated we shall always take it to be the identity function $f(t, x) \equiv x$. For example, in the 1-dimensional case $d = m = 1$ for $f(t, x) \equiv x$ we have

$$f_{(0)} = a, \quad f_{(1)} = b, \quad f_{(1,1)} = bb'$$

and

$$f_{(0,1)} = ab' + \frac{1}{2} b^2 b''.$$

Here the prime $'$ denotes the ordinary or partial derivative with respect to the $x$ variable, depending on whether or not the function being differentiated depends only on $x$ or on both $t$ and $x$.

**Exercise 5.3.1**     *For the 1-dimensional case with $f(x) \equiv x$ determine the Ito coefficient functions $f_{(1,0)}$ and $f_{(1,1,1)}$*

To facilitate the construction of Ito-Taylor expansions we shall now list the Ito coefficient functions in the autonomous case $d = 1$ with $f(x) \equiv x$ for all multi-indices $\alpha$ of length $l(\alpha) \leq 3$. To simplify our presentation we shall use the abbreviation

$$(3.4) \qquad\qquad \sigma = \frac{1}{2} \sum_{j=1}^{m} \left(b^j\right)^2.$$

From (3.3) we can show for $j_1, j_2, j_3 \in \{1, \ldots, m\}$ that

$$(3.5) \qquad\qquad f_{(0)} = a, \quad f_{(j_1)} = b^{j_1},$$

$$f_{(0,0)} = aa' + \sigma a'', \quad f_{(0,j_1)} = ab^{j_1\prime} + \sigma b^{j_1\prime\prime},$$

$$f_{(j_1,0)} = b^{j_1} a', \quad f_{(j_1,j_2)} = b^{j_1} b^{j_2\prime},$$

$$\begin{aligned}
f_{(0,0,0)} &= a\left(aa'' + (a')^2 + \sigma'a'' + \sigma a'''\right) \\
&\quad + \sigma\left(aa''' + 3a'a'' + \sigma''a'' + 2\sigma'a''' + \sigma a^{(4)}\right),
\end{aligned}$$

$$\begin{aligned}
f_{(0,0,j_1)} &= a\left(a'b^{j_1\prime} + ab^{j_1\prime\prime} + \sigma'b^{j_1\prime\prime} + \sigma b^{j_1\prime\prime\prime}\right) \\
&\quad + \sigma\left(a''b^{j_1\prime} + 2a'b^{j_1\prime\prime} + ab^{j_1\prime\prime\prime} + \sigma''b^{j_1\prime\prime} + 2\sigma'b^{j_1\prime\prime\prime} + \sigma b^{j_1\,(4)}\right),
\end{aligned}$$

$$f_{(0,j_1,0)} = a\left(a'b^{j_1\prime} + a''b^{j_1}\right) + \sigma\left(a'b^{j_1\prime\prime} + 2a''b^{j_1\prime} + a'''b^{j_1}\right),$$

$$f_{(0,j_1,j_2)} = a\left(b^{j_1\prime}b^{j_2\prime} + b^{j_1}b^{j_2\prime\prime}\right) + \sigma\left(b^{j_1\prime\prime}b^{j_2\prime} + 2b^{j_1\prime}b^{j_2\prime\prime} + b^{j_1}b^{j_2\prime\prime\prime}\right),$$

$$f_{(j_1,0,0)} = b^{j_1}\left((a')^2 + aa'' + \sigma'a'' + \sigma a'''\right),$$

$$f_{(j_1,0,j_2)} = b^{j_1}\left(a'b^{j_2\prime} + ab^{j_2\prime\prime} + \sigma'b^{j_2\prime\prime} + \sigma b^{j_2\prime\prime\prime}\right),$$

$$f_{(j_1,j_2,0)} = b^{j_1}\left(a'b^{j_2\prime} + a''b^{j_2}\right), \quad f_{(j_1,j_2,j_3)} = b^{j_1}\left(b^{j_2\prime}b^{j_3\prime} + b^{j_2}b^{j_3\prime\prime}\right).$$

Thus the coefficients $a$ and $b$ here must be at least 4 times continuously differentiable.

**Exercise 5.3.2**     *Verify the formulae in (3.5) and determine $f_{(j_1,j_2,j_3,j_4)}$ for $j_1, \ldots, j_4 \in \{1, \ldots, m\}$.*

## Stratonovich Coefficient Functions

Here we shall need the operators

(3.6)
$$L^0 = \frac{\partial}{\partial t} + \sum_{k=1}^{d} a^k \frac{\partial}{\partial x^k}$$

and

(3.7)
$$L^j = \sum_{k=1}^{d} b^{k,j} \frac{\partial}{\partial x^k}.$$

for $j \in \{1, \ldots, m\}$, where

(3.8)
$$\underline{a} = a - \frac{1}{2} \sum_{j=1}^{m} L^j b^j$$

For each $\alpha = (j_1, \ldots, j_l)$ and function $f \in C^h(\Re^+ \times \Re^d, \Re)$ with $h = l(\alpha)$ we define recursively the *Stratonovich coefficient function*

(3.9)
$$\underline{f}_\alpha = \begin{cases} f & : \quad l = 0 \\ \underline{L}^{j_1} \underline{f}_{-\alpha} & : \quad l \geq 1. \end{cases}$$

When the function $f$ is not explicitly stated in the text we shall always take it to be the identity function $f(t, x) \equiv x$. We remark that the Stratonovich coefficient functions generally do not contain as many higher order derivatives as the corresponding Ito coefficient functions. For example, in the 1-dimensional case $d = m = 1$ for the identity function $f(t, x) \equiv x$ we have

$$\underline{f}_{(0)} = \underline{a}, \quad \underline{f}_{(1)} = b, \quad \underline{f}_{(1,1)} = bb'$$

and

$$\underline{f}_{(0,1)} = \underline{a}b' = \left( a - \frac{1}{2}bb' \right) b'.$$

In contrast, we saw earlier that the Ito coefficient function $f_{(0,1)}$ corresponding to the last of these is equal to $ab' + \frac{1}{2}b^2 b''$.

**Exercise 5.3.3**   *For the 1-dimensional case with $f(t, x) \equiv x$ determine the Stratonovich coefficient functions $\underline{f}_{(1,0)}$ and $\underline{f}_{(1,1,1)}$.*

We shall now list the Stratonovich coefficient functions in the autonomous case $d = 1$ with $f(x) \equiv x$ for all multi- indices $\alpha$ of length $l(\alpha) \leq 3$. These follow from (3.9) and will be useful later when we construct Stratonovich -Taylor expansions:

(3.10)
$$\underline{f}_{(0)} = \underline{a}, \quad \underline{f}_{(j_1)} = b^{j_1}, \quad \underline{f}_{(0,0)} = \underline{a}\underline{a}',$$

$$\underline{f}_{(0,j_1)} = \underline{a}b^{j_1}{}', \quad \underline{f}_{(j_1,0)} = \underline{a}'b^{j_1}, \quad \underline{f}_{(j_1,j_2)} = b^{j_1}b^{j_2}{}',$$

$$\underline{f}_{(0,0,0)} = \underline{a}\left(\underline{a}\underline{a}'' + (\underline{a}')^2\right), \quad \underline{f}_{(0,0,j_1)} = \underline{a}\left(\underline{a}b^{j_1}{}'' + \underline{a}'b^{j_1}{}'\right),$$

$$\underline{f}_{(0,j_1,0)} = \underline{a}\left(\underline{a}''b^{j_1} + \underline{a}'b^{j_1}{}'\right), \quad \underline{f}_{(j_1,0,0)} = b^{j_1}\left(\underline{a}\underline{a}'' + (\underline{a}')^2\right),$$

$$\underline{f}_{(0,j_1,j_2)} = \underline{a}\left(b^{j_1}b^{j_2}{}'' + b^{j_1}{}'b^{j_2}{}'\right), \quad \underline{f}_{(j_1,0,j_2)} = b^{j_1}\left(\underline{a}b^{j_1}{}'' + \underline{a}'b^{j_1}{}'\right),$$

$$\underline{f}_{(j_1,j_2,0)} = b^{j_1}\left(\underline{a}''b^{j_2} + \underline{a}'b^{j_2}{}'\right), \quad \underline{f}_{(j_1,j_2,j_3)} = b^{j_1}\left(b^{j_2}b^{j_3}{}'' + b^{j_2}{}'b^{j_3}{}'\right),$$

where $j_1$, $j_2$, $j_3 \in \{1,\ldots, m\}$. We note that the functions $\underline{a}$ and $b$ here need only be three times continuously differentiable, compared with four times for their counterparts in the corresponding Ito coefficient functions.

**Exercise 5.3.4**     *Verify the formulae in (3.10) and determine* $\underline{f}_{(j_1,j_2,j_3,j_4)}$ *for* $j_1, \ldots, j_4 \in \{1, \ldots, m\}$.

## 5.4    Hierarchical and Remainder Sets

The multiple stochastic integrals appearing in a stochastic Taylor expansion with constant integrands cannot be chosen completely arbitrarily. Rather, the set of corresponding multi-indices must form an hierarchical set.
    We call a subset $\mathcal{A} \subset \mathcal{M}$ an *hierarchical set* if $\mathcal{A}$ is nonempty:

$$(4.1) \qquad\qquad\qquad \mathcal{A} \neq \emptyset;$$

if the multi-indices in $\mathcal{A}$ are uniformly bounded in length:

$$(4.2) \qquad\qquad\qquad \sup_{\alpha \in \mathcal{A}} l(\alpha) < \infty;$$

and if

$$(4.3) \qquad\qquad -\alpha \in \mathcal{A} \quad \text{for each} \quad \alpha \in \mathcal{A} \setminus \{v\},$$

where $v$ is the multi-index of length zero (see (2.4)). Thus, if a multi-index $\alpha$ belongs to an hierarchical set, then so does the multi-index $-\alpha$ obtained by deleting the first component of $\alpha$.
    For example, the sets

$$\{v\}, \quad \{v, (0), (1)\}, \quad \{v, (0), (1), (1, 1)\}$$

are hierarchical sets.
    If we form a stochastic Taylor expansion for a given hierarchical set, then the remainder term involves only those multiple stochastic integrals with multi-

indices which belong to the corresponding remainder set. For any given hierarchical set $\mathcal{A}$ we define the *remainder set* $\mathcal{B}(\mathcal{A})$ of $\mathcal{A}$ by

(4.4) $$\mathcal{B}(\mathcal{A}) = \{\alpha \in \mathcal{M} \setminus \mathcal{A} : -\alpha \in \mathcal{A}\}.$$

This means that the remainder set consists of all of the next following multi-indices with respect to the given hierarchical set. It is constructed by adding a further component taking all possible values at the beginning of the "maximal" multi-indices of the hierarchical set. For example, when $m = 1$ we have the remainder sets

$$\mathcal{B}(\{v\}) = \{(0),(1)\}, \quad \mathcal{B}(\{v,(0),(1)\}) = \{(0,0),(0,1),(1,0),(1,1)\}$$

and

$$\mathcal{B}(\{v,(0),(1),(1,1)\}) = \{(0,0),(0,1),(1,0),(0,1,1),(1,1,1)\}.$$

**Exercise 5.4.1** *Which of the following sets are hierarchical sets:*

$$\emptyset,\ \{(0)\},\ \{v,(1)\},\ \{v,(0),(0,1)\},\ \{v,(0),(1),(0,1)\}?$$

**Exercise 5.4.2** *Determine the remainder sets corresponding to the hierarchical sets in Exercise 5.4.1.*

**Exercise 5.4.3** *Are the sets*

(4.5) $$\Gamma_r = \{\alpha \in \mathcal{M} : l(\alpha) \le r\},$$

*where $r = 1,\ 2,\ \ldots$, hierarchical sets?*

**Exercise 5.4.4** *Determine the remainder set $\mathcal{B}(\Gamma_r)$ for the $\Gamma_r$ in Exercise 5.4.3 which are hierarchical sets.*

**Exercise 5.4.5** *Are the sets*

(4.6) $$\Lambda_r = \{\alpha \in \mathcal{M} : l(\alpha) + n(\alpha) \le r\},$$

*where $r = 1,\ 2,\ \ldots$, hierarchical sets?*

## 5.5 Ito–Taylor Expansions

We shall now state and prove the Ito–Taylor expansion for a $d$-dimensional Ito process

(5.1) $$X_t = X_{t_0} + \int_{t_0}^t a(s, X_s)\, ds + \sum_{j=1}^m \int_{t_0}^t b^j(s, X_s)\, dW_s^j,$$

where $t \in [t_0, T]$, using the notation introduced in the preceding sections.

**Theorem 5.5.1** *Let $\rho$ and $\tau$ be two stopping times with*

(5.2) $$t_0 \leq \rho(\omega) \leq \tau(\omega) \leq T,$$

*w.p.1; let $\mathcal{A} \subset \mathcal{M}$ be an hierarchical set; and let $f : \Re^+ \times \Re^d \to \Re$. Then the* Ito-Taylor expansion

$$(5.3) \qquad f(\tau, X_\tau) = \sum_{\alpha \in \mathcal{A}} I_\alpha \left[ f_\alpha(\rho, X_\rho) \right]_{\rho, \tau} + \sum_{\alpha \in \mathcal{B}(\mathcal{A})} I_\alpha \left[ f_\alpha(\cdot, X_\cdot) \right]_{\rho, \tau},$$

*holds, provided all of the derivatives of $f$, $a$ and $b$ and all of the multiple Ito integrals appearing in (5.3) exist.*

We shall prove Theorem 5.5.1 at the end of this section by an iterated application of the Ito formula.

If we apply the Ito-Taylor expansion (5.3) in the case $d = m = 1$ for $f(t, x) \equiv x$, $\rho = t_0$, $\tau = t$ and the hierarchical set

$$\mathcal{A} = \{ \alpha \in \mathcal{M} l(\alpha) \leq 3 \}$$

to the autonomous Ito process $X_t$ with drift $a(x)$ and diffusion coefficient $b(x)$, then we obtain the following expansion:

$$
\begin{aligned}
(5.4) \quad X_t \;=\; & X_{t_0} + a\,I_{(0)} + b\,I_{(1)} + \left( aa' + \frac{1}{2}b^2 a'' \right) I_{(0,0)} \\[2mm]
& + \left( ab' + \frac{1}{2}b^2 b'' \right) I_{(0,1)} + ba'\,I_{(1,0)} + bb'\,I_{(1,1)} \\[2mm]
& + \left[ a \left( aa'' + (a')^2 + bb'a'' + \frac{1}{2}b^2 a''' \right) + \frac{1}{2}b^2 \left( aa''' + 3a'a'' \right) \right. \\[2mm]
& \qquad \left. + \left( (b')^2 + bb'' \right) a'' + 2bb'a''' \right) + \frac{1}{4}b^4 a^{(4)} \Big] I_{(0,0,0)} \\[2mm]
& + \left[ a \left( a'b' + ab'' + bb'b'' + \frac{1}{2}b^2 b''' \right) + \frac{1}{2}b^2 \left( a''b' + 2a'b'' \right. \right. \\[2mm]
& \qquad \left. \left. + ab''' + \left( (b')^2 + bb'' \right) b'' + 2bb'b''' + \frac{1}{2}b^2 b^{(4)} \right) \right] I_{(0,0,1)} \\[2mm]
& + \left[ a \left( b'a' + ba'' \right) + \frac{1}{2}b^2 \left( b''a' + 2b'a'' + ba''' \right) \right] I_{(0,1,0)} \\[2mm]
& + \left[ a \left( (b')^2 + bb'' \right) + \frac{1}{2}b^2 \left( b''b' + 2bb'' + bb''' \right) \right] I_{(0,1,1)} \\[2mm]
& + b \left( aa'' + (a')^2 + bb'a'' + \frac{1}{2}b^2 a''' \right) I_{(1,0,0)} \\[2mm]
& + b \left( ab'' + a'b' + bb'b'' + \frac{1}{2}b^2 b''' \right) I_{(1,0,1)} \\[2mm]
& + b \left( a'b' + a''b \right) I_{(1,1,0)} + b \left( (b')^2 + bb'' \right) I_{(1,1,1)} + R.
\end{aligned}
$$

Here we have used the abbreviations introduced in the previous sections.

The following example shows how the Ito-Taylor expansion reduces to the Ito formula. We take the hierarchical set $\mathcal{A} = \{v\}$, where $v$ denotes the multi-index of length zero (see (2.4)), which has the remainder set

(5.5) $$\mathcal{B}(\{v\}) = \{(0), (1), \ldots, (m)\}.$$

Then (5.3) takes the form

(5.6) $$\begin{aligned} f(\tau, X_\tau) &= I_v [f_v(\rho, X_\rho)]_{\rho, \tau} + \sum_{\alpha \in \mathcal{B}(\{v\})} I_\alpha [f_\alpha(\cdot, X.)]_{\rho, \tau} \\ &= f(\rho, X_\rho) + \int_\rho^\tau L^0 f(s, X_s)\, ds + \sum_{j=1}^m \int_\rho^\tau L^j f(s, X_s)\, dW_s^j, \end{aligned}$$

which is obviously the Ito formula. In this case the expansion part contains only the single term $f(\rho, X_\rho)$ and the remainder only the multiple Ito integrals of multiplicity one.

The Ito-Taylor expansion can, in a sense, be considered as a generalization of the deterministic Taylor formula. To illustrate this fact we shall consider the 1-dimensional case with $d = 1$, $f(t, x) = f(x)$, $a \equiv 1$, $b \equiv 0$, $\rho = 0$ and $\tau = t \in [0, T]$. This means that we are considering the situation where the Ito process reduces to the time variable, that is with

(5.7) $$X_t \equiv t \quad \text{for all} \quad t \in [0, T].$$

From (3.1)–(3.3), the Ito coefficient functions with multi- indices $\alpha = (j_1, \ldots, j_l)$ thus vanish if any $j_i \geq 1$ and the others are given by

(5.8) $$f_\alpha = \begin{cases} f & : \quad \alpha = v \\ f^{(l)} & : \quad l \geq 1 \text{ and } j_1 = \cdots = j_l = 0 \end{cases}$$

Here we assume that $f \in C^\infty$ and write $f^{(l)}$ for its $l$th derivative. For each $l = 0, 1, \ldots$ we take the hierarchical set

(5.9) $$\Gamma_l = \{\alpha \in \mathcal{M} : l(\alpha) \leq l\}$$

and the corresponding remainder set

(5.10) $$\mathcal{B}(\Gamma_l) = \{\alpha \in \mathcal{M} : l(\alpha) = l + 1\}$$

(see Exercises 5.4.3 and 5.4.4). Then, from the definition of multiple Ito integrals (2.12), we have

$$I_\alpha [f(X.)]_{t_0, t} = \int_{t_0}^t \cdots \int_{t_0}^{s_2} f(s_1)\, ds_1 \ldots ds_k$$

for $0 \leq t_0 \leq t < \infty$ and multi-indices $\alpha = \{j_1, \ldots, j_k\}$ with $j_1 = \ldots = j_k = 0$, $k = 1, 2, \ldots$ and $f \in \mathcal{H}_\alpha$. Using (5.7)–(5.10), for each $k = 0, 1, \ldots$ the Ito-Taylor expansion (5.3) here is then

$$(5.11) \quad f(t) \;=\; f(X_{t_0}) + \sum_{\alpha \in \Gamma_k \setminus \{v\}} I_\alpha \left[ f_\alpha(X_{t_0}) \right]_{t_0, t} + \sum_{\alpha \in \mathcal{B}(\Gamma_k)} I_\alpha \left[ f_\alpha(X_\cdot) \right]_{t_0, t}$$

$$= \sum_{i=0}^{k} \int_{t_0}^{t} \cdots \int_{t_0}^{s_2} f^{(i)}(X_{t_0}) \, ds_1 \ldots ds_i$$

$$+ \int_{t_0}^{t} \cdots \int_{t_0}^{s_2} f^{(k+1)}(X_{s_1}) \, ds_1 \ldots ds_{k+1}$$

$$= f(t_0) + \sum_{i=1}^{k} \frac{1}{i!} f^{(i)}(t_0)(t - t_0)^i + \int_{t_0}^{t} \cdots \int_{t_0}^{s_2} f^{(k+1)}(s_1) \, ds_1 \ldots ds_{k+1}.$$

This is obviously a version of the usual deterministic Taylor expansion.

The most important property of the Ito-Taylor expansion is the fact that it allows a sufficiently smooth function of an Ito process to be expanded as the sum of a finite number of terms represented by multiple Ito integrals with constant integrands and a remainder which consists of a finite number of other multiple Ito integrals with nonconstant integrands. This expansion is characterized by the particular choice of the hierarchical set. In practical situations there are usually sufficiently many degrees of freedom to choose the hierarchical set in an appropriate way. We shall use the Ito-Taylor expansion later to construct time discrete approximations of an Ito process.

**Exercise 5.5.2**    *Determine the truncated Ito-Taylor expansion, that is without the remainder term, in the autonomous case with $d = m = 1$, $f(t, x) \equiv x$, $\rho = 0$ and $\tau = t \in [0, T]$ for the hierarchical set $\mathcal{A} = \{v, (0), (1), (1,1)\}$.*

**Exercise 5.5.3**    *Determine the truncated Ito-Taylor expansion for the Ornstein-Uhlenbeck process (that is with $d = m = 1$, $a(t, x) = -x$ and $b(t, x) \equiv 1$, see Exercise 1.7.2) for $f(t, x) \equiv x$, $\rho = 0$ and $\tau = t \in [0, T]$ and the hierarchical set $\Gamma_l$ for each $l = 1, 2, \ldots$ defined in (5.9).*

We shall now prove the Ito-Taylor expansion by an iterated application of a slightly more general version of the Ito formula than we derived in Section 3 of Chapter 3. To begin we shall formulate this generalized Ito formula in terms of the notation of the present chapter.

**Lemma 5.5.4**    *Let $\rho$ and $\tau$ be two stopping times with*

$$(5.12) \qquad\qquad t_0 \leq \rho(\omega) \leq \tau(\omega) \leq T,$$

*w.p.1, and let $f : \Re^+ \times \Re^d \to \Re$ belong to the class $C^{1,2}$. Then the following version of the Ito formula holds*

$$(5.13) \qquad f(\tau, X_\tau) = f(\rho, X_\rho) + \sum_{j=0}^{m} I_{(j)} \left[ L^j f(\cdot, X_\cdot) \right]_{\rho, \tau}.$$

**Proof**   For each stopping time $\rho$ with $t_0 \leq \rho \leq T < \infty$, w.p.1, the process $\{X_{t \wedge \rho}, \ t \geq t_0\}$ is a semi-martingale (see Ikeda and Watanabe (1989)). Thus from the Ito formula for semi-martingales and from (2.12) and (3.1)–(3.2) it follows for all $t \geq T$ that

$$
(5.14) \qquad f(\rho, X_\rho) = f(t \wedge \rho, X_{t \wedge \rho})
$$

$$
= f(t_0, X_{t_0}) + \sum_{j=0}^{m} I_{(j)} \left[ L^j f(\cdot, X_\cdot) \right]_{t_0, t \wedge \rho}
$$

$$
= f(t_0, X_{t_0}) + \sum_{j=0}^{m} I_{(j)} \left[ L^j f(\cdot, X_\cdot) \right]_{t_0, \rho}.
$$

We subtract this from (5.13) to obtain

$$
f(\tau, X_\tau) - f(\rho, X_\rho) = \sum_{j=0}^{m} I_{(j)} \left[ L^j f(\cdot, X_\cdot) \right]_{t_0, \tau} - \sum_{j=0}^{m} I_{(j)} \left[ L^j f(\cdot, X_\cdot) \right]_{t_0, \rho}.
$$

$$
= \sum_{j=0}^{m} I_{(j)} \left[ L^j f(\cdot, X_\cdot) \right]_{\rho, \tau},
$$

which proves the lemma. $\square$

   We recall from (2.6) the concatenation operation $*$ which adjoins two multi-indices.

**Lemma 5.5.5**   *Let $\rho$ and $\tau$ be two stopping times with*

$$
(5.15) \qquad\qquad t_0 \leq \rho(\omega) \leq \tau(\omega) \leq T,
$$

*w.p.1; let $f : \Re^+ \times \Re^d \to \Re$; and let $\alpha, \ \beta \in \mathcal{M}$ with $l(\beta) \geq 1$. Then*

$$
(5.16) \qquad I_\alpha \left[ f_\beta(\cdot, X_\cdot) \right]_{\rho, \tau} = I_\alpha \left[ f_\beta(\rho, X_\rho) \right]_{\rho, \tau} + \sum_{j=0}^{m} I_{(j)*\alpha} \left[ f_{(j)*\beta}(\cdot, X_\cdot) \right]_{\rho, \tau},
$$

*provided $f_\beta \in \mathcal{H}_\alpha$ and all of the derivatives and multiple stochastic integrals appearing in (5.16) exist.*

**Proof**   We shall prove this formula by induction on $l(\alpha)$.

   For $l(\alpha) = 0$ we have $\alpha = v$. Hence it follows from (2.12), Lemma 5.5.4, (3.3) and (2.6) that

$$
(5.17) \quad I_\alpha \left[ f_\beta(\cdot, X_\cdot) \right]_{\rho, \tau} = f_\beta(\tau, X_\tau)
$$

$$
= f_\beta(\rho, X_\rho) + \sum_{j=0}^{m} I_{(j)} \left[ L^j f_\beta(\cdot, X_\cdot) \right]_{\rho, \tau}
$$

$$
= I_\alpha \left[ f_\beta(\rho, X_\rho) \right]_{\rho, \tau} + \sum_{j=0}^{m} I_{(j)*\alpha} \left[ f_{(j)*\beta}(\cdot, X_\cdot) \right]_{\rho, \tau}.
$$

Now let $l(\alpha) = k \geq 1$, where $\alpha = (j_1, \ldots, j_k)$. Then, from (2.12) and the inductive assumption we have

$$(5.18) \quad I_\alpha \left[f_\beta(\cdot, X.)\right]_{\rho,\tau} = I_{(j_k)} \left[I_{\alpha-}\left[(f_\beta(\cdot, X.))\right]_{\rho,\cdot}\right]_{\rho,\tau}$$

$$= I_{(j_k)} \left[I_{\alpha-}\left[(f_\beta(\rho, X_\rho))\right]_{\rho,\cdot}\right]_{\rho,\tau}$$

$$+ \sum_{j=0}^{m} I_{(j_k)} \left[I_{(j)*\alpha-}\left[f_{(j)*\beta}(\cdot, X.)\right]_{\rho,\cdot}\right]_{\rho,\tau}$$

$$= I_\alpha \left[f_\beta(\rho, X_\rho)\right]_{\rho,\tau}$$

$$+ \sum_{j=0}^{m} I_{(j)*\alpha}\left[f_{(j)*\beta}(\cdot, X.)\right]_{\rho,\tau}. \quad \Box$$

**Proof of Theorem 5.5.1** We shall prove this theorem by induction on

$$(5.19) \qquad\qquad l_1(\mathcal{A}) = \sup_{\alpha \in \mathcal{A}} l(\alpha).$$

For $l_1(\mathcal{A}) = 0$ we have $\mathcal{A} = \{v\}$ with the remainder set

$$\mathcal{B}(\mathcal{A}) = \{(0), (1), \cdots, (m)\}.$$

It follows then from Lemma 5.5.4 with (2.12) and (3.3) that

$$f(\tau, X_\tau) = \sum_{\alpha \in \mathcal{A}} I_\alpha \left[f_\alpha(\rho, X_\rho)\right]_{\rho,\tau} + \sum_{\alpha \in \mathcal{B}(\mathcal{A})} I_\alpha \left[f_\alpha(\cdot, X.)\right]_{\rho,\tau}.$$

Now let $l_1(\mathcal{A}) = k \geq 1$. If we set

$$\mathcal{E} = \{\alpha \in \mathcal{A} : l(\alpha) \leq k - 1\},$$

which is an hierarchical set, then by the inductive assumption we obtain

$$(5.20) \qquad f(\tau, X_\tau) = \sum_{\alpha \in \mathcal{E}} I_\alpha \left[f_\alpha(\rho, X_\rho)\right]_{\rho,\tau} + \sum_{\alpha \in \mathcal{B}(\mathcal{E})} I_\alpha \left[f_\alpha(\cdot, X.)\right]_{\rho,\tau}.$$

Since $\mathcal{A}$ is an hierarchical set with $l_1(\alpha) = k$, it follows from (4.3) and (4.4) that

$$(5.21) \qquad\qquad \mathcal{A} \setminus \mathcal{E} \subseteq \mathcal{B}(\mathcal{E}).$$

For $\beta = \alpha \in \mathcal{A} \setminus \mathcal{E}$ the assumptions of Lemma 5.5.4 hold, so we can rewrite (5.20) as

$$f(\tau, X_\tau) = \sum_{\alpha \in \mathcal{E}} I_\alpha \left[f_\alpha(\rho, X_\rho)\right]_{\rho,\tau} + \sum_{\alpha \in \mathcal{A} \setminus \mathcal{E}} I_\alpha \left[f_\alpha(\cdot, X.)\right]_{\rho,\tau}$$

$$+ \sum_{\alpha \in \mathcal{B}(\mathcal{E}) \setminus (\mathcal{A} \setminus \mathcal{E})} I_\alpha \left[f_\alpha(\cdot, X.)\right]_{\rho,\tau}$$

$$= \sum_{\alpha \in \mathcal{E}} I_\alpha \left[ f_\alpha \left( \rho, X_\rho \right) \right]_{\rho, \tau}$$

$$+ \sum_{\alpha \in \mathcal{A} \setminus \mathcal{E}} \left[ I_\alpha \left[ f_\alpha \left( \rho, X_\rho \right) \right]_{\rho, \tau} + \sum_{j=0}^{m} I_{(j)*\alpha} \left[ f_{(j)*\alpha}(\cdot, X_\cdot) \right]_{\rho, \tau} \right]$$

$$+ \sum_{\alpha \in \mathcal{B}(\mathcal{E}) \setminus (\mathcal{A} \setminus \mathcal{E})} I_\alpha \left[ f_\alpha(\cdot, X_\cdot) \right]_{\rho, \tau}$$

$$= \sum_{\alpha \in \mathcal{A}} I_\alpha \left[ f_\alpha \left( \rho, X_\rho \right) \right]_{\rho, \tau} + \sum_{\alpha \in \mathcal{B}_1} I_\alpha \left[ f_\alpha(\cdot, X_\cdot) \right]_{\rho, \tau}.$$

Now because of (4.4) we have

$$\mathcal{B}_1 = \left[ \mathcal{B}(\mathcal{E}) \setminus (\mathcal{A} \setminus \mathcal{E}) \right] \bigcup \left[ \bigcup_{j=0}^{m} \{(j) * \alpha \in \mathcal{M} : \alpha \in \mathcal{A} \setminus \mathcal{E}\} \right]$$

$$= \left[ \{\alpha \in \mathcal{M} \setminus \mathcal{E} : -\alpha \in \mathcal{E}\} \setminus \{\alpha \in \mathcal{M} \setminus \mathcal{E} : \alpha \in \mathcal{A}\} \right]$$

$$\bigcup \{\alpha \in \mathcal{M} : -\alpha \in \mathcal{A} \setminus \mathcal{E}\}$$

$$= \{\alpha \in \mathcal{M} \setminus \mathcal{A} : -\alpha \in \mathcal{E}\} \bigcup \{\alpha \in \mathcal{M} \setminus \mathcal{A} : -\alpha \in \mathcal{A} \setminus \mathcal{E}\}$$

$$= \{\alpha \in \mathcal{M} \setminus \mathcal{A} : -\alpha \in \mathcal{A}\}$$

$$= \mathcal{B}(\mathcal{A}).$$

This completes the proof of Theorem 5.5.1. $\square$

## 5.6 Stratonovich-Taylor Expansions

We shall now formulate the Stratonovich-Taylor expansion for the $d$- dimensional Ito process

(6.1) $$X_t = X_{t_0} + \int_{t_0}^{t} a(s, X_s) \, ds + \sum_{j=1}^{m} \int_{t_0}^{t} b^j(s, X_s) \, dW_s^j$$

$$= X_{t_0} + \int_{t_0}^{t} \underline{a}(s, X_s) \, ds + \sum_{j=1}^{m} \int_{t_0}^{t} b^j(s, X_s) \circ dW_s^j,$$

where $t \in [t_0, T]$. For this we shall use the terminology that was introduced in the preceding sections.

**Theorem 5.6.1** *Let $\rho$ and $\tau$ be two stopping times with*

(6.2) $$t_0 \leq \rho \leq \tau \leq T < \infty,$$

*w.p.1; let $f : \Re^+ \times \Re^d \to \Re$; and let $\mathcal{A} \subset \mathcal{M}$ be an hierarchical set. Then the*
Stratonovich-Taylor expansion

$$(6.3) \qquad f(\tau, X_\tau) = \sum_{\alpha \in \mathcal{A}} J_\alpha \left[ \underline{f}_\alpha (\rho, X_\rho) \right]_{\rho, \tau} + \sum_{\alpha \in \mathcal{B}(\mathcal{A})} J_\alpha \left[ \underline{f}_\alpha (\cdot, X.) \right]_{\rho, \tau},$$

*holds, provided all of the derivatives of $f$, $a$ and $b$ and all of the multiple Stratonovich integrals appearing in (6.3) exist.*

We shall give an indication of the proof of the theorem at the end of the section.

Let us apply the Stratonovich-Taylor expansion in the case $d = m = 1$ for $f(t, x) \equiv x$, $\rho = t_0$, $\tau = t$ and the hierarchical set

$$\mathcal{A} = \{\alpha \in \mathcal{M} : l(\alpha) \leq 3\}$$

to an autonomous Ito process. Then we obtain the expansion

$$(6.4) \quad X_t \;\; = \;\; X_{t_0} + \underline{a} J_{(0)} + b J_{(1)} + \underline{a} a' J_{(0,0)} + \underline{a} b' J_{(0,1)} + b \underline{a}' J_{(1,0)}$$

$$+ bb' J_{(1,1)} + \underline{a} \left( \underline{a} a'' + (\underline{a}')^2 \right) J_{(0,0,0)} + \underline{a} \left( \underline{a} b'' + \underline{a}' b' \right) J_{(0,0,1)}$$

$$+ \underline{a} \left( a'' b + \underline{a}' b' \right) J_{(0,1,0)} + b \left( \underline{a} a'' + (\underline{a}')^2 \right) J_{(1,0,0)}$$

$$+ \underline{a} \left( bb'' + (b')^2 \right) J_{(0,1,1)} + b \left( \underline{a} b'' + \underline{a}' b' \right) J_{(1,0,1)}$$

$$+ b ( \underline{a}'' b + \underline{a}' b' ) J_{(1,1,0)} + b \left( bb'' + (b')^2 \right) J_{(1,1,1)} + R,$$

where we have used previously introduced abbreviations. Compare this with the corresponding Ito-Taylor expansion (5.4).

If we take the hierarchical set $\mathcal{A} = \{v\}$ and its remainder set

$$\mathcal{B}(\{v\}) = \{(0), \cdots, (m)\},$$

then we obtain from (6.3)

$$(6.5) \qquad f(\tau, X_\tau) \;\; = \;\; J_v \left[ \underline{f}_v (\rho, X_\rho) \right]_{\rho, \tau} + \sum_{\alpha \in \mathcal{B}(\{v\})} J_\alpha \left[ \underline{f}_\alpha (\cdot, X.) \right]_{\rho, \tau}$$

$$= \;\; f(\rho, X_\rho) + \int_\rho^\tau \underline{L}^0 f(s, X_s) \, ds$$

$$+ \sum_{j=1}^m \int_{t_0}^t \underline{L}^j f(s, X_s) \circ dW_s^j,$$

which is the Stratonovich counterpart to the Ito formula and is similar to the deterministic chain rule. It is easy to see that (5.3) reduces to the deterministic Taylor formula, as in (5.11), in the 1-dimensional case where $a \equiv 1$ and all of the $b^j \equiv 0$. Thus the Stratonovich-Taylor expansion can also be regarded as a generalization of the deterministic Taylor formula. It allows us to expand a sufficiently smooth function of an Ito process as the sum of a finite number of terms involving multiple Stratonovich integrals with constant integrands and a remainder term consisting of a finite number of other multiple Stratonovich integrals with nonconstant integrands. Like the Ito-Taylor expansion, it is also characterized by the choice of the hierarchical set. However, for the same hierarchical set the Ito-Taylor expansion usually involves higher order derivatives of $f$, $a$ and $b$ than does the Stratonovich- Taylor expansion. Because of its analogous structure to the deterministic Taylor expansion, the Stratonovich-Taylor expansion seems to be a more natural generalization of it than is the Ito-Taylor expansion. Consequently, the Stratonovich-Taylor expansion may sometimes be more convenient for, amongst other things, the investigation of numerical schemes.

**Exercise 5.6.2**    *Determine the truncated Stratonovich-Taylor expansion for the hierarchical set $\mathcal{A} = \{v, (0), (1), (1,1)\}$ in the case $d = m = 1$, $f(t,x) \equiv x$, $\rho = 0$ and $\tau = t \in [0, T]$.*

**Exercise 5.6.3**    *Determine the truncated Stratonovich-Taylor expansions for the hierarchical sets $\Gamma_l$ as defined in (5.9), where $l = 1, 2, \ldots$ for the function $f(t,x) \equiv x$ and the Ornstein-Uhlenbeck process (for which $d = m = 1$, $a(t,x) = -x$ and $b(t,x) \equiv 1$). Compare the result with the corresponding Ito-Taylor expansions from Exercise 5.5.3.*

The Stratonovich-Taylor expansion is proved by an iterated application of the Stratonovich formula (6.5) using the following lemma. The concatenation operation $*$ here simply adjoins the two multi-indices (see (2.5)).

**Lemma 5.6.4**    *Let $\rho$ and $\tau$ be two stopping times with $t_0 \leq \rho \leq \tau \leq T < \infty$, w.p.1; let $f : \Re^+ \times \Re^d \to \Re$; and let $\alpha$, $\beta \in \mathcal{M}$ with $l(\beta) \geq 1$. Then*

$$(6.6) \qquad J_\alpha \left[\underline{f}_\beta(\cdot, X_\cdot)\right]_{\rho,\tau} = J_\alpha \left[\underline{f}_\beta(\rho, X_\rho)\right]_{\rho,\tau} + \sum_{j=0}^{m} J_{(j)*\alpha} \left[\underline{f}_{(j)*\beta}(\cdot, X_\cdot)\right]_{\rho,\tau}$$

*holds, provided that $\underline{f}_\beta \in \mathcal{H}_\alpha$ and that all of the multiple Stratonovich integrals and all of the derivatives appearing in (6.6) exist.*

**Proof**    The proof is analogous to the proof of Lemma 5.5.5 for Ito-Taylor expansions, except that here we need to use the definitions of multiple Stratonovich integrals (2.31) and of Stratonovich coefficient functions (3.9). □

The proof of Theorem 5.6.1 is then an easy consequence of Lemma 5.6.4. It mimics that of Theorem 5.5.1 for the corresponding Ito-Taylor expansion, so is omitted.

**Exercise 5.6.5**    *Write out full proofs for Lemma 5.6.4 and Theorem 5.6.1.*

## 5.7   Moments of Multiple Ito Integrals

For applications of stochastic Taylor expansions it is often necessary to estimate the multiple stochastic integrals appearing in the remainder terms. In this
section we shall derive several estimates of moments of multiple Ito stochastic
integrals, which we shall use later in the construction of time discrete approximations of Ito processes. These estimates may also be useful in other
applications of stochastic Taylor expansions.

### Multiple Integrals with Zero First Moments

To begin we shall show that the first moment of a multiple Ito integral vanishes
if it has at least one integration with respect to a component of the Wiener
process. This means that not all of the components of the multi-index $\alpha$ are
equal to 0, which is equivalent to the condition

$$l(\alpha) \neq n(\alpha).$$

This property is thus a generalization of property (3.2.9) of the Ito integral.

**Lemma 5.7.1**   $Let\ \alpha \in \mathcal{M} \setminus \{v\}$ $with\ l(\alpha) \neq n(\alpha),$ $let\ f \in \mathcal{H}_\alpha$ $and\ let\ \rho\ and$
$\tau\ be\ two\ stopping\ times\ with\ t_0 \leq \rho \leq \tau \leq T < \infty,\ w.p.1.$ $Then$

$$(7.1) \qquad E\left(I_\alpha\left[f(\cdot)\right]_{\rho,\tau} \Big| \mathcal{A}_\rho\right) = 0, \quad w.p.1.$$

**Proof**   In the proof we shall use of the properties of local martingales, details
of which the reader can find in Ikeda and Watanabe (1989). In view of (2.12)
we can write

$$(7.2) \qquad E\left(I_\alpha\left[f(\cdot)\right]_{\rho,\tau} \Big| \mathcal{A}_\rho\right) = E\left(I_\alpha\left[f(\cdot)I_{\{\cdot \leq \tau\}}\right]_{\rho,T} \Big| \mathcal{A}_\rho\right).$$

When $\alpha = (j_1, \ldots, j_k)$ with $j_k \in \{1, \ldots, m\}$ the process

$$\left\{I_\alpha\left[f(\cdot)I_{\{\cdot \leq \tau\}}\right]_{\rho,t}, \rho \leq t \leq T\right\}$$

is a local martingale, so (7.1) holds. On the other hand when $\alpha = (j_1, \ldots, j_k)$
with $j_k = 0$ we apply (2.12) and obtain

$$(7.3) \quad E\left(I_\alpha\left[f(\cdot)\right]_{\rho,\tau} \Big| \mathcal{A}_\rho\right) = E\left(\int_\rho^\tau I_{\alpha-}\left[f(\cdot)I_{\{\cdot \leq \tau\}}\right]_{\rho,s} ds \Big| \mathcal{A}_\rho\right)$$

$$= \int_\rho^\tau E\left(I_{\alpha-}\left[f(\cdot)I_{\{\cdot \leq \tau\}}\right]_{\rho,s} \Big| \mathcal{A}_\rho\right) ds.$$

Then, if $j_{k-1} \in \{1, \ldots, m\}$, the process

$$\left\{I_{\alpha-}\left[f(\cdot)I_{\{\cdot \leq \tau\}}\right]_{\rho,s}, \rho \leq s \leq \tau\right\}$$

is a local martingale and thus (7.1) holds. However, if $j_{k-1} = 0$ we proceed as
above until we reach a component of $\alpha$ which is not equal to 0, and then (7.1)
follows.   $\square$

## A Second Moment Estimate

We shall now derive an estimate for the second moment of a multiple Ito integral and the covariance of two such integrals. In order to present the result and proof in a clear and compact form we shall first introduce some additional terminology.

Given a multi-index $\alpha \in \mathcal{M}$, we define $\alpha^+$ to be the multi-index obtained from $\alpha$ by deleting all of the components equal to 0. For example, if $\alpha = (1,0,2,1)$, then we have

$$\alpha^+ = (1,0,2,1)^+ = (1,2,1).$$

We shall denote by $k_0(\alpha)$ the number of components of $\alpha$ equal to 0 preceding the first nonzero component of $\alpha$ or until the end of $\alpha$ if all of its components are zeros. In addition, we shall denote by $k_i(\alpha)$, for $i = 1, \ldots, l(\alpha^+)$, the number of components of $\alpha$ between the $i$th nonzero component and the $(i+1)$th nonzero component or the end of $\alpha$ if $i = l(\alpha^+)$. Thus, for $\alpha = (0,1,2,0)$ we have $\alpha^+ = (1,2)$, $l(\alpha^+) = 2$ and

$$k_0(\alpha) = 1, \quad k_1(\alpha) = 0, \quad k_2(\alpha) = 1.$$

We shall also use the combinatorial symbol

$$(7.4) \qquad C_i^k = \frac{i!}{k!(i-k)!},$$

for $k = 0, 1, \ldots, i$ and $i = 0, 1, \ldots$ with the convention that $0! = 1$. Finally, for any $\alpha, \beta \in \mathcal{M}$ we shall write

$$(7.5) \qquad w(\alpha, \beta) = l(\alpha^+) + \sum_{i=0}^{l(\alpha^+)} (k_i(\alpha) + k_i(\beta)).$$

The following lemma is valid for both scalar and vector functions. In the latter case $(f, g)$ is the usual Euclidean scalar product.

**Lemma 5.7.2** Let $\alpha, \beta \in \mathcal{M}$, let $f \in \mathcal{H}_\alpha$, $g \in \mathcal{H}_\beta$ and let $\rho$ and $\tau$ be two stopping times with $t_0 \leq \rho \leq \tau \leq T < \infty$, w.p.1, where $\tau$ is $\mathcal{A}_\rho$-measurable. Then

$$(7.6) \quad E\left( \left( I_\alpha\left[ f(\cdot) \right]_{\rho,\tau}, I_\beta\left[ g(\cdot) \right]_{\rho,\tau} \right) \big| \mathcal{A}_\rho \right)$$

$$\begin{cases} = 0 & : \ \alpha^+ \neq \beta^+ \\[2mm] \leq K_{f,g} \dfrac{(\tau - \rho)^{w(\alpha,\beta)}}{w(\alpha,\beta)!} \prod_{i=0}^{l(\alpha^+)} C_{k_i(\alpha)+k_i(\beta)}^{k_i(\alpha)} & : \ \alpha^+ = \beta^+, \end{cases}$$

with

$$(7.7) \qquad K_{f,g} = \sup_{s_1,s_2 \in [\rho,\tau]} E\left( |(f(s_1), g(s_2))| \, \big| \mathcal{A}_\rho \right) \qquad where$$

$$C_{k_i(\alpha)+k_i(\beta)}^{k_i(\alpha)} = \frac{(k_i(\alpha) + k_i(\beta))!}{k_i(\alpha)! k_i(\beta)!}.$$

Moreover, (7.6) holds with equality when $f \equiv g \equiv 1$.

**Proof** For all $\alpha \in \mathcal{M}$ with $\alpha = \alpha^+ = (j_1, \ldots, j_l)$, all $g \in \mathcal{H}_\alpha$ and $l$, $k_i = 0$, $1, \ldots$ with $i \in \{0, \ldots, l\}$ we define recursively the multiple stochastic integrals

(7.8) $\qquad H_{\alpha^+}[k_0, \ldots, k_l; g(\cdot)]_{\rho,\tau}$

$$= \begin{cases} g(\tau) & : \; l = 0, k_0 = 0 \\ \int_\rho^\tau H_v[k_0 - 1; g(\cdot)]_{\rho,u} \, du & : \; l = 0, k_0 \geq 1 \\ \int_\rho^\tau \frac{(\tau - u)^{k_l}}{k_l!} H_{\alpha^+-}[k_0, \ldots, k_{l-1}; g(\cdot)]_{\rho,u} \, dW_u^{j_l} & : \; l \geq 1 \end{cases}$$

Now for any $g(\cdot, t) \in \mathcal{H}_{(0,1)}$ and $j \in \{1, \ldots, m\}$ we have

(7.9) $\qquad \zeta_1 := \int_\rho^\tau \int_\rho^\tau g(s,t) \, dW_s^j \, dt = \int_\rho^\tau \int_\rho^\tau g(s,t) \, dt \, dW_s^j =: \zeta_2,$

w.p.1, since from the properties of the Ito integral (see Lemma 3.2.2) it can be shown that

$$E\left(|\zeta_1 - \zeta_2|^2\right) = E\left(|\zeta_1|^2\right) - 2E\left((\zeta_1, \zeta_2)\right) + E\left(|\zeta_2|^2\right) = 0.$$

The multiple integrals defined in (7.8) satisfy the relation

(7.10) $\qquad \zeta_3 \;\; := \;\; \int_\rho^\tau H_{\alpha^+}[k_0, \ldots, k_{l-1}, k_l; g(\cdot)]_{\rho,u} \, du$

$$= \;\; H_{\alpha^+}[k_0, \ldots, k_{l-1}, k_l + 1; g(\cdot)]_{\rho,\tau}.$$

This follows directly from the definition when $l = 0$. For the case $l \geq 1$, using (7.5) and (7.8), we have

(7.11) $\quad \zeta_3 \;\; = \;\; \int_\rho^\tau \int_\rho^\tau I_{\{u \in [\rho,z]\}} \frac{(z-u)^{k_l}}{k_l!} H_{\alpha^+-}[0, \ldots, k_{l-1}; g(\cdot)]_{\rho,u} \, dW_u^{j_l} \, dz$

$$= \;\; \int_\rho^\tau H_{\alpha^+-}[k_0, \ldots, k_{l-1}; g(\cdot)]_{\rho,u} \int_\rho^\tau I_{\{u \in [\rho,z]\}} \frac{(z-u)^{k_l}}{k_l!} \, dz \, dW_u^{j_l}$$

$$= \;\; \int_\rho^\tau H_{\alpha^+-}[k_0, \ldots, k_{l-1}; g(\cdot)]_{\rho,u} \frac{(\tau - u)^{k_l+1}}{(k_l + 1)!} \, dW_u^{j_l}$$

$$= \;\; H_{\alpha^+}[k_0, \ldots, k_{l-1}, k_l + 1); g(\cdot)]_{\rho,\tau}.$$

In addition, we can show by induction for any $\alpha \in \mathcal{M}$ and $f \in \mathcal{H}_\alpha$ that

(7.12) $\qquad I_\alpha[f(\cdot)]_{\rho,\tau} = H_{\alpha^+}[k_0(\alpha), \ldots, k_{l(\alpha^+)}(\alpha); f(\cdot)]_{\rho,\tau}.$

When $l(\alpha) = 0$ we have $\alpha = \alpha^+ = v$, so (7.10) follows from (2.12) and (7.5). When $l(\alpha) = l \geq 1$ for $\alpha = (j_1, \ldots, j_l)$ with $j_l \in \{1, \ldots, m\}$ we have $(\alpha-)^+ = (\alpha^+)-$ and $k_{l(\alpha+)} = 0$. From (2.12), the inductive assumption and (7.5) we then obtain

$$
\begin{aligned}
I_\alpha \left[ f(\cdot) \right]_{\rho,\tau} &= \int_\rho^\tau I_{\alpha-} \left[ f(\cdot) \right]_{\rho,u} dW_u^j \\
&= \int_\rho^\tau H_{(\alpha+)-} \left[ k_0(\alpha), \ldots, k_{l(\alpha+)-1}(\alpha); f(\cdot) \right]_{\rho,u} dW_u^{j_l} \\
&= H_{\alpha+} \left[ k_0(\alpha), \ldots, k_{l(\alpha+)-1}(\alpha), 0; f(\cdot) \right]_{\rho,\tau}
\end{aligned}
$$

Finally, when $l(\alpha) = l \geq 1$ for $\alpha = (j_1, \ldots, j_l)$ with $j_l = 0$ we have $\alpha^+ = (\alpha-)^+$ and $k_{l(\alpha+)} \geq 1$. Then from (2.12), the inductive assumption and (7.10) we have

$$
\begin{aligned}
I_\alpha \left[ f(\cdot) \right]_{\rho,\tau} &= \int_\rho^\tau I_{\alpha-} \left[ f(\cdot) \right]_{\rho,u} du \\
&= \int_\rho^\tau H_{\alpha+} \left[ k_0(\alpha), \ldots, k_{l(\alpha+)}(\alpha) - 1; f(\cdot) \right]_{\rho,u} du \\
&= H_{\alpha+} \left[ k_0(\alpha), \ldots, k_{l(\alpha+)}(\alpha); f(\cdot) \right]_{\rho,\tau} .
\end{aligned}
$$

This completes the proof of (7.10)

We shall now prove (7.6) by induction on $l(\alpha^+)$ for the case that $\alpha^+ = \beta^+$. For this we shall require the identity

$$
(7.13) \qquad \int_\rho^\tau (\tau - u)^k (u - \rho)^i \, du = \frac{k! \, i!}{(k+i+1)!} (\tau - \rho)^{k+i+1},
$$

which can be easily shown by integrating by parts. When $l(\alpha^+) = l(\beta^+) = 0$ we have $\alpha^+ = \beta^+ = v$. Hence when $l(\alpha^+) \geq 1$ and $l(\beta^+) \geq 1$ it follows from (7.5) and (7.8) that

$$
(7.14) \qquad E \left( \left( I_\alpha \left[ f(\cdot) \right]_{\rho,\tau}, I_\beta \left[ g(\cdot) \right]_{\rho,\tau} \right) \Big| \mathcal{A}_\rho \right)
$$

$$
= E \left( \left( \int_\rho^\tau \cdots \int_\rho^{u_2} f(u_1) \, du_1 \ldots du_{l(\alpha)}, \int_\rho^\tau \cdots \int_\rho^{s_2} g(s_1) \, ds_1 \ldots ds_{l(\beta)} \right) \Big| \mathcal{A}_\rho \right)
$$

$$
= \int_\rho^\tau \cdots \int_\rho^{u_2} \int_\rho^\tau \cdots \int_\rho^{s_2} E \left( (f(u_1), g(s_1)) \Big| \mathcal{A}_\rho \right) ds_1 \ldots ds_{l(\beta)} du_1 \ldots du_{l(\alpha)}
$$

$$
\leq K_{f,g} \frac{(\tau - \rho)^{l(\alpha)+l(\beta)}}{l(\alpha)! \, l(\beta)!}
$$

$$
= K_{f,g} \frac{(\tau - \rho)^{w(\alpha,\beta)}}{w(\alpha,\beta)!} \prod_{i=0}^{l(\alpha^+)} C_{k_i(\alpha)+k_i(\beta)}^{k_i(\alpha}.
$$

The same result follows in an analogous manner when $l(\alpha) = l(\beta) = 0$. Finally, when $l(\alpha^+) = l(\beta^+) = l \geq 1$ with $\alpha^+ = \beta^+ = (j_1, \ldots, j_l)$ we obtain from (7.5), (7.8) and the inductive assumption

$$E\left(\left(I_\alpha\left[f(\cdot)\right]_{\rho,\tau}, I_\beta\left[g(\cdot)\right]_{\rho,\tau}\right) \Big| \mathcal{A}_\rho\right)$$

$$= E\left(\left(\int_\rho^\tau \frac{(\tau - u)^{k_l(\alpha)}}{k_l(\alpha)!} H_{(\alpha^+)-}\left[k_0(\alpha), \ldots, k_{l-1}(\alpha); f(\cdot)\right]_{\rho,u} dW_u^{j_l},\right.\right.$$

$$\left.\left.\int_\rho^\tau \frac{(\tau - u)^{k_l(\beta)}}{k_l(\beta)!} H_{(\beta^+)-}\left[k_0(\beta), \ldots, k_{l-1}(\beta); g(\cdot)\right]_{\rho,u} dW_u^{j_l}\right) \Big| \mathcal{A}_\rho\right)$$

$$= \int_\rho^\tau \frac{(\tau - u)^{k_l(\alpha)+k_l(\beta)}}{k_l(\alpha)!k_l(\beta)!} E\left(\left(H_{(\alpha^+)-}\left[k_0(\alpha), \ldots, k_{l-1}(\alpha); f(\cdot)\right]_{\rho,u},\right.\right.$$

$$\left.\left. H_{(\beta^+)-}\left[k_0(\beta), \ldots, k_{l-1}(\beta); g(\cdot)\right]_{\rho,u}\right) \Big| \mathcal{A}_\rho\right) du$$

$$\leq K_{f,g} \int_\rho^\tau \frac{(\tau - u)^{k_l(\alpha)+k_l(\beta)}(u - \rho)^{\bar{w}(\alpha,\beta)}}{k_l(\alpha)!k_l(\beta)!\bar{w}(\alpha,\beta)!} du \prod_{i=0}^{l-1} C_{k_i(\alpha)+k_i(\beta)}^{k_i(\alpha}),$$

where

$$\bar{w}(\alpha,\beta) = l - 1 + \sum_{i=0}^{l-1}(k_i(\alpha) + k_i(\beta)).$$

Then, using the identity (7.9), we have

$$E\left(\left(I_\alpha\left[f(\cdot)\right]_{\rho,\tau}, I_\beta\left[g(\cdot)\right]_{\rho,\tau}\right) \Big| \mathcal{A}_\rho\right)$$

$$\leq K_{f,g} \frac{(k_l(\alpha) + k_l(\beta))\,\bar{w}(\alpha,\beta)!(\tau - \rho)^{w(\alpha,\beta)}}{k_l(\alpha)!k_l(\beta)!\bar{w}(\alpha,\beta)!w(\alpha,\beta)!} \prod_{i=0}^{l-1} C_{k_i(\alpha)+k_i(\beta)}^{k_i(\alpha})$$

$$\leq K_{f,g} \frac{(\tau - \rho)^{w(\alpha,\beta)}}{w(\alpha,\beta)!} \prod_{i=0}^{l-1} C_{k_i(\alpha)+k_i(\beta)}^{k_i(\alpha}),$$

where equality holds if $f \equiv g \equiv 1$.

Finally, it remains to observe that when $\alpha^+ \neq \beta^+$ the result follows from (7.5), (7.8) and the properties of the Ito integral. $\square$

## A Uniform Mean-Square Estimate

We shall now establish a uniform mean-square estimate for multiple Ito integrals, which we shall formulate in a way that is convenient for later purposes.

**Lemma 5.7.3**    *Let $\alpha \in \mathcal{M} \setminus \{v\}$, let $g \in \mathcal{H}_\alpha$, let $\delta > 0$ and let $\rho$ and $\tau$ be two stopping times with $\tau$ $\mathcal{A}_\rho$-measurable and $t_0 \le \rho \le \tau \le \rho + \delta \le T < \infty$, w.p.1. Then*

$$(7.15) \qquad H_\tau = E \left( \sup_{\rho \le s \le \tau} \left| I_\alpha \left[ g(\cdot) \right]_{\rho,s} \right|^2 \Big| \mathcal{A}_\rho \right)$$

$$\le \; 4^{l(\alpha)-n(\alpha)} \delta^{l(\alpha)+n(\alpha)-1} \int_\rho^\tau R_{\rho,s} \, ds$$

*where*

$$(7.16) \qquad R_{\rho,s} = E \left( \sup_{\rho \le t \le s} |g(t)|^2 \Big| \mathcal{A}_\rho \right) < \infty$$

*for $s \in [\rho, \tau]$.*

**Proof**  We shall prove the assertion by induction on $l(\alpha)$. For $l(\alpha) = 1$ and $\alpha = (0)$ it follows from (2.14) that

$$(7.17) \qquad H_\tau = E \left( \sup_{\rho \le s \le \tau} \left| \int_\rho^s g(z) \, dz \right|^2 \Big| \mathcal{A}_\rho \right)$$

$$\le \; E \left( \sup_{\rho \le s \le \tau} (s - \rho) \int_\rho^s |g(z)|^2 \, dz \Big| \mathcal{A}_\rho \right)$$

$$\le \; E \left( \delta \int_\rho^s |g(z)|^2 \, dz \Big| \mathcal{A}_\rho \right)$$

$$\le \; 4^{l(\alpha)-n(\alpha)} \delta^{l(\alpha)+n(\alpha)-1} \int_\rho^\tau R_{\rho,z} \, dz.$$

On the other hand, when $\alpha = (j)$ for $j \in \{1, \ldots, m\}$ it is easy, because of (2.12), to see that the process

$$\left\{ I_\alpha \left[ g(\cdot) \right]_{\rho,t} , \, t \in [\rho, T] \right\}$$

is a martingale. Hence with the help of the Doob inequality (2.3.7) we obtain

$$(7.18) \qquad H_\tau = E \left( \sup_{\rho \le s \le \tau} \left| \int_\rho^s g(z) \, dW_z^j \right|^2 \Big| \mathcal{A}_\rho \right)$$

$$\le \; 4 \sup_{\rho \le s \le \tau} E \left( \left| \int_\rho^s g(z) \, dW_z^j \right|^2 \Big| \mathcal{A}_\rho \right)$$

$$\le \; 4 \sup_{\rho \le s \le \tau} \int_\rho^\tau E \left( |g(z)|^2 \Big| \mathcal{A}_\rho \right) dz$$

$$\le \; 4^{l(\alpha)-n(\alpha)} \delta^{l(\alpha)+n(\alpha)-1} \int_\rho^\tau R_{\rho,z} \, dz.$$

Now let $l(\alpha) = k + 1$ where $\alpha = (j_1, \ldots, j_{k+1})$ and $k \geq 1$. If $j_{k+1} = 0$, then (2.12) implies that

$$(7.19) \qquad H_\tau = E \left( \sup_{\rho \leq s \leq \tau} \left| \int_\rho^s I_{\alpha-} [g(\cdot)]_{\rho,z} \, dz \right|^2 \Big| \mathcal{A}_\rho \right)$$

$$\leq E \left( \delta \int_\rho^\tau \left| I_{\alpha-} [g(\cdot)]_{\rho,z} \right|^2 \, dz \Big| \mathcal{A}_\rho \right)$$

$$\leq \delta^2 E \left( \sup_{\rho \leq s \leq \tau} \left| I_{\alpha-} [g(\cdot)]_{\rho,s} \right|^2 \Big| \mathcal{A}_\rho \right).$$

Using the inductive assumption it then follows that

$$H_\tau \leq \delta^2 \delta^{l(\alpha-)+n(\alpha-)-1} 4^{l(\alpha-)-n(\alpha-)} \int_\rho^\tau R_{\rho,z} \, dz$$

$$= 4^{l(\alpha)-n(\alpha)} \delta^{l(\alpha)+n(\alpha)-1} \int_\rho^\tau R_{\rho,z} \, dz.$$

Finally, let $j_{k+1} \in \{1, \ldots, m\}$. Then, because of (2.12), the process

$$\left\{ I_\alpha [g(\cdot)]_{\rho,t}, \, t \in [\rho, T] \right\}$$

is a martingale. Hence, using the Doob inequality (2.3.7), we obtain

$$(7.20) \qquad H_\tau = E \left( \sup_{\rho \leq s \leq \tau} \left| \int_\rho^s I_{\alpha-} [g(\cdot)]_{\rho,z} \, dW_z^{j_{k+1}} \right|^2 \Big| \mathcal{A}_\rho \right)$$

$$\leq 4 \sup_{\rho \leq s \leq \tau} E \left( \left| \int_\rho^s I_{\alpha-} [g(\cdot)]_{\rho,z} \, dW_z^{j_{k+1}} \right|^2 \Big| \mathcal{A}_\rho \right)$$

$$\leq 4 \int_\rho^\tau E \left( \sup_{\rho \leq s \leq \tau} \left| I_{\alpha-} [g(\cdot)]_{\rho,s} \right|^2 \Big| \mathcal{A}_\rho \right) \, dz$$

$$\leq 4\delta E \left( \sup_{\rho \leq s \leq \tau} \left| I_{\alpha-} [g(\cdot)]_{\rho,s} \right|^2 \Big| \mathcal{A}_\rho \right).$$

With the inductive assumption we then have

$$H_\tau \leq 4\delta \, 4^{l(\alpha-)-n(\alpha-)} \delta^{l(\alpha-)+n(\alpha-)-1} \int_\rho^\tau R_{\rho,z} \, dz$$

$$= 4^{l(\alpha)-n(\alpha)} \delta^{l(\alpha)+n(\alpha)-1} \int_\rho^\tau R_{\rho,z} \, dz.$$

This completes the proof of Lemma 5.7.3. $\square$

With the same ideas and using the same notation as in the preceding proof it is easy to derive the following estimate which is a slight variation of (7.15).

**Lemma 5.7.4**  *Let $\alpha \in \mathcal{M} \setminus \{v\}$, let $g \in \mathcal{H}_\alpha$, let $\delta > 0$ and let $\rho$ and $\tau$ be two stopping times with $\tau$ $\mathcal{A}_\rho$-measurable and $t_0 \leq \rho \leq \tau \leq \rho + \delta \leq T < \infty$, w.p.1. Then*

$$(7.21) \quad E\left(\sup_{\rho \leq s \leq \tau} \left|I_\alpha\left[g(\cdot)\right]_{\rho,s}\right|^2 \Big| \mathcal{A}_\rho\right) \leq 4^{l(\alpha)-n(\alpha)} \delta^{l(\alpha)+n(\alpha)} \frac{1}{l(\alpha)!} R_{\rho,\tau}.$$

## Estimates of Higher Moments

The next lemma provides an estimate for the higher moments of a multiple Ito integral.

**Lemma 5.7.5**  *Let $\alpha \in \mathcal{M}$, let $g \in \mathcal{H}_\alpha$, let $q = 1, 2, \ldots$, let $\delta > 0$ and let $\rho$ and $\tau$ be two stopping times with $\tau$ $\mathcal{A}_\rho$- measurable and $t_0 \leq \rho \leq \tau \leq \rho + \delta \leq T < \infty$, w.p.1. Then*

$$(7.22) \quad \left(E\left(\left|I_\alpha\left[g(\cdot)\right]_{\rho,\tau}\right|^{2q} \Big| \mathcal{A}_\rho\right)\right)^{1/q}$$

$$\leq \left(2(2q-1)e^T\right)^{l(\alpha)-n(\alpha)} (\tau - \rho)^{l(\alpha)+n(\alpha)} R,$$

*where*

$$(7.23) \quad R = \left(E\left(\sup_{\rho \leq s \leq \tau} |g(s)|^{2q} \Big| \mathcal{A}_\rho\right)\right)^{1/q}.$$

**Proof**  Let $\alpha = (j_1, \ldots, j_l)$, with $l(\alpha) = l$, and let $W_t^0 = t$. Then from (2.12) and an assertion on page 78 of Krylov (1980) we have

$$H_{\alpha,q}(\rho,\tau) = \left(E\left(\left|I_\alpha\left[g(\cdot)\right]_{\rho,\tau}\right|^{2q} \Big| \mathcal{A}_\rho\right)\right)^{1/q}$$

$$= \left(E\left(\left|\int_\rho^\tau I_{\alpha-}\left[g(\cdot)\right]_{\rho,s_1} dW_{s_1}^{j_{l(\alpha)}}\right|^{2q} \Big| \mathcal{A}_\rho\right)\right)^{1/q}$$

$$\leq \left(2(2q-1)e^T\right)^{1-n((j_{l(\alpha)}))} (\tau - \rho)^{n((j_{l(\alpha)}))}$$

$$\times \int_\rho^\tau \left(E\left(\left|I_{\alpha-}\left[g(\cdot)\right]_{\rho,s_1}\right|^{2q} \Big| \mathcal{A}_\rho\right)\right)^{1/q} ds_1$$

$$\leq \left(2(2q-1)e^T\right)^{1-n((j_{l(\alpha)}))} (\tau - \rho)^{n((j_{l(\alpha)}))} \int_\rho^\tau H_{\alpha-,q}(\rho,s_1)\, ds_1$$

$$\leq \left(2(2q-1)e^T\right)^{l(\alpha)-n(\alpha)} (\tau - \rho)^{l(\alpha)+n(\alpha)} R,$$

which is (7.22). $\square$

## 5.8    Strong Approximation of Multiple Stochastic Integrals

Multiple stochastic integrals of higher multiplicity cannot always be expressed in terms of simpler stochastic integrals, especially when the Wiener process is multi-dimensional. Nevertheless it is still possible to represent them in an efficient way. Here we shall present one such method for multiple Stratonovich integrals based on a Kahunen-Loève or Fourier series expansion of a Wiener process similar to that discussed in Section 4 of Chapter 2.

Our starting point is the Brownian bridge process (1.8.7)

$$\left\{ W_t - \frac{t}{\Delta} W_\Delta, \, 0 \leq t \leq \Delta \right\}$$

formed from the given $m$-dimensional Wiener process $W_t = (W_t^1, \ldots, W_t^m)$ on the time interval $[0, \Delta]$. The componentwise Fourier expansion of this process is

$$(8.1) \quad W_t^j - \frac{t}{\Delta} W_\Delta^j = \frac{1}{2} a_{j,0} + \sum_{r=1}^{\infty} \left( a_{j,r} \cos \left( \frac{2r\pi t}{\Delta} \right) + b_{j,r} \sin \left( \frac{2r\pi t}{\Delta} \right) \right)$$

with random coefficients

$$(8.2) \quad a_{j,r} = \frac{2}{\Delta} \int_0^\Delta \left( W_s^j - \frac{s}{\Delta} W_\Delta^j \right) \cos \left( \frac{2r\pi s}{\Delta} \right) \, ds$$

and

$$(8.3) \quad b_{j,r} = \frac{2}{\Delta} \int_0^\Delta \left( W_s^j - \frac{s}{\Delta} W_\Delta^j \right) \sin \left( \frac{2r\pi s}{\Delta} \right) \, ds$$

for $j = 1, \ldots, m$ and $r = 0, 1, 2, \ldots$. The series in equation (8.1) is understood to converge in the mean-square sense. As linear transformations of Gaussian random variables the coefficients here are Gaussian. Using Exercise 3.2.7 and property 3.2.10 of Ito integrals it can be shown that $a_{j,r}$ and $b_{j,r}$ are $N(0; \Delta/2\pi^2 r^2)$ distributed. Similarly, the coefficients are pairwise independent.

We can then truncate the series in (8.1) to obtain an approximation of the Brownian bridge process. For each $p = 1, 2, \ldots$ we write

$$(8.4) \quad W_t^{j,p} = \frac{t}{\Delta} W_\Delta^j + \frac{1}{2} a_{j,0} + \sum_{r=1}^{p} \left( a_{j,r} \cos \left( \frac{2r\pi t}{\Delta} \right) + b_{j,r} \sin \left( \frac{2r\pi t}{\Delta} \right) \right).$$

This process has differentiable sample paths on $[0, \Delta]$. As we shall see in Section 1 of Chapter 6, Riemann-Stieltjes integrals with respect to such a process will converge to Stratonovich stochastic integrals rather than to Ito stochastic integrals.

## Derivation of Multiple Stratonovich Integrals

Using the relationship (2.39) and the Fourier expansion (8.1) for the Brownian bridge we can systematically derive formulae for multiple Stratonovich integrals $J_{\alpha,t}$ on an interval $t \in [0, \Delta]$ for multi-indices $\alpha$ of increasing length. In what follows we assume $j, j_1, j_2 \in \{1, \ldots, m\}$ and $t \in [0, \Delta]$. In addition, we shall write

$$\gamma = \frac{\pi}{\Delta}.$$

From (2.31) we have

$$(8.5) \qquad J_{(0),t} = t, \qquad J_{(0,0),t} = \frac{1}{2} t^2,$$

$$J_{(j),t} = \frac{1}{\Delta} W_\Delta^j J_{(0),t} + \frac{1}{2} a_{j,0} + \sum_{r=1}^{\infty} \left( a_{j,r} \cos(2\gamma rt) + b_{j,r} \sin(2\gamma rt) \right),$$

and hence

$$(8.6) \qquad J_{(j,0),t} \;=\; \int_0^t J_{(j),s}\, ds$$

$$=\; \frac{1}{\Delta} W_\Delta^j J_{(0,0),t} + \frac{1}{2} a_{j,0} J_{(0),t}$$

$$+ \frac{\Delta}{2\pi} \sum_{r=1}^{\infty} \frac{1}{r} \left( a_{j,r} \sin(2\gamma rt) - b_{j,r} \left[ \cos(2\gamma rt) - 1 \right] \right).$$

In addition, from (2.39) we have

$$(8.7) \qquad J_{(0,j),t} = J_{(j),t} J_{(0),t} - J_{(j,0),t}.$$

Then we find that

$$(8.8) \qquad J_{(j_1,j_2),t} \;=\; \int_0^t J_{(j_1),s} \circ dW_s^{j_2}$$

$$=\; \frac{1}{\Delta} W_\Delta^{j_1} J_{(0,j_2),t} + \frac{1}{2} a_{j_1,0} J_{(j_2),t}$$

$$+ \int_0^t \sum_{r=1}^{\infty} \left( a_{j_1,r} \cos(2\gamma rs) + b_{j_1,r} \sin(2\gamma rs) \right) \circ dW_s^{j_2}$$

$$=\; \frac{1}{\Delta} W_\Delta^{j_1} J_{(0,j_2),t} + \frac{1}{2} a_{j_1,0} J_{(j_2),t} + A_{(j_1,j_2)} J_{(0),t} + F_{(j_1,j_2),t}$$

where

$$A_{(j_1,j_2)} = \frac{\pi}{\Delta} \sum_{r=1}^{\infty} r \left( a_{j_1,r} b_{j_2,r} - b_{j_1,r} a_{j_2,r} \right) \qquad \text{and}$$

$$F_{(j_1,j_2),t} = \frac{1}{4} \sum_{r=1}^{\infty} F_r(t) + \sum_{\substack{k,r=1 \\ k \neq r}}^{\infty} k \, F_{k,r}(t)$$

with

$$
\begin{aligned}
F_r(t) \;=\; & (a_{j_1,r} a_{j_2,r} - b_{j_1,r} b_{j_2,r}) \left[1 - \cos(4\gamma rt)\right] \\
& + (a_{j_1,r} b_{j_2,r} + b_{j_1,r} a_{j_2,r}) \sin(4\gamma rt) \\
& + \frac{2}{\pi r} W_{\Delta}^{j_2} \left(a_{j_1,r} \sin(2\gamma rt) + b_{j_1,r} [\cos(2\gamma rt) - 1]\right)
\end{aligned}
$$

and

$$
\begin{aligned}
F_{k,r}(t) \;=\; & a_{j_1,r} a_{j_2,k} \left\{ \frac{\cos(2\gamma(k+r)t)}{2(k+r)} + \frac{\cos(2\gamma(k-r)t)}{2(k-r)} - \frac{k}{k^2 - r^2} \right\} \\
& + a_{j_1,r} b_{j_2,k} \left\{ \frac{\sin(2\gamma(k+r)t)}{2(k+r)} + \frac{\sin(2\gamma(k-r)t)}{2(k-r)} \right\} \\
& + b_{j_1,r} b_{j_2,k} \left\{ \frac{\cos(2\gamma(k-r)t)}{2(k-r)} - \frac{\cos(2\gamma(k+r)t)}{2(k+r)} - \frac{r}{k^2 - r^2} \right\} \\
& + \frac{\Delta}{2\pi} b_{j_1,r} a_{j_2,k} \left\{ \frac{\sin(2\gamma(k+r)t)}{2(k+r)} - \frac{\sin(2\gamma(k-r)t)}{2(k-r)} \right\}.
\end{aligned}
$$

Continuing in this way we can also obtain, in principle at least, further higher order multiple Stratonovich integrals.

**Exercise 5.8.1**     *Show that*

$$
\begin{aligned}
J_{(j,0,0),t} \;=\; & \frac{1}{\Delta} W_{\Delta}^{j} J_{(0,0,0),t} + \frac{1}{2} a_{j,0} J_{(0,0),t} \\
& - \frac{1}{2\gamma} \sum_{r=1}^{\infty} \frac{1}{r} \left( \frac{a_{j,r}}{2\gamma r} \left[\cos(2\gamma rt) - 1\right] + b_{j,r} \left[ \frac{1}{2\gamma r} \sin(2\gamma rt) - t \right] \right).
\end{aligned}
$$

## Representation of Multiple Stratonovich Integrals

Taking $t = \Delta$ in (8.5)–(8.8) and corresponding expressions for the multiple Stratonovich integrals $J_{\alpha,t}$ over the interval $[0, \Delta]$ we obtain the following formulae for the $J_{\alpha,\Delta}$, which we shall write simply as $J_\alpha$, for multi-indices $\alpha$ with length $l(\alpha) \leq 3$. Here $j, j_1, j_2, j_3 \in \{1, \ldots, m\}$.

$$(8.9) \qquad J_{(0)} \;=\; \Delta, \qquad J_{(j)} = W_{\Delta}^{j}, \qquad J_{(0,0)} = \frac{1}{2}\Delta^2,$$

$$J_{(j,0)} \;=\; \frac{1}{2}\Delta \left( W_{\Delta}^{j} + a_{j,0} \right), \qquad J_{(0,j)} = \frac{1}{2}\Delta \left( W_{\Delta}^{j} - a_{j,0} \right);$$

$$J_{(j_1,j_2)} = \frac{1}{2} W_{\Delta}^{j_1} W_{\Delta}^{j_2} - \frac{1}{2}\left( a_{j_2,0} W_{\Delta}^{j_1} - a_{j_1,0} W_{\Delta}^{j_2} \right) + \Delta A_{j_1 j_2} \qquad \text{with}$$

$$A_{j_1,j_2} = \frac{\pi}{\Delta} \sum_{r=1}^{\infty} r \left( a_{j_1,r} b_{j_2,r} - b_{j_1,r} a_{j_2,r} \right);$$

$$J_{(0,0,0)} = \frac{1}{3!} \Delta^3, \qquad J_{(0,j,0)} = \frac{1}{3!} \Delta^2 W_\Delta^j - \frac{1}{\pi} \Delta^2 b_j,$$

$$J_{(j,0,0)} = \frac{1}{3!} \Delta^2 W_\Delta^j + \frac{1}{4} \Delta^2 a_{j,0} + \frac{1}{2\pi} \Delta^2 b_j,$$

$$J_{(0,0,j)} = \frac{1}{3!} \Delta^2 W_\Delta^j - \frac{1}{4} \Delta^2 a_{j,0} + \frac{1}{2\pi} \Delta^2 b_j$$

with

$$b_j = \sum_{r=1}^{\infty} \frac{1}{r} b_{j,r};$$

$$J_{(j_1,0,j_2)} = \frac{1}{3!} \Delta W_\Delta^{j_1} W_\Delta^{j_2} + \frac{1}{2} a_{j_1,0} J_{(0,j_2)} + \frac{1}{2\pi} \Delta W_\Delta^{j_2} b_{j_1} - \Delta^2 B_{j_1,j_2}$$
$$- \frac{1}{4} \Delta a_{j_2,0} W_\Delta^{j_1} + \frac{1}{2\pi} \Delta W_\Delta^{j_1} b_{j_2},$$

$$J_{(0,j_1,j_2)} = \frac{1}{3!} \Delta W_\Delta^{j_1} W_\Delta^{j_2} - \frac{1}{\pi} \Delta W_\Delta^{j_2} b_{j_1} + \Delta^2 B_{j_1,j_2} - \frac{1}{4} \Delta a_{j_2,0} W_\Delta^{j_1}$$
$$+ \frac{1}{2\pi} \Delta W_\Delta^{j_1} b_{j_2} + \Delta^2 C_{j_1,j_2} + \frac{1}{2} \Delta^2 A_{j_1,j_2}$$

with

$$B_{j_1,j_2} = \frac{1}{2\Delta} \sum_{r=1}^{\infty} \left( a_{j_1,r} a_{j_2,r} + b_{j_1,r} b_{j_2,r} \right)$$

and

$$C_{j_1,j_2} = -\frac{1}{\Delta} \sum_{\substack{r,l=1 \\ r \neq l}}^{\infty} \frac{r}{r^2 - l^2} \left( r a_{j_1,r} a_{j_2,l} + l b_{j_1,r} b_{j_2,l} \right);$$

$$J_{(j_1,j_2,0)} = \frac{1}{2} \Delta W_\Delta^{j_1} W_\Delta^{j_2} - \frac{1}{2} \Delta \left( a_{j_2,0} W_\Delta^{j_1} - a_{j_1,0} W_\Delta^{j_2} \right) + \Delta^2 A_{j_1,j_2}$$
$$- J_{(j_1,0,j_2)} - J_{(0,j_1,j_2)};$$

$$J_{(j_1,j_2,j_3)} = \frac{1}{\Delta} W_\Delta^{j_1} J_{(0,j_2,j_3)} + \frac{1}{2} a_{j_1,0} J_{(j_2,j_3)} + \frac{1}{2\pi} b_{j_1} W_\Delta^{j_2} W_\Delta^{j_3}$$
$$- \Delta W_\Delta^{j_2} B_{j_1,j_3} + \Delta W_\Delta^{j_3} \left( \frac{1}{2} A_{j_1,j_2} - C_{j_2,j_1} \right) + \Delta^{3/2} D_{j_1,j_2,j_3}$$

with

$$
D_{j_1,j_2,j_3} = -\frac{\pi}{2\Delta^{3/2}} \sum_{r,l=1}^{\infty} l \left[ a_{j_2,l} \left( a_{j_3,l+r} b_{j_1,r} - a_{j_1,r} b_{j_3,l+r} \right) \right.
$$

$$
\left. + b_{j_2,l} \left( a_{j_1,r} a_{j_3,l+r} + b_{j_1,r} b_{j_3,l+r} \right) \right]
$$

$$
+ \frac{\pi}{2\Delta^{3/2}} \sum_{l=1}^{\infty} \sum_{r=1}^{l-1} l \left[ a_{j_2,l} \left( a_{j_1,r} b_{j_3,l-r} + a_{j_3,l-r} b_{j_1,r} \right) \right.
$$

$$
\left. - b_{j_2,l} \left( a_{j_1,r} a_{j_3,l-r} - b_{j_1,r} b_{j_3,l-r} \right) \right]
$$

$$
+ \frac{\pi}{2\Delta^{3/2}} \sum_{l=1}^{\infty} \sum_{r=l+1}^{\infty} l \left[ a_{j_2,l} \left( a_{j_3,r-l} b_{j_1,r} - a_{j_1,r} b_{j_3,r-l} \right) \right.
$$

$$
\left. + b_{j_2,l} \left( a_{j_1,r} a_{j_3,r-l} + b_{j_1,r} b_{j_3,r-l} \right) \right]
$$

Similar formulae can also be derived for the multiple Stratonovich integrals of higher multiplicity.

**Exercise 5.8.2**    *Verify formula (8.9) for the Stratonovich integrals of multiplicity 1 and 2.*

## Approximate Multiple Stratonovich Integrals

For each $j = 1, \ldots, m$ and $r = 1, \ldots, p$ with $p = 1, 2, \ldots$ we shall define independent standard Gaussian random variables $\xi_j$, $\zeta_{j,r}$, $\eta_{j,r}$, $\mu_{j,p}$ and $\phi_{j,p}$ by

$$
(8.10) \quad \xi_j = \frac{1}{\sqrt{\Delta}} W_\Delta^j, \quad \zeta_{j,r} = \sqrt{\frac{2}{\Delta}} \pi r a_{j,r}, \quad \eta_{j,r} = \sqrt{\frac{2}{\Delta}} \pi r b_{j,r},
$$

$$
\mu_{j,p} = \frac{1}{\sqrt{\Delta \rho_p}} \sum_{r=p+1}^{\infty} a_{j,r}, \quad \phi_{j,p} = \frac{1}{\sqrt{\Delta \alpha_p}} \sum_{r=p+1}^{\infty} \frac{1}{r} b_{j,r}
$$

where

$$
\rho_p = \frac{1}{12} - \frac{1}{2\pi^2} \sum_{r=1}^{p} \frac{1}{r^2} \qquad \alpha_p = \frac{\pi^2}{180} - \frac{1}{2\pi^2} \sum_{r=1}^{p} \frac{1}{r^4}.
$$

Using these random variables we can approximate each of the above multiple Stratonovich integrals $J_{(j_1,\ldots,j_l)}$ by expressions $J_{(j_1,\ldots,j_l)}^p$ for $p = 1, 2, \ldots$, which are similar in form to $J_{(j_1,\ldots,j_l)}$. Here we also have $j, j_1, j_2, j_3 \in \{1, \ldots, m\}$.

$$
(8.11) \qquad J_{(0)}^p = \Delta, \quad J_{(j)}^p = \sqrt{\Delta} \, \xi_j, \quad J_{(0,0)}^p = \frac{1}{2} \Delta^2,
$$

$$
J_{(j,0)}^p = \frac{1}{2} \Delta \left( \sqrt{\Delta} \xi_j + a_{j,0} \right), \quad J_{(0,j)}^p = \frac{1}{2} \Delta \left( \sqrt{\Delta} \xi_j - a_{j,0} \right) \qquad \text{where}
$$

$$a_{j,0} = -\frac{1}{\pi}\sqrt{2\Delta}\sum_{r=1}^{p}\frac{1}{r}\zeta_{j,r} - 2\sqrt{\Delta\rho_p}\,\mu_{j,p};$$

$$J^p_{(j_1,j_2)} = \frac{1}{2}\Delta\xi_{j_1}\xi_{j_2} - \frac{1}{2}\sqrt{\Delta}\,(a_{j_2,0}\xi_{j_1} - a_{j_1,0}\xi_{j_2}) + \Delta\,A^p_{j_1,j_2}$$

with

$$A^p_{j_1,j_2} = \frac{1}{2\pi}\sum_{r=1}^{p}\frac{1}{r}\left(\zeta_{j_1,r}\eta_{j_2,r} - \eta_{j_1,r}\zeta_{j_2,r}\right);$$

$$
\begin{aligned}
J^p_{(0,0,0)} &= \frac{1}{3!}\Delta^3, & J^p_{(0,j,0)} &= \frac{1}{3!}\Delta^{5/2}\xi_j - \frac{1}{\pi}\Delta^2 b_j, \\
J^p_{(j,0,0)} &= \frac{1}{3!}\Delta^{5/2}\xi_j + \frac{1}{4}\Delta^2 a_{j,0} + \frac{1}{2\pi}\Delta^2 b_j, \\
J^p_{(0,0,j)} &= \frac{1}{3!}\Delta^{5/2}\xi_j - \frac{1}{4}\Delta^2 a_{j,0} + \frac{1}{2\pi}\Delta^2 b_j
\end{aligned}
$$

with

$$b_j = \sqrt{\frac{\Delta}{2}}\sum_{r=1}^{p}\frac{1}{r^2}\eta_{j,r} + \sqrt{\Delta\alpha_p}\,\phi_{j,p};$$

$$
\begin{aligned}
J^p_{(j_1,0,j_2)} &= \frac{1}{3!}\Delta^2\xi_{j_1}\xi_{j_2} + \frac{1}{2}a_{j_1,0}J^p_{(0,j_2)} + \frac{1}{2\pi}\Delta^{3/2}\xi_{j_2}b_{j_1} - \Delta^2 B^p_{j_1,j_2} \\
&\quad -\frac{1}{4}\Delta^{3/2}a_{j_2,0}\xi_{j_1} + \frac{1}{2\pi}\Delta^{3/2}\xi_{j_1}b_{j_2}, \\
J^p_{(0,j_1,j_2)} &= \frac{1}{3!}\Delta^2\xi_{j_1}\xi_{j_2} - \frac{1}{\pi}\Delta^{3/2}\xi_{j_2}b_{j_1} + \Delta^2 B^p_{j_1,j_2} - \frac{1}{4}\Delta^{3/2}a_{j_2,0}\xi_{j_1} \\
&\quad +\frac{1}{2\pi}\Delta^{3/2}\xi_{j_1}b_{j_2} + \Delta^2\,C^p_{j_1,j_2} + \frac{1}{2}\Delta^2\,A^p_{j_1,j_2},
\end{aligned}
$$

with

$$B^p_{j_1,j_2} = \frac{1}{4\pi^2}\sum_{r=1}^{p}\frac{1}{r^2}\left(\zeta_{j_1,r}\zeta_{j_2,r} + \eta_{j_1,r}\eta_{j_2,r}\right)$$

and

$$C^p_{j_1,j_2} = -\frac{1}{2\pi^2}\sum_{\substack{r,l=1 \\ r\neq l}}^{p}\frac{r}{r^2-l^2}\left(\frac{1}{l}\zeta_{j_1,r}\zeta_{j_2,l} - \frac{l}{r}\eta_{j_1,r}\eta_{j_2,l}\right);$$

$$
\begin{aligned}
J^p_{(j_1,j_2,0)} &= \frac{1}{2}\Delta^2\xi_{j_1}\xi_{j_2} - \frac{1}{2}\Delta^{3/2}\left(a_{j_2,0}\xi_{j_1} - a_{j_1,0}\xi_{j_2}\right) + \Delta^2 A^p_{j_1,j_2} \\
&\quad -J^p_{(j_1,0,j_2)} - J^p_{(0,j_1,j_2)};
\end{aligned}
$$

$$J^p_{(j_1,j_2,j_3)} \;=\; \frac{1}{\sqrt{\Delta}}\,\xi_{j_1}\,J^p_{(0,j_2,j_3)} + \frac{1}{2}\,a_{j_1,0}\,J^p_{(j_2,j_3)} + \frac{1}{2\pi}\,\Delta b_{j_1}\,\xi_{j_2}\,\xi_{j_3}$$

$$-\Delta^{3/2}\,\xi_{j_2}\,B^p_{j_1,j_3} + \Delta^{3/2}\,\xi_{j_3}\left(\frac{1}{2}A^p_{j_1,j_2} - C^p_{j_2,j_1}\right) + \Delta^{3/2}\,D^p_{j_1,j_2,j_3}$$

with

$$D^p_{j_1,j_2,j_3} \;=\; -\frac{1}{\pi^2 2^{5/2}}\sum_{r,l=1}^{p}\frac{1}{l(l+r)}\Big[\zeta_{j_2,l}\,(\zeta_{j_3,l+r}\eta_{j_1,r} - \zeta_{j_1,r}\eta_{j_1,l+r})$$

$$+\eta_{j_2,l}\,(\zeta_{j_1,r}\zeta_{j_3,l+r} + \eta_{j_1,r}\eta_{j_3,l+r})\Big]$$

$$+\frac{1}{\pi^2 2^{5/2}}\sum_{l=1}^{p}\sum_{r=1}^{l-1}\frac{1}{r(l-r)}\Big[\zeta_{j_2,l}\,(\zeta_{j_1,r}\eta_{j_3,l-r} + \zeta_{j_3,l-r}\eta_{j_1,r})$$

$$-\eta_{j_2,l}\,(\zeta_{j_1,r}\zeta_{j_3,l-r} - \eta_{j_1,r}\eta_{j_3,l-r})\Big]$$

$$+\frac{1}{\pi^2 2^{5/2}}\sum_{l=1}^{p}\sum_{r=l+1}^{2p}\frac{1}{r(r-l)}\Big[\zeta_{j_2,l}\,(\zeta_{j_3,r-l}\eta_{j_1,r} - \zeta_{j_1,r}\eta_{j_3,r-l})$$

$$+\eta_{j_2,l}\,(\zeta_{j_1,r}\zeta_{j_3,r-l} + \eta_{j_1,r}\eta_{j_3,r-l})\Big],$$

where for $r > p$ we set $\eta_{j,r} = 0$ and $\zeta_{j,r} = 0$ for all $j = 1, \ldots, m$.

**Exercise 5.8.3**   *Verify formula (8.11) for the approximate Stratonovich integrals of multiplicity 1 and 2.*

## Mean-Square Convergence of Approximate Multiple Stratonovich Integrals

We shall now examine the mean-square error between $J^p_\alpha$ and $J_\alpha$. The most sensitive approximation is $J^p_{(j_1,j_2)}$ because the others are either identical to $J_\alpha$ or their mean-square error can be estimated by a constant times $\Delta^\gamma$ for some $\gamma \geq 3$. We have

$$(8.12) \qquad E\left(\left|J^p_{(j_1,j_2)} - J_{(j_1,j_2)}\right|^2\right)$$

$$= \; \Delta^2 E\left(A^p_{j_1,j_2} - A_{j_1,j_2}\right)^2$$

$$= \; \Delta^2 E\left(\frac{\pi}{\Delta}\sum_{r=p+1}^{\infty} r\left(a_{j_1,r}b_{j_2,r} - b_{j_1,r}a_{j_2,r}\right)\right)^2.$$

Now it follows from (8.1)–(8.2) that

$$E\left(a_{j,r}b_{j,r}\right) = E\left(a_{j,r}b_{j,k}\right) = 0,$$
$$E\left(a_{j,r}a_{j,k}\right) = E\left(b_{j,r}b_{j,k}\right) = 0,$$
$$E\left(a_{j_1,r}a_{j_2,r}\right) = E\left(b_{j_1,r}b_{j_2,r}\right) = 0$$

and

$$E\left(a_{j,r}^2\right) = E\left(b_{j,r}^2\right) = \frac{\Delta}{2r^2\pi^2}$$

for $j, j_1, j_2 = 1, \ldots, m$ with $r \neq k$ and $j_1 \neq j_2$. Hence

(8.13) $$E\left(\left|J_{(j_1,j_2)}^p - J_{(j_1,j_2)}\right|^2\right) = \frac{\Delta^2}{2\pi^2}\sum_{r=p+1}^{\infty}\frac{1}{r^2} = \Delta^2\rho_p$$

where

$$\rho_p = \frac{1}{2\pi^2}\sum_{r=p+1}^{\infty}\frac{1}{r^2}.$$

From this and the remark at the beginning of the subsection we can conclude that

(8.14) $$E\left(\left|J_\alpha^p - J_\alpha\right|^2\right) \leq \Delta^2\rho_p$$

for multi-indices $\alpha$ with $l(\alpha) \leq 3$ provided $\Delta$ is sufficiently small. We note that

(8.15) $$\rho_p = \frac{1}{2\pi^2}\sum_{r=p+1}^{\infty}\frac{1}{r^2} \leq \frac{1}{2\pi^2}\int_p^{\infty}\frac{1}{u^2}\,du = \frac{1}{2\pi^2 p},$$

which provides us with an easily calculated upper bound for the error. Some more precise values of $\rho_p$ are given in Table 5.8.1.

| $p$ | 1 | 2 | 3 | 4 | 5 | 10 | 20 |
|---|---|---|---|---|---|---|---|
| $\rho_p$ | 0.033 | 0.020 | 0.014 | 0.011 | 0.009 | 0.005 | 0.003 |

**Table 5.8.1** Some values of the error bound $\rho_p$.

We note that there exists a finite constant $C$ such that for sufficiently small $\Delta$ we have the estimate

(8.16) $$E\left(\left|I_\alpha^p - I_\alpha\right|^2\right) \leq C\frac{\Delta^2}{p}$$

for the approximate multiple Ito integrals $I_\alpha^p$ obtained from the corresponding approximate multiple Stratonovich integrals $J_\alpha^p$ as described in (2.35).

# 5.9   Strong Convergence
## of Truncated Ito-Taylor Expansions

In this section we shall use the estimates derived in the last section to investigate the mean-square error of truncated Ito-Taylor expansions. We shall prove that they also converge with probability one.

For $k = 0, 1, \ldots$ we take the truncated Ito-Taylor expansion

$$(9.1) \qquad X_k(t) = \sum_{\alpha \in \Lambda_k} I_\alpha \left[ f_\alpha(t_0, X_{t_0}) \right]_{t_0, t}$$

for $t \in [t_0, T]$, the function $f(t, x) \equiv x$ and the hierarchical set

$$(9.2) \qquad \Lambda_k = \{ \alpha \in \mathcal{M} : l(\alpha) + n(\alpha) \le k \}$$

(see Exercise 5.4.5). We shall assume that the necessary derivatives and multiple integrals exist for all $\alpha \in \Lambda_k \cup \mathcal{B}(\Lambda_k)$. Here we shall always use $[a]$ to denote the largest integer not exceeding $a$.

**Propostion 5.9.1**   *Suppose that $f_\alpha(t_0, X_{t_0}) \in \mathcal{H}_\alpha$ for all $\alpha \in \Lambda_k$ and that $f_\alpha(\cdot, X.) \in \mathcal{H}_\alpha$ with*

$$(9.3) \qquad \sup_{t_0 \le t \le T} E\left( |f_\alpha(t, X_t)|^2 \right) \le C_1 C_2^{l(\alpha) + n(\alpha)} \left[ \frac{1}{2} (l(\alpha) + n(\alpha)) \right]!$$

*for all $\alpha \in \mathcal{B}(\Lambda_k)$. Then*

$$(9.4) \qquad E\left( |X_t - X_k(t)|^2 \right) \le C_3 \frac{(C_4(t - t_0))^{k+1}}{[\frac{1}{2}(k+1)]!}$$

*for all $t \in [t_0, T]$, so the truncated Ito-Taylor expansion (9.1) converges to the Ito process $X_t$ in the mean-square sense.*

**Proof**   From the Ito-Taylor formula (5.3) and Lemma 5.7.1 we obtain

$$(9.5) \qquad \left( E\left( |X_t - X_k(t)|^2 \right) \right)^{1/2}$$

$$= \left( E\left( \left| \sum_{\alpha \in \mathcal{B}(\Lambda_k)} I_\alpha \left[ f_\alpha(\cdot, X.) \right]_{t_0, t} \right|^2 \right) \right)^{1/2}$$

$$\le \sum_{\alpha \in \mathcal{B}(\Lambda_k)} \left( E\left( \left| I_\alpha \left[ f_\alpha(\cdot, X.) \right]_{t_0, t} \right|^2 \right) \right)^{1/2}$$

$$\le \sum_{\alpha \in \mathcal{B}(\Lambda_k)} \left( K_{f_\alpha, f_\alpha} \frac{(t - t_0)^{l(\alpha) + n(\alpha)}}{(l(\alpha) + n(\alpha))!} \prod_{i=0}^{l(\alpha^+)} C_{2k_i(\alpha)}^{k_i(\alpha)} \right)^{1/2}.$$

It follows from Lemma 5.7.2 and the definitions of $\Lambda_k$ and $\mathcal{B}(\Lambda_k)$ that we have to sum only over those $\alpha \in \mathcal{B}(\Lambda_k)$ for which

$$(9.6) \qquad 0 \leq l(\alpha^+) \leq k+1.$$

Further, it can be shown that

$$(9.7) \qquad l(\alpha) + n(\alpha) = l(\alpha^+) + 2n(\alpha) \in \{k+1, k+2\}$$

for all $\alpha \in \mathcal{B}(\Lambda_k)$. Therefore we have

$$n(\alpha) \in \left\{ \frac{1}{2} \left( k + 1 - l(\alpha^+) \right), \frac{1}{2} \left( k + 2 - l(\alpha^+) \right) \right\},$$

and, since $n(\alpha)$ is an integer, we must have

$$(9.8) \qquad n(\alpha) = \frac{1}{2} \left( k + 2 - l(\alpha^+) \right).$$

Moreover, the number of elements in $\mathcal{B}(\Lambda_k)$ is obviously finite. In fact, it has the upper bound

$$(9.9) \qquad \#\mathcal{B}(\Lambda_k) \leq (m+1)^{k+1}.$$

For $k_i$, $r_i = 0, 1, \ldots$ where $i \in \{0, 1, \ldots, l\}$ and $l = 0, 1, \ldots$ it follows by induction on $k_0 + r_0 + k_1 + r_1$ that

$$C_{k_0+r_0}^{k_0} C_{k_1+r_1}^{k_1} \leq C_{k_0+r_0+k_1+r_1}^{k_0+k_1}$$

and, hence, that

$$(9.10) \qquad \prod_{i=0}^{l} C_{k_i+r_i}^{k_i} \leq C_{k_0+r_0+\cdots+k_l+r_l}^{k_0+\cdots+k_l}.$$

Thus from (9.5) with (7.7), (9.3), (9.7), (9.9) and (9.10) we obtain

$$E\left( |X_t - X_k(t)|^2 \right)$$

$$\leq C_5 \left( \sum_{\alpha \in \mathcal{B}(\Lambda_k)} \left( C_2^{l(\alpha)+n(\alpha)} \left[ \frac{1}{2}(l(\alpha) + n(\alpha)) \right]! C_{2n(\alpha)}^{n(\alpha)} \right)^{1/2} \right)^2$$

$$\times \frac{(t - t_0)^{k+1}}{(k+1)!}$$

$$\leq \max_{l=0,\ldots,k+2} \left( C_5 \left[ \frac{1}{2}l + \left[ \frac{1}{2}(k+2-l) \right] \right]! \frac{2[\frac{1}{2}(k+2-l)]!}{([\frac{1}{2}(k+2-l)]!)^2} \right)$$

$$\times \frac{(C_4(t-t_0))^{k+1}}{(k+1)!}$$

$$\leq C_5 \frac{(C_4(t-t_0))^{k+1}}{(k+1)!} \left( \left[ \frac{1}{2}(k+2) \right]! \left( 2 \left[ \frac{1}{2}(k+1) \right]! \left[ \frac{1}{2}(k+2) \right]! \right) \right)^2$$

$$\leq C_3 \frac{(C_4(t-t_0))^{k+1}}{[\frac{1}{2}(k+1)]!}. \quad \square$$

We note from the upper bound (9.4) that the mean-square error of the truncated Ito-Taylor expansion (9.1) vanishes as $k \to \infty$, so the expansion converges in the mean-square sense to the Ito process. Under additional assumptions this can be strengthened to convergence with probability one uniformly on the interval $[t_0, T]$.

**Propostion 5.9.2**    *Suppose in addition to the assumptions of Proposition 5.9.1 that $f_\alpha(t_0, X_{t_0})$, $f_\alpha(\cdot, X.) \in \mathcal{H}_\alpha$ and*

$$(9.11) \qquad \sup_{t_0 \leq t \leq T} E\left(|f_\alpha(t, X_t)|^2\right) \leq C_1 \, C_2^{l(\alpha)+n(\alpha)}$$

*for all $\alpha \in \mathcal{M}$. Then the truncated Ito-Taylor expansion $X_k(t)$ converges with probability one as $k \to \infty$ to $X_t$ uniformly in $t \in [t_0, T]$ and*

$$(9.12) \qquad X_t = \lim_{k \to \infty} X_k(t) = \sum_{\alpha \in \mathcal{M}} I_\alpha \left[f_\alpha(t_0, X_{t_0})\right]_{t_0, t},$$

*w.p.1, for all $t \in [t_0, T]$.*

**Proof**    We have already noted in (9.9) that the number of elements in $\mathcal{B}(\Lambda_k)$ does not exceed $(m+1)^{k+1}$. Using this in the truncated Ito-Taylor expansion (9.1) and starting from the Ito-Taylor expansion (5.3) we obtain

$$V^{(k)} := E\left(\sup_{t_0 \leq t \leq T} |X_t - X_k(t)|^2\right)$$

$$= E\left(\sup_{t_0 \leq t \leq T} \left|\sum_{\alpha \in \mathcal{B}(\Lambda_k)} I_\alpha \left[f_\alpha(\cdot, X.)\right]_{t_0, t}\right|^2\right)$$

$$\leq (m+1)^{k+1} \sum_{\alpha \in \mathcal{B}(\Lambda_k)} E\left(\sup_{t_0 \leq t \leq T} \left|I_\alpha \left[f_\alpha(\cdot, X.)\right]_{t_0, t}\right|^2\right).$$

From Lemma 5.7.4 and the assumptions of the proposition it then follows that

$$(9.13) \qquad V^{(k)} \leq (m+1)^{k+1} \sum_{\alpha \in \mathcal{B}(\Lambda_k)} C_1 \, C_2^{l(\alpha)+n(\alpha)} 4^{l(\alpha)-n(\alpha)}$$

$$\times \frac{(T-t_0)^{l(\alpha)+n(\alpha)}}{l(\alpha)!}$$

$$\leq (m+1)^{k+1} \sum_{\alpha \in \mathcal{B}(\Lambda_k)} C_1 \frac{(4C_2(T-t_0))^{l(\alpha)+n(\alpha)}}{l(\alpha)!}.$$

For each $\alpha \in \mathcal{B}(\Lambda_k)$ we can combine (9.7) with (9.8) to conclude that

$$l(\alpha) = l(\alpha^+) + n(\alpha)$$

$$\begin{aligned}
&= l(\alpha^+) + \left[\frac{1}{2}\left(k + 2 - l(\alpha^+)\right)\right] \\
&\geq \frac{1}{2}l(\alpha^+) + \frac{1}{2}(k+1) \\
&\geq \frac{1}{2}(k+1).
\end{aligned}$$

Inserting this in (9.13) we obtain

$$(9.14) \qquad \begin{aligned}
V^{(k)} &\leq C_3 \frac{(4(m+1)C_2(T-t_0))^{k+1}}{[\frac{1}{2}(k+1)]!} \\
&\leq C_4 \frac{C_5^{[\frac{1}{2}(k+1)]}}{[\frac{1}{2}(k+1)]!}.
\end{aligned}$$

The Chebyshev inequality (1.4.14) thus gives

$$\sum_{k=1}^{\infty} P\left(\sup_{t_0 \leq t \leq T} |X_t - X_k(t)| > \epsilon\right)$$

$$\begin{aligned}
&\leq \frac{1}{\epsilon^2} \sum_{k=1}^{\infty} V^{(k)} \\
&\leq 2C_4 \frac{1}{\epsilon^2} \sum_{k=1}^{\infty} \frac{(C_5)^k}{k!} \\
&\leq 2C_4 \frac{1}{\epsilon^2} \exp(C_5) < \infty
\end{aligned}$$

for any $\epsilon > 0$. The desired result then follows by an application of the Borel-Cantelli Lemma 2.1.4. □

In conclusion we remark that the uniform mean-square error estimate

$$(9.15) \qquad E\left(\sup_{t_0 \leq t \leq T} |X_t - X_k(t)|^2\right) \leq C_4 \frac{(C_5)^{[\frac{1}{2}(k+1)]}}{[\frac{1}{2}(k+1)]!}$$

can be deduced from the the preceding proof.

# 5.10   Strong Convergence of Truncated Stratonovich-Taylor Expansions

We shall now estimate the mean-square error of truncated Stratonovich-Taylor expansions. For $k = 0, 1, \ldots$ we introduce the truncated Stratonovich-Taylor expansion

$$(10.1) \qquad \underline{Z}_k(t) = \sum_{\alpha \in \Lambda_k} J_\alpha \left[ \left( \underline{f}_\alpha (t_0, X_{t_0}) \right) \right]_{t_0, t}$$

for $t \in [t_0, T]$, the function $f(t, x) \equiv x$ and the hierarchical set

$$(10.2) \qquad \Lambda_k = \{ \alpha \in \mathcal{M} : l(\alpha) + n(\alpha) \leq k \}$$

for $k = 0, 2, 4, \ldots$. Here the $\underline{f}_\alpha$ are the corresponding Stratonovich coefficient functions (see (3.9)) and we assume that the necessary derivatives and integrals exist for all $\alpha \in \Lambda_k \cup \mathcal{B}(\Lambda_k)$.

**Proposition 5.10.1**   *Let $k = 0, 2, 4, \ldots$. Suppose that $\underline{f}_\alpha(t_0, X_{t_0}) \in \mathcal{H}_\alpha$ for all $\alpha \in \Lambda_k$ and that $\underline{f}_\alpha(\cdot, X.) \in \mathcal{H}_\alpha$ with*

$$(10.3) \qquad \sup_{t_0 \leq t \leq T} E \left( \left| \underline{f}_\alpha(t, X_t) \right|^2 \right) \leq C_1 \, C_2^{l(\alpha) + n(\alpha)} \left[ \frac{1}{2} \left( l(\alpha) + n(\alpha) \right) \right]!$$

*for all $\alpha \in \mathcal{B}(\Lambda_k)$. Then*

$$(10.4) \qquad E \left( |X_t - \underline{Z}_k(t)|^2 \right) \leq C_3 \frac{(C_4(t - t_0))^{k+1}}{\left[ \frac{1}{2}(k+1) \right]!}$$

*for each $t \in [t_0, T]$.*

The proof is completely analogous to that of the corresponding result for truncated Ito-Taylor expansions, Proposition 5.9.1, if we use the fact noted in Remark 5.2.8 that every multiple Stratonovich integral $J_\alpha$ can be represented as a finite sum of multiple Ito integrals $I_\beta$ with multi-indices $\beta$ such that

$$l(\beta) + n(\beta) \leq l(\alpha) + n(\alpha).$$

The above result can be strengthened to almost sure convergence.

**Proposition 5.10.2**   *Suppose in addition to the assumptions of Proposition 5.10.1 that $\underline{f}_\alpha(t_0, X_{t_0}), \underline{f}_\alpha(\cdot, X.) \in \mathcal{H}_\alpha$ and*

$$(10.5) \qquad \sup_{t_0 \leq t \leq T} E \left( \left| \underline{f}_\alpha(t, X_t) \right|^2 \right) \leq C_1 \, C_2^{l(\alpha) + n(\alpha)}$$

*for all $\alpha \in \mathcal{M}$. Then the truncated Stratonovich-Taylor expansion $\underline{Z}_k(t)$ converges with probability one as $k \to \infty$ to $X_t$ uniformly in $t \in [t_0, T]$ and*

$$(10.6) \qquad X_t = \lim_{k \to \infty} \underline{Z}_k(t) = \sum_{\alpha \in \mathcal{M}} J_\alpha \left[ \underline{f}_\alpha (t_0, X_{t_0}) \right]_{t_0, t},$$

*w.p.1, for $t \in [t_0, T]$.*

The proof is similar to that of Proposition 5.9.2, so it will be omitted.

**Exercise 5.10.3**  *Prove Propositions 5.10.1 and 5.10.2.*

# 5.11 Weak Convergence of Truncated Ito-Taylor Expansions

Here we shall determine an estimate for the weak error of truncated Ito-Taylor expansions. For this we consider the truncated *weak Ito-Taylor expansion*

$$(11.1) \qquad \eta_\beta(t) = \sum_{\alpha \in \Gamma_\beta} I_\alpha \left[ f_\alpha \left( t_0, X_{t_0} \right) \right]_{t_0,t},$$

for $t \in [t_0, T]$, where $\Gamma_\beta$ is the hierarchical set

$$(11.2) \qquad \Gamma_\beta = \{ \alpha \in \mathcal{M} : l(\alpha) \le \beta \}$$

and the $f_\alpha$ are the Ito coefficient functions (3.3) corresponding to $f(t,x) \equiv x$. We assume here that the necessary derivatives and integrals exist for all $\alpha \in \Gamma_\beta \cup B(\Gamma_\beta)$.

We recall from Section 8 of Chapter 4 that $C_P^l(\Re^d, \Re)$ denotes the space of $l$ times continuously differentiable functions $g : \Re^d \to \Re$ for which $g$ and all of its partial derivatives of order up to and including $l$ have polynomial growth. For notational simplicity we shall restrict attention to the autonomous case and denote by $X_t^{X_0}$, or simply $X_t$ if no misunderstanding is possible, the value at time $t$ of the diffusion process $X$ which starts at $X_0$ at time $t_0 = 0$. We note that the coefficient functions $f_\alpha$ do not then depend on $t$, which we shall thus omit along with $t_0$.

**Proposition 5.11.1**  *Let $\beta \in \{1, 2, \ldots\}$ and $T \in (0, \infty)$ be given and suppose that the drift and diffusion components $a^k = a^k(x)$ and $b^{kj} = b^{kj}(x)$ belong to the space $C_P^{2(\beta+1)}(\Re^d, \Re)$ and satisfy Lipschitz conditions and linear growth bounds for $k = 1, 2, \ldots, d$ and $j = 1, 2, \ldots, m$. Then for each $g \in C_P^{2(\beta+1)}(\Re^d, \Re)$ there exist constants $K \in (0, \infty)$ and $r \in \{1, 2, \ldots\}$ such that*

$$(11.3) \qquad \sup_{0 \le t \le T} \left| E\left( g\left( X_t \right) - g\left( \eta_\beta(t) \right) \Big| \mathcal{A}_0 \right) \right| \le K \left( 1 + |X_0|^{2r} \right) T^{\beta+1}.$$

We shall prove this Proposition at the end of the section after stating and proving a series of lemmata. First, we note in comparing Proposition 5.11.1 with Proposition 5.9.1 that we do not need to include the same terms in a truncated weak Ito-Taylor expansion as in a truncated strong Ito-Taylor expansion in order to obtain a given order of convergence.

In what follows we shall fix $\beta$ at the value given in Proposition 5.11.1. We shall also denote by $K \in (0, \infty)$ and $r \in \{1, 2, \ldots\}$ constants, but these will generally differ in value from one occurrence to the next.

For $t \in [0, T]$ we shall write

$$(11.4) \qquad \eta_t = \eta_\beta(t) = \sum_{\alpha \in \Gamma_\beta} I_\alpha [f_\alpha (X_0)]_t ,$$

omitting the superfluous $t_0 = 0$ from (11.1). Then, from the definition (5.2.12) of multiple Ito integrals we obtain

$$(11.5) \qquad \eta_t = X_0 + \sum_{j=0}^{m} \int_0^t \bar{b}_s^j \, dW_s^j$$

with $dW_s^0 = ds$ and

$$(11.6) \qquad \bar{b}_t^j = \sum_{\substack{\alpha \in \Gamma_\beta \backslash \{v\} \\ j_{l(\alpha)} = j}} I_{\alpha-} [f_\alpha (X_0)]_t$$

for $j = 0, 1, \ldots, m$, where $\alpha = (j_1, \ldots, j_{l(\alpha)})$.

**Lemma 5.11.2**    *Under the assumptions of Proposition 5.11.1 for each $p = 1, 2, \ldots$ there exist constants $K \in (0, \infty)$ and $r \in \{1, 2, \ldots\}$ such that*

$$(11.7) \qquad \hat{T} := E \left( \sup_{0 \le t \le T} \left| \bar{b}_t^j \right|^{2q} \Big| \mathcal{A}_0 \right) \le K \left( 1 + |X_0|^{2r} \right)$$

*for each $q = 1, \ldots, p$ and $j = 0, \ldots, m$.*

**Proof**    Let us fix $p$ and $j$. From (11.6), the Doob inequality (2.3.7), (11.2) and Lemma 5.7.5 we obtain

$$\hat{T} = E \left( \sup_{0 \le t \le T} \left| \sum_{\substack{\alpha \in \Gamma_\beta \backslash \{v\} \\ j_{l(\alpha)} = j}} I_{\alpha-} [f_\alpha (X_0)]_t \right|^{2q} \Big| \mathcal{A}_0 \right)$$

$$\le K \sum_{\substack{\alpha \in \Gamma_\beta \backslash \{v\} \\ j_{l(\alpha)} = j}} E \left( \sup_{0 \le t \le T} |I_{\alpha-} [f_\alpha (X_0)]_t|^{2q} \Big| \mathcal{A}_0 \right)$$

$$\le K \sum_{\alpha \in \Gamma_\beta \backslash \{v\}} E \left( \sup_{0 \le t \le T} |I_{\alpha-} [f_\alpha (X_0)]_t|^{2q} \Big| \mathcal{A}_0 \right)$$

$$\le K \sum_{\alpha \in \Gamma_\beta \backslash \{v\}} E \left( |f_\alpha (X_0)|^{2q} \Big| \mathcal{A}_0 \right) .$$

From the polynomial growth bound on each $f_\alpha$ we then have

$$\hat{T} \le K \left( 1 + |X_0|^{2r} \right)$$

for some $r \in \{1, 2, \ldots\}$.  $\square$

**Lemma 5.11.3** *Under the assumptions of Proposition 5.11.1 for each $p = 1, 2, \ldots$ there exist constants $K \in (0, \infty)$ and $r \in \{1, 2, \ldots\}$ such that*

$$(11.8) \qquad E \left( \sup_{0 \le t \le T} |\eta_t|^{2q} \,\Big|\, \mathcal{A}_0 \right) \le K \left( 1 + |X_0|^{2r} \right)$$

*for each $q = 1, \ldots, p$.*

**Proof** Let us fix $p$. Using estimates on page 85 of Krylov (1977), which are similar to those in the proof of Theorem 4.5.6, we obtain from (11.5)

$$
\begin{aligned}
H_t \;&:=\; E \left( \sup_{0 \le s \le t} |\eta_t|^{2q} \,\Big|\, \mathcal{A}_0 \right) \\
&\le\; K \left( |X_0|^{2q} + \int_0^t E \left( \sum_{j=0}^m |\bar{b}_s^j|^{2q} \,\Big|\, \mathcal{A}_0 \right) ds \right) \\
&\le\; K \left( |X_0|^{2q} + \int_0^t \sum_{j=0}^m E \left( |\bar{b}_s^j|^{2q} \,\Big|\, \mathcal{A}_0 \right) ds \right) \\
&\le\; K \left( |X_0|^{2q} + \int_0^t \sum_{j=0}^m E \left( \sup_{0 \le u \le T} |\bar{b}_u^j|^{2q} \,\Big|\, \mathcal{A}_0 \right) ds \right)
\end{aligned}
$$

for each $t \in [0, T]$. It then follows from Lemma 5.11.2 that

$$
\begin{aligned}
H_t \;&\le\; K \left( |X_0|^{2q} + \int_0^t \left( 1 + |X_0|^{2r} \right) ds \right) \\
&\le\; K \left( 1 + |X_0|^{2r} \right). \;\square
\end{aligned}
$$

For more compact notation we shall write

$$(11.9) \qquad P_l = \{1, 2, \ldots, d\}^l$$

and

$$(11.10) \qquad F_{\vec{p}}(y) = \prod_{h=1}^l y^{p_h}$$

for all $y = (y^1, \ldots, y^d)$ and $\vec{p} = (p_1, \ldots, p_l) \in P_l$ where $l = 1, 2, \ldots$. Starting from (11.5) and (11.10), by a generalization of the Ito formula (4.8.3) to semimartingales (see Ikeda and Watanabe (1989)) we obtain

$$(11.11) \qquad F_{\vec{p}} (\eta_t - X_0) = \sum_{j=0}^m \int_0^t \hat{b}_{\vec{p}}^j(s) \, dW_s^j$$

for all $t \in [0, T]$, $l = 1, 2, \ldots$ and $\vec{p} \in P_l$, where

(11.12)
$$\hat{b}_{\vec{p}}^j(t) = \sum_{k=1}^{d} \bar{b}_t^{k,j} \frac{\partial}{\partial y^k} F_{\vec{p}}(\eta_t - X_0)$$

for $j = 1, \ldots, m$ and

(11.13)  $\hat{b}_{\vec{p}}^0(t) = \sum_{k=1}^{d} \bar{b}_t^{k,0} \frac{\partial}{\partial y^k} F_{\vec{p}}(\eta_t - X_0)$

$$+ \frac{1}{2} \sum_{k,l=1}^{d} \sum_{j=1}^{m} \bar{b}_t^{k,j} \bar{b}_t^{l,j} \frac{\partial^2}{\partial y^k \partial y^l} F_{\vec{p}}(\eta_t - X_0).$$

**Lemma 5.11.4**    *Under the assumptions of Proposition 5.11.1 for each $p = 1, 2, \ldots$ there exist constants $K \in (0, \infty)$ and $r \in \{1, 2, \ldots\}$ such that*

(11.14)   $\hat{Z}(t) := E\left(\sup_{0 \le s \le t} |F_{\vec{p}}(\eta_s - X_0)|^{2q} \Big| \mathcal{A}_0\right) \le K\left(1 + |X_0|^{2r}\right) t^{ql}$

*for all $t \in [0, T]$, $q = 0, 1, \ldots, p \cdot 2^{2(\beta+1)-l}$ and $\vec{p} \in P_l$ where $l = 1, \ldots, 2(\beta+1)$.*

**Proof**  Let us fix $p$. We shall prove (11.14) by induction on $l = 1, \ldots, 2(\beta+1)$.
**1.**  Let $l = 1$. It follows from (11.10)–(11.13) and the Doob inequality (2.3.7) that

$$\hat{Z}(t) \le K E\left(\sup_{0 \le s \le t} \sum_{j=0}^{m} \left|\int_0^s \hat{b}_{\vec{p}}^j(u)\, dW_u^j\right|^{2q} \Big| \mathcal{A}_0\right)$$

$$\le K \sum_{j=0}^{m} \sum_{k=1}^{d} E\left(\sup_{0 \le s \le t} \left|\int_0^s \bar{b}_u^{k,j}\, dW_u^j\right|^{2q} \Big| \mathcal{A}_0\right)$$

$$\le K \sum_{j=0}^{m} \sum_{k=1}^{d} E\left(\left|\int_0^t \bar{b}_s^{k,j}\, dW_s^j\right|^{2q} \Big| \mathcal{A}_0\right).$$

Then, using Lemma 5.7.4 and Lemma 5.11.2 we obtain

$$\hat{Z}(t) \le K \sum_{j=0}^{m} \sum_{k=1}^{d} E\left(\sup_{0 \le s \le t} |\bar{b}_s^{k,j}|^{2q} \Big| \mathcal{A}_0\right) t^q \le K\left(1 + |X_0|^{2r}\right) t^q.$$

**2.**  Let $l = 2, \ldots, 2(\beta+1)$. It follows analogously with the preceding step that

$$\hat{Z}(t) \le K E\left(\sup_{0 \le s \le t} \sum_{j=0}^{m} \sum_{k=1}^{d} \left|\int_0^s \bar{b}_u^{k,j} \frac{\partial}{\partial y^k} F_{\vec{p}}(\eta_u - X_0)\, dW_u^j\right|^{2q} \Big| \mathcal{A}_0\right)$$

$$+ K E\left(\sup_{0 \le s \le t} \sum_{k,r=1}^{d} \sum_{j=1}^{m} \left|\int_0^s \bar{b}_u^{k,j} \bar{b}_u^{r,j} \frac{\partial^2}{\partial y^k \partial y^r} F_{\vec{p}}(\eta_u - X_0)\, du\right|^{2q} \Big| \mathcal{A}_0\right)$$

$$\leq \ K \sum_{j=0}^{m} \sum_{k=1}^{d} E \left( \left| \int_0^t \bar{b}_s^{k,j} \frac{\partial}{\partial y^k} F_{\vec{p}}(\eta_s - X_0)\, dW_s^j \right|^{2q} \Big| \mathcal{A}_0 \right)$$

$$+ K \sum_{k,r=1}^{d} \sum_{j=1}^{m} E \left( \left| \int_0^t \bar{b}_s^{k,j} \bar{b}_s^{r,j} \frac{\partial^2}{\partial y^k \partial y^r} F_{\vec{p}}(\eta_s - X_0)\, ds \right|^{2q} \Big| \mathcal{A}_0 \right).$$

Using Lemma 5.7.5 we then find that

$$\hat{Z}(t) \ \leq \ K \sum_{j=0}^{m} \sum_{k=1}^{d} E \left( \sup_{0 \leq s \leq t} \left| \bar{b}_s^{k,j} \frac{\partial}{\partial y^k} F_{\vec{p}}(\eta_s - X_0) \right|^{2q} \Big| \mathcal{A}_0 \right) t^q$$

$$+ K \sum_{k,r=1}^{d} \sum_{j=1}^{m} E \left( \sup_{0 \leq s \leq t} \left| \bar{b}_s^{k,j} \bar{b}_s^{r,j} \frac{\partial^2}{\partial y^k \partial y^r} F_{\vec{p}}(\eta_s - X_0) \right|^{2q} \Big| \mathcal{A}_0 \right) t^{2q}$$

$$\leq \ K \sum_{j=0}^{m} \sum_{k=1}^{d} \Bigg[ E \left( \sup_{0 \leq s \leq t} \left| \bar{b}_s^{k,j} \right|^{4q} \Big| \mathcal{A}_0 \right)$$

$$\times E \left( \sup_{0 \leq s \leq t} \left| \frac{\partial}{\partial y^k} F_{\vec{p}}(\eta_s - X_0) \right|^{4q} \Big| \mathcal{A}_0 \right) \Bigg]^{1/2} t^q$$

$$+ K \sum_{k,r=1}^{d} \sum_{j=1}^{m} \Bigg[ E \left( \sup_{0 \leq s \leq t} \left| \bar{b}_s^{k,j} \right|^{8q} \Big| \mathcal{A}_0 \right) E \left( \sup_{0 \leq s \leq t} \left| \bar{b}_s^{r,j} \right|^{8q} \Big| \mathcal{A}_0 \right) \Bigg]^{1/4}$$

$$\times \Bigg[ E \left( \sup_{0 \leq s \leq t} \left| \frac{\partial^2}{\partial y^k \partial y^r} F_{\vec{p}}(\eta_s - X_0) \right|^{4q} \Big| \mathcal{A}_0 \right) \Bigg]^{1/2} t^{2q}.$$

From Lemma 5.11.2 and the fact that each differentiation reduces the product defined by (11.10) by one factor we obtain

$$\hat{Z}(t) \ \leq \ K \left( 1 + |X_0|^{2r} \right) \left( \max_{\vec{p} \in P_{l-1}} \left[ E \left( \sup_{0 \leq s \leq t} |F_{\vec{p}}(\eta_s - X_0)|^{4q} \Big| \mathcal{A}_0 \right) \right]^{1/2} \right) t^q$$

$$+ K \left( 1 + |X_0|^{4r} \right) \left( \max_{\vec{p} \in P_{l-2}} \left[ E \left( \sup_{0 \leq s \leq t} |F_{\vec{p}}(\eta_s - X_0)|^{4q} \Big| \mathcal{A}_0 \right) \right]^{1/2} \right) t^{2q}.$$

Concluding with the induction assumption we finally have

$$\hat{Z}(t) \ \leq \ K \left( 1 + |X_0|^{2r} \right) \left\{ t^{q(l-1)} t^q + t^{q(l-2)} t^{2q} \right\}$$

$$\leq \ K \left( 1 + |X_0|^{2r} \right) t^{ql}. \ \square$$

The proof of the next lemma follows immediately from the estimates (4.5.16) and (4.5.17).

**Lemma 5.11.5** *Under the assumptions of Proposition 5.11.1 for each* $p = 1, 2, \ldots$ *there exists a constant* $K \in (0, \infty)$ *such that*

$$(11.15) \qquad E\left(\sup_{0 \le t \le T} |X_t^y|^{2q} \Big| \mathcal{A}_0\right) \le K\left(1 + |y|^{2q}\right)$$

*and*

$$(11.16) \qquad E\left(|X_t^y - y|^{2q} \Big| \mathcal{A}_0\right) \le K\left(1 + |y|^{2q}\right) t^q$$

*for all* $y \in \Re^d$, $t \in [0, T]$ *and* $q = 0, 1, \ldots, p$.

**Lemma 5.11.6** *Under the assumptions of Proposition 5.11.1 we have*

$$(11.17) \qquad X_t^{X_0} - \eta_t = \sum_{\alpha \in B(\Gamma_\beta)} I_\alpha \left[f_\alpha\left(X_\cdot^{X_0}\right)\right]_t$$

*for all* $t \in [0, T]$ *and* $\beta = 1, 2, \ldots$, *with the remainder set*

$$(11.18) \qquad B(\Gamma_\beta) = \{\alpha \in \mathcal{M} : l(\alpha) = \beta + 1\}.$$

**Proof** Using (11.4) we have

$$
\begin{aligned}
X_t^{X_0} - \eta_t &= X_t^{X_0} - X_0 - (\eta_t - X_0) \\
&= X_t^{X_0} - X_0 - \sum_{\alpha \in \Gamma_\beta \setminus \{v\}} I_\alpha \left[f_\alpha\left(X_0\right)\right]_t.
\end{aligned}
$$

Hence by the Ito-Taylor formula (5.5.3) the increment $X_t^{X_0} - X_0$ satisfies

$$X_t^{X_0} - X_0 = \sum_{\alpha \in \Gamma_\beta \setminus \{v\}} I_\alpha \left[f_\alpha\left(X_0\right)\right]_t + \sum_{\alpha \in B(\Gamma_\beta)} I_\alpha \left[f_\alpha\left(X_\cdot^{X_0}\right)\right]_t.$$

Inserting this expression into our first equation gives (11.7). Finally, the assertion (11.18) follows from (5.4.4) and (11.2). □

**Lemma 5.11.7** *Under the assumptions of Proposition 5.11.1 there exist constants* $K \in (0, \infty)$ *and* $r \in \{1, 2, \ldots\}$ *such that*

$$(11.19) \qquad \hat{F}(t) := \left|E\left(F_{\vec{p}}(X_t^{X_0} - X_0) - F_{\vec{p}}(\eta_t - X_0)\Big|\mathcal{A}_0\right)\right|$$

$$\le K\left(1 + |X_0|^r\right) t^{\beta+1}$$

*for all* $t \in [0, T]$, $\vec{p} \in P_l$ *and* $l = 1, \ldots, 2\beta + 1$.

**Proof**

**1.** Let $l = 1$. Then $\vec{p} = (k)$ where $k \in \{1, \ldots, d\}$, so by (11.10) and Lemma 5.11.6 we have

$$\hat{F}(t) = \left| E\left(X_t^k - \eta_t^k \middle| \mathcal{A}_0\right) \right| = \left| \sum_{\alpha \in B(\Gamma_\beta)} E\left(I_\alpha \left[f_\alpha^k\left(X_\cdot^{X_0}\right)\right]_t \middle| \mathcal{A}_0\right) \right|.$$

According to Lemma 5.7.1 the expectations here are zero for those $\alpha$ with $l(\alpha) \neq n(\alpha)$. Examining the remainder set $B(\Gamma_\beta)$ given by (11.18) we see that only the term with $\alpha$ satisfying $l(\alpha) = n(\alpha) = \beta + 1$ remains, that is for the multi-index $\alpha^* = (j_1, \ldots, j_{\beta+1})$ where $j_1 = \cdots = j_{\beta+1} = 0$. Thus

$$\hat{F}(t) \leq \int_0^t \cdots \int_0^{s_2} \left| E\left(f_{\alpha^*}^k\left(X_{s_1}^{X_0}\right) \middle| \mathcal{A}_0\right) \right| ds_1 \ldots ds_{\beta+1}.$$

From the polynomial growth bound on $f_{\alpha^*}^k$ we then have

$$\hat{F}(t) \leq \int_0^t \cdots \int_0^{s_2} K\left(1 + E\left(\sup_{0 \leq s \leq T} \left|X_s^{X_0}\right|^{2r} \middle| \mathcal{A}_0\right)\right) ds_1 \ldots ds_{\beta+1}.$$

Finally, using the moment estimate (4.5.16) we obtain

$$(11.20) \qquad \hat{F}(t) \leq K\left(1 + |X_0|^{2r}\right) t^{\beta+1}.$$

**2.** For $l = 2, \ldots, 2\beta + 1$ we take the deterministic Taylor expansion of the function $F_{\vec{p}}$ about $\eta_t - X_0$ obtaining

$$(11.21) \qquad \hat{F}(t) \leq \sum_{r=1}^l \frac{1}{r!} \sum_{k_1=1}^d \cdots \sum_{k_r=1}^d \hat{F}_{k_1, \ldots, k_r}(t)$$

where

$$(11.22) \qquad \hat{F}_{k_1, \ldots, k_r}(t) = \left| E\left(\left(X_t^{k_1} - \eta_t^{k_1}\right) \cdots \left(X_t^{k_r} - \eta_t^{k_r}\right) \right.\right.$$
$$\left.\left. \times \frac{\partial^d}{\partial y^{k_1} \ldots \partial y^{k_r}} F_{\vec{p}}\left(\eta_t - X_0\right) \middle| \mathcal{A}_0\right) \right|.$$

**3.** We shall first estimate $\hat{F}_k$ for $k = 1, \ldots, d$. From the definition (11.10) of $F_{\vec{p}}$ there exists a $\vec{p}' \in P_{l-1}$ and a $q \in \{1, \ldots, l\}$ such that

$$\frac{\partial}{\partial y^k} F_{\vec{p}'}\left(\eta_t - X_0\right) = q F_{\vec{p}'}\left(\eta_t - X_0\right).$$

Hence in view of (11.11)

$$\frac{\partial}{\partial y^k} F_{\vec{p}}\left(\eta_t - X_0\right) = q \sum_{j=0}^m \int_0^t \hat{b}_{\vec{p}'}^j(s)\, dW_s^j.$$

We shall use this relation in (11.17), (11.22) and (5.2.12) to estimate

$$\hat{F}_k(t) = \left| E\left( \sum_{\alpha \in B(\Gamma_\beta)} I_\alpha \left[ f_\alpha^k \left( X^{X_0} \right) \right]_t q \sum_{j=0}^m \int_0^t \hat{b}^j_{p'}(s)\, dW^j_s \Big| \mathcal{A}_0 \right) \right|$$

$$\leq \sum_{\alpha \in B(\Gamma_\beta)} \sum_{j=0}^m q \left| E\left( I_\alpha \left[ f_\alpha^k \left( X^{X_0} \right) \right]_t I_{(j)} \left[ \hat{b}^j_{p'}(\cdot) \right]_t \Big| \mathcal{A}_0 \right) \right|$$

Applying Lemma 5.7.2 we have

$$\hat{F}_k(t) \leq \sum_{\alpha \in B(\Gamma_\beta)} \sum_{j=0}^m K\, t^{l(\alpha)} \left( \tilde{K}_1 \tilde{K}_2^j \right)^{1/2}$$

with $l(\alpha) = \beta + 1$ (see (11.18)),

$$\tilde{K}_1 = E\left( \sup_{0 \leq s \leq T} \left| f_\alpha^k \left( X^{X_0}_s \right) \right|^2 \Big| \mathcal{A}_0 \right)$$

and

$$\tilde{K}_2^j = E\left( \sup_{0 \leq s \leq T} \left| \hat{b}^j_{p'}(s) \right|^2 \Big| \mathcal{A}_0 \right).$$

From (11.15) and the polynomial growth bound on $f_\alpha$ we have

$$\tilde{K}_1 \leq K \left( 1 + |X_0|^{2r} \right).$$

Then, using (11.11)–(11.14), (11.6) and (11.7) analogously as in the proof of Lemma 5.11.4 we obtain

$$\tilde{K}_2^j \leq K \left( 1 + |X_0|^{2r} \right).$$

Combining these last estimates we thus have

$$(11.23) \qquad \hat{F}_k(t) \leq K \left( 1 + |X_0|^{2r} \right) t^{\beta+1}.$$

**4.** We shall now estimate the $\hat{F}_{k_1,\ldots,k_r}(t)$ for $r = 2, \ldots, l$. For each $(k_1, \ldots, k_l) \in P_r$ there is a finite $q \in \{1, 2, \ldots\}$ and $\vec{p}' \in P_{l-r}$ such that

$$\frac{\partial^r}{\partial y^{k_1} \ldots \partial y^{k_r}} F_{\vec{p}'} (\eta_t - X_0) = q\, F_{\vec{p}'} (\eta_t - X_0).$$

From (11.22) we thus obtain

$$(11.24) \qquad \hat{F}_{k_1,\ldots,k_r}(t) = \left| E\left( \left( X^{k_1}_t - \eta^{k_1}_t \right) \cdots \left( X^{k_r}_t - \eta^{k_r}_t \right) \right.\right.$$

$$\left.\left. \times \frac{\partial^d}{\partial y^{k_1} \ldots \partial y^{k_r}} F_{\vec{p}'} (\eta_t - X_0) \Big| \mathcal{A}_0 \right) \right|$$

$$\leq \; q \left[ E\left( |F_{\tilde{p}'}(\eta_t - X_0)|^2 \Big| \mathcal{A}_0 \right) E\left( \left( X_t^{k_1} - \eta_t^{k_1} \right)^2 \cdots \left( X_t^{k_r} - \eta_t^{k_r} \right)^2 \Big| \mathcal{A}_0 \right) \right]^{1/2}$$

$$\leq \; q \left[ E\left( |F_{\tilde{p}'}(\eta_t - X_0)|^2 \Big| \mathcal{A}_0 \right) \right]^{1/2} \left[ E\left( \left| X_t^{k_1} - \eta_t^{k_1} \right|^4 \Big| \mathcal{A}_0 \right) \right]^{1/4}$$

$$\times \cdots \times \left[ E\left( \left| X_t^{k_r} - \eta_t^{k_r} \right|^{2^{r+1}} \Big| \mathcal{A}_0 \right) \right]^{1/2^{r+1}}.$$

From (11.17), Lemma 5.7.5, the polynomial growth bound on $f_\alpha$, (11.18) and Lemma 5.11.5 for each $k = 1, \ldots, d$ and $l = 2, 4, \ldots, 2^{r+1}$ we have

$$\left[ E\left( \left| X_t^k - \eta_t^k \right|^{2l} \Big| \mathcal{A}_0 \right) \right]^{1/2l}$$

$$= \; \left[ E\left( \Big| \sum_{\alpha \in B(\Gamma_\beta)} I_\alpha \left[ f_\alpha^k (X_\cdot^{X_0}) \right]_t \Big|^{2l} \Big| \mathcal{A}_0 \right) \right]^{1/2l}$$

$$\leq \; \sum_{\alpha \in B(\Gamma_\beta)} \left[ E\left( \Big| I_\alpha \left[ f_\alpha^k (X_\cdot^{X_0}) \right]_t \Big|^{2l} \Big| \mathcal{A}_0 \right) \right]^{1/2l}$$

$$\leq \; \sum_{\alpha \in B(\Gamma_\beta)} K \, t^{l(\alpha)/2} \left[ E\left( \sup_{0 \leq s \leq T} \left| f_\alpha^k (X_s^{X_0}) \right|^{2l} \Big| \mathcal{A}_0 \right) \right]^{1/2l}$$

$$\leq \; \sum_{\alpha \in B(\Gamma_\beta)} K \, t^{(\beta+1)/2} \left[ 1 + E\left( \sup_{0 \leq s \leq T} \left| X_s^{X_0} \right|^{2l} \Big| \mathcal{A}_0 \right) \right]^{1/2l}$$

$$\leq \; K \left( 1 + |X_0|^{2r} \right) t^{(\beta+1)/2}.$$

Using this with (11.14) in (11.24) we then obtain

$$(11.25) \qquad \hat{F}_{k_1,\ldots,k_r}(t) \; \leq \; K \left( 1 + |X_0|^{2r} \right) t^{l-r} t^{(\beta+1)/2} t^{(\beta+1)/2}$$

$$\leq \; K \left( 1 + |X_0|^{2r} \right) t^{\beta+1}.$$

**5.** Finally, using (11.23) and (11.25) in (11.21), we have

$$\hat{F}_{k_1,\ldots,k_r}(t) \leq K \left( 1 + |X_0|^{2r} \right) t^{\beta+1}$$

for $l = 2, \ldots, 2\beta + 1$ too. $\square$

**Proof of Proposition 5.11.1** The function $g$ is $2(\beta+1)$ times differentiable so we use the deterministic Taylor expansion to obtain

$$(11.26) \qquad H_0(t) := \left| E\left( g(X_t) - g(\eta_t) \right) \Big| \mathcal{A}_0 \right|$$

$$= \left| E\left(\{g\left(X_t\right) - g\left(X_0\right)\} - \{g\left(\eta_t\right) - g\left(X_0\right)\}\right) \middle| \mathcal{A}_0 \right|$$

$$= \left| E\left( \sum_{l=1}^{2\beta+1} \frac{1}{l!} \sum_{\vec{p}\in P_l} [\partial_y^{\vec{p}} g\left(X_0\right)] F_{\vec{p}}\left(X_t - X_0\right) + R_0\left(X_t\right) \right.\right.$$

$$\left.\left. - \sum_{l=1}^{2\beta+1} \frac{1}{l!} \sum_{\vec{p}\in P_l} [\partial_y^{\vec{p}} g\left(X_0\right)] F_{\vec{p}}\left(\eta_t - X_0\right) + R_0\left(\eta_t\right) \middle| \mathcal{A}_0 \right)\right|$$

for each $t \in [0, T]$. Here the remainder terms are of the form

$$(11.27)\quad R_0(Z) = \frac{1}{(2(\beta+1))!} \sum_{\vec{p}\in P_{2(\beta+1)}} [\partial_y^{\vec{p}} g\left(X_0 + \Theta_{\vec{p}}(Z)\left(Z - X_0\right)\right)]$$

$$\times F_{\vec{p}}\left(Z - X_0\right)$$

for $Z = X_t$ and $\eta_t$, respectively, where $\Theta_{\vec{p}}(Z)$ is a $d \times d$ diagonal matrix with diagonal components

$$(11.28)\qquad\qquad \Theta_{\vec{p}}^{k,k}(Z) = \hat{\Theta}_{\vec{p}}^{k}(Z) \in [0,1]$$

for $k = 1, \ldots, d$.

Now, from (11.26) with (11.19) and the polynomial growth bound for $g$ and its derivatives we obtain

$$(11.29)\qquad H_0(t) \leq \sum_{l=1}^{2\beta+1} \frac{1}{l!} \sum_{\vec{p}\in P_l} |\partial_y^{\vec{p}} g\left(X_0\right)|$$

$$\times \left| E\left( F_{\vec{p}}\left(X_t - X_0\right) - F_{\vec{p}}\left(\eta_t - X_0\right) \middle| \mathcal{A}_0 \right)\right|$$

$$+ \left| E\left( R_0\left(X_t\right) \middle| \mathcal{A}_0\right)\right| + \left| E\left( R_0\left(\eta_t\right) \middle| \mathcal{A}_0\right)\right|$$

$$\leq K \left( 1 + |X_0|^{2r} \right) t^{\beta+1} + E\left( |R_0\left(X_t\right)| \middle| \mathcal{A}_0\right)$$

$$+ E\left( |R_0\left(\eta_t\right)| \middle| \mathcal{A}_0\right).$$

Using (11.27), (11.28), (11.16) and the polynomial growth bound on the derivatives of $g$ we obtain

$$(11.30)\qquad E\left( |R_0\left(X_t\right)| \middle| \mathcal{A}_0\right) \leq K \left( 1 + |X_0|^{2r} \right) t^{\beta+1}.$$

Applying Lemma 5.11.4, by similar arguments we also have

$$(11.31)\qquad E\left( |R_0\left(\eta_t\right)| \middle| \mathcal{A}_0\right) \leq K \left( 1 + |X_0|^{2r} \right) t^{\beta+1}.$$

Summarizing (11.26), (11.29), (11.30) and (11.31), we finally have

$$H_0(t) \leq K \left(1 + |X_0|^{2r}\right) T^{\beta+1}$$

uniformly in $t \in [0, T]$. This completes the proof of Proposition 5.11.1. $\square$

# 5.12 Weak Approximations of Multiple Ito Integrals

The main assertion of Proposition 5.11.1 still holds true if we replace the multiple Ito integrals $I_\alpha$ by other random variables $\hat{I}_\alpha$ which satisfy corresponding moment conditions. We shall write

$$(12.1) \qquad U_\beta(t) = \sum_{\alpha \in \Gamma_\beta} f_\alpha\left(t_0, X_{t_0}\right) \hat{I}_{\alpha, t_0, t}$$

for $t \in [t_0, T]$ and $\beta \in \{1, 2, \ldots\}$.

**Corollary 5.12.1** *Suppose under the assumptions of Proposition 5.11.1 that for $t \in [t_0, T]$ and $\beta \in \{1, 2, \ldots\}$ there exists a constant $K \in (0, \infty)$ such that the moment condition*

$$(12.2) \qquad \left| E\left( \prod_{k=1}^{l} I_{\alpha_k, t_0, t} - \prod_{k=1}^{l} \hat{I}_{\alpha_k, t_0, t} \middle| \mathcal{A}_{t_0} \right) \right| \leq K \, (t - t_0)^{\beta+1}$$

*holds for all choices of multi-indices $\alpha_k \in \Gamma_\beta \setminus \{v\}$ with $k = 1, \ldots, l$ and $l = 1, \ldots, 2\beta + 1$. Then for each $g \in C_P^{2(\beta+1)}(\Re^d, \Re)$ there exists constants $K_g \in (0, \infty)$ and $r \in \{1, 2, \ldots\}$ such that*

$$(12.3) \qquad \left| E\left( g\left(X_t\right) - g\left(U_\beta(t)\right) \middle| \mathcal{A}_{t_0} \right) \right| \leq K \left(1 + |X_{t_0}|^{2r}\right) (t - t_0)^{\beta+1},$$

*where $X_t = X_t^{t_0, X_{t_0}}$.*

We leave the proof as an exercise since it is a straightforward extension of the proof of Proposition 5.11.1.

**Exercise 5.12.2** *Prove Corollary 5.12.1.*

To evaluate the conditional expectation of the first product in (12.2) we shall need the following lemma.

**Lemma 5.12.3** *Let $\alpha = (j_1, \ldots, j_l)$, $\beta = (j'_1, \ldots, j'_p) \in \mathcal{M}$ with $l, p \in \{1, 2, \ldots\}$. Then*

$$(12.4) \qquad I_{\alpha, t} \, I_{\beta, t} = \int_0^t I_{\alpha, s} \, I_{\beta-, s} \, dW_s^{j'_p} + \int_0^t I_{\alpha-, s} \, I_{\beta, s} \, dW_s^{j_l}$$

$$+ \int_0^t I_{\alpha-,s} \, I_{\beta-,s} \, I_{\{j_l = j'_p \neq 0\}} \, ds$$

for $t \geq 0$.

We shall leave the proof as an exercise too as it is similar to that of Proposition 5.2.3, but here we need to start with a multi-dimensional linear Ito process with components representing the multiple Ito integrals up to $I_\alpha$ and $I_\beta$. An application of the Ito formula (3.4.6) then gives the desired result.

**Exercise 5.12.4**    *Verify (12.4)*.

Using Lemma 5.12.3 and the moment estimates (7.1) and (7.6) for multiple Ito integrals we shall now state the values of the conditional expectations

$$E\left(I_{\alpha_1} \cdots I_{\alpha_k}\right) := E\left(\prod_{k=1}^l I_{\alpha_k, t_0, t_0 + \Delta} \Big| \mathcal{A}_0\right)$$

for $\Delta \in [0, T - t_0]$ and $\alpha_1, \alpha_2, \ldots, \alpha_k \in \Gamma_\beta \setminus \{v\}$ which are relevant in the moment conditions (12.2) for the cases $\beta = 1, 2, 3$. We have

$$(12.5) \qquad E\left(I_{\alpha_1} \cdots I_{\alpha_k}\right) = 0$$

when the number of nonzero components of the multi-indices involved,

$$\lambda := \sum_{k=1}^l \left(l\left(\alpha_k\right) - n\left(\alpha_k\right)\right),$$

is odd. Furthermore, when $\lambda$ is even we find that

$$(12.6) \qquad \left|E\left(I_{\alpha_1} \cdots I_{\alpha_k}\right)\right| \leq K \, \Delta^\rho$$

where

$$\rho := \frac{1}{2}\lambda + \sum_{k=1}^l n\left(\alpha_k\right).$$

We shall say that the expectation in (12.6) has order $\rho$ in the time increment $\Delta$. Excluding those already given in (12.5) and those of the constants

$$I_{(0)} = \Delta \qquad I_{(0,0)} = \frac{1}{2}\Delta^2 \qquad I_{(0,0,0)} = \frac{1}{3!}\Delta^3,$$

the expectations of products of order $\rho = 1$, 2 and 3 are as follows. Here $j_1$, $\ldots$, $j_l = 1, \ldots, m$ and $\delta_{i,j}$ is the Kronecker delta symbol, that is

$$\delta_{i,j} = \begin{cases} 1 & : \quad i = j \\ 0 & : \quad \text{otherwise.} \end{cases}$$

$$(12.7) \qquad E\left(I_{(j_1)} I_{(j_2)}\right) = \Delta \, \delta_{j_1, j_2};$$

$$E\left(I_{(j_1)} I_{(0,j_2)}\right) = E\left(I_{(j_1)} I_{(j_2,0)}\right) = \frac{1}{2}\Delta^2 \delta_{j_1,j_2};$$

$$E\left(I_{(j_1,j_2)} I_{(j_3,j_4)}\right) = \frac{1}{2}\Delta^2 \delta_{j_1,j_3} \delta_{j_2,j_4};$$

$$E\left(I_{(j_1)} I_{(j_2)} I_{(j_3,j_4)}\right) = \left\{ \begin{array}{lll} \Delta^2 & : & j_1 = j_2 = j_3 = j_4 \\ \frac{1}{2}\Delta^2 & : & j_3 \neq j_4 \text{ and } j_1 = j_3, j_2 = j_4 \\ & & \text{or } j_1 = j_4, j_2 = j_3 \\ 0 & : & \text{otherwise} \end{array} \right.$$

$$E\left(I_{(j_1)} I_{(j_2)} I_{(j_3)} I_{(j_4)}\right) = \left\{ \begin{array}{lll} 3\Delta^2 & : & j_1 = j_2 = j_3 = j_4 \\ \Delta^2 & : & \{j_1, j_2, j_3, j_4\} \text{ consists of 2 distinct} \\ & & \text{pairs of identical numbers} \\ 0 & : & \text{otherwise} \end{array} \right.$$

$$E\left(I_{(j_1)} I_{(j_2,j_3,j_4)}\right) = 0, \qquad E\left(I_{(j_1,0)} I_{(0,j_2)}\right) = \frac{1}{3!}\Delta^3 \delta_{j_1,j_2};$$

$$E\left(I_{(j_1,0)} I_{(j_2,0)}\right) = E\left(I_{(0,j_1)} I_{(0,j_2)}\right) = \frac{1}{3}\Delta^3 \delta_{j_1,j_2};$$

$$E\left(I_{(0,j_1)} I_{(j_2,j_3,j_4)}\right) = E\left(I_{(j_1,0)} I_{(j_2,j_3,j_4)}\right) = 0;$$

$$\begin{aligned} E\left(I_{(0,j_1,j_2)} I_{(j_3,j_4)}\right) &= E\left(I_{(j_1,0,j_2)} I_{(j_3,j_4)}\right) \\ &= E\left(I_{(j_1,j_2,0)} I_{(j_3,j_4)}\right) = \frac{1}{3!}\Delta^3 \delta_{j_1,j_3} \delta_{j_2,j_4}; \end{aligned}$$

$$E\left(I_{(0,0,j_1)} I_{(j_2)}\right) = E\left(I_{(0,j_1,0)} I_{(j_2)}\right) = E\left(I_{(j_1,0,0)} I_{(j_2)}\right) = \frac{1}{3!}\Delta^3 \delta_{j_1,j_2};$$

$$E\left(I_{(j_1,0,0)} I_{(j_2,0)}\right) = E\left(I_{(0,0,j_1)} I_{(0,j_2)}\right) = \frac{1}{8}\Delta^4 \delta_{j_1,j_2};$$

$$E\left(I_{(0,j_1,0)} I_{(j_2,0)}\right) = E\left(I_{(0,j_1,0)} I_{(0,j_2)}\right) = \frac{1}{3!}\Delta^4 \delta_{j_1,j_2};$$

$$E\left(I_{(j_1,0,0)} I_{(0,j_2)}\right) = E\left(I_{(0,0,j_1)} I_{(j_2,0)}\right) = \frac{1}{4!}\Delta^4 \delta_{j_1,j_2};$$

$$E\left(I_{(j_1,j_2,0)} I_{(j_3,0)}\right) = E\left(I_{(j_1,j_2,0)} I_{(0,j_3)}\right) = E\left(I_{(j_1,0,j_2)} I_{(j_3,0)}\right) = 0;$$

$$E\left(I_{(j_1,0,j_2)}I_{(0,j_3)}\right) = E\left(I_{(0,j_1,j_2)}I_{(j_3,0)}\right) = E\left(I_{(0,j_1,j_2)}I_{(0,j_3)}\right) = 0;$$

$$E\left(I_{(j_1,j_2,j_3)}I_{(j_2,j_3)}\right) = 0;$$

$$E\left(I_{(0,j_1)}I_{(j_2)}I_{(j_3)}I_{(j_4)}\right) = E\left(I_{(j_1,0)}I_{(j_2)}I_{(j_3)}I_{(j_4)}\right)$$

$$= \begin{cases} \frac{3}{2}\Delta^3 & : \quad j_1 = j_2 = j_3 = j_4 \\ \frac{1}{2}\Delta^3 & : \quad \{j_1,j_2,j_3,j_4\} \text{ consists of 2 distinct} \\ & \qquad \text{pairs of identical numbers} \\ 0 & : \quad \text{otherwise} \end{cases}$$

$$E\left(I_{(0,j_1)}I_{(j_2)}I_{(j_3,j_4)}\right) = \frac{1}{3!}\Delta^3\delta_{j_1,j_3}\delta_{j_2,j_4};$$

$$E\left(I_{(j_1,0)}I_{(j_2)}I_{(j_3,j_4)}\right) = \frac{1}{3}\Delta^3\delta_{j_1,j_3}\delta_{j_2,j_4};$$

$$E\left(I_{(0,j_1,j_2)}I_{(j_3)}I_{(j_4)}\right) = E\left(I_{(j_1,0,j_2)}I_{(j_3)}I_{(j_4)}\right) = E\left(I_{(j_1,j_2,0)}I_{(j_3)}I_{(j_4)}\right)$$

$$= \begin{cases} \frac{1}{3!}\Delta^3 & : \quad j_1 = j_2, j_3 = j_4 \text{ or } j_1 = j_4, j_2 = j_3 \\ 0 & : \quad \text{otherwise} \end{cases}$$

$$E\left(I_{(j_1,j_2,j_3)}I_{(j_4,j_5,j_6)}\right) = \frac{1}{3!}\Delta^3\delta_{j_1,j_4}\delta_{j_2,j_5}\delta_{j_3,j_6};$$

$$E\left(I_{(j_1,)}\cdots I_{(j_6)}\right) = \begin{cases} 15\Delta^3 & : \quad j_1 = \cdots = j_6 \\ 3\Delta^3 & : \quad \text{1 pair and 1 quadruple of identical } j_i \\ \Delta^3 & : \quad \text{3 different pairs of identical } j_i \\ 0 & : \quad \text{otherwise} \end{cases}$$

$$E\left(I_{(j_1,j_2,j_3)}I_{(j_4)}I_{(j_5,j_6)}\right) = \frac{1}{3!}\Delta^3\left[\delta_{j_1,j_4}\delta_{j_2,j_5}\delta_{j_3,j_6} + \delta_{j_1,j_5}\delta_{j_2,j_4}\delta_{j_3,j_6} + \delta_{j_1,j_5}\delta_{j_2,j_6}\delta_{j_3,j_4}\right]$$

$$\begin{aligned} E\left(I_{(j_1,j_2,j_3)}I_{(j_4)}I_{(j_5)}I_{(j_6)}\right) &= \frac{1}{3!}\Delta^3\Big[\delta_{j_1,j_4}\left(\delta_{j_2,j_5}\delta_{j_3,j_6} + \delta_{j_2,j_6}\delta_{j_3,j_5}\right) \\ &\qquad + \delta_{j_1,j_5}\left(\delta_{j_2,j_4}\delta_{j_3,j_6} + \delta_{j_2,j_6}\delta_{j_3,j_4}\right) \\ &\qquad + \delta_{j_1,j_6}\left(\delta_{j_2,j_5}\delta_{j_3,j_4} + \delta_{j_2,j_4}\delta_{j_3,j_5}\right)\Big]; \end{aligned}$$

$$E\left(I_{(j_1,j_2)} I_{(j_3,j_4)} I_{(j_5,j_6)}\right) = \frac{1}{3!}\Delta^3 \big[\delta_{j_2,j_4}\left(\delta_{j_1,j_5}\delta_{j_3,j_6} + \delta_{j_1,j_6}\delta_{j_3,j_5}\right)$$

$$+\delta_{j_2,j_6}\left(\delta_{j_1,j_3}\delta_{j_4,j_5} + \delta_{j_1,j_4}\delta_{j_3,j_5}\right)$$

$$+\delta_{j_4,j_6}\left(\delta_{j_1,j_3}\delta_{j_2,j_5} + \delta_{j_2,j_3}\delta_{j_1,j_5}\right)\big];$$

$$E\left(I_{(j_1,j_2)} I_{(j_3,j_4)} I_{(j_5)} I_{(j_6)}\right) = \frac{1}{3!}\Delta^3 \big[2\delta_{j_1,j_3}\left(\delta_{j_2,j_5}\delta_{j_4,j_6} + \delta_{j_2,j_6}\delta_{j_4,j_5}\right)$$

$$+\delta_{j_2,j_3}\left(\delta_{j_1,j_5}\delta_{j_4,j_6} + \delta_{j_1,j_6}\delta_{j_4,j_5}\right)$$

$$+\delta_{j_1,j_4}\left(\delta_{j_3,j_5}\delta_{j_2,j_6} + \delta_{j_3,j_6}\delta_{j_2,j_5}\right)$$

$$+2\delta_{j_2,j_4}\left(\delta_{j_1,j_5}\delta_{j_3,j_6} + \delta_{j_3,j_5}\delta_{j_1,j_6}\right)\big];$$

$$E\left(I_{(j_1,j_2)} I_{(j_3)} I_{(j_4)} I_{(j_5)} I_{(j_6)}\right) = \frac{1}{2}\big[E\left(I_{(j_1)}\cdots I_{(j_6)}\right) - \Delta\delta_{j_1,j_2}E\left(I_{(j_3)}\cdots I_{(j_6)}\right)\big].$$

We shall now use the above relations to propose some weak approximations $\hat{I}_\alpha$ of the multiple Ito integrals $I_\alpha$ which satisfy the condition (12.2) for each of the orders $\beta = 1, 2$ or 3. In all of these cases we set:

$$\hat{I}_{(0)} = \Delta, \qquad \hat{I}_{(0,0)} = \frac{1}{2}\Delta^2, \qquad \hat{I}_{(0,0,0)} = \frac{1}{3!}\Delta^3.$$

For $\beta = 1$, $m = 1, 2, \ldots$ and $j \in \{1, \ldots, m\}$ we can set:

$$(12.8) \qquad \hat{I}_{(j)} = \Delta\hat{W}^j,$$

where the $\Delta\hat{W}^j$ are independent $N(0, \Delta)$ Gaussian random variables or independent two-point distributed random variables with

$$P\left(\Delta\hat{W}^j = \pm\sqrt{\Delta}\right) = \frac{1}{2}.$$

For $\beta = 2$, $m = 1, 2, \ldots$ and $j_1, j_2 \in \{1, \ldots, m\}$ we can set:

$$(12.9) \qquad \hat{I}_{(j)} = \Delta\hat{W}^j, \qquad \hat{I}_{(0,j_1)} = \hat{I}_{(j_1,0)} = \frac{1}{2}\Delta\,\Delta\hat{W}^j,$$

$$\hat{I}_{(j_1,j_2)} = \frac{1}{2}\left(\Delta\hat{W}^{j_1}\Delta\hat{W}^{j_2} + V_{j_1,j_2}\right),$$

where the $\Delta\hat{W}^j$ are independent $N(0, \Delta)$ Gaussian or three-point distributed random variables with

$$P\left(\Delta\hat{W}^j = \pm\sqrt{3\Delta}\right) = \frac{1}{6}, \qquad P\left(\Delta\hat{W}^j = 0\right) = \frac{2}{3},$$

and the $V_{j_1,j_2}$ are independent two-point distributed random variables with

$$P\left(V_{j_1,j_2} = \pm\Delta\right) = \frac{1}{2}$$

for $j_2 = 1, \ldots, j_1 - 1$,

$$V_{j_1,j_1} = -\Delta$$

and

$$V_{j_2,j_1} = -V_{j_1,j_2}$$

for $j_2 = j_1 + 1, \ldots, m$ and $j_1 = 1, \ldots, m$. Obviously the $V_{j_1,j_2}$ are not required when $m = 1$.

For $\beta = 3$ and $m = 1$ we can set:

(12.10)
$$\hat{I}_{(j)} = \Delta\hat{W}^j,$$

$$\hat{I}_{(1,0)} = \Delta\hat{Z}, \qquad \hat{I}_{(0,1)} = \Delta\,\Delta\hat{W} - \Delta\hat{Z},$$

$$\hat{I}_{(1,1)} = \frac{1}{2}\left(\left(\Delta\hat{W}\right)^2 - \Delta\right),$$

$$\hat{I}_{(0,0,1)} = \hat{I}_{(0,1,0)} = \hat{I}_{(1,0,0)} = \frac{1}{6}\Delta^2\,\Delta\hat{W},$$

$$\hat{I}_{(1,1,0)} = \hat{I}_{(1,0,1)} = \hat{I}_{(0,1,1)} = \frac{1}{6}\Delta\left(\left(\Delta\hat{W}\right)^2 - \Delta\right),$$

$$\hat{I}_{(1,1,1)} = \frac{1}{6}\Delta\hat{W}\left(\left(\Delta\hat{W}\right)^2 - 3\Delta\right)$$

where the $\Delta\hat{W}$ and $\Delta\hat{Z}$ are correlated Gaussian random variables with $\Delta\hat{W} \sim N(0;\Delta)$, $\Delta\hat{Z} \sim N(0;\frac{1}{3}\Delta^3)$ and $E\left(\Delta\hat{W}\,\Delta\hat{Z}\right) = \frac{1}{2}\Delta^2$.

For many stochastic differential equations not all of the multiple stochastic integrals in an Ito-Taylor expansion may actually appear because their coefficient functions are zero due to some special structural feature of the equation. In such situations it is sometimes feasible, practically, to find weak approximations of multiple stochastic integrals when $m > 1$ and $\beta \geq 3$ too.

# Chapter 6

# Modelling with Stochastic Differential Equations

Important issues which arise when stochastic differential equations are used in applications are discussed in this chapter, in particular the appropriateness of the Ito or Stratonovich version of an equation. Stochastic stability, parametric estimation, stochastic control and filtering are also considered.

## 6.1 Ito or Stratonovich?

The differing Ito and Stratonovich interpretations of stochastic integrals and stochastic differential equations results from the peculiar property that the sample paths of a Wiener process are, almost surely, not differentiable or even of bounded variation. This is, perhaps, not so strange if we remember that such equations arise from an attempt to add to ordinary differential equations random fluctuations described by a Gaussian white noise, which can be considered formally to be the (nonexistent) derivative of a Wiener process. In reality white noise processes are often meant to be tractable idealizations of real coloured noise processes, for which the autocorrelation at different time instants is made arbitrarily small. The stochastic calculi of Ito and Stratonovich provide us with mathematically valid formulations of stochastic differential equations, but leave unanswered the question of which interpretation we should use. In fact, the answer depends on how exactly we intend the white noise processes to approximate the real noise processes and on how the stochastic differential equation itself approximates the real situation being modelled.

   Physically realizable processes are often smooth with at least a small degree of autocorrelation. If $R_t^{(n)}$ is such a process, close in some way to a Wiener process $W_t$, the differential equation

$$(1.1) \qquad dX_t^{(n)} = a\left(t, X_t^{(n)}\right) \, dt + b\left(t, X_t^{(n)}\right) \, dR_t^{(n)}$$

is a random differential equation, that is an ordinary differential equation in each of its sample paths involving Riemann-(or Lebesgue-) Stieltjes integrals in its integral equation form

$$(1.2) \qquad X_t^{(n)} = X_{t_0} + \int_{t_0}^{t} a\left(s, X_s^{(n)}\right) \, ds + \int_{t_0}^{t} b\left(s, X_s^{(n)}\right) \, dR_s^{(n)}.$$

It can be solved, in principle at least, by the methods of classical calculus. Since these methods are also valid for the Stratonovich calculus, this suggests that the Stratonovich interpretation may be the appropriate one for the limiting stochastic differential equation obtained by replacing the real process $R_t^{(n)}$ in (1.1) by a Wiener process $W_t$. This can be seen explicitly in the linear equation

$$(1.3) \qquad\qquad dX_t^{(n)} = aX_t^{(n)} \, dt + bX_t^{(n)} \, dR_t^{(n)},$$

where $R_t^{(n)}$ is the piecewise differentiable linear interpolation of a Wiener process $W_t$ on a partition $0 = t_1^{(n)} < t_2^{(n)} < \ldots < t_{n+1}^{(n)} = T$ defined by

$$R_t^{(n)} = W_{t_j^{(n)}} + \left( W_{t_{j+1}^{(n)}} - W_{t_j^{(n)}} \right) \frac{t - t_j^{(n)}}{t_{j+1}^{(n)} - t_j^{(n)}}$$

for $t_j^{(n)} \leq t \leq t_{j+1}^{(n)}$ and $j = 1, 2, \ldots, n$. This process converges sample pathwise, almost surely, on $t_0 \leq t \leq T$ to $W_t$ as $n \to \infty$ if the partition length

$$\delta^{(n)} = \max_{1 \leq j \leq n} \left| t_{j+1}^{(n)} - t_j^{(n)} \right| \longrightarrow 0 \quad \text{as} \quad n \to \infty.$$

By classical calculus the solution of (1.3) is

$$X_t^{(n)} = X_{t_0} \exp\left( a(t - t_0) + b \left( R_t^{(n)} - R_{t_0}^{(n)} \right) \right),$$

and this converges sample pathwise to

$$X_t = X_{t_0} \exp\left( a(t - t_0) + b(W_t - W_{t_0}) \right),$$

which is the solution of the Stratonovich SDE

$$dX_t = aX_t \, dt + bX_t \circ dW_t.$$

A theorem of Wong and Zakai establishes the same result under quite broad assumptions on the smooth approximating noise processes $R_t^{(n)}$ and the coefficients $a(t, x)$ and $b(t, x)$. From this we can assert that the Stratonovich interpretation of a stochastic differential equation is the appropriate one when the white noise is used as an idealization of a smooth real noise process. In such cases the Ito counterpart of the Stratonovich SDE is a useful artifice which, for example, allows access to the appropriate moment equations or the Fokker-Planck equation.

In engineering and the physical sciences many stochastic differential equations are obtained by including random fluctuations in ordinary differential equations, which have been deduced from phenomological or physical laws. The underlying systems being modelled here are usually continuous in both time and state. In contrast, many biological systems are intrinsically discrete in either time or state, or both. For example, in genetics and population dynamics the population size is integer valued, successive generations may not overlap in time, breeding may occur in separated seasons and environmental parameters may only change at discrete instants. In these cases diffusion processes satisfying Ito stochastic differential equations are a convenient and mathematically tractable approximation to the actual process. For example, the exponential growth equation

$$\dot{x} = \frac{dx}{dt} = ax$$

and its noisy counterpart

$$dX_t = aX_t\, dt + bX_t\, \xi_t\, dt,$$

resulting from a noisy growth coefficient $a + b\xi_t$, are often only convenient continuous time approximations of the discrete time systems

$$x(t_{n+1}) - x(t_n) = a\, x(t_n)\Delta t_n$$

and

(1.4) $$X_{t_{n+1}} - X_{t_n} = a\, X_{t_n}\Delta t_n + b\, X_{t_n}\, \xi_{t_n}\, \Delta t_n.$$

In turn these may be continuous state approximations of a discrete state system. In view of the Cenral Limit Theorem, as the result of a large number of small discrete events we can suppose that the $\xi_{t_n}$ in (1.4) are independent standard Gaussian random variables. Thus we can take

$$\xi_{t_n}\Delta t_n = W_{t_{n+1}} - W_{t_n} = \Delta W_{t_n}$$

for a standard Wiener process $W_t$; this makes the noise term in (1.4) equal to $bX_{t_n}\Delta W_{t_n}$. If the noise is due to external environmental effects, it can be argued that the intensity $bX_{t_n}$ and the noise $\Delta W_{t_n}$ in the transition from the $n$th to the $(n+1)$th time instant should be independent. In other words, the intensity should be nonanticipative. If we approximate (1.4) by a continuous time process, that is by a stochastic differential equation, then the Ito interpretation

$$dX_t = aX_t\, dt + bX_t\, dW_t$$

is the appropriate one as we shall seewhen we consider the Euler scheme in Section 2 of Chapter 10.

From a purely mathematical viewpoint both the Ito and Stratonovich calculi are correct. Which one we should use in a particular context depends on extraneous circumstances. Nevertheless, once a choice has been made the other calculus can be applied when advantageous to the appropriately modified stochastic differential equation. In this way the strengths of both calculi can be used to maximum benefit.

## 6.2 Diffusion Limits of Markov Chains

Many problems in biology and other fields are intrinsically discrete and are often modelled by Markov chains. When small increments are involved these Markov chains are usually approximated by mathematically more tractable diffusion processes, the solutions of appropriate Ito stochastic differential equations. Conversely, when a stochastic differential equation is solved numerically, Markov chains are constructed to approximate its diffusion process solutions. In both cases the validity and accuracy of such approximations are obviously matters of some importance.

In PC-Exercise 1.8.2 we examined the convergence in distribution of a sequence of random walks to a standard Wiener process on the time interval $[0, 1]$. This is one of the simplest examples of a sequence of Markov chains approximating a diffusion process. For each $N = 1, 2, 3, \ldots$ we partitioned the time interval $[0, 1]$ at the instants $t_k^{(n)} = k/N$ for $k = 0, 1, \ldots, N$ and generated a random walk starting at $X_0^{(N)} = 0$ with equally probable spatial steps of length $N^{-1/2}$ to the left or right of the current position $X_k^{(N)}$ at the time instant $t_{k+1}^{(N)}$ to the next position $X_{k+1}^{(N)}$. That is, we evaluated

$$(2.1) \qquad X_{k+1}^{(N)} = X_k^{(N)} + \frac{1}{\sqrt{N}} \zeta_k$$

for $k = 0, 1, \ldots, N$, where the $\zeta_k$ are independent 2-point distributed random variables taking the values $+1$ and $-1$ with equal probabilities $1/2$. We interpolated the random variables $X_k^{(N)}$ in a piecewise constant manner on the time intervals $[t_k^{(N)}, t_{k+1}^{(N)})$ for $k = 0, 1, \ldots, N - 1$ to define a process $X^{(N)}$ on the entire time interval $[0, 1]$, and then used the Central Limit Theorem to conclude that the $X^{(N)}$ converged in distribution to a standard Wiener process on the time interval $[0, 1]$ as $N \to \infty$. It is obvious here that the $X_k^{(N)}$ generated by (2.1) for $k = 1, 2, \ldots, N$ are the realizations of a Markov chain on the countably infinite state space

$$R_{N^{-1/2}}^1 = \left\{ x_i = i \frac{1}{\sqrt{N}} : i = 0, \pm 1, \pm 2, \ldots \right\},$$

which is called the $N^{-1/2}$-grid, with transition probabilities $p^{i-1,i} = p^{i,i+1} = 1/2$ and $p^{i,j} = 0$ otherwise, for all $i, j = 0, \pm 1, \pm 2, \ldots$.

In any of the above random walks the step size was the same at every point of the lattice $R_{N^{-1/2}}^1$. A natural generalization is to allow it to depend on the current location $X_k^{(N)}$ with the next position being determined by

$$(2.2) \qquad X_{k+1}^{(N)} = X_k^{(N)} + a_N \left( X_k^{(N)} \right) \frac{1}{N} + b_N \left( X_k^{(N)} \right) \frac{1}{\sqrt{N}} \zeta_k$$

for given functions $a_N$ and $b_N$ defined on $R_{N^{-1/2}}^1$ such that

$$x + a_N(x) \frac{1}{N} \pm b_N(x) \frac{1}{\sqrt{N}} \in R_{N^{-1/2}}^1$$

for all $x \in R_{N^{-1/2}}^1$. The functions $a_N$ and $b_N$ may be derived from certain functions $a$ and $b$ defined on all of $\Re$ or may converge to such functions. In either case we can form an Ito stochastic differential equation

$$(2.3) \qquad dX_t = a(X_t)\, dt + b(X_t)\, dW_t,$$

the solutions of which will be diffusion processes if the coefficient functions are sufficiently regular. In general the increments $X_{k+1}^{(N)} - X_k^{(N)}$ of the Markov chains defined by (2.2) will not be i.i.d. random variables, so we cannot use the Central Limit Theorem simply to conclude that they converge in distribution to a diffusion process satisfying (2.3). Such convergence can nevertheless still be established, but only if additional properties are satisfied by the Markov chains. A necessary condition is that the Markov chains are consistent with the diffusion processes satisfying (2.3) in the sense that their conditioned first and second order moments and probable jump sizes satisfy, asymptotically in $N \to \infty$ at least, similar properties to the three limit conditions (1.7.9)–(1.7.11) which define diffusion processes. That is, the Markov chains $\{X_k^{(N)},\, k = 0, 1,$ ..., $\}$ should satisfy *consistency conditions* such as

$$(2.4) \qquad \lim_{N \to \infty} N\, E\left(X_{k+1}^{(N)} - X_k^{(N)} \,\big|\, X_k^{(N)} = x\right) = \lim_{N \to \infty} E\left(a_N(x)\right) = a(x),$$

$$(2.5) \qquad \lim_{N \to \infty} N\, E\left(\left(X_{k+1}^{(N)} - X_k^{(N)}\right)^2 \,\big|\, X_k^{(N)} = x\right) = \lim_{N \to \infty} E\left(b_N^2(x)\right)$$
$$= b^2(x)$$

and

$$(2.6) \qquad \lim_{N \to \infty} N\, E\left(\left|X_{k+1}^{(N)} - X_k^{(N)}\right|^3 \,\big|\, X_k^{(N)} = x\right) = 0$$

for all $x \in \Re$.

These remarks also apply to more complicated types of driving noises than the 2-point process considered so far, and for other relationships between the spatial and temporal step sizes. In such cases it may, however, be difficult to write down explicitly the state spaces and the transition matrices of the Markov chains.

**Example 6.2.1**    *A population has $2N$ genes with two alleles a and A. In the current generation there are $i$ genes of type a and $2N - i$ of type A. The new generation, also of size $2N$, is selected by $2N$ trials with replacement, with each trial yielding an a allele with probability $p_i = i/2N$ and an A allele with probability $q_i = 1 - p_i$. The proportion of genes of type a in successive generations forms a Markov chain with transition probabilities*

$$p^{i,j} = \frac{(2N)!}{j!(2N - j)!}\, (p_i)^j\, (q_i)^{(2N-j)}$$

*for $i, j = 1, 2, \ldots, 2N$. If $X_k^{(N)}$ denotes the proportion in the kth generation, that is $X_k^{(N)} = i/2n$, it can be shown that*

$$\lim_{N \to \infty} N E \left( X_{k+1}^{(N)} - X_k^{(N)} \big| X_k^{(N)} = x \right) = 0$$

*and*

$$\lim_{N \to \infty} N E \left( \left( X_{k+1}^{(N)} - X_k^{(N)} \right)^2 \big| X_k^{(N)} = x \right) = x^2 (1 - x)^2.$$

*In addition, if the chain is interpolated piecewise constant on time intervals of equal length* $\Delta = N^{-2}$, *then the interpolated process converges in distribution to a solution of the Ito stochastic differential equation*

$$dX_t = X_t (1 - X_t) \, dW_t.$$

**Exercise 6.2.2**    *Construct Markov chains driven by the random telegraphic noise process (see Exercise 1.6.8) which are consistent with a diffusion process* $X_t = at + bW_t$ *where* $a$ *and* $b$ *are nonzero constants. Then use the Central Limit Theorem to verify that the appropriate piecewise constant interpolations of these chains converge in distribution to the diffusion process on any finite time interval.*

## 6.3    Stochastic Stability

Most differential equations, deterministic or stochastic, cannot be solved explicitly. Nevertheless we can often deduce alot of useful information, usually qualitative, about the behaviour of their solutions from the functional form of their coefficients. Of particular interest in applications is the long term asymptotic behaviour and sensitivity of the solutions to small changes, for example measurement errors, in the initial values. From existence and uniqueness theory we know that the solutions of a differential equation are continuous in their initial values, at least over a finite time interval. Extending this idea to an infinite time interval leads to the concept of stability.

For an ordinary differential equation

(3.1)                      $$\dot{x} = \frac{dx}{dt} = a(t, x)$$

we usually talk about the stability of an *equilibrium point* or *steady state* $x = c$, where $a(t, c) = 0$ for all $t$, which we can assume without loss of generality to equal 0. We say that $x = 0$ is *stable* for the differential equation (3.1) if for every $\epsilon > 0$ there exists a $\delta = \delta(t_0, \epsilon) > 0$ such that

(3.2)          $$|x(t; t_0, x_0)| < \epsilon \quad \text{for all} \quad t \geq t_0 \quad \text{and} \quad |x_0| < \delta,$$

where $x(t; t_0, x_0)$ is the solution of (3.1) with initial value $x(t_0; t_0, x_0) = x_0$. If, in addition, there is a $\delta_0 = \delta_0(t_0) > 0$ such that

(3.3)                  $$\lim_{t \to \infty} x(t; t_0, x_0) = 0 \quad \text{for all} \quad |x_0| < \delta_0$$

we say that $x = 0$ is *asymptotically stable* for (3.1). We include the qualifier *uniformly* if $\delta$ and $\delta_0$ do not depend on $t_0$ and *global* if (3.3) holds for any $\delta_0$. These definitions are due to Lyapunov who introduced a test for the stability of an equilibrium point in terms of a function $V$, now called a *Lyapunov function*, resembling the potential energy in a mechanical system. Essentially, the equilibrium lies at the bottom of a potential energy well and the potential energy decreases monotonically along the solutions of the differential equation, at least in a small neighbourhood of the equilibrium point.

In particular, for a Lyapunov function $V$ it is required that

$$V(t,0) = 0, \qquad V(t,x) > 0$$

for all $x \neq 0$ and that

(3.4) $$\frac{d}{dt} V\left(t, x(t; t_0, x_0)\right) < 0$$

for $t \geq t_0$ and all $x_0$ sufficiently small. By the chain rule

$$\frac{d}{dt} V\left(t, x(t; t_0, x_0)\right)$$

$$= \frac{\partial V}{\partial t}\left(t, x(t; t_0, x_0)\right) + \frac{\partial V}{\partial x}\left(t, x(t; t_0, x_0)\right) \frac{dx}{dt}$$

$$= \frac{\partial V}{\partial t}\left(t, x(t; t_0, x_0)\right) + \frac{\partial V}{\partial x}\left(t, x(t; t_0, x_0)\right) a\left(t, x(t; t_0, x_0)\right),$$

so (3.4) follows if

$$\frac{\partial V}{\partial t}(t, x) + \frac{\partial V}{\partial x}(t, x) a(t, x) < 0$$

for all $x$ in a neighbourhood of 0 and all $t \geq t_0$. We do not need to know the solutions of (3.1) explicitly to check the validity of this inequality.

**Example 6.3.1**    *The function* $V(x) = x^2$ *is a Lyapunov function for the ordinary differential equation*

(3.5) $$\dot{x} = \frac{dx}{dt} = -x - x^3,$$

*which has a unique equilibrium point* $x = 0$. *Here*

$$\frac{dV}{dx}(x) a(x) = -2x^2 - 2x^4 \leq -2V(x)$$

*for all* $x \in \Re$, *so along the solutions of (3.5) we have*

$$\frac{d}{dt} V\left(t, x(t; t_0, x_0)\right) \leq -2V\left(t, x(t; t_0, x_0)\right).$$

*Hence*

$$V(t, x(t; t_0, x_0)) \leq V(x_0)e^{-2(t-t_0)}$$

*for all $t \geq t_0$ and $x_0$, from which we can conclude that $x = 0$ is globally uniformly asymptotically stable for (3.5).*

This simple example gives the gist of Lyapunov's method for investigating the stability of an equilibrium point. It is a powerful method when an appropriate Lyapunov function can be found, but that is not always easily done. An extensive theory has been developed, listing necessary or sufficient conditions on the Lyapunov function for the stability, asymptotic stability and instability of an equilibrium point. We shall indicate some of these for stochastic differential equations.

In view of the variety of convergences for stochastic processes there are many different ways of defining stability concepts for stochastic differential equations. We shall discuss some of these in relation to a scalar Ito equation

$$(3.6) \qquad\qquad dX_t = a(t, X_t)\, dt + b(t, X_t)\, dW_t$$

with a steady solution $X_t \equiv 0$, so $a(t, 0) = 0$ and $b(t, 0) = 0$. Let us assume that a unique solution $X_t = X_t^{t_0, x_0}$ exists for all $t \geq t_0$ and each nonrandom initial value $X_{t_0} = x_0$ under consideration. A widely accepted definition of stochastic stability is a probabilistic one due to Hasminski. We call the steady solution $X_t \equiv 0$ *stochastically stable* if for any $\epsilon > 0$ and $t_0 \geq 0$

$$\lim_{x_0 \to 0} P\left( \sup_{t \geq t_0} \left| X_t^{t_0, x_0} \right| \geq \epsilon \right) = 0,$$

and *stochastically asymptotically stable* if, in addition,

$$\lim_{x_0 \to 0} P\left( \lim_{t \to \infty} \left| X_t^{t_0, x_0} \right| \to 0 \right) = 1$$

or *stochastically asymptotically stable in the large* when

$$P\left( \lim_{t \to \infty} \left| X_t^{t_0, x_0} \right| \to 0 \right) = 1 \quad \text{for all} \quad x_0 \in \Re.$$

Another definition involving $p$th-moments is also widely used. In this case $X_t \equiv 0$ is called *stable in $p$th-mean* if for every $\epsilon > 0$ and $t_0 \geq 0$ there exists a $\delta = \delta(t_0, \epsilon) > 0$ such that

$$E\left( \left| X_t^{t_0, x_0} \right|^p \right) < \epsilon \quad \text{for all} \quad t \geq t_0 \quad \text{and} \quad |x_0| < \delta$$

and *asymptotically stable in $p$th-mean* if, in addition, there exists a $\delta_0 = \delta_0(t_0) > 0$ such that

$$\lim_{t \to \infty} E\left( \left| X_t^{t_0, x_0} \right|^p \right) = 0 \quad \text{for all} \quad |x_0| < \delta_0.$$

The qualifiers uniform and global are used in the same way here as for their deterministic stability counterparts. Of particular interest in applications are the

$p = 1$ and $p = 2$ cases, *stability in mean* and *mean-square stability*, respectively.

Lyapunov functions can also be used to test for the stochastic stability or $p$th-mean stability of the steady solution $X_t \equiv 0$ of an Ito SDE (3.6). Let $V(t, x)$ be sufficiently smooth so that the Ito formula can be used. Then $V(t, X_t)$ has the Ito stochastic differential

$$dV(t, X_t) = LV(t, X_t) \, dt + b(t, X_t) \frac{\partial V}{\partial x}(t, X_t) \, dW_t$$

or the equivalent integral representation

$$(3.7) \qquad V(t, X_t) - V(t_0, X_{t_0})$$

$$= \int_{t_0}^{t} LV(s, X_s) \, ds + \int_{t_0}^{t} b(s, X_s) \frac{\partial V}{\partial x}(s, X_s) \, dW_s,$$

where the operator $L$ is defined by

$$(3.8) \qquad LV = \frac{\partial V}{\partial t} + a \frac{\partial V}{\partial x} + \frac{1}{2} b^2 \frac{\partial^2 V}{\partial x^2}.$$

If we can find a function $V$ such that

$$(3.9) \qquad LV(t, x) \leq 0$$

for all $x$ and $t \geq t_0$, which is easily checked and does not require explicit knowledge of the solutions $X_t$ of (3.6), then the equality (3.7) can be replaced by the inequality

$$(3.10) \qquad V(t, X_t) - V(t_0, X_{t_0}) \leq \int_{t_0}^{t} b(s, X_s) \frac{\partial V}{\partial x}(s, X_s) \, dW_s.$$

Taking conditional expectations and using property (3.2.9) of Ito integrals we obtain

$$E\left(V(t, X_t) \,|\, \mathcal{A}_{t_0}\right) \leq V(t_0, X_{t_0}),$$

w.p.1, where $\mathcal{A}_t$ is the $\sigma$-algebra generated by $X_s$ for $t_0 \leq s \leq t$. This inequality says that the Lyapunov function $V(t, X_t)$ evaluated along the solutions of (3.6) is a supermartingale. Hence we can use the maximal martingale inequality (2.3.6) to obtain

$$P\left(\sup_{t_0 \leq t \leq T} V(t, X_t) \geq \epsilon\right) \leq \frac{1}{\epsilon} V(t_0, x_0)$$

for all $\epsilon > 0$ and all $T > t_0$, and thus

$$(3.11) \qquad P\left(\sup_{t \geq t_0} V(t, X_t) \geq \epsilon\right) \leq \frac{1}{\epsilon} V(t_0, x_0).$$

If, in addition, there are monotonically increasing continuous functions $\alpha = \alpha(r)$ and $\beta = \beta(r)$ of $r > 0$ with $\alpha(0) = \beta(0) = 0$ such that

$$(3.12) \qquad\qquad \alpha\left(|x|\right) \leq V(t, x) \leq \beta\left(|x|\right)$$

for all $x$ and $t \geq 0$, we can conclude from (3.11) that the steady solution $X_t \equiv 0$ of (3.6) is uniformly stochastically stable. Similarly, if instead of (3.9) we have for all $x$ and $t \geq 0$

$$(3.13) \qquad\qquad LV(t, x) \leq -c\left(|x|\right)$$

for some continuous positive function $c = c(r)$ of $r > 0$ with $c(0) = 0$, then we can prove that the steady solution $X_t \equiv 0$ is uniformly stochastically stable in the large. We remark that in many problems it is convenient not to require all of the necessary partial derivates of $V$ to be continuous at $x = 0$. The Ito formula will then not be valid in a neighbourhood of 0, but refinements of the proofs yield the same results as above.

**Example 6.3.2**    *Consider the linear Ito SDE*

$$(3.14) \qquad\qquad dX_t = aX_t \, dt + bX_t \, dW_t$$

*and the Lyapunov function* $V(x) = x^2$. *Then*

$$LV(x) = \left(a + \frac{1}{2} b^2\right) x^2,$$

*so (3.13) holds with* $c(r) = -(a + \frac{1}{2}b^2)r^2$ *provided* $a + \frac{1}{2}b^2 < 0$. *For this parameter range the zero solution of (3.14) is uniformly stochastically asymptotically stable in the large. This result is sharp because the exact solution of (3.14) is*

$$X_t = X_0 \exp\left(\left(a + \frac{1}{2} b^2\right)(t - t_0) + b\left(W_t - W_{t_0}\right)\right)$$

*and converges to 0 with probability one as* $t \to \infty$ *if and only if* $a + \frac{1}{2}b^2 < 0$ *holds, since*

$$\lim_{t \to \infty} \frac{W_t - W_{t_0}}{t - t_0} = 0, \quad \text{w.p.1.}$$

**Exercise 6.3.3**    *Use an appropriate Lyapunov function to show that the steady solution* $X_t \equiv 0$ *of the Ito SDE*

$$dX_t = aX_t(1 + X_t^2) \, dt + bX_t \, dW_t$$

*is stochastically asymptotically stable if* $a + \frac{1}{2}b^2 < 0$.

Similar results can be obtained for the $p$th-mean stabilities, and for $d$-dimensional vector stochastic differential equations

$$dX_t = a(t, X_t) \, dt + b(t, X_t) \, dW_t.$$

In the latter case the operator $L$ takes the form

$$LV = \frac{\partial V}{\partial t} + \sum_{i=1}^{d} a^i \frac{\partial V}{\partial x_i} + \frac{1}{2} \sum_{i,j=1}^{d} (bb^\mathsf{T})^{i,j} \frac{\partial^2 V}{\partial x_i \partial x_j}.$$

Another method for determining the stability of a steady solution is to linearize the differential equation about this solution and to analyse the stability of the zero solution of the resulting linear differential equation. In many cases its stability or asymptotic stability will imply the corresponding property of the steady solution of the nonlinear equation. For the scalar differential equation (3.1) the linear differential equation obtained by linearizing about a steady solution $x = c$ is

(3.15)
$$\frac{dz}{dt} = \bar{a}(t)z,$$

where $z = x - c$ and $\bar{a}(t) = \frac{\partial a}{\partial x}(t, c)$, and has the solution

$$z(t; t_0, x_0) = z_0 \exp\left(\int_{t_0}^{t} \bar{a}(s) \, ds\right).$$

Thus the zero solution $z \equiv 0$ of (3.15) is asymptotically stable if and only if

(3.16)
$$\lambda = \limsup_{t \to t_0} \frac{1}{t - t_0} \int_{t_0}^{t} \bar{a}(s) \, ds < 0.$$

The limit superior has been used because the usual limit may not exist, as is the case of $\bar{a}(t) = -2 + t \sin t$. When the original differential equation (3.1) is autonomous, the coefficient $\bar{a}(t)$ in (3.15) is a constant, say $\bar{a}$ and $\lambda = \bar{a}$ is just the eigenvalue of the 1×1 matrix $[\bar{a}]$. For a vector differential equation the linearized equation has the form

(3.17)
$$\frac{dz}{dt} = A(t)z,$$

where $A(t)$ is a $d \times d$ matrix. In the autonomous case, with $A(t) \equiv A$ a constant matrix, the solutions of (3.17) are

$$z(t; t_0, z_0) = z_0 \exp\left((t - t_0)A\right).$$

and the zero solution of (3.17) is asymptotically stable if and only if the eigenvalues of the matrix $A$ all have negative real parts. For the nonautonomous case matters are much more complicated. The counterparts of the real parts of the eigenvalues are the *Lyapunov exponents* defined by

(3.18)
$$\lambda(t_0, z_0) = \limsup_{t \to \infty} \frac{1}{t - t_0} \ln |z(t; t_0, z_0)|,$$

of which (3.16) is a simple example. Asymptotic stability of the zero solution thus follows if and only if the Lyapunov exponents are negative for all $t_0$ and $z_0$

$\neq 0$. However, in practice, it is generally not easy to determine the Lyapunov exponents explicitly.

We can also use Lyapunov exponents for stochastic differential equations. For this it is more convenient to use Stratonovich SDEs. We shall suppose that $\bar{X}_t$ is a stochastically stationary solution, not necessarily a point like 0, of a $d$-dimensional Stratonovich SDE

$$(3.19) \qquad dX_t = \underline{a}(t, X_t)\, dt + \sum_{k=1}^{m} b^k(t, X_t) \circ dW_t^k,$$

where $W = (W^1, \ldots, W^m)$ is an $m$-dimensional standard Wiener process. With $Z_t = X_t - \bar{X}_t$ we obtain the linearized system

$$(3.20) \qquad dZ_t = A(t, \omega) Z_t\, dt + \sum_{k=1}^{m} B^k(t, \omega) Z_t \circ dW_t^k$$

where $A$, $B^1$, $B^2$, ..., $B^m$ are $d \times d$ matrices defined componentwise by

$$A(t, \omega)^{i,j} = \frac{\partial \underline{a}^i}{\partial x_j}\left(t, \bar{X}_t(\omega)\right)$$

and

$$B^k(t, \omega)^{i,j} = \frac{\partial b^{k,i}}{\partial x_j}\left(t, \bar{X}_t(\omega)\right)$$

for $i$, $j = 1, 2, \ldots, d$ and $k = 1, 2, \ldots, m$.

**Exercise 6.3.4**    *Linearize the 2-dimensional Ito SDE*

$$d\begin{pmatrix} X_t^1 \\ X_t^2 \end{pmatrix} = \begin{pmatrix} X_t^2 \\ -bX_t^2 - \sin X_t^1 - c\sin 2X_t^1 \end{pmatrix} dt$$

$$+ \begin{pmatrix} 0 \\ -a\left(X_t^2\right)^2 + \sin X_t^1 \end{pmatrix} dW_t,$$

*where a, b and c are constants and W is a scalar Wiener process, about the steady solution $(\bar{X}_t^1, \bar{X}_t^1) \equiv (0, 0)$. Determine the corresponding Stratonovich SDE and linearize it about $(0,0)$ too. What is the relationship between the two linearized SDEs?*

In the scalar case $d = 1$ we can solve (3.20) explicitly (see (4.4.11)). Writing its solution as

$$\ln |Z_t| = \ln |Z_0| + \int_{t_0}^{t} A(s, \omega)\, ds + \int_{t_0}^{t} B(s, \omega)\, dW_s,$$

and using the fact that

$$\lim_{t \to \infty} \frac{1}{t - t_0} \int_{t_0}^{t} B(s, \omega)\, dW_s = 0, \quad \text{w.p.1,}$$

we obtain a unique Lyapunov exponent

$$\lambda = \limsup_{t\to\infty} \frac{1}{t-t_0} \ln|Z_t| = \limsup_{t\to\infty} \frac{1}{t-t_0} \int_{t_0}^{t} A(s,\omega)\,ds, \quad \text{w.p.1.}$$

The zero solution of the linearized equation is thus stochastically asymptotically stable if and only if this number $\lambda$ is negative. If in the autonomous case the stationary solution $\bar{X}_t$ has a stationary probabilitity density $\bar{p}(x)$, found by solving the stationary Fokker-Planck equation (1.7.16) with the appropriately modified drift coefficient $a(x)$, then we can use ergodicity (4.4.18) and Theorem 4.8.8 to conclude that

$$(3.21) \qquad \limsup_{t\to\infty} \frac{1}{t-t_0} \int_{t_0}^{t} A(s,\omega)\,ds = \int_{\alpha}^{\beta} \underline{a}'(x)\bar{p}(x)\,dx.$$

Here the interval $\alpha < x < \beta$, which may be infinite, is the support of the density function $\bar{p}$. In the degenerate case that $\bar{X}_t \equiv 0$ the density $\bar{p}(x) = \delta(x)$, the Dirac delta function, and the integral on the right reduces to the eigenvalue $\underline{a}'(0)$ of the scalar equation. We thus have an effective means of evaluating Lyapunov exponents in the scalar autonomous case.

**Exercise 6.3.5** *Evaluate the Lyapunov exponent of the zero solution of the linear Stratonovich SDE*

$$dX_t = aX_t\,dt + bX_t \circ dW_t.$$

*For which values of the parameters a and b is the Lyapunov exponent negative?*

For the general $d$-dimensional case (3.20), we can use the *Multiplicative Ergodic Theorem of Oseledec* to assert the existence, w.p.1, of $d$ nonrandom Lyapunov exponents

$$\lambda_d \leq \lambda_{d-1} \leq \cdots \leq \lambda_1$$

and a partitioning of $\Re^d$ into random subsets $E_d(\omega)$, $E_{d-1}(\omega)$, $\cdots$, $E_1(\omega)$. For solutions of (3.20) starting in these sets the limits (3.18) take values $\lambda_d$, $\lambda_{d-1}$, $\cdots$, $\lambda_1$, respectively. The stochastic asymptotic stability of the zero solution of (3.20) thus follows if and only if $\lambda_1 < 0$. For $\bar{X}_t \equiv 0$ and an autonomous system there is a formula like (3.21) for this top Lyapunov exponent $\lambda_1$. To determine it we change to spherical coordinates $r = |z|$ and $s = z/|z|$ for $z \in \Re^d \setminus \{0\}$, in which case the linear Stratonovich system (3.20) transforms into the system

$$(3.22) \qquad dR_t = R_t\,q(S_t)\,dt + \sum_{k=1}^{m} R_t\,q^k(S_t) \circ dW_t^k$$

$$(3.23) \qquad dS_t = h(S_t, A)\,dt + \sum_{k=1}^{m} h(S_t, B^k) \circ dW_t^k$$

on $\Re^+ \times S^{d-1}$, where $S^{d-1}$ is the unit sphere in $\Re^d$ and

$$q(s) = s^\top As + \sum_{k=1}^{m} \left( \frac{1}{2} s^\top \left( B^k + \left( B^k \right)^\top \right) s - \left( s^\top B^k s \right)^2 \right),$$

$$q^k(s) = s^\top B^k s, \quad h(s, A) = \left( A - (s^\top As) I \right) s.$$

Equation (3.23) does not involve the radial variable $R_t$ and is an example of a stochastic differential equation on a compact manifold. The advantage of using the Stratonovich interpretation here is that the transformation laws of classical calculus apply, and are in fact used to derive (3.22) and (3.23) from (3.20). From (3.22) we have

$$\ln R_t = \ln |Z_t| = \int_{t_0}^{t} q(S_u) \, du + \sum_{k=1}^{m} \int_{t_0}^{t} q^k(S_u) \circ dW_u^k,$$

and hence

(3.24) $$\limsup_{t \to \infty} \frac{1}{t} \ln |Z_t| = \limsup_{t \to \infty} \frac{1}{t - t_0} \int_{t_0}^{t} q(S_u) \, du.$$

Under appropriate conditions on the matrices $A$, $B^1$, $B^2$, ..., $B^m$ there exists an ergodic solution $\bar{S}_t$ of (3.23) with invariant probability distribution $\bar{\mu}(s)$ on $S^{d-1}$ such that the Ergodic Theorem 4.8.8 applies to the limit on the right hand side of (3.24) to give the following expression for the top Lyapunov exponent

$$\lambda_1 = \int_{S^{d-1}} q(s) \, d\bar{\mu}(s).$$

While of considerable theoretical benefit, this formula is difficult to use in practice and usually must be evaluated either numerically as in Section 3 of Chapter 17 or in terms of asymptotic expansions of the noise parameters. There is also a formula, which is more complicated, for $\lambda_1$ in the case of a general ergodic solution $\bar{X}_t \not\equiv 0$ of the original nonlinear SDE (3.19).

**Exercise 6.3.6**    *Determine the equations (3.22) and (3.23) for the linearized Stratonovich SDE derived in Exercise 6.3.4.*

To conclude this section on stochastic stability we return briefly to the $p$th-moment stabilities, in particular mean-square stability. For a $d$-dimensional linear Ito system

(3.25) $$dX_t = A(t)X_t \, dt + \sum_{k=1}^{m} B^k(t)X_t \, dW_t^k$$

we saw in Section 8 of Chapter 4 that that the $d \times d$ matrix valued second moment $P(t) = E(X_t X_t^\top)$ satisfies the deterministic matrix differential equation

$$\frac{dP}{dt} = A(t)P + PA(t)^\top + \sum_{k=1}^{m} B^k(t)PB^k(t)^\top,$$

which is linear in $P$. On account of the symmetry of the matrix $P$, we can write this equation as a linear system of the from

$$(3.26) \qquad \frac{d\tilde{p}}{dt} = \mathcal{A}(t)\tilde{p}$$

where $\tilde{p}$ is an $\frac{1}{2}d(d+1)$-dimensional vector consisting of the free components of $P$ and $\mathcal{A}(t)$ is a square matrix. The mean-square stability of the zero solution of (3.25) can be determined from that of (3.26). Moreover, since

$$|E(X_t)| \leq E(|X_t|) \leq \sqrt{E(X_t X_t^T)},$$

the mean-square stability will imply the stability of the first moment and, hence, that of the mean.

**Exercise 6.3.7**   *Derive the linear system (3.26) for the second moment of the solution of the linearized Ito SDE in Exercise 6.3.4. Hence determine the parameter values for which the steady solution of the original Ito SDE is mean-square asymptotically stable.*

## 6.4   Parametric Estimation

Often a modeller can use phenomological or theoretical arguments to justify the use of a differential equation with a particular structure, but needs to estimate the appropriate parameter values from observations of the actual system being modelled. Such parameter estimates are, strictly speaking, random variables, so something needs to be known about the probability distribution of their deviations from the true values of the parameters. For example, if we have a stochastic differential equation with an asymptotically stable ergodic stationary solution, we might expect that the parameter values obtained from the observation of a single trajectory over a finite time interval would converge to the true parameter values as the length of the observation interval increases without bound. Then we could use various limit theorems to obtain an indication of the reliability of such estimates.

   To illustrate the basic ideas of parametric estimation we shall consider a scalar stochastic differential equation with additive noise

$$(4.1) \qquad dX_t = \alpha\, a(X_t)\, dt + dW_t$$

where $\alpha$ is the parameter to be estimated and the function $a = a(x)$ is possibly nonlinear. The maximum likelihood estimate $\hat{\alpha}(T)$ determined from observations of a trajectory of a solution process (4.1) over the time interval $0 \leq t \leq T$ is the value of $\alpha$ which maximizes the *likelihood ratio*

$$(4.2) \qquad L(\alpha, T) = \exp\left( -\frac{1}{2}\alpha^2 \int_0^T a^2(X_t)\, dt + \alpha \int_0^T a(X_t)\, dX_t \right)$$

of the process $X = \{X_t, 0 \le t \le T\}$ with respect to the Wiener process $W$ $= \{W_t, 0 \le t \le T\}$. This is derived using the *Girsanov Theorem*, already mentioned in Section 8 of Chapter 4, which says that the right side of (4.2) is equal to the Radon-Nikodym derivative $dP_X/dP_W$ of the probability measures $P_X$ and $P_W$, corresponding to the processes $X$ and $W$, on the function space $\mathcal{C}([0,T], \Re)$. It can be justified heuristically by approximating (4.1) with the Euler difference scheme

$$(4.3) \qquad\qquad Y_{i+1} = Y_i + \alpha\, a(Y_i)\Delta + \Delta W_i$$

for $i = 0, 1, \ldots, N - 1$ where $\Delta = T/N$. The increments $\Delta W_0$, $\Delta W_1$, $\ldots$, $\Delta W_{N-1}$ of the Wiener process are independent $N(0; \Delta)$ distributed random variables, so their joint probability density is given by

$$
(4.4) \qquad p_W^{(N)} = \prod_{i=0}^{N-1} \frac{1}{\sqrt{2\pi\Delta}} \exp\left(-\frac{1}{2\Delta}(\Delta W_i)^2\right)
$$

$$
= \frac{1}{(2\pi\Delta)^{N/2}} \exp\left(-\frac{1}{2\Delta}\sum_{i=0}^{N-1}(\Delta W_i)^2\right).
$$

Writing $\Delta Y_i = Y_{i+1} - Y_i$ for $i = 0, 1, \ldots, N - 1$ we can determine the joint probability density $p_Y^{(N)}$ for $\Delta Y_0$, $\Delta Y_1$, $\ldots$, $\Delta Y_{N-1}$ by substituting $\Delta Y_i - \alpha\, a(Y_i)\Delta$ for $\Delta W_i$ in (4.4) to obtain

$$
(4.5) \qquad p_Y^{(N)} = \frac{1}{(2\pi\Delta)^{N/2}} \exp\left(-\frac{1}{2\Delta}\sum_{i=0}^{N-1}(\Delta Y_i)^2\right)
$$

$$
\times \exp\left(-\frac{1}{2}\alpha^2\sum_{i=0}^{N-1} a^2(Y_i)\,\Delta + \alpha\sum_{i=0}^{N-1} a(Y_i)\,\Delta Y_i\right).
$$

The Radon-Nikodym derivative of the discrete process $Y = \{Y_i, i = 0, 1, \ldots, N-1\}$ with respect to the Wiener process $W$ is simply the ratio

$$
(4.6) \qquad \frac{p_Y^{(N)}}{p_W^{(N)}} = \exp\left(-\frac{1}{2\Delta}\sum_{i=0}^{N-1}\left((\Delta Y_i)^2 - (\Delta W_i)^2\right)\right)
$$

$$
\times \exp\left(-\frac{1}{2}\alpha^2\sum_{i=0}^{N-1} a^2(Y_i)\,\Delta + \alpha\sum_{i=0}^{N-1} a(Y_i)\,\Delta Y_i\right).
$$

Taking limits as $N \to \infty$, we see that the exponent of the first term vanishes and that of the other term converges to the difference of the integrals in the formula (4.2).

We differentiate the likelihood ratio (4.2) with respect to $\alpha$ and solve the equation

$$(4.7) \qquad\qquad \frac{\partial L}{\partial \alpha}(\alpha, T) = 0$$

to obtain the *maximum likelihood estimator*

$$(4.8) \qquad \hat{\alpha}(T) = \int_0^T a(X_t)\, dX_t \Big/ \int_0^T a^2(X_t)\, dt.$$

This is a random variable depending on the particular sample path of the process $X$ that is observed over the time interval $[0, T]$. From (4.1) and (4.8) we have

$$(4.9) \qquad \hat{\alpha}(T) - \alpha = \int_0^T a(X_t)\, dW_t \Big/ \int_0^T a^2(X_t)\, dt,$$

where $\alpha$ is the true value of the parameter. If

$$(4.10) \qquad E\left(\int_0^T a^2(X_t)\, dt\right) < \infty$$

for all $T$ and if the stochastic differential equation (4.1) has a stationary solution with density $\bar{p}$ found by solving the stationary Fokker-Planck equation (1.7.16), then it follows from the Ergodicity Theorem 4.8.8 that

$$(4.11) \qquad \frac{1}{T}\int_0^T a(X_t)\, dW_t \longrightarrow 0$$

and

$$(4.12) \qquad \frac{1}{T}\int_0^T a^2(X_t)\, dt \longrightarrow \int_{\Re} a^2(x)\bar{p}(x)\, dx$$

with probability one as $T \to \infty$. We can thus conclude from (4.9) that

$$(4.13) \qquad \hat{\alpha}(T) \longrightarrow \alpha$$

with probability one as $T \to \infty$. Moreover, a version of the Central Limit Theorem tells us that $T^{1/2}\left(\hat{\alpha}(T) - \alpha\right)$ converges in distribution as $T \to \infty$ to an $N(0; \sigma^2)$-distributed random variable with variance

$$(4.14) \qquad \sigma^2 = \left(\int_{\Re} a^2(x)\bar{p}(x)\, dx\right)^{-1}.$$

We can use this information to determine confidence intervals for $\alpha$ and to estimate an appropriate value of $T$ for a desired confidence level.

**Exercise 6.4.1** *Consider the Langevin equation*

$$(4.15) \qquad dX_t = \alpha\, X_t\, dt + dW_t$$

*and suppose that the true value of $\alpha$ is $-1$. Determine the maximum likelihood estimate $\hat{\alpha}(T)$ and the variance $\sigma^2$.*

The preceding discussion pertains to the simplest situations of parametric estimation. In the literature many generalizations have been considered, such as to multiplicative noise, partial observations of the solution process, including time-discrete observations, and coloured noise. In those cases where the original equation should be interpreted as a Stratonovich stochastic differential equation, the maximum likelihood estimate should be determined for the equivalent Ito equation in order to obtain a correct estimate of the true parameter value. We note that matters are much more complicated when the drift is nonlinear in the parameter. In Section 2 of Chapter 13 we shall test some parametric estimators using numerical methods.

## 6.5  Optimal Stochastic Control

In many applications the state of the system is described by a stochastic differential equation involving parameters which can be adjusted so that some task can be achieved in an optimal manner. Typical situations are the regulation of a chemical reactor or of a satellite orbit with minimum energy expenditure, or the selection of an investment portfolio with maximum income. Often we can describe these mathematically in terms of a state equation

$$(5.1) \qquad dX_t = a(t, X_t, u)\, dt + b(t, X_t, u)\, dW_t$$

with a control parameter $u \in \Re^k$ which is to be chosen so as to minimize a cost criterion

$$(5.2) \qquad J(s, x; u) = E\left( K(\tau, X_\tau) + \int_s^\tau F(t, X_t, u)\, dt \,\bigg|\, X_s = x \right),$$

where $K$ and $F$ are given functions and $\tau$ is a specified Markov time, which is often a constant.

In choosing the control parameters we can usually take into account whatever information we have about the state of the system, that is we implement controls which are functionals of $\{X_z(\omega), s \le z \le t\}$ at each instant $t$. We can use *Markovian feedback controls* of the form $u(t, X_t)$, where $u$ is a nonrandom Lebesgue measurable function, when we have perfect information about the state. If we have or make use of no information, then we are restricted to the *openloop controls* of the form $u(t, \omega)$ which are nonanticipative with respect to the Wiener process in (5.1). On inserting a control from either of these classes into (5.1), under appropriate assumptions, we obtain an Ito stochastic differential equation. We then minimize the cost functional (5.2) with respect to one of these classes of controls. For Markovian feedback controls the minimum value of the cost functional

$$(5.3) \qquad H(s, x) = \min_{u(\cdot)} J\left(s, x; u(\cdot)\right)$$

satisfies the *Hamilton-Jacobi-Bellman (HJB) equation*

(5.4) $$\min_{u \in \Re^k} \{F(s,x,u) + L_u H(s,x)\} = 0$$

with the final time condition

(5.5) $$H(T,x) = K(T,x),$$

where the Markov time $\tau = T$ and $L_u$ is the operator

(5.6) $$L_u = \frac{\partial}{\partial s} + \sum_{i=1}^{d} a^i(s,x,u)\frac{\partial}{\partial x_i} + \frac{1}{2}\sum_{i,j=1}^{d} D^{i,j}(s,x,u)\frac{\partial^2}{\partial x_i \partial x_j}$$

with $D = bb^\mathsf{T}$. (Compare this with the operator of the Kolmogorov backward equation (2.4.7)). This is only a necessary condition which the minimum must satisfy if it exists. However, various mild regularity assumptions imply both the existence of the minimum and the sufficiency of the condition. The problem of finding the optimal stochastic control thus reduces to a deterministic minimization problem, that of solving the HJB equation (5.4). We remark that an optimal Markovian feedback control found by this means gives a cost functional value that is at least as good, if not better, than that for an optimal openloop control. Rather than present a mathematical justification of these statements here, we shall show how the method works by using it to solve the linear-quadratic regulator problem.

We consider the $d \times d$ matrices $A(t)$, $C(t)$ and $R$, the $d \times m$ matrix $\sigma(t)$, the $k \times k$ matrix $G(t)$ and the $d \times k$ matrix $M(t)$, where all of the matrices are continuous in $t$ and $R$ is a constant matrix. In addition, we assume that the $C(t)$ and $R$ are symmetric and positive semi-definite and that the $G(t)$ are symmetric and positive definite. The *linear-quadratic regulator problem* is to minimize the quadratic cost criterion

(5.7) $$J(s,x;u) = E\left(X_T^\mathsf{T} R X_T + \int_s^T (X_t^\mathsf{T} C(t)X_t + u^\mathsf{T} G(t)u)\, dt \,\Big|\, X_s = x\right),$$

where $X_t$ satisfies the linear stochastic differential equation

(5.8) $$dX_t = (A(t)X_t + M(t)u)\, dt + \sigma(t)\, dW_t,$$

over the class of Markovian feedback controls. (The state equation (5.8) is linear in $X_t$ for constant $u$, but will become nonlinear if a nonlinear feedback control $u(t, X_t)$ is used). The HJB equation (5.4) here is

(5.9) $$\frac{\partial H}{\partial s} + x^\mathsf{T} C(s)x + \sum_{i=1}^{d}(A(s)x)^i\frac{\partial H}{\partial x_i} + \frac{1}{2}\sum_{i,j=1}^{d} D^{i,j}\frac{\partial^2 H}{\partial x_i \partial x_j}$$

$$+ \min_{u \in \Re^k}\left\{u^\mathsf{T} G(t)u + \sum_{i=1}^{d}(M(s)u)^i\frac{\partial H}{\partial x_i}\right\} = 0,$$

where $D = \sigma\sigma^\mathsf{T}$, with the final time condition

$$(5.10) \qquad H(T, x) = x^{\mathsf{T}} R x.$$

To solve this we guess a solution of the form

$$(5.11) \qquad H(s, x) = x^{\mathsf{T}} S(s) x + a(s),$$

where the $S(t)$ are symmetric positive definite $d \times d$ matrices and $a(t) \in \Re^d$, with both continuously differentiable in $t$. Obviously (5.10) holds if we have

$$(5.12) \qquad S(T) = R \quad \text{and} \quad a(T) = 0.$$

We substitute (5.11) into the left side of (5.9) without the minimization, then differentiate the expression obtained with respect to $x$ and find that the derivative vanishes for

$$(5.13) \qquad u = u^*(s, x) = -G(s)^{-1} M(s) x.$$

For these values of $u$ the term being minimized in the HJB equation (5.9) also vanishes and the resulting partial differential equation reduces to a matrix ordinary differential *Riccati* equation

$$(5.14) \qquad \frac{dS}{ds} = -A(s)^{\mathsf{T}} S - S A(s) + S M(s) G(s)^{-1} M(s) S - C(s)$$

for $S = S(s)$, provided $a(s)$ is chosen so that

$$(5.15) \qquad \frac{da}{ds} = -\mathrm{tr}\left\{\sigma(s)\sigma(s)^{\mathsf{T}} S(s)\right\}$$

where tr denotes the trace of the matrix. Thus we need to solve (5.14) and then evaluate $a(s)$ from (5.15) subject to the final time conditions (5.12). Using this expression for $S(s)$, the optimal control $u^*(t, X_t)$ is given by (5.13) and the minimum cost by

$$(5.16) \qquad H(s, x) = x^{\mathsf{T}} S(s) x + \int_s^T \mathrm{tr}\left\{\sigma(t)\sigma(t)^{\mathsf{T}} S(t)\right\} dt.$$

Since equation (5.14) does not involve the noise coefficient $\sigma(s)$, the integral term in (5.16) is thus the total additional cost due to the presence of noise in the control system. The optimal control $u^*(t, X_t)$ is linear in $X_t$, so the corresponding state equation (5.8) is a linear SDE; in the scalar case the optimal control corresponds a scaled response in the opposite direction to the displacement. In general, the Riccati equation (5.14) must be solved numerically, but this can be done off-line and stored for later use since (5.14) only involves known coefficients.

**Exercise 6.5.1** *Derive (5.13)–(5.16) for the scalar case $d = k = m = 1$.*

**Exercise 6.5.2** *State explicitly the optimal control, the optimal cost and the optimal trajectory of the linear-quadratic regulator problem with state equation*

$$dX_t = (-4X_t + 2u) \, dt + 4 \, dW_t$$

*and cost functional*

$$J(s, x; u) = E\left(X_1^2 + \int_s^1 \left(8X_t^2 + u^2\right) dt \,\Big|\, X_s = x\right).$$

Usually for nonlinear state equations (5.1) explicit solutions cannot be found for the HJB equation. In the case of openloop controls alternative necessary conditions for optimality can be derived using martingale theory. Let $u(\cdot)$ be an openloop control and let $X_t^{u(\cdot)}$ be the corresponding solution of the state equation (5.1) with initial value $X_0^{u(\cdot)} = x$. Then the stochastic process $M_t^{u(\cdot)}$ defined by

$$(5.17) \qquad M_t^{u(\cdot)} = \int_0^t F\left(s, X_s^{u(\cdot)}, u(s)\right) ds + H\left(t, X_t^{u(\cdot)}\right)$$

is a submartingale with respect to the probability measure $P^{u(\cdot)}$ determined from the original probabilty measure $P$ by the Girsanov transformation (see Section 8 of Chapter 4). The crucial point here is that $M_t^{u(\cdot)}$ is a martingale if and only if the control is optimal. Martingale methods can then be used to prove a stochastic version of the Pontryagin Maximum Principle.

Situations of partial information often occur because we can only make noise contaminated observations of some of the state components of the control system. For instance, we may only be able to observe a process $Y_t$, usually of lower dimension than $X_t$, which satisfies

$$(5.18) \qquad dY_t = H(t)X_t \, dt + \Gamma(t) \, dW_t^*,$$

where $W_t^*$ is a Wiener process independent of that in the state equation. For an estimate $\hat{X}_t$ of the state $X_t$ we could use the conditional expectation

$$\hat{X}_t = E\left(X_t \,|\, \mathcal{Y}_t\right),$$

where $\mathcal{Y}_t$ is the $\sigma$-algebra generated by $Y_s$ for $s \le t$. As indicated by (2.2.19) such an estimate is the best mean-square approximation of the state $X_t$ over all $\mathcal{Y}_t$- measurable $Z_t$, that is

$$E\left(\left|X_t - \hat{X}_t\right|^2\right) \le E\left(|X_t - Z_t|^2\right).$$

It seems natural to use this estimate $\hat{X}_t$ in a feedback control. In fact, for a linear-quadratic regulator problem (5.7)–(5.8) with the observation equation 5.18, the optimal control $u^*(t, \hat{X}_t)$ takes the same form (5.13) as for perfect information when $\hat{X}_t$ is determined by the Kalman-Bucy filter, which will be considered in the next section. This is an example of the *Separation Principle* of linear stochastic control theory, which says that a stochastic control problem with linear state and observation equations reduces to a deterministic control problem and a linear filtering problem. The Kalman-Bucy filter requires a deterministic matrix Riccati equation to be solved. In general, partial information in stochastic control results in an infinite dimensional problem for which a numerical method must be used.

To conclude this brief sketch of optimal stochastic control we remark that the optimal controls $u^*(t, x)$ may only be piecewise continuous or measurable in their variables. Consequently basic existence and uniqueness theorems such as Theorem 4.5.3 do not apply, but the relevant generalizations have been developed. The derivation of efficient numerical methods for solving stochastic control problems is of important practical consequence.

## 6.6 Filtering

The procedure of determining an estimate of the state of a system from noise contaminated observations is known as *filtering*. As indicated in the previous section, filtering is often a crucial step in solving optimal stochastic control problems. It is also important in many other situations, such as tracking an aircraft or reconstructing a radio signal. To be of practical use a filter should be robust and, usually, implementable on-line. In general, this leads to demanding computational problems and requires some deep underlying mathematical theory. There are nevertheless some exceptions, notably the Kalman-Bucy filter for linear Gaussian systems.

For simplicity we shall give a description of the *Kalman-Bucy filter* for state and observation equations with constant coefficients. Suppose that the state process $X_t$ is a $d$-dimensional Gaussian process satisfying the (narrow-sense) linear stochastic differential equation

$$(6.1) \qquad dX_t = AX_t\, dt + B\, dW_t$$

where $A$ is a $d \times d$ matrix, $B$ a $d \times m$ matrix and $W = \{W_t,\ t \geq 0\}$ an $m$-dimensional standard Wiener process. To ensure that the solution $X_t$ of (6.1) is Gaussian, the initial value $X_{t_0}$ should be a Gaussian random variable. Then the mean and covariance matrices of $X_t$ satisfy ordinary differential equations; see (4.6.9) and (4.6.10). Suppose also that the observed process $Y_t$ is an $e$-dimensional process, where $1 \leq e \leq d$, and is related to $X_t$ by the equation

$$(6.2) \qquad dY_t = HX_t\, dt + \Gamma\, dW_t^*$$

with $Y_0 = 0$, where $H$ is a $d \times e$ matrix, $\Gamma$ an $e \times n$ matrix and $W^* = \{W_t^*, t \geq 0\}$ an $n$-dimensional standard Wiener process which is independent of the Wiener process $W$. Finally, for each $t \geq 0$, let $\mathcal{A}_t$ be the $\sigma$-algebra generated by $X_{t_0}$, $Y_s$ and $W_s$ for $0 \leq s \leq t$ and $\mathcal{Y}_t$ the $\sigma$-algebra generated by the observations $Y_s$ for $0 \leq s \leq t$, so $\mathcal{Y}_t \subset \mathcal{A}_t$. The Kalman-Bucy filter uses the conditional expectation

$$(6.3) \qquad \hat{X}_t = E\left(X_t \,|\, \mathcal{Y}_t\right)$$

as its estimate of the state $X_t$, this being the best mean-square estimate of $X_t$ with

(6.4)
$$E\left(\left|X_t - \hat{X}_t\right|^2\right) \le E\left(\left|X_t - Z_t\right|^2\right)$$

for all $e$-dimensional $\mathcal{Y}_t$-measurable processes $Z_t$. This estimate $\hat{X}_t$ is a $\mathcal{Y}_t$-measurable, and hence $\mathcal{A}_t$-measurable, Gaussian process with the same mean as the state process $X_t$, but with a different covariance matrix. In fact, the error covariance matrix

(6.5)
$$S(t) = E\left(\left(X_t - \hat{X}_t\right)\left(X_t - \hat{X}_t\right)^{\mathsf{T}}\right)$$

satisfies the matrix Riccati equation

(6.6)
$$\frac{dS}{dt} = AS + SA^{\mathsf{T}} + BB^{\mathsf{T}} - SH^{\mathsf{T}}\left(\Gamma\Gamma^{\mathsf{T}}\right)^{-1}HS$$

with the initial value $S(0) = E(X_0 X_0^{\mathsf{T}})$; this differs from the linear equation (4.8.10) for the second moment $P(t) = E(X_t X_t^{\mathsf{T}})$ by the quadratic correction term. In addition, the estimate $\hat{X}_t$ satisfies the stochastic differential equation

(6.7)
$$d\hat{X}_t = \left(A - SH^{\mathsf{T}}\left(\Gamma\Gamma^{\mathsf{T}}\right)^{-1}H\right)\hat{X}_t\,dt + SH^{\mathsf{T}}\left(\Gamma\Gamma^{\mathsf{T}}\right)^{-1}dY_t,$$

where the observation process $Y_t$ appears instead of a Wiener process. The stochastic integral with respect to $Y_t$ here is defined in a similar way to the Ito integral. The derivation of equations (6.6) and (6.7) is based on the fact that the best mean-square estimate of an element of a Hilbert space $\mathcal{H}$ with respect to a subspace $\mathcal{S}$ of $\mathcal{H}$ is its orthogonal projection onto $\mathcal{S}$. In this case the appropriate Hilbert space $\mathcal{H}$ consists of the mean-square integrable processes $f$ $= \{f_s, 0 \le s \le t\}$ adapted to the family of $\sigma$-algebras $\{\mathcal{A}_s, 0 \le s \le t\}$ and the subspace $\mathcal{S}$ consists of those processes adapted to the observation $\sigma$-algebras $\{\mathcal{Y}_s, 0 \le s \le t\}$. The inner product here is

$$(f,g) = E\left(\int_0^t f_s^{\mathsf{T}} g_s\,ds\right),$$

with $f$ and $g$ orthogonal if $(f,g) = 0$. The estimate $\hat{X}_t$ and the error $X_t - \hat{X}_t$ are orthogonal in this sense, and also in the stronger pointwise sense that

$$E\left(\left(X_t - \hat{X}_t\right)^{\mathsf{T}} X_t\right) = 0$$

for each $t \ge 0$. We note that the Riccati equation (6.6) must be solved numerically, but it only involves the known coefficients of the state and observation equations (6.1) and (6.2), so this can be done off-line. The coefficients of the estimate equation (6.7) are then known, so this equation can be solved on-line as the observations become available; generally this must be done numerically too. Finally, it has been shown that the Kalman-Bucy filter is robust in the sense that the estimate changes only slightly with slight changes in the coefficients and the nature of the noise processes. This is particularly important in practice since the actual noises will often be broad-band approximations of Gaussian white noise.

**Exercise 6.6.1**    *Solve the Riccati equation (6.6) for the scalar case $d = e = m = n = 1$.*

**Exercise 6.6.2**    *Show that the Kalman-Bucy filter in the scalar case with $X_t \equiv X_0$ where $E(X_0) = 0$ and $E(X_0^2) = \sigma^2$, that is with*

$$dX_t \equiv 0, \qquad dY_t = X_t \, dt + dW_t^*$$

*and $Y_0 = 0$, gives the estimate*

$$\hat{X}_t = \frac{\sigma^2}{1 + \sigma^2 t} Y_t.$$

In using the Kalman-Bucy filter we are essentially determining the conditional probability distribution of the state $X_t$ in the estimate $\hat{X}_t$. This observation has been used as a starting point for investigating nonlinear filtering problems. To begin, we suppose that the state process $X = \{X_t, \, t \geq 0\}$ is a continuous time Markov chain on the finite state space $\{1, 2, ..., N\}$ with its $N$-dimensional probability vectors $p(t)$, with components $p^i(t) = P(X_t = i)$ for $i = 1, 2, ..., N$, satisfying a vector ordinary differential equation

(6.8) $$\frac{dp}{dt} = A(t)p$$

where the $A(t)$ are the time-dependent intensity matrices of the Markov chain at time $t$; see Example 1.6.8. Further, we suppose that the observations $Y_t$ represent an $e$-dimensional process satisfying

(6.9) $$dY_t = h(X_t) \, dt + dW_t^*$$

with $Y_0 = 0$, where $W^* = \{W_t^*, \, t \geq 0\}$ is an $e$-dimensional Wiener process which is independent of $X$. As before we denote the $\sigma$-algebra generated by the observations $Y_s$ for $0 \leq s \leq t$ by $\mathcal{Y}_t$ and in addition denote the $N \times N$ diagonal matrix with $ii$th component $h_k(i)$ by $H_k$ for $k = 1, 2, ..., N$. Then the conditional probabilities of $X_t$ given $\mathcal{Y}_t$ are

(6.10) $$\hat{p}^i(t) = P(X_t = i | \mathcal{Y}_t) = Q_t^i \Big/ \sum_{j=1}^{N} Q_t^j$$

for $i = 1, 2, ..., N$, which follows from the Fujisaki-Kallianpur-Kunita formula. Here the $N$-dimensional process $Q_t = (Q_t^1, Q_t^2, ..., Q_t^N)$ satisfies the linear stochastic differential equation, called the *Zakai equation*,

$$(6.11) \qquad dQ_t = A(t)Q_t \, dt + \sum_{k=1}^{N} H_k \, Q_t \, dY_t^k$$

with respect to the observation process $Y_t$. The appropriate initial condition here is $Q_0 = p(0)$.

**Exercise 6.6.3** *Write down the SDE (6.11) for 1-dimensional observations $Y_t$ of random telegraphic noise with $h(\pm 1) = \pm 1$ in equation (6.9).*

An analogous result holds when the state process $X_t$ is a nondegenerate diffusion process satisfying an Ito stochastic differential equation. For simplicity we suppose that the drift and diffusion matrix are independent of time and are sufficiently smooth. Then the probability densities $p(t, x)$ satisfy the Fokker-Planck equation (2.4.5), which we write in operator form as

$$(6.12) \qquad \frac{\partial p}{\partial t} = \mathcal{L}^* p.$$

In addition suppose that the $e$-dimensional observation process satisfies an equation of similar form to (6.9), except now the domain of $h$ is $\Re^d$ for an $d$-dimensional diffusion process $X_t$. Then the conditional probability densities of $X_t$ given $\mathcal{Y}_t$ are

$$(6.13) \qquad \bar{p}(t, x) = Q_t(x) \Big/ \int_{\Re^d} Q_t(x) \, dx,$$

where the unnormalized densities $Q_t(x)$ satisfy the *stochastic partial differential equation*, called the *Wong-Zakai equation*,

$$(6.14) \qquad dQ_t(x) = \mathcal{L}Q_t(x) \, dt + h(x) \, Q_t(x) \, dY_t,$$

with respect to the observation process $Y_t$ where $Q_0(x) = p(0, x)$.

For both the Markov chain and diffusion state processes the conditional probabilities provide information about an estimate of the state in terms of the observations, but unlike the Kalman-Bucy filter do not give a simple sample path representation of the estimate. In the Markov chain case the unnormalized conditional probabilities satisfy an Ito type stochastic differential equation (6.11) for which the numerical methods discussed in this book can be applied. See in particular Section 3 of Chapter 13. Appropriate numerical methods for stochastic partial differential equations like (6.14) are yet to be developed. One possible method would be to approximate the diffusion process by a Markov chain and then to solve the corresponding ordinary stochastic differential equation (6.11).

Another approach is based on the Kallianpur-Striebel formula. This refers to the general nonlinear filtering problem with both the state process $X = \{X_t, t \in [0, T]\}$ and the observation process $Y = \{Y_t, t \in [0, T]\}$ satisfying nonlinear stochastic differential equations, which we write in integral form as

$$(6.15) \qquad X_t \;=\; X_0 + \int_0^t a\left(X_s\right) ds + \int_0^t b\left(X_s\right) dW_s^1,$$

$$Y_t \;=\; \int_0^t h\left(X_s\right) ds + W_t^2$$

for $t \in [0,T]$. Here the given probability space is $(\Omega,\; \mathcal{A},\; P)$, $W = (W^1,\; W^2)$ $= \{(W_t^1,\; W_t^2),\; t \in [0,T]\}$ is a 2-dimensional standard Wiener process with respect to a family of nondecreasing sub-$\sigma$-algebras $\{\mathcal{A}_t,\; t \in [0,T]\}$ and the initial state $X_0$ is $\mathcal{A}_0$-measurable. As before we denote the $\sigma$-algebra generated by the observations $Y_s$ for $0 \le s \le t$ by $\mathcal{Y}_t$. Using the Girsanov transformation (4.8.17) we introduce a new probability measure $\dot{P} = L_T^{-1} P$ with the Radon-Nikodym derivative

$$(6.16) \qquad L_t = \exp\left(-\frac{1}{2}\int_0^t h^2\left(X_s\right) ds + \int_0^t h\left(X_s\right) dY_s\right)$$

for $t \in [0,T]\}$. The observation process $Y$ is then a Wiener process with respect to $\dot{P}$ and is independent of $X$ under $\dot{P}$. Furthermore, for each sufficiently smooth function $f$ the *Kallianpur-Striebel formula* for the nonlinear filter

$$(6.17) \qquad E\left(f\left(X_t\right) \mid \mathcal{Y}_t\right) = \frac{\dot{E}\left(f\left(X_t\right) L_t \mid \mathcal{Y}_t\right)}{\dot{E}\left(L_t \mid \mathcal{Y}_t\right)},$$

holds for all $t \in [0,T]\}$, where $\dot{E}$ denotes the expectation with respect to the probability measure $\dot{P}$.

# Chapter 7

# Applications of Stochastic Differential Equations

This chapter consists of a selection of examples from the literature of applications of stochastic differential equations. These are taken from a wide variety of disciplines with the aim of stimulating the readers' interest to apply stochastic differential equations in their own particular fields of interest and of providing an indication of how others have used models described by stochastic differential equations. Here we simply describe the equations and refer readers to the original papers for the justification and analysis of the models.

## 7.1 Population Dynamics, Protein Kinetics and Genetics

**Population Dynamics**  The simplest deterministic model of population growth is the exponential equation $\dot{x} = ax$ where $a$ is the Malthusian growth coefficient, which is usually a positive constant, but may vary in sign and magnitude with time $t$ to cater for seasonal variations. The vagaries of the environment can be modelled by allowing $a$ to vary randomly as $a + \sigma \xi_t$ for a zero mean process $\xi_t$. We saw in Section 1 of Chapter 6 that when $\xi_t$ is a Gaussian white noise process we obtain a linear stochastic differential equation, either an Ito equation or a Stratonovich equation depending on which interpretation we choose. Both the deterministic and stochastic models here admit unbounded exponential growth, which is untenable in an environment of finite resources. Under such circumstances a finite supportable carrying capacity $K$ is appropriate, with the population decreasing whenever it exceeds this value. We can incorporate this feature easily into the deterministic model by replacing the growth constant $a$ by the linear factor $a(K - x)$. Then we obtain the linear-quadratic *Verhulst equation*

$$(1.1) \qquad \dot{x} = a(K - x)x,$$

which is often written as

$$(1.2) \qquad \dot{x} = \lambda x - x^2$$

with $aK$ replaced by $\lambda$ and $ax$ by $x$.

On randomizing the parameter $\lambda$ in (1.2) to $\lambda + \sigma \xi_t$, we obtain a stochastic differential equation

$$(1.3) \qquad dX_t = \left(\lambda X_t - X_t^2\right) dt + \sigma X_t \, dW_t$$

using the Ito interpretation, which can be solved explicitly by the reduction method in Section 3 of Chapter 4; for the explicit solution see equation (4.4.51).

Some caution is needed in randomizing the parameter $\lambda$, or more precisely the original parameters $a$ and $K$ in (1.1), in this fashion. Since negative as well as positive values make sense for a growth rate, there are no conceptual difficulties in randomizing the growth parameter $a$ as above. In contrast, by its very definition the carrying capacity $K$ must be positive, so randomizing it as $K + \sigma\xi_t$ for white noise $\xi_t$ will lead to unrealistic stochastic models. One way around this is to begin with Schoener's deterministic equation

$$(1.4) \qquad \dot{x} = R\,x\,\left(E(T - \Lambda x) - C - \Gamma x\right)$$

instead of the simpler looking Verhulst equation (1.1), to which it reduces with parameters $a$ and $K$ defined by

$$a = R\,(E\Lambda + \Gamma), \quad K = \frac{ET - C}{E\Lambda + \Gamma}.$$

We shall not elaborate here on the meaning of the parameters in (1.4) other than to remark that some may naturally take both positive and negative values and thus be meaningfully randomized as above with white noise fluctuations.

Apart from contrived situations such as laboratory experiments, single species population dynamics models are usually unrealistic since in nature most species coexist with others and are affected by their presence in one way or another. Such interactions may be benevolent, neutral or malevolent, as in symbiotic, predator-prey or competitive relationships, respectively. These can all be incorporated in population dynamics models with terms coupling together separate single species models. A frequently studied deterministic model of multi-species interaction is the *Volterra-Lotka system*

$$(1.5) \qquad \dot{x}^i = x^i \left(a^i + \sum_{j=1}^{d} b^{i,j} x^j\right)$$

for $i = 1, 2, \ldots, d$ in the case of $d$ different species. Randomizing the growth parameters $a^i$ as $a^i + \sigma^i \xi_t^i$ leads to a system of stochastic differential equations, which we shall interpret as Ito equations with independent Wiener processes,

$$(1.6) \qquad dX_t^i = X_t^i \left(a^i + \sum_{j=1}^{d} b^{i,j} X_t^j\right) dt + \sigma^i X_t^i \, dW_t^i$$

for $i = 1, 2, \ldots, d$. More generally, we might obtain noise terms of the form $\sum_{j=1}^{d} \sigma^{i,j}(X_t^1, X_t^2, \ldots, X_t^d) \, dW_t^j$ for $i = 1, 2, \ldots, d$ if, for example, the coupling parameters are also randomized. Explicit solutions are not known for equations (1.6), so they must be solved numerically or a qualitative investigation made of the boundedness and stability of their solutions.

**Protein Kinetics** Stochastic counterparts of many ordinary differential equations modelling chemical kinetics, such as the Brusselator equations, can be derived by randomizing coefficients. For example, the kinetics of the proportion $x$ of one of two possible forms of certain proteins can be modelled by an ordinary differential equation of the form

$$(1.7) \qquad \dot{x} = \alpha - x + \lambda x(1 - x),$$

where $0 \leq x \leq 1$ and the other form has proportion $y = 1 - x$. For random fluctuations of the interaction coefficient $\lambda$ of the form $\lambda + \sigma \xi_t$ with white noise $\xi_t$, the appropriate stochastic version of (1.7) is given by the Stratonovich stochastic differential equation

$$(1.8) \qquad dX_t = (\alpha - X_t + \lambda X_t (1 - X_t)) \, dt + \sigma X_t (1 - X_t) \circ dW_t.$$

The solutions of (1.8) are not known explicitly, but remain in the interval $[0, 1]$, w.p.1, since at both endpoints the diffusion coefficient vanishes and the drift is directed into the interval for positive values of the $\alpha$, which are necessary for (1.7) to be meaningful.

**Genetics** The Ito equation equivalent to (1.8),

$$(1.9) \qquad dX_t = a(X_t) \, dt + \sigma X_t (1 - X_t) \, dW_t$$

where

$$a(x) = \left( \alpha - x + \lambda x(1 - x) + \frac{1}{2}\sigma^2 x(1 - x)(1 - 2x) \right),$$

also has a genetical application, with $X_t$ representing the proportion at time $t$ of one of two possible alleles of a certain gene. A discrete-time Markov process can be constructed to model the changes from generation to generation in the allele proportions due to natural selection, which favours the allele most suited to the current state of a randomly fluctuating environment, and to mutations which, in this case, transform one allele form into the other. Essentially, this discrete-time Markov process converges to a diffusion process $X_t$ satisfying the equation (1.9) as the total number of alleles becomes arbitrarily large. Similar arguments have been used to derive a system of Ito equations from Kimura's discrete-time Markov model of the distributional dynamics of such an allele amongst $d$ different geographical sites. For example, Shiga obtained the system of equations

$$(1.10) \qquad dX_t^i = a^i \left( X_t^1, \ldots, X_t^d \right) dt + \sqrt{\frac{1}{2}X_t^i \left( 1 - X_t^i \right)} \, dW_t^i,$$

where

$$a^i \left( x^1, \ldots, x^d \right) = \left( v^i - (u^i + v^i) \, x^i + s^i x^i (1 - x_i) + \sum_{i,j=1}^{d} m^{j,i} \left( x^j - x^i \right) \right)$$

for $i = 1, 2, \ldots, d$. Here $W_t^1, W_t^2, \ldots, W_t^d$ are independent Wiener processes, $s_i$ the relative fitness of the specified gene in the $i$th region, $u^i$ and $v^i$ the mutation rates between the two alleles in the $i$th region, and $m^{j,i}$ the relative migration rate from the $j$th to the $i$th region.

# 7.2   Experimental Psychology and Neuronal Activity

**Experimental Psychology**   The coordination of human movement, particularly of periodically repeated movement, has been extensively investigated by experimental psychologists with the objective of gaining a deeper understanding of neurological control mechanisms. The neurological system is without doubt extremely complicated, yet in some situations a single characteristic appears to dominate and a satisfactory phenomological model can be constructed to describe its dynamics. One such example is the following experiment carried out by Kelso. A subject sits at a table with his wrists fixed vertically on the table and moves the index finger on each hand periodically to the left and right. Depending on the frequency of this movement and on whether it is chosen freely by the subject or synchronized with a metronome, just one of two stable steady states is observed: a symmetric steady state, in which the two index fingers move inwards and outwards together, and an asymmetric steady state in which one moves inwards while the other moves outwards. The dominant characteristic here is the phase difference $\phi$ of the two fingers, with the symmetric state corresponding to $\phi = 0$ and the asymmetric one $\phi_0 = \pi$ ( or $-\pi$, since $\phi$ takes values modulo $2\pi$). It was always found that the asymmetric state suddenly gives way to the symmetric state when the frequency of oscillations exceeds a certain critical value. Moreover, the ratio of this critical frequency to the frequency freely chosen by the subject was the same for all subjects.

Haken and his coworkers proposed the ordinary differential equation

$$(2.1) \qquad \dot{\phi} = -a \sin \phi - 2b \sin 2\phi$$

as a deterministic model of the phase dynamics, this being the simplest periodic one with steady states and stability characteristics consistent with the experimental observations. Here $a$ and $b$ are positive parameters that must be determined from experimental data. In order to describe fluctuations in this data, Schöner included an additive noise term $\sigma \xi_t$ in (2.1), obtaining a stochastic differential equation

$$(2.2) \qquad dX_t = -\left(a \sin X_t + 2b \sin 2X_t\right) dt + \sigma \, dW_t,$$

with $X_t$ interpreted modulo $2\pi$. This form of noise is meant to model noisy fluctuations in a very large number of weakly coupled neuronal cells.

While it does not make the equation any easier to solve, we can express the required $2\pi$-periodicity of $X_t$ directly by writing (2.2) as a Stratonovich stochastic differential equation on the unit circle, namely

$$dX_t^1 = \left(a + 4bX_t^1\right)\left(X_t^2\right)^2 dt - \sigma X_t^2 \circ dW_t$$
$$dX_t^2 = -\left(a + 4bX_t^1\right)X_t^1 X_t^2 dt + \sigma X_t^1 \circ dW_t,$$

where $X_t^1 = \cos X_t$ and $X_t^2 = \sin X_t$.

**Neuronal Activity**  Many stochastic models have been proposed to describe the spontaneous firing activity of a single neuron. These are usually based on jump processes and allow arbitrarily large hyperpolarization values for the membrane potential. Attempts have been made to avoid the latter difficulty by requiring the depolarization to be state dependent. In addition, diffusion approximations are often sought in order to simplify the subsequent mathematical analysis. A model incorporating these features has been derived by Kallianpur. It involves the Ito stochastic differential equation for the membrane potential

$$(2.3) \qquad dX_t = \left(-\frac{1}{\tau}X_t + \alpha\left(V_E - X_t\right) + \beta\left(X_t - V_I\right)\right) dt$$
$$+ \sqrt{\gamma\sigma_E^2\left(V_E - X_t\right)^2 + \epsilon\sigma_I^2\left(X_t - V_I\right)^2}\, dW_t$$

with $V_I \leq X_t \leq V_E$, for inhibitory and excitatory membrane potentials $V_I < V_E$. Here $W = \{W_t,\, t \geq 0\}$ is a standard Wiener process and the constants $\alpha$, $\beta$, $\gamma$ and $\tau$ are positive. A related model due to Lansky and Lanska, which can be obtained from (2.3) essentially by setting $\gamma = 0$, has been analysed in detail in a paper by Giorno, Lansky, Nobile and Ricciardi.

# 7.3  Investment Finance and Option Pricing

**Investment Finance**  Given the apparent random fluctuations of share prices on the stock exchange, it seems natural to use stochastic differential equations in models of share price dynamics or, more generally, in models of investment finance. One of the first to do this was Merton, whose simple model contains the basic ideas that have been used in recent, more sophisticated models. Merton considered an investor who chooses between two different types of investment, one risky and the other safe. The investor must implement an investment strategy which will maximize some utility function, such as his net wealth or cash flow, while avoiding bankruptcy. Merton supposed that the price $p_s$ of the safe investment increased steadily according to the exponential growth ordinary diffferential equation

$$(3.1) \qquad\qquad \dot{p}_s = a\, p_s$$

for some constant rate $a > 0$. In addition, he supposed that the price $p_r$ of the risky investment satisfied a similar equation, but including noisy fluctuations with intensity proportional to the price. Thus

(3.2)                                   $\dot{p}_r = b\,p_r + \beta\,p_r\,\xi_t,$

where $\xi_t$ is a Gaussian white noise process, which he interpreted as an Ito
stochastic differential equation

(3.3)                                   $dP_t^r = bP_t^r\,dt + \beta P_t^r\,dW_t$

where $\{W_t,\ t \ge 0\}$ is a standard Wiener process. Here $b$ and $\beta$ are positive
constants and $a < b$ since the risky investment is, potentially at least, more
profitable than the safe one. At each instant of time the investor must select
the fraction $f$ of his wealth that he will put into the risky investment, with the
remaining fraction $1 - f$ going into the safe one. If his current consumption
rate is $c \ge 0$, then it follows from (3.1) and (3.3) that his wealth $X_t$ satisfies
the stochastic differential equation

$$dX_t = f\,(bX_t\,dt + \beta X_t\,dW_t) + (1 - f)aX_t\,dt - c\,dt,$$

which can be rewritten as the Ito stochastic differential equation

(3.4)                     $dX_t = (\{(1 - f)a + fb\}\,X_t - c)\,dt + f\beta X_t\,dW_t.$

When the investor has perfect information about his current wealth, Markovian
feedback controls of the form $u(t, X_t) = (f(t, X_t),\ c(t, X_t))$ provide a natural
way for choosing his current investment mixture and consumption rate. If $X_t^{u(\cdot)}$
is the corresponding solution of (3.4) he may want to choose $u(\cdot)$ to maximize
the expected value of some utility function $U$ at time $T$. This gives rise to an
optimal stochastic control problem with profit functional

$$J\left(s, x; u(\cdot)\right) = E\left(U\left(X_T^{u(\cdot)}\right) \mid X_s^{u(\cdot)} = x\right)$$

which is to be maximized. The problem is complicated by the presence of a
nonnegative consumption rate in (3.4), which may result in bankruptcy at a
random first exit time

$$\tau\left(s, x; u(\cdot); \omega\right) = \inf\left\{t \ge s : X_t^{u(\cdot)}(\omega) = 0 \mid X_s^{u(\cdot)} = x\right\}.$$

**Option Pricing**    Suppose that the price of a risky asset, for example a stock
or an exchange rate, evolves according to the Ito stochastic differential equation

(3.5)                           $X_t = X_0 + \displaystyle\int_0^t b\,(s, X_s)\,dW_s$

for $t \in [0, T]$, where $W$ is a Wiener process with respect to an underlying
probability measure $P$ which can be interpeted as the probability measure of
the risk-neutral world.

For simplicity we shall consider only options on this single risky asset, assuming that there are no dividends and that the interest rate is zero. A Euro-
pean call option with striking price $c$, for example, gives the right to buy the
stock at time $T$ at the fixed price $c$. The resulting payoff is then given by

(3.6)
$$f(X_T) = (X_T - c)^+ = \begin{cases} X_T - c & : \quad X_T > c \\ 0 & : \quad X_T \le c. \end{cases}$$

Suppose that we apply a dynamical portfolio strategy or hedging strategy

$$(\zeta_t, \eta_t)_{t \in [0,T]},$$

where at time $t \in [0, T]$ we hold the amount $\eta_t$ in a riskless asset of constant value, say 1, and the amount $\zeta_t$ in the risky asset. Then the value $V_t$ of the portfolio at time $t$ is

(3.7)
$$V_t = \zeta_t X_t + \eta_t.$$

An important problem is to determine the fair price of the option. It follows from the well known formula of Black and Scholes that

(3.8)
$$V_0 = E(f(X_T)).$$

The corresponding self-financing hedging strategy, in quite general situations, leads to a perfect replication of the claim, that is

(3.9)
$$V_T = f(X_T).$$

On the other hand, if there is some intrinsic risk, then the situation is more complicated. This case has been considered by Föllmer, Sondermann and Schweizer.

# 7.4  Turbulent Diffusion and Radio-Astronomy

**Turbulent Diffusion**  Stochastic differential equations have long been used to model turbulent diffusion and related phenomena, dating back to Langevin's equations for Brownian motion. Let $X_t \in \Re^3$ represent the position of a fluid particle at time $t$ and $V_t$ its velocity. As a simple model for the Lagrangian dynamics of such a particle, Obukhov proposed the following 6-dimensional system of stochastic differential equations:

(4.1)
$$dX_t = V_t \, dt, \qquad dV_t = \sigma \, dW_t,$$

where $\sigma$ is a scalar diffusion coefficient and $W_t$ is a 3-dimensional standard Wiener process. To account for weak frictional forces acting on the particle, he replaced the equations by

(4.2)
$$dX_t = V_t \, dt, \qquad dV_t = -\frac{1}{T} V_t \, dt + \sigma \, dW_t,$$

where $T$ is a rather large relaxation time for the process $V_t$, the components of which are now Ornstein-Uhlenbeck processes. Variations of equations (4.1) and (4.2) have since been considered, for example with coloured noise instead

of white noise. In some instances Poisson processes have also been used as the
driving process.

The spectral density of atmospheric turbulent fluctuations contains a wide
gap in the vicinity of the frequency 1 per hour, that is of period 1 hour. The
large-scale synoptic fluctuations $V_t^{(1)}$ have a period much larger than 1 hour,
whereas the micro-scale fluctuations $V_t^{(2)}$ have a much smaller period and are
superimposed on the synoptic fluctuations, but are independent of them. Atmospheric scientists have thus used two different relaxation times $T_1$ and $T_2 \ll$
$T_1$ for these two types of fluctuations. Consequently, two coupled systems of
stochastic differential equations are needed for a statistical model of turbulent
diffusion with two such time-scales. A typical example, due to Bywater and
Chung in a slightly different context, has the form

$$dV_t^{(1)} = \left(-\frac{1}{T_1} V_t^{(1)} - \beta \left(V_t^{(1)} - V_t^{(2)}\right)\right) dt + \sigma_1 \, dW_t^{(1)}$$

$$dV_t^{(2)} = \left(-\frac{1}{T_2} V_t^{(2)} + \beta \left(V_t^{(1)} - V_t^{(2)}\right)\right) dt + \sigma_2 \, dW_t^{(2)}$$

where $W_t^{(1)}$ and $W_t^{(2)}$ are two independent 3-dimensional standard Wiener
processes.

**Radio-Astronomy**  In radio-astronomy signals from a star are analysed to
obtain estimates of certain characteristic parameters of the star. According
to Le Gland such a signal can be represented as a complex-valued stochastic
process

(4.3) $$\eta_t = a \, \exp\left(\imath \left(b + X_t\right)\right) + r \, \xi_t$$

where $\imath = \sqrt{-1}$, $a$ is the signal amplitude and $b$ the mean phase, with $X_t$ a real-
valued zero-mean process representing the effects of atmospheric turbulence on
the signal. In addition, $\xi_t$ is a complex-valued Gaussian white noise process
modelling measurement errors and $r$ is a nonzero number or an invertible matrix
for vector valued signals. For a given process $X = \{X_t, t \geq 0\}$, equation (4.3)
can be written as a complex-valued stochastic differential equation

(4.4) $$dY_t = a \, \exp\left(\imath \left(b + X_t\right)\right) dt + r \, dW_t$$

where $\{W_t, t \geq 0\}$ is a complex-valued Wiener process, that is $W_t = W_t^1 +$
$\imath W_t^2$ for independent Wiener processes $\{W_t^1, t \geq 0\}$ and $\{W_t^2, t \geq 0\}$. With
$Y_t = Y_t^1 + \imath Y_t^2$, equation (4.4) can be written as two real-valued equations

(4.5) $$dY_t^1 = a \, \cos\left(b + X_t\right) dt + r \, dW_t^1$$

and

(4.6) $$dY_t^2 = a \, \sin\left(b + X_t\right) dt + r \, dW_t^2.$$

Le Gland used an Ornstein-Uhlenbeck process $X_t$ satisfying the stochastic differential equation

(4.7) $$dX_t = -\beta X_t \, dt + \sigma \sqrt{2\beta} \, dW_t^3,$$

for certain parameters $\beta$ and $\sigma$, to model the atmospheric turbulence. Here $\{W_t^3, t \geq 0\}$ is another Wiener process which is independent of those in (4.5) and (4.6).

Given the model consisting of equations (4.5)–(4.7) and the observed signal $\{Y_t, t \geq 0\}$, the problem confronting the astronomer is to determine estimates of the characteristic parameters $a$ and $b$ of the star. In addition, appropriate values of the parameters $\beta$, $\sigma$ and $r$ are required and must either be provided from other sources or also estimated. Assuming that these latter parameters are known, one possible method of estimating $a$ and $b$ is to consider them as time-independent processes satisfying the degenerate stochastic differential equations

$$da_t = 0, \qquad db_t = 0$$

and then to use a nonlinear filter to determine estimates $\hat{a}_t$ and $\hat{b}_t$ for them, subject to the nonlinear observation equations (4.5) and (4.6)..

# 7.5  Helicopter Rotor and Satellite Orbit Stability

**Helicopter Rotor Stability**  The possible destabilization of a helicopter by turbulence in the vicinity of a rotating rotor blade is a matter of obvious concern. To investigate this problem Pardoux, Pignol and Talay proposed a model for the dynamics of a rotor blade with two degrees of freedom, in which the velocity of the helicopter, the geometric characteristics of the rotor blade and the statistical characteristics of the turbulence around the blade appeared as parameters. This resulted in a 4-dimensional system of differential equations. As a first approximation they considered a linear deterministic system of the form

$$\dot{x} = A(t)x + f(t)$$

where the matrices $A(t)$ and the vectors $f(t)$ are periodic functions with the period of rotation of the rotor blade. To this they added noise terms, either coloured or white. In the latter case they obtained the linear Stratonovich stochastic differential equation

(5.1) $$dX_t = (A(t)X_t + f(t)) \, dt + (B(t)X_t + g(t)) \, \sigma(t) \circ dW_t.$$

Here $W_t$ is a 1-dimensional Wiener process and $\sigma(t)$ is the intensity of the noise, which, like the matrices $B(t)$ and the vectors $g(t)$, is also periodic with the same period as that of the rotor blades.

The appropriate type of stability here is that of a unique periodic in law solution $\bar{X}_t$ such that

$$\lim_{t \to \infty} |X_t - \bar{X}_t| = 0, \quad \text{w.p.1},$$

for any other solution $X_t$ starting at an arbitrary deterministic initial value. This stability is also characterized by the negativeness of all of the corresponding Lyapunov exponents of the solution $\bar{X}_t$. These have been evaluated by Talay using a stochastic numerical procedure (see Section 3 of Chapter 17).

Pardoux and Pignol also considered the more realistic case of coloured noise, in particular noise of the form $\xi^\epsilon(t) = \epsilon^{-1/2} Z(t/\epsilon)$ for $\epsilon > 0$ sufficiently small, where $Z(t)$ is a stationary Ornstein-Uhlenbeck process. For $\epsilon \to 0$, Pardoux showed that the top Lyapunov exponent of the counterpart of system (5.1) with this coloured noise converges to that of the Stratonovich stochastic differential equation (5.1), which suggests that the behaviour of the model is robust to changes in the type of noise.

**Satellite Orbital Stability**   The rapid fluctuations on the earth's atmospheric density and other disturbances in the upper atmosphere must be taken into account when modelling the dynamics of satellites. Usually these are incorporated into deterministic models based on Newtonian mechanics in the form of random forcing or randomized coefficients. One such model that arose from the problem of stabilizing a satellite in a circular orbit is the following due to Sagirow:

$$(5.2) \qquad \ddot{x} + b\left(1 + a\,\xi_t\right)\dot{x} + \left(1 + a\,\xi_t\right)\sin x - c\sin 2x = 0$$

where $x$ is the radial perturbation about the given orbit, $\xi_t$ is a Gaussian white noise and $a$, $b$ and $c$ are constants, with $b$ and $c$ positive. With $X_t = (X_t^1, X_t^2) = (x, \dot{x})$, this can be written as a 2-dimensional Ito stochastic differential equation

$$(5.3) \qquad d\begin{pmatrix} X_t^1 \\ X_t^2 \end{pmatrix} = \begin{pmatrix} X_t^2 \\ -b\,X_t^2 - \sin X_t^1 - c\sin 2X_t^1 \end{pmatrix} dt$$
$$+ \begin{pmatrix} 0 \\ -ab\,X_t^2 - b\sin X_t^1 \end{pmatrix} dW_t$$

where $\{W_t,\ t \geq 0\}$ is a standard Wiener process. Alternatively, it may be argued that the Gaussian white noise is only an approximation of a real noise, so (5.3) should be written as a Stratonovich stochastic differential equation. The choice of interpretation will, for instance, have an effect on the outcome of a stability analysis.

**Satellite Attitude Dynamics**   Another model, which has been investigated by Balakrishnan, is for the attitude dynamics of a satellite. The basic equations are the kinematic (Euler) attitude motion equations for a rigid body subject to random torques. These are the 3-dimensional first order system

$$(5.4) \qquad\qquad M\dot{x} + x \otimes Mx + Kx = D\xi_t$$

where $x$ is the 3-dimensional state (attitude) vector, $M$, $K$ and $D$ $3 \times 3$ real symmetric positive definite matrices, $\xi_t$ a 3-dimensional vector Gaussian white noise process and $\otimes$ the vector cross product. Here $M$ is the moment of inertia matrix, $K$ a friction matrix and $D$ the noise intensity correlation matrix. This can be written as a 3-dimensional vector Ito stochastic differential equation

$$(5.5) \qquad dX_t = f(X_t)\,dt + M^{-1}D\,dW_t$$

where $\{W_t, t \geq 0\}$ is a 3-dimensional standard Wiener process and

$$f(x) = -M^{-1}Kx - M^{-1}x \otimes Mx.$$

The cross product term here is quadratic in the state variables, so this drift vector does not satisfy a global Lipschitz condition or a global growth bound and the conventional existence and uniqueness theorems such as Theorem 4.5.3 do not apply. However the dot product

$$x^{\mathsf{T}} \cdot x \otimes Mx = 0$$

holds for all $x \in \Re^3$, so

$$x^{\mathsf{T}} \cdot M f(x) = -x^{\mathsf{T}}Kx \leq -\lambda\,|x|^2$$

for some $\lambda > 0$ and this ensures the global existence of strong solutions of (5.5). This example shows that models which do not satisfy the commonly assumed regularity properties may still have well behaved solutions on account of special structural features of the model.

# 7.6 Biological Waste Treatment, Hydrology and Indoor Air Quality

**Biological Waste Treatment**  The continuous flow cultivation of micro-organisms for the biological treatment of urban waste water has attracted widespread attention. Most of the mathematical models of microbial growth in current usage trace back to a model of Monod for single strain bacterial growth in a single substrate. Harris proposed a model of an anaerobic digester of sewage sludge based on a perfect mixing model of a continuously stirred tank reactor with noisy fluctuations in the concentrations and feedrates, and with flow equalization prior to the reaction being used to stabilize the process. He considered five state variables $(X^1, X^2, X^3, X^4, X^5)$ with $X^1$ the equalization tank volume, $X^2$ the incoming sludge flow rate, $X^3$ the incoming substrate (pollutant) concentration, $X^4$ the substrate concentration leaving the equalization tank for the reactor tank, and $X^5$ the substrate concentration leaving the reactor tank.

To account for random fluctuations in the influent flow rate and pollutant concentration due to rainfall and industrial and domestic waste discharge, Harris used the pair of Ito stochastic differential equations

$$(6.1) \qquad dX_t^2 = \left(\alpha_1 + 1 - \beta_1 \frac{X_t^2}{X_t^1}\right) dt + \sqrt{2\gamma_1 \frac{X_t^2}{X_t^1}}\, dW_t^1$$

$$(6.2) \qquad dX_t^3 = \frac{\left(\alpha_2 + 1 - \beta_2 X_t^3\right)}{X_t^1} \, dt + \sqrt{2\gamma_2 \frac{X_t^3}{X_t^1}} \, dW_t^2$$

to model the turbulent flow and concentration pattern. Here the $\alpha_i$, $\beta_i$ and $\gamma_i$ are positive parameters characterizing the noisy fluctuations. Then, using mass balance for the equalization tank, he obtained a pair of ordinary differential equations for the equalization tank volume and the substrate concentration leaving the equalization tank:

$$(6.3) \qquad dX_t^1 = \left(\frac{\alpha_1 + 1}{\beta_1} - X_t^2\right) dt$$

$$(6.4) \qquad dX_t^4 = \left(X_t^3 - X_t^4\right) \frac{X_t^2}{X_t^1} \, dt.$$

Finally, the mass balance for the reactor tank gave an ordinary differential equation for the substrate concentration leaving the reactor tank:

$$(6.5) \qquad dX_t^5 = \left(a + b\left(X_t^4 - X_t^5\right)\right) dt$$

where $a$ and $b$ are factors depending on the growth rate of the micro-organisms, the yield constant and the noise parameters.

The five equations (6.1)–(6.5) form a nonlinear 5-dimensional vector stochastic differential equation with no noise terms appearing explicitly in three of the five component equations, so the corresponding rows in the diffusion matrix have zero entries. Harris presented a numerical method for this system, and used it to plot the time evolution of the expected gain in concentration of the biological oxygen demanding states, that is $E(X_t^5/X_t^3)$ against time $t$ for various values of inputs and parameters.

**Hydrology**  The outflow $x$ from a lake or reservoir depends on the precipitation $P$ over its catchment area, the evaporation from the water surface, the seepage flow and other factors which can be combined and represented abstractly by a parameter $\Phi$. Hydrologists have modelled the outflow from a single reservoir by a first order ordinary differential equation

$$(6.6) \qquad \dot{x} = F(x)$$

where the function on the right is given by

$$(6.7) \qquad F(x) = AP\,x^\alpha - AP\,\Phi\left(1 - \psi(x)\right)x^\alpha$$
$$-AE\,\psi(x)x^\alpha - Ax^\beta - Bx^\gamma.$$

The last term in (6.7) is the seepage flow and $\psi(x) = C + Dx^\delta$ is determined by the ratio of the surface areas of the catchment region and the reservoir. The constants $A$, $B$, $C$ and $D$ are nonnegative, whereas the indices $\alpha$, $\beta$, $\gamma$ and $\delta$ may also be negative.

Since the precipitation $P$, evaporation $E$ and catchment abstractions $\Phi$ are subject to fluctuations from many sources, it is reasonable to consider them to be random variables. Unny and Karmeshu argued that they could be approximated by independent Gaussian white noise fluctuations about positive mean values, namely

$$P + \sigma_1 \xi_t^1, \quad E + \sigma_2 \xi_t^2 \quad \text{and} \quad \Phi + \sigma_3 \xi_t^3.$$

They substituted these into (4.2), using $P\Phi + P\sigma_3 \xi_t^3 + \Phi\sigma_1 \xi_t^1$ for the product of $P + \sigma_1 \xi_t^1$ and $\Phi + \sigma_3 \xi_t^3$ in the second term, and obtained the Stratonovich stochastic differential equation

$$(6.8) \qquad dX_t \;=\; F(X_t)\,dt + \sigma_1 \left(1 - \Phi + A\Phi\psi(X_t)(X_t)^\alpha\right) \circ dW_t^1$$
$$-\sigma_2 A\psi(X_t) \circ dW_t^2$$
$$-\sigma_3 AP\left(1 - \psi(X_t)\right)(X_t)^\alpha \circ dW_t^3$$

as a stochastic counterpart of (6.6). Here $W_t^1$, $W_t^2$ and $W_t^3$ are independent standard Wiener processes. Unny and Karmeshu simulated numerically some sample paths of various solutions of (6.8). They also considered a cascade of lakes or reservoirs, such as the Great Lakes of North America, with the output of one being an additional input into the next. This resulted in a system of Stratonovich equations of the form (6.8) with additional coupling terms.

**Air Quality** Air conditioning units are used not only to regulate the temperature within a room, but also to filter the air to remove pollutants. Traditionally determinsitic models, based on mass and heat balances, have been used to help calibrate such units. Restricting attention here to the air quality in a single room, an established deterministic model for the concentration of a pollutant such as $CO_2$ is

$$(6.9) \qquad v\dot{C}^i \;=\; -k\left(q^0 + f^1 q^1 + q^2\right)C^i + S - R$$
$$+k\left(q^0(1 - f^0) + q^2\right)C^0$$

where $C^i$ is the indoor concentration and $C^0$ the outdoor concentration; $q$ is the volumetric flow rate for the make-up air $q^0$, the recirculation $q^1$, the infiltration $q^2$, the exfiltration $q^3$, and the exhaust $q^4$; $f$ is the filter efficiency for the make-up $f^0$ and the recirculation $f^1$ air; $v$ is the volume of the room; $S$ is the indoor source emission rate; $R$ is the indoor sink removal rate; and $k$ is a factor which accounts for the inefficiency of the mixing.

In reality, many of the above parameters cannot be predicted with certainty due to random fluctuations, which can be quite large. To obtain a more realistic model, Haghighat, Fazio and Unny replaced $S$, $C^0$, $q^0$ and $q^2$ by random fluctuations about deterministic values $\bar{S}$, $\bar{C}^0$, $\bar{q}^0$ and $\bar{q}^2$ (which may vary in time), that is by

$$S_t = \bar{S} + \xi_t^{(1)}, \qquad C_t^0 = \bar{C}^0 + \xi_t^{(2)},$$
$$q_t^0 = \bar{q}^0 + \xi_t^{(3)}, \qquad q_t^2 = \bar{q}^2 + \xi_t^{(4)},$$

where $\xi_t^{(1)}, \xi_t^{(2)}, \xi_t^{(3)}$ and $\xi_t^{(4)}$ are noise terms. They substituted these into (6.9), neglecting the product $\xi_t^{(2)}\xi_t^{(3)}$, to obtain

(6.10)                    $$v\, dC_t^i = g(C^i, t)\, dt + H(C^i, t)\, d\xi_t$$

where $\xi_t = (\xi_t^{(1)}, \xi_t^{(2)}, \xi_t^{(3)}, \xi_t^{(4)})$, $g$ is the scalar function

$$
\begin{aligned}
g(C^i, t) \;=\; & -k\left(\bar{q}^0 + f^1 q^1 + \bar{q}^2\right) C^i + \bar{S} - R \\
& + k(\bar{q}^0\left(1 - f^0\right) + \bar{q}^2)\bar{C}^0,
\end{aligned}
$$

and $H$ is the (row) vector function with

$$
H(C^i, t)^{\mathsf{T}} = \begin{pmatrix}
1 \\
k\left(\bar{q}^0\left(1 - f^0\right) + \bar{q}^2\right) \\
k(\bar{q}^0\left(1 - f^0\right) - C^i)\,\bar{C}^0 \\
k\left(\bar{C}^0 - C^i\right)
\end{pmatrix}.
$$

Using independent Wiener processes for the noise terms, they interpreted (6.10) as an Ito stochastic differential equation. They also considered coupled systems of such equations to model the pollutant concentrations in several different rooms.

Haghighat, Chandrashekar and Unny proposed a model, with a similar mathematical structure, of the thermal variations within a building subject to external temperature fluctuations.

## 7.7  Seismology and Structural Mechanics

**Seismology**  The vertical motion of the ground level during an earthquake, particularly during the period of strong movement, has been extensively modelled by Ito or Ito-like stochastic differential equations. For example, Bolotin assumed that the acceleration $\ddot{z}(t)$ of the ground level $z(t)$ at time $t$ has the form

(7.1)                          $$\ddot{z}(t) = I(t)\,\xi_t$$

for some random process $\xi_t$ with a deterministic intensity function $I$. In particular, for Gaussian white noise he suggested that $I(t) = hte^{-\alpha t}$ or

(7.2)                          $$I(t) = e^{-\alpha t} - e^{-\beta t},$$

where $0 < \alpha < \beta$ and $h > 0$ are known constants, are appropriate representations of the noise intensity. In this case (7.1) can be written as a 2-dimensional vector Ito stochastic differential equation

(7.3)            $$d\begin{pmatrix} Z_t^1 \\ Z_t^2 \end{pmatrix} = \begin{pmatrix} Z_t^2 \\ 0 \end{pmatrix} dt + \begin{pmatrix} 0 \\ 1 \end{pmatrix} I(t)\, dW_t$$

where $Z_t^1 = z(t)$ is the vertical displacement, $Z_t^2 = \dot{z}(t)$ the velocity in the vertical direction and $\{W_t, \, t \geq 0\}$ a standard Wiener process.

A variation of (7.1) was proposed by Shinozuka and Sato, with the Gaussian white noise passing through a linear filter with impulse response $h(t)$ after the application of an intensity function $I(t)$. This resulted in an integro-differential relationship

$$(7.4) \qquad \ddot{z}(t) = \int_0^t h(t-s)I(s)\, \xi_s \, ds$$

instead of (7.1). A typical choice here for the intensity function $I(t)$ is (7.2) and for the impulse response

$$h(t) = e^{-\delta \omega t} \frac{\sin\left(\omega\sqrt{1-\delta^2}\, t\right)}{\omega\sqrt{1-\delta^2}},$$

where $\delta$ and $\omega$ are known constants. Equation (7.4) can then be written as a 2-dimensional vector stochastic differential equation like (7.3), but with $I(t)\, dW_t$ replaced by $dY_t$ for

$$Y_t = \int_0^t h(t-s)I(s)\, dW_s.$$

In another variation of (7.1) Kozin replaced the acceleration by a time dependent linear combination of the acceleration and its derivatives. For example, with $a(t) = \ddot{z}(t)$ he considered equations of the form

$$(7.5) \qquad \ddot{a}(t) + c_1(t)\dot{a}(t) + c_0(t)a(t) = I(t)\, \xi_t$$

where $c_1(t) = c_{10}$ and $c_0(t) = c_{00} + c_{01}t + c_{02}t^2 + c_{03}t^3$ for known coefficients $c_{10}, c_{00}, c_{01}, c_{02}$ and $c_{03}$. For Gaussian white noise (7.5) can be written as a 2-dimensional vector Ito equation in $A_t^1 = a(t)$ and $A_t^2 = \dot{a}(t)$ or a 4-dimensional equation in $Z_t^1 = z(t)$, $Z_t^2 = \dot{z}(t)$, $Z_t^3 = \ddot{z}(t) = a(t)$ and $Z_T^4 = z^{(3)}(t) = \dot{a}(t)$, the latter being

$$(7.6) \qquad d\begin{pmatrix} Z_t^1 \\ Z_t^2 \\ Z_t^3 \\ Z_t^4 \end{pmatrix} = \begin{pmatrix} Z_t^1 \\ Z_t^2 \\ Z_t^3 \\ -c_1(t)Z_t^4 - c_0(t)Z_t^3 \end{pmatrix} dt + \begin{pmatrix} 0 \\ 0 \\ 0 \\ 1 \end{pmatrix} I(t)\, dW_t$$

where $\{W_t, \, t \geq 0\}$ is once again a standard Wiener process.

An important use of these models of earthquake dynamics is as excitation input into a system modelling the dynamics of a structure such as a building during an earthquake. Typically, a forced nonlinear oscillator of the form

$$(7.7) \qquad \ddot{x} + g(x, \dot{x})\dot{x} + \omega^2 x = -\ddot{z}(t)$$

is used to describe the dynamics of the structure, where $\ddot{z}(t)$ is the vertical acceleration of the ground level obtained from (7.1), (7.4) or (7.5). Equation (7.7) can be written as a 2-dimensional Ito-like stochastic differential equation

$$(7.8) \qquad d \begin{pmatrix} X_t^1 \\ X_t^2 \end{pmatrix} = \begin{pmatrix} X_t^2 \\ -\omega^2 X_t^1 - g(X_t^1, X_t^2) X_t^2 \end{pmatrix} dt - \begin{pmatrix} 0 \\ 1 \end{pmatrix} dZ_t^2$$

where $Z_t^2$ is the speed component of the earthquake model, for example (7.1) or (7.6).

**Structural Mechanics**   An Euler-Bernoulli beam under axial loading satisfies a fourth order nonlinear partial integro-differential equation under the Kirchhoff assumption that the axial extension depends only on time $t$ and not on the distance $x$ along the beam. An axial forcing $P(t)$ produces an end displacement $u(t)$. If all but the first spatial mode are disregarded, it follows that the corresponding time-dependent coefficient satisfies the second order nonlinear ordinary differential equation

$$(7.9) \qquad \ddot{T} + 2D\omega_1 \dot{T} + \omega_1^2 (1 + \epsilon u(t)) T + \gamma T^3 = 0$$

where $\omega_1$ is the first natural frequency of the beam. Random fluctuations in the external forcing $P(t)$, due for example to wind gusts or earth vibrations, result in a randomly fluctuating end displacement term $u(t)$. Writing this process as $U_t = U_0 + \sigma \xi_t$ for a white noise $\xi_t$, Wedig derived a vector Stratonovich stochastic differential equation

$$d \begin{pmatrix} T_t^1 \\ T_t^2 \end{pmatrix} = \begin{pmatrix} T_t^2 \\ -2D\omega_1 T_t^2 - (\omega_1^2 + \epsilon U_0) T_t^1 + \gamma (T_t^1)^3 \end{pmatrix} dt$$

$$+ \begin{pmatrix} 0 \\ \epsilon \omega_1^2 \sigma T_t^1 \end{pmatrix} \circ dW_t$$

where $(T^1, T^2) = (T, \dot{T})$ and $W_t$ is a scalar standard Wiener process. He investigated the effect of changes in the parameters on the stability of the null solution of (7.9).

Similar types of models have been constructed for other kinds of mechanical problems when randomly fluctuating parameters or input forces are included in an originally deterministic model for greater realism. This may be done directly in terms of ordinary differential equations or may, as above, involve the approximation of more complicated equations by ordinary differential equations. Wedig also considered a similar system of the latter type to model the effects of a rough surface on the flow of a fluid through a canal. As an example of the former, Hennig, Grunwald and Platen have developed a model of the flow-induced oscillations in a pressurized water reactor control mechanism consisting of two coupled rods.

# 7.8  Fatigue Cracking, Optical Bistability and Nemantic Liquid Crystals

**Fatigue Cracking**  It has been recognized that fatigue failures in solid materials such as aluminum or steel result from the nucleation and propagation of cracks. During their course of propagation the cracks encounter various types of metallurgical structures and imperfections, so in general their rate of propagation varies with time. Crack growth experiments usually show marked statistical variation, suggesting that randomness is a characteristic feature of crack growth. Both laboratory experiments on the fatigue life of test samples and observations of damage to real structures indicate that fatigue damage, which is usually measured by the length $L$ of the dominant crack, is affected by such factors as imposed stress, the properties of the material and environmental conditions. Various parameters have been used to describe the effects of random loading on crack growth, in particular the stress intensity range $S$, the stress ratio $Q$ and frequency $\mu$. Let $m$ be the average of external influences such as temperature, $D$ the intensity of the noise, and $c$ and $\rho$ material parameters. Sobczyk proposed the following Stratonovich stochastic differential equation for the time evolution of the length $L_t$ of the dominant crack in the material:

$$(8.1) \qquad dL_t = m\,f\,(L_t)^p\,dt + f\,(L_t)^p \circ dW_t$$

where $p > 0$, $f = mcg(Q)S^{2p}$ for some function $g$ of the stress ratio $Q$, and $\{W_t,\ t \geq 0\}$ is a standard Wiener process. The exponent $p$ is determined from experimental data. When $p = 1$ equation (8.1) is linear and its solutions exist for all time $t \geq 0$. Often materials are found to have values of $p > 1$, in which case (8.1) is a nonlinear stochastic differential equation and its solutions may explode in finite time. This explosion time $\tau$ is generally a random time depending on the particular sample path and corresponds to the time of ultimate damage or fatigue failure in the material.

**Optical Bistability**  A single state variable, the total cavity field amplitude $x$, has been found adequate to model the absorption of a single-mode laser beam injected into a Fabry-Perot cavity filled with, say, sodium vapour. Assuming that the gas atoms with two levels are driven incoherently by optical pumping and coherently by the injected laser beam, the frequency of which coincides exactly with the resonance frequency of the atoms, Gragg derived the ordinary differential equation

$$(8.2) \qquad \dot{x} = a - x - 2c\,\frac{x}{1 + x^2}$$

for the field amplitude, after adiabatically eliminating its phase. Here $a$ is the amplitude of the injected laser beam and $c$ is the cooperativity parameter, which is proportional to the negative of the population inversion of the atoms, so $c > 0$ corresponds to absorption and $c < 0$ to stimulated emission. Both $a$ and $c$ can fluctuate randomly, with the latter being more interesting because it appears multiplicatively in equation (8.2). In particular, Gragg considered Gaussian white noise fluctuations $c_t = c + \sigma \xi_t$ in the cooperativity parameter, about a positive mean $c$, and obtained the Stratonovich stochastic differential equation

$$(8.3) \qquad dX_t = \left( a - X_t - 2c\, \frac{X_t}{1 + X_t^2} \right) dt - 2\sigma\, \frac{X_t}{1 + X_t^2} \circ dW_t.$$

He converted this into the equivalent Ito stochastic differential equation and solved the corresponding Fokker-Planck equation for the stationary probability density. This has extrema close to the steady states of the deterministic equation (8.2) for sufficiently small noise intensity. For $0 < c < 4$ and large noise intensity there are two additional maxima not present in the original deterministic model, thus corresponding to a noise induced optical bistability.

**Nemantic Liquid Crystal**   A nemantic liquid crystal is a liquid consisting of elongated molecules which, on average, align themselves in a preferred direction. For example, when the liquid is enclosed between two parallel glass plates this direction is prescribed by the plates. In contrast, with a solid crystal the position of the molecules is not fixed, but can fluctuate in response to, say, an imposed magnetic field. If $\theta$ denotes the angle of maximum deformation of the first mode from the preferred direction, it can be shown that

$$(8.4) \qquad \dot{\theta} = -a\,\theta + b\,H^2 \left( \theta - \frac{1}{2}\theta^3 \right),$$

for certain positive constants $a$ and $b$, where $H$ is the amplitude of the magnetic field.

Horsthemke and Lefever have considered the effect of random fluctuations in the magnetic field. Since $H$ appears quadratically in (8.4) rather than linearly, they argued that it is inappropriate to simply replace $H^2$ by $H^2 + \sigma \xi_t$ with Gaussian white noise in (8.4). Instead they proposed that $H$ should be replaced by $H + \sigma \xi_t$ for a stationary Ornstein-Uhlenbeck process $\xi_t$ and they then used a method based on the expansion of solutions of a Fokker-Planck like equation to derive the Stratonovich stochastic differential equation

$$(8.5) \qquad d\Theta_t = \left( -a\,\Theta_t + b \left( H^2 + \frac{1}{2}\sigma^2 \right) \left( \Theta_t - \frac{1}{2}\Theta_t^3 \right) \right) dt$$
$$+ b \sqrt{\frac{1}{2}\sigma^2 + 4H^2\sigma^2} \left( \Theta_t - \frac{1}{2}\Theta_t^3 \right) \circ dW_t,$$

where $W = \{W_t, t \geq 0\}$ is a scalar standard Wiener process. The parameters $H$ and $\sigma$ appear here quite differently than they would have if $H^2$ had been randomized in a linear manner. In particular, the parameter values in (8.5) are increased by the nonlinear randomization.

Other models arising from the randomization of parameters which occur nonlinearly in a preliminary deterministic model can also be found in the book by Horshemke and Lefever.

# 7.9 Blood Clotting Dynamics and Cellular Energetics

**Blood Clotting Dynamics** The repair of small blood vessels and the pathological growth of internal blood clots involve the formation of platelet aggregates adhering to portions of the vessel wall. Fogelson has proposed a microsopic model in which blood is represented by a suspension of discrete massless platelets in a viscous incompressible fluid. The platelets are initially noncohesive; however, if stimulated by an above-threshhold concentration of the chemical adenosine di phosphate (ADP) or by contact with the adhesive injured proportion of the vessel wall, they become cohesive and secrete more ADP into the fluid. Cohesion between the platelets and adhesion of a platelet to the injured wall are modelled by creating elastic links, whereas repulsive forces are used to prevent a platelet from coming too close to another platelet or to the wall. These forces effect the fluid only in the neighbourhood of an aggregate. The platelets and the secreted ADP both move by fluid advection and diffusion.

In particular, Fogelson considered a steady 2-dimensional flow with negligible inertial effects, so the velocity $u$ satisfies the Stokes' equations

(9.1) $$\Delta u - \nabla p + f = 0, \qquad \nabla \cdot u = 0$$

where $p$ is the pressure and $f$ the force density due to the interactions between the platelets. He denoted by $X_t^{(k)}$ the 2-dimensional position of the $k$th platelet at time $t$ and assumed that its velocity was the superposition of advection of the blood flow and a random effect resulting from the local stirring in the blood induced by the tumbling and colliding of the platelets amongst and with the much larger red blood cells. He thus obtained the 2-dimensional stochastic differential equation

(9.2) $$dX_t^{(k)} = u\left(X_t^{(k)}\right) dt + \sigma \, dW_t,$$

where $W_t$ is a standard 2-dimensional Wiener process and $\sigma$ is the diffusion coefficient of the platelets. The number of platelets, and hence of such stochastic differential equations is extremely large and could vary in time in a neighbourhood of an injury. The reader is referred to Fogelson's paper for the remaining details of the model. He had to solve the Stokes' equations (9.1) numerically to determine the velocity $u$ and then used the numerical values in each of the stochastic differential equations (9.2).

**Cellular Energetics** Cellular energetics is a complex combination of catabolism and anabolism, the breaking down of certain molecules and the synthesis of others. The chemical adenosine tri phosphate (ATP) plays a central role in the transfer of the energy released by catabolism to the anabolic sites within a cell. A typical catabolic reaction is the aerobic breakdown of carbohydrates via glycolysis, the Krebs cycle and, ultimately, oxidative phosphorylation. The irreversible ATP-utilizing anabolic reactions are subject to natural random fluctuations in the living cell. Their efficiency, and that of oxidative phosphorylation, is known to be maintained by the buffering ability of the adenylate kinase reaction, a reversible reaction between the three adenine nucleotides ATP, ADP and AMP (adenosine mono phosphate). However, with current experimental techniques it is not possible to determine the instantaneous intracellular variations in these efficiencies. To gain some insight into what may be happening, Veuthey and Stucki proposed a system of stochastic differential equations modelling these reactions, which they then investigated by means of numerical simulations.

Veuthey and Stucki used rate laws to obtain the system of differential equations

(9.3)
$$\begin{aligned}
\dot{T} &= J_p + J_1 + J_a \\
\dot{D} &= -J_p - J_1 - 2J_a \\
\dot{M} &= J_a
\end{aligned}$$

for the concentrations $T$, $D$ and $M$ of ATP, ADP and AMP, respectively. Here the phenomenological linear relation

(9.4)
$$J_p = L_p X_p + L_{po} X_o$$

is the rate of ATP production and

$$J_o = L_{po} X_p + L_o X_o$$

is the rate of oxygen consumption, with $L_p$, $L_{po}$ and $L_o$ the phenomenological Onsager coefficients summarizing the overall kinetic properties of the process. $X_o$ is the redox potential of the oxidizable substrate and $X_p$ is the phosphate potential which has the form

(9.5)
$$X_p = -\alpha - \beta \ln \left( \frac{T}{D} \right)$$

for positive parameters $\alpha$ and $\beta$. In addition, $J_1 = L_1 X_p$ is the load flow of the energy conversion, with the load conductance $L_1$ summarizing all of the irreversible ATP-utilizing reactions in the cell. Finally, the adenylate kinase reaction is represented by the relationship $J_a = L_a X_a$, with $L_a$ a constant proportional to the activity of the adenylate kinase and $X_a$ the adenylate kinase potential which has the form

(9.6)
$$X_a = -\gamma - \delta \ln \left( \frac{T M}{D^2} \right)$$

for positive parameters $\gamma$ and $\delta$.

For oxidative phosphorylation to occur it is necessary that $L_1$ be matched to $L_p$ according to $L_1/L_p = \sqrt{1 - q^2}$, where $q$ is the degree of coupling of oxidative phosphorylation and is a dimensionless, normalized measure of the cross-coupling ratio $L_{po}/\sqrt{L_p L_o}$. The model is completed by assuming that $L_1$ is subject to real noise fluctuations about a mean $\bar{L}_1$, that is

(9.7)
$$L_1 = \bar{L}_1 + \rho_t$$

where $\rho_t$ is an Ornstein-Uhlenbeck process satisfying the stochastic differential equation

(9.8)
$$d\rho_t = \gamma \rho_t + \sigma \, dW_t.$$

Equations (9.2)–(9.8) can be rewritten as a 4-dimensional vector stochastic differential equation in $T$, $D$, $M$ and $\rho$.

# 7.10 Josephson Junctions, Communications and Stochastic Annealing

**Josephson Tunneling Junctions** A Josephson tunneling junction consists of two superconductors separated by a thin oxide layer. When the ratio of the amplitudes of the wave functions of the Cooper pairs of electrons in the two superconductors remains constant, their phase difference $\phi$ satisfies the Josephson equation

(10.1)
$$\dot{\phi} = \frac{2e}{\hbar} V,$$

where $e$ is the unit electron charge, $\hbar$ is Planck's constant and $V$ is the potential difference across the oxide layer. If the external resistance is very large, the total current $I$ is constant and consists of a noisy current due to the resistance $R$ across the oxide layer, a current due to the capacitance $C$ of the junction and a current due to the tunneling of Cooper pairs through the junction with maximum amplitude $I_{max}$, the maximum Josephson current. Thus $I$ is given by

(10.2)
$$I = \frac{V}{R} - \zeta_t + C\dot{V} + I_{max} \sin \phi,$$

where $\zeta_t$ is Gaussian white noise with intensity $2kT/R$, with $k$ being Boltzmann's constant and $T$ the absolute temperature. We can combine (10.1) and (10.2) to obtain a second order differential equation with additive noise for the phase difference $\phi$

(10.3)
$$\frac{\hbar}{2e} C \ddot{\phi} + \frac{\hbar}{2e R} \frac{1}{R} \dot{\phi} + I_{max} \sin \phi = I + \zeta_t,$$

which we interpret as a 2-dimensional vector stochastic differential equation with additive noise. Of primary interest here is the relationship between current $I$ and the mean voltage $E(V) = \hbar E(\dot{\phi})/2e$.

**Communications**    Noise also plays a central role in modern communication theory. An example is in the stable tuning of radio and television sets, which is achieved with a device called a phase-locked loop or PLL. A PLL ties the phase of an oscillator signal to that of a reference signal. Ideally the phases are locked together, but noisy fluctuations in, say, the input voltage often cause the phases to slip.

For a basic PLL device containing a linear filter it can be assumed that the phase difference $\phi$ of the input and reference signals satisfies a second order differential equation with additive white noise $\zeta_t$ of the form

(10.4) $$a \ddot{\phi} + \dot{\phi} + \alpha b \sin \phi = \omega - \omega_0 - \alpha \zeta_t,$$

which, in turn, can be written as a 2-dimensional vector Ito stochastic differential equation. Here $a = RC$, the series effect of a resistor and capacitor in the linear filter of the PLL, $b$ is related to the product of the amplitudes of the two signals and $\alpha$ is the intensity of the white noise. An important task, then, is to use (10.4) to predict the relationship between the mean rate of change of the phase difference $E(\dot{\phi})$ and the detuning $\omega - \omega_0$. The cycle slip rate per second, that is the average number of times per second that the phases differ by $2\pi$, provides a measure of quality of a PLL.

**Stochastic Annealing**    The global minimum of a function $V : \Re^d \to \Re$ can often be located from the asymptotic behaviour of the solutions of the gradient ordinary differential equation

(10.5) $$\dot{x} = -\nabla V(x)$$

for which it is a steady state. This method is known as the simulated annealing procedure. It has, however, the serious shortcoming that a computed solution of (10.5) may become trapped at a local minimum of $V$ rather than converge to the desired, global minimum $\bar{X}$. One way to circumvent this difficulty is to consider the solutions of the related stochastic differential equation with additive noise

(10.6) $$dX_t = -\nabla V(X_t)\, dt + \sigma(t)\, dW_t,$$

where $\{W_t,\ t \geq 0\}$ is an $d$-dimensional standard Wiener process, for an appropriate choice of scalar diffusion coefficient $\sigma(t)$. Suppose that $\nabla V$ is uniformly Lipschitz and satisfies the growth bound

$$|\nabla V(x)|^2 \leq K \left(1 + |x|^2\right)$$

for some constant $K > 0$ and all $x \in \Re^d$. Then, for $\sigma(t) = c/\sqrt{\log(t + 2)}$ with $c > 0$, it can be shown that the distribution of $X_t$ converges to the limit of the Gibbs densities proportional to $\exp(-V(x)/T))$ as the "absolute temperature" $T = \sigma(t)^2 \to 0$ for $t \to \infty$, which is concentrated at the global minimum $\bar{X}$ of $V$. Other choices of $\sigma(t)$ may lead to convergence to a nonglobal local minimum with higher probability than to $\bar{X}$. In fact, the stronger result

(10.7) $$\gamma \leq E \left( \left|X_t - \bar{X}\right|^2 \log t \right)$$

can be established for some $\gamma > 0$ and all $t$ sufficiently large. This provides a lower bound for the rate of mean-square convergence of $X_t$ to $\bar{X}$. An upper bound in (10.7) also holds when $V$ has no local minima other than $\bar{X}$. This procedure of using the solutions $X_t$ of (10.6) to locate $\bar{X}$ is called *stochastic annealing*. Generally, a numerical method is needed to solve stochastic differential equation (10.6).

# Chapter 8

# Time Discrete Approximation of Deterministic Differential Equations

In this chapter we summarize the basic concepts and assertions of the numerical analysis of initial value problems for deterministic ordinary differential equations. The material is presented so as to facilitate generalizations to the stochastic setting and to highlight the differences between the deterministic and stochastic cases.

## 8.1  Introduction

In general it is not possible to find explicitly the solution $x = x(t; t_0, x_0)$ of an initial value problem (IVP)

$$(1.1) \qquad \dot{x} = \frac{dx}{dt} = a(t, x), \qquad x(t_0) = x_0$$

for the deterministic differential equations that occur in many scientific and technological models. Even when such a solution can be found, it may be only in implicit form or too complicated to visualize and evaluate numerically. Necessity has thus lead to the development of methods for calculating numerical approximations to the solutions of such initial value problems. The most widely applicable and commonly used of these are the *time discrete approximation* or *difference methods,* in which the continuous time differential equation is replaced by a discrete- time difference equation generating values $y_1$, $y_2$, ..., $y_n$, ... to approximate $x(t_1; t_0, x_0)$, $x(t_2; t_0, x_0)$, ..., $x(t_n; t_0, x_0)$, ... at given discretization times $t_0 < t_1 < t_2 < \cdots < t_n < \cdots$. These approximations should be quite accurate, one hopes, if the time increments $\Delta_n = t_{n+1} - t_n$ for $n = 0, 1, 2, \ldots$ are sufficiently small. As a background for the development of discretization methods for stochastic differential equations, in this chapter we shall review the basic difference methods used for ordinary differential equations and consider some related issues such as their convergence and stability.

The simplest difference method for the IVP (1.1) is the *Euler method*

$$(1.2) \qquad y_{n+1} = y_n + a(t_n, y_n)\, \Delta_n$$

for a given time discretization $t_0 < t_1 < t_2 < \cdots < t_n < \cdots$ with increments $\Delta_n = t_{n+1} - t_n$ where $n = 0, 1, 2, \ldots$. Once the initial value $y_0$ has been specified, usually $y_0 = x_0$, the approximations $y_1$, $y_2$, ..., $y_n$, ... can be calculated by recursively applying formula (1.2). We can derive (1.2) by freezing the right hand side of the differential equation over the time interval $t_n \leq t < t_{n+1}$ at the value $a(t_n, y_n)$ and then integrating to obtain the tangent to the solution $x(t; t_n, y_n)$ of the differential equation with the initial value $x(t_n) = y_n$. The difference

$$(1.3) \qquad l_{n+1} = x(t_{n+1}; t_n, y_n) - y_{n+1},$$

which is generally not zero, is called the *local discretization error* for the nth time step. This is usually not the same as the *global discretization error*

$$(1.4) \qquad e_{n+1} = x(t_{n+1}; t_0, x_0) - y_{n+1}$$

for the same time step, which is the error with respect to the sought solution of the original IVP (1.1). Nevertheless, we can use the local discretization error to estimate the global discretization error. It must be emphasized that (1.3) and (1.4) assume that we can perform all arithmetic calculations exactly. In practice, both we and digital computers are restricted to a finite number of decimal places when doing calculations and roundoff all excess decimal places, thus introducing *roundoff error*. We shall denote it by $r_{n+1}$ for the nth time step.

The key to estimating the size of discretization errors is the Taylor formula with remainder, which for a twice continuously differentiable function $x = x(t)$ is

$$(1.5) \qquad x(t_{n+1}) = x(t_n) + \dot{x}(t_n)\,\Delta_n + \frac{1}{2!}\,\ddot{x}(\theta_n)\,\Delta_n^2$$

for some $\theta_n$ satisfying $t_n < \theta_n < t_{n+1}$. For $x(t) \equiv x(t; t_n, y_n)$, the solution of the differential equation with $x(t_n) = y_n$, we thus have

$$(1.6) \qquad x(t_{n+1}) = x(t_n) + a(t_n, x(t_n))\,\Delta_n + \frac{1}{2!}\,\ddot{x}(\theta_n)\,\Delta_n^2.$$

Since $x(t_n) = y_n$ here, on subtracting (1.5) from (1.6) we find that the local discretization error (1.3) has the form

$$l_{n+1} = \frac{1}{2!}\,\ddot{x}(\theta_n)\,\Delta_n^2.$$

If we knew that $|\ddot{x}(t)| < M$ for all $t$ in some interval $[t_0, T]$ of interest, then we would have the estimate

$$(1.7) \qquad |l_{n+1}| \leq \frac{1}{2!}\,M\,\Delta_n^2$$

for any discretization time subinterval $[t_n, t_{n+1}]$ with $t_0 \leq t_n < t_{n+1} \leq T$. We can obtain such a bound on $\ddot{x}$ using the fact that

$$\ddot{x} = \frac{d}{dt}\dot{x} = \frac{d}{dt}a(t, x(t))$$

$$= \frac{\partial a}{\partial t}(t, x(t)) + \frac{\partial a}{\partial x}(t, x(t))\frac{dx}{dt}$$

$$= \frac{\partial a}{\partial t}(t, x(t)) + \frac{\partial a}{\partial x}(t, x(t))\, a(t, x(t)).$$

If $a$, $\frac{\partial a}{\partial t}$ and $\frac{\partial a}{\partial x}$ are continuous and if we knew that all solutions $x(t)$ under consideration remained in some closed and bounded set $C$ for all $t_0 < t < T$, we could use

$$M = \max\left|\frac{\partial a}{\partial t}(t, x)\right| + \max\left|\frac{\partial a}{\partial x}(t, x)\, a(t, x)\right|,$$

where the maxima are taken over $(t, x) \in [t_0, T] \times C$, in the inequality (1.7). This particular value of $M$ will usually give a gross overestimate, but from (1.7) we can see that the local discretization error for the Euler method (1.2) is of order $\Delta_n^2$.

To estimate the global discretization error we shall assume, for simplicity, that $a = a(t, x)$ satisfies a uniform Lipschitz condition

$$|a(t, x) - a(t, y)| \leq K\, |x - y|$$

and that the time discretization involves equidistant time instants $t_n = t_0 + n\Delta$ for $n = 0, 1, 2, \ldots$. Applying the Taylor formula (1.5) to the solution $x(t) \equiv x(t; t_0, x_0)$ we have (1.6) with $\Delta_n \equiv \Delta$, but now $x(t_n) \neq y_n$ in general. Subtracting (1.2) then gives

$$(1.8) \qquad e_{n+1} = e_n + \{a(t_n, x(t_n)) - a(t_n, y_n)\}\,\Delta + \frac{1}{2}\ddot{x}(\theta_n)\,\Delta^2$$

and, using the Lipschitz condition on $a$ and a bound on $\ddot{x}$, thus

$$|e_{n+1}| \leq |e_n| + K\,|e_n|\,\Delta + \frac{1}{2}M\,\Delta^2.$$

We can then show by induction that the difference inequality

$$(1.9) \qquad |e_{n+1}| \leq (1 + K\,\Delta)\,|e_n| + \frac{1}{2}M\,\Delta^2$$

with $e_0 = x_0 - y_0 = 0$ implies that

$$|e_{n+1}| \leq \frac{1}{2}\left(\frac{(1 + K\Delta)^n - 1}{(1 + K\Delta) - 1}\right)M\,\Delta^2 \leq \frac{1}{2}\left(e^{nK\Delta} - 1\right)\frac{M}{K}\,\Delta,$$

since $(1 + K\Delta)^n \leq e^{nK\Delta}$. Hence the global discretization error for the Euler method (1.2) satisfies

$$(1.10) \qquad |e_{n+1}| \leq \frac{1}{2}\left(e^{K(T - t_0)} - 1\right)\frac{M}{K}\,\Delta$$

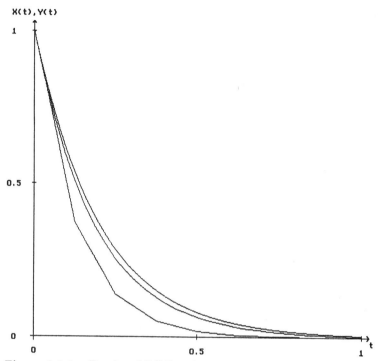

**Figure 8.1.1**    Results of PC-Exercise 8.1.1.

for discretization times $t_n = t_0 + n\Delta \leq T$. It is obviously one power of $\Delta$ less than the local discretization error.

**PC-Exercise 8.1.1**    *Apply the Euler method (1.2) to the IVP*

$$\frac{dx}{dt} = -5\,x, \qquad x(0) = 1,$$

*with time steps of equal length* $\Delta = 2^{-3}$ *and* $2^{-5}$ *over the time interval* $0 \leq t \leq 1$. *Plot the results and the exact solution* $x(t) = e^{-5t}$ *against* $t$.

In Figure 8.1.1 the upper curve corresponds to the exact solution and the lower and middle ones to the Euler method with step sizes $\Delta = 2^{-3}$ and $2^{-5}$, respectively. We see that the global discretization error is smaller for the smaller step size.

In the next PC-Exercise we shall look more closely at the dependence of the global truncation error on the step size. For this we recall that a function $f(\Delta) = A\,\Delta^\gamma$ becomes linear in logarithmic coordinates, that is

$$\log_a f(\Delta) = \log_a A + \gamma \log_a \Delta$$

for logarithms to the base $a \neq 1$. In comparative studies we shall take time steps of the form $\Delta = a^{-n}$ for $n = 1, 2, \ldots$ and $a > 1$. We shall usually halve the time step successively, in which case logarithms to the base $a = 2$ will be

Ld(f(Delta))

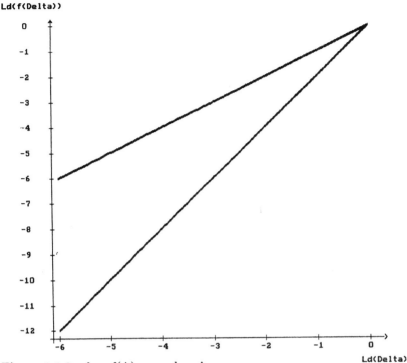

**Figure 8.1.2**    $\log_2 f(\Delta)$ versus $\log_2 \Delta$.

appropriate. In Figure 8.1.2 we plot the values of $\log_2 f(\Delta)$ against $\log_2 \Delta$ for the two functions $f(\Delta) = \Delta$ and $f(\Delta) = \Delta^2$ with $\Delta = 2^{-n}$ for $n = 0, 1, \ldots, 6$.

**PC-Exercise 8.1.2**    *For the IVP in PC-Exercise 8.1.1 calculate the global discretization error at time $t = 1$ for the Euler method with time steps of equal length $\Delta = 1$, $2^{-1}$, $2^{-2}$, $\ldots$, $2^{-13}$, rounding off to 5 significant digits. Plot the logarithm to the base 2 of these errors against $\log_2 \Delta$ and determine the slope of the resulting curve.*

From Figure 8.1.3 we see that the calculated global discretization error $\tilde{e}_{n+1}$ for the Euler method is proportional to the step size $\Delta$ for $\Delta \leq 2^{-3}$, provided $\Delta$ is not too small. For $\Delta \leq 2^{-11}$ the error $\tilde{e}_{n+1}$ begins to increase here as $\Delta$ is further decreased. This does not contradict (1.10), but occurs because $\tilde{e}_{n+1}$ also includes the roundoff error. To estimate $\tilde{e}_{n+1}$ we must add the roundoff error $r_n$ to the right hand side of (1.9). For $|r_n| \leq R$ for each $n$ we then obtain the estimate

$$|\tilde{e}_{n+1}| \leq \frac{1}{2} \left( e^{K(T-t_0)} - 1 \right) \left( \frac{2R}{K} \frac{1}{\Delta} + \frac{M}{K} \Delta \right)$$

instead of (1.10). For very small $\Delta$ the reciprocal term dominates the bound. While it represents the worst case scenario, this bound is still indicative of the cummulative effect of roundoff error, since for smaller $\Delta$ more calculations are

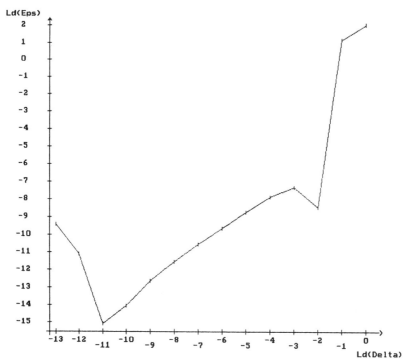

**Figure 8.1.3**    Results of PC-Exercise 8.1.2.

required to reach the given end time. We shall look more closely at randomly distributed roundoff errors in Section 4.

The presence of roundoff error means there is a minimum step size $\Delta_{min}$ for each initial value problem, below which we cannot improve the accuracy of the approximations calculated by means of the Euler method. To obtain a more accurate approximation we need to use another method with a higher order discretization error. The Taylor expansion provides a systematic framework for developing and investigating such schemes. For the rest of this section we shall, however, continue with a more heuristic approach.

For the Euler method we simply froze the right hand side of the differential equation at the value $a(t_n, y_n)$ at the beginning of each discretization subinterval $t_n < t < t_{n+1}$. We should obtain a more accurate approximation if we included more information from elsewhere in the subinterval. For instance, we could use the average of the values at both end points, in which case we have the *trapezoidal method*

$$(1.11) \qquad y_{n+1} = y_n + \frac{1}{2} \left\{ a(t_n, y_n) + a(t_{n+1}, y_{n+1}) \right\} \Delta_n.$$

This is called an *implicit* scheme because the unknown quantity $y_{n+1}$ appears in both sides of (1.11) and, in general, cannot be isolated algebraically. To circumvent this difficulty we could use the Euler method (1.2) to approximate

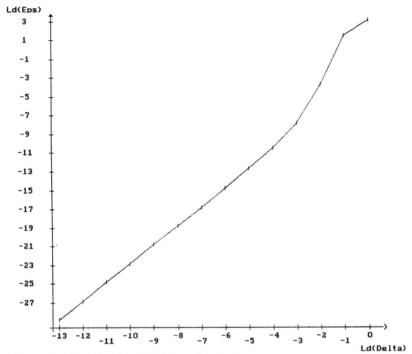

**Figure 8.1.4**    Results of PC-Exercise 8.1.3.

the $y_{n+1}$ term on the right hand side of (1.11). Then we obtain the *modified trapezoidal method*

$$\bar{y}_{n+1} = y_n + a(t_n, y_n)\, \Delta_n$$

$$y_{n+1} = y_n + \frac{1}{2} \left\{ a(t_n, y_n) + a(t_{n+1}, \bar{y}_{n+1}) \right\} \Delta_n,$$

or

$$(1.12) \quad y_{n+1} = y_n + \frac{1}{2} \left\{ a(t_n, y_n) + a\left(t_{n+1}, y_n + a(t_n, y_n)\Delta_n\right) \right\} \Delta_n,$$

which is also known as the *improved Euler* or *Heun method*. It is a simple example of a *predictor-corrector method* with the predictor $\bar{y}_{n+1}$ inserted into the corrector equation to give the next iterate $y_{n+1}$.

Both the trapezoidal and the modified trapezoidal methods have local discretization errors of third order in $\Delta_n$. This can be verified by comparing the Taylor formula with third order remainder of the solution $x(t; t_0, x_0)$ of the differential equation and (1.11) or (1.12) with $a(t_{n+1}, y_{n+1})$ or $a(t_{n+1}, \bar{y}_{n+1})$ expanded about $(t_n, y_n)$. The global discretization error for both methods is of second order in $\Delta = \max_n \Delta_n$, which is again one order less than the local discretization error.

**PC-Exercise 8.1.3**    *Repeat PC-Exercise 8.1.2 with the usual arithmetic of the PC for the modified trapezoidal method (1.12). Compare the results with those for the Euler method.*

Figure 8.1.4 indicates that the global truncation error for the modified trapezoidal method is proportional to $\Delta^2$ for $\Delta \leq 2^{-3}$, whereas that of the Euler method is proportional to $\Delta$.

**Exercise 8.1.4**     *Show that the local discretization errors of the trapezoidal and modified trapezoidal methods are of third order in $\Delta_n$.*

Even higher order difference methods can be derived by using more accurate approximations of the right hand side of the differential equation over each discretization subinterval $t_n < t < t_{n+1}$. These are called *one-step methods* if they involve only the values $y_n$ and $y_{n+1}$ in addition, of course, to $t_n$ and $\Delta_n$. Explicit one-step methods are usually written in the general form

$$(1.13) \qquad y_{n+1} = y_n + \Psi(t_n, y_n, \Delta_n)\,\Delta_n,$$

for some function $\Psi = \Psi(t, x, \Delta)$, which is called the *increment function*. We say that it is a *pth order* method if its global discretization error is bounded by the $p$th power of $\Delta = \max_n \Delta_n$. If the functions $a$ and $\Psi$ are sufficiently smooth it can be shown that a $(p+1)$th order local discretization error implies a $p$th order global discretization error; see Theorem 8.3.2 in the next section. For example, the Euler method (1.2) is a 1st order one-step scheme with $\Psi(t, x, \Delta) = a(t, x)$ and the Heun method (1.12) is a 2nd order one-step scheme with

$$\Psi(t, x, \Delta) = \frac{1}{2}\left\{a(t, x) + a\left(t + \Delta, x + a(t, x)\,\Delta\right)\right\}.$$

The function $\Psi$ cannot be chosen completely arbitrarily. For instance, it should be *consistent* with the differential equation, that is satisfy

$$\lim_{\Delta \downarrow 0} \Psi(t, x, \Delta) = a(t, x),$$

if the values calculated from (1.13) are to converge to the desired solution of the differential equation. The Euler method is obviously consistent and the Heun method is consistent when $a(t, x)$ is continuous in both variables.

Some difference methods achieve higher accuracy by using information from previous discretization subintervals when calculating $y_{n+1}$ on $t_n < t < t_{n+1}$. In these *multi-step* methods $y_{n+1}$ depends on the previous $k$ values $y_n, y_{n-1}, \ldots, y_{n-k}$ for some $k > 1$. An example involving equal time steps $\Delta$ is the *3-step Adams-Bashford method*

$$(1.14) \quad y_{n+1} = y_n + \frac{1}{12}\left\{23\,a(t_n, y_n) - 16\,a(t_{n-1}, y_{n-1}) + 5\,a(t_{n-2}, y_{n-2})\right\}\Delta,$$

which turns out to have third order global discretization error. It is derived by replacing the right hand side of the differential equation on the time interval $t_n < t < t_{n+1}$ by the unique cubic polynomial passing through the points $(t_n, a(t_n, y_n))$, $(t_{n-1}, a(t_{n-1}, y_{n-1}))$ and $(t_{n-2}, a(t_{n-2}, y_{n-2}))$. Notice that these are the only points where the function $a$ has to be evaluated. Since the evaluations from the previous steps can be saved for use in the current step, this method essentially requires only the value of $a(t_n, y_n)$ to be calculated in the $n$th step, once the procedure has been started. In order to start it we must specify the first 3 values $y_0$, $y_1$ and $y_2$, which is usually done be calculating $y_1$ and $y_2$ with a one-step method starting at $y_0$.

**PC-Exercise 8.1.5** *Repeat PC-Exercise 8.1.3 using the 3-step Adams-Bashford method (1.14) with the Heun method (1.12) as its starting routine.*

**Exercise 8.1.6** *Show that the 3-step Adams-Bashford method (1.14) has fourth order local discretization error.*

Finally, we note that we can sometimes obtain higher order accuracy from a one-step scheme by the method of extrapolation. For example, suppose we use the Euler scheme (1.2) with $N$ equal time steps $\Delta = T/N$ on the interval $0 \le t \le T$. If $x(T)$ is the true value at time $T$ and $y_N(\Delta)$ the corresponding value from the Euler scheme, then we have

$$(1.15) \qquad y_N(\Delta) = x(T) + e(T)\Delta + O(\Delta^2),$$

where we have written the global truncation error as $e(T)\Delta + O(\Delta^2)$. If, instead, we use the Euler scheme with $2N$ time steps of equal length $\Delta/2$, then we have

$$(1.16) \qquad y_{2N}\left(\frac{1}{2}\Delta\right) = x(T) + \frac{1}{2}e(T)\Delta + O(\Delta^2).$$

We can eliminate $e(T)$ from (1.15) and (1.16) to obtain

$$x(T) = 2y_{2N}\left(\frac{1}{2}\Delta\right) - y_N(\Delta) + O(\Delta^2).$$

Thus we have a second order approximation

$$(1.17) \qquad Z_N(\Delta) = 2y_{2N}\left(\frac{1}{2}\Delta\right) - y_N(\Delta)$$

for $x(t)$ from the first order Euler scheme. Of course, this requires our repeating the Euler scheme calculations for half the original time step, but for complicated differential equations it may involve fewer and simpler calculations than a second order one-step scheme. This method is known as *Richardson* or *Romberg extrapolation*. It can also be applied to more general one-step schemes and to multi-step schemes.

**PC-Exercise 8.1.7** *Compare the error of the Euler and Richardson extrapolation approximations of $x(1)$ for the solution of the initial value problem*

$$\frac{dx}{dt} = -x, \qquad x(0) = 1$$

*for equal time steps $\Delta = 2^{-3}$, $2^{-4}$,..., $2^{-10}$. Plot $\log_2$ of the errors against $\log_2 \Delta$.*

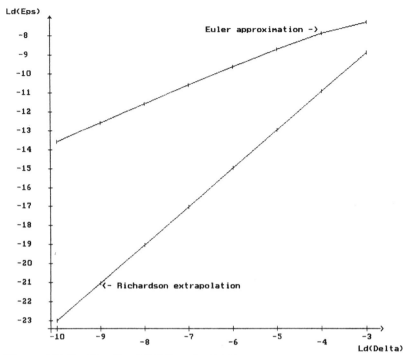

Figure 8.1.5    Results of PC-Exercise 8.1.7.

## 8.2   Taylor Approximations and Higher Order Methods

A one-step difference method with global discretization error of order $p$ is readily suggested locally by the *Taylor formula with* $(p + 1)$th *order remainder*

$$(2.1) \qquad x(t_{n+1}) \quad = \quad x(t_n) + \frac{dx}{dt}(t_n)\,\Delta_n + \cdots + \frac{1}{p!}\frac{d^p x}{dt^p}(t_n)\,\Delta_n^p$$

$$+\frac{1}{(p+1)!}\frac{d^{p+1}x}{dt^{p+1}}(\theta_n)\,\Delta_n^{p+1},$$

where $t_n < \theta_n < t_{n+1}$ and $\Delta_n = t_{n+1} - t_n$, for any $p = 1, 2, 3, \ldots$. We can apply this formula to a solution $x(t)$ of the differential equation

$$(2.2) \qquad \frac{dx}{dt} = a(t, x)$$

if the function $a = a(t, x)$ and its partial derivatives of orders up to and including $p$ are continuous as this assures that $x(t)$ is $p + 1$ times continuously differentiable. Indeed, from (2.2) and the chain rule, by repeatedly differentiating $a(t, x(t))$ we have

$$\frac{dx}{dt} = a, \qquad \frac{d^2 x}{dt^2} = a_t + a_x\, a,$$

$$\frac{d^3 x}{dt^3} = a_{tt} + 2a_{tx}\, a + a_{xx}\, a^2 + a_t\, a_x + a_x^2\, a,$$

and so on, where we have used subscripts to indicate partial derivatives. Evaluating these expressions at $(t_n, y_n)$ for the solution $x(t) = x(t; t_n, y_n)$ of (2.2) and omitting the remainder term in (2.1) we obtain a one-step method for $y_{n+1}$, which we shall call the $p$th *order truncated Taylor method*. This method obviously has local discretization error of order $p+1$ and can be shown to have global discretization error of order $p$. The 1st order truncated Taylor method is just the Euler method (1.2), the 2nd *order truncated Taylor method* is

$$(2.3) \quad y_{n+1} = y_n + a(t_n, y_n)\, \Delta_n + \frac{1}{2!}\left\{a_t(t_n, y_n) + a_x(t_n, y_n)a(t_n, y_n)\right\}\, \Delta_n^2$$

and the 3rd *order truncated Taylor method*

$$(2.4) \quad y_{n+1} = y_n + a\, \Delta_n + \frac{1}{2!}\left\{a_t + a_x\, a\right\}\, \Delta_n^2$$
$$+ \frac{1}{3!}\left\{a_{tt} + 2a_{tx}\, a + a_{xx}\, a^2 + a_t\, a_x + a_x^2\, a\right\}\, \Delta_n^3,$$

where $a$ and its partial derivatives are evaluated at $(t_n, y_n)$.

**PC-Exercise 8.2.1** *Use the 2nd order truncated Taylor method (2.3) with equal length time steps $\Delta = 2^{-3}, \ldots, 2^{-10}$ to calculate approximations to the solution $x(t) = 2/(1 + e^{-t^2})$ of the initial value problem*

$$\frac{dx}{dt} = t\, x\, (2 - x), \qquad x(0) = 1$$

*over the interval $0 \le t \le 0.5$. Repeat the calculations using the 3rd order truncated Taylor method (2.4). Plot $\log_2$ of the global discretization errors at time $t = 0.5$ against $\log_2 \Delta$.*

In Figure 8.2.1 the upper curve with slope 2 corresponds to the 2nd order truncated Taylor method and the lower one with slope 3 to the 3rd order truncated Taylor method.

The coefficients in higher order truncated Taylor methods soon become unwieldy and error prone to determine for all but the simplest differential equations. Moreover, considerable computational effort is needed for the evaluation of these coefficients, so these methods are not particularly efficient. They are almost never used in practice, except to provide a reference point for the development and analysis of other, more efficient higher order difference schemes.

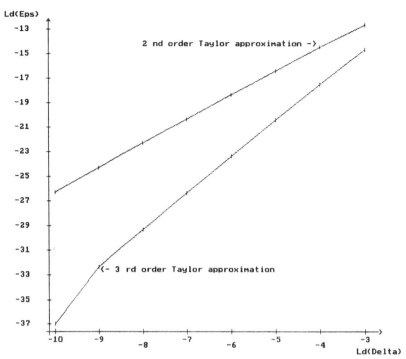

**Figure 8.2.1**   Results of PC-Exercise 8.2.1.

One way of simplifying the coefficients in a truncated Taylor method is to replace the partial derivatives by their forward difference quotients, for example replacing $a_t(t_n, y_n)$ and $a_x(t_n, y_n)$ by

$$\frac{a(t_{n+1}, y_n) - a(t_n, y_n)}{\Delta_n} \quad \text{and} \quad \frac{a(t_n, y_{n+1}) - a(t_n, y_n)}{y_{n+1} - y_n},$$

respectively. This will lead to an implicit scheme because $y_{n+1}$ appears on both sides of the recursion formula and, generally, cannot be solved for algebraically. As in the trapezoidal method (1.11) we could use the Euler method (1.2), say, to predict a value of $y_{n+1}$ to use in the terms on the right hand side of the formula, thus obtaining an explicit method. For the 2nd order truncated Taylor method this results first in the trapezoidal method (1.11) and then in the Heun method (1.12). The higher order coefficients will usually be considerably more complicated, but are at least derivative free.

The standard procedure with most one-step methods

$$(2.5) \qquad\qquad y_{n+1} = y_n + \Psi(t_n, y_n, \Delta_n)\, \Delta_n$$

is first to derive the function $\Psi = \Psi(t, x, \Delta)$ by an heuristic argument and then to compare the method with a truncated Taylor method or expansion to determine the order of its discretization error. The *Runge-Kutta methods* are typical of this approach. For what will turn out to be the 2nd order methods of this type, $\Psi$ is chosen with the form

(2.6) $\qquad \Psi(t, x, \Delta) = \alpha \, a(t, x) + \beta \, a\left(t + \gamma \, \Delta, x + \gamma \, a(t, x) \, \Delta\right),$

for certain constants $\alpha$, $\beta$ and $\gamma$, which represents an weighted averaging of the right hand side of the differential equation (2.2) over two points. Expanding the second term about $(t, x)$, we obtain

$$
\begin{aligned}
\Psi \;=\; & (\alpha + \beta) \, a + \gamma \beta \left(a_t + a_x \, a\right) \Delta \\
& + \frac{1}{2} \gamma^2 \beta \left(a_{tt} + 2 a_{tx} \, a + a_{xx} \, a^2\right) \Delta^2 \\
& + \text{ higher order terms,}
\end{aligned}
$$

where $a$ and its partial derivatives are all evaluated at $(t, x)$. Hence, subtracting (2.5) with this expansion for $\Psi$ evaluated at $(t_n, y_n, \Delta_n)$ from the 3rd order truncated Taylor method (2.4) we get

$$
\begin{aligned}
(1 - \alpha - \beta) \, a \, \Delta_n \;+\;\; & \left(\frac{1}{2!} - \gamma \beta\right) \left(a_t + a_x \, a\right) \Delta_n^2 \\
+ \;\; & \frac{1}{2} \left(\frac{1}{3} - \gamma^2 \beta\right) \left(a_{tt} + 2 a_{tx} \, a + a_{xx} \, a^2\right) \Delta_n^3 \\
+ \;\; & \frac{1}{6} \left(a_t \, a_x + a_x^2 \, a\right) \Delta_n^3 \;+\; \text{ higher order terms,}
\end{aligned}
$$

where everything is now evaluated at $(t_n, y_n)$. The first two terms here drop out if we choose the weighting paramters $\alpha$, $\beta$ and $\gamma$ so that

(2.7) $\qquad\qquad\qquad \alpha + \beta = 1, \qquad \gamma \beta = \frac{1}{2}.$

In general it will not be possible to eliminate both of the $\Delta^3$ terms by a judicious choice of parameters $\beta$ and $\gamma$ because the second of these terms need not vanish identically. The parameter constraints (2.7) assure that a difference method with $\Psi$ given by (2.6) will have local discretization error of order 3 and hence global discretization error of order 2. Since one of the parameters in (2.7) can be chosen arbitrarily, this gives an infinite number of 2nd order difference schemes. Note that the first constraint in (2.7) assures that all of these methods are consistent, as defined in Section 1. The choice $\alpha = \beta = 1/2$, $\gamma = 1$ gives the Heun method (1.12), which is also called *the* 2nd order Runge-Kutta method.

We can use an analogous derivation for the 4th order Runge-Kutta method, starting with a weighted average over four points to approximate the right hand side of the differential equation. Now the comparison is made with the 5th order truncated Taylor method. The classical 4th *order Runge-Kutta method* is an explicit method given by

(2.8) $\qquad\qquad y_{n+1} = y_n + \frac{1}{6} \left\{ k_n^{(1)} + 2 \, k_n^{(2)} + 2 \, k_n^{(3)} + k_n^{(4)} \right\} \Delta_n$

where

$$k_n^{(1)} = a(t_n, y_n),$$

$$k_n^{(2)} = a\left(t_n + \frac{1}{2}\Delta_n, y_n + \frac{1}{2}k_n^{(1)}\Delta_n\right),$$

$$k_n^{(3)} = a\left(t_n + \frac{1}{2}\Delta_n, y_n + \frac{1}{2}k_n^{(2)}\Delta_n\right),$$

$$k_n^{(4)} = a\left(t_{n+1}, y_n + k_n^{(3)}\Delta_n\right).$$

When $a = a(t)$, a function of $t$ only, the increment $y_{n+1} - y_n$ in (2.8) is just a Simpson rule approximation of the definite integral

$$\int_{t_n}^{t_{n+1}} a(t)\, dt.$$

**PC-Exercise 8.2.2**    *Repeat PC-Exercise 8.2.1 using the 4th order Runge-Kutta method (2.8) with equal length time steps $\Delta = 2^{-2}, \ldots, 2^{-7}$*

Even higher order Runge-Kutta schemes have been derived. It turns out that the number of evaluations of the function $a$ needed for a $p$th order Runge-Kutta method is $p$ for $2 \leq p \leq 4$, $p + 1$ for $5 \leq p \leq 7$ and $p + 2$ for $p \geq 8$. The 4th order Runge-Kutta methods are the most commonly used, representing a good compromise between accuracy and computational effort.

Often multi-step methods do not require as many evaluations of the function $a$ per time step as one-step methods of the same order. An example is the 3-step Adams-Bashford method (1.14), which, essentially, requires the function $a$ to be evaluated only at a single point, namely $(t_n, y_n)$, for each iteration once the recursion procedure has been got going. This contrasts with the two evaluations per iteration needed by the Heun method (1.12), which is also a second order method. These considerations were of some importance before digital computers came into widespread usage and the calculations had to be done manually. They must still be borne in mind today, particularly for lengthy calculations, both for efficiency and to reduce roundoff error. Most explicit multi-step methods express $y_{n+1}$ as a linear combination of the values $y_i$ and $a(t_i, y_i)$ at the previous $k$ discretization times, where $k$ is fixed and denotes the number of steps of the method. In implicit methods the term $a(t_{n+1}, y_{n+1})$ also appears. For time steps of equal length $\Delta$ we write such multi-step methods in the general form

$$(2.9) \qquad y_{n+1} = \sum_{j=1}^{k} \alpha_j\, y_{n+1-j} + \sum_{j=0}^{k} \beta_j\, a\left(t_{n+1-j}, y_{n+1-j}\right) \Delta,$$

where the $\alpha_j$ and $\beta_j$ are given constants, with $\beta_0 = 0$ for an explicit scheme and $\beta_0 \neq 0$ for an implicit scheme. Most of these methods are derived by replacing the right hand side of the differential equation over $t_n \leq t \leq t_{n+1}$ by a polynomial passing through the points $(t_j, a(t_j, y_j))$ under consideration. Their local discretization error can be determined by comparison with truncated Taylor methods. The global discretization error also depends on the order of the starting routine, which should be, preferably, at least the same as that of the multi-step method itself.

Examples of multi-step methods are the *midpoint method*

$$(2.10) \qquad y_{n+1} = y_{n-1} + 2\,a(t_n, y_n)\,\Delta,$$

the *Milne method*

$$(2.11) \quad y_{n+1} = y_{n-3} + \frac{4}{3}\left\{2a(t_n, y_n) - a(t_{n-1}, y_{n-1}) + 2a(t_{n-2}, y_{n-2})\right\}\Delta$$

and the *Adams-Moulton method*

$$(2.12) \quad y_{n+1} = y_n + \frac{1}{12}\left\{5a(t_{n+1}, y_{n+1}) + 8a(t_n, y_n) - a(t_{n-1}, y_{n-1})\right\}\Delta.$$

The first two of these are explicit methods and the third is implicit. They have local discretization errors of orders 3, 5 and 4, respectively.

Note that an arbitrary choice of coefficients $\alpha_j$ and $\beta_j$ in (2.9) may result in an inconsistent method. Also, even if the coefficients are determined by an interpolating polynomial, a multi-step method may have some undesirable properties, such as being susceptible to numerical instabilities.

**PC-Exercise 8.2.3** *Calculate the discretization errors in using the Euler method (1.2) and the midpoint method (2.10) started with the Euler method to approximate the solution $x(t) = \frac{2}{3}e^{-3t} + \frac{1}{3}$ of the initial value problem*

$$\frac{dx}{dt} = -3x + 1, \qquad x(0) = 1$$

*over the interval $0 \le t \le 1$. Use time steps of equal length $\Delta = 0.1$ and plot on $x$ versus $t$ axes.*

The trapezoidal method (1.11) and the Adams-Moulton method (2.12) are examples of implicit difference methods. These are often more stable than their explicit counterparts. A root finding method such as the Newton method could be used at each step to calculate an approximation of the unknown value $y_{n+1}$. Another approach is to use an explicit method to predict an approximation $y_{n+1}^0$ to $y_{n+1}$, which is then inserted into the right hand side of the implicit method to calculate another approximation $y_{n+1}^1$. This correction procedure can then be repeated $l \ge 0$ times to produce a final approximation $y_{n+1}^{l+1}$. The resulting method is called a *predictor-corrector method*. A very simple example with $l = 0$ is the Heun or modified trapezoidal method (1.12), which uses the Euler method as its predictor. Besides providing the desired approximate value at each time step, a predictor- corrector method also gives an easy indication of the local dicretization error. This could be useful in choosing an appropriate, possibly varying, step size.

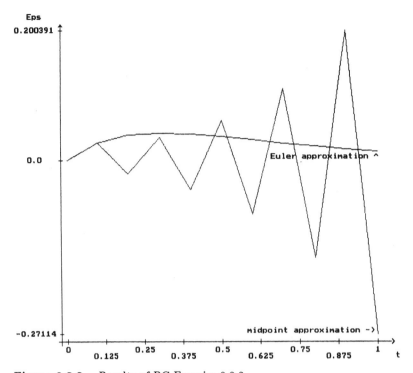

**Figure 8.2.2**     Results of PC-Exercise 8.2.3.

## 8.3     Consistency, Convergence and Stability

We usually not know the exact solution of an initial value problem that we are trying to approximate by a finite difference method. Then, to assure that an approximation will be reasonably accurate, we have to be able to keep the unknown discretization and roundoff errors under control and sufficiently small. We can use certain a priori information about the difference method, that is information obtainable without explicit knowledge of the exact solution, to tell us whether this is possible. In particular, we can check if the method is *consistent* with the differential equation, if the estimates of the global discretization error *converge* to zero with the maximum time step, and if the method is *stable*, that is if propagated errors remain bounded.

We shall assume for any differential equation

$$(3.1) \qquad \frac{dx}{dt} = a(t, x)$$

under consideration that the function $a = a(t, x)$ and its partial derivatives of sufficiently high order are continuous everywhere. For most of the common one-step methods

$$(3.2) \qquad y_{n+1} = y_n + \Psi(t_n, y_n, \Delta_n)\, \Delta_n$$

the increment function $\Psi = \Psi(t, x, \Delta)$ will then be continuous in all three variables and satisfy a local Lipschitz condition in $x$. Such methods are generally also *consistent* with the differential equation (3.1), that is they satisfy

$$(3.3) \qquad \Psi(t, x, 0) = a(t, x)$$

everywhere. Comparing (3.2) with a truncated Taylor method, we can then establish that the local discretization error (1.3) has order $p + 1$ for some $p \geq 1$; when (3.3) is violated we can only get a local discretization error of order 1.

By *convergence* of a one-step method (3.2) we mean that the global discretization error (1.4) converges to zero with the maximum time step $\Delta = \max_n \Delta_n$, that is

$$(3.4) \qquad \lim_{\Delta \downarrow 0} |e_{n+1}| = \lim_{\Delta \downarrow 0} |x(t_{n+1}; t_0, x_0) - y_{n+1}| = 0,$$

where $y_0 = x_0$, on any finite time interval $[t_0, T]$. Of more practical significance than convergence itself is the rate of convergence, which is provided by the order of the global discretization error. Since the local discretization error effects the global discretization error, we cannot expect to have convergence if (3.2) is not consistent. The following two theorems indicate the precise link between consistency and convergence, and between the orders of the local and global discretization errors. To simplify the proofs we shall assume that the increment function $\Psi$ of (3.2) satisfies a global Lipschitz condition

$$(3.5) \qquad |\Psi(t', x', \Delta') - \Psi(t, x, \Delta)| \leq K \left( |t' - t| + |x' - x| + |\Delta' - \Delta| \right)$$

in $(t, x, \Delta)$ and a global bound of the form

$$(3.6) \qquad |\Psi(t, x, 0)| \leq L$$

for all $(t, x)$, although it is possible to weaken these assumptions.

**Theorem 8.3.1** *A one-step method (3.2) with increment function $\Psi$ satisfying conditions (3.5) and (3.6) is convergent if and only if it is consistent.*

**Proof** It follows from the Lipschitz condition (3.5) that the differential equation

$$(3.7) \qquad \frac{dz}{dt} = \Psi(t, z, 0)$$

has a unique continuously differentiable solution $z(t) = z(t; t_0, x_0)$ with the initial value $z(t_0) = x_0$. Hence by the Mean Value Theorem there exists a $\theta_n$ with $0 < \theta_n < 1$ such that

$$(3.8) \qquad z(t_{n+1}) - z(t_n) = \Psi(t_n + \theta_n \Delta_n, z(t_n + \theta_n \Delta_n), 0) \, \Delta_n.$$

Writing $\bar{e}_n = y_n - z(t_n)$, where $y_n$ satisfies (3.2) with $y_0 = x_0$, we have

$$
\begin{aligned}
\bar{e}_{n+1} &= \bar{e}_n + \{\Psi(t_n, y_n, \Delta_n) - \Psi(t_n + \theta_n \Delta_n, z(t_n + \theta_n \Delta_n), 0)\} \Delta_n \\
&= \bar{e}_n + \{\Psi(t_n, y_n, \Delta_n) - \Psi(t_n, z(t_n), 0)\} \Delta_n \\
&\quad + \{\Psi(t_n, z(t_n), 0) - \Psi(t_n + \theta_n \Delta_n, z(t_n + \theta_n \Delta_n), 0)\} \Delta_n,
\end{aligned}
$$

from which we obtain

$$
(3.9) \qquad |\bar{e}_{n+1}| \leq |\bar{e}_n| + K \left(|\bar{e}_n| + \Delta_n\right) \Delta_n
$$
$$
+ K \left(\theta_n \Delta_n + |z(t_n) - z(t_n + \theta_n \Delta_n)|\right) \Delta_n
$$

by means of the Lipschitz condition (3.5). Using the Mean Value Theorem again and the bound (3.6) we have

$$
|z(t_n) - z(t_n + \theta_n \Delta_n)| = \left|\Psi(t_n + \bar{\theta}_n \theta_n \Delta_n, z(t_n + \bar{\theta}_n \theta_n \Delta_n), 0)\right| \theta_n \Delta_n \leq L \theta_n \Delta_n
$$

for some $0 < \bar{\theta}_n < 1$. Inserting this into (3.9), we then get

$$
(3.10) \qquad |\bar{e}_{n+1}| \leq (1 + K\Delta) |\bar{e}_n| + K (L + 2) \Delta^2,
$$

where $\Delta = \max_n \Delta_n$. We can use induction to show that

$$
|\bar{e}_n| \leq (L + 2) \left(e^{K(T - t_0)} - 1\right) \Delta
$$

on an interval $[t_0, T]$, from which we conclude that the approximations $y_n$ generated by the one-step method (3.2) converge to the solution $z(t) = z(t; t_0, x_0)$ of (3.7) on $[t_0, T]$.

Assuming consistency, the differential equations (3.1) and (3.7) are the same, so by the uniqueness of solutions of initial value problems $z(t) \equiv x(t)$ for $t_0 \leq t \leq T$. From the above considerations we have thus established convergence of the one-step method (3.2).

Assuming convergence, we have $z(t) \equiv x(t)$ for $t_0 \leq t \leq T$. If there were a point $(t_0, x_0)$ where $a(t_0, x_0) \neq \Psi((t_0, x_0, 0))$, we would have

$$
\frac{dx}{dt}(t_0) = a(t_0, x_0) \neq \Psi((t_0, x_0, 0)) = \frac{dz}{dt}(t_0),
$$

which contradicts the fact that $z(t) \equiv x(t)$. Hence the consistency condition (3.3) must hold.

This completes the proof of Theorem 8.3.1. $\square$

**Theorem 8.3.2**    *A one-step method (3.2) with increment function $\Psi$ satisfying the global Lipschitz condition (3.5) and with local discretization error of order $p + 1$ has global discretization error of order $p$.*

**Proof**    Let $x(t) = x(t; t_0, x_0)$ be the solution of the initial value problem (3.1) and let $y_n$ be generated by (3.2) with $y_0 = x_0$. Then the global discretization error (1.4) satisfies

$$
\begin{aligned}
e_{n+1} &= y_{n+1} - x(t_{n+1}) \\
&= e_n + \Psi(t_n, y_n, \Delta_n)\,\Delta_n + x(t_n) - x(t_{n+1}) \\
&= e_n + \{\Psi(t_n, y_n, \Delta_n) - \Psi(t_n, x(t_n), \Delta_n)\}\,\Delta_n \\
&\quad + \{\Psi(t_n, x(t_n), \Delta_n)\,\Delta_n + x(t_n) - x(t_{n+1})\}\,,
\end{aligned}
$$

where the very last term is the local discretization error. From the global Lipschitz condition (3.5) and the assumption that the local discretization error is of order $p+1$ we obtain

$$
\begin{aligned}
|e_{n+1}| &\le |e_n| + K\,|e_n|\,\Delta_n + D\,\Delta_n^{p+1} \\
&\le (1 + K\Delta)\,|e_n| + D\,\Delta^{p+1},
\end{aligned}
$$

where $\Delta = \max_n \Delta_n$ and $D$ is some positive constant, from which it follows that

$$
|\bar{e}_n| \le \frac{D}{K}\left(e^{K(T-t_0)} - 1\right)\Delta^p
$$

on the interval $[t_0, T]$. The global discretization error is thus of order $p$. $\square$

**Exercise 8.3.3**    *Show that the increment function $\Psi(t, x, \Delta)$ of the Heun method (1.12) satisfies a global Lipschitz condition (3.5) in $(t, x, \Delta)$ when $a(t, x)$ satisfies a global Lipschitz condition in $(t, x)$. Also, show that the Heun method is consistent and hence convergent with global discretization error of order 2.*

We may still encounter difficulties when trying to implement a difference method which is known to be convergent. For example, the differential equation

$$
(3.11) \qquad\qquad \frac{dx}{dt} = -16\,x
$$

has exact solutions $x(t) = x_0 e^{-16t}$, which all converge very rapidly to zero. For this differential equation the Euler method with constant time step $\Delta$,

$$
y_{n+1} = (1 - 16\Delta)\,y_n,
$$

has exact iterates $y_n = (1 - 16\Delta)^n\,y_0$. If we choose $\Delta > 2^{-3}$ these iterates oscillate with increasing amplitude instead of converging to zero like the exact solutions of (3.11). This is a simple example of a *numerical instability*, which, in this particular case, we can overcome simply by taking the time step $\Delta < 2^{-3}$. For some other methods, such as the midpoint method (2.10) investigated in PC-Exercise 8.2.3, the numerical instabilities persist no matter how small we take $\Delta$. The structure of these methods can make them intrinsically unstable, causing small errors such as roundoff errors to grow rapidly and ultimately rendering the calculations useless.

The idea of numerical stability of a one-step method is that errors will remain bounded with respect to an initial error for any differential equation (3.1) with right hand side $a(t, x)$ satisfying a Lipschitz condition. To be specific, we say that a one-step method (3.2) is *numerically stable* if for each interval $[t_0, T]$ and differential equation (3.1) with $a(t, x)$ satisfying a Lipschitz condition there exist positive constants $\Delta_0$ and $M$ such that

$$(3.12) \qquad\qquad |y_n - \tilde{y}_n| \le M \, |y_0 - \tilde{y}_0|$$

for $n = 0, 1, \ldots, n_T$ and any two solutions $y_n$, $\tilde{y}_n$ of (3.2) corresponding to any time discretizations with $\max_n \Delta_n < \Delta_0$. The constants $\Delta_0$ and $M$ here may also depend on the particular time interval $t_0 \le t \le T$ in addition to the differential equation under consideration. (3.12) is analogous to the continuity in initial conditions, uniformly on finite time intervals, of the solutions of the differential equation (3.1). The following result thus comes as no surprise.

**Theorem 8.3.4**    *A one-step method (3.2) is numerically stable if the increment function $\Psi$ satisfies a global Lipschitz condition (3.5).*

The commonly used one-step methods are numerically stable. However, the constant $M$ in (3.12) may be quite large. For example, if we replace the minus sign by a plus sign in the differential equation (3.11), we obtain

$$y_n - \tilde{y}_n = (1 + 16\Delta)^n \, (y_0 - \tilde{y}_0)$$

for the Euler method. The numerical stability condition (3.12) requires a bound like $e^{16(T-t_0)}$ for $M$, in contrast with $M \le 1$, provided $\Delta_0 < 2^{-3}$, for the original differential equation. The difference is due to the fact that the solutions of the modified differential equation are diverging exponentially fast, whereas those of the original are converging exponentially fast. In both cases the Euler method keeps the error under control, but in the former case the initial error must be considerably smaller if it is to remain small.

**Exercise 8.3.5**    *Prove Theorem 8.3.4.*

To ensure that the errors in the Euler method for (3.11) do not grow, that is the bound $M \le 1$ in (3.12), we need to take step sizes less than $2^{-3}$. This may seem inordinately small given that the differential equation itself is very stable. The situation does not improve if we use the higher order Heun method (1.12). However, the implicit trapezoidal method (1.11) offers a substantial improvement. In this case it is

$$(3.13) \qquad\qquad y_{n+1} = y_n + \frac{1}{2} \left\{ -16 \, y_n - 16 \, y_{n+1} \right\} \Delta,$$

which we can solve explicitly to get

$$y_{n+1} = \left( \frac{1 - 8\Delta}{1 + 8\Delta} \right) y_n.$$

Here

$$(3.14) \qquad\qquad \left| \frac{1 - 8\Delta}{1 + 8\Delta} \right| < 1$$

for any $\Delta > 0$. For a nonlinear differential equation we usually cannot solve an implicit method algebraically for $y_{n+1}$ as in (3.13). Nevertheless this example highlights a significant advantage of implicit methods, which sometimes makes the additional work needed to solve numerically for $y_{n+1}$ worthwhile.

In the preceding discussion we tried to ensure that the error would not grow over an infinite time horizon. This leads to the idea of asymptotic numerical stability. We shall say that a one-step method (3.1) is *asymptotically numerically stable* for a given differential equation if there exist positive constants $\Delta_a$ and $M$ such that

$$(3.15) \qquad \lim_{n \to \infty} |y_n - \bar{y}_n| \le M |y_0 - \bar{y}_0|$$

for any two solutions $y$ and $\bar{y}$ of (3.2) corresponding to any time discretization with $\max_n \Delta_n < \Delta_a$.

It is easy to see that the Euler method is asymptotically numerically stable for the differential equation (3.11) with $\Delta_a \le 2^{-3}$, whereas the implicit trapezoidal method (3.13) is asymptotically numerically stable for this differential equation without any restriction on $\Delta_a$. On the other hand the Euler method is not asymptotically numerically stable for the differential equation $\dot{x} = 16x$ for any $\Delta_a > 0$.

Knowing just that a one-step method is numerically stable does not tell us how to pick an appropriate step size $\Delta$. In fact, the answer will depend very much on the particular differential equation under consideration. To obtain an indication of suitable values of $\Delta$ we consider a class of test equations. These are the complex-valued linear differential equations

$$(3.16) \qquad \frac{dx}{dt} = \lambda\, x,$$

with $\lambda = \lambda_r + \imath\lambda_i$, which have oscillating solutions when $\lambda_i \neq 0$. We can obviously write (3.16) equivalently as a 2-dimensional differential equation

$$\frac{d}{dt} \left( \begin{array}{c} x^1 \\ x^2 \end{array} \right) = \left[ \begin{array}{cc} \lambda_r & -\lambda_i \\ \lambda_i & \lambda_r \end{array} \right] \left( \begin{array}{c} x^1 \\ x^2 \end{array} \right)$$

where $x = x^1 + \imath x^2$. The suitable values of the step size $\Delta > 0$ are expressed in terms of the *region of absolute stability* for the method, consisting of the complex numbers $\lambda\Delta$ for which an error in $y_0$ at $t_0$ will not grow in subsequent iterations of the method applied to the differential equation (3.16). Essentially, these are the values of $\lambda$ and $\Delta$ producing a bound $M \le 1$ in (3.12). For the Euler method we thus require

$$|1 + \lambda\,\Delta| \le 1,$$

so its region of absolute stability is the unit disc in the complex plane centered on $z = -1 + 0\imath$.

**Exercise 8.3.6** *Determine and sketch the region of absolute stability for the trapezoidal method (1.11).*

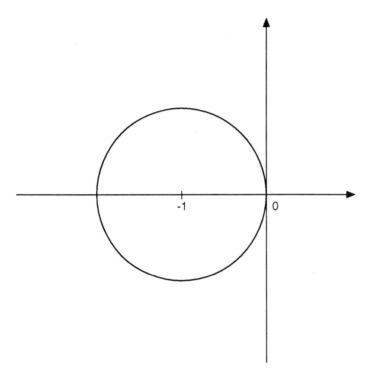

**Figure 8.3.1**   Stability region for the Euler method.

We shall now consider a 2-dimensional linear differential equation

$$(3.17) \qquad \frac{d}{dt}\begin{pmatrix} x^1 \\ x^2 \end{pmatrix} = \begin{bmatrix} -\alpha_1 & 0 \\ 0 & -\alpha_2 \end{bmatrix}\begin{pmatrix} x^1 \\ x^2 \end{pmatrix}$$

with initial value $(x_0^1, x_0^2) = (1, 1)$, where the two eigenvalues of the coefficient matrix are negative and very different, that is with

$$0 \le \alpha_2 \ll \alpha_1.$$

The components of (3.17) are uncoupled, so they can be solved separately to give

$$(3.18) \qquad x^1(t) = e^{-\alpha_1 t}, \qquad x^2(t) = e^{-\alpha_2 t}.$$

Since $\alpha_1$ is much larger than $\alpha_2$ the first component shows a very fast exponential decay in comparison with the second, that is the relaxation time of the first component is very much smaller than that of the second. In other words the two components have widely differing time scales. In the literature such a system of equations is often called a *stiff system*. In the general $d$-dimensional case we shall say that a linear system is *stiff* if the real parts of the eigenvalues $\lambda_1, \ldots, \lambda_d$ of the coefficient matrix satisfy

$$\max_{k=1,\ldots,d} \mathrm{Re}\,(\lambda_k) \gg \min_{k=1,\ldots,d} \mathrm{Re}\,(\lambda_k).$$

Now, if we apply the Euler method (1.2) to (3.17), for the first component to remain within the region of absolute stability we need a step size

$$\Delta \leq \frac{2}{\alpha_1}.$$

We saw in Figure 8.1.3 of Section 1 that there is a lower bound on the step size for which the influence of the roundoff error to remain acceptable. But the upper bound $2/\alpha_1$ might already be too small to allow for the control over roundoff errors in the second component. Thus, the Euler scheme may not be applicable for a stiff system.

A much more stable result is shown when we apply the *implicit Euler scheme*

(3.19) $$y_{n+1} = y_n + a\,(t_{n+1}, y_{n+1})\,\Delta$$

to the test equation (3.16). Using similar notation as above we obtain

$$y_n = (1 - \lambda\Delta)^{-n}\,y_0$$

and, hence, for $\lambda$ with $\mathrm{Re}(\lambda) = \lambda_r < 0$ and all $\Delta > 0$ we have

$$|y_n - \bar{y}_n| \leq |y_0 - \bar{y}_0|$$

for all $n = 0, 1, \ldots$ and any two solutions $y_n$, $\bar{y}_n$ of (3.19). Thus, the implicit Euler method (3.19) applied to the stiff system (3.17) would still behave stably in its first component when $\Delta > 2/\alpha_1$.

We shall say that a numerical method is *A-stable* if its region of absolute stability contains all of the left half of the complex plane, that is all $\lambda\Delta$ with $\mathrm{Re}(\lambda) < 0$ and $\Delta > 0$. Hence the implicit Euler method (3.19) is A-stable, whereas the Euler method (1.2) is not.

**Exercise 8.3.7** *Check whether or not the trapezoidal method (1.11) is A-stable.*

We shall conclude this section with a few remarks on consistency, convergence and stability of multi-step methods, for which matters are somewhat more complicated than for one-step methods. The definition of convergence for a multi-step method assumes that the starting values are exact, although in practice these will be calculated approximately by means of a one-step method. The idea of stability is similar to that for one-step methods, but now some new phenomena can occur. Suppose we have a $k$-step method

(3.20) $$\sum_{j=0}^{k} (\alpha_{n+1-j}\,y_{n+1-j} + \beta_{n+1-j}\,a(t_{n+1-j}, y_{n+1-j})\Delta) = 0.$$

For the linear differential equation (3.15) this is a linear recursion

$$(3.21) \qquad \sum_{j=0}^{k} (\alpha_{n+1-j} + \beta_{n+1-j} \lambda \Delta) \, y_{n+1-j} = 0,$$

which is also satisfied by iterated errors $e_n = y_n - \tilde{y}_n$. To solve (3.21) we try solutions of the form $y_n = \xi^n$ and find that $\xi$ must be a root, possibly complex valued, of the polynomial equation

$$(3.22) \qquad \sum_{i=0}^{k} (\alpha_i + \beta_i \lambda \Delta) \, \xi^i = 0.$$

Any errors introduced will thus die out if all of the roots lie within the unit circle in the complex plane, that is have modulus $|\xi| < 1$. One of the roots will be approximately equal to $e^{\lambda \Delta}$ and the corresponding iterates $y_n = x_0 \, \xi^n$ correspond to the differential equation solution values $x(t_n) = x_0 e^{\lambda t_n}$. The problem now is that (3.22) may have other roots lying outside of the unit circle and these may lead to iterates of the multi-step method increasing in magnitude when the differential equation has no such solutions. We say that the multi-step method (3.20) is *stable* when all of the roots of (3.22) lie within the unit complex circle for $\Delta$ sufficiently small. A necessary and sufficient condition for this is that the roots of the polynomial

$$\sum_{i=0}^{k} \alpha_i \, \xi^i = 0$$

lie within the unit complex circle, or possibly also on the unit circle if a root is simple. The term *strong stability* is used if all roots except $\xi = 1$ lie inside the unit circle and *weak stability* if other roots also lie on the circle. For example, the Adams-Bashford method (1.14) is strongly stable, whereas the midpoint (2.10) and the Milne (2.11) methods are weakly stable. The presence of the extra roots on the unit circle means that (3.22) may have roots lying outside the unit circle no matter how small $\Delta$ is taken, which can lead to numerical instabilities. For example, the midpoint method (2.10) has roots

$$\xi = \frac{1}{2} \lambda \Delta \pm \sqrt{1 + \left(\frac{1}{2} \lambda \Delta\right)^2},$$

one of which has modulus greater than 1. Finally, as a partial analogue of Theorem 8.3.1, we remark that it can be shown that a multi-step method is convergent if it is consistent and stable.

**Exercise 8.3.8**    *Determine the polynomials (3.22) for the Adams-Bashford method (1.14) and the Adams-Moulton method (2.12).*

# 8.4  Roundoff Error

Roundoff errors occur because, in practice, arithmetic operations can only be carried out to a finite number of significant decimal places. In principle we could determine the roundoff error of each calculation, and hence the accumulated roundoff error, exactly, though this is infeasible in all but the simplest situations and we have to use estimates instead. Assuming constant roundoff error $r$ at each step, in Section 1 we derived a theoretical upper bound proportional to $r/\Delta$, where $\Delta$ is the maximum time step, for the Euler method. This assumption is certainly not true, but the implication from it that there is a minimum time step $\Delta_{min}$ below which the total error will begin to increase is consistent with what actually happens in numerical calculations, as we saw in PC-Exercise 8.1.2.

More realistic estimates of the accumulated roundoff error can be determined from a statistical analysis, assuming that the local roundoff errors are independent, identically distributed random variables. It is commonly assumed that they are uniformly distributed over the interval

(4.1) $$\left[-5 \times 10^{-(s+1)},\ 5 \times 10^{-(s+1)}\right],$$

where $s$ is the number of significant decimal places used. To check the appropriateness of this distribution we could repeat the calculations using double precision arithmetic and use the difference of single and double precision results to represent the roundoff error. If double precision arithmetic is not available we can simulate the same effect by using arithmetic to $s$ decimal places, say $s = 4$, instead of single precision and the computer's prescribed precision instead of double precision.

**PC-Exercise 8.4.1**  *Calculate* 300 *iterates of*

$$y_{n+1} = \frac{\pi}{3}\, y_n$$

*with initial value $y_0 = 0.1$ using the prescribed arithmetic of the PC, at each step rounding the value of $y_{n+1}$ obtained to four significant figures. Plot the relative frequencies of the roundoff errors in a histogram on the interval*

$$[-5 \times 10^{-5}, 5 \times 10^{-5}]$$

*using* 40 *equal subintervals.*

If the local roundoff errors $r_n$ take values in the interval (4.1), then after $N$ calculations the accumulated roundoff error

$$R_N = \sum_{n=1}^{N} r_n$$

would lie in the interval

$$\left[-5N \times 10^{-(s+1)}, 5N \times 10^{-(s+1)}\right].$$

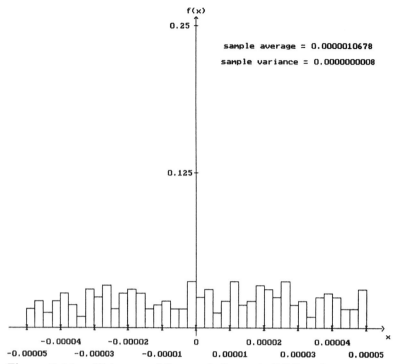

**Figure 8.4.1**    Histogram of the roundoff error in PC-Exercise 8.4.1.

After $N = 10^s$ calculations this is $[-0.5, +0.5]$, so all decimal places of accuracy may be lost. However, this worst case scenario is highly unlikely to occur. If the $r_n$ are uniformly distributed over the interval (4.1) they have mean and variance

$$\mu = E(r_n) = 0, \qquad \sigma^2 = \text{Var}(r_n) = \frac{1}{12} 10^{-2s}.$$

Thus, if they are also independent, the accumulated roundoff error has mean and variance

$$E(R_N) = 0, \qquad \text{Var}(R_N) = N \sigma^2.$$

By the Central Limit Theorem (see (1.5.9)) the normalized random variables $Z_N = R_N / \sigma \sqrt{N}$ are approximately standard Gaussian for large $N$. From this, as in Section 9 of Chapter 1, we can conclude that the values of $R_N$ lie with probability 0.95 in the interval

$$\left[ -1.96 \times 10^{-s} \sqrt{N/12}, \ 1.96 \times 10^{-s} \sqrt{N/12} \right]$$

when $N$ is large. The ratio of $1.96 \times 10^{-s} \sqrt{N/12}$ to $5N \times 10^{-(s+1)}$ is approximately $1/\sqrt{N}$, so for large $N$ the accuracy is in fact considerably better than predicted by the worst case scenario above. Of course, it may be much worse in some instances, but these occur with small probabilities.

**PC-Exercise 8.4.2** *Use the Euler method with equal time steps $\Delta = 2^{-2}$ for the differential equation*

$$\frac{dx}{dt} = x$$

*over the interval $0 \leq t \leq 1$ with $N = 10^3$ different initial values $x(0)$ between $0.4$ and $0.6$. Use both four significant figure arithmetic and the prescribed arithmetic of the PC and determine the final accumulative roundoff error $R_{1/\Delta}$ in each case, plotting them in a histogram on the interval $[-5 \times 10^{-4}, 5 \times 10^{-4}]$ with 40 equal subintervals. In addition, calculate the sample mean and sample variance of the $R_{1/\Delta}$ values.*

As a final comment we remark that the roundoff error may be considered as being independent of the discretization error.

**PC-Exercise 8.4.3** *Repeat PC-Exercise 8.4.2 with $N = 200$ and with time steps $\Delta = 2^{-2}$, $2^{-3}$, $2^{-4}$ and $2^{-5}$, determining $R_{1/\Delta}$ in each case. Plot the 90% confidence intervals for the mean value of the error against $\Delta$.*

# Chapter 9

# Introduction to Stochastic Time Discrete Approximation

To introduce the reader to the main issues concerning the time discrete approximation of Ito processes, we shall examine the stochastic Euler scheme in some detail in this chapter. We shall consider some examples of typical problems that can be handled by the simulation of approximating time discrete trajectories. In addition, general definitions for time discretizations and time discrete approximations will be given, and the strong and weak convergence criteria for time discrete approximations introduced. These concepts will all be developed more extensively in the subsequent chapters.

## 9.1 The Euler Approximation

One of the simplest time discrete approximations of an Ito process is the *Euler approximation*, or the *Euler-Maruyama approximation* as it is sometimes called. We shall consider an Ito process $X = \{X_t, t_0 \leq t \leq T\}$ satisfying the scalar stochastic differential equation

$$(1.1) \qquad dX_t = a(t, X_t)\, dt + b(t, X_t)\, dW_t$$

on $t_0 \leq t \leq T$ with the initial value

$$(1.2) \qquad X_{t_0} = X_0.$$

For a given discretization $t_0 = \tau_0 < \tau_1 < \cdots < \tau_n < \cdots < \tau_N = T$ of the time interval $[t_0, T]$, an *Euler approximation* is a continuous time stochastic process $Y = \{Y(t), t_0 \leq t \leq T\}$ satisfying the iterative scheme

$$(1.3) \quad Y_{n+1} = Y_n + a(\tau_n, Y_n)(\tau_{n+1} - \tau_n) + b(\tau_n, Y_n)\left(W_{\tau_{n+1}} - W_{\tau_n}\right),$$

for $n = 0, 1, 2, \ldots, N-1$ with initial value

$$(1.4) \qquad Y_0 = X_0,$$

where we have written

$$(1.5) \qquad Y_n = Y(\tau_n)$$

for the value of the approximation at the discretization time $\tau_n$. We shall also write

(1.6) $$\Delta_n = \tau_{n+1} - \tau_n$$

for the $n$th time increment and call

(1.7) $$\delta = \max_n \Delta_n$$

the *maximum time step*. For much of this chapter we shall consider equidistant discretization times

(1.8) $$\tau_n = t_0 + n\,\delta$$

with $\delta = \Delta_n \equiv \Delta = (T - t_0)/N$ for some integer $N$ large enough so that $\delta \in (0,1)$.

When the diffusion coefficient is identically zero, that is when $b \equiv 0$, the stochastic iterative scheme (1.3) reduces to the deterministic Euler scheme (8.1.2) for the ordinary differential equation

$$\frac{dx}{dt} = a(t,x).$$

The sequence $\{Y_n,\ n = 0,\ 1,\dots,N\}$ of values of the Euler approximation (1.3) at the instants of the time discretization $(\tau)_\delta = \{\tau_n,\ n = 0,\ 1,\ \dots,\ N\}$ can be computed in a similar way to those of the deterministic case. The main difference is that we now need to generate the random increments

(1.9) $$\Delta W_n = W_{\tau_{n+1}} - W_{\tau_n},$$

for $n = 0,\ 1,\ \dots,\ N-1$, of the Wiener process $W = \{W_t,\ t \geq 0\}$. From Chapters 1 and 2 we know that these increments are independent Gaussian random variables with mean

(1.10) $$E\left(\Delta W_n\right) = 0$$

and variance

(1.11) $$E\left((\Delta W_n)^2\right) = \Delta_n.$$

We can use a sequence of independent Gaussian pseudo-random numbers generated by one of the random number generators introduced in Section 3 of Chapter 1 for the increments (1.9) of the Wiener process.

For simpler notation we shall often write

(1.12) $$f = f(\tau_n, Y_n)$$

for each function $f$ defined on $\Re^+ \times \Re^d$ and $n = 0,\ 1,\ \dots,\ N-1$ when no misunderstanding is possible. We can then rewrite the Euler scheme (1.3) in the abbreviated form

(1.13) $$Y_{n+1} = Y_n + a\,\Delta_n + b\,\Delta W_n,$$

for $n = 0, 1, \ldots, N - 1$. Usually, we shall leave unstated the initial condition (1.4).

The recursive structure of the Euler scheme, which evaluates approximate values to the Ito process at the discretization instants only, is the key to its successful implementation on a digital computer. In this book we shall focus on time discrete approximations with such a recursive structure. We shall use the term *scheme* to denote a recursive algorithm which provides the values of a time discrete approximation at the given discretization instants. We emphasize that we shall always consider a time discrete approximation to be a continuous time stochastic process defined on the whole interval $[t_0, T]$, although we shall mainly be interested in its values at the discretization times.

For a given time discretization the Euler scheme (1.3) determines values of the approximating process at the discretization times only. If required, values can then be determined at the intermediate instants by an appropriate interpolation method. The simplest is the *piecewise constant interpolation* with

$$(1.14) \qquad Y(t) = Y_{n_t}$$

for $t \in \Re^+$, where $n_t$ is the integer defined by

$$(1.15) \qquad n_t = \max\{n = 0, 1, \ldots, N : \tau_n \leq t\},$$

that is the largest integer $n$ for which $\tau_n$ does not exceed $t$. However, the *linear interpolation*

$$(1.16) \qquad Y(t) = Y_{n_t} + \frac{t - \tau_{n_t}}{\tau_{n_t+1} - \tau_{n_t}} (Y_{n_t+1} - Y_{n_t})$$

is often used because it is continuous and simple.

In general, the sample paths of an Ito process inherit the irregularity of the sample paths of its driving Wiener process, in particular their nondifferentiability. It will not be possible to reproduce the finer structure of such paths on a computer, so we shall concentrate on the values of a time discrete approximation at the given discretization instants.

**Exercise 9.1.1**    *Derive the distribution of the random variable $Y_{n_t}$ at any time $t \geq 0$ for the Euler approximation with equidistant discretization times of a scalar Ito process with constant drift and diffusion coefficients.*

## 9.2   Example of a Time Discrete Simulation

To illustrate various aspects of the simulation of a time discrete approximation of an Ito process we shall examine a simple example in some detail. We shall consider the Ito process $X = \{X_t, t \geq 0\}$ satisfying the linear stochastic differential equation

(2.1)                         $$dX_t = aX_t\, dt + bX_t\, dW_t$$

for $t \in [0, T]$ with the initial value $X_0 \in \Re^1$.

This is an Ito process with drift

(2.2)                               $$a(t, x) = ax$$

and diffusion coefficient

(2.3)                               $$b(t, x) = bx.$$

We know from (4.4.6) that (2.1) has the explicit solution

(2.4)               $$X_t = X_0 \exp\left(\left(a - \frac{1}{2}b^2\right)t + b\,W_t\right)$$

for $t \in [0, T]$ and the given Wiener process $W = \{W_t,\ t \geq 0\}$. Knowing the solution (2.4) explicitly gives us the possibility of comparing the Euler approximation with the exact solution and to calculate the error.

To simulate a trajectory of the Euler approximation for a given time discretization we simply start from the initial value $Y_0 = X_0$, and proceed recursively to generate the next value

(2.5)                   $$Y_{n+1} = Y_n + a\,Y_n\,\Delta_n + b\,Y_n\,\Delta W_n$$

for $n = 0, 1, 2, \ldots$ according to the Euler scheme (1.13) with drift and diffusion coefficients (2.2) and (2.3), respectively. Here $\Delta W_n$ is the $N(0; \Delta_n)$ distributed Gaussian increment of the Wiener process $W$ over the subinterval $\tau_n \leq t \leq \tau_{n+1}$.

For comparison, we can use (2.4) to determine the corresponding values of the exact solution for the same sample path of the Wiener process, obtaining

(2.6)           $$X_{\tau_n} = X_0 \exp\left(\left(a - \frac{1}{2}b^2\right)\tau_n + b\sum_{i=1}^{n}\Delta W_{i-1}\right).$$

**PC-Exercise 9.2.1**    *Generate equidistant Euler approximations on the time interval $[0, 1]$ with equal step size $\Delta = 2^{-2}$ for the Ito process $X$ satisfying (2.1) with $X_0 = 1.0$, $a = 1.5$ and $b = 1.0$. Plot both the linearly interpolated approximation and the exact solution for the same sample path of the Wiener process. See PC-Exercise 4.4.2.*

We need to be careful when writing down a time discrete scheme such as (2.5) to make sure that the resulting expressions are meaningful. For instance, difficulties may arise because the increments in the noise can take extremely large values of either sign, even though this can occur only with very small probability. This will be more serious for nonlinear equations such as (4.4.37) and (4.4.38) than for the linear equation (2.1).

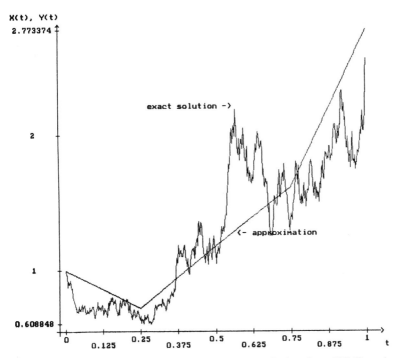

**Figure 9.2.1**    Euler approximation and exact solution from PC-Exercise 9.2.1.

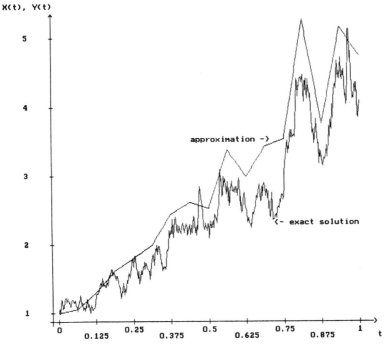

**Figure 9.2.2**    The Euler approximation for the smaller step size $\Delta = 2^{-4}$.

Figure 9.2.1 illustrates a typical output for PC-Exercise 9.2.1.

It is of no surprise that the Euler approximation differs from the Ito process. However, we may expect a closer resemblence if we use a smaller step size.

**PC-Exercise 9.2.2**    *Repeat PC-Exercise 9.2.1 with step size* $\Delta = 2^{-4}$.

From Figure 9.2.2 we note that the Euler approximation is closer to the Ito process at the endpoint $T = 1$ when the step size is smaller. We would hope that the Euler approximation for a finer time discretization would be closer to the Ito process in some useful sense.

So far we have not specified a criterion to judge the quality, that is the accuracy, of a time discrete approximation. Obviously, such a criterion should reflect the main goal of a practical simulation. It turns out that there are two basic types of tasks connected with the simulation of solutions of stochastic differential equations. The first occurs in situations where a good *pathwise approximation* is required, for instance in direct simulations, filtering problems or testing statistical estimators. In the second type interest focuses on *approximating expectations* of functionals of the Ito process, such as its probability distribution and its moments. This is relevant in many practical problems because, usually, such functionals cannot be determined analytically. In the following sections we shall look at both of these types of objectives in the context of the above example.

However, we emphasize that one should always be careful in interpreting the results of individual numerical simulations as they may differ significantly from the true solution. A statistical analysis of many different simulations is required for a meaningful comparison.

## 9.3   Pathwise Approximations

Usually we do not know the solutions of a stochastic differential equation explicitly, so we use simulations to try to discover something about them. If we do happen to know a solution explicitly, then, as in the PC-Exercises 9.2.1 and 9.2.2, we can calculate the error of an approximation using the *absolute error criterion*. This is simply the expectation of the absolute value of the difference between the approximation and the Ito process at the time $T$, that is

$$(3.1) \qquad\qquad \epsilon = E\left(|X_T - Y(T)|\right),$$

which gives a measure of the pathwise closeness at the end of the time interval $[0, T]$. When the diffusion coefficient $b \equiv 0$ and the initial value is deterministic, randomness has no effect and the expectation in (3.1) is superfluous. The criterion (3.1) then reduces to the deterministic absolute error criterion, that is the global truncation error (8.1.4).

We shall use the example of the preceding section to examine the absolute error criterion more closely. Rather than derive a theoretical estimate for the absolute error here, we shall try to estimate it statistically using computer experiments. To this end we shall repeat $N$ different simulations of sample paths of the Ito process and their Euler approximation corresponding to the same sample paths of the Wiener process. We shall denote the values at time $T$ of the $k$th simulated trajectories by $X_{T,k}$ and $Y_{T,k}$, respectively, and estimate the absolute error by the statistic

$$(3.2) \qquad \hat{\epsilon} = \frac{1}{N} \sum_{k=1}^{N} |X_{T,k} - Y_{T,k}|.$$

**PC-Exercise 9.3.1** *Simulate $N = 25$ trajectories of the Ito process $X$ satisfying (2.1) with $X_0 = 1.0$, $a = 1.5$, $b = 1.0$ and their Euler approximations with equidistant time steps of step size $\Delta = 2^{-4}$ corresponding to the same sample paths of the Wiener process on the time interval $[0, T]$ for $T = 1$. Evaluate the statistic $\hat{\epsilon}$ defined by (3.2). Repeat this for step sizes $\Delta = 2^{-5}$, $2^{-6}$ and $2^{-7}$, and form a table of the corresponding $\Delta$ and $\hat{\epsilon}$ values.*

| $\Delta$ | $2^{-4}$ | $2^{-5}$ | $2^{-6}$ | $2^{-7}$ |
|---|---|---|---|---|
| $\hat{\epsilon}$ | 0.5093 | 0.4446 | 0.3265 | 0.2295 |

**Table 9.3.1** Absolute errors $\hat{\epsilon}$ for different step lengths $\Delta$.

We list our results for PC-Exercise 9.3.1 in Table 9.3.1 and note the improvement in the estimate $\hat{\epsilon}$ of the absolute error with decreasing step size $\Delta$.

**PC-Exercise 9.3.2** *Repeat PC-Exercise 9.3.1 using a different seed, that is initial value, for the random number generator. (This may be done automatically by the PC, but can be easily programed if not).*

We combine our results from PC-Exercises 9.3.1 and 9.3.2 in Table 9.3.2, writing $\hat{\epsilon}_1$ and $\hat{\epsilon}_2$, respectively, for the corresponding absolute error statistic (3.2).

| $\Delta$ | $2^{-4}$ | $2^{-5}$ | $2^{-6}$ | $2^{-7}$ |
|---|---|---|---|---|
| $\hat{\epsilon}_1$ | 0.5093 | 0.4446 | 0.3265 | 0.2292 |
| $\hat{\epsilon}_2$ | 0.4692 | 0.3788 | 0.2234 | 0.1477 |

**Table 9.3.2** Absolute errors $\hat{\epsilon}_1$ and $\hat{\epsilon}_2$ for different step lengths $\Delta$.

Comparing the results in Table 9.3.2 we see in both cases that the estimate of the absolute error decreases with decreasing step size. However, these estimates are random variables and take different values in the two batches. For large $N$ we know from the Central Limit Theorem, mentioned in Section 5 of Chapter 1, that the error $\hat{\epsilon}$ becomes asymptotically a Gaussian random variable and converges in distribution to the nonrandom expectation $\epsilon$ of the absolute value of the error as $N \to \infty$.

In practice it is impossible to generate an infinite number of trajectories.

However, we can estimate the variance $\sigma_\epsilon^2$ of $\hat{\epsilon}$ and then use it to construct a confidence interval for the absolute error $\epsilon$. To do this we arrange the simulations into $M$ batches of $N$ simulations each and estimate the variance of $\hat{\epsilon}$ in the following way. We denote by $Y_{T,k,j}$ the value of the $k$th generated Euler trajectory in the $j$th batch at time $T$ and by $X_{T,k,j}$ the corresponding value of the Ito process. The average errors

$$(3.3) \qquad \hat{\epsilon}_j = \frac{1}{N} \sum_{k=1}^{N} |X_{T,k,j} - Y_{T,k,j}|$$

of the $M$ batches $j = 1, 2, \ldots, M$ are then independent and approximately Gaussian for large $N$. We have arranged the errors into batches because, as explained in Section 9 of Chapter 1, we can then use the Student t-distribution to construct confidence intervals for a sum of independent Gaussian, or in this case approximately Gaussian, random variables with unknown variance. In particular, we estimate the mean of the batch averages

$$(3.4) \qquad \hat{\epsilon} = \frac{1}{M} \sum_{j=1}^{M} \hat{\epsilon}_j = \frac{1}{NM} \sum_{j=1}^{M} \sum_{k=1}^{N} |X_{T,k,j} - Y_{T,k,j}|$$

and then use the formula

$$(3.5) \qquad \hat{\sigma}_\epsilon^2 = \frac{1}{M-1} \sum_{j=1}^{M} (\hat{\epsilon}_j - \hat{\epsilon})^2$$

to estimate the variance $\sigma_\epsilon^2$ of the batch averages. Experience has shown that the batch averages can be interpreted as being Gaussian for batch sizes $N \geq 15$; we shall usually take $N = 100$. For the Student $t$-distribution with $M - 1$ degrees of freedom an $100(1 - \alpha)\%$ confidence interval for $\epsilon$ has the form

$$(3.6) \qquad (\hat{\epsilon} - \Delta\hat{\epsilon}, \; \hat{\epsilon} + \Delta\hat{\epsilon})$$

with

$$(3.7) \qquad \Delta\hat{\epsilon} = t_{1-\alpha,M-1} \sqrt{\frac{\hat{\sigma}_\epsilon^2}{M}},$$

where $t_{1-\alpha,M-1}$ is determined from the Student t-distribution with $M - 1$ degrees of freedom. For $M = 20$ and $\alpha = 0.1$ we have $t_{1-\alpha,M-1} \approx 1.73$ from Table 1.9.1. In this case the absolute error $\epsilon$ will lie in the corresponding confidence interval (3.6) with probability $1 - \alpha = 0.9$.

**PC-Exercise 9.3.3**    *Simulate $M = 10$ batches each with $N = 100$ trajectories of the Ito process $X$ satisfying (2.1) with $X_0 = 1.0$, $a = 1.5$, $b = 0.1$ and their Euler approximations with equidistant time steps of step size $\Delta = 2^{-4}$ corresponding to the same sample paths of the Wiener process on the time interval $[0, T]$ for $T = 1$. Evaluate the 90% confidence interval for the absolute error $\epsilon$. Repeat this for $M = 20$, $40$ and $100$ batches, in each case using the batches already simulated, and plot the confidence intervals on $\epsilon$ versus $M$ axes.*

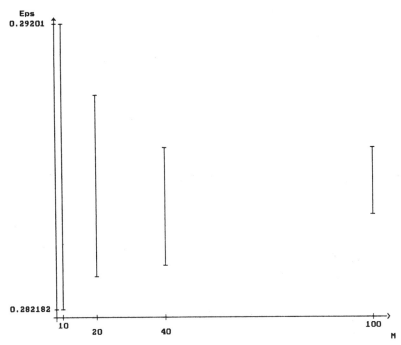

**Figure 9.3.1**   Confidence intervals for increasing numbers of batches.

Our results for PC-Exercise 9.3.3, plotted in Figure 9.3.1, indicate that the length of the confidence interval for the absolute error decreases as the number of batches increases. In fact, this is predicted by formula (3.7), which says that we need to increase the number of batches fourfold in order to halve the length of the confidence interval. As is evident from the PC-Exercise, it can be very time consuming computationally to achieve this additional accuracy.

The preceding computations provide us with a method for determining the number of simulations needed to obtain a confidence interval of specified length for the absolute error $\epsilon$.  Since the length $2\Delta\hat{\epsilon}$ of the confidence interval is inversely proportional to the square root of the number of batches $M$ only, the required number of batches for a chosen confidence interval of sufficiently small length may be very large. Consequently, some thought should be given to decide how much accuracy is really needed in the answer of a given problem.

We shall now look more closely at the relationship between the absolute error of Euler approximations and the step size.

**PC-Exercise 9.3.4**    *Simulate $M = 20$ batches each with $N = 100$ trajectories of the Ito process $X$ satisfying (2.1) with $X_0 = 1.0$, $a = 1.5$, $b = 0.1$ and their Euler approximations with equidistant time steps of step size $\Delta = 2^{-2}$ corresponding to the same sample paths of the Wiener process on the time interval $[0, T]$ for $T = 1$. Evaluate the 90% confidence interval for the absolute error $\epsilon$. Repeat this for step sizes $\Delta = 2^{-3}$, $2^{-4}$ and $2^{-5}$, and plot the confidence intervals on $\epsilon$ versus $\Delta$ axes.*

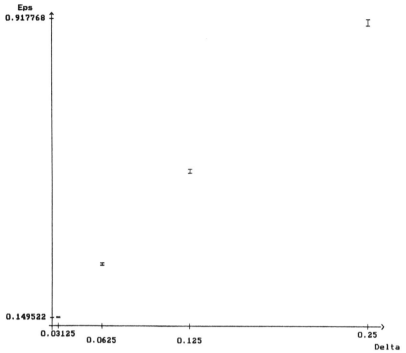

**Figure 9.3.2**   Confidence intervals for increasing step size.

Figure 9.3.2 shows that the step size $\Delta$ has a definite effect on the magnitude of the absolute error $\epsilon$ and on the length of the confidence interval. We could include in Figure 9.3.2 the graph of a function

$$(3.8) \qquad\qquad \tilde{\epsilon}(\Delta) = K\,\Delta^{1/2}$$

for an appropriate constant $K$ which would suggest that the absolute error is proportional to the square root of the step size. We can see this more clearly if we plot the results using $\log_2$ versus $\log_2$ coordinates for which the graph of (3.8) becomes a straight line with slope $1/2$.

**PC-Exercise 9.3.5**   *Replot the results of PC-Exercise 9.3.4 on $\log_2 \epsilon$ versus $\log_2 \Delta$ axes.*

The confidence intervals in Figure 9.3.3 follow closely a straight line with slope $1/2$. In fact, we shall prove in Theorem 10.2.2 of Chapter 10 that this is true for an Euler approximation of a general Ito process and shall call the exponent $1/2$ of $\Delta$ in (3.8) the corresponding *order of strong convergence*.

We can decompose the random variable $\hat{\epsilon}$, the mean of the batch averages, into two parts, that is as

$$(3.9) \qquad\qquad \hat{\epsilon} = \epsilon_{sys} + \epsilon_{stat}, \qquad \text{where}$$

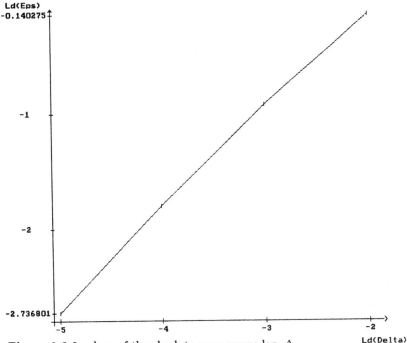

**Figure 9.3.3**    $\log_2$ of the absolute error versus $\log_2 \Delta$.

(3.10)                                  $\epsilon_{sys} = E(\hat{\epsilon})$

denotes the *systematic error* and $\epsilon_{stat}$ the *statistical error*. From (3.1) and (3.4) we have

$$
\begin{aligned}
(3.11) \qquad \epsilon_{sys} &= E(\hat{\epsilon}) \\
&= E\left( \frac{1}{NM} \sum_{j=1}^{M} \sum_{k=1}^{N} |X_{T,k,j} - Y_{T,k,j}| \right) \\
&= E(|X_T - Y(T)|) \\
&= \epsilon,
\end{aligned}
$$

that is, the systematic error coincides with the absolute error. In Chapters 10, 11 and 12 we shall introduce time discrete approximations with smaller systematic errors as $\Delta \to 0$, that is with higher orders of strong convergence than the Euler approximations.

From (3.9) and (3.10) the statistical error obviously satisfies

(3.12)                                  $\epsilon_{stat} = \hat{\epsilon} - \epsilon.$

For a large number $NM$ of independent simulations the Central Limit Theorem says that the statistical error is asymptotically Gaussian with mean zero and and variance

(3.13)                $\text{Var}(\epsilon_{stat}) = \text{Var}\left(\hat{\epsilon} - \epsilon\right) = E\left((\hat{\epsilon} - \epsilon)^2\right)$

$$= \frac{1}{(NM)^2} \sum_{j=1}^{M} \sum_{k=1}^{N} E\left((|X_{T,k,j} - Y_{T,k,j}| - \epsilon)^2\right)$$

$$= \frac{1}{NM} E\left((|X_T - Y(T)| - \epsilon)^2\right)$$

$$= \frac{1}{NM} \text{Var}\left(|X_T - Y(T)|\right).$$

Thus the statistical error depends on the total number $NM$ of simulations and not separately on either the number $N$ of simulations in each batch or on the number $M$ of batches used.

Finally, we mention that roundoff errors must also be taken into account in practical simulations. They are certainly present in the errors calculated in the above PC-Exercises. Usually, however, we consider idealized estimates of discretization errors assuming that no roundoff error occurs, just as we did in Chapter 8 for deterministic numerical schemes, and then investigate the roundoff error separately.

# 9.4   Approximation of Moments

In many practical situations we do not need so strong a convergence as the pathwise approximation considered in the previous section. For instance, we may only be interested in the computation of moments, probabilities or other functionals of the Ito process. Since the requirements for their simulation are not as demanding as for pathwise approximations, it is natural and convenient to classify them as a separate class of problems.

To help the reader understand this weaker type of convergence, we shall carry out some computer experiments to investigate the mean error

(4.1)                $\mu = E(Y(T)) - E(X_T)$

for the same linear stochastic differential equation as in Section 2

(4.2)                $dX_t = a\, X_t\, dt + b\, X_t\, dW_t$

for $t \in [0, T]$ and its Euler approximation

$$Y_{n+1} = Y_n + a\, Y_n \Delta_n + b\, Y_n\, \Delta W_n$$

for $n = 0, 1, 2, \ldots, N-1$. Here, as before, $\Delta_n = \tau_{n+1} - \tau_n$ denotes the step size and $\Delta W_n = W_{\tau_{n+1}} - W_{\tau_n}$ the increment of the Wiener process. In Section 7 we shall generalize (4.1) to the approximation of polynomial, and more general, functionals of the process, including its higher moments. We note that $\mu$ can take negative as well as positive values.

An important feature of this type of approximation is that we do not need to use the same sample path of the Wiener process when generating $X$ and $Y$ in order to obtain a small mean error. Rather, we require only that the probability distributions of $X_T$ and $Y(T)$ are sufficiently close to each other, but not necessarily the actual realizations of the random variables. This is implied by a much weaker form of convergence than that needed for pathwise approximations.

For the example (4.2) it is easy to show (see (4.2.10)) that the mean of the Ito process is

$$(4.3) \qquad E(X_T) = E(X_0) \exp(aT).$$

As in the last section, we shall arrange the simulated trajectories of the Euler approximation into $M$ batches with $N$ trajectories in each. Then, we shall estimate the mean error of the $j$th batch by the statistic

$$(4.4) \qquad \hat{\mu}_j = \frac{1}{N} \sum_{k=1}^{N} Y_{T,k,j} - E(X_T),$$

for $j = 1, 2, \ldots, M$, and their average by the statistic

$$(4.5) \qquad \hat{\mu} = \frac{1}{M} \sum_{j=1}^{M} \hat{\mu}_j = \frac{1}{MN} \sum_{j=1}^{M} \sum_{k=1}^{N} Y_{T,k,j} - E(X_T).$$

Similarly, we shall estimate the variance of the batch averages $\hat{\mu}_j$ by

$$(4.6) \qquad \hat{\sigma}_\mu^2 = \frac{1}{M-1} \sum_{j=1}^{M} (\hat{\mu}_j - \hat{\mu})^2.$$

The $100(1 - \alpha)\%$ confidence interval of the Student $t$-distribution with $M - 1$ degrees of freedom for the mean error $\mu$ is

$$(4.7) \qquad (\hat{\mu} - \Delta\hat{\mu}, \, \hat{\mu} + \Delta\hat{\mu})$$

where

$$(4.8) \qquad \Delta\hat{\mu} = t_{1-\alpha, M-1} \sqrt{\frac{\hat{\sigma}_\mu^2}{M}}.$$

The mean error $\mu$ will thus lie in this confidence interval with at least probability $1 - \alpha$.

**PC-Exercise 9.4.1** *Generate $M = 10$ batches of $N = 100$ trajectories of the Euler approximation for the Ito process (4.2) with $X_0 = 1.0$, $a = 1.5$, $b = 0.1$ for step length $\Delta = 2^{-4}$ and terminal time $T = 1$. Determine the 90% confidence interval for the mean error $\mu$. Then repeat this for $M = 20$, $40$ and $100$ batches using the batches already simulated and plot the intervals on $\mu$ versus $M$ axes.*

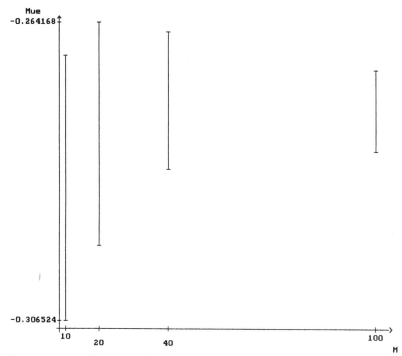

**Figure 9.4.1**    Confidence intervals for increasing number of batches.

Figure 9.4.1 shows that a fourfold increase in the number of batches halves the length of the confidence interval, just as was needed in the previous section for pathwise approximations. More interesting is the dependence of the mean error on the step size.

**PC-Exercise 9.4.2**    *Generate $M = 20$ batches of $N = 100$ trajectories of the Euler approximation as in PC-Exercise 9.4.1. Determine the 90% confidence interval for the mean error $\mu$. Then repeat this for step sizes $\Delta = 2^{-3}$, $2^{-4}$ and $2^{-5}$, and plot the intervals on $\mu$ versus $\Delta$ axes.*

We can see from Figure 9.4.2, which contains our results for PC-Exercise 9.4.2, that the choice of step size $\Delta$ has a clear effect on the mean error. It appears that $\hat{\mu}$ is proportional to $\Delta$. To highlight this impression we could include the graph of a linear function

$$(4.9) \qquad\qquad \tilde{\mu}(\Delta) = K\,\Delta$$

with an appropriate constant $K$ in the figure. As in the last section it is useful here to plot the results in $\log_2$-$\log_2$ coordinates. For convenience we shall also call $|\mu|$ the mean error.

**PC-Exercise 9.4.3**    *Replot the results of PC-Exercise 9.4.2 on $\log_2 |\mu|$ versus $\log_2 \Delta$ axes.*

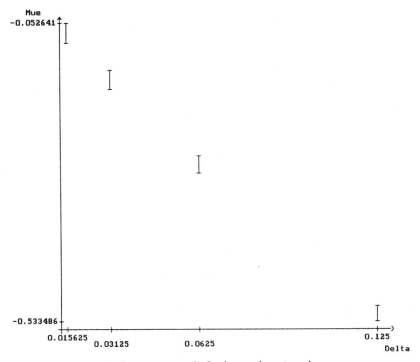

**Figure 9.4.2**    Confidence intervals for increasing step size.

From Figure 9.4.3 we see that the $\log_2$ of the mean error closely follows a straight line of slope 1 in $\log_2 \Delta$. This contrasts with the slope $1/2$ for the strong pathwise approximations in the previous section. In Chapter 14 we shall examine this functional dependence theoretically for the Euler approximation and also for other approximations.

We can also decompose the random estimate $\hat{\mu}$ for the mean error $\mu$ into a *systematic error* $\mu_{sys}$ and a *statistical error* $\mu_{stat}$, with

$$(4.10) \qquad \hat{\mu} = \mu_{sys} + \mu_{stat} \qquad \text{where}$$

$$(4.11) \qquad \mu_{sys} = E(\hat{\mu}).$$

Then

$$
\begin{aligned}
(4.12) \qquad \mu_{sys} &= E(\hat{\mu}) \\
&= E\left(\frac{1}{MN}\sum_{j=1}^{M}\sum_{k=1}^{N}Y_{T,k,j}\right) - E(X_T) \\
&= E(Y(T)) - E(X_T) \\
&= \mu,
\end{aligned}
$$

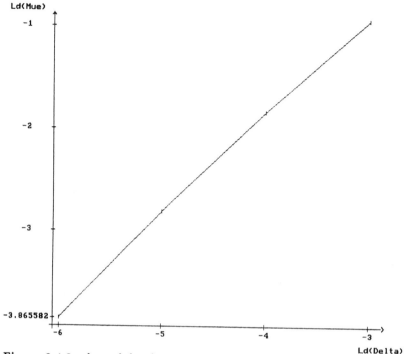

**Figure 9.4.3**    $\log_2$ of the absolute mean error versus $\log_2 \Delta$.

so the systematic error is the same as the mean error. In Chapters 14 and 15 we shall introduce more complicated time discrete approximations with the objective of decreasing the mean error for a given step size.

For a large number $MN$ of independent simulations we can conclude from the Central Limit Theorem that the statistical error $\hat{\mu}_{stat}$ becomes asymptotically Gaussian with mean zero and variance

$$(4.13) \qquad \text{Var}\,(\mu_{stat}) = \text{Var}\,(\hat{\mu}) = \frac{1}{MN}\,\text{Var}(Y(T)).$$

This depends on the total number $MN$ of simulations and not separately on the number $M$ of batches or number $N$ of simulations in each batch. The comment in the previous section that the proposed degree of accuracy should be chosen with care applies here too, since a fourfold increase in the number of simulations is also required here to halve the length of a confidence interval. The successful implementation of weak approximations is often in direct conflict with the size of this variance of the estimated functional.

In contrast with pathwise approximations, we can use variance reduction techniques to construct approximations which give the same first moment of a general functional of the Ito process with a much smaller variance. In this way we can achieve the same accuracy as the Euler approximation with a substantial reduction in computational effort. We shall discuss variance reduction techniques in Chapter 16.

In conclusion, we remark that roundoff errors also affect weak approximations. We shall take them into account later in the context of the numerical stability of the corresponding numerical schemes.

**Exercise 9.4.4**  *Show (4.13).*

# 9.5  General Time Discretizations and Approximations

A general class of time discretizations and approximations is introduced in this section, which can be omitted on the first reading.

For simplicity we have so far concentrated on the equidistant time discretization

$$(5.1) \qquad (\tau)_\delta = \{\tau_n : n = 0, 1, \ldots\}$$

of a bounded interval $[t_0, T]$ with discretization times

$$(5.2) \qquad \tau_n = t_0 + n\,\Delta$$

for $n = 0, 1, \ldots$ and constant step size $\delta \equiv \Delta \in (0, \delta_0)$ for some finite $\delta_0 > 0$.

In some applications it is desirable to have a more flexible time discretization, for instance, to allow step size control where the next step size depends on the current value of the time discrete approximation. In such cases the step size will be random. There are many other situations in which a variable step size, random or deterministic, may be useful. In addition, while we do not consider the approximation of Ito processes with a Poisson jump component in this book, it is obvious that the discretization times of such approximations could include the random jump times.

We recall (1.15) where we defined the integer $n_t$ as the largest integer $n$ for which $\tau_n$ does not exceed $t$, that is

$$n_t = \max\{n = 0, 1, \ldots : \tau_n \leq t\}.$$

Then, for a given maximum step size $\delta \in (0, \delta_0)$ we define a *time discretization*

$$(5.3) \qquad (\tau)_\delta = \{\tau_n : n = 0, 1, \ldots\}$$

as a sequence of time instants $\{\tau_n : n = 0, 1, \ldots\}$, which may be random, satisfying

$$(5.4) \qquad 0 \leq \tau_0 < \tau_1 < \cdots < \tau_n < \cdots < \infty,$$

$$(5.5) \qquad \sup_n (\tau_{n+1} - \tau_n) \leq \delta \qquad \text{and}$$

(5.6)                                  $n_t < \infty$,

w.p.1, for all $t \in \Re^+$, where $\tau_{n+1}$ is $\mathcal{A}_{\tau_n}$-measurable for each $n = 0$, 1, ....
Here $\{\mathcal{A}_t, \, t \geq 0\}$ is a preassigned increasing family of $\sigma$-algebras, generally
associated with the Ito or Wiener process under consideration. Also, when we
have a time discretization $(\tau)_\delta$ on a bounded interval $[t_0, \, T]$ we shall usually
choose $\tau_0 = t_0$ and $\tau_{n_T} = T$.

While the discretization points may be random, they cannot be completely
arbitrary. The restriction that $\tau_{n+1}$ be $\mathcal{A}_{\tau_n}$-measurable for each $n = 0$, 1, ...
means that the step size $\Delta_n = \tau_{n+1} - \tau_n$ can depend only on the information
available at the discretization point $\tau_n$ for each $n = 0$, 1, .... In addition,
(5.5) says that each step size $\Delta_n$ can be no larger than the specified maximum
allowable step size $\delta$, whereas condition (5.6) ensures that there can be only a
finite number of discretization instants in any bounded interval.

In the sequel we shall introduce more complicated stochastic schemes than
the Euler scheme. Like the Euler scheme they will only generate discrete time
processes, but we can then construct continuous time processes from them by
interpolation, for example.

We shall call a process $Y = \{Y(t), \, t \geq 0\}$, which is right continuous with left
hand limits, a *time discrete approximation* with maximum step size $\delta \in (0, \delta_0)$
if it is based on a time discretization $(\tau)_\delta$ such that $Y(\tau_n)$ is $\mathcal{A}_{\tau_n}$-measurable
and $Y(\tau_{n+1})$ can be expressed as a function of $Y(\tau_0)$, ..., $Y(\tau_n)$, $\tau_0$, ..., $\tau_n$,
$\tau_{n+1}$ and a finite number $l$ of $\mathcal{A}_{\tau_{n+1}}$-measurable random variables $Z_{n+1,j}$ for $j$
$= 1$, ..., $l$ and each $n = 0$, 1, ....

This definition allows the recursive computation of the values of the approximation at the given discretization times. Since the computation of $Y(\tau_n)$
should not involve more information than is available at time $\tau_n$ we restrict
$Y(\tau_n)$ to be $\mathcal{A}_{\tau_n}$-measurable. The value of $Y(\tau_{n+1})$ may then depend on the
values of $Y$ at earlier discretization times, on the step size and on a finite number of random variables which generate the noise mainly within the current
time step. We note that the Ito process itself is a time discrete approximation. A Markov chain on a discrete state space is another example, since we
simply take $Y(\tau_{n+1}) = Y(\tau_n) + Z_{n+1,1}$ where $Z_{n+1,1}$ is a random variable
characterized by the transition probabilities of the chain.

Various kinds of interpolation methods are covered by this definition. Since
it asks only for right continuous approximations with left hand limits, it includes the right continuous piecewise constant interpolations (1.14) as well as
the linear interpolations (1.16) of the values of the approximation at the discretization times. A time discrete approximation thus defined is a continuous
time stochastic process on $\Re^+$. It corresponds to a scheme, which describes a
recursive algorithm for the generation of the values at the discretization points,
and a prescribed interpolation method.

## 9.6   Strong Convergence and Consistency

In Section 3 we considered the pathwise approximation of an Ito process $X$ by an Euler approximation $Y$ and introduced the absolute error criterion

(6.1) $$\epsilon = E\left(|X_T - Y(T)|\right),$$

the expectation of the absolute value of the difference between the Ito process and the approximation at a finite terminal time $T$.

We shall say that a general time discrete approximation $Y^\delta$ with maximum step size $\delta$ *converges strongly* to $X$ at time $T$ if

(6.2) $$\lim_{\delta\downarrow 0} E\left(\left|X_T - Y^\delta(T)\right|\right) = 0.$$

While the Euler approximation is the simplest useful time discrete approximation, it is, generally, not particularly efficient numerically. We shall thus derive and investigate other time discrete approximations in the following chapters. In order to assess and compare different time discrete approximations, we need to know their rates of strong convergence.

We shall say that a time discrete approximation $Y^\delta$ *converges strongly with order* $\gamma > 0$ at time $T$ if there exists a positive constant $C$, which does not depend on $\delta$, and a $\delta_0 > 0$ such that

(6.3) $$\epsilon(\delta) = E\left(\left|X_T - Y^\delta(T)\right|\right) \le C\,\delta^\gamma$$

for each $\delta \in (0, \delta_0)$.

We note that (6.3) is a straightforward generalization of the usual deterministic convergence criterion (8.3.4) and reduces to it when the diffusion coefficient vanishes and the initial value is deterministic. Various other criteria have also been suggested in the literature, but (6.3) is a natural generalization of the deterministic one and allows mathematically sharp orders of convergence to be derived. In fact, we can also establish stronger versions of (6.3) involving uniform convergence on the interval $[t_0, T]$ (see Theorem 10.6.3). In Chapters 10 and 11 we shall investigate the strong convergence of a number of different time discrete approximations. We shall see, in particular, that the Euler approximation has strong order of convergence $\gamma = 0.5$, as suggested by the computer experiments in Section 3.

**Exercise 9.6.1**   *Does the Euler approximation of an Ito process with constant drift and diffusion coefficients converge with some strong order $\gamma > 0.5$?*

As with deterministic numerical schemes, the concept of consistency of a stochastic time discrete approximation is closely entwined with that of convergence and is often easier to verify. We shall say that a discrete time approximation $Y^\delta$ corresponding to a time discretization $(\tau)_\delta = \{\tau_n\colon n = 0, 1, \ldots\}$ with maximum step size $\delta$ is *strongly consistent* if there exists a nonnegative function $c = c(\delta)$ with

(6.4) $$\lim_{\delta \downarrow 0} c(\delta) = 0$$

such that

(6.5) $$E\left(\left|E\left(\left.\frac{Y_{n+1}^{\delta} - Y_n^{\delta}}{\Delta_n}\right|\mathcal{A}_{\tau_n}\right) - a(\tau_n, Y_n^{\delta})\right|^2\right) \leq c(\delta)$$

and

(6.6) $$E\left(\frac{1}{\Delta_n}\left|Y_{n+1}^{\delta} - Y_n^{\delta} - E\left(Y_{n+1}^{\delta} - Y_n^{\delta}\big|\mathcal{A}_{\tau_n}\right) - b(\tau_n, Y_n^{\delta})\Delta W_n\right|^2\right) \leq c(\delta)$$

for all fixed values $Y_n^{\delta} = y$ and $n = 0, 1, \ldots$.

Condition (6.5) requires the mean of the increment of the approximation to converge to that of the Ito process. In the absence of noise it is equivalent to the definition of consistency (8.3.3) of a deterministic one-step scheme. From condition (6.6) it follows that the variance of the difference between the random parts of the approximation and the Ito process also converges to zero. Thus strong consistency gives an indication of the pathwise closeness. In fact, it implies the strong convergence of the time discrete approximation to the Ito process, which we shall now prove in a simple context.

We shall consider the 1-dimensional case $d = m = 1$ with the Ito process $X$ satisfying the autonomous stochastic differential equation

(6.7) $$dX_t = a(X_t)\, dt + b(X_t)\, dW_t,$$

together with a time discrete approximation $Y^{\delta}$ corresponding to an equidistant time discretization $(\tau)_{\delta}$ with time step $\Delta_n \equiv \delta$. Let us suppose that the assumptions of the strong existence and uniqueness theorem, Theorem 4.5.3, are satisfied, in particular that the coefficients $a$ and $b$ satisfy a uniform Lipschitz condition and a growth bound. We remark that the following theorem holds in much more general cases too.

**Theorem 9.6.2**   *Under the assumptions of Theorem 4.5.3 a strongly consistent equidistant time discrete approximation $Y^{\delta}$ of a 1-dimensional autonomous Ito process $X$ with $Y^{\delta}(0) = X_0$ converges strongly to $X$.*

**Proof**   For $0 \leq t \leq T$ we set

$$Z(t) = \sup_{0 \leq s \leq t} E\left(\left|Y_{n_s}^{\delta} - X_s\right|^2\right)$$

and obtain

$$Z(t) = \sup_{0 \leq s \leq t} E\left(\left|\sum_{n=0}^{n_s - 1}\left(Y_{n+1}^{\delta} - Y_n^{\delta}\right) - \int_0^s a(X_r)\, dr - \int_0^s b(X_r)\, dW_r\right|^2\right)$$

$$\leq \ C_1 \sup_{0 \leq s \leq t} \left\{ E \left( \left| \sum_{n=0}^{n_s-1} \left( E \left( Y_{n+1}^\delta - Y_n^\delta \,\middle|\, \mathcal{A}_{\tau_n} \right) - a(Y_n^\delta) \Delta_n \right) \right|^2 \right) \right.$$

$$+ E \left( \left| \sum_{n=0}^{n_s-1} \left( Y_{n+1}^\delta - Y_n^\delta - E \left( Y_{n+1}^\delta - Y_n^\delta \,\middle|\, \mathcal{A}_{\tau_n} \right) - b(Y_n^\delta) \Delta W_n \right) \right|^2 \right)$$

$$+ E \left( \left| \int_0^{\tau_{n_s}} \left( a(Y_{n_r}^\delta) - a(X_r) \right) dr \right|^2 \right)$$

$$+ E \left( \left| \int_0^{\tau_{n_s}} \left( b(Y_{n_r}^\delta) - b(X_r) \right) dW_r \right|^2 \right)$$

$$\left. + E \left( \left| \int_{\tau_{n_s}}^s a(X_r) \, dr \right|^2 \right) + E \left( \left| \int_{\tau_{n_s}}^s b(X_r) \, dW_r \right|^2 \right) \right\}.$$

Using the conditional independence of the summands in the first sum, the Lipschitz condition and growth bound for the coefficients $a$ and $b$ and an estimate for the second moment of $X_t$ we derive

$$Z(t) \leq$$

$$C_1 \sup_{0 \leq s \leq t} \left\{ \sum_{n=0}^{n_s-1} E \left( \left| Y_{n+1}^\delta - Y_n^\delta - E \left( Y_{n+1}^\delta - Y_n^\delta \,\middle|\, \mathcal{A}_{\tau_n} \right) - b(Y_n^\delta) \Delta W_n \right|^2 \right) \right.$$

$$+ T \delta \sum_{n=0}^{n_s-1} E \left( \left| E \left( \frac{Y_{n+1}^\delta - Y_n^\delta}{\Delta_n} \,\middle|\, \mathcal{A}_{\tau_n} \right) - a(Y_n^\delta) \right|^2 \right)$$

$$\left. + K^2(1+T) \int_0^{\tau_{n_s}} Z(r) \, dr + K^2 (1+T)(1+C_2)\delta \right\}.$$

In view of strong consistency it follows from (6.5) and (6.6) that

$$Z(t) \leq C_3 \int_0^t Z(r) \, dr + C_4 \left( \delta + c(\delta) \right)$$

and, thus, by the Gronwall inequality (Lemma 4.5.1) that

$$Z(t) \leq C_5 \left( \delta + c(\delta) \right).$$

Using the Lyapunov inequality (2.2.14) we can then conclude that

$$E\left(\left|Y^{\delta}(T) - X_T\right|\right) \le \sqrt{Z(T)} \le \sqrt{C_5\left(\delta + c(\delta)\right)},$$

from which the assertion of the proposition follows.  □

**Exercise 9.6.3**    *Show that the Euler scheme (1.3) is strongly consistent with*
$c(\delta) \equiv 0$ *and hence from (6.8) deduce that it has strong order of convergence
at least* $\gamma = 0.5$ *under the assumptions of Theorem 9.6.2.*

**Exercise 9.6.4**    *A formal generalization of the Heun scheme (8.1.12) for
ordinary differential equations to the stochastic differential equation (6.7) is*

$$Y_{n+1} \;=\; Y_n + \frac{1}{2}\Big\{a(Y_n) + a\left(Y_n + a(Y_n)\,\Delta_n + b(Y_n)\,\Delta W_n\right)\Big\}\Delta_n$$

$$+ \frac{1}{2}\Big\{b(Y_n) + b\left(Y_n + a(Y_n)\,\Delta_n + b(Y_n)\,\Delta W_n\right)\Big\}\Delta W_n.$$

*Show that it is generally not strongly consistent. For what types of coefficients
is it strongly consistent?*

**Exercise 9.6.5**    *Does an Euler approximation of the Ito process in Exercise
9.6.1 based on another Wiener process which is independent of that driving the
Ito process also converge strongly to the Ito process?*

## 9.7    Weak Convergence and Consistency

In Section 4 we examined the approximation of the first moment of a particular
Ito process $X$ by the Euler approximation with respect to the mean error

$$(7.1) \qquad\qquad \mu = E\left(Y(T)\right) - E\left(X_T\right).$$

In particular, we saw that this criterion differs in its properties from the strong
convergence criterion. To some extent (7.1) is special and not appropriate for
applications where the approximation of some higher moment

$$E\left(\left|X_T\right|^q\right)$$

with $q = 2, 3, \ldots$ or of some functional

$$(7.2) \qquad\qquad E\left(g\left(X_T\right)\right)$$

is of interest. Like (7.1) these do not require the pathwise approximation of the
Ito process, but only an approximation of the probability distribution of $X_T$.

    We shall say that a general time discrete approximation $Y^{\delta}$ corresponding
to a time discretization $(\tau)_{\delta}$ converges weakly to $X$ at time $T$ as $\delta \downarrow 0$ with
respect to a class $\mathcal{C}$ of test functions $g: \Re^d \to \Re$ if we have

(7.3) $$\lim_{\delta \downarrow 0} \left| E\left(g\left(X_T\right)\right) - E\left(g\left(Y^\delta(T)\right)\right) \right| = 0$$

for all $g \in C$. If $C$ contains all polynomials this definition implies the convergence of all moments, so theoretical investigations involving it will require the existence of all moments. In the deterministic case with a zero diffusion coefficient and a nonrandom initial value, (7.3) with $g(x) \equiv x$ reduces to the usual deterministic convergence criterion, just as the strong convergence criterion (6.2) does.

To compare different time discrete approximations we need to consider the rate of weak convergence. We recall from Section 8 of Chapter 4 that $C_P^l(\Re^d, \Re)$ denotes the space of $l$ times continuously differentiable functions $g \colon \Re^d \to \Re$ which, together with their partial derivatives of orders up to and including order $l$, have polynomial growth. We shall use such a space as the class of test functions. Since it contains all of the polynomials it will suffice for most practical purposes.

We shall say that a time discrete approximation $Y^\delta$ *converges weakly with order* $\beta > 0$ to $X$ at time $T$ as $\delta \downarrow 0$ if for each $g \in C_P^{2(\beta+1)}(\Re^d, \Re)$ there exists a positive constant $C$, which does not depend on $\delta$, and a finite $\delta_0 > 0$ such that

(7.4) $$\left| E\left(g\left(X_T\right)\right) - E\left(g\left(Y^\delta(T)\right)\right) \right| \le C\,\delta^\beta$$

for each $\delta \in (0, \delta_0)$.

In Chapters 14 and 15 we shall investigate the order of weak convergence of various time discrete approximations theoretically. In particular, we shall prove that the Euler approximation usually converges with weak order $\beta = 1$, in contrast with the strong order $\gamma = 0.5$. We shall see that the strong and weak convergence criteria lead to the development of different time discrete approximations which are only efficient with respect to one of the two criteria. This fact makes it important to clarify the aim of a simulation before choosing an approximation scheme:

Is a good pathwise approximation of the Ito process required or is the approximation of some functional of the Ito process the real objective?

**Exercise 9.7.1** *Consider an Ito process with constant drift and diffusion coefficients. Does the Euler approximation (1.3) based on another Wiener process which is independent of that driving the Ito process also converge with some weak order $\beta > 0$?*

The utility of a property that is more easily verified than weak convergence leads us to weak consistency. We shall say that a time discrete approximation $Y^\delta$ with maximum step size $\delta$ is *weakly consistent* if there exists a nonnegative function $c = c(\delta)$ with

(7.5) $$\lim_{\delta \downarrow 0} c(\delta) = 0$$

such that

$$(7.6) \qquad E\left(\left|E\left(\frac{Y^\delta_{n+1} - Y^\delta_n}{\Delta_n}\bigg| \mathcal{A}_{\tau_n}\right) - a(\tau_n, Y^\delta_n)\right|^2\right) \le c(\delta)$$

and

$$(7.7) \qquad E\left(\left|E\left(\frac{1}{\Delta_n}\left(Y^\delta_{n+1} - Y^\delta_n\right)\left(Y^\delta_{n+1} - Y^\delta_n\right)^{\mathsf{T}} \bigg| \mathcal{A}_{\tau_n}\right)\right.\right.$$
$$\left.\left. - b(\tau_n, Y^\delta_n)b(\tau_n, Y^\delta_n)^{\mathsf{T}}\right|^2\right) \le c(\delta)$$

for all fixed values $Y^\delta_n = y$ and $n = 0, 1, \ldots$.

We note that condition (7.6) involving the mean of the increment of the approximation is the same as condition (6.5) in the definition of strong consistency. However, (7.7) differs considerably from the second condition (6.6) for strong consistency. It is much weaker because only the variance of the increment of the approximation has to be close to that of the Ito process, whereas for strong consistency the variance of the difference between the increments of the approximation and the Ito process must vanish.

**Exercise 9.7.2**   *Show that the Euler approximation (1.3) is weakly consistent.*

It is not difficult to see that an Euler approximation based on a Wiener process different from that driving the Ito process is still weakly consistent, although it is not strongly consistent; see Exercises 9.6.5 and 9.7.1. The following exercise illustrates the flexibility that we have in generating the noise increments in weakly consistent schemes.

**Exercise 9.7.3**   *Show in the 1-dimensional case $d = m = 1$ that the Euler scheme*

$$Y_{n+1} = Y_n + a(\tau_n, Y_n)\Delta_n + b(\tau_n, Y_n)\xi_n \Delta_n^{1/2},$$

*where the $\xi_n$ are independent two-point random variables with $P(\xi_n = \pm 1) = \frac{1}{2}$, is weakly consistent.*

Under quite natural assumptions a weakly consistent scheme is weakly convergent. As with Theorem 9.6.2 for strong consistency and convergence, we prove it here for the special case of an Ito process satisfying the 1-dimensional autonomous stochastic differential equation (6.7). The result also holds in much more general situations. The proof contains the basic ideas and methods associated with the weak convergence of time discrete approximations, which we shall later apply repeatedly.

**Theorem 9.7.4**   *Suppose that the drift coefficient $a = a(x)$ and the diffusion coefficient $b = b(x)$ of an Ito process satisfying (6.7) are four times continuously differentiable with polynomial growth and uniformly bounded derivatives. Let $Y^\delta$ be a weakly consistent time discrete approximation with equidistant time steps $\Delta_n \equiv \delta$ and initial value $Y^\delta(0) = X_0$ which satisfies the moment bounds*

(7.8) $$E\left(\max_n |Y_n^\delta|^{2q}\right) \le K\left(1 + E\left(|X_0|^{2q}\right)\right)$$

*for* $q = 1, 2, \ldots,$ *and*

(7.9) $$E\left(\frac{1}{\Delta_n}|Y_{n+1}^\delta - Y_n^\delta|^6\right) \le c(\delta)$$

*for* $n = 0, 1, 2, \ldots,$ *where* $c(\delta)$ *is as in (7.5). Then* $Y^\delta$ *converges weakly to the given Ito process.*

**Proof**   We know from Theorem 4.8.6 that the functional

(7.10) $$u(s, x) = E\left(g(X_T)\big| X_s = x\right)$$

is a solution of the final value problem

(7.11) $$\frac{\partial u}{\partial s} + \mathcal{L}u = \frac{\partial u}{\partial s} + a\frac{\partial u}{\partial x} + \frac{1}{2}b^2\frac{\partial^2 u}{\partial x^2} = 0$$

(7.12) $$u(T, x) = g(x),$$

with $\frac{\partial u}{\partial s}(s, x)$ continuous and $u(s, x)$ four times continuously differentiable in $x$, where all of these partial derivatives have polynomial growth.

  We shall denote by $X^{s,x}$ the Ito process starting at $x$ at time $s$, so

(7.13) $$X_t^{s,x} = x + \int_s^t a\left(X_r^{s,x}\right)dr + \int_s^t b\left(X_r^{s,x}\right)dW_r.$$

From (7.11) and the Ito formula (3.3.6) we obtain

(7.14) $$E\left(u\left(\tau_{n+1}, X_{\tau_{n+1}}^{\tau_n,x}\right) - u(\tau_n, x)\big| \mathcal{A}_{\tau_n}\right) = 0.$$

We shall define

(7.15) $$H_\delta = \left|E\left(g\left(Y^\delta(T)\right)\right) - E\left(g\left(X_T\right)\right)\right|,$$

which by means of (7.10) and (7.12) we can rewrite as

$$\begin{aligned} H_\delta &= \left|E\left(u\left(T, Y^\delta(T)\right) - u\left(0, Y_0^\delta\right)\right)\right| \\ &= \left|E\left(\sum_{n=0}^{n_T-1}\left\{u\left(\tau_{n+1}, Y_{n+1}^\delta\right) - u\left(\tau_n, Y_n^\delta\right)\right\}\right)\right|. \end{aligned}$$

For notational simplicity we shall henceforth omit the $\delta$ superscript on the $Y_n^\delta$.

From (7.14) we thus have

$$
\begin{aligned}
H_\delta \;=\;& \left| E\left( \sum_{n=0}^{n_T-1} \Big[ \{ u(\tau_{n+1}, Y_{n+1}) - u(\tau_n, Y_n) \} \right.\right. \\
& \left.\left. - \Big\{ u\left(\tau_{n+1}, X^{\tau_n, Y_n}_{\tau_{n+1}}\right) - u\left(\tau_n, X^{\tau_n, Y_n}_{\tau_n}\right) \Big\} \Big] \right) \right| \\
=\;& \left| E\left( \sum_{n=0}^{n_T-1} \Big[ \{ u(\tau_{n+1}, Y_{n+1}) - u(\tau_{n+1}, Y_n) \} \right.\right. \\
& \left.\left. - \Big\{ u\left(\tau_{n+1}, X^{\tau_n, Y_n}_{\tau_{n+1}}\right) - u(\tau_{n+1}, Y_n) \Big\} \Big] \right) \right| .
\end{aligned}
$$

In view of its differentiability we can expand $u$ in $x$, thus obtaining

$$
\begin{aligned}
H_\delta \;=\;& \left| E\left( \sum_{n=0}^{n_T-1} \left[ \frac{\partial u}{\partial x}(\tau_{n+1}, Y_n) \left\{ (Y_{n+1} - Y_n) - \left( X^{\tau_n, Y_n}_{\tau_{n+1}} - Y_n \right) \right\} \right.\right.\right. \\
& + \frac{1}{2} \frac{\partial^2 u}{\partial x^2}(\tau_{n+1}, Y_n) \left\{ (Y_{n+1} - Y_n)^2 - \left( X^{\tau_n, Y_n}_{\tau_{n+1}} - Y_n \right)^2 \right\} \\
& \left.\left.\left. + R_n(Y_{n+1}) - R_n(X^{\tau_n, Y_n}_{\tau_{n+1}}) \right] \right) \right| ,
\end{aligned}
$$

where

$$
R_n(Z) = \frac{1}{3!} \frac{\partial^3 u}{\partial x^3} \left(\tau_{n+1}, Y_n + \theta_{n,Z}\left(Z - Y_n\right)\right)(Z - Y_n)^3
$$

for some $\theta_{n,Z} \in (0,1)$. We can bound this expression using (7.5)–(7.9), the polynomial growth of $u$, $a$, $b$ and their derivatives and the moment estimates of an Ito process given in Theorem 4.5.4. With all of the partial derivatives in what follows evaluated at $(\tau_{n+1}, Y_n)$ we have

(7.16)    $\displaystyle \lim_{\delta \downarrow 0} H_\delta$

$$
\begin{aligned}
\leq\;& \lim_{\delta \downarrow 0} \sum_{n=0}^{n_T-1} E\left( \left| \frac{\partial u}{\partial x} \right| \left| E\left( (Y_{n+1} - Y_n) - \left( X^{\tau_n, Y_n}_{\tau_{n+1}} - Y_n \right) \Big| \mathcal{A}_{\tau_n} \right) \right| \right. \\
& \left. + \frac{1}{2} \left| \frac{\partial^2 u}{\partial x^2} \right| \left| E\left( (Y_{n+1} - Y_n)^2 - \left( X^{\tau_n, Y_n}_{\tau_{n+1}} - Y_n \right)^2 \Big| \mathcal{A}_{\tau_n} \right) \right| + K \Delta_n^{3/2} \right) \\
\leq\;& \lim_{\delta \downarrow 0} \sum_{n=0}^{n_T-1} \Delta_n\, E\left( \left| \frac{\partial u}{\partial x} \right| \left| E\left( \frac{Y_{n+1} - Y_n}{\Delta_n} \Big| \mathcal{A}_{\tau_n} \right) - a(\tau_n, Y_n) \right| \right.
\end{aligned}
$$

$$+\frac{1}{2}\left|\frac{\partial^2 u}{\partial x^2}\right|\left|E\left(\frac{(Y_{n+1}-Y_n)^2}{\Delta_n}\Big|\mathcal{A}_{\tau_n}\right)-b(\tau_n,Y_n)^2\right|\right)$$

$$\leq \lim_{\delta\downarrow 0}\sum_{n=0}^{n_T-1}\Delta_n\left[\left\{E\left(\left|\frac{\partial u}{\partial x}\right|^2\right)E\left(\left|E\left(\frac{Y_{n+1}-Y_n}{\Delta_n}\Big|\mathcal{A}_{\tau_n}\right)-a(\tau_n,Y_n)\right|^2\right)\right\}^{1/2}\right.$$

$$\left.+\frac{1}{2}\left\{E\left(\left|\frac{\partial^2 u}{\partial x^2}\right|^2\right)E\left(\left|E\left(\frac{(Y_{n+1}-Y_n)^2}{\Delta_n}\Big|\mathcal{A}_{\tau_n}\right)-b(\tau_n,Y_n)^2\right|^2\right)\right\}^{1/2}\right]$$

$$\leq \lim_{\delta\downarrow 0}\sum_{n=0}^{n_T-1}\Delta_n\,K\,\sqrt{c(\delta)}\leq \lim_{\delta\downarrow 0}T\,K\,\sqrt{c(\delta)}=0.$$

Thus the time discrete approximation $Y^\delta$ converges weakly to the given Ito process. $\square$

**Exercise 9.7.5**    *Show that weak consistency reduces, roughly speaking, to the consistency of a deterministic one-step scheme defined by (8.3.3) in the absence of noise.*

**Exercise 9.7.6**    *Use conditional expectations and Theorem 4.8.6 to verify that the correlation functions*

$$C(t,t+h)=E\left(X_t\,X_{t+h}\right)$$

*can also be approximated by weak approximations of $X$ under appropriate assumptions.*

## 9.8    Numerical Stability

We saw in Section 3 of Chapter 8 that the consistency and convergence of a deterministic numerical scheme alone are no guarantee that the scheme can be used effectively for a given stiff differential equation. In addition, the propagation of initial errors and roundoff errors must be kept under control. This problem also arises for *stiff stochastic differential equations*, which are characterized in the $d$-dimensional linear case (6.3.20)

$$dZ_t = AZ_t\,dt + \sum_{k=1}^m B^k Z_t \circ dW_t^k$$

by the relationship

$$\lambda_1 \gg \lambda_d,$$

where $\lambda_d \leq \lambda_{d-1} \leq \ldots \leq \lambda_1$ are the Lyapunov exponents for this equation defined in Section 3 of Chapter 6. This generalizes the deterministic notion of stiffness in Section 3 of Chapter 8 to the stochastic context because the real parts of the eigenvalues of the coefficient matrix of a deterministic linear differential equation are its Lyapunov exponents. Thus, stochastic stiffness also refers to the presence of two or more widely differing time scales in the solutions. We note that a stiff linear ordinary differential equation is also stiff in the stochastic sense.

More generally, we shall say that a stochastic differential equation such as (6.3.19)

$$dX_t = \underline{a}(t, X_t) \, dt + \sum_{k=1}^{m} b^k(t, X_t) \circ dW_t^k$$

is *stiff* if its linearized system (6.3.20) is stiff in the above sense. To handle stiff stochastic differential equations, in particular, and error propagation, in general, we need a counterpart to the deterministic concept of numerical stability for stochastic numerical schemes.

Let $Y^\delta$ denote a time discrete approximation with maximum step size $\delta > 0$ starting at time $t_0$ at $Y_0^\delta$, with $\bar{Y}^\delta$ denoting the corresponding approximation starting at $\bar{Y}_0^\delta$. We shall say that a time discrete approximation $Y^\delta$ is *stochastically numerically stable* for a given stochastic differential equation if for any finite interval $[t_0, T]$ there exists a positive constant $\Delta_0$ such that for each $\epsilon > 0$ and each $\delta \in (0, \Delta_0)$

$$(8.1) \qquad \lim_{|Y_0^\delta - \bar{Y}_0^\delta| \to 0} \sup_{t_0 \leq t \leq T} P\left(\left|Y_{n_t}^\delta - \bar{Y}_{n_t}^\delta\right| \geq \epsilon\right) = 0.$$

In addition, we shall say that a time discrete approximation is *stochastically numerically stable* if it is stochastically numerically stable for the class of stochastic differential equations for which the approximation converges to the corresponding solution of the equation. For brevity we shall usually refer to stochastic numerical stability as just *numerical stability* in the sequel.

It will turn out that nearly all the one step stochastic schemes proposed in this book are numerically stable under sufficient smoothness and regularity conditions on the drift and diffusion coefficients. By adapting the proof of Theorem 4.5.3, the existence and uniqueness theorem, or of Theorem 9.6.2 it is straight forward to show that the Euler method (1.3) is numerically stable under the assumptions of Theorem 4.5.3.

**Exercise 9.8.1**   *Show under the assumptions of Theorem 4.5.3 that the Euler method (1.3) is numerically stable.*

**PC-Exercise 9.8.2**   *Simulate the exact solution and the Euler approximation with step size $\delta = \Delta = 2^{-4}$ corresponding to the same realization of the Wiener process for the 1-dimensional stochastic differential equation*

$$dX_t = 5\,X_t\,dt + dW_t$$

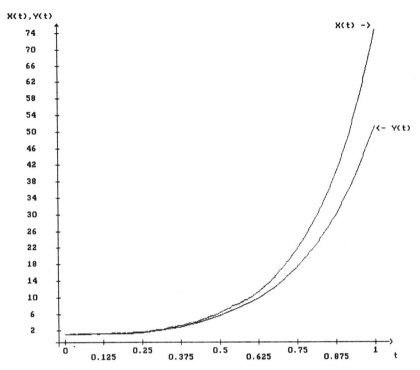

**Figure 9.8.1** The exact solution and Euler trajectory from PC-Exercise 9.8.2.

with $X_0 = 1$ on the interval $[0, T]$, where $T = 1$. Plot both paths on the same $X$ versus $t$ axes. Note that the exact solution here can be simulated using the correlated Gaussian random variables

$$\int_{\tau_n}^{\tau_{n+1}} \left(e^{-5(s-\tau_n)} - 1\right) dW_s \quad and \quad \int_{\tau_n}^{\tau_{n+1}} dW_s.$$

The propagation of an initial error will thus remain bounded on any bounded interval for a numerically stable scheme. We emphasize that the numerical stability criterion applies only to step sizes $\delta > 0$ that are less than some critical value $\Delta_0$, which will usually depend on the time interval $[t_0, T]$, and the differential equation under consideration. This critical value may be extremely small in some cases. From the above PC-Exercise we can see that as the time interval $[t_0, T]$ becomes relatively large, the propagated error of a numerically stable scheme, which is theoretically still under control, may, in fact, become so unrealistically large as to make the approximation useless for some practical purposes. For instance, when simulating first exit times we do not know the appropriate time interval in advance and so must allow for an arbitrarily large interval. In effect, we then need to control the error propagation over the infinite time interval $[t_0, \infty)$, which will only be possible for particular stochastic differential equations.

To cover such situations we shall say that a time discrete approximation $Y^\delta$ is *asymptotically numerically stable* for a given stochastic differential equation if it is numerically stable and there exists a positive constant $\Delta_a$ such that for each $\epsilon > 0$ and $\delta \in (0, \Delta_a)$

$$(8.2) \qquad \lim_{|Y_0^\delta - \bar{Y}_0^\delta| \to 0} \lim_{T \to \infty} P\left( \sup_{t_0 \le t \le T} |Y_{n_t}^\delta - \bar{Y}_{n_t}^\delta| \ge \epsilon \right) = 0,$$

where we have used the same notation as in (8.1).

**PC-Exercise 9.8.3**    *Repeat  PC-Exercise  9.8.2  for  the  1-dimensional stochastic differential equation*

$$dX_t = -5\,X_t\,dt + dW_t$$

*with $X_0 = 1$ on the interval $[0,\,T]$, where $T = 1$.*

**Exercise 9.8.4**    *Show that the Euler method is numerically asymptotically stable for the stochastic differential equation in PC-Exercise 9.8.3.*

As with the A-stability of deterministic differential equations, we can also consider asymptotical numerical stability of a stochastic scheme with respect to an appropriately restricted class of stochastic differential equations. We shall choose the class of complex-valued linear test equations

$$(8.3) \qquad\qquad dX_t = \lambda\,X_t\,dt + dW_t,$$

where the parameter $\lambda$ is a complex number with real part $\text{Re}(\lambda) < 0$ and $W$ is a real-valued standard Wiener process. This represents a simple stochastic generalization by including additive noise in the deterministic test equations (8.3.16) used to test for the A-stability of deterministic schemes. In the stochastic case the critical value $\Delta_a$ will also depend on the parameter $\lambda$. Obviously, we can write (8.3) as a 2-dimensional Ito stochastic differential equation with linear drift and constant diffusion coefficients in terms of the components $(X_1, X_2)$ where $X = X_1 + \imath X_2$. From Theorem 4.8.8 we know that (8.3) has an ergodic solution when $\text{Re}(\lambda) < 0$, which makes these equations a good choice of test equations for situations involving additive noise.

**Exercise 9.8.5**    *Write the complex valued equation (8.3) as a 2-dimensional Ito stochastic differential equation in terms of the real and imaginary parts of $X$ and $\lambda$.*

Let us suppose that we can write a given scheme with equidistant step size $\Delta \equiv \delta$ applied to the test equations (8.3) with $\text{Re}(\lambda) < 0$ in the recursive form

$$(8.4) \qquad\qquad Y_{n+1}^\Delta = Y_n^\Delta G(\lambda\Delta) + Z_n^\Delta,$$

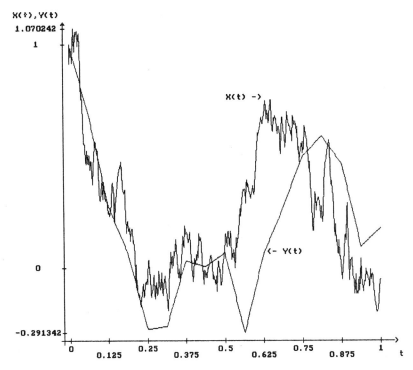

**Figure 9.8.2**    The exact solution and Euler trajectory from PC-Exercise 9.8.3.

for $n = 0, 1, \ldots$, where $G$ is a mapping of the complex plane $C$ into itself and the $Z_0^\Delta$, $Z_1^\Delta$, ... are random variables which do not depend on $\lambda$ or on the $Y_0^\Delta$, $Y_1^\Delta$, .... Then, we shall call the set of complex numbers $\lambda\Delta$ with

(8.5)  $$\mathrm{Re}(\lambda) < 0 \quad \text{and} \quad |G(\lambda\Delta)| < 1$$

the *region of absolute stability* of the scheme. From this region we can determine the appropriate equidistant step size $\Delta$ such that an error in the approximation by this scheme of a particular test equation from the class (8.3) will not grow in subsequent iterations. Obviously, the scheme is asymptotically numerically stable for such a test equation if $\lambda\Delta$ belongs to the region of absolute stability.

The Euler scheme (1.3) with equidistant step size $\Delta > 0$ for the stochastic differential equation (8.3) is

$$Y_{n+1}^\Delta = Y_n^\Delta (1 + \lambda\Delta) + \Delta W_n.$$

Thus

$$\left| Y_n^\Delta - \bar{Y}_n^\Delta \right| \le |1 + \lambda\Delta|^n \left| Y_0^\Delta - \bar{Y}_0^\Delta \right|,$$

where $\bar{Y}_n^\Delta$ is the solution starting at $\bar{Y}_0^\Delta$. The additive noise terms cancel out here, so we obtain the same region of absolute stability as in the deterministic case. See (8.3.16) and Figure 8.3.1. The *implicit Euler scheme*

$$(8.6) \qquad Y_{n+1}^{\Delta} = Y_n^{\Delta} + a\left(\tau_{n+1}, Y_{n+1}^{\Delta}\right)\Delta + b\left(\tau_n, Y_n^{\Delta}\right)\Delta W_n$$

takes the form

$$Y_{n+1}^{\Delta} = Y_n^{\Delta} + \lambda Y_{n+1}^{\Delta}\Delta + \Delta W_n$$

for the test equation (8.3), from which we obtain

$$\left|Y_n^{\Delta} - \bar{Y}_n^{\Delta}\right| \le |1 - \lambda\Delta|^{-n}\left|Y_0^{\Delta} - \bar{Y}_0^{\Delta}\right|$$

and the same region of absolute stability as for the deterministic implicit Euler scheme (8.3.18). Thus, for any $\lambda$ with $\mathrm{Re}(\lambda) < 0$ the step size $\Delta > 0$ can be chosen arbitrarily large.

Generalizing the deterministic definition, we shall say that a stochastic scheme is *A-stable* if its region of absolute stability is the whole of the left half of the complex plane, that is if it consists of all $\lambda\Delta$ with $\mathrm{Re}(\lambda) < 0$ and $\Delta > 0$. Hence the implicit Euler scheme (8.6) is A-stable, but not the Euler scheme (1.3). It is apparent that an A-stable stochastic scheme is also A-stable in the deterministic sense.

The test equation (8.3) can be interpreted as a linearization of a damped oscillator with noise added. In this sense A-stability is related to additive noise. Various problems arise in the case of multiplicative noise. For instance, if we apply the *fully implicit Euler scheme*

$$(8.7) \qquad Y_{n+1}^{\Delta} = Y_n^{\Delta} + a\left(\tau_{n+1}, Y_{n+1}^{\Delta}\right)\Delta + b\left(\tau_{n+1}, Y_{n+1}^{\Delta}\right)\Delta W_n$$

with Gaussian distributed $\Delta W_n = W_{\tau_{n+1}} - W_{\tau_n}$ to the 1-dimensional homogeneous linear Ito stochastic differential equation

$$dX_t = a\,X_t\,dt + b\,X_t\,dW_t,$$

then we obtain

$$Y_n^{\Delta} = Y_0^{\Delta}\prod_{k=0}^{n-1}\frac{1}{1 - a\,\Delta - b\,\Delta W_k}.$$

However, this expression is not suitable as an approximation because one of its factors may become infinite. In fact, the first absolute moment $E(|Y_n^{\Delta}|)$ does not exist, as can be easily seen from Exercise 1.4.7. It seems then that fully implicit methods involving unbounded random variables, such as (8.7), are not practicable, except perhaps in special cases such as for a linear equation with a strongly attracting drift and a very weak noise intensity. As we shall see later, we need to use random variables such as $\Delta W_n$ in strong approximations so in this book we shall concentrate on implicit strong approximations that have implicit coefficients only in the nonrandom terms. In contrast, for weak approximations we need only use bounded random variables to obtain convergence, so fully implicit weak approximations are generally possible. An example is the scheme

(8.8) $$Y_{n+1}^\Delta = Y_n^\Delta + \bar{a}\left(\tau_{n+1}, Y_{n+1}^\Delta\right)\Delta + b\left(\tau_{n+1}, Y_{n+1}^\Delta\right)\Delta\hat{W}_n$$

where $\bar{a} = a - bb'$ and the $\Delta\hat{W}_n$, $n = 0, 1, \ldots$ are independent two-point distributed random variables with

(8.9) $$P\left(\Delta\hat{W}_n = \pm\sqrt{\Delta}\right) = \frac{1}{2}.$$

Obviously, the $|\Delta\hat{W}_n|$ are bounded by $\sqrt{\Delta}$, whereas the Gaussian distributed increments $\Delta W_n$ are unbounded. We note that the drift $\bar{a}$ in the scheme (8.8) has been adjusted so that the resulting scheme is consistent.

# Chapter 10

# Strong Taylor Approximations

In this chapter we shall use stochastic Taylor expansions to derive time discrete approximations with respect to the strong convergence criterion, which we shall call strong Taylor approximations. We shall mainly consider the corresponding strong Taylor schemes, and shall see that the desired order of strong convergence determines the truncation to be used. To establish the appropriate orders of various schemes we shall make frequent use of a technical lemma estimating multiple Ito integrals. This is Lemma 10.8.1, which is stated and proved in Section 8 at the end of the chapter, although it will be used earlier.

## 10.1  Introduction

To simplify our notation in this and in the following chapters, we shall use the following operators, which were introduced in Chapter 5:

$$(1.1) \qquad L^0 = \frac{\partial}{\partial t} + \sum_{k=1}^{d} a^k \frac{\partial}{\partial x^k} + \frac{1}{2} \sum_{k,l=1}^{d} \sum_{j=1}^{m} b^{k,j} b^{l,j} \frac{\partial^2}{\partial x^k \partial x^l},$$

$$(1.2) \qquad \underline{L}^0 = \frac{\partial}{\partial t} + \sum_{k=1}^{d} \underline{a}^k \frac{\partial}{\partial x^k}$$

and

$$(1.3) \qquad L^j = \underline{L}^j = \sum_{k=1}^{d} b^{k,j} \frac{\partial}{\partial x^k}$$

for $j = 1, 2, \ldots, m$, where

$$(1.4) \qquad \underline{a}^k = a^k - \frac{1}{2} \sum_{j=1}^{m} \underline{L}^j b^{k,j}$$

for $k = 1, 2, \ldots, d$. In addition, as in Chapter 5, we shall abbreviate multiple Ito integrals by

$$(1.5) \qquad I_{(j_1,\ldots,j_l)} = \int_{\tau_n}^{\tau_{n+1}} \cdots \int_{\tau_n}^{s_2} dW_{s_1}^{j_1} \cdots dW_{s_l}^{j_l}$$

and multiple Stratonovich integrals by

$$(1.6) \qquad J_{(j_1,\ldots,j_l)} = \int_{\tau_n}^{\tau_{n+1}} \cdots \int_{\tau_n}^{s_2} odW_{s_1}^{j_1} \cdots o\, dW_{s_l}^{j_l}$$

for $j_1, \ldots, j_l \in \{0, 1, \ldots, m\}$, $l = 1, 2, \ldots$ and $n = 0, 1, \ldots$ with the convention that

$$(1.7) \qquad W_t^0 = t$$

for all $t \in \Re^+$. We shall also use the abbreviation

$$f = f(\tau_n, Y_n),$$

for $n = 0, 1, \ldots$, in the schemes for any given function $f$ defined on $\Re^+ \times \Re^d$, and usually not explicitly mention the initial value $Y_0$ or the step numbers $n = 0, 1, \ldots$.

If we refer to the order of convergence of a scheme in this chapter, we shall mean the order of strong convergence (9.6.3) of a corresponding time discrete approximation. Finally, we shall always suppose that the Ito process $X$ under consideration satisfies the, in general, nonautonomous stochastic differential equation

$$(1.8) \qquad X_t = X_0 + \int_0^t a(s, X_s)\, ds + \sum_{j=1}^m \int_0^t b^j(s, X_s)\, dW_s^j$$

in Ito form or

$$(1.9) \qquad X_t = X_0 + \int_0^t \underline{a}(s, X_s)\, ds + \sum_{j=1}^m \int_0^t b^j(s, X_s) \circ dW_s^j$$

in its equivalent Stratonovich form, for $t \in [0, T]$.

## 10.2  The Euler Scheme

We shall begin with the Euler scheme, also called the Euler–Maruyama scheme, which we have already looked at in Chapter 9. It represents the simplest strong Taylor approximation and, generally, as we shall see, attains the order of strong convergence $\gamma = 0.5$.

In the 1-dimensional case $d = m = 1$ the *Euler scheme* has the form

$$(2.1) \qquad Y_{n+1} = Y_n + a\, \Delta + b\, \Delta W,$$

where

$$(2.2) \qquad \Delta = \tau_{n+1} - \tau_n = I_{(0)} = J_{(0)}$$

is the length of the time discretization subinterval $[\tau_n, \tau_{n+1}]$ and

$$(2.3) \qquad \Delta W = W_{\tau_{n+1}} - W_{\tau_n}$$

is the $N(0; \Delta)$ increment of the Wiener process $W$ on $[\tau_n, \tau_{n+1}]$.

In the multi-dimensional case with scalar noise, $d = 1, 2, \ldots$ and $m = 1$, the $k$th component of the *Euler scheme* is given by

(2.4) $$Y_{n+1}^k = Y_n^k + a^k \Delta + b^k \Delta W,$$

for $k = 1, 2, \ldots, d$, where the drift and diffusion coefficients are $d$-dimensional vectors $a = (a^1, \ldots, a^d)$ and $b = (b^1, \ldots, b^d)$. We note that the different components of the time discrete approximation $Y$ are coupled through the drift and diffusion coefficients, as in the corresponding stochastic differential equation (1.8).

For the general multi-dimensional case with $d, m = 1, 2, \ldots$ the $k$th component of the *Euler scheme* has the form

(2.5) $$Y_{n+1}^k = Y_n^k + a^k \Delta + \sum_{j=1}^{m} b^{k,j} \Delta W^j.$$

Here

(2.6) $$\Delta W^j = W_{\tau_{n+1}}^j - W_{\tau_n}^j = I_{(j)} = J_{(j)}$$

is the $N(0; \Delta)$ distributed increment of the $j$th component of the $m$-dimensional standard Wiener process $W$ on $[\tau_n, \tau_{n+1}]$, and $\Delta W^{j_1}$ and $\Delta W^{j_2}$ are independent for $j_1 \neq j_2$. The diffusion coefficient $b = [b^{k,j}]$ is a $d \times m$-matrix.

The Euler scheme (2.5) obviously corresponds to the truncated Ito-Taylor expansion containing only the time and Wiener integrals of multiplicity one, so an Euler approximation can be interpreted as an order 0.5 strong Ito-Taylor approximation. See (5.5.3).

**PC-Exercise 10.2.1** *Determine explicitly the 1-dimensional Ito process $X$ satisfying*

$$dX_t = -\frac{1}{2} X_t \, dt + X_t \, dW_t^1 + X_t \, dW_t^2$$

*on the time interval $[0, T]$ with $T = 1$ for the initial value $X_0 = 1$, where $W^1$ and $W^2$ are two independent standard Wiener processes (see (4.4.59)). Then simulate $M = 20$ batches each of $N = 100$ trajectories of $X$ and their Euler approximations corresponding to the same sample paths of the Wiener processes with equidistant time steps of step size $\Delta = 2^{-3}$. Determine the 90%-confidence interval for the absolute error $\epsilon$ at time $T$; see PC-Exercise 9.3.5. Repeat the calculations for step sizes $2^{-4}$, $2^{-5}$ and $2^{-6}$, and plot $\log_2 \epsilon$ against $\log_2 \Delta$.*

In Theorem 10.2.2 below, we shall show, assuming Lipschitz and linear growth conditions on the coefficients $a$ and $b$, that the Euler approximation has the order of strong convergence $\gamma = 0.5$, as is suggested by the slope 0.5 of the curve in Figure 10.2.1. In special cases the Euler scheme may actually achieve a higher order of strong convergence. For example, when the noise is *additive*, that is when the diffusion coefficient has the form

(2.7) $$b(t, x) \equiv b(t)$$

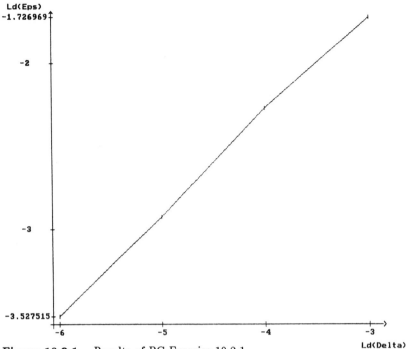

**Figure 10.2.1**    Results of PC-Exercise 10.2.1.

for all $(t, x) \in \Re^+ \times \Re^d$ and under appropriate smoothness assumptions on $a$ and $b$ it turns out that the Euler scheme has order of strong convergence $\gamma = 1.0$. See Theorem 10.3.4.

Usually the Euler scheme gives good numerical results when the drift and diffusion coefficients are nearly constant. In general, however, it is not particularly satisfactory and the use of higher order schemes is recommended.

**Theorem 10.2.2**    *Suppose that*

$$(2.8) \qquad\qquad E\left(|X_0|^2\right) < \infty,$$

$$(2.9) \qquad\qquad E\left(\left|X_0 - Y_0^\delta\right|^2\right)^{1/2} \leq K_1\, \delta^{1/2},$$

$$(2.10) \qquad |a(t, x) - a(t, y)| + |b(t, x) - b(t, y)| \leq K_2\, |x - y|,$$

$$(2.11) \qquad\qquad |a(t, x)| + |b(t, x)| \leq K_3\, (1 + |x|)$$

*and*

$$(2.12) \qquad |a(s, x) - a(t, x)| + |b(s, x) - b(t, x)| \leq K_4\, (1 + |x|)\, |s - t|^{1/2}$$

*for all $s$, $t \in [0, T]$ and $x$, $y \in \Re^d$, where the constants $K_1, \ldots, K_4$ do not depend on $\delta$. Then, for the Euler approximation $Y^\delta$ the estimate*

$$(2.13) \qquad E\left(\left|X_T - Y^\delta(T)\right|\right) \le K_5\,\delta^{1/2}$$

*holds, where the constant $K_5$ does not depend on $\delta$.*

**Proof** In view of the Lipschitz condition (2.10) and the linear growth condition (2.11), it follows from estimate (4.5.16) for the Ito process $X$ that

$$(2.14) \qquad E\left(\sup_{0 \le s \le T} |X_s|^2 \,\Big|\, \mathcal{A}_0\right) \le C_1\left(1 + |X_0|^2\right).$$

By arguments analogous to those in the proof of (4.5.16) and Lemma 10.8.1 we can also show

$$(2.15) \qquad E\left(\sup_{0 \le s \le T} |Y^\delta(s)|^2 \,\Big|\, \mathcal{A}_0\right) \le C_2\left(1 + |Y_0^\delta|^2\right),$$

where the constant $C_2$ does not depend on $\delta$, for the Euler approximation $Y^\delta$ interpolated continuously by

$$(2.16) \qquad Y^\delta(t) = Y_n^\delta + \int_{\tau_n}^t a\left(\tau_n, Y_n^\delta\right)\,ds + \sum_{j=1}^m \int_{\tau_n}^t b^j\left(\tau_n, Y_n^\delta\right)\,dW_s^j$$

for $t \in [\tau_n, \tau_{n+1}]$, $n = 0, 1, \ldots$.
   Then, from (1.8) and (2.16) we obtain

$$(2.17) \qquad Z(t) = E\left(\sup_{0 \le s \le t} |X_s - Y^\delta(s)|^2 \,\Big|\, \mathcal{A}_0\right)$$

$$\le C_3\left\{|X_0 - Y_0^\delta|^2 + \sum_{j=0}^m \left(R_t^{(j)} + S_t^{(j)} + T_t^{(j)}\right)\right\},$$

where the terms being summed will be defined below and upper bounds determined for them. From (2.10), (5.2.6), (5.3.3) and Lemma 10.8.1 we have

$$(2.18) \quad R_t^{(j)} := E\left(\sup_{0 \le s \le t}\left|\sum_{n=0}^{n_s-1} I_{(j)}\left[f_{(j)}\left(\tau_n, X_{\tau_n}\right) - f_{(j)}\left(\tau_n, Y_n^\delta\right)\right]_{\tau_n, \tau_{n+1}}\right.\right.$$

$$\left.\left. + I_{(j)}\left[f_{(j)}\left(\tau_{n_s}, X_{\tau_{n_s}}\right) - f_{(j)}\left(\tau_{n_s}, Y_{n_s}^\delta\right)\right]_{\tau_{n_s}, s}\right|^2 \,\Big|\, \mathcal{A}_0\right)$$

$$\le C_4 \int_0^t E\left(\sup_{0 \le s \le u}\left|f_{(j)}\left(\tau_{n_s}, X_{\tau_{n_s}}\right) - f_{(j)}\left(\tau_{n_s}, Y_{n_s}^\delta\right)\right|^2 \,\Big|\, \mathcal{A}_0\right)\,du$$

$$\le C_4 K_2^2 \int_0^t Z(u)\,du$$

for each $j = 0, \ldots, m$ and $t \in [0, T]$. Further, from Theorem 4.5.4 and Lemma 10.8.1 we obtain

$$(2.19) \quad S_t^{(j)} := E\left(\sup_{0 \le s \le t}\left|\sum_{n=0}^{n_s - 1} I_{(j)}\left[f_{(j)}(\tau_n, X_{\tau_n}) - f_{(j)}(\tau_n, X.)\right]_{\tau_n, \tau_{i+1}}\right.\right.$$

$$\left.\left. + I_{(j)}\left[f_{(j)}(\tau_{n_s}, X_{\tau_{n_s}}) - f_{(j)}(\tau_{n_s}, X.)\right]_{\tau_{n_s}, s}\right|^2 \Big| A_0\right)$$

$$\le C_5 \int_0^t E\left(\sup_{0 \le s \le u}\left|X_{\tau_{n_s}} - X_s\right|^2 \Big| A_0\right) du$$

$$\le C_6 \left(1 + |X_0|^2\right) \delta$$

for each $j = 0, \ldots, m$ and $t \in [0, T]$. Finally, it follows analogously that

$$(2.20) \quad T_t^{(j)} := E\left(\sup_{0 \le s \le t}\left|\sum_{n=0}^{n_s - 1} I_{(j)}\left[f_{(j)}(\tau_n, X.) - f_{(j)}(\cdot, X.)\right]_{\tau_n, \tau_{n+1}}\right.\right.$$

$$\left.\left. + I_{(j)}\left[f_{(j)}(\tau_{n_s}, X.) - f_{(j)}(\cdot, X.)\right]_{\tau_{n_s}, s}\right|^2 \Big| A_0\right)$$

$$\le C_7 \left(1 + |X_0|^2\right) \delta$$

for each $j = 0, \ldots, m$ and $t \in [0, T]$.

Combining (2.18) to (2.20) we have

$$(2.21) \qquad Z(t) \le C_3 \left|X_0 - Y_0^\delta\right|^2 + C_8 \left(1 + |X_0|^2\right) \delta + C_9 \int_0^t Z(u)\, du,$$

to which we apply the Gronwall inequality (Lemma 4.5.1) to obtain

$$(2.22) \qquad Z(t) \le C_{10} \left\{\left|X_0 - Y_0^\delta\right|^2 + \left(1 + |X_0|^2\right) \delta\right\}$$

for each $t \in [0, T]$. The conclusion of the theorem then follows immediately from the definition (2.17) of $Z(t)$. $\square$

**Remark 10.2.3**    In the above proof we have, in fact, proven a stronger result than that asserted in the theorem, namely we have established a uniform error bound over the whole time interval $[0, T]$ rather than an error bound at just the final instant $T$.

## 10.3 The Milstein Scheme

We shall now examine a scheme proposed by Milstein, which turns out to be an order 1.0 strong Taylor scheme.

If, in the 1-dimensional case with $d = m = 1$, we add to the Euler scheme (2.1) the term

$$bb' I_{(1,1)} = \frac{1}{2} bb' \left\{ (\Delta W)^2 - \Delta \right\}$$

from the Ito-Taylor expansion (5.5.3), then we obtain the *Milstein scheme*

$$(3.1) \qquad Y_{n+1} = Y_n + a \Delta + b \Delta W + \frac{1}{2} bb' \left\{ (\Delta W)^2 - \Delta \right\}$$

We note that we obtain the same scheme from the truncated Stratonovich-Taylor expansion (5.6.3) with the hierarchical set $\mathcal{A} = \{v, (0), (1), (1, 1)\}$, that is

$$Y_{n+1} = Y_n + \underline{a} \Delta + b \Delta W + \frac{1}{2} bb' (\Delta W)^2$$

where

$$\underline{a} = a - \frac{1}{2} bb'.$$

**Exercise 10.3.1**   *Show that the Milstein scheme (3.1) is strongly consistent for bounded $bb'$.*

In Theorem 10.6.3 we shall prove that the Milstein scheme has the order of strong convergence $\gamma = 1.0$ under the assumption that $a \in C^{1,1}(\Re^+ \times \Re^d)$ and $b \in C^{1,2}(\Re^+ \times \Re^d)$. Thus, with the addition of just one more term to the Euler scheme to form the Milstein scheme we increase the strong convergence order from $\gamma = 0.5$ to $\gamma = 1.0$. The strong order $\gamma = 1.0$ of the Milstein scheme corresponds to that of the Euler scheme in the deterministic case without any noise, that is with $b \equiv 0$. See Section 1 of Chapter 8. The additional term in the Milstein scheme marks the point of divergence of stochastic numerical analysis from the deterministic. In this sense we can regard the Milstein scheme as the proper generalization of the deterministic Euler scheme for the strong convergence criterion because it gives the same order of strong convergence as for the deterministic case.

**PC-Exercise 10.3.2**   *Consider the Ito process $X$ satisfying the linear stochastic differential equation*

$$dX_t = a X_t \, dt + b X_t \, dW_t$$

*on the time interval $[0, T]$ with $T = 1$ with $X_0 = 1.0$, $a = 1.5$, $b = 0.1$, which was investigated in Section 2 of Chapter 9. Generate $M = 20$ batches each of $N = 100$ simulations of $X_T$ and of the values $Y^\delta(T)$ of the Milstein approximation (3.1) corresponding to the same sample path of the Wiener process for equidistant time steps with $\delta = \Delta = 2^{-3}$. Then evaluate the 90%-confidence*

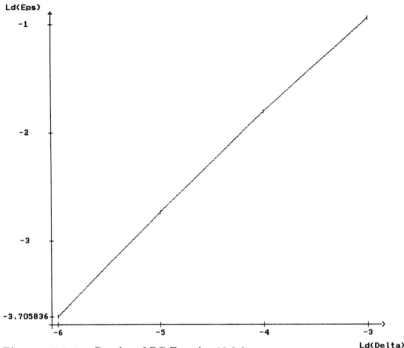

**Figure 10.3.1**    Results of PC-Exercise 10.3.2.

interval for the absolute error $\epsilon$. Repeat for step sizes $2^{-4}$, $2^{-5}$ and $2^{-6}$, and plot $\log_2 \epsilon$ against $\log_2 \Delta$.

Figure 10.3.1 suggests a linear dependence of the absolute error $\epsilon$ on the step size $\Delta$.

In the multi-dimensional case with $m = 1$ and $d = 1, 2, \ldots$ the $k$th component of the *Milstein scheme* is given by

$$(3.2) \qquad Y_{n+1}^k = Y_n^k + a^k\, \Delta + b^k\, \Delta W + \frac{1}{2}\left(\sum_{l=1}^{d} b^l \frac{\partial b^k}{\partial x^l}\right)\left\{(\Delta W)^2 - \Delta\right\}.$$

In the general multi-dimensional case with $d, m = 1, 2, \ldots$ the $k$th component of the *Milstein scheme* has the form

$$(3.3) \qquad Y_{n+1}^k = Y_n^k + a^k\, \Delta + \sum_{j=1}^{m} b^{k,j}\, \Delta W^j + \sum_{j_1,j_2=1}^{m} L^{j_1} b^{k,j_2} I_{(j_1,j_2)}$$

in terms of multiple Ito integrals $I_{(j_1,j_2)}$, or

$$(3.4) \qquad Y_{n+1}^k = Y_n^k + \underline{a}^k\, \Delta + \sum_{j=1}^{m} b^{k,j}\, \Delta W^j + \sum_{j_1,j_2=1}^{m} \underline{L}^{j_1} b^{k,j_2} J_{(j_1,j_2)}$$

if multiple Stratonovich integrals $J_{(j_1,j_2)}$ are used. The Milstein approximation thus represents both the order 1.0 strong Ito-Taylor approximation and the order 1.0 strong Stratonovich-Taylor approximation.

We remark that for $j_1 \neq j_2$ with $j_1$, $j_2 = 1$, ..., $m$ the multiple stochastic integrals

$$(3.5) \qquad J_{(j_1,j_2)} = I_{(j_1,j_2)} = \int_{\tau_n}^{\tau_{n+1}} \int_{\tau_n}^{s_1} dW_{s_2}^{j_1} \, dW_{s_1}^{j_2}$$

appearing in the scheme (3.4) cannot be so easily expressed in terms of the increments $\Delta W^{j_1}$ and $\Delta W^{j_2}$ of the components of the Wiener process as in the case $j_1 = j_2$ for which, from (5.2.21) and (5.2.44), we have

$$(3.6) \qquad I_{(j_1,j_1)} = \frac{1}{2} \left\{ (\Delta W^{j_1})^2 - \Delta \right\} \quad \text{and} \quad J_{(j_1,j_1)} = \frac{1}{2} \left( \Delta W^{j_1} \right)^2 .$$

The application of the Milstein scheme in the multi-dimensional case becomes practicable with the approximation of the multiple Stratonovich integrals proposed in Section 8 of Chapter 5. For $j_1 \neq j_2$ with $j_1$, $j_2 = 1$, ..., $m$ we can use (5.8.11) to approximate the integral $J_{(j_1,j_2)}$ by

$$(3.7) \qquad J_{(j_1,j_2)}^p = \Delta \left( \frac{1}{2} \xi_{j_1} \xi_{j_2} + \sqrt{\rho_p} \left( \mu_{j_1,p} \xi_{j_2} - \mu_{j_2,p} \xi_{j_1} \right) \right)$$

$$+ \frac{\Delta}{2\pi} \sum_{r=1}^{p} \frac{1}{r} \left( \zeta_{j_1,r} \left( \sqrt{2} \xi_{j_2} + \eta_{j_2,r} \right) - \zeta_{j_2,r} \left( \sqrt{2} \xi_{j_1} + \eta_{j_1,r} \right) \right)$$

where

$$(3.8) \qquad \rho_p = \frac{1}{12} - \frac{1}{2\pi^2} \sum_{r=1}^{p} \frac{1}{r^2}$$

and $\xi_j$, $\mu_{j,p}$, $\eta_{j,r}$ and $\zeta_{j,r}$ are independent $N(0; 1)$ Gaussian random variables with

$$(3.9) \qquad \xi_j = \frac{1}{\sqrt{\Delta}} \Delta W^j$$

for $j = 1$, ..., $m$, $r = 1$, ... $p$ and $p = 1, 2, ...$. The size of $p$ obviously influences the accuracy of $J_{(j_1,j_2)}^p$ as an approximation of $J_{(j_1,j_2)}$. We can see from (5.8.16) that we must choose

$$(3.10) \qquad p = p(\Delta) \geq \frac{K}{\Delta}$$

for some positive constant $K$ to obtain the order of strong convergence $\gamma = 1.0$ for the Milstein scheme with these approximations of the multiple Stratonovich integrals.

In many important practical problems the diffusion coefficients have special properties which allow the Milstein scheme to be simplified in a way that avoids the use of double stochastic integrals involving different components of the

Wiener process. For instance, with additive noise (2.7) the diffusion coefficients depend at most on time $t$ and not on the $x$ variable and the Milstein scheme reduces to the Euler scheme, which involves no double stochastic integrals.

Another important special case is that of *diagonal noise*, where $d = m$ and each component $X^k$ of the Ito process $X$ is disturbed only by the corresponding component $W^k$ of the Wiener process $W$ and the diagonal diffusion coefficient $b^{k,k}$ depends only on $x^k$, that is

$$(3.11) \qquad b^{k,j}(t,x) \equiv 0 \quad \text{and} \quad \frac{\partial b^{j,j}}{\partial x^k}(t,x) \equiv 0$$

for each $(t,x) \in \Re^+ \times \Re^d$ and $j, k = 1, \ldots, m$ with $j \neq k$. Thus for diagonal noise the components of the Ito process are coupled only through the drift term. It is easy to see that the *Milstein scheme for diagonal noise* reduces to

$$(3.12) \qquad Y_{n+1}^k = Y_n^k + a^k \Delta + b^{k,k} \Delta W^k + \frac{1}{2} b^{k,k} \frac{\partial b^{k,k}}{\partial x^k} \left\{ (\Delta W^k)^2 - \Delta \right\}.$$

A more general, but important special case is that of commutative noise in which the diffusion matrix satisfies the *commutativity condition*

$$(3.13) \qquad \underline{L}^{j_1} b^{k,j_2} = \underline{L}^{j_2} b^{k,j_1}$$

for all $j_1, j_2 = 1, \ldots, m$, $k = 1, \ldots, d$ and $(t,x) \in \Re^+ \times \Re^d$. For instance, additive noise, diagonal noise and *linear noise*, the last being

$$(3.14) \qquad b^{k,j}(t,x) = b^{k,j}(t) x^k$$

for all $x = (x^1, \ldots, x^d) \in \Re^d$, $t \in \Re^+$ and $j = 1, \ldots, d$, $k = 1, \ldots, m$, all satisfy the commutativity condition (3.13).

A closer inspection of the applications in Chapter 7 reveals that many of them involve commutative noise in one form or another. In formulating a model in terms of stochastic differential equations it is well worth checking whether noncommutative noise is essential, and avoiding it if possible.

In view of (3.6) and Proposition 5.2.3 we have

$$(3.15) \qquad I_{(j_1,j_2)} + I_{(j_2,j_1)} = \Delta W^{j_1} \Delta W^{j_2}$$

for $j_1, j_2 = 1, \ldots, m$ with $j_1 \neq j_2$. Inserting this into (3.4), we see that the *Milstein scheme for commutative noise* can be written as

$$(3.16) \qquad Y_{n+1}^k = Y_n^k + \underline{a}^k \Delta + \sum_{j=1}^m b^{k,j} \Delta W^j + \frac{1}{2} \sum_{j_1,j_2=1}^m \underline{L}^{j_1} b^{k,j_2} \Delta W^{j_1} \Delta W^{j_2}.$$

**PC-Exercise 10.3.3**    *Consider the scalar Ito process $X$ in PC-Exercise 10.2.1 with $d = 1$ and $m = 2$ on the time interval $[0,T]$ with $T = 1$. Compute $M = 20$ batches each of $N = 100$ simulations of $X_T$ and of the values $Y^\delta(T)$*

*of the Milstein scheme corresponding to the same sample paths of the Wiener processes for equidistant time steps with step size $\delta = \Delta = 2^{-3}$. Evaluate the 90% confidence interval for the absolute error $\epsilon$. Repeat for step sizes $2^{-4}$, $2^{-5}$ and $2^{-6}$, and plot $\log_2 \epsilon$ against $\log_2 \Delta$. Approximate the muliple integral $J_{(1,2)}$ by $J^p_{(1,2)}$ with $p = 2$. Repeat with $p = 10$.*

Our results for PC-Exercise 10.3.3, plotted in Figure 10.3.2, indicate a smaller error for the smaller step sizes when $p = 10$.

The following example due to Clark and Cameron shows that in the non-commutative case we need the multiple stochastic integrals $I_{(j_1,j_2)} = J_{(j_1,j_2)}$ for $j_1 \neq j_2$ to obtain the order of strong convergence $\gamma = 1.0$.

**Example 10.3.4**    *We consider in the two-dimensional case $d = m = 2$ an Ito process $X = (X^1, X^2)$ with components given by*

$$X^1_t = \int_0^t dW^1_s \quad and \quad X^2_t = \int_0^t X^1_s \, dW^2_s.$$

*Since*

$$\underline{L}^{j_1} b^{2,2} = 1 \neq 0 = \underline{L}^{j_2} b^{2,1}$$

*here, the commutativity condition (3.13) is not satisfied. Nevertheless, we can apply the numerical scheme (3.16) with equidistant time steps $\Delta$, obtaining*

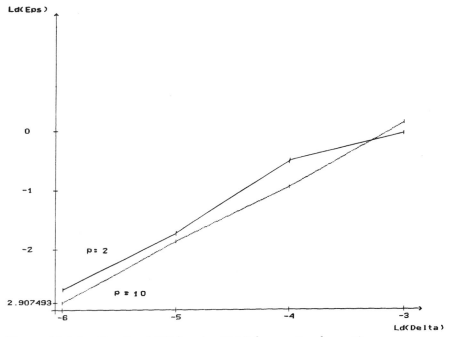

**Figure 10.3.2**    Results of PC-Exercise 10.3.3 for $p = 2$ and $p = 10$.

$$(3.17) \qquad \begin{aligned} Y_{n+1}^1 &= Y_n^1 + \Delta W^1 \\ Y_{n+1}^2 &= Y_n^2 + Y_n^1 \, \Delta W^2 + \frac{1}{2} \Delta W^1 \Delta W^2 \end{aligned}$$

for $n = 0, 1, \ldots$ with the initial value $Y_0^1 = Y_0^2 = 0$. Clark and Cameron showed that the mean-square error of the second component has the form

$$(3.18) \qquad \left( E \left( |Y^2(T) - X_T^2|^2 \right) \right)^{1/2} = \frac{1}{2} T^{1/2} \Delta^{1/2}.$$

Hence, for this particular example the scheme (3.16) attains only the strong order $\gamma = 0.5$, which is the same as for the Euler scheme. However, if we replace the term $\frac{1}{2} \Delta W^1 \Delta W^2$ in (3.17) by the double Ito integral $I_{(1,2)}$, then we have the general Milstein scheme (3.3) for our example and this has strong order $\gamma = 1.0$.

We shall now state conditions assuring that the Milstein scheme (3.3) has the strong order of convergence $\gamma = 1.0$.

**Theorem 10.3.5**   *Suppose that*

$$(3.19) \qquad E\left( |X_0|^2 \right) < \infty,$$

$$(3.20) \qquad E\left( |X_0 - Y_0^\delta|^2 \right)^{1/2} \leq K_1 \, \delta^{1/2},$$

$$(3.21) \qquad \begin{aligned} |\underline{a}(t,x) - \underline{a}(t,y)| &\leq K_2 \, |x - y| \\ \left| b^{j_1}(t,x) - b^{j_1}(t,y) \right| &\leq K_2 \, |x - y| \\ \left| \underline{L}^{j_1} b^{j_2}(t,x) - \underline{L}^{j_1} b^{j_2}(t,y) \right| &\leq K_2 \, |x - y| \end{aligned}$$

$$(3.22) \qquad \begin{aligned} |\underline{a}(t,x)| + \left| \underline{L}^j \underline{a}(t,x) \right| &\leq K_3 \, (1 + |x|) \\ \left| b^{j_1}(t,x) \right| + \left| \underline{L}^j b^{j_2}(t,x) \right| &\leq K_3 \, (1 + |x|) \\ \left| \underline{L}^j \underline{L}^{j_1} b^{j_2}(t,x) \right| &\leq K_3 \, (1 + |x|) \end{aligned}$$

*and*

$$(3.23) \qquad \begin{aligned} |\underline{a}(s,x) - \underline{a}(t,x)| &\leq K_4 \, (1 + |x|) \, |s - t|^{1/2} \\ \left| b^{j_1}(s,x) - b^{j_1}(t,x) \right| &\leq K_4 \, (1 + |x|) \, |s - t|^{1/2} \\ \left| \underline{L}^{j_1} b^{j_2}(s,x) - \underline{L}^{j_1} b^{j_2}(t,x) \right| &\leq K_4 \, (1 + |x|) \, |s - t|^{1/2} \end{aligned}$$

for all $s, t \in [0, T]$, $x, y \in \Re^d$, $j = 0, \ldots, m$ and $j_1, j_2 = 1, \ldots, m$, where the constants $K_1, \ldots, K_4$ do not depend on $\delta$.

*Then for the Milstein approximation $Y^\delta$ the estimate*

(3.24) $$E\left(\left|X_T - Y^\delta(T)\right|\right) \le K_5\,\delta$$

*holds, where the constant $K_5$ does not depend on $\delta$.*

The proof of Theorem 10.3.5 follows directly from the much stronger and more general assertion of Theorem 10.6.3, which we shall state and prove in Section 6.

**Exercise 10.3.6** *Derive (3.18) using the stochastic Ito-Taylor expansion (5.5.3).*

## 10.4 The Order 1.5 Strong Taylor Scheme

Generally said, we can obtain more accurate strong Taylor schemes by including further multiple stochastic integrals from the stochastic Taylor expansion in the scheme. These multiple stochastic integrals contain additional information about the sample path of the Wiener process. The necessity of their inclusion is a fundamental difference between the numerical analysis of stochastic differential equations and that of deterministic differential equations.

We shall now examine a Taylor scheme which has strong order $\gamma = 1.5$. By adding more terms from the Ito-Taylor expansion (5.5.3) to the Milstein scheme (3.1), in the autonomous 1-dimensional case $d = m = 1$ Platen and Wagner obtained the *order 1.5 strong Ito-Taylor scheme*

$$
\begin{aligned}
(4.1)\qquad Y_{n+1} \;=\; & Y_n + a\,\Delta + b\,\Delta W + \frac{1}{2}\,bb'\left\{(\Delta W)^2 - \Delta\right\} \\[2mm]
& + a'b\,\Delta Z + \frac{1}{2}\left(aa' + \frac{1}{2}b^2 a''\right)\Delta^2 \\[2mm]
& + \left(ab' + \frac{1}{2}b^2 b''\right)\left\{\Delta W\,\Delta - \Delta Z\right\} \\[2mm]
& + \frac{1}{2}b\left(bb'' + (b')^2\right)\left\{\frac{1}{3}(\Delta W)^2 - \Delta\right\}\Delta W.
\end{aligned}
$$

Here the additional random variable $\Delta Z$ is required to represent the double integral

(4.2) $$\Delta Z = I_{(1,0)} = \int_{\tau_n}^{\tau_{n+1}} \int_{\tau_n}^{s_2} dW_{s_1}\,ds_2.$$

We know from Exercise 5.2.7 that $\Delta Z$ is normally distributed with mean $E(\Delta Z) = 0$, variance $E((\Delta Z)^2) = \frac{1}{3}\Delta^3$ and covariance $E(\Delta Z \Delta W) = \frac{1}{2}\Delta^2$. In addition, as in PC-Exercise 1.4.12, the pair of correlated normally distributed random variables $(\Delta W, \Delta Z)$ can be determined from two independent $N(0;1)$ distributed random variables $U_1$ and $U_2$ by means of the transformation

(4.3)        $\Delta W = U_1 \sqrt{\Delta}, \qquad \Delta Z = \frac{1}{2} \Delta^{3/2} \left( U_1 + \frac{1}{\sqrt{3}} U_2 \right).$

All of the other multiple stochastic integrals appearing in the truncated Ito-Taylor expansion used to derive (4.1) can be expressed in terms of $\Delta$, $\Delta W$ and $\Delta Z$. In particular, the last term in (4.1) contains the triple Ito integral

(4.4)        $I_{(j,j,j)} = \frac{1}{2} \left\{ \frac{1}{3} \left( \Delta W^j \right)^2 - \Delta \right\} \Delta W^j,$

where, in this case, $j = 1$.

**Exercise 10.4.1**    *Derive the terms of the scheme (4.1) from the stochastic Taylor expansion (5.5.3) with the hierarchical set*

$$\{ v, (0), (1), (1,1), (0,1), (1,0), (0,0), (1,1,1) \}$$

*and show that it is strongly consistent if a, b and their derivatives of first and second order are bounded.*

In the deterministic case where $b \equiv 0$, the order 1.5 strong Ito-Taylor scheme (4.1) reduces to the deterministic 2nd order truncated Taylor method (8.2.3). There is obviously no counterpart to the order 1.5 Ito-Taylor scheme in the deterministic setting. In the next section we shall see that the 2.0 order strong

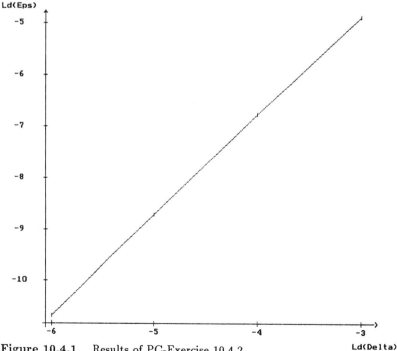

**Figure 10.4.1**    Results of PC-Exercise 10.4.2.

Ito-Taylor scheme, which turns out to be equivalent to the 2.0 order strong Stratonovich-Taylor scheme, is a better stochastic generalization of the deterministic 2nd order truncated Taylor method.

**PC-Exercise 10.4.2**     *Repeat PC-Exercise 10.3.2 with the order 1.5 strong Taylor approximation.*

In the multi-dimensional case with $d = 1, 2, \ldots$ and $m = 1$, that is with *scalar noise*, the $k$th component of the *order 1.5 strong Taylor scheme* is given by

$$(4.5) \qquad Y_{n+1}^k = Y_n^k + a^k \Delta + b^k \Delta W$$

$$+ \frac{1}{2} L^1 b^k \left\{ (\Delta W)^2 - \Delta \right\} + L^1 a^k \Delta Z$$

$$+ L^0 b^k \left\{ \Delta W \Delta - \Delta Z \right\} + \frac{1}{2} L^0 a^k \Delta^2$$

$$+ \frac{1}{2} L^1 L^1 b^k \left\{ \frac{1}{3} (\Delta W)^2 - \Delta \right\} \Delta W.$$

In the general multi-dimensional case with $d, m = 1, 2, \ldots$ the $k$th component of the *order 1.5 strong Taylor scheme* takes the form

$$(4.6) \qquad Y_{n+1}^k = Y_n^k + a^k \Delta + \frac{1}{2} L^0 a^k \Delta^2$$

$$+ \sum_{j=1}^{m} \left( b^{k,j} \Delta W^j + L^0 b^{k,j} I_{(0,j)} + L^j a^k I_{(j,0)} \right)$$

$$+ \sum_{j_1, j_2 = 1}^{m} L^{j_1} b^{k,j_2} I_{(j_1, j_2)}$$

$$+ \sum_{j_1, j_2, j_3 = 1}^{m} L^{j_1} L^{j_2} b^{k,j_3} I_{(j_1, j_2, j_3)}.$$

As already happened in the general multi-dimensional version of the Milstein scheme (3.3), here we also have multiple Ito integrals with respect to different components of the Wiener process. To implement such a scheme we can represent these multiple Ito integrals in terms of multiple Stratonovich integrals by means of the relations (5.2.36) and then use the approximations (5.8.11) for these multiple Stratonovich integrals:

Let $\xi_j, \zeta_{j,1}, \ldots, \zeta_{j,p}, \eta_{j,1}, \ldots, \eta_{j,p}, \mu_{j,p}$ and $\phi_{j,p}$ be independent standard Gaussian random variables. Then for $j, j_1, j_2, j_3 = 1, \ldots, m$ and $p = 1, 2, \ldots$ we have

$$(4.7) \qquad I_{(j)} = \Delta W^j = \sqrt{\Delta}\, \xi_j, \qquad I_{(j,0)} = \frac{1}{2} \Delta \left( \sqrt{\Delta}\, \xi_j + a_{j,0} \right) \qquad \text{with}$$

$$a_{j,0} = -\frac{\sqrt{2\Delta}}{\pi} \sum_{r=1}^{p} \frac{1}{r} \zeta_{j,r} - 2\sqrt{\Delta\rho_p} \, \mu_{j,p}$$

where

$$\rho_p = \frac{1}{12} - \frac{1}{2\pi^2} \sum_{r=1}^{p} \frac{1}{r^2};$$

$$I_{(0,j)} = \Delta W^j \, \Delta - I_{(j,0)}, \qquad I_{(j,j)} = \frac{1}{2} \left\{ (\Delta W^j)^2 - \Delta \right\},$$

$$I_{(j,j,j)} = \frac{1}{2} \left\{ \frac{1}{3} (\Delta W^j)^2 - \Delta \right\} \Delta W^j;$$

$$I_{(j_1,j_2)}^p = \frac{1}{2} \Delta \, \xi_{j_1} \xi_{j_2} - \frac{1}{2} \sqrt{\Delta} \, (\xi_{j_1} a_{j_2,0} - \xi_{j_2} a_{j_1,0}) + A_{j_1,j_2}^p \, \Delta$$

for $j_1 \neq j_2$ where

$$A_{j_1,j_2}^p = \frac{1}{2\pi} \sum_{r=1}^{p} \frac{1}{r} \left( \zeta_{j_1,r} \, \eta_{j_2,r} - \eta_{j_1,r} \, \zeta_{j_2,r} \right);$$

and

$$I_{(j_1,j_2,j_3)}^p = J_{(j_1,j_2,j_3)}^p - \frac{1}{2} \left( I_{\{j_1=j_2\}} I_{(0,j_3)} + I_{\{j_2=j_3\}} I_{(j_1,0)} \right)$$

with $J_{(j_1,j_2,j_3)}^p$ as defined in (5.8.11). From the estimate (5.8.16) we thus have

$$(4.8) \qquad E\left( |I_\alpha^p - I_\alpha|^2 \right) \leq C \frac{\Delta^2}{p}$$

for the above approximations of the multiple Ito integrals. We shall see from (10.6.16) that we need to choose $p$ so that

$$(4.9) \qquad p = p(\Delta) \geq \frac{K}{\Delta^2}$$

for an appropriate positive constant $K$ to ensure that the Taylor scheme (4.6) with these approximations of the multiple Ito integrals really does have order 1.5 of strong convergence.

In several important practical situations the Taylor scheme (4.6) reduces to a form in which the multiple stochastic integrals with respect to different components of the Wiener process do not appear. In the case of *additive noise* (2.7), where $b$ is a constant or depends only on time $t$, the order 1.5 strong Taylor scheme reduces to the form

$$(4.10) \quad Y_{n+1}^k = Y_n^k + a^k \, \Delta + \sum_{j=1}^{m} b^{k,j} \, \Delta W^j + \frac{1}{2} L^0 a^k \, \Delta^2$$

$$+ \sum_{j=1}^{m} \left[ L^j a^k \, \Delta Z^j + \frac{\partial}{\partial t} b^{k,j} \left\{ \Delta W^j \Delta - \Delta Z^j \right\} \right] \qquad \text{with}$$

(4.11) $$\Delta Z^j = I_{(j,0)},$$

for $j = 1, \ldots, m$, as in (4.2).

Another special case is that of *diagonal noise* (3.11) for which (4.6) reduces to

$$(4.12) \quad Y_{n+1}^k = Y_n^k + a^k \Delta + b^{k,k} \Delta W^k$$

$$+ \frac{1}{2} L^0 a^k \Delta^2 + \frac{1}{2} L^k b^{k,k} \left\{ (\Delta W^k)^2 - \Delta \right\}$$

$$+ L^0 b^{k,k} \left\{ \Delta W^k \Delta - \Delta Z^k \right\} + L^k a^k \Delta Z^k$$

$$+ \frac{1}{2} L^k L^k b^{k,k} \left\{ \frac{1}{3} (\Delta W^k)^2 - \Delta \right\} \Delta W^k.$$

More general than diagonal noise is what we shall call *commutative noise of the second kind*, that is for which the diffusion matrix $b$ satisfies the *second commutativity condition*

$$(4.13) \quad L^{j_1} L^{j_2} b^{k,j_3}(t, x) = L^{j_2} L^{j_1} b^{k,j_3}(t, x)$$

for $k = 1, \ldots, d$ and $j_1, j_2, j_3 = 1, \ldots, m$ for all $(t, x) \in \Re \times \Re^d$, as well as the commutativity condition (3.13). Now, in addition to (3.6), (3.15) and (4.4) we can derive from (5.2.16) the relations

$$(4.14) \quad I_{(j_1,j_2,j_3)} + I_{(j_2,j_1,j_3)} + I_{(j_2,j_3,j_1)} + I_{(j_3,j_2,j_1)} + I_{(j_3,j_1,j_2)} + I_{(j_1,j_3,j_2)}$$

$$= \begin{cases} \Delta W^{j_1} \Delta W^{j_2} \Delta W^{j_3} & : \quad j_1 \neq j_2, j_1 \neq j_3, j_2 \neq j_3 \\ \Delta W^{j_1} \left\{ (\Delta W^{j_2})^2 - \Delta \right\} & : \quad j_1 \neq j_2, j_1 \neq j_3, j_2 = j_3 \\ \left\{ (\Delta W^{j_1})^2 - 3\Delta \right\} \Delta W^{j_1} & : \quad j_1 = j_2 = j_3 \end{cases}$$

for $j_1, j_2, j_3 = 1, \ldots, m$. Using this we obtain the following version of the order 1.5 *strong Taylor scheme for commutative noise of the second kind*:

$$(4.15) \quad Y_{n+1}^k = Y_n^k + a^k \Delta + \sum_{j=1}^m b^{k,j} \Delta W^j$$

$$+ \frac{1}{2} L^0 a^k \Delta^2 + \frac{1}{2} \sum_{j=1}^m L^j b^{k,j} \left\{ (\Delta W^j)^2 - \Delta \right\}$$

$$+ \sum_{j_1=1}^m \sum_{j_2=1}^{j_1-1} L^{j_1} b^{k,j_2} \Delta W^{j_1} \Delta W^{j_2}$$

$$+ \sum_{j=1}^m \left( L^0 b^{k,j} \left\{ \Delta W^j \Delta - \Delta Z^j \right\} + L^j a^k \Delta Z^j \right)$$

$$+\frac{1}{2}\sum_{\substack{j_1,j_2=1\\j_1\neq j_2}}^{m} L^{j_1}L^{j_2}b^{k,j_2}\,\Delta W^{j_1}\left\{\left(\Delta W^{j_2}\right)^2-\Delta\right\}$$

$$+\sum_{j_1=1}^{m}\sum_{j_2=1}^{j_1-1}\sum_{j_3=1}^{j_2-1} L^{j_1}L^{j_2}b^{k,j_3}\,\Delta W^{j_1}\Delta W^{j_2}\Delta W^{j_3}$$

$$+\frac{1}{2}\sum_{j=1}^{m} L^j L^j b^{k,j}\left\{\frac{1}{3}\left(\Delta W^j\right)^2-\Delta\right\}\Delta W^j.$$

We note that it is not necessary here to generate the $\Delta Z^j$ when the drift and diffusion coefficients satisfy

(4.16) $$L^0 b^{k,j}(t,x)=L^j a^k(t,x)$$

for all $j=1,\ldots,m$ with $k=1,\ldots,d$ and all $(t,x)\in\Re\times\Re^d$.

Conditions under which the order 1.5 strong Taylor approximation actually achieves the order 1.5 of strong convergence will be given in Theorem 10.6.3, which we shall state and prove in Section 6. Finally, we remark that we could similarly derive an order 1.5 strong Stratonovich-Taylor scheme from the corresponding truncated Stratonovich-Taylor expansion. We shall not do so here because such a scheme would already contain most of the terms needed for an order 2.0 strong Taylor scheme and, in general, is not as efficient numerically as the order 1.5 strong Ito-Taylor scheme considered above.

**Exercise 10.4.3**     *Derive (4.14).*

## 10.5   The Order 2.0 Strong Taylor Scheme

In contrast with the last section, we shall now use the Stratonovich-Taylor expansion and retain those terms which are needed to obtain an order 2.0 strong scheme.

In the autonomous 1-dimensional case $d=m=1$ the *order 2.0 strong Taylor scheme*, due to the authors, has the form

(5.1) $$\begin{aligned}
Y_{n+1} &= Y_n+\underline{a}\,\Delta+b\,\Delta W+\frac{1}{2!}bb'(\Delta W)^2+b\underline{a}'\,\Delta Z\\[4pt]
&\quad+\frac{1}{2}\underline{a}\,\underline{a}'\,\Delta^2+\underline{a}b'\{\Delta W\,\Delta-\Delta Z\}\\[4pt]
&\quad+\frac{1}{3!}b\left(bb'\right)'(\Delta W)^3+\frac{1}{4!}b\left(b\left(bb'\right)'\right)'(\Delta W)^4\\[4pt]
&\quad+\underline{a}\left(bb'\right)'J_{(0,1,1)}+b\left(\underline{a}b'\right)'J_{(1,0,1)}\\[4pt]
&\quad\quad+b\left(b\underline{a}'\right)'J_{(1,1,0)}.
\end{aligned}$$

Here the Gaussian random variables $\Delta W$ and $\Delta Z$ are the same as in the preceding section. In particular, see (4.2) for the definition of $\Delta Z$. The next exercise shows how they can be generated approximately together with the multiple Stratonovich integrals of multiplicity 3 appearing in the above scheme

**Exercise 10.5.1** *Derive the following approximate multiple Stratonovich integrals:*

$$\Delta W = J^p_{(1)} = \sqrt{\Delta}\,\zeta_1, \qquad \Delta Z = J^p_{(1,0)} = \frac{1}{2}\Delta\left(\sqrt{\Delta}\,\zeta_1 + a_{1,0}\right),$$

$$J^p_{(1,0,1)} = \frac{1}{3!}\Delta^2\zeta_1^2 - \frac{1}{4}\Delta\,a_{1,0}^2 + \frac{1}{\pi}\Delta^{3/2}\,\zeta_1\,b_1 - \Delta^2\,B^p_{1,1},$$

$$J^p_{(0,1,1)} = \frac{1}{3!}\Delta^2\zeta_1^2 - \frac{1}{2\pi}\Delta^{3/2}\,\zeta_1\,b_1 + \Delta^2\,B^p_{1,1} - \frac{1}{4}\Delta^{3/2}\,a_{1,0}\,\zeta_1 + \Delta^2\,C^p_{1,1},$$

$$J^p_{(1,1,0)} = \frac{1}{3!}\Delta^2\zeta_1^2 + \frac{1}{4}\Delta\,a_{1,0}^2 - \frac{1}{2\pi}\Delta^{3/2}\,\zeta_1\,b_1 + \frac{1}{4}\Delta^{3/2}\,a_{1,0}\,\zeta_1 - \Delta^2\,C^p_{1,1}$$

*with*

$$a_{1,0} = -\frac{1}{\pi}\sqrt{2\Delta}\sum_{r=1}^{p}\frac{1}{r}\xi_{1,r} - 2\sqrt{\Delta\rho_p}\,\mu_{1,p}, \qquad \rho_p = \frac{1}{12} - \frac{1}{2\pi^2}\sum_{r=1}^{p}\frac{1}{r^2}$$

$$b_1 = \sqrt{\frac{\Delta}{2}}\sum_{r=1}^{p}\frac{1}{r^2}\eta_{1,r} + \sqrt{\Delta\alpha_p}\,\phi_{1,p}, \qquad \alpha_p = \frac{\pi^2}{180} - \frac{1}{2\pi^2}\sum_{r=1}^{p}\frac{1}{r^4}$$

$$B^p_{1,1} = \frac{1}{4\pi^2}\sum_{r=1}^{p}\frac{1}{r^2}\left(\xi_{1,r}^2 + \eta_{1,r}^2\right),$$

$$C^p_{1,1} = -\frac{1}{2\pi^2}\sum_{\substack{r,l=1 \\ r\neq l}}^{p}\frac{r}{r^2 - l^2}\left(\frac{1}{l}\xi_{1,r}\xi_{1,l} - \frac{l}{r}\eta_{1,r}\eta_{1,l}\right)$$

*where $\zeta_1$, $\xi_{1,r}$ $\eta_{1,r}$, $\mu_{1,p}$ and $\phi_{1,p}$ for $r = 1$, ..., $p$ and $p = 1, 2, \ldots$ denote independent standard Gaussian random variables; see Chapter 5.*

**PC-Exercise 10.5.2** *Repeat PC-Exercise 10.3.2 with the order 2.0 strong Taylor approximation (5.1). Recall from (5.2.39) that*

$$J_{(1,1,0)} + J_{(1,0,1)} + J_{(0,1,1)} = J_{(1,1)}\,\Delta, \qquad J_{(0,1)} + J_{(1,0)} = J_{(1)}\,\Delta$$

*and use Exercise 10.5.1 with $p = 5$ to approximate the required integrals.*

For the multi-dimensional case $d = 1, 2, \ldots$ with *scalar noise*, that is with $m = 1$, the $k$th component of the *order 2.0 strong Taylor scheme* is given by

$$(5.2) \qquad Y^k_{n+1} = Y^k_n + \underline{a}^k\,\Delta + b^k\,\Delta W + \frac{1}{2!}L^1 b^k\,(\Delta W)^2 + \underline{L}^1\underline{a}^k\,\Delta Z$$

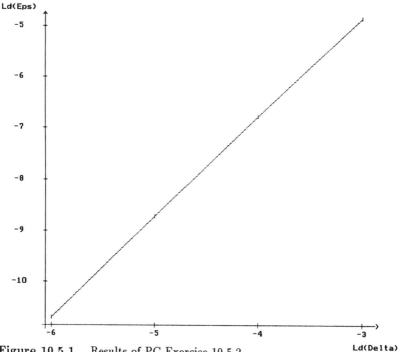

**Figure 10.5.1**     Results of PC-Exercise 10.5.2.

$$+\frac{1}{2}\underline{L}^0\underline{a}^k\,\Delta^2+\underline{L}^0 b^k\,\{\Delta W\,\Delta-\Delta Z\}$$

$$+\frac{1}{3!}\,\underline{L}^1\underline{L}^1 b^k\,(\Delta W)^3+\frac{1}{4!}\,\underline{L}^1\underline{L}^1\underline{L}^1 b^k\,(\Delta W)^4$$

$$+\underline{L}^0\underline{L}^1 b^k\,J_{(0,1,1)}+\underline{L}^1\underline{L}^0 b^k\,J_{(1,0,1)}$$

$$+\underline{L}^1\underline{L}^1\underline{a}^k\,J_{(1,1,0)},$$

where the operators $\underline{L}^j$ for $j=0,\,1,\,\ldots,\,m$ were defined in (1.2) and (1.3).

In the general multi-dimensional case with $d,\,m=1,\,2,\,\ldots$ the $k$th component of the *order* 2.0 *strong Taylor scheme* satisfies

$$(5.3)\quad Y^k_{n+1}\;=\;Y^k_n+\underline{a}^k\,\Delta+\frac{1}{2}\,\underline{L}^0\underline{a}^k\,\Delta^2$$

$$+\sum_{j=1}^m\left(b^{k,j}\,\Delta W^j+\underline{L}^0 b^{k,j}\,J_{(0,j)}+\underline{L}^j\underline{a}^k\,J_{(j,0)}\right)$$

$$+\sum_{j_1,j_2=1}^m\left(\underline{L}^{j_1}b^{k,j_2}J_{(j_1,j_2)}+\underline{L}^0\underline{L}^{j_1}b^{k,j_2}J_{(0,j_1,j_2)}\right.$$

$$\left.+\underline{L}^{j_1}\underline{L}^0 b^{k,j_2}J_{(j_1,0,j_2)}+\underline{L}^{j_1}\underline{L}^{j_2}\underline{a}^k\,J_{(j_1,j_2,0)}\right)$$

$$+ \sum_{j_1,j_2,j_3=1}^{m} \underline{L}^{j_1} \underline{L}^{j_2} b^{k,j_3} J_{(j_1,j_2,j_3)}$$

$$+ \sum_{j_1,j_2,j_3,j_4=1}^{m} \underline{L}^{j_1} \underline{L}^{j_2} \underline{L}^{j_3} b^{k,j_4} J_{(j_1,j_2,j_3,j_4)}.$$

Conditions under which the order 2.0 strong Taylor approximation converges strongly with order $\gamma = 2.0$ will be given in Theorem 10.6.3. These are smoothness and growth conditions on the drift and diffusion coefficients of the stochastic differential equations.

**Exercise 10.5.3**    *Derive the order 2.0 strong Taylor scheme (5.3) from a truncated Stratonovich-Taylor expansion and show that it is strongly consistent if $\underline{a}$, b and their derivatives of orders up to and including order 4 are bounded.*

The multiple Stratonovich integrals appearing in (5.3) can be represented or approximated similarly as in (5.8.11) and Exercise 10.5.1. In several special cases the scheme (5.3) reduces to simpler ones which avoid the use of some of the multiple Stratonovich integrals. In the autonomous case with *additive noise* (2.7) the *order 2.0 strong Taylor scheme* takes the form

$$(5.4) \qquad Y_{n+1}^k = Y_n^k + \underline{a}^k \Delta + \frac{1}{2} \underline{L}^0 \underline{a}^k \Delta^2 + \sum_{j=1}^{m} b^{k,j} \Delta W^j$$

$$+ \sum_{j=1}^{m} \underline{L}^j \underline{a}^k J_{(j,0)} + \sum_{j_1,j_2=1}^{m} \underline{L}^{j_1} \underline{L}^{j_2} \underline{a}^k J_{(j_1,j_2,0)}.$$

**Exercise 10.5.4**    *Derive the order 2.0 strong Taylor scheme with additive noise (5.4) from (5.3).*

As with the Milstein scheme, we would also obtain the above order 2.0 strong Taylor scheme if we started with an Ito-Taylor expansion instead of a Stratonovich -Taylor expansion. However, the Stratonovich approach is more convenient because of the simpler form of the coefficient functions and the simpler way in which the multiple Stratonovich integrals can be approximated. Moreover, in the proof that the scheme converges strongly with order $\gamma = 2.0$, the remainder term in the Ito-Taylor expansion involves higher order derivatives of the coefficients of the stochastic differential equation than in the Stratonovich-Taylor expansion.

Finally, we emphasize that it is well worth checking whether or not a substantial simplification of the strong Taylor scheme is possible when the stochastic differential equation has some special structure. We shall see an example of this in Section 4 of Chapter 11 for second order stochastic differential equations with additive noise, which commonly occur in noisy electrical and mechanical systems.

# 10.6    General Strong
## Ito-Taylor Approximations

In the preceding sections we considered strong Taylor schemes of up to order
two. Continuing in the same way by adding more terms from the stochastic
Taylor expansions, we can obtain higher order schemes. In this section we shall
describe, using more compact notation, the terms of the Ito-Taylor expansion
which are required for the corresponding strong Taylor scheme to achieve a
desired order of strong convergence. For this we shall use the notation of
Chapter 5 where we derived the stochastic Taylor expansions. We remind the
reader that $v$ denotes the multi-index of zero length $l(v) = 0$ and $I_v \equiv 1$. In
addition, writing $b^0 = a$, we have the coefficient functions

$$(6.1) \qquad f_\alpha(t, x) = L^{j_1} \cdots L^{j_{l-1}} b^{j_l}(t, x)$$

for all $(t, x) \in \Re \times \Re^d$ and all multi-indices $\alpha = (j_1, \ldots, j_l) \in \mathcal{M}$. We shall find
that the strong Ito-Taylor scheme of order $\gamma = 0.5, 1.0, 1.5, \ldots$ is associated
with the set of multi-indices

$$(6.2) \quad \mathcal{A}_\gamma = \left\{ \alpha \in \mathcal{M} : l(\alpha) + n(\alpha) \le 2\gamma \ \text{ or } \ l(\alpha) = n(\alpha) = \gamma + \frac{1}{2} \right\},$$

where $l(\alpha)$ denotes the length of the multi-index $\alpha$ and $n(\alpha)$ the number of its
zero components.

**Exercise 10.6.1**    *Show that $\mathcal{A}_\gamma$ for $\gamma = 0.5, 1.0, 1.5, \ldots$ is an hierarchical
set.*

**Exercise 10.6.2**    *Determine $\mathcal{A}_\gamma$ explicitly for $\gamma = 0, 0.5, 1.0, 1.5$ and $2.0$
when $m = 1$.*

Suppose we have a time discretization $(\tau)_\delta$ as defined in (9.5.3). For $\gamma = $
0.5, 1.0, 1.5, 2.0, ... in the general multi-dimensional case $d, m = 1, 2, \ldots$ we
define the *order $\gamma$ strong Ito-Taylor scheme* by the vector equation

$$(6.3) \qquad Y_{n+1} = \sum_{\alpha \in \mathcal{A}_\gamma} f_\alpha(\tau_n, Y_n) I_\alpha,$$

where $I_\alpha$ is the multiple Ito integral for the index $\alpha$ over the time interval $[\tau_n,
\tau_{n+1}]$. In view of Exercise 10.6.2, the strong Taylor schemes (6.3) of order $\gamma = $
0.5, 1.0, 1.5 and 2.0 coincide with those discussed earlier in this chapter.

The strong Ito-Taylor schemes (6.3) generate recursively approximate values
of the given Ito process at the discretization times. We can interpolate these
schemes in various ways to obtain general time discrete approximations as
defined in Section 5 of Chapter 9. We shall introduce a specific interpolation
to define the corresponding strong Ito-Taylor approximations in a way that
will allow us to prove a strong convergence result which holds uniformly on a
finite time interval. To do this we shall use the definition (5.2.12) of a multiple

Ito integral on a varying time interval and the definition of a general time discretization from Section 5 of Chapter 9. Let $\gamma = 0.5,\ 1.0,\ 1.5,\ 2.0,\ \ldots$ and $d$, $m = 1, 2, \ldots$. Then, for a given time discretization $(\tau)_\delta$ we define the general multi-dimensional *order $\gamma$ strong Ito-Taylor approximation* $Y = \{Y(t),\ t \geq 0\}$ by the vector equation

$$(6.4) \qquad Y(t) \;=\; Y_{n_t} + \sum_{\alpha \in \mathcal{A}_\gamma \setminus \{v\}} I_\alpha \left[ f_\alpha \left( \tau_{n_t}, Y_{n_t} \right) \right]_{\tau_{n_t}, t}$$

$$= \sum_{\alpha \in \mathcal{A}_\gamma} I_\alpha \left[ f_\alpha \left( \tau_{n_t}, Y_{n_t} \right) \right]_{\tau_{n_t}, t},$$

where $n_t$ was defined in (9.1.15), starting from a given $\mathcal{A}_0$-measurable random variable $Y(0)$.

It is obvious that the values of a strong Ito-Taylor approximation defined in this way coincide at the discretization times with those of the corresponding strong Ito-Taylor scheme. Such strong Ito-Taylor approximations are continuous processes and their sample paths display an irregular behaviour similar to that of the driving Wiener process. This would not be so if we had, for instance, used a piecewise constant or piecewise linear interpolation.

We shall now state a theorem for the convergence of strong Ito-Taylor approximations, which we shall prove at the end of the section.

**Theorem 10.6.3**   *Let* $Y^\delta = \{Y^\delta(t),\ t \in [0,T]\}$ *be the order $\gamma$ strong Ito-Taylor approximation, for a given* $\gamma = 0.5,\ 1.0,\ 1.5,\ 2.0,\ \ldots$, *corresponding to a time discretization* $(\tau)_\delta$, *where* $\delta \in (0,1)$. *Suppose that the coefficient functions* $f_\alpha$ *satisfy*

$$(6.5) \qquad |f_\alpha(t,x) - f_\alpha(t,y)| \leq K_1 |x - y|$$

*for all* $\alpha \in \mathcal{A}_\gamma$, $t \in [0,T]$ *and* $x,\ y \in \Re^d$;

$$(6.6) \qquad f_{-\alpha} \in C^{1,2} \quad and \quad f_\alpha \in \mathcal{H}_\alpha$$

*for all* $\alpha \in \mathcal{A}_\gamma \cup \mathcal{B}(\mathcal{A}_\gamma)$; *and*

$$(6.7) \qquad |f_\alpha(t,x)| \leq K_2 (1 + |x|)$$

*for all* $\alpha \in \mathcal{A}_\gamma \cup \mathcal{B}(\mathcal{A}_\gamma)$, $t \in [0,T]$ *and* $x \in \Re^d$. *Then*

$$(6.8) \qquad E\left( \sup_{0 \leq t \leq T} |X_t - Y^\delta(t)|^2 \Big| \mathcal{A}_0 \right)$$

$$\leq K_3 \left( 1 + |X_0|^2 \right) \delta^{2\gamma} + K_4 \left| X_0 - Y^\delta(0) \right|^2.$$

*The constants* $K_1$, $K_2$, $K_3$, *and* $K_4$ *here do not depend on* $\delta$.

As an immediate consequence of Theorem 10.6.3 and the Lyapunov inequality (1.4.12) we have

**Corollary 10.6.4** *in addition to the assumptions of Theorem 10.6.3 that*

$$(6.9) \qquad E\left(|X_0|^2\right) < \infty$$

*and*

$$(6.10) \qquad \sqrt{E\left(|X_0 - Y^\delta(0)|^2\right)} \le K_5\,\delta^\gamma.$$

*Then*

$$(6.11) \qquad E\left(\sup_{0 \le t \le T} |X_t - Y^\delta(t)|\right) \le K_6\,\delta^\gamma.$$

Thus, for the strong Ito-Taylor approximations defined by means of the above interpolation we have strong convergence of order $\gamma$ not only at the endpoint $T$, but also uniformly within the whole time interval $[0, T]$.

In fact, we could generalize Theorem 10.6.3 and Corollary 10.6.4 from the second moment to any $p$th moment for $p \ge 2$ if, instead of (6.9) and (6.10), we have

$$E\left(|X_0|^p\right) < \infty, \quad \left(E\left(|X_0 - Y^\delta(0)|^p\right)\right)^{1/p} \le K_5^*\,\delta^\gamma.$$

Then, using Jensen's inequality (1.4.10) and the Doob inequality (2.3.7) it can eventually be shown that

$$\left(E\left(\sup_{0 \le t \le T} |X_t - Y^\delta(t)|^p\right)\right)^{1/p} \le K_6^*\,\delta^\gamma.$$

In Section 4 we suggested a way of approximating the multiple Ito integrals appearing in the schemes there. This leads to the problem of determining the order of strong convergence of a strong Ito-Taylor scheme with the multiple Ito integrals $I_\alpha$ replaced by the approximate multiple Ito integrals $I_\alpha^p$. To be specific, for $\gamma = 1.0, 1.5, 2.0, \ldots$ we shall consider the scheme

$$(6.12) \qquad Y_{n+1} = \sum_{\alpha \in \mathcal{A}_\gamma} f_\alpha(\tau_n, Y_n) I_\alpha^p,$$

with $I_v^p \equiv 1$, where the parameter $p$ may depend on the maximum step size $\delta$.

**Corollary 10.6.5**   *Suppose in addition to the assumptions of Corollary 10.6.4 that*

$$(6.13) \qquad E\left(|I_\alpha - I_\alpha^p|^2\right) \le K_7\,\delta^{2\gamma+1}$$

*on each time interval $[\tau_n, \tau_{n+1}]$ for each $n = 0, 1, \ldots$ and all $\alpha \in \mathcal{A}_\gamma$. Then*

$$(6.14) \qquad E\left(|X_T - Y^\delta(T)|\right) \le K_8\,\delta^\gamma,$$

*where $Y^\delta$ is defined by (6.12).*

That is, the approximate scheme (6.12) also converges with strong order $\gamma$. Corollary 10.6.5 can be proved by an easy application of the arguments which will be used in the proof of Theorem 10.6.3, so we leave the proof to the reader.

In (5.8.16) we obtained the estimate

$$(6.15) \qquad E\left(|I_\alpha - I_\alpha^p|^2\right) \le C \frac{\delta^2}{p}$$

for sufficiently small maximum step size $\delta$ and all multi-indices $\alpha \in \mathcal{M}$ with length $l(\alpha) \le 3$. Thus, to obtain a strong scheme of order $\gamma = 1.0$, $1.5$ or $2.0$ we need to choose $p$ so that

$$(6.16) \qquad p \ge p(\delta) = \frac{C}{K_7} \delta^{1-2\gamma}.$$

This result can be easily generalized to higher orders too.

**Proof of Theorem 10.6.3** The proof will use the assertion of Lemma 10.8.1 which is stated and proved at the end of the chapter.

From the estimate (4.5.16) we have

$$(6.17) \qquad E\left(\sup_{0 \le s \le T} |X_s|^2 \,\big|\, \mathcal{A}_0\right) \le C_1\left(1 + |X_0|^2\right).$$

By similar arguments to those in the proof of Lemma 10.8.1 we can show that

$$(6.18) \qquad E\left(\sup_{0 \le s \le T} |Y^\delta(s)|^2 \,\big|\, \mathcal{A}_0\right) \le C_2\left(1 + |Y^\delta(0)|^2\right),$$

where the constant $C_2$ does not depend on the maximum step size $\delta$. In addition, from the Ito-Taylor expansion (5.5.3) we can represent the Ito process $X$ as

$$(6.19) \qquad X_\tau = \sum_{\alpha \in \mathcal{A}_\gamma} I_\alpha \left[f_\alpha\left(\rho, X_\rho\right)\right]_{\rho,\tau} + \sum_{\alpha \in \mathcal{B}(\mathcal{A}_\gamma)} I_\alpha \left[f_\alpha\left(\cdot, X.\right)\right]_{\rho,\tau}$$

for any two stopping times $\rho$ and $\tau$ with $0 \le \rho \le \tau \le T$, w.p.1. Thus, we can write

$$X_t = X_0 + \sum_{\alpha \in \mathcal{A}_\gamma \setminus \{v\}} \left\{ \sum_{n=0}^{n_t-1} I_\alpha \left[f_\alpha\left(\tau_n, X_{\tau_n}\right)\right]_{\tau_n, \tau_{n+1}} + I_\alpha \left[f_\alpha\left(\tau_{n_t}, X_{\tau_{n_t}}\right)\right]_{\tau_{n_t}, t} \right\}$$

$$(6.20)$$

$$+ \sum_{\alpha \in \mathcal{B}(\mathcal{A}_\gamma)} \left\{ \sum_{n=0}^{n_t-1} I_\alpha \left[f_\alpha(\cdot, X.)\right]_{\tau_n, \tau_{n+1}} + I_\alpha \left[f_\alpha(\cdot, X.)\right]_{\tau_{n_t}, t} \right\}.$$

From (6.20) and (6.4) we obtain

$$(6.21) \qquad Z(t) = E\left(\sup_{0 \le s \le t} |X_s - Y^\delta(s)|^2 \,\big|\, \mathcal{A}_0\right)$$

$$\le C_3\left(|X_0 - Y^\delta(0)|^2 + \sum_{\alpha \in \mathcal{A}_\gamma \setminus \{v\}} R_t^\alpha + \sum_{\alpha \in \mathcal{B}(\mathcal{A}_\gamma)} U_t^\alpha\right)$$

for all $t \in [0, T]$, where $R_t^\alpha$ and $U_t^\alpha$ will be defined below when they are estimated. In particular, from Lemma 10.8.1 and the Lipschitz condition (6.5) we have

$$(6.22) \quad R_t^\alpha := E\left( \sup_{0 \leq s \leq t} \left| \sum_{n=0}^{n_s - 1} I_\alpha \left[ f_\alpha \left( \tau_n, X_{\tau_n} \right) - f_\alpha \left( \tau_n, Y_n^\delta \right) \right]_{\tau_n, \tau_{n+1}} \right. \right.$$

$$\left. \left. + I_\alpha \left[ f_\alpha \left( \tau_{n_t}, X_{\tau_{n_t}} \right) - f_\alpha \left( \tau_{n_s}, Y_{n_t}^\delta \right) \right]_{\tau_{n_s}, s} \right|^2 \Bigg| \mathcal{A}_0 \right)$$

$$\leq C_4 \int_0^t E\left( \sup_{0 \leq s \leq u} \left| f_\alpha \left( \tau_{n_t}, X_{\tau_{n_t}} \right) - f_\alpha \left( \tau_{n_s}, Y_{n_t}^\delta \right) \right|^2 \Bigg| \mathcal{A}_0 \right) du$$

$$\leq C_4 K_1^2 \int_0^t Z(u) \, du$$

for all $\alpha \in \mathcal{A}_\gamma$.

In addition, for all $\alpha \in \mathcal{B}(\mathcal{A}_\gamma)$ from Lemma 10.8.1, (6.7) and (6.18) we have

$$(6.23) \quad U_t^\alpha := E\left( \sup_{0 \leq s \leq t} \left| \sum_{n=0}^{n_s - 1} I_\alpha \left[ f_\alpha \left( \cdot, X. \right) \right]_{\tau_n, \tau_{n+1}} \right. \right.$$

$$\left. \left. + I_\alpha \left[ f_\alpha \left( \cdot, X. \right) \right]_{\tau_{n_s}, s} \right|^2 \Bigg| \mathcal{A}_0 \right)$$

$$\leq C_5 \left( 1 + |X_0|^2 \right) \delta^{\phi(\alpha)},$$

where

$$(6.24) \qquad \phi(\alpha) = \begin{cases} 2(l(\alpha) - 1) & : \quad l(\alpha) = n(\alpha) \\ l(\alpha) + n(\alpha) - 1 & : \quad l(\alpha) \neq n(\alpha) \end{cases}.$$

For all such $\alpha$ we have $l(\alpha) \geq \gamma + 1$ when $l(\alpha) = n(\alpha)$ and $l(\alpha) + n(\alpha) \geq 2\gamma + 1$ when $l(\alpha) \neq n(\alpha)$. Thus, for all $\alpha \in \mathcal{B}(\mathcal{A}_\gamma)$, (6.23) gives

$$(6.25) \qquad\qquad U_t^\alpha \leq C_5 \left( 1 + |X_0|^2 \right) \delta^{2\gamma}.$$

Combining (6.21), (6.22) and (6.25) we obtain

$$Z(t) \leq C_7 \left| X_0 - Y^\delta(0) \right|^2 + C_8 \left( 1 + |X_0|^2 \right) \delta^{2\gamma} + C_9 \int_0^t Z(u) \, du$$

for all $t \in [0, T]$. By the assumed bounds (6.17) and (6.18) $Z(t)$ is bounded, so by the Gronwall inequality (Lemma 4.5.1) we obtain

$$Z(T) \leq K_3 \left( 1 + |X_0|^2 \right) \delta^{2\gamma} + K_4 \left| X_0 - Y^\delta(0) \right|^2,$$

which is the assertion of Theorem 10.6.3. □

# 10.7 General Strong Stratonovich-Taylor Approximations

We have seen in Sections 3 and 5 that the order 1.0 and 2.0 strong Taylor schemes have a simple structure when derived from Stratonovich-Taylor expansions. We shall now describe the terms that must be retained in such an expansion in order to obtain a scheme with an arbitrary desired order of strong convergence. In doing so we shall only consider schemes with integer order $\gamma$ = 1.0, 2.0, ..., since those of fractional order already contain almost all of the terms of the following integer order strong Taylor scheme.

For $\gamma$ = 1.0, 2.0, 3.0, ... we shall use the hierarchical set

$$(7.1) \qquad \mathcal{A}_\gamma = \{\alpha \in \mathcal{M} : l(\alpha) + n(\alpha) \le 2\gamma\},$$

which, for such integer values of $\gamma$ is equivalent to the set defined by (6.2). We remind the reader that the multiple Stratonovich integrals $J_\alpha$ and the Stratonovich coefficient functions $\underline{f}_\alpha$ were defined in (5.2.31) and (5.3.9), respectively, of Chapter 5.

In the general multi-dimensional case $d$, $m = 1, 2, \ldots$ we define the *order $\gamma$ strong Stratonovich-Taylor scheme* for $\gamma$ = 1.0, 2.0, 3.0, ... by the vector equation

$$(7.2) \qquad Y_{n+1} = \sum_{\alpha \in \mathcal{A}_\gamma} \underline{f}_\alpha(\tau_n, Y_n) J_\alpha.$$

Obviously, the schemes (7.2) for $\gamma$ = 1.0 and 2.0 coincide with those proposed in Sections 3 and 5.

Then, analogously with the general Ito-Taylor approximations considered in the previous section, for a given time discretization $(\tau)_\delta$ we define the general multi-dimensional *strong Stratonovich-Taylor approximation* $Y = \{Y(t), t \ge 0\}$ of order $\gamma$ = 1.0, 2.0, 3.0, ... by the vector equation

$$(7.3) \qquad Y(t) = \sum_{\alpha \in \mathcal{A}_\gamma} J_\alpha \left[ \underline{f}_\alpha(\tau_n, Y_n) \right]_{\tau_n, t}$$

for $t \in (\tau_n, \tau_{n+1}]$. Here $d$, $m = 1, 2, \ldots$ and $n = 0, 1, \ldots$. In addition, we have written, as usual, $Y_n = Y(\tau_n)$ and start from a given $\mathcal{A}_0$-measurable random variable $Y_0$. By interpolating the strong Stratonovich-Taylor scheme (7.2) in this way we also obtain a continuous approximation with sample paths which mimic the irregular behaviour of the underlying Wiener process.

We shall now state a theorem, the proof of which we shall sketch at the end of the section, for the convergence of the strong Stratonovich-Taylor approximations (7.3).

**Theorem 10.7.1** *Let $Y^\delta = \{Y^\delta(t), t \in [0, T]\}$ be the order $\gamma$ strong Stratonovich -Taylor approximation, for a given $\gamma$ = 1.0, 2.0, 3.0, ..., corresponding to a time discretization $(\tau)_\delta$, where $\delta \in (0, 1)$. Suppose that the coefficient* *functions $\underline{f}_\alpha$ satisfy*

(7.4)
$$\left| \underline{f}_\alpha(t, x) - \underline{f}_\alpha(t, y) \right| \leq K_1 \left| x - y \right|$$

*for all $\alpha \in \mathcal{A}_\gamma$, $t \in [0, T]$ and $x$, $y \in \Re^d$;*

(7.5)
$$\underline{f}_{-\alpha} \in \mathcal{C}^{1,1} \quad and \quad \underline{f}_\alpha \in \mathcal{H}_\alpha$$

*for all $\alpha \in \mathcal{A}_\gamma \cup \mathcal{B}(\mathcal{A}_\gamma)$; and*

(7.6)
$$\left| \underline{f}_\alpha(t, x) \right| \leq K_2 \left( 1 + |x| \right)$$

*for all $\alpha \in \mathcal{A}_\gamma \cup \mathcal{B}(\mathcal{A}_\gamma)$, all $t \in [0, T]$ and all $x \in \Re^d$. Then*

(7.7)
$$E \left( \sup_{0 \leq t \leq T} \left| X_t - Y^\delta(t) \right|^2 \Big| \mathcal{A}_0 \right)$$

$$\leq K_3 \left( 1 + |X_0|^2 \right) \delta^{2\gamma} + K_4 \left| X_0 - Y^\delta(0) \right|^2 .$$

*The constants $K_1$, $K_2$, $K_3$, and $K_4$ here do not depend on $\delta$.*

Comparing the strong Ito- and Stratonovich-Taylor approximations for $\gamma = 1.0, 2.0, \ldots$ and the assumptions of Theorems 10.6.3 and 10.7.1 we can see that the Stratonovich-Taylor approximation has a simpler structure and requires less smoothness of the drift and diffusion coefficients than the Ito-Taylor approximation.

As an immediate consequence of Theorem 10.7.1 we have the following corollary on strong convergence of order $\gamma$ uniformly on the time interval $[0, T]$ for the order $\gamma$ strong Stratonovich-Taylor approximation (7.3).

**Corollary 10.7.2**    *Suppose in addition to the assumptions of Theorem 10.7.1 that*

(7.8)
$$E \left( |X_0|^2 \right) < \infty$$

*and*

(7.9)
$$\sqrt{E \left( |X_0 - Y^\delta(0)|^2 \right)} \leq K_5 \, \delta^\gamma .$$

*Then*

(7.10)
$$E \left( \sup_{0 \leq t \leq T} \left| X_t - Y^\delta(t) \right| \right) \leq K_6 \, \delta^\gamma .$$

For computational purposes we can also replace the multiple Stratonovich integrals $J_\alpha$ appearing in (7.3 ) by the approximate multiple Stratonovich integrals $J_\alpha^p$ introduced in Section 8 of Chapter 5. That is, for $\gamma = 1.0, 2.0, 3.0,$ $\ldots$ we can use the scheme

(7.11)
$$Y_{n+1} = \sum_{\alpha \in \mathcal{A}_\gamma} \underline{f}_\alpha \left( \tau_n, Y_n \right) J_\alpha^p .$$

Here we write $J_v^p \equiv 1$ and, if necessary, can choose the parameter $p$ to depend on the maximum step size $\delta$.

**Corollary 10.7.3**   *Suppose in addition to the assumptions of Corollary 10.7.2 that*

$$(7.12) \qquad E\left(|J_\alpha - J_\alpha^p|^2\right) \le K_7 \delta^{2\gamma+1}$$

*on each time interval $[\tau_n, \tau_{n+1}]$ for each $n = 0, 1, \ldots$ and all $\alpha \in \mathcal{A}_\gamma$. Then*

$$(7.13) \qquad E\left(|X_T - Y^\delta(T)|\right) \le K_8 \delta^\gamma,$$

*where $Y^\delta$ is defined by (7.11).*

Thus, the approximate scheme (7.11) will also converge with strong order $\gamma$ if the multiple Stratonovich integrals are approximated appropriately.

Now, from (5.8.14) and (5.8.15) we have the estimate

$$(7.14) \qquad E\left(|J_\alpha - J_\alpha^p|^2\right) \le C \frac{\delta^2}{p}$$

for sufficiently small maximum step size $\delta$ and all multi-indices $\alpha \in \mathcal{M}$ with length $l(\alpha) \le 3$. We can thus obtain a strong scheme (7.11) of order $\gamma = 1.0$ if we choose $p$ such that

$$(7.15) \qquad p \ge p(\delta) = \frac{C}{K_7} \delta^{1-2\gamma}.$$

A similar result holds for higher orders too.

**Proof of Theorem 10.7.1**   The proof is similar to that of Theorem 10.6.3 for the Ito-Taylor approximation, so we shall just indicate its salient features.

From the Stratonovich-Taylor expansion (5.6.3) and equation (7.3) we obtain the following expression for the difference between the Ito process $X$ and the strong Stratonovich-Taylor approximation $Y^\delta$:

$$(7.16) \qquad X_t - Y^\delta(t) = X_0 - Y^\delta(0)$$

$$+ \sum_{\alpha \in \mathcal{A}_\gamma} \left\{ \sum_{n=1}^{n_t-1} J_\alpha \left[ \underline{f}_\alpha(\tau_n, X_{\tau_n}) - \underline{f}_\alpha(\tau_n, Y_n^\delta) \right]_{\tau_n, \tau_{n+1}} \right.$$

$$\left. + J_\alpha \left[ \underline{f}_\alpha(\tau_{n_t}, X_{\tau_{n_t}}) - \underline{f}_\alpha(\tau_{n_t}, Y_{\tau_{n_t}}^\delta) \right]_{\tau_{n_t}, t} \right\}$$

$$+ \sum_{\alpha \in \mathcal{B}(\mathcal{A}_\gamma)} \left\{ \sum_{n=1}^{n_t-1} J_\alpha \left[ \underline{f}_\alpha(\cdot, X\cdot) \right]_{\tau_n, \tau_{n+1}} + J_\alpha \left[ \underline{f}_\alpha(\cdot, X\cdot) \right]_{\tau_{n_t}, t} \right\}$$

for all $t \in [0, T]$.

To determine the uniform mean-square error between $X_t$ and $Y^\delta(t)$ we need to estimate

$$(7.17) \quad \underline{F}_t^\alpha := E\left(\sup_{0 \le s \le t}\left|\sum_{n=1}^{n_s-1} J_\alpha\left[g(\cdot)\right]_{\tau_n,\tau_{n+1}} + J_\alpha\left[g(\cdot)\right]_{\tau_{n_s},s}\right|^2 \Big| \mathcal{A}_0\right)$$

for all $\alpha \in \mathcal{A}_\gamma \cup \mathcal{B}(\mathcal{A}_\gamma)$. We can do this as in the proof of Lemma 10.8.1, except here we now have multiple Stratonovich integrals. Using the relation (5.2.34) between multiple Ito and Stratonovich integrals, together with Lemma 10.8.1 and Remark 5.2.8, we have

$$\underline{F}_t^\alpha \le K \, \delta^{\phi(\alpha)} \int_{t_0}^t E\left(\sup_{t_0 \le s \le u} |g(s)|^2 \Big| \mathcal{A}_0\right) du,$$

where $\phi$ was defined in (6.24) and $K$ does not depend on $\delta$. The rest of the proof is then similar to that of Theorem 10.6.3. □

## 10.8 A Lemma on Multiple Ito Integrals

We now state and prove a uniform mean-square estimate for multiple Ito integrals with respect to a given time discretization, which we have already referred to earlier in this chapter and shall use frequently in the sequel. The terminology used here is from Chapter 5.

**Lemma 10.8.1** *Suppose for a multi-index $\alpha \in \mathcal{M} \setminus \{v\}$, time discretization $(\tau)_\delta$ with $\delta \in (0,1)$ and right continuous adapted process $g \in H_\alpha$ that*

$$(8.1) \qquad R_{t_0,u} := E\left(\sup_{t_0 \le s \le u} |g(s)|^2 \Big| \mathcal{A}_{t_0}\right) < \infty,$$

*and let*

$$F_t^\alpha := E\left(\sup_{t_0 \le z \le t}\left|\sum_{n=0}^{n_z-1} I_\alpha\left[g(\cdot)\right]_{\tau_n,\tau_{n+1}} + I_\alpha\left[g(\cdot)\right]_{\tau_{n_z},z}\right|^2 \Big| \mathcal{A}_{t_0}\right).$$

*Then*

$$(8.2) \qquad F_t^\alpha \le \begin{cases} (T - t_0)\delta^{2(l(\alpha)-1)} \int_{t_0}^t R_{t_0,u}\, du & : \quad l(\alpha) = n(\alpha) \\ 4^{l(\alpha)-n(\alpha)+2}\delta^{l(\alpha)+n(\alpha)-1} \int_{t_0}^t R_{t_0,u}\, du & : \quad l(\alpha) \ne n(\alpha), \end{cases}$$

*w.p.1, for each $t \in [t_0, T]$.*

**Proof** When $l(\alpha) = n(\alpha)$ (5.2.12) implies that

$$F_t^\alpha = E\left(\sup_{t_0 \le z \le t}\left|\int_{t_0}^z I_{\alpha-}\left[g(\cdot)\right]_{\tau_{n_u},u}\, du\right|^2 \Big| \mathcal{A}_{t_0}\right)$$

$$\leq\ E\left(\sup_{t_0\leq z\leq t}(z-t_0)\int_{t_0}^{z}\left|I_{\alpha-}\left[g(\cdot)\right]_{\tau_{n_u},u}\right|^2 du\,\Big|\mathcal{A}_{t_0}\right)$$

$$\leq\ (T-t_0)\int_{t_0}^{t}E\left(E\left(\sup_{\tau_{n_u}\leq s\leq t}\left|I_{\alpha-}\left[g(\cdot)\right]_{\tau_{n_u},s}\right|^2\Big|\mathcal{A}_{\tau_{n_u}}\right)\Big|\mathcal{A}_{t_0}\right)du.$$

Using Lemma 5.7.3 we then obtain

(8.3) $\qquad F_t^{\alpha}\ \leq\ (T-t_0)4^{l(\alpha-)-n(\alpha-)}\delta^{l(\alpha-)+n(\alpha-)}$

$$\times\int_{t_0}^{t}E\left(\int_{\tau_{n_u}}^{u}R_{\tau_{n_u},s}\,ds\,\Big|\mathcal{A}_{t_0}\right)du$$

$$\leq\ (T-t_0)\delta^{l(\alpha-)+n(\alpha-)-1}\int_{t_0}^{t}E\left(R_{\tau_{n_u},u}\,\Big|\mathcal{A}_{t_0}\right)du$$

$$\leq\ (T-t_0)\delta^{2(l(\alpha)-1)}\int_{t_0}^{t}R_{t_0,u}\,du,$$

which is the desired result in this case.

Now suppose that $l(\alpha)\neq n(\alpha)$ where $\alpha=(j_1,\ \ldots,\ j_l)$ and $j_l=0$. Then

(8.4) $\qquad F_t^{\alpha}\ \leq\ 2E\left(\sup_{t_0\leq z\leq t}\left|\sum_{n=0}^{n_z-1}I_{\alpha}\left[g(\cdot)\right]_{\tau_n,\tau_{n+1}}\right|^2\Big|\mathcal{A}_{t_0}\right)$

$$+2E\left(\sup_{t_0\leq z\leq t}\left|I_{\alpha}\left[g(\cdot)\right]_{\tau_{n_z},z}\right|^2\Big|\mathcal{A}_{t_0}\right).$$

By Lemma 5.7.1 the discrete time process

$$\left\{\sum_{n=0}^{r}I_{\alpha}\left[g(\cdot)\right]_{\tau_n,\tau_{n+1}},r=0,1,\ldots\right\}$$

is a discrete square-integrable martingale, so with the aid of the Doob inequality (2.3.7), Lemma 5.7.1 and Lemma 5.7.3 we obtain

(8.5) $\qquad E\left(\sup_{t_0\leq z\leq t}\left|\sum_{n=0}^{n_z-1}I_{\alpha}\left[g(\cdot)\right]_{\tau_n,\tau_{n+1}}\right|^2\Big|\mathcal{A}_{t_0}\right)$

$$\leq\ \sup_{t_0\leq z\leq t}4E\left(\left|\sum_{n=0}^{n_z-1}I_{\alpha}\left[g(\cdot)\right]_{\tau_n,\tau_{n+1}}\right|^2\Big|\mathcal{A}_{t_0}\right)$$

$$\leq\ \sup_{t_0\leq z\leq t}4E\left(\left|\sum_{n=0}^{n_z-2}I_{\alpha}\left[g(\cdot)\right]_{\tau_n,\tau_{n+1}}\right|^2\right.$$

$$+2 \sum_{n=0}^{n_z-2} I_\alpha \left[g(\cdot)\right]_{\tau_n,\tau_{n+1}} E \left( I_\alpha \left[g(\cdot)\right]_{\tau_{n_z-1},\tau_{n_z}} \Big| \mathcal{A}_{\tau_{n_z-1}} \right)$$

$$\left. +E \left( \left| I_\alpha \left[g(\cdot)\right]_{\tau_{n_z-1},\tau_{n_z}} \right|^2 \Big| \mathcal{A}_{\tau_{n_z-1}} \right) \Big| \mathcal{A}_{t_0} \right)$$

$$\leq \sup_{t_0 \leq z \leq t} 4E \left( \left| \sum_{n=0}^{n_z-2} I_\alpha \left[g(\cdot)\right]_{\tau_n,\tau_{n+1}} \right|^2 \right.$$

$$\left. +4^{l(\alpha)-n(\alpha)} \delta^{l(\alpha)+n(\alpha)-1} \int_{\tau_{n_z-1}}^{\tau_{n_z}} R_{\tau_{n_z-1},u} \, du \Big| \mathcal{A}_{t_0} \right)$$

$$\leq \sup_{t_0 \leq z \leq t} 4E \left( \left| \sum_{n=0}^{n_z-3} I_\alpha \left[g(\cdot)\right]_{\tau_n,\tau_{n+1}} \right|^2 \right.$$

$$+4^{l(\alpha)-n(\alpha)} \delta^{l(\alpha)+n(\alpha)-1} \int_{\tau_{n_z-2}}^{\tau_{n_z-1}} R_{\tau_{n_z-2},u} \, du$$

$$\left. +4^{l(\alpha)-n(\alpha)} \delta^{l(\alpha)+n(\alpha)-1} \int_{\tau_{n_z-1}}^{z} R_{\tau_{n_z-2},u} \, du \Big| \mathcal{A}_{t_0} \right)$$

$$\leq \sup_{t_0 \leq z \leq t} 4E \left( 4^{l(\alpha)-n(\alpha)} \delta^{l(\alpha)+n(\alpha)-1} \int_{t_0}^{z} R_{t_0,u} \, du \Big| \mathcal{A}_{t_0} \right)$$

$$\leq 4^{l(\alpha)-n(\alpha)+1} \delta^{l(\alpha)+n(\alpha)-1} \int_{t_0}^{t} R_{t_0,u} \, du.$$

To bound the second term on the right of (8.4) we apply (5.2.12) and Lemma 5.7.3 to obtain

$$(8.6) \qquad E \left( \sup_{t_0 \leq z \leq t} \left| I_\alpha \left[g(\cdot)\right]_{\tau_{n_z},z} \right|^2 \Big| \mathcal{A}_{t_0} \right)$$

$$= E \left( \sup_{t_0 \leq z \leq t} \left| \int_{\tau_{n_z}}^{z} I_{\alpha-} \left[g(\cdot)\right]_{\tau_{n_z},u} \, du \right|^2 \Big| \mathcal{A}_{t_0} \right)$$

$$\leq E \left( \sup_{t_0 \leq z \leq t} (z - \tau_{n_z}) \int_{\tau_{n_z}}^{z} \left| I_{\alpha-} \left[g(\cdot)\right]_{\tau_{n_z},u} \right|^2 \, du \Big| \mathcal{A}_{t_0} \right)$$

$$\leq \delta \int_{t_0}^{t} E \left( E \left( \sup_{\tau_{n_u} \leq s \leq u} \left| I_{\alpha-} \left[g(\cdot)\right]_{\tau_{n_u},s} \right|^2 \Big| \mathcal{A}_{\tau_{n_u}} \right) \Big| \mathcal{A}_{t_0} \right) \, du$$

$$\leq \delta 4^{l(\alpha-)-n(\alpha-)} \int_{t_0}^{t} E \left( \int_{\tau_{n_u}}^{u} R_{\tau_{n_u},s} \, ds \, \delta^{l(\alpha-)+n(\alpha-)-1} \Big| \mathcal{A}_{t_0} \right) \, du$$

$$\leq 4^{l(\alpha)-n(\alpha)} \delta^{l(\alpha)+n(\alpha)-1} \int_{t_0}^{t} R_{t_0,u} \, du.$$

Combining (8.4), (8.6) and (8.7) gives the desired result in this case.

Finally, in the case that $l(\alpha) \neq n(\alpha)$ where $\alpha = (j_1, \ldots, j_l)$ and $j_l = 1, \ldots, m$, we also have a martingale and can use the Doob inequality. With the help of Lemma 5.7.3 we get

$$(8.7) \quad F_t^\alpha \;=\; E\left( \sup_{t_0 \leq z \leq t} \left| \int_{t_0}^z I_{\alpha-}[g(\cdot)]_{\tau_{n_u},u} \, dW_u^{j_l} \right|^2 \bigg| \mathcal{A}_{t_0} \right)$$

$$\leq \; 4 \sup_{t_0 \leq z \leq t} E\left( \left| \int_{t_0}^z I_{\alpha-}[g(\cdot)]_{\tau_{n_u},u} \, dW_u^{j_l} \right|^2 \bigg| \mathcal{A}_{t_0} \right)$$

$$\leq \; 4 \sup_{t_0 \leq z \leq t} \int_{t_0}^z E\left( E\left( \left| I_{\alpha-}[g(\cdot)]_{\tau_{n_u},u} \right|^2 \bigg| \mathcal{A}_{\tau_{n_u}} \right) \bigg| \mathcal{A}_{t_0} \right) \, du$$

$$\leq \; 4 \int_{t_0}^t E\left( E\left( \sup_{\tau_{n_u} \leq s \leq u} \left| I_{\alpha-}[g(\cdot)]_{\tau_{n_u},s} \right|^2 \bigg| \mathcal{A}_{\tau_{n_u}} \right) \bigg| \mathcal{A}_{t_0} \right) \, du$$

$$\leq \; 4 \, 4^{l(\alpha-)-n(\alpha-)} \int_{t_0}^t E\left( \int_{\tau_{n_u}}^u R_{\tau_{n_u},s} \, ds \, \delta^{l(\alpha-)+n(\alpha-)-1} \bigg| \mathcal{A}_{t_0} \right) \, du$$

$$\leq \; 4^{l(\alpha)-n(\alpha)} \delta^{l(\alpha)+n(\alpha)-1} \int_{t_0}^t R_{t_0,u} \, du.$$

This is the desired result in the final case, and thus completes the proof of the lemma. $\square$

# Chapter 11

# Explicit Strong Approximations

In this chapter we shall propose and examine strong schemes which avoid the use of derivatives in much the same way that Runge-Kutta schemes do in the deterministic setting. We shall also call these Runge-Kutta schemes, but it must be emphasized that they are not simply heuristic generalizations of deterministic Runge-Kutta schemes to stochastic differential equations. The notation and abbreviations of the last chapter will continue to be used, often without direct reference.

## 11.1 Explicit Order 1.0 Strong Schemes

A disadvantage of the strong Taylor approximations is that the derivatives of various orders of the drift and diffusion coefficients must be evaluated at each step, in addition to the coefficients themselves. This can make the implementation of such schemes a complicated undertaking. There are stochastic schemes which avoid the use of derivatives of the drift and diffusion coefficients. However, in view of the differences between deterministic and stochastic calculi, heuristic generalizations to stochastic differential equations of the widely used deterministic numerical schemes such as the Runge-Kutta schemes have limited value. Exercise 9.6.4 and the following PC-Exercise illustrate this problem.

**PC-Exercise 11.1.1**    *Consider the Ito process $X$ satisfying the stochastic differential equation*

(1.1) $$dX_t = a\, X_t\, dt + b\, X_t\, dW_t,$$

*with $X_0 = 1.0$, $a = 1.5$ and $b = 0.1$ on the time interval $[0, T]$, where $T = 1$, and the following heuristic generalization of the Heun method (8.1.12)*

(1.2) $$Y_{n+1} = Y_n + \frac{1}{2}\left\{a\left(\bar{\Upsilon}_n\right) + a\left(Y_n\right)\right\}\Delta_n + \frac{1}{2}\left\{b\left(\bar{\Upsilon}_n\right) + b\left(Y_n\right)\right\}\Delta W_n,$$

*with supporting value*

$$\bar{\Upsilon}_n = Y_n + a\left(Y_n\right)\Delta_n + b\left(Y_n\right)\Delta W_n$$

*and initial value $Y_0 = 1.0$; here $a(x) = ax$ and $b(x) = bx$. Generate $M = 20$ batches each of $N = 100$ simulations of $X_T$ and of the values $Y^\delta(T)$ of (1.2) corresponding to the same sample paths of the Wiener process for equidistant*

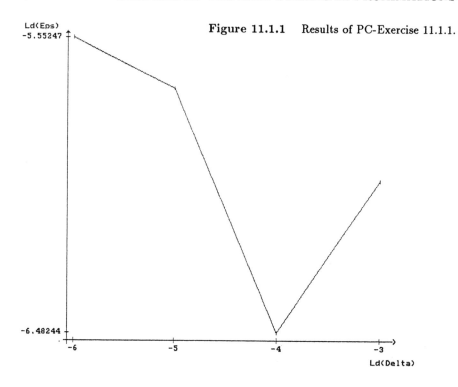

**Figure 11.1.1**    Results of PC-Exercise 11.1.1.

*time discretizations with step size $\delta = \Delta = 2^{-3}$, $2^{-4}$, $2^{-5}$ and $2^{-6}$. Plot $\log_2$ of the absolute error against $\log_2 \Delta$.*

While the formally generalized Heun method (1.2) may at first seem acceptable, the results of PC-Exercise 11.1.1 plotted in Figure 11.1.1 indicate that it does not converge. This should not be surprising, since we saw in Exercise 9.6.4 that such a scheme is generally not strongly consistent.

**Exercise 11.1.2**    *Show that the scheme*

$$Y_{n+1} = Y_n + \frac{1}{2} \left\{ a \left( \bar{\Upsilon}_n \right) + a \right\} \Delta_n + b \left( Y_n \right) \Delta W_n$$

*with $\bar{\Upsilon}_n$ as in (1.2) is strongly consistent.*

Several strongly consistent first order *derivative free schemes* can be derived from the Milstein scheme (10.3.3) simply by replacing the derivatives there by the corresponding difference ratios. However, these differences require the use of supporting values of the coefficients at additional points. An example is the following scheme, which we shall call the *explicit order* 1.0 *strong scheme*. We shall first consider some special cases.

In the 1-dimensional case with $d = m = 1$ an *explicit order* 1.0 *strong scheme* proposed by Platen is given by

$$(1.3) \qquad Y_{n+1} = Y_n + a\,\Delta + b\,\Delta W + \frac{1}{2\sqrt{\Delta}} \left\{ b(\tau_n, \bar{\Upsilon}_n) - b \right\} \left\{ (\Delta W)^2 - \Delta \right\}$$

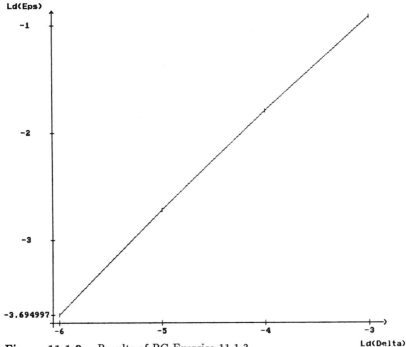

**Figure 11.1.2**    Results of PC-Exercise 11.1.3.

with the supporting value

$$(1.4) \qquad \bar{\Upsilon}_n = Y_n + a\,\Delta + b\,\sqrt{\Delta}.$$

Using the deterministic Taylor expasion it is easy to show that the ratio

$$\frac{1}{\sqrt{\Delta}} \left\{ b\left(\tau_n, Y_n + a\,\Delta + b\,\sqrt{\Delta}\right) - b(\tau_n, Y_n) \right\}$$

is a forward difference approximation for $b\,\frac{\partial b}{\partial x}$ at $(\tau_n, Y_n)$ if we neglect higher terms.

**PC-Exercise 11.1.3**    *Repeat PC-Exercise 11.1.1 using the explicit order 1.0 strong scheme (1.3)-(1.4). Compare the results with those for the Milstein scheme (10.3.1) in PC-Exercise 10.3.2.*

In the multi-dimensional case $d = 1, 2, \ldots$ with *scalar noise*, that is with $m = 1$, the $k$th component of the explicit *order* 1.0 *strong scheme* has the form

$$(1.5) \quad Y_{n+1}^k = Y_n^k + a^k\,\Delta + b^k\,\Delta W + \frac{1}{2\sqrt{\Delta}} \left\{ b^k(\tau_n, \bar{\Upsilon}_n) - b^k \right\} \left\{ (\Delta W)^2 - \Delta \right\}$$

with the vector supporting value

$$(1.6) \qquad \bar{\Upsilon}_n = Y_n + a\,\Delta + b\,\sqrt{\Delta}.$$

In the general multi-dimensional case $d$, $m = 1, 2, \ldots$ the $k$th component of the *explicit order 1.0 strong scheme* is

$$(1.7) \quad Y_{n+1}^k = Y_n^k + a^k\,\Delta + \sum_{j=1}^{m} b^{k,j}\,\Delta W^j$$

$$+ \frac{1}{\sqrt{\Delta}} \sum_{j_1,j_2=1}^{m} \left\{ b^{k,j_2}\left(\tau_n, \bar{\Upsilon}_n^{j_1}\right) - b^{k,j_2} \right\} I_{(j_1,j_2)}$$

with the vector supporting values

$$(1.8) \qquad \bar{\Upsilon}_n^j = Y_n + a\,\Delta + b^j\,\sqrt{\Delta}$$

for $j = 1, 2, \ldots$. We note that $b$ has to be evaluated $m + 1$ times here for each time step. There is also a Stratonovich version of (1.7) with $\underline{a}$ and $J_{(j_1,j_2)}$ substituted for $a$ and $I_{(j_1,j_2)}$.

In Section 5 we shall show that the explicit order 1.0 strong scheme converges with strong order $\gamma = 1.0$ under conditions similar to those for the Milstein scheme (10.3.3). In actual computations the double Ito integrals with respect to different components of the Wiener process in (1.7) can be approximated by the method proposed in Section 3 of Chapter 10.

In the most general case of the scheme (1.7)–(1.8) each component $b^{k,j}$ of the diffusion matrix must be evaluated at the $m + 1$ vector valued points $Y_n$, $\bar{\Upsilon}_n^1, \ldots, \bar{\Upsilon}_n^m$ for each discretization time. The number of such evalutions may be less in special cases. For example, in the autonomous case with additive noise (10.2.7) the scheme (1.7)–(1.8) is just the Euler scheme, in which the $b^{k,j}$ need to be evaluated at $Y_n$ only. For *diagonal noise* (10.3.11) the scheme (1.7)–(1.8) reduces to

$$(1.9) \quad Y_{n+1}^k = Y_n^k + a^k\,\Delta + b^{k,k}\,\Delta W^k$$

$$+ \frac{1}{2\sqrt{\Delta}} \left\{ b^{k,k}\left(\tau_n, \bar{\Upsilon}_n^k\right) - b^{k,k} \right\} \left\{ (\Delta W^k)^2 - \Delta \right\}$$

with the vector supporting value

$$(1.10) \qquad \bar{\Upsilon}_n^k = Y_n + a\,\Delta + b^k\,\sqrt{\Delta}.$$

Here we require only two evaluations for each $b^{k,k}$. When the noise is *commutative*, as in (10.3.13), the scheme (1.7)–(1.8) has the Stratonovich analogue

$$(1.11) \qquad Y_{n+1}^k = Y_n^k + \underline{a}^k\,\Delta + \frac{1}{2} \sum_{j=1}^{m} \left\{ b^{k,j}\left(\tau_n, \bar{\Upsilon}_n\right) + b^{k,j} \right\} \Delta W^j$$

with the vector supporting values

$$(1.12) \qquad \bar{\Upsilon}_n = Y_n + \underline{a}\,\Delta + \sum_{j=1}^{m} b^j\,\Delta W^j,$$

where $\underline{a}$ denotes the corrected Stratonovich drift (10.1.4). Here we need to evaluate each $b^{k,j}$ at two different points for each time step. This scheme turns out to be rather convenient.

**Exercise 11.1.4**    *Write out the explicit order 1.0 strong scheme (1.11) for the scalar stochastic differential equation*

$$(1.13) \qquad dX_t = -\frac{1}{2}\,X_t\,dt + X_t\,dW_t^1 + X_t\,dW_t^2,$$

*where $W^1$ and $W^2$ are independent scalar Wiener processes. Does (1.13) have commutative noise?*

**PC-Exercise 11.1.5**    *Repeat PC-Exercise 10.2.1 for equation (1.13) with the scheme (1.11).*

In fact, we could omit the $\underline{a}\,\Delta$ term in (1.12) and (1.14) without losing the strong order of convergence $\gamma = 1.0$.

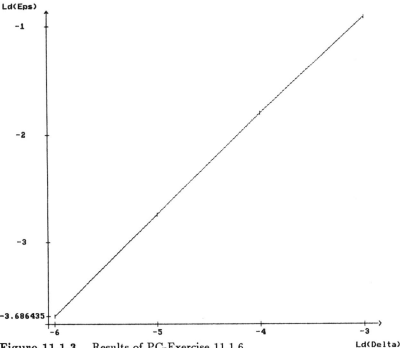

**Figure 11.1.3**    Results of PC-Exercise 11.1.6.

**PC-Exercise 11.1.6**    *Repeat PC-Exercise 11.1.1 for the explicit order* 1.0
*strong scheme (1.3) using the supporting value*

$$(1.14) \qquad \bar{\Upsilon}_n = Y_n + b\sqrt{\Delta}$$

*instead of (1.4). Compare the results with those of PC-Exercise 10.3.2 and
PC-Exercise 11.1.3.*

The modified scheme in the preceding PC-Exercise shows a slightly worse
result than for the other schemes. This is because some higher order terms
in the stochastic Taylor expansion are better approximated by the supporting
value (1.4) than by (1.14).

In Theorem 11.5.1 we shall see that many other derivative free schemes can
also be constructed. For certain types of stochastic differential equations spe-
cially constructed explicit order 1.0 schemes may sometimes be more efficient.

## 11.2    Explicit Order 1.5 Strong Schemes

As in the preceding section, we can also derive derivative free schemes of order
1.5 by replacing the derivatives in the order 1.5 strong Taylor scheme (10.4.1)
by corresponding finite differences. For notational simplicity we shall mainly
state the schemes for the autonomous case.

In the autonomous 1-dimensional case $d = m = 1$ such an *explicit order* 1.5
*strong scheme* due to Platen has the form

$$(2.1) \quad Y_{n+1} \; = \; Y_n + b\,\Delta W + \frac{1}{2\sqrt{\Delta}}\left\{a(\bar{\Upsilon}_+) - a(\bar{\Upsilon}_-)\right\}\Delta Z$$

$$+\frac{1}{4}\left\{a(\bar{\Upsilon}_+) + 2a + a(\bar{\Upsilon}_-)\right\}\Delta$$

$$+\frac{1}{4\sqrt{\Delta}}\left\{b(\bar{\Upsilon}_+) - b(\bar{\Upsilon}_-)\right\}\left\{(\Delta W)^2 - \Delta\right\}$$

$$+\frac{1}{2\Delta}\left\{b(\bar{\Upsilon}_+) - 2b + b(\bar{\Upsilon}_-)\right\}\left\{\Delta W\Delta - \Delta Z\right\}$$

$$+\frac{1}{4\Delta}\left[b(\bar{\Phi}_+) - b(\bar{\Phi}_-) - b(\bar{\Upsilon}_+) + b(\bar{\Upsilon}_-)\right]$$

$$\times\left\{\frac{1}{3}(\Delta W)^2 - \Delta\right\}\Delta W$$

with

$$(2.2) \qquad \bar{\Upsilon}_\pm = Y_n + a\,\Delta \pm b\sqrt{\Delta}$$

and

$$(2.3) \qquad \bar{\Phi}_\pm = \bar{\Upsilon}_+ \pm b(\bar{\Upsilon}_+)\sqrt{\Delta}.$$

Here $\Delta Z$ is the multiple Ito integral $I_{(1,0)}$ defined in (10.4.2). We see that $a$ must be evaluated at three points and $b$ at five supporting values for each time step.

**PC-Exercise 11.2.1** *Repeat PC-Exercise 11.1.1 using the explicit order 1.5 strong scheme (2.1)–(2.3) and compare the results with those of PC-Exercise 10.4.2.*

In the general multi-dimensional autonomous case with $d, m = 1, 2, \ldots$ the $k$th component of the *explicit order 1.5 strong scheme* satisfies

$$(2.4) \quad Y_{n+1}^k = Y_n^k + a^k \Delta + \sum_{j=1}^m b^{k,j} \Delta W^j$$

$$+ \frac{1}{2\sqrt{\Delta}} \sum_{j_2=0}^m \sum_{j_1=1}^m \left\{ b^{k,j_2}\left(\bar{\Upsilon}_+^{j_1}\right) - b^{k,j_2}\left(\bar{\Upsilon}_-^{j_1}\right) \right\} I_{(j_1,j_2)}$$

$$+ \frac{1}{2\Delta} \sum_{j_2=0}^m \sum_{j_1=1}^m \left\{ b^{k,j_2}\left(\bar{\Upsilon}_+^{j_1}\right) - 2b^{k,j_2} + b^{k,j_2}\left(\bar{\Upsilon}_-^{j_1}\right) \right\} I_{(0,j_2)}$$

$$+ \frac{1}{2\Delta} \sum_{j_1,j_2,j_3=1}^m \left[ b^{k,j_3}\left(\bar{\Phi}_+^{j_1,j_2}\right) - b^{k,j_3}\left(\bar{\Phi}_-^{j_1,j_2}\right) \right.$$

$$\left. - b^{k,j_3}\left(\bar{\Upsilon}_+^{j_1}\right) + b^{k,j_3}\left(\bar{\Upsilon}_-^{j_1}\right) \right] I_{(j_1,j_2,j_3)}$$

with

$$(2.5) \qquad \bar{\Upsilon}_\pm^j = Y_n + \frac{1}{m} a \Delta \pm b^j \sqrt{\Delta}$$

and

$$(2.6) \qquad \bar{\Phi}_\pm^{j_1,j_2} = \bar{\Upsilon}_+^{j_1} \pm b^{j_2}\left(\bar{\Upsilon}_+^{j_1}\right) \sqrt{\Delta},$$

where we have written $b^{k,0}$ for $a^k$ in the summation terms.

We shall see in Theorem 11.5.1 that this scheme converges with strong order $\gamma = 1.5$ under analogous conditions on the drift and diffusion coefficients as for the order 1.5 strong Ito-Taylor scheme (10.4.6). For actual calculations here we can also approximate the multiple Ito integrals in (2.4) as in (10.4.7). In general, each step of the scheme (2.4) requires $2m + 1$ evaluations of the drift coefficient $a^k$ and $2m(m+1)+1$ evaluations of the diffusion coefficient $b^{k,j}$ for $k = 1, \ldots, d$ and $j = 1, \ldots, m$. These numbers may be reduced in special cases. For instance, with *additive noise* (10.2.7) the scheme (2.4)–(2.6) takes the form

$$(2.7) \qquad Y_{n+1}^k = Y_n^k + \sum_{j=1}^m b^{k,j} \Delta W^j$$

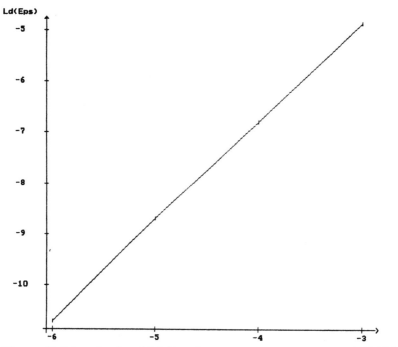

**Figure 11.2.1** Results of PC-Exercise 11.2.1.

$$+\frac{1}{2\sqrt{\Delta}}\sum_{j=1}^{m}\left\{a^k\left(\bar{\Upsilon}_+^j\right)-a^k\left(\bar{\Upsilon}_-^j\right)\right\}\Delta Z^j$$

$$+\frac{1}{4}\Delta\sum_{j=1}^{m}\left\{a^k\left(\bar{\Upsilon}_+^j\right)-\frac{2(m-2)}{m}a^k+a^k\left(\bar{\Upsilon}_-^j\right)\right\}$$

with

$$(2.8) \qquad \bar{\Upsilon}_\pm^j = Y_n + \frac{1}{m}a\,\Delta \pm b^j\,\sqrt{\Delta}$$

and

$$(2.9) \qquad \Delta Z^j = I_{(j,0)}.$$

Here we still need $m+1$ evaluations of the drift coefficient $a^k$ at each time step.

For *diagonal noise* (10.3.11) the *autonomous explicit order* 1.5 *strong scheme* has $k$th component

$$(2.10) \qquad Y_{n+1}^k = Y_n^k + b^{k,k}\,\Delta W^k$$

$$+\frac{1}{2\sqrt{\Delta}}\sum_{j=1}^{m}\left\{a^k\left(\bar{\Upsilon}_+^j\right)-a^k\left(\bar{\Upsilon}_-^j\right)\right\}\Delta Z^k$$

$$+\frac{1}{4}\Delta\sum_{j=1}^{m}\left\{a^k\left(\bar{\Upsilon}_+^j\right)-\frac{2(m-2)}{m}a^k+a^k\left(\bar{\Upsilon}_-^j\right)\right\}$$

$$+\frac{1}{4\sqrt{\Delta}}\left\{b^{k,k}\left(\bar{\Upsilon}^k_+\right)-b^{k,k}\left(\bar{\Upsilon}^k_-\right)\right\}\left\{(\Delta W^k)^2-\Delta\right\}$$

$$+\frac{1}{2\Delta}\left\{b^{k,k}\left(\bar{\Upsilon}^k_+\right)-2b^{k,k}+b^{k,k}\left(\bar{\Upsilon}^k_-\right)\right\}\left\{\Delta W^k\,\Delta-\Delta Z^k\right\}$$

$$+\frac{1}{4\Delta}\left[b^{k,k}\left(\bar{\Phi}^k_+\right)-b^{k,k}\left(\bar{\Phi}^k_-\right)-b^{k,k}\left(\bar{\Upsilon}^k_+\right)+b^{k,k}\left(\bar{\Upsilon}^k_-\right)\right]$$

$$\times\left\{\frac{1}{3}(\Delta W^k)^2-\Delta\right\}\Delta W^k$$

with

(2.11)
$$\bar{\Upsilon}^k_\pm=Y_n+\frac{1}{m}a\,\Delta\pm b^k\,\sqrt{\Delta}$$

and

(2.12)
$$\bar{\Phi}^k_\pm=\bar{\Upsilon}^k_+\pm b^k\left(\bar{\Upsilon}^k_+\right)\sqrt{\Delta}.$$

This scheme requires $2m+1$ evaluations of $a^k$ and $4m+1$ evaluations of $b^{k,k}$ at each time step.

For the autonomous case of the explicit order 1.5 strong scheme for *commutative noise of the second kind* (10.4.13) in vector form we have

(2.13)
$$Y_{n+1}=Y_n+\sum_{j=1}^{m}b^j\,\Delta W^j$$

$$+\frac{1}{2\sqrt{\Delta}}\sum_{j=1}^{m}\left\{a\left(\bar{\Upsilon}^j_+\right)-a\left(\bar{\Upsilon}^j_-\right)\right\}\Delta Z^j$$

$$+\frac{1}{4}\Delta\sum_{j=1}^{m}\left\{a\left(\bar{\Upsilon}^j_+\right)-\frac{2(m-2)}{m}a+a\left(\bar{\Upsilon}^j_-\right)\right\}$$

$$+\frac{1}{4\sqrt{\Delta}}\sum_{j=1}^{m}\left\{b^j\left(\bar{\Upsilon}^j_+\right)-b^j\left(\bar{\Upsilon}^j_-\right)\right\}\left\{(\Delta W^j)^2-\Delta\right\}$$

$$+\frac{1}{4\Delta}\sum_{j=1}^{m}\left[b^j\left(\bar{\Phi}^j_+\right)-b^j\left(\bar{\Phi}^j_-\right)-b^j\left(\bar{\Upsilon}^j_+\right)+b^j\left(\bar{\Upsilon}^j_-\right)\right]$$

$$\times\left\{\frac{1}{3}\left(\Delta W^j\right)^2-\Delta\right\}\Delta W^j$$

$$+\frac{1}{2\Delta}\sum_{j_2=1}^{m}\sum_{j_1=1}^{m}\left\{b^{j_2}\left(\bar{\Upsilon}^{j_1}_+\right)-2b^{j_2}+b^{j_2}\left(\bar{\Upsilon}^{j_1}_-\right)\right\}\left\{\Delta W^{j_2}\,\Delta-\Delta Z^{j_2}\right\}$$

$$+\frac{1}{2\sqrt{\Delta}}\sum_{j_2=1}^{m}\sum_{j_1=1}^{j_2-1}\left\{b^{j_2}\left(\bar{\Upsilon}^{j_1}_+\right)-b^{j_2}\left(\bar{\Upsilon}^{j_1}_-\right)\right\}\Delta W^{j_1}\Delta W^{j_2}$$

$$+\frac{1}{4\Delta} \sum_{\substack{j_1,j_2=1 \\ j_1 \neq j_2}}^{m} \left[ b^{j_2} \left( \bar{\Phi}_+^{j_1,j_2} \right) - b^{j_2} \left( \bar{\Phi}_-^{j_1,j_2} \right) - b^{j_2} \left( \bar{\Upsilon}_+^{j_1} \right) + b^{j_2} \left( \bar{\Phi}_-^{j_1,j_2} \right) \right]$$

$$\times \left\{ \left( \Delta W^{j_2} \right)^2 - \Delta \right\} \Delta W^{j_1}$$

$$+\frac{1}{2\Delta} \sum_{j_1=1}^{m} \sum_{j_2=1}^{j_1-1} \sum_{j_3=1}^{j_2-1} \left[ b^{j_3} \left( \bar{\Phi}_+^{j_1,j_2} \right) - b^{j_3} \left( \bar{\Phi}_-^{j_1,j_2} \right) - b^{j_3} \left( \bar{\Upsilon}_+^{j_1} \right) + b^{j_3} \left( \bar{\Upsilon}_-^{j_1} \right) \right]$$

$$\times \Delta W^{j_1} \Delta W^{j_2} \Delta W^{j_3}$$

with

(2.14)
$$\bar{\Upsilon}_\pm^j = Y_n + \frac{1}{m} a \Delta \pm b^j \sqrt{\Delta}$$

and

(2.15)
$$\bar{\Phi}_\pm^{j_1,j_2} = \bar{\Upsilon}_+^{j_1} \pm b^{j_2} \left( \bar{\Upsilon}_+^{j_1} \right) \sqrt{\Delta}.$$

The above schemes can often be simplified even more by making use of any other special structure of a particular stochastic differential equation under consideration. Nevertheless, the order 1.5 schemes remain quite complicated.

For completeness we state the *nonautonomous* general multi-dimensional version of the *explicit order* 1.5 *strong scheme* in vector form

(2.16)  $\displaystyle Y_{n+1} = Y_n + a\Delta + \sum_{j=1}^{m} b^j \Delta W^j$

$$+\frac{1}{2\sqrt{\Delta}} \sum_{j_2=0}^{m} \sum_{j_1=1}^{m} \left\{ b^{j_2} \left( \tau_n, \bar{\Upsilon}_+^{j_1} \right) - b^{j_2} \left( \tau_n, \bar{\Upsilon}_-^{j_1} \right) \right\} I_{(j_1,j_2)}$$

$$+\frac{1}{\Delta} \sum_{j=0}^{m} \left\{ b^j \left( \tau_{n+1}, Y_n \right) - b^j \right\} I_{(0,j)}$$

$$+\frac{1}{2\Delta} \sum_{j_2=0}^{m} \sum_{j_1=1}^{m} \left\{ b^{j_2} \left( \tau_n, \bar{\Upsilon}_+^{j_1} \right) - 2b^{j_2} + b^{j_2} \left( \tau_n, \bar{\Upsilon}_-^{j_1} \right) \right\} I_{(0,j_2)}$$

$$+\frac{1}{2\Delta} \sum_{j_1,j_2,j_3=1}^{m} \left[ b^{j_3} \left( \tau_n, \bar{\Phi}_+^{j_1,j_2} \right) - b^{j_3} \left( \tau_n, \bar{\Phi}_-^{j_1,j_2} \right) \right.$$

$$\left. -b^{j_3} \left( \tau_n, \bar{\Upsilon}_+^{j_1} \right) + b^{j_3} \left( \tau_n, \bar{\Upsilon}_-^{j_1} \right) \right] I_{(j_1,j_2,j_3)}$$

with

(2.17)
$$\bar{\Upsilon}_\pm^j = Y_n + \frac{1}{m} a \Delta \pm b^j \sqrt{\Delta}$$

and

(2.18)
$$\bar{\Phi}_\pm^{j_1,j_2} = \bar{\Upsilon}_+^{j_1} \pm b^{j_2} \left( \tau_n, \bar{\Upsilon}_+^{j_1} \right) \sqrt{\Delta},$$

where we have written $b^0$ for $a$ in the summation terms. We can again use (10.4.7) to approximate the multiple Ito integrals appearing here.

Finally, we state the explicit order 1.5 strong scheme, in vector form, for the nonautonomous case with *additive noise*

$$(2.19) \quad Y_{n+1} \;=\; Y_n + a\,\Delta + \sum_{j=1}^{m} b^j\,\Delta W^j$$

$$+\frac{1}{4}\sum_{j=1}^{m}\left\{ a\left(\tau_{n+1},\bar{\Upsilon}_{+}^{j}\right) - 2a + a\left(\tau_{n+1},\bar{\Upsilon}_{-}^{j}\right)\right\}\Delta$$

$$+\frac{1}{2\sqrt{\Delta}}\sum_{j=1}^{m}\left\{ a\left(\tau_{n+1},\bar{\Upsilon}_{+}^{j}\right) - a\left(\tau_{n+1},\bar{\Upsilon}_{-}^{j}\right)\right\}\Delta Z^j$$

$$+\frac{1}{\Delta}\sum_{j=1}^{m}\left\{ b^j\left(\tau_{n+1}\right) - b^j\right\}\left\{\Delta W^j\Delta - \Delta Z^j\right\}$$

with

$$\bar{\Upsilon}_{\pm}^{j} = Y_n + \frac{1}{m}a\,\Delta \pm b^j\sqrt{\Delta}$$

for $j = 1, \ldots, m$.

## 11.3  Explicit Order 2.0 Strong Schemes

In the previous two sections we obtained explicit schemes from Taylor schemes basically by replacing derivatives by their corresponding finite differences. This procedure works well for low order explicit schemes, but leads to evermore complicated formulae as the order is increased. However, we can often take advantage of some special structure of the equations under consideration to derive relatively simple higher order explicit schemes which do not involve derivatives of the drift and diffusion cofficients.

To avoid technical difficulties and complicated expressions we shall restrict our attention here at first to the case of *additive noise*, that is with diffusion coefficient satisfying

$$(3.1) \qquad\qquad b(t,x) \equiv b(t)$$

for all $t$ and $x$. We recall from the examples in Chapter 7 that additive noise arises naturally and is appropriate in a wide variety of situations. Moreover, the Ito and Stratonovich representations coincide for additive noise. However, we shall use the Stratonovich notation in presenting an explicit order 2.0 strong scheme here for the additive noise case to facilitate comparison with its counterparts for more complicated types of noise.

For the autonomous 1-dimensional case $d = m = 1$ an *explicit order* 2.0 *strong scheme for additive noise* due to Chang has the form

(3.2)          $$Y_{n+1} = Y_n + \frac{1}{2} \left\{ \underline{a} \left( \bar{\Upsilon}_+ \right) + \underline{a} \left( \bar{\Upsilon}_- \right) \right\} \Delta + b \, \Delta W$$

with

$$\bar{\Upsilon}_\pm = Y_n + \frac{1}{2} \underline{a} \Delta + \frac{1}{\Delta} b \left\{ \Delta Z \pm \sqrt{2 J_{(1,1,0)} \Delta - (\Delta Z)^2} \right\},$$

where the random variables $\Delta W$, $\Delta Z$ and $J_{(1,1,0)}$ can be approximated as described in Exercise 10.5.1. We note that $\underline{a}$ must be evaluated at three different points for each iteration of the scheme (3.2) and that

$$2 J_{(1,1,0)} \Delta = \int_0^\Delta (W_s)^2 \, ds \int_0^\Delta ds \geq \left( \int_0^\Delta W_s \, ds \right)^2 = (\Delta Z)^2 .$$

In the *nonautonomous* multi-dimensional case $d = 1, 2, \ldots$ with $m = 1$ we propose a similar *explicit order* 2.0 *strong scheme for additive noise* with $k$th component

(3.3)   $$Y_{n+1}^k = Y_n^k + \frac{1}{2} \left\{ \underline{a}^k \left( \tau_n + \frac{1}{2}\Delta, \bar{\Upsilon}_+ \right) + \underline{a}^k \left( \tau_n + \frac{1}{2}\Delta, \bar{\Upsilon}_- \right) \right\} \Delta$$

$$+ b^k \, \Delta W + \frac{1}{\Delta} \left\{ b^k \left( \tau_{n+1} \right) - b^k \right\} \left\{ \Delta W \Delta - \Delta Z \right\}$$

with

$$\bar{\Upsilon}_\pm = Y_n + \frac{1}{2} \underline{a} \Delta + \frac{1}{\Delta} b \left\{ \Delta Z \pm \sqrt{2 J_{(1,1,0)} \Delta - (\Delta Z)^2} \right\},$$

for $k = 1, \ldots, d$.

It can be shown with the help of Theorem 11.5.2, which will be stated and proved in Section 5, that the scheme (3.3) has order of strong convergence $\gamma = 2.0$ under suitable assumptions on the drift and diffusion coefficients.

**Exercise 11.3.1**     *Verify that the nonautonomous linear stochastic differential equation with additive noise (4.4.4)*

$$dX_t = \left( \frac{2}{1+t} X_t + (1+t)^2 \right) dt + (1+t)^2 \, dW_t$$

*with initial value $X_0 = 1$ has, for $t \geq t_0 = 0$, the exact solution*

$$X_t = (1+t)^2 (1 + W_t + t) .$$

**PC-Exercise 11.3.2**     *Generate $M = 20$ batches each of $N = 100$ simulations of $X_T$ for the Ito process in Exercise 11.3.1, with $T = 0.5$, and of the value $Y^\delta(T)$ of the explicit order 2.0 strong scheme (3.3) corresponding to the same sample paths of the Wiener process for equal step sizes $\delta = \Delta = 2^{-1}, 2^{-2}, 2^{-3}$ and $2^{-4}$. See Exercise 10.5.1 for approximating $J_{(1,1,0)}$ by $J_{(1,1,0)}^p$ with $p = 15$. Plot $\log_2$ of the absolute error against $\log_2 \Delta$.*

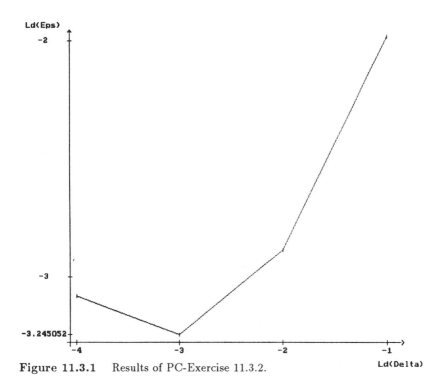

**Figure 11.3.1** Results of PC-Exercise 11.3.2.

Figure 11.3.1 shows an increase in the error for step sizes $\Delta \leq 2^{-3}$. This effect is delayed to even smaller $\Delta$ if we use an approximation $J^p_{(1,1,0)}$ with a larger $p$.

Explicit order 2.0 strong schemes for stochastic differential equations with other kinds of special structure, such as with linearly occuring noise (10.3.14), can also be derived from the order 2.0 Stratonovich-Taylor expansion in a similar fashion. These schemes still require those multiple stochastic integrals which survive in the corresponding Taylor expansion.

## 11.4 Multistep Schemes

We saw in Chapter 8 that deterministic multi-step methods are often more efficient computationally than one-step methods of the same order because they require, essentially, only one new evalution of the right hand side of the differential equation for each iteration. In addition, such multi-step schemes are sometimes more stable for larger time steps, although there are also unstable multi-step schemes as we saw in Figure 8.2.2 for the midpoint method (8.2.10). Stochastic simulations typically require the calculation of many different realizations of the approximating process, so efficiency and stability are crucial factors to be taken into account. In this section we shall briefly introduce some stochastic multi-step schemes, concentrating for simplicity on two-step schemes. As with deterministic multi-step schemes a one-step scheme will be used to generate the initial steps needed to start the multi-step scheme. For a two-step scheme just one iteration of the starting routine is needed.

We can obtain stochastic multi-step schemes by heuristically generalizing known deterministic multi-step schemes to stochastic differential equations as we did with the deterministic Runge-Kutta schemes in Section 2, but these turn out to be of limited value for much the same reasons.

## A Two-Step Order 1.0 Strong Scheme

It is not an easy task to write out and investigate higher order multi-step schemes in the most general case. However, we can take advantage of the structure of certain types of stochastic differential equations to obtain relatively simple multi-step schemes. Ilustrative of this is the 2-dimensional Ito system

$$(4.1) \qquad dX_t^1 \;=\; X_t^2\, dt$$

$$dX_t^2 \;=\; \left\{-a(t)\, X_t^2 + b\left(t, X_t^1\right)\right\}\, dt + \sum_{j=1}^m c^j\left(t, X_t^1\right)\, dW_t^j,$$

which is typical of the noisy electrical and mechanical systems considered in Chapter 7. The Milstein scheme (10.3.3) for system (4.1) takes the form

$$(4.2) \qquad Y_{n+1}^1 \;=\; Y_n^1 + Y_n^2\, \Delta$$

$$Y_{n+1}^2 \;=\; Y_n^2 - a(\tau_n)\, Y_n^2\, \Delta + b\left(\tau_n, Y_n^1\right)\, \Delta$$

$$+ \sum_{j=1}^m c^j\left(\tau_n, Y_n^1\right)\, \Delta W_n^j$$

and has strong order 1.0. A useful simplifying feature of (4.2) is the absence of the double Ito integrals $I_{(j_1, j_2)}$. Moreover, we can solve the first equation for

$$Y_n^2 = \frac{1}{\Delta}\left(Y_{n+1}^1 - Y_n^1\right)$$

and insert it into the second to obtain a two-step scheme for the first component $Y^1$, provided we use an equidistant discretization. This resulting two-step scheme,

$$(4.3) \qquad Y_{n+2}^1 \;=\; \left\{2 - a(\tau_n)\, \Delta\right\}\, Y_{n+1}^1 - \left\{1 - a(\tau_n)\, \Delta\right\}\, Y_n^1$$

$$+ b\left(\tau_n, Y_n^1\right)\, \Delta^2 + \sum_{j=1}^m c^j\left(\tau_n, Y_n^1\right)\, \Delta W_n^j\, \Delta,$$

is due to Lépingle and Ribémont.

Thus we have a two-step scheme for the first component of the approximation, which is equivalent to the 2-dimensional Milstein scheme (4.2) for the system (4.1). We can use the first equation of (4.2) both as a starting routine for (4.3) and to calculate approximations of the second component if they are required.

**PC-Exercise 11.4.1**   *Use the two-point scheme (4.3) with time step $\delta = \Delta$ $= 2^{-4}$ to simulate and plot a linearly interpolated approximate trajectory of the first component of system (4.6) with*

$$a(t) \equiv 1.0, \quad b(t, x) = x^1 \left(1 - \left(x^1\right)^2\right), \quad c(t, x) \equiv 0.01$$

*and $X_0^1 = -2.0$, $X_0^2 = 0$ on the interval $[0, 8]$. Use the first equation of the 2-dimensional Milstein scheme (4.1) as the starting routine.*

## Two-Step Order 1.5 Strong Schemes

For the 1-dimensional case $d = m = 1$ we propose the *two-step order 1.5 strong scheme*

$$(4.4) \qquad Y_{n+1} = Y_{n-1} + 2a\,\Delta - a'\,(Y_{n-1})\,b\,(Y_{n-1})\,\Delta W_{n-1}\Delta$$
$$+ V_n + V_{n-1}$$

with

$$V_n = b\,\Delta W_n + \left(ab' + \frac{1}{2}b^2 b''\right)\{\Delta W_n \Delta - \Delta Z_n\}$$

$$+ a'b\,\Delta Z_n + \frac{1}{2}\,bb'\left\{(\Delta W_n)^2 - \Delta\right\}$$

$$+ \frac{1}{2}\,b\,(bb')'\left\{\frac{1}{3}\,(\Delta W_n)^2 - \Delta\right\}\Delta W_n,$$

where the random variables $\Delta W_n$ and $\Delta Z_n$ are taken as usual; see (10.4.3). We note that (4.4) can be interpreted as a higher strong order stochastic generalization of the deterministic midpoint method (8.2.10). In particular, the scheme (4.4) does not require the second derivative $a''$ to be determined or evaluated.

In the general multi-dimensional case $d, m = 1, 2, \ldots$ we have in vector form the *two-step order 1.5 scheme*

$$(4.5) \qquad Y_{n+1} = Y_{n-1} + 2a\,\Delta - \sum_{j=1}^{m} L^j a\,(\tau_{n-1}, Y_{n-1})\,\Delta W_{n-1}^j \Delta$$

$$+ V_n + V_{n-1}$$

with

$$V_n = \sum_{j=1}^{m}\left[b^j\,\Delta W_n^j + L^0 b^j\,\{\Delta W_n^j \Delta - \Delta Z_n^j\} + L^j a\,\Delta Z_n^j\right]$$

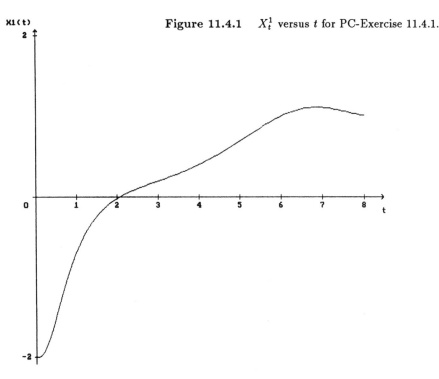

**Figure 11.4.1**    $X_t^1$ versus $t$ for PC-Exercise 11.4.1.

$$+ \sum_{j_1,j_2=1}^{m} L^{j_1} b^{j_2} \, I_{(j_1,j_2),\tau_n,\tau_{n+1}}$$

$$+ \sum_{j_1,j_2,j_3=1}^{m} L^{j_1} L^{j_2} b^{j_3} \, I_{(j_1,j_2,j_3),\tau_n,\tau_{n+1}}.$$

We can approximate the Ito integrals here by the previously described methods. We also have a *derivative free counterpart* of the above scheme

$$(4.6) \qquad Y_{n+1} \;=\; Y_{n-1} + 2a\,\Delta$$

$$- \frac{\sqrt{\Delta}}{2} \sum_{j=1}^{m} \left\{ a\left(\tau_{n-1}, \tilde{\Upsilon}_{n-1}^{j+}\right) - a\left(\tau_{n-1}, \tilde{\Upsilon}_{n-1}^{j-}\right) \right\} \Delta W_{n-1}^{j}$$

$$+ V_n + V_{n-1}$$

with

$$V_n \;=\; \sum_{j=1}^{m} \left[ b^{j}\,\Delta W_n^{j} + \frac{1}{\Delta}\left\{ b^{j}\left(\tau_{n+1}, Y_n\right) - b^{j} \right\} I_{(0,j),\tau_n,\tau_{n+1}} \right]$$

$$+ \frac{1}{2\sqrt{\Delta}} \sum_{j_2=0}^{m} \sum_{j_1=1}^{m} \left\{ b^{j_2}\left(\tau_n, \tilde{\Upsilon}_n^{j_1+}\right) - b^{j_2}\left(\tau_n, \tilde{\Upsilon}_n^{j_1-}\right) \right\} I_{(j_1,j_2),\tau_n,\tau_{n+1}}$$

$$+\frac{1}{2\Delta}\sum_{j_1,j_2=1}^{m}\left\{b^{j_2}\left(\tau_n,\bar{\Upsilon}_n^{j_1+}\right)-2b^{j_2}+b^{j_2}\left(\tau_n,\bar{\Upsilon}_n^{j_1-}\right)\right\}I_{(0,j_2),\tau_n,\tau_{n+1}}$$

$$+\frac{1}{2\Delta}\sum_{j_1,j_2,j_3=1}^{m}\left[b^{j_3}\left(\tau_n,\bar{\Phi}_n^{j_1,j_2+}\right)-b^{j_3}\left(\tau_n,\bar{\Phi}_n^{j_1,j_2-}\right)\right.$$

$$\left.-b^{j_3}\left(\tau_n,\bar{\Upsilon}_n^{j_1+}\right)+b^{j_3}\left(\tau_n,\bar{\Upsilon}_n^{j_1-}\right)\right]I_{(j_1,j_2,j_3),\tau_n,\tau_{n+1}}$$

with

$$\bar{\Upsilon}_n^{j\pm}=Y_n+\frac{1}{m}\,a\,\Delta\pm b^j\,\sqrt{\Delta}$$

and

$$\bar{\Phi}_n^{j_1,j_2,\pm}=\bar{\Upsilon}_n^{j_1+}\pm b^{j_2}\left(\tau_n,\bar{\Upsilon}_n^{j_1+}\right)\sqrt{\Delta},$$

where we have written $b^0$ for the drift $a$ in the second summation term. This scheme becomes much simpler for *additive noise*, reducing to

$$(4.7)\qquad Y_{n+1}\;=\;Y_{n-1}+2a\,\Delta-\sum_{j=1}^{m}A_{n-1}^j\,\Delta W_{n-1}^j\sqrt{\Delta}$$

$$+V_n+V_{n-1}$$

with

$$V_n\;=\;\frac{1}{\Delta}\sum_{j+1}^{m}\left[b^j\,\Delta W_n^j\Delta+\left\{b^j\left(\tau_{n+1},Y_n\right)-b^j\right\}\left\{\Delta W_n^j\Delta-\Delta Z_n^j\right\}\right.$$

$$\left.+A_n^j\,\Delta Z_n^j\,\sqrt{\Delta}\right]$$

where

$$A_n^j=\frac{1}{2}\left\{a\left(\tau_n,\bar{\Upsilon}_n^{j+}\right)-a\left(\tau_n,\bar{\Upsilon}_n^{j-}\right)\right\}$$

and

$$\bar{\Upsilon}_n^{j\pm}=Y_n+\frac{1}{m}\,a\,\Delta\pm b^j\,\sqrt{\Delta}.$$

**PC-Exercise 11.4.2** *Repeat PC-Exercise 10.3.2 for the scheme (4.5) with the explicit order 1.5 strong scheme as the starting routine.*

We shall discuss the proof of convergence of the above two-step schemes at the end of Section 6 of Chapter 12 in connection with implicit two-step schemes.

## 11.5    General Strong Schemes

We shall now introduce and examine two classes of strong schemes which generalize all of the explicit strong schemes considered so far and, in addition, contain
other strong schemes of all orders $\gamma = 0.5, 1.0, 1.5, 2.0, \ldots$. These two classes
will be defined in relation to the Ito-Taylor and Stratonovich-Taylor approximations, respectively. In doing so we shall use the hierarchical set introduced
in (10.6.2), that is

$$\mathcal{A}_\gamma = \left\{ \alpha \in \mathcal{M} : l(\alpha) + n(\alpha) \leq 2\gamma \quad \text{or} \quad l(\alpha) = n(\alpha) = \gamma + \frac{1}{2} \right\}$$

for $\gamma = 0.5, 1.0, 1.5, 2.0, \ldots$.

### Strong Ito Approximations

We shall call a recursive relation of the form

$$(5.1) \qquad\qquad Y_{n+1} = Y_n + \sum_{\alpha \in \mathcal{A}_\gamma \backslash \{v\}} g_{\alpha,n} \, I_\alpha + R_n$$

a *strong Ito scheme of order* $\gamma$  for some $\gamma = 0.5, 1.0, 1.5, 2.0, \ldots$ if the
coefficient functions $g_{\alpha,n}$ are $\mathcal{A}_{\tau_n}$-measurable for $n = 0, 1, \ldots, n_T - 1$ and
satisfy the estimates

$$(5.2) \qquad\qquad E\left( \max_{0 \leq n \leq n_T} |g_{\alpha,n} - f_\alpha(\tau_n, Y_n)|^2 \right) \leq K \, \delta^{2\gamma - \phi(\alpha)}$$

for all $\alpha \in \mathcal{A}_\gamma \backslash \{v\}$, where

$$(5.3) \qquad\qquad \phi(\alpha) = \begin{cases} 2l(\alpha) - 2 & : \quad l(\alpha) = n(\alpha) \\ l(\alpha) + n(\alpha) - 1 & : \quad l(\alpha) \neq n(\alpha), \end{cases}$$

with the $R_n$ satisfying the estimate

$$(5.4) \qquad\qquad E\left( \max_{1 \leq n \leq n_T} \left| \sum_{0 \leq k \leq n-1} R_k \right|^2 \right) \leq K \, \delta^{2\gamma}.$$

The class of strong Ito schemes is quite broad because the coefficient functions $g_{\alpha,n}$ and the remainders $R_n$ of quite diverse structures can often easily
satisfy the required measurability condition and the estimates (5.2) and (5.4).
For instance, it includes many multistep schemes in addition to the explicit
strong schemes so far considered. However, a strong Ito scheme of order $\gamma$
must contain the multiple Ito integral $I_\alpha$ for every multi-index $\alpha \in \mathcal{A}_\gamma \backslash \{v\}$,
as it may not attain the order $\gamma$ if some of these are missing.

**Theorem 11.5.1** *Let $Y^\delta$ be a time discrete approximation generated by a strong Ito scheme of order $\gamma \in \{\, 0.5,\ 1.0,\ 1.5,\ 2.0,\ \ldots\,\}$ satisfying the assumptions of Corollary 10.6.4 for the strong Ito-Taylor approximation of order $\gamma$. Then*

$$(5.5) \qquad E\left(\max_{0\le n\le n_T} \left|Y_n^\delta - X_{\tau_n}\right|^2\right) \le K\,\delta^{2\gamma}$$

*and, hence, $Y^\delta$ converges strongly with order $\gamma$.*

**Proof** We shall compare $Y^\delta$ with the strong Ito-Taylor approximation of order $\gamma$ defined in (10.6.3), which we denote here by $\bar{Y}^\delta$.

It follows from (5.1) and (10.6.3) that

$$(5.6) \qquad H_t \;:=\; E\left(\max_{0\le n\le n_t}\left|\bar{Y}_n^\delta - Y_n^\delta\right|^2\right)$$

$$= E\left(\max_{0\le n\le n_t}\left|\sum_{k=0}^{n-1}\sum_{\alpha\in\mathcal{A}_\gamma\backslash\{v\}} f_\alpha\left(\tau_k,\bar{Y}_k^\delta\right)I_\alpha \right.\right.$$

$$\left.\left. -\sum_{k=0}^{n-1}\left(\sum_{\alpha\in\mathcal{A}_\gamma\backslash\{v\}} g_{\alpha,k}\,I_\alpha + R_k\right)\right|^2\right)$$

$$\le K_1 \sum_{\alpha\in\mathcal{A}_\gamma\backslash\{v\}}\left\{ E\left(\max_{0\le n\le n_t}\left|\sum_{k=0}^{n-1}\left[f_\alpha\left(\tau_k,Y_k^\delta\right)-f_\alpha\left(\tau_k,\bar{Y}_n^\delta\right)\right]I_\alpha\right|^2\right)\right.$$

$$\left. +E\left(\max_{0\le n\le n_t}\left|\sum_{k=0}^{n-1}\left[f_\alpha\left(\tau_k,Y_k^\delta\right)-g_{\alpha,k}\right]I_\alpha\right|^2\right)\right\}$$

$$+K_1\,E\left(\max_{0\le n\le n_t}\left|\sum_{k=0}^{n-1} R_k\right|^2\right)$$

for all $t\in[0,T]$. Then, from Lemma 10.8.1, (5.2)–(5.4), (10.6.5) and (5.6) we obtain

$$(5.7)\quad H_t \;\le\; K_2 \sum_{\alpha\in\mathcal{A}_\gamma\backslash\{v\}}\left\{\left(\int_0^t E\left(\max_{0\le n\le n_u}\left|f_\alpha\left(\tau_n,Y_n^\delta\right)-f_\alpha\left(\tau_n,\bar{Y}_n^\delta\right)\right|^2 du\right)\right.\right.$$

$$\left.\left. +\int_0^t E\left(\max_{0\le n\le n_u}\left|f_\alpha\left(\tau_n,Y_n^\delta\right)-g_{\alpha,n}\right|^2 du\right)\right\}\delta^{\phi(\alpha)}\right.$$

$$+K_2\,\delta^{2\gamma}$$

$$\le\; K_3 \int_0^t H_u\,du + K_4\,\delta^{2\gamma}.$$

We apply the Gronwall inequality (Lemma 4.5.1) to (5.7) to obtain the desired estimate (5.5), and then apply the Lyapunov inequality (1.4.12)

$$E\left(|Y_{n_T}^\delta - X_T|\right) \le \left(E\left(|Y_{n_T}^\delta - X_T|^2\right)\right)^{1/2} \le K_5\,\delta^\gamma,$$

from which we conclude that $Y^\delta$ converges strongly with order $\gamma$. □

As in Corollary 10.6.5, we can replace the multiple Ito integrals $I_\alpha$ in the scheme (5.1) by the approximate multiple Ito integrals $I_\alpha^p$. The resulting scheme will also have strong order $\gamma$ if the assumptions of Theorem 11.5.1 and the assumed bound (10.6.13) in Corollary 10.6.5 are satisfied.

As an illustration of the above theorem we shall apply it to show that the explicit strong scheme

$$(5.8) \quad Y_{n+1} = Y_n + a\,\Delta + b\,\Delta W_n + \frac{1}{2\sqrt{\Delta}}\left\{b\left(\tau_n, \bar{\Upsilon}_n\right) - b\right\}\left\{(\Delta W_n)^2 - \Delta\right\}$$

with

$$\bar{\Upsilon}_n = Y_n + a\,\Delta + b\,\sqrt{\Delta}$$

proposed in (11.1.3) converges strongly with order $\gamma = 1.0$ under suitable assumptions on the drift and diffusion coefficients. We shall first rewrite (5.8) with the help of the usual Taylor expansion and the relation (5.2.12) in the form

$$(5.9) \qquad Y_{n+1} = Y_n + g_{(0),n}\,I_{(0)} + g_{(1),n}\,I_{(1)} + g_{(1,1),n}\,I_{(1,1)} + R_n$$

with

$$(5.10) \qquad g_{(0),n} = a, \qquad g_{(1),n} = b, \qquad g_{(1,1),n} = bb'$$

and

$$(5.11) \quad R_n = \left\{ab'\,\Delta + b''\left(Y_n + \theta\left(a\,\Delta + b\,\sqrt{\Delta}\right)\right)\left(a\,\Delta + b\,\sqrt{\Delta}\right)^2\right\}$$

$$\times \frac{1}{2\sqrt{\Delta}}\left\{(\Delta W_n)^2 - \Delta\right\}$$

for some $\theta \in (0,1)$. Obviously, from (5.10) and (5.3.5), we have

$$f_{(0)} = g_{(0),n} \qquad f_{(1)} = g_{(1),n} \qquad f_{(1,1)} = g_{(1,1),n}$$

so condition (5.2) of a strong Ito scheme is automatically satisfied. For simplicity we shall use an equidistant time discretization and assume that $a$, $b$, $b'$ and $b''$ are uniformly bounded. Then, in view of (5.11), the sum $\sum_{k=0}^{n-1} R_k$ has zero mean and bounded finite variance. Moreover, it satisfies the martingale property (2.3.5), so we can use a discrete version of the Doob inequality (2.3.7) to obtain the estimate

$$E\left(\max_{0\le n\le n_T}\left|\sum_{k=0}^{n-1} R_k\right|^2\right)$$

$$\leq \quad K \Delta E \left( \max_{0 \leq n \leq n_T} \left| \sum_{k=0}^{n-1} \{ (\Delta W_k)^2 - \Delta \} \right|^2 \right)$$

$$\leq \quad 4K \Delta \max_{0 \leq n \leq n_T} E \left( \left| \sum_{k=0}^{n-1} \{ (\Delta W_k)^2 - \Delta \} \right|^2 \right)$$

$$\leq \quad 4K \Delta \sum_{k=0}^{n_T-1} E \left( \left| (\Delta W_k)^2 - \Delta \right|^2 \right)$$

$$\leq \quad 8KT \Delta^2.$$

The scheme (5.8) thus also satisfies condition (5.4) with equal step sizes $\delta = \Delta$, so is a strong Ito scheme. Under the assumptions of Theorem 11.5.1 we can then conclude that (5.8) is of strong order $\gamma = 1.0$.

## Strong Stratonovich Approximations

Our second class of strong schemes is defined in relation to Stratonovich-Taylor approximations. We shall call a recursive relation of the form

$$(5.12) \qquad Y_{n+1} = Y_n + \sum_{\alpha \in \mathcal{A}_\gamma \backslash \{v\}} \underline{g}_{\alpha,n} J_\alpha + \underline{R}_n$$

a *strong Stratonovich scheme of order* $\gamma$ for some $\gamma = 1.0, 2.0, \ldots$ if the coefficient functions $\underline{g}_{\alpha,n}$ are $\mathcal{A}_{\tau_n}$-measurable for $n = 0, 1, \ldots, n_T - 1$ and satisfy the estimates

$$(5.13) \qquad E \left( \max_{0 \leq n \leq n_T} \left| \underline{g}_{\alpha,n} - \underline{f}_\alpha(\tau_n, Y_n) \right|^2 \right) \leq K \, \delta^{2\gamma - \phi(\alpha)},$$

for all $\alpha \in \mathcal{A}_\gamma \setminus \{v\}$, where $\phi(\alpha)$ was defined in (5.3), with the $R_n$ satisfying the estimate

$$(5.14) \qquad E \left( \max_{1 \leq n \leq n_T} \left| \sum_{0 \leq k \leq n-1} \underline{R}_k \right|^2 \right) \leq K \, \delta^{2\gamma}.$$

The class of strong Stratonovich schemes encompasses a wide range of one-step and multistep schemes, like their Ito counterparts. The following theorem provides conditions under which a strong Stratonovich scheme of order $\gamma$ actually converges strongly with order $\gamma$.

**Theorem 11.5.2** $Y^\delta$ *be a time discrete approximation generated by a strong Stratonovich scheme of order* $\gamma \in \{1.0, 2.0, \ldots \}$ *satisfying the assumptions of Corollary 10.7.2 for the strong Stratonovich-Taylor approximation of order* $\gamma$. *Then*

$$(5.15) \qquad E \left( \max_{0 \leq n \leq n_t} \left| Y_n^\delta - X_{\tau_n} \right|^2 \right) \leq K \, \delta^{2\gamma}$$

*and, hence,* $Y^\delta$ *converges strongly with order* $\gamma$.

The proof of Theorem 11.5.2 is very similar to that of Theorem 11.5.1, so will be omitted. It is just necessary to recall that the multiple Stratonovich integrals appearing in it can be expressed in terms of multiple Ito integrals of the corresponding mean-square order.

We can also replace the multiple Stratonovich integrals $J_\alpha$ in the scheme (5.12) by the approximate multiple Stratonovich integrals $J_\alpha^p$. The resulting scheme will retain the strong order $\gamma$ if the assumptions of Theorem 11.5.2 and the assumed bound (10.7.12) in Corollary 10.7.3 are satisfied.

To conclude this section we emphasize yet again the diversity of one-step and multistep strong schemes which fall under the classification of the general strong schemes introduced here. It will be a worthwhile challenge to derive efficient strong schemes for different special classes of problems. Here we have tried to provide some guiding principles and a few illustrative examples.

**Exercise 11.5.3**    *Under suitable assumptions show that the scheme (2.1) is an order 1.5 strong scheme.*

**Exercise 11.5.4**    *Show that the scheme (3.2) is an order 2.0 strong scheme under appropriate assumptions.*

# Chapter 12

# Implicit Strong Approximations

In this chapter we shall consider implicit strong schemes which are necessary for the simulation of the solutions of stiff stochastic differential equations. The regions of absolute stability of several of these implicit strong schemes and other explicit strong schemes will also be investigated.

## 12.1 Introduction

In our discussion of numerical stability in Section 3 of Chapter 8 and in Section 8 of Chapter 9 we mentioned that it is often necessary to use implicit schemes to simulate the solutions of stiff differential equations. These schemes usually have a wide range of step sizes suitable for the approximation of dynamical behaviour, in particular with vastly different time scales, without the excessive accumulation of unavoidable initial and roundoff errors.

We also saw in Section 8 of Chapter 9 that difficulties can arise in applying fully implicit schemes to obtain strong approximations of solutions of stochastic differential equations, because they usually involve reciprocals of Gaussian random variables which do not have finite absolute moments. Consequently, finite absolute moments generally will not exist for fully implicit strong approximations and a strong convergence analysis would not make sense. For mainly this reason we shall restrict our attention here to "semi-implicit" strong approximations, which we shall call implicit, with implicit terms obtained from the corresponding Taylor approximation by suitably modifying the coefficient functions $a$ and $aa'$ of the nonrandom multiple stochastic integrals $I_{(0)} = \Delta$ and $I_{(0,0)} = \frac{1}{2}\Delta^2$, respectively. For this we use the stochastic Taylor formula in the form

$$f(X_t) = f(X_{t+\Delta}) - \left\{ a(X_t) f'(X_t) + \frac{1}{2} b^2(X_t) f''(X_t) \right\} \Delta$$

$$- b(X_t) f'(X_t) \Delta W - \cdots,$$

for $f = a$ and $aa'$, respectively.

In implementing an implicit scheme we need to solve an additional algebraic equation at each time step, which can usually be done with standard numerical methods such as the Newton-Raphson method. We shall see that we can improve the stability of simulations considerably without too much additional computational effort by using implicit schemes.

The orders of strong convergence of the implicit schemes that we shall propose in this chapter follow under corresponding assumptions from Theorems 11.5.1 and 11.5.2, respectively, and will be discussed in the last section of the chapter.

We shall continue to use the notation introduced in Chapter 10 without further explanation.

## 12.2   Implicit Strong Taylor Schemes

In this section we shall systematically examine implicit strong schemes obtained by adapting corresponding strong Taylor schemes.

### The Implicit Euler Scheme

The simplest implicit strong Taylor scheme has strong order $\gamma = 0.5$. This is the *implicit Euler scheme*, which, in the 1-dimensional case $d = m = 1$, has the form

$$(2.1) \qquad Y_{n+1} = Y_n + a\left(\tau_{n+1}, Y_{n+1}\right) \Delta + b\,\Delta W,$$

where we follow our convention in writing $b = b(\tau_n, Y_n)$. From this and the explicit Euler scheme (10.2.1) we can easily construct a *family of implicit Euler schemes*

$$(2.2) \qquad Y_{n+1} = Y_n + \left\{\alpha a\left(\tau_{n+1}, Y_{n+1}\right) + (1 - \alpha)\,a\right\} \Delta + b\,\Delta W,$$

where the parameter $\alpha \in [0, 1]$ characterizes the degree of implicitness. We note that for $\alpha = 0$ we have the explicit Euler scheme (10.2.1) and for $\alpha = 1$ the scheme (2.1), whereas for $\alpha = 0.5$ we obtain from (2.2) a generalization of the deterministic trapezoidal method (8.1.11).

In the general multi-dimensional case $d, m = 1, 2, \ldots$ the *family of implicit Euler schemes* has $k$th component

$$(2.3)\ \ Y_{n+1}^k = Y_n^k + \left\{\alpha_k a^k\left(\tau_{n+1}, Y_{n+1}\right) + (1 - \alpha_k)\,a^k\right\} \Delta + \sum_{j=1}^{m} b^{k,j}\,\Delta W^j$$

with parameters $\alpha_k \in [0, 1]$ for $k = 1, \ldots, d$.

We shall use the following 2-dimensional linear Ito stochastic differential equations in PC-Exercises to test the above family of implicit schemes and others to be introduced later.

**Example 12.2.1**     *Let $W$ be a 1-dimensional Wiener process and*

$$A = \begin{bmatrix} a^{1,1} & a^{1,2} \\ a^{2,1} & a^{2,2} \end{bmatrix} \quad and \quad B = bI = \begin{bmatrix} b & 0 \\ 0 & b \end{bmatrix}$$

*for real numbers $a^{1,1}$ $a^{1,2}$, $a^{2,1}$, $a^{2,2}$ and $b$. Then*

(2.4) $$dX_t = AX_t \, dt + BX_t \, dW_t$$

*is a 2-dimensional, homogeneous linear Ito stochastic differential equation in the terminology of Section 8 of Chapter 4. Since the matrices A and B commute, it follows from (4.8.7) that (2.4) has the explicit solution*

(2.5) $$X_t = \exp\left(\left(A - \frac{1}{2}B^2\right)t + BW_t\right)X_0.$$

**Exercise 12.2.2** *Use (6.3.22)–(6.3.24) and the fact that the sum of the Lyapunov exponents equals the trace of the matrix $A - \frac{1}{2}B^2$, that is the sum of its main diagonal components, to show that the Lyapunov exponents of (2.4) are the real parts of the eigenvalues of the matrix $A - \frac{1}{2}B^2$.*

For the matrices

(2.6) $$A = \begin{bmatrix} -a & a \\ a & -a \end{bmatrix} \quad \text{and} \quad B = \begin{bmatrix} b & 0 \\ 0 & b \end{bmatrix}$$

the Lyapunov exponents of the stochastic differential equation (2.4) are

(2.7) $$\lambda_1 = -\frac{1}{2}b^2, \qquad \lambda_2 = -\frac{1}{2}b^2 - 2a.$$

With $a = 25$ and $b = 2$, for example, we have $\lambda_1 = -2 \gg \lambda_2 = -52$. In accordance with Section 8 of Chapter 9, we can thus consider equation (2.4) with the corresponding coefficient matrices $A$ and $B$ to be an example of a stiff stochastic differential equation. We note that the matrix exponential in the expression (2.5) can be evaluated by diagonalizing the matrix

$$\left(A - \frac{1}{2}B^2\right)t + BW_t$$

at each instant $t \geq 0$. In this way we obtain

(2.8) $$X_t = P\begin{bmatrix} \exp\left(\rho^+(t)\right) & 0 \\ 0 & \exp\left(\rho^-(t)\right) \end{bmatrix}P^{-1}X_0$$

where

$$\rho^{\pm}(t) = \left(-a - \frac{1}{2}b^2 \pm a\right)t + bW_t$$

and

$$P = \frac{1}{\sqrt{2}}\begin{bmatrix} 1 & 1 \\ 1 & -1 \end{bmatrix} \quad \text{with} \quad P^{-1} = P.$$

**PC-Exercise 12.2.3** *Consider the 2-dimensional Ito process X satisfying the stochastic differential equation (2.4) with coefficient matrices (2.6) with a $= 5$ and $b = 0.01$ on the time interval $[0, T]$ with initial value $X_0 = (1, 0)$ and $T = 1$. Compute $M = 20$ batches each of $N = 100$ simulations of $X_T$ and of the values $Y^\delta(T)$ of the implicit Euler approximation with $\alpha_1 = \alpha_2 = 0$, $0.5$ and $1.0$*

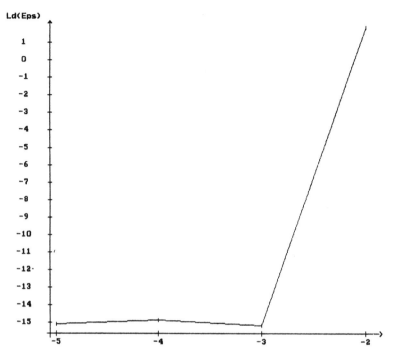

Figure 12.2.1     Results of PC-Exercise 12.2.3 with $\alpha_1 = \alpha_2 = 0$.

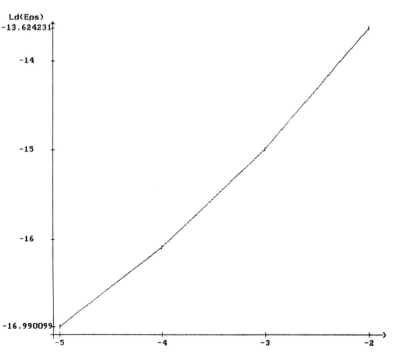

Figure 12.2.2     Results of PC-Exercise 12.2.3 with $\alpha_1 = \alpha_2 = 0.5$.

*corresponding to the same sample paths of the Wiener process for equidistant time steps with step size $\delta = \Delta = 2^{-2}$. Evaluate the 90% confidence intervals for the absolute errors $\epsilon$. Repeat for step sizes $\delta = 2^{-3}$, $2^{-4}$ and $2^{-5}$ and plot $\log_2 \epsilon$ versus $\log_2 \delta$ for the three cases $\alpha_1 = \alpha_2 = 0$, 0.5 and 1.0, respectively.*

We can see from Figure 12.2.1 that the explicit Euler scheme, that is with $\alpha_1 = \alpha_2 = 0$, does not give an acceptable result for the stiff system under consideration. On the other hand Figure 12.2.2 indicates that the generalized trapezoidal scheme with $\alpha_1 = \alpha_2 = 0.5$ is reliable and, from the slope, that it is an order $\gamma = 0.5$ strong scheme. Figure 12.2.3 suggests similar behaviour for the implicit Euler scheme with $\alpha_1 = \alpha_2 = 1.0$.

## The Implicit Milstein Scheme

The Milstein scheme (10.3.1) is the order 1.0 strong Taylor scheme. We shall call its implicit counterpart an *implicit Milstein scheme*. In the 1-dimensional case $d = m = 1$ this has the *Ito version*

$$(2.9) \qquad Y_{n+1} = Y_n + a\left(\tau_{n+1}, Y_{n+1}\right) \Delta + b\,\Delta W + \frac{1}{2} bb' \left\{(\Delta W)^2 - \Delta\right\}$$

and the *Stratonovich version*

$$(2.10) \qquad Y_{n+1} = Y_n + \underline{a}\left(\tau_{n+1}, Y_{n+1}\right) \Delta + b\,\Delta W + \frac{1}{2} bb'(\Delta W)^2,$$

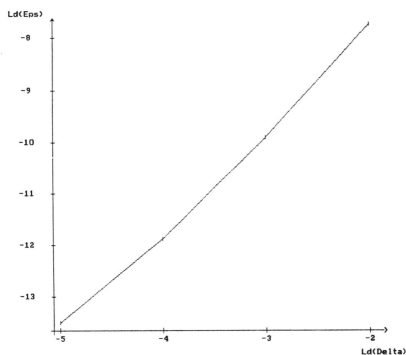

**Figure 12.2.3**    Results of PC-Exercise 12.2.3 with $\alpha_1 = \alpha_2 = 1.0$.

where $\underline{a} = a - \frac{1}{2}bb'$. These will generally be different, so we have two distinct types of implicit Milstein schemes.

As with the Euler schemes, we can also interpolate between the explicit and implicit Milstein schemes of the same type. In the general multi-dimensional case $d$, $m = 1$, $2$, $\ldots$ with *commutative noise* (see (10.3.13)) we obtain an *Ito version* of a *family of implicit Milstein schemes* with $k$th component

$$(2.11) \quad Y_{n+1}^k = Y_n^k + \left\{ \alpha_k a^k \left( \tau_{n+1}, Y_{n+1} \right) + (1 - \alpha_k) a^k \right\} \Delta$$

$$+ \sum_{j=1}^m b^{k,j} \, \Delta W^j$$

$$+ \frac{1}{2} \sum_{j_1, j_2 = 1}^m L^{j_1} b^{k,j_2} \left\{ \Delta W^{j_1} \Delta W^{j_2} - \delta_{j_1, j_2} \Delta \right\},$$

where $\delta_{j_1, j_2}$ is the Kronecker delta symbol

$$\delta_{j_1, j_2} = \begin{cases} 1 & : \quad j_1 = j_2 \\ 0 & : \quad \text{otherwise,} \end{cases}$$

and a *Stratonovich version* with $k$th component

$$(2.12) \quad Y_{n+1}^k = Y_n^k + \left\{ \alpha_k \underline{a}^k \left( \tau_{n+1}, Y_{n+1} \right) + (1 - \alpha_k) \underline{a}^k \right\} \Delta$$

$$+ \sum_{j=1}^m b^{k,j} \, \Delta W^j + \frac{1}{2} \sum_{j_1, j_2 = 1}^m L^{j_1} b^{k,j_2} \Delta W^{j_1} \Delta W^{j_2}.$$

Here the parameter $\alpha_k \in [0, 1]$ indicates the degree of implicitness of the $k$th component for $k = 1, \ldots, d$. When $\alpha_k = 0$ we have the explicit Milstein scheme in the $k$th component, and the implicit Milstein scheme when it equals 1.0. The case $\alpha_k = 0.5$ gives an order $\gamma = 1.0$ generalization of the deterministic trapezoidal method (8.1.11).

In the general case the *family of implicit Milstein schemes* has $k$th component

$$(2.13) \quad Y_{n+1}^k = Y_n^k + \left\{ \alpha_k a^k \left( \tau_{n+1}, Y_{n+1} \right) + (1 - \alpha_k) a^k \right\} \Delta$$

$$+ \sum_{j=1}^m b^{k,j} \, \Delta W^j + \sum_{j_1, j_2 = 1}^m L^{j_1} b^{k,j_2} I_{(j_1, j_2)}$$

in its *Ito version* and

$$(2.14) \quad Y_{n+1}^k = Y_n^k + \left\{ \alpha_k \underline{a}^k \left( \tau_{n+1}, Y_{n+1} \right) + (1 - \alpha_k) \underline{a}^k \right\} \Delta$$

$$+ \sum_{j=1}^m b^{k,j} \, \Delta W^j + \sum_{j_1, j_2 = 1}^m L^{j_1} b^{k,j_2} J_{(j_1, j_2)}$$

in its *Stratonovich version*, where $\alpha_k \in [0,1]$ for $k = 1,\ldots, d$. The multiple stochastic integrals $I_{(j_1,j_2)}$ and $J_{(j_1,j_2)}$ here can be approximated by the method proposed in Section 3 of Chapter 10.

**PC-Exercise 12.2.4**    *Repeat PC-Exercise 12.2.3 for the Ito version of the implicit Milstein schemes with $\alpha_1 = \alpha_2 = 0.5$ and 1.0.*

Figures 12.2.4 and 12.2.5 suggest that the implicit Milstein schemes with $\alpha_1 = \alpha_2 = 0.5$ and 1.0 have strong order $\gamma = 1.0$.

## The Implicit Order 1.5 Strong Taylor Scheme

For the 1-dimensional autonomous case $d = m = 1$ the *implicit order 1.5 strong Taylor scheme* in its simplest version has the form

$$(2.15) \quad Y_{n+1} = Y_n + \frac{1}{2}\left\{a\left(Y_{n+1}\right) + a\right\} \Delta$$

$$+ \left(ab' + \frac{1}{2}b^2b''\right)\left\{\Delta W \Delta - \Delta Z\right\}$$

$$+ a'b\left\{\Delta Z - \frac{1}{2}\Delta W \Delta\right\} + \frac{1}{2}bb'\left\{(\Delta W)^2 - \Delta\right\}$$

$$+ \frac{1}{2}b\left(bb'\right)'\left\{\frac{1}{3}(\Delta W)^2 - \Delta\right\}\Delta W,$$

where $\Delta W$ and $\Delta Z$ are Gaussian random variables with zero means and $E((\Delta W)^2) = \Delta$, $E((\Delta Z)^2) = \frac{1}{3}\Delta^3$ and $E(\Delta W \Delta Z) = \frac{1}{2}\Delta^2$, as in (10.4.3).

In the general multi-dimensional case we obtain the *family of implicit order 1.5 strong Taylor schemes* with $k$th component

$$(2.16) \quad Y_{n+1}^k = Y_n^k + \left\{\alpha_k a^k\left(\tau_{n+1}, Y_{n+1}\right) + (1 - \alpha_k)a^k\right\} \Delta$$

$$+ \left(\frac{1}{2} - \alpha_k\right)\left\{\beta_k L^0 a^k\left(\tau_{n+1}, Y_{n+1}\right) + (1 - \beta_k)L^0 a^k\right\} \Delta^2$$

$$+ \sum_{j=1}^m \left(b^{k,j}\,\Delta W^j + L^0 b^{k,j} I_{(0,j)} + L^j a^k\left\{I_{(j,0)} - \alpha_k \Delta W^j \Delta\right\}\right)$$

$$+ \sum_{j_1,j_2=1}^m L^{j_1} b^{k,j_2} I_{(j_1,j_2)} + \sum_{j_1,j_2,j_3=1}^m L^{j_1} L^{j_2} b^{k,j_3} I_{(j_1,j_2,j_3)},$$

where the parameters $\alpha_k$, $\beta_k \in [0,1]$ for $k = 1, \ldots, d$. The multiple stochastic integrals here can also be approximated as in Section 4 of Chapter 10.

The preceding scheme has a simpler structure in some important practical cases. For instance, for *additive noise* where $b(t,x) \equiv b(t)$, the scheme (2.16) reduces to

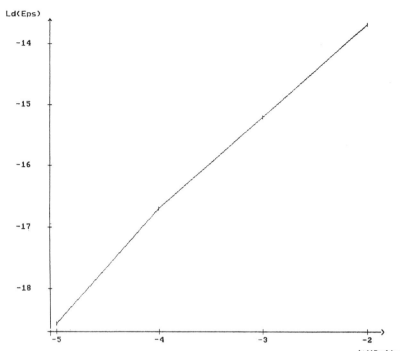

**Figure 12.2.4** Results of PC-Exercise 12.2.4 with $\alpha_1 = \alpha_2 = 0.5$.

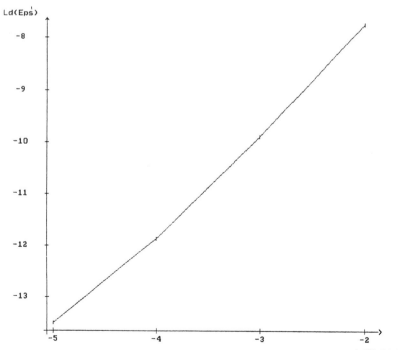

**Figure 12.2.5** Results of PC-Exercise 12.2.4 with $\alpha_1 = \alpha_2 = 1.0$.

$$(2.17) \quad Y_{n+1}^k = Y_n^k + \left\{ \alpha_k a^k \left( \tau_{n+1}, Y_{n+1} \right) + (1 - \alpha_k) a^k \right\} \Delta$$

$$+ \left( \frac{1}{2} - \alpha_k \right) \left\{ \beta_k L^0 a^k \left( \tau_{n+1}, Y_{n+1} \right) + (1 - \beta_k) L^0 a^k \right\} \Delta^2$$

$$+ \sum_{j=1}^m \left[ b^{k,j} \Delta W^j + \frac{\partial b^{k,j}}{\partial t} \left\{ \Delta W^j \Delta - \Delta Z^j \right\} \right.$$

$$\left. + L^j a^k \left\{ \Delta Z^j - \alpha_k \Delta W^j \Delta \right\} \right]$$

with $\alpha_k$, $\beta_k \in [0, 1]$ for $k = 1, \ldots, d$. For the choice $\alpha_1 = \cdots = \alpha_d = 0.5$ we obtain a particularly simple scheme of strong order $\gamma = 1.5$, since the second line in (2.17) vanishes.

Another interesting special case is that of *commutative noise of the second kind* (10.4.13). Then, in the general multi-dimensional case with $d$, $m = 1, 2,$ $\ldots$, the scheme (2.16) takes the form

$$(2.18) \quad Y_{n+1}^k = Y_n^k + \left\{ \alpha_k a^k \left( \tau_{n+1}, Y_{n+1} \right) + (1 - \alpha_k) a^k \right\} \Delta$$

$$+ \left( \frac{1}{2} - \alpha_k \right) \left\{ \beta_k L^0 a^k \left( \tau_{n+1}, Y_{n+1} \right) + (1 - \beta_k) L^0 a^k \right\} \Delta^2$$

$$+ \sum_{j=1}^m \left[ b^{k,j} \Delta W^j + \frac{1}{2} L^j b^{k,j} \left\{ \left( \Delta W^j \right)^2 - \Delta \right\} \right.$$

$$\left. + L^0 b^{k,j} \left\{ \Delta W^j \Delta - \Delta Z^j \right\} + L^j a^k \left\{ \Delta Z^j - \alpha_k \Delta W^j \Delta \right\} \right]$$

$$+ \sum_{j_1=1}^m \sum_{j_2=1}^{j_1-1} L^{j_1} b^{k,j_2} \Delta W^{j_1} \Delta W^{j_2}$$

$$+ \frac{1}{2} \sum_{\substack{j_1, j_2 = 1 \\ j_1 \neq j_2}}^m L^{j_1} L^{j_2} b^{k,j_2} \Delta W^{j_1} \left\{ \left( \Delta W^{j_2} \right)^2 - \Delta \right\}$$

$$+ \sum_{j_1=1}^m \sum_{j_2=1}^{j_1-1} \sum_{j_3=1}^{j_2-1} L^{j_1} L^{j_2} b^{k,j_3} \Delta W^{j_1} \Delta W^{j_2} \Delta W^{j_3}$$

$$+ \frac{1}{2} \sum_{j=1}^m L^j L^j b^{k,j} \left\{ \frac{1}{3} \left( \Delta W^j \right)^2 - \Delta \right\} \Delta W^j,$$

where $\alpha_k$, $\beta_k \in [0, 1]$ for $k = 1, \ldots, d$. We note that this scheme involves only the Gaussian random variables $\Delta W^j$ and $\Delta Z^j$, $j = 1, \ldots, m$.

**PC-Exercise 12.2.5** *Repeat PC-Exercise 12.2.3 for the implicit order 1.5 strong Taylor scheme (2.16) with $\alpha_1 = \alpha_2 = 0.5$.*

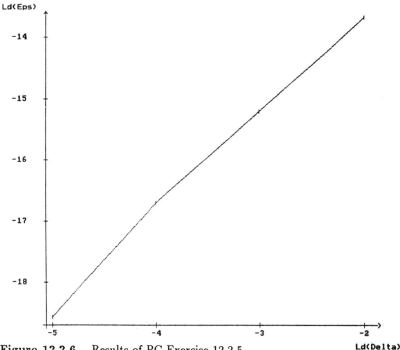

**Figure 12.2.6**    Results of PC-Exercise 12.2.5.

Figure 12.2.6 suggests that the implicit order 1.5 strong Taylor scheme with $\alpha_1 = \alpha_2 = 0.5$ does indeed have strong order $\gamma = 1.5$.

## The Implicit Order 2.0 Strong Taylor Scheme

Analogously as in Section 5 of Chapter 10, we can also derive implicit schemes with strong order $\gamma = 2.0$ from the Stratonovich-Taylor expansion.

In the 1-dimensional autonomous case $d = m = 1$ we have the following Stratonovich version of the *implicit order 2.0 strong Taylor scheme:*

$$(2.19) \quad Y_{n+1} \;=\; Y_n + \frac{1}{2}\{\underline{a}\,(Y_{n+1}) + \underline{a}\}\,\Delta + b\,\Delta W$$

$$+\underline{a}b'\,\{\Delta W \Delta - \Delta Z\} + \underline{a}'b\left\{\Delta Z - \frac{1}{2}\Delta W\,\Delta\right\}$$

$$+\frac{1}{2}\,bb'\,(\Delta W)^2 + \frac{1}{3!}\,b\,(bb')'\,(\Delta W)^3$$

$$+\frac{1}{4!}\,b\left(b\,(bb')'\right)'\,(\Delta W)^4$$

$$+\underline{a}\,(bb')'\,J_{(0,1,1)} + b\,(\underline{a}b')'\,J_{(1,0,1)}$$

$$+b\,(\underline{a}'b)'\left\{J_{(1,1,0)} - \frac{1}{4}(\Delta W)^2\,\Delta\right\}.$$

The multiple integrals appearing here can be approximated as in Exercise 10.5.1.

In the general multi-dimensional case $d$, $m = 1$, $2$, ... the $k$th component of the *implicit order* 2.0 *strong Taylor scheme* is given by

$$(2.20) \qquad Y_{n+1}^k = Y_n^k + \left\{ \alpha_k \underline{a}^k \left( \tau_{n+1}, Y_{n+1} \right) + (1 - \alpha_k) \underline{a}^k \right\} \Delta$$

$$+ \left( \frac{1}{2} - \alpha_k \right) \left\{ \beta_k \underline{L}^0 \underline{a}^k \left( \tau_{n+1}, Y_{n+1} \right) + (1 - \beta_k) \underline{L}^0 \underline{a}^k \right\} \Delta^2$$

$$+ \sum_{j=1}^m \left[ b^{k,j} \Delta W^j + \underline{L}^0 b^{k,j} J_{(0,j)} + \underline{L}^j \underline{a}^k \left\{ J_{(j,0)} - \alpha_k \Delta W^j \Delta \right\} \right]$$

$$+ \sum_{j_1,j_2=1}^m \left[ \underline{L}^{j_1} b^{k,j_2} J_{(j_1,j_2)} + \underline{L}^0 \underline{L}^{j_1} b^{k,j_2} J_{(0,j_1,j_2)} \right.$$

$$+ \underline{L}^{j_1} \underline{L}^0 b^{k,j_2} J_{(j_1,0,j_2)}$$

$$\left. + \underline{L}^{j_1} \underline{L}^{j_2} \underline{a}^k \left\{ J_{(j_1,j_2,0)} - \alpha_k \Delta J_{(j_1,j_2)} \right\} \right]$$

$$+ \sum_{j_1,j_2,j_3=1}^m \underline{L}^{j_1} \underline{L}^{j_2} b^{k,j_3} J_{(j_1,j_2,j_3)}$$

$$+ \sum_{j_1,j_2,j_3,j_4=1}^m \underline{L}^{j_1} \underline{L}^{j_2} \underline{L}^{j_3} b^{k,j_4} J_{(j_1,j_2,j_3,j_4)},$$

where $\alpha_k$, $\beta_k \in [0,1]$ for $k = 1$, ..., $d$.

In special cases the scheme (2.20) also simplifies to ones which do not involve all of the multiple Stratonovich integrals. For instance, in the autonomous case with *additive noise* (10.2.7) the $k$th component of the order 2.0 implicit strong Taylor scheme reduces to

$$(2.21) \qquad Y_{n+1}^k = Y_n^k + \left\{ \alpha_k \underline{a}^k \left( Y_{n+1} \right) + (1 - \alpha_k) \underline{a}^k \right\} \Delta$$

$$+ \left( \frac{1}{2} - \alpha_k \right) \left\{ \beta_k \underline{L}^0 \underline{a}^k \left( Y_{n+1} \right) + (1 - \beta_k) \underline{L}^0 \underline{a}^k \right\} \Delta^2$$

$$+ \sum_{j=1}^m \left[ b^{k,j} \Delta W^j + \underline{L}^j \underline{a}^k \left\{ J_{(j,0)} - \alpha_k \Delta W^j \Delta \right\} \right]$$

$$+ \sum_{j_1,j_2=1}^m \underline{L}^{j_1} \underline{L}^{j_2} \underline{a}^k \left\{ J_{(j_1,j_2,0)} - \alpha_k \Delta J_{(j_1,j_2)} \right\},$$

with $\alpha_k$, $\beta_k \in [0,1]$ for $k = 1$, ..., $d$. This scheme becomes particularly simple when $\alpha_k = 0.5$ for $k = 1$, ..., $d$, as it then differs from the strong order 1.5 scheme (2.17) only by the inclusion of the last term.

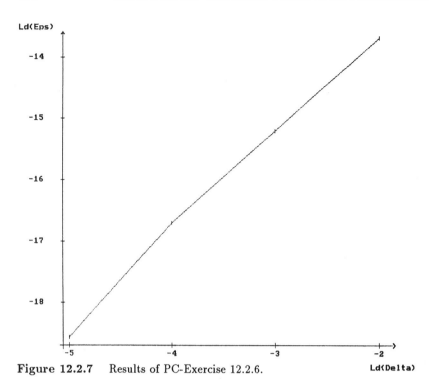

**Figure 12.2.7**    Results of PC-Exercise 12.2.6.

There are also Ito counterparts of the above implicit order 2.0 strong schemes, but the Stratonovich versions are more convenient, as we have already seen for the explicit schemes themselves.

**PC-Exercise 12.2.6**    *Repeat PC-Exercise 12.2.3 for the implicit order* 2.0 *strong Taylor scheme (2.20) with* $\alpha_1 = \alpha_2 = 0.5$.

Figure 12.2.7 gives the impression that the scheme (2.20) really is an order 2.0 strong scheme.

## 12.3    Implicit Strong Runge-Kutta Schemes

In this section we shall discuss implicit schemes which avoid the use of derivatives in the terms involving non-deterministic multiple stochastic integrals. They are obtained from the corresponding implicit strong Taylor schemes by replacing the derivatives there by finite differences expressed in terms of appropriate supporting values. For this reason we shall call them implicit strong Runge-Kutta schemes, but we emphasize that they are not simply heuristic adaptations to stochastic differential equations of the deterministic Runge-Kutta schemes.

## The Implicit Order 1.0 Strong Runge-Kutta Scheme

In the 1-dimensional case $d = m = 1$ the *implicit order* 1.0 *strong Runge-Kutta scheme* is

$$(3.1) \qquad Y_{n+1} = Y_n + a(\tau_{n+1}, Y_{n+1})\Delta + b\Delta W$$

$$+ \frac{1}{2\sqrt{\Delta}} (b(\tau_n, \tilde{\Upsilon}_n) - b)\{(\Delta W)^2 - \Delta\}$$

with supporting value

$$\tilde{\Upsilon} = Y_n + a\Delta + b\sqrt{\Delta}.$$

By interpolating between this implicit scheme and the corresponding explicit scheme (11.1.7), we can form a *family of implicit order* 1.0 *strong Runge-Kutta schemes*. In the general multi-dimensional case $d, m = 1, 2, \ldots$ these have $k$th component

$$(3.2) \quad Y_{n+1}^k = Y_n^k + \{\alpha_k a(\tau_{n+1}, Y_{n+1}^k) + (1 - \alpha_k)a^k\}\Delta + \sum_{j=1}^m b^{k,j}\Delta W^j$$

$$+ \frac{1}{\sqrt{\Delta}} \sum_{j_1,j_2=1}^m \{b^{k,j_2}(\tau_n, \tilde{\Upsilon}_n^{j_1}) - b^{k,j_2}\} I_{(j_1,j_2)}$$

with vector supporting values

$$\tilde{\Upsilon}_n^j = Y_n + a\Delta + b^j\sqrt{\Delta},$$

for $j = 1, \ldots, m$, and parameters $\alpha_k \in [0, 1]$ for $k = 1, \ldots, d$. The multiple Ito integrals $I_{(j_1,j_2)}$ here can be approximated as in Section 3 of Chapter 10.

There is also a *Stratonovich version* of (3.2). For *commutative noise* (10.3.13) it reduces to

$$(3.3) \qquad Y_{n+1}^k = Y_n^k + \{\alpha_k \underline{a}^k(\tau_{n+1}, Y_{n+1}^k) + (1 - \alpha_k)\underline{a}^k\}\Delta$$

$$+ \frac{1}{2} \sum_{j=1}^m \{b^{k,j}(\tau_n, \bar{\Psi}_n) + b^{k,j}\}\Delta W^j$$

with vector supporting values

$$\bar{\Psi}_n = Y_n + \underline{a}\Delta + \sum_{j=1}^m b^j\Delta W^j$$

and parameters $\alpha_k \in [0, 1]$ for $k = 1, \ldots, d$. We remark that the term $\underline{a}\Delta$ can be omitted from $\bar{\Psi}_n$ and note that the diffusion coefficient $b$ must be evaluated at two points for each time step.

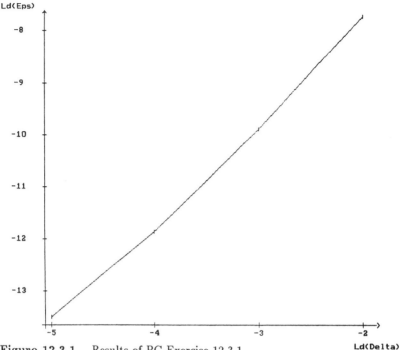

**Figure 12.3.1**     Results of PC-Exercise 12.3.1.

**PC-Exercise 12.3.1**     *Repeat PC-Exercise 12.2.3 for the scheme (3.3) with*
$\alpha_1 = \alpha_2 = 1.0$. *Then repeat the calculation with the supporting value*

$$\bar{\Psi}_n = Y_n + b\,\Delta W.$$

*Compare the results with those of PC-Exercise 12.2.4.*

## The Implicit Order 1.5 Strong Runge-Kutta Scheme

As above, we can also derive implicit order 1.5 strong schemes by replacing derivatives by their corresponding finite differences.

In the autonomous 1-dimensional case $d = m = 1$ an *implicit order* 1.5 *strong Runge-Kutta scheme* obtained in this way has the form

$$(3.4) \qquad Y_{n+1} \;=\; Y_n + \frac{1}{2}\left\{a\left(Y_{n+1}\right) + a\right\}\Delta + b\,\Delta W$$

$$+\frac{1}{4\sqrt{\Delta}}\left\{b\left(\bar{\Upsilon}_+\right) - b\left(\bar{\Upsilon}_-\right)\right\}\left\{(\Delta W)^2 - \Delta\right\}$$

$$+\frac{1}{2\Delta}\left\{b\left(\bar{\Upsilon}_+\right) - 2b + b\left(\bar{\Upsilon}_-\right)\right\}\left\{\Delta W\Delta - \Delta Z\right\}$$

$$+\frac{1}{2\sqrt{\Delta}}\left\{a\left(\bar{\Upsilon}_+\right) - a\left(\bar{\Upsilon}_-\right)\right\}\left\{\Delta Z - \frac{1}{2}\Delta W\Delta\right\}$$

$$+\frac{1}{4\Delta}\left\{b\left(\bar{\Phi}_+\right)-b\left(\bar{\Phi}_-\right)-b\left(\bar{\Upsilon}_+\right)+b\left(\bar{\Upsilon}_-\right)\right\}$$

$$\times\left\{\frac{1}{3}(\Delta W)^2-\Delta\right\}\Delta W$$

with supporting values

$$\bar{\Upsilon}_\pm=Y_n+a\,\Delta\pm b\,\sqrt{\Delta}$$

and

$$\bar{\Phi}_\pm=\bar{\Upsilon}_+\pm b\left(\bar{\Upsilon}_+\right)\,\sqrt{\Delta}.$$

For the general multi-dimensional case $d$, $m=1,2,\ldots$ we have the *implicit order 1.5 strong Runge-Kutta scheme* in vector form

$$(3.5)\qquad Y_{n+1}=Y_n+\frac{1}{2}\left\{a\left(\tau_{n+1},Y_{n+1}\right)+a\right\}\Delta$$

$$+\sum_{j=1}^{m}\left[b^j\,\Delta W^j+\frac{1}{\Delta}\left\{b^j\left(\tau_{n+1},Y_n\right)-b^j\right\}I_{(0,j)}\right.$$

$$\left.+\frac{1}{2\sqrt{\Delta}}\left\{a\left(\tau_n,\bar{\Upsilon}_+^j\right)-a\left(\tau_n,\bar{\Upsilon}_-^j\right)\right\}\left\{I_{(j,0)}-\frac{1}{2}\Delta W^j\Delta\right\}\right]$$

$$+\frac{1}{2\sqrt{\Delta}}\sum_{j_1,j_2=1}^{m}\left\{b^{j_2}\left(\tau_n,\bar{\Upsilon}_+^{j_1}\right)-b^{j_2}\left(\tau_n,\bar{\Upsilon}_-^{j_1}\right)\right\}I_{(j_1,j_2)}$$

$$+\frac{1}{2\Delta}\sum_{j_1,j_2=1}^{m}\left\{b^{j_2}\left(\tau_n,\bar{\Upsilon}_+^{j_1}\right)-2b^{j_2}+b^{j_2}\left(\tau_n,\bar{\Upsilon}_-^{j_1}\right)\right\}I_{(0,j_2)}$$

$$+\frac{1}{2\Delta}\sum_{j_1,j_2,j_3=1}^{m}\left[b^{j_3}\left(\tau_n,\bar{\Phi}_+^{j_1,j_2}\right)-b^{j_3}\left(\tau_n,\bar{\Phi}_-^{j_1,j_2}\right)\right.$$

$$\left.-b^{j_3}\left(\tau_n,\bar{\Upsilon}_+^{j_1}\right)+b^{j_3}\left(\tau_n,\bar{\Upsilon}_-^{j_1}\right)\right]I_{(j_1,j_2,j_3)}$$

with vector supporting values

$$\bar{\Upsilon}_\pm^j=Y_n+\frac{1}{m}a\,\Delta\pm b^j\,\sqrt{\Delta}$$

and

$$\bar{\Phi}_\pm^{j_1,j_2}=\bar{\Upsilon}_+^{j_1}\pm b^{j_1}\left(\tau_n,\bar{\Upsilon}_+^{j_2}\right)\,\sqrt{\Delta}.$$

To avoid too many terms here we have restricted our attention to the degree of implicitness $\alpha_k=0.5$. We note that the diffusion coefficient $b$ must be evaluated at $2m+1$ points for each time step. This scheme also simplifies considerably in special cases. For instance, with *additive noise* it becomes

$$(3.6)\qquad Y_{n+1}=Y_n+\frac{1}{2}\left\{a\left(\tau_{n+1},Y_{n+1}\right)+a\right\}\Delta$$

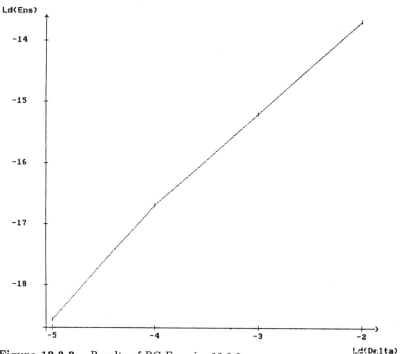

**Figure 12.3.2**    Results of PC-Exercise 12.3.2.

$$+ \sum_{j=1}^{m} \left[ b^j \, \Delta W^j + \frac{1}{\Delta} \left\{ b^j \left( \tau_{n+1} \right) - b^j \right\} \left\{ \Delta W^j \Delta - \Delta Z^j \right\} \right.$$

$$\left. + \frac{1}{2\sqrt{\Delta}} \left\{ a \left( \tau_n, \tilde{\Upsilon}_+^j \right) - a \left( \tau_n, \tilde{\Upsilon}_-^j \right) \right\} \left\{ \Delta Z^j - \frac{1}{2} \Delta W^j \Delta \right\} \right]$$

with

$$\tilde{\Upsilon}_\pm^j = Y_n + \frac{1}{m} \, a \, \Delta \pm b^j \, \sqrt{\Delta}.$$

**PC-Exercise 12.3.2**    *Repeat PC-Exercise 12.2.3 with the scheme (3.5).*

## An Implicit Order 2.0 Strong Runge-Kutta Scheme

To avoid cumbersome notation we shall restrict ourselves to the case of scalar additive noise and, as before, to the Stratonovich version of the order 2.0 schemes.

In the autonomous 1-dimensional case $d = m = 1$ with *scalar additive noise* we propose the *implicit order* 2.0 *strong Runge-Kutta scheme*

$$(3.7) \quad Y_{n+1} = Y_n + \left\{ \underline{a} \left( \tilde{\Upsilon}_+ \right) + \underline{a} \left( \tilde{\Upsilon}_- \right) - \frac{1}{2} \left( \underline{a} \left( Y_{n+1} \right) + \underline{a} \right) \right\} \Delta + b \, \Delta W$$

with

$$\bar{\Upsilon}_\pm = Y_n + \frac{1}{2} \underline{a} \Delta + \frac{1}{\Delta} b \left( \Delta \tilde{Z} \pm \tilde{\zeta} \right),$$

where

$$\Delta \tilde{Z} = \frac{1}{2} \Delta Z + \frac{1}{4} \Delta W \Delta$$

and

$$\tilde{\zeta} = \sqrt{ J_{(1,1,0)} \Delta - \frac{1}{2} (\Delta Z)^2 + \frac{1}{8} \left( (\Delta W)^2 + \frac{1}{2} (2 \Delta Z \, \Delta^{-1} - \Delta W)^2 \right) \Delta^2 }.$$

This generalizes in the multi-dimensional case $d = 1, 2, \ldots$ and $m = 1$ with *scalar additive noise* $b(t, x) \equiv b(t)$ to the implicit order 2.0 strong Runge-Kutta scheme in vector form

$$(3.8) \quad Y_{n+1} = Y_n + \left\{ \underline{a} (\bar{\Upsilon}_+) + \underline{a} (\bar{\Upsilon}_-) - \frac{1}{2} (\underline{a} (Y_{n+1}) + \underline{a}) \right\} \Delta + b \, \Delta W$$

$$+ \frac{1}{\Delta} \left\{ b (\tau_{n+1}) - b \right\} \left\{ \Delta W \Delta - \Delta Z \right\}$$

with

$$\bar{\Upsilon}_\pm = Y_n + \frac{1}{2} \underline{a} \Delta + \frac{1}{\Delta} b \left( \Delta \tilde{Z} \pm \tilde{\zeta} \right),$$

where $\Delta \tilde{Z}$ and $\tilde{\zeta}$ are as above.

**PC-Exercise 12.3.3** *Repeat PC-Exercise 11.3.2 with the scheme (3.8) with $\underline{a}$ determined by interpreting the time $t$ as the first component of a 2-dimensional Ito process $\{(t, X_t), t \geq 0\}$.*

**Exercise 12.3.4** *Use Theorem 11.5.2 to show under appropriate assumptions that (3.8) is an order 2.0 strong scheme.*

## 12.4 Implicit Two-Step Strong Schemes

We saw in Section 4 of Chapter 11 that the number of evaluations of derivatives of the drift and diffusion coefficients can be reduced by using multi-step schemes. Here we shall describe some implicit two-step schemes which can be used for the efficient simulation of stiff stochastic differential equations.

### Implicit Two-Step Order 1.0 Strong Schemes

For the autonomous 1-dimensional case $d = m = 1$ we have the following *Ito version* of an *implicit two-step order 1.0 strong scheme*:

$$(4.1) \quad Y_{n+1} = Y_{n-1} + \{ a (\tau_{n+1}, Y_{n+1}) + a \} \Delta + V_n + V_{n-1} \quad \text{with}$$

$$V_n = b\,\Delta W_n + \frac{1}{2}\,bb'\left\{(\Delta W_n)^2 - \Delta\right\}.$$

The *Stratonovich version* of this scheme is

(4.2)        $$Y_{n+1} = Y_{n-1} + \{\underline{a}\,(\tau_{n+1}, Y_{n+1}) + \underline{a}\}\,\Delta + \underline{V}_n + \underline{V}_{n-1}$$

with

$$\underline{V}_n = b\,\Delta W_n + \frac{1}{2}\,bb'\,(\Delta W_n)^2,$$

and a *derivative free Stratonovich version* is

(4.3)        $$Y_{n+1} = Y_{n-1} + \{\underline{a}\,(\tau_{n+1}, Y_{n+1}) + \underline{a}\}\,\Delta + \tilde{V}_n + \tilde{V}_{n-1}$$

with

$$\tilde{V}_n = b\,\Delta W_n + \frac{1}{2\sqrt{\Delta}}\left\{b\left(\tau_n, \tilde{\Upsilon}_n\right) - b\right\}(\Delta W_n)^2$$

where

$$\tilde{\Upsilon}_n = Y_n + \underline{a}\,\Delta + b\,\sqrt{\Delta}.$$

There is a *family of implicit order 1.0 two-step strong schemes* for which the $k$th component of the *Ito version* in the general multi-dimensional case $d$, $m = 1, 2,\ldots$ is given by

(4.4)   $$\begin{aligned}
Y_{n+1} &= (1 - \gamma_k)\,Y_n^k + \gamma_k Y_{n-1}^k \\
&\quad + \big[\alpha_{2,k}a^k\,(\tau_{n+1}, Y_{n+1}) + \{\gamma_k\alpha_{1,k} + (1 - \alpha_{2,k})\}\,a^k \\
&\qquad + \gamma_k\,(1 - \alpha_{1,k})\,a^k\,(\tau_{n-1}, Y_{n-1})\big]\,\Delta \\
&\quad + V_n^k + \gamma_k V_{n-1}^k
\end{aligned}$$

with

$$V_n^k = \sum_{j=1}^{m} b^{k,j}\,\Delta W_n^j + \sum_{j_1,j_2=1}^{m} L^{j_1} b^{k,j_2}\,I_{(j_1,j_2),\tau_n,\tau_{n+1}}$$

and parameters $\alpha_{1,k}$, $\alpha_{2,k}$, $\gamma_k \in [0,1]$ for $k = 1, \ldots, d$.

The *Stratonovich version* of (4.4) is

(4.5)   $$\begin{aligned}
Y_{n+1} &= (1 - \gamma_k)\,Y_n^k + \gamma_k Y_{n-1}^k \\
&\quad + \big[\alpha_{2,k}\underline{a}^k\,(\tau_{n+1}, Y_{n+1}) + \{\gamma_k\alpha_{1,k} + (1 - \alpha_{2,k})\}\,\underline{a}^k \\
&\qquad + \gamma_k\,(1 - \alpha_{1,k})\,\underline{a}^k\,(\tau_{n-1}, Y_{n-1})\big]\,\Delta \\
&\quad + \underline{V}_n^k + \gamma_k\,\underline{V}_{n-1}^k
\end{aligned}$$

with

(4.6)        $$\underline{V}_n^k = \sum_{j=1}^{m} b^{k,j}\,\Delta W_n^j + \sum_{j_1,j_2=1}^{m} \underline{L}^{j_1} b^{k,j_2}\,J_{(j_1,j_2),\tau_n,\tau_{n+1}}$$

and parameters $\alpha_{1,k}$, $\alpha_{2,k}$, $\gamma_k \in [0,1]$ for $k = 1, \ldots, d$. We can obtain a *derivative free version* of this scheme if, instead of (4.6), we use

$$(4.7) \quad \underline{V}_n^k = \sum_{j=1}^m b^{k,j} \Delta W_n^j + \frac{1}{\sqrt{\Delta}} \sum_{j_1,j_2=1}^m \left\{ b^{k,j_2}\left(\tau_n, \tilde{\Upsilon}_n^{j_1}\right) - b^{k,j_2} \right\} J_{(j_1,j_2),\tau_n,\tau_{n+1}}$$

with

$$\tilde{\Upsilon}_n^j = Y_n + \underline{a}\,\Delta + b^j\,\sqrt{\Delta}.$$

For *commutative noise* we can substitute $\frac{1}{2}\Delta W_n^{j_1}\Delta W_n^{j_2}$ for the Stratonovich double integral $J_{(j_1,j_2),\tau_n,\tau_{n+1}}$ in (4.6) and (4.7) or we may set

$$\underline{V}_n^k = \frac{1}{2}\sum_{j=1}^m \left\{ b^{k,j}\left(\tau_n, \tilde{\Upsilon}_n\right) + b^{k,j} \right\} \Delta W_n^j$$

with

$$\tilde{\Upsilon}_n = Y_n + \underline{a}\,\Delta + \sum_{j=1}^m b^j\,\Delta W_n^j.$$

We note that if we set $\gamma_k = 0$ in the above schemes, then we obtain our implicit one-step schemes. On the other hand, if we set $\alpha_{2,k} = 0$, then we have our explicit two-step schemes. We can control the stability of these schemes by an appropriate choice of parameters $\gamma_k$, $\alpha_{1,k}$ and $\alpha_{2,k}$. In Section 5 we shall check their regions of absolute stability.

**PC-Exercise 12.4.1**    *Repeat PC-Exercise 12.2.3 for the scheme (4.4) with parameters $\gamma_k = \alpha_{1,k} = \alpha_{2,k} = 1.0$ using the implicit Milstein scheme (2.11) as the starting routine.*

## Implicit Two-Step Order 1.5 Strong Schemes

In the autonomous 1-dimensional case $d = m = 1$ an *implicit two-step order 1.5 strong scheme* is given by

$$(4.8) \quad Y_{n+1} = Y_{n-1} + \frac{1}{2}\left\{ a\left(Y_{n+1}\right) + 2a + a\left(Y_{n-1}\right) \right\} \Delta$$

$$+ V_n + V_{n-1}$$

with

$$V_n = b\,\Delta W_n + \left( ab' + \frac{1}{2}b^2 b'' \right) \left\{ \Delta W_n\,\Delta - \Delta Z_n \right\}$$

$$+ a'b \left\{ \Delta Z_n - \frac{1}{2}\Delta W_n\,\Delta \right\} + \frac{1}{2}\,bb'\left\{ \left(\Delta W_n\right)^2 - \Delta \right\}$$

$$+ \frac{1}{2}\,b\,(bb')'\left\{ \frac{1}{3}\left(\Delta W_n\right)^2 - \Delta \right\} \Delta W_n.$$

We have in the general multi-dimensional case $d, m = 1, 2, \ldots$ the following *family of implicit two-step order* 1.5 *strong schemes* with $k$th component:

$$(4.9) \qquad Y_{n+1}^k = (1 - \gamma_k) Y_n^k + \gamma_k Y_{n-1}^k$$

$$+ \frac{1}{2} \left\{ a^k \left( \tau_{n+1}, Y_{n+1} \right) + (1 + \gamma_k) a^k + \gamma_k a^k \left( \tau_{n-1}, Y_{n-1} \right) \right\} \Delta$$

$$- \frac{1}{2} (1 - \gamma_k) \sum_{j=1}^m L^j a^k \left( \tau_{n-1}, Y_{n-1} \right) \Delta W_{n-1}^j \Delta$$

$$+ V_n^k + \gamma_k V_{n-1}^k$$

with

$$V_n^k = \sum_{j=1}^m \left[ b^{k,j} \, \Delta W_n^j + L^0 b^{k,j} \left\{ \Delta W_n^j \Delta - \Delta Z_n^j \right\} \right.$$

$$\left. + L^j a^k \left\{ \Delta Z_n^j - \frac{1}{2} \Delta W_n^j \Delta \right\} \right]$$

$$+ \sum_{j_1, j_2 = 1}^m L^{j_1} b^{k,j_2} I_{(j_1, j_2), \tau_n, \tau_{n+1}}$$

$$+ \sum_{j_1, j_2, j_3 = 1}^m L^{j_1} L^{j_2} b^{k,j_3} I_{(j_1, j_2, j_3), \tau_n, \tau_{n+1}}$$

and $\gamma_k \in [0, 1]$ for $k = 1, \ldots, d$.

A *derivative free* version of the above family of implicit two-step order 1.5 strong schemes (4.9) has $k$th component

$$(4.10) \qquad Y_{n+1}^k = (1 - \gamma_k) Y_n^k + \gamma_k Y_{n-1}^k$$

$$+ \frac{1}{2} \left\{ a^k \left( \tau_{n+1}, Y_{n+1} \right) + (1 + \gamma_k) a^k + \gamma_k a^k \left( \tau_{n-1}, Y_{n-1} \right) \right\} \Delta$$

$$- \frac{1}{4} (1 - \gamma_k) \sqrt{\Delta} \sum_{j=1}^m \left\{ a^k \left( \tau_{n-1}, \bar{\Upsilon}_{n-1}^{j+} \right) - a^k \left( \tau_{n-1}, \bar{\Upsilon}_{n-1}^{j-} \right) \right\} \Delta W_{n-1}^j$$

$$+ V_n^k + \gamma_k V_{n-1}^k$$

with

$$V_n^k = \sum_{j=1}^m \left[ b^{k,j} \, \Delta W_n^j + \frac{1}{\Delta} \left\{ b^{k,j} \left( \tau_{n+1}, Y_n \right) - b^{k,j} \right\} I_{(0,j), \tau_n, \tau_{n+1}} \right.$$

$$\left. + \frac{1}{2\sqrt{\Delta}} \left\{ a^k \left( \tau_n, \bar{\Upsilon}_n^{j+} \right) - a^k \left( \tau_n, \bar{\Upsilon}_n^{j-} \right) \right\} \left\{ \Delta Z_n^j - \frac{1}{2} \Delta W_n^j \Delta \right\} \right]$$

$$+\frac{1}{2}\sum_{j_1,j_2=1}^{m}\left[\frac{1}{\sqrt{\Delta}}\left\{b^{k,j_2}\left(\tau_n,\bar{\Upsilon}_n^{j_1+}\right)-b^{k,j_2}\left(\tau_n,\bar{\Upsilon}_n^{j_1-}\right)\right\}I_{(j_1,j_2),\tau_n,\tau_{n+1}}\right.$$

$$\left.+\frac{1}{\Delta}\left\{b^{k,j_2}\left(\tau_n,\bar{\Upsilon}_n^{j_1+}\right)-2b^{k,j_2}+b^{k,j_2}\left(\tau_n,\bar{\Upsilon}_n^{j_1-}\right)\right\}I_{(0,j_2),\tau_n,\tau_{n+1}}\right]$$

$$+\frac{1}{2\Delta}\sum_{j_1,j_2,j_3=1}^{m}\left[b^{k,j_3}\left(\tau_n,\bar{\Phi}_n^{j_1,j_2+}\right)-b^{k,j_3}\left(\tau_n,\bar{\Phi}_n^{j_1,j_2-}\right)\right.$$

$$\left.-b^{k,j_3}\left(\tau_n,\bar{\Upsilon}_n^{j_1+}\right)+b^{k,j_3}\left(\tau_n,\bar{\Upsilon}_n^{j_1-}\right)\right]I_{(j_1,j_2,j_3),\tau_n,\tau_{n+1}}$$

with

$$\bar{\Upsilon}_n^{j\pm}=Y_n+\frac{1}{m}a\,\Delta\pm b^j\,\sqrt{\Delta},$$

$$\bar{\Phi}_n^{j_1,j_2\pm}=\bar{\Upsilon}_n^{j_1+}\pm b^{j_2}\left(\tau_n,\bar{\Upsilon}_n^{j_1+}\right)\sqrt{\Delta}$$

and $\gamma_k\in[0,1]$ for $k=1,\ldots,d$.

If we set $\gamma_k=0$ in the above two-step schemes we obtain implicit one-step schemes. We shall see later from (6.7) that it is possible to derive other schemes with different degrees of implicitness than the $\alpha_k=0.5$ values used above.

**PC-Exercise 12.4.2**  *Repeat PC-Exercise 12.2.3 for the scheme (4.9) with* $\gamma_k=1.0$ *using the implicit order 1.5 strong Taylor scheme (2.16) as the starting routine.*

## The Implicit Two-Step Order 2.0 Strong Scheme

As before, we shall consider only the Stratonovich versions of order 2.0 schemes.

For the autonomous 1-dimensional case $d=m=1$ with *additive noise* we have the *implicit two-step order 2.0 strong scheme*

$$(4.11)\qquad Y_{n+1}=Y_{n-1}+\frac{1}{2}\left\{\underline{a}\left(Y_{n+1}\right)+2\underline{a}+\underline{a}\left(Y_{n-1}\right)\right\}\Delta$$

$$+V_n+V_{n-1}$$

with

$$V_n=b\,\Delta W_n+\underline{a}'b\left\{\Delta Z_n-\frac{1}{2}\Delta W_n\,\Delta\right\}$$

$$+\underline{a}''b^2\left\{J_{(1,1,0),\tau_n,\tau_{n+1}}-\frac{1}{4}\left(\Delta W_n\right)^2\Delta\right\}.$$

We note that if $\underline{a}''\equiv0$, as in the linear case, then the final term in $V_n$ with the multiple integral $J_{(1,1,0)}$ vanishes.

In the nonautonomous multi-dimensional case with *scalar additive noise d*
$= 1, 2, \ldots$ and $m = 1$ there is a *family of implicit two-step order 2.0 strong
schemes* for which the $k$th component is

(4.12) $\qquad Y_{n+1}^k = (1 - \gamma_k)\, Y_n^k + \gamma_k Y_{n-1}^k$

$$+ \frac{1}{2} \left\{ \underline{a}^k \left( \tau_{n+1}, Y_{n+1} \right) + (1 + \gamma_k)\, \underline{a}^k + \gamma_k \underline{a}^k \left( \tau_{n-1}, Y_{n-1} \right) \right\} \Delta$$

$$- \frac{1}{2} (1 - \gamma_k)\, \underline{L}^1 \underline{a}^k \left( \tau_{n-1}, Y_{n-1} \right) \Delta W_{n-1}\, \Delta$$

$$- \frac{1}{4} (1 - \gamma_k)\, \underline{L}^1 \underline{L}^1 \underline{a}^k \left( \Delta W_n \right)^2 \Delta$$

$$+ V_n^k + \gamma_k V_{n-1}^k \qquad \text{with}$$

$$V_n^k \;\; = \;\; b^k \Delta W_n + \frac{\partial b^k}{\partial t} \left\{ \Delta W_n\, \Delta - \Delta Z_n \right\}$$

$$+ \underline{L}^1 \underline{a}^k \left\{ \Delta Z_n - \frac{1}{2} \Delta W_n\, \Delta \right\}$$

$$+ \underline{L}^1 \underline{L}^1 \underline{a}^k \left\{ J_{(1,1,0), \tau_n, \tau_{n+1}} - \frac{1}{4} \left( \Delta W_n \right)^2 \Delta \right\},$$

where $\gamma_k \in [0, 1]$ for $k = 1, \ldots, d$.

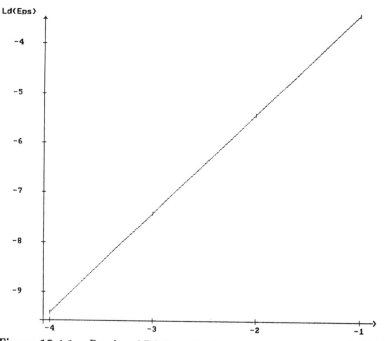

**Figure 12.4.1**    Results of PC-Exercise 12.4.3.

All of the schemes introduced in this section follow from formula (6.7), which will be given in Section 6. Other two-step strong schemes, including derivative free ones can also be derived from (6.7). However, this is more conveniently done for special classes of stochastic differential equations which allow simplifications in the formulation of the schemes.

**PC-Exercise 12.4.3**    *Repeat PC-Exercise 11.3.2 for the scheme (4.12) with* $\gamma_k = 1.0$ *using the implicit order 2.0 strong Taylor scheme (2.20) with* $\alpha_k = 0.5$ *as the starting routine.*

## 12.5   A-Stability of Strong One-Step Schemes

Here we shall investigate the region of absolute stability, which we defined in Section 8 of Chapter 9, for some of the explicit and implicit strong schemes presented in Chapters 10, 11 and 12.

We recall that we introduced complex-valued test equations (9.8.3)

$$(5.1) \qquad dX_t = \lambda X_t \, dt + dW_t$$

where $\lambda$ is a complex number with real part $Re(\lambda) < 0$ and $W$ is a real-valued standard Wiener process. When we apply a one-step scheme to (5.1) we often obtain a recursive expression of the form

$$(5.2) \qquad Y_{n+1} = G(\lambda\Delta) Y_n + Z_n,$$

where the $Z_n$ are random variables which do not depend on $\lambda$ or on the $Y_0$, $Y_1$, …… The *region of absolute stability* of the scheme is then the subset of the complex plane consisting of those $\lambda\Delta$ with $Re(\lambda) < 0$ and real-valued $\Delta > 0$ which are mapped by $G$ into the unit circle, that is with

$$(5.3) \qquad |G(\lambda\Delta)| < 1.$$

If this region coincides with the left half of the complex plane, we say the scheme is *A-stable*.

In Section 8 of Chapter 9 we found that the region of absolute stability of the explicit Euler scheme (10.2.4) is the open disc of unit radius centered on the point $(-1, 0)$, that is the complex number $-1 + 0\imath$. It is easy to see that this set is also the region of absolute stability for the Milstein schemes (10.3.3) and (10.3.4). In addition we saw that the implicit Euler scheme (9.8.6) is A-stable. For the families of implicit Euler and Milstein schemes, (2.3) and (2.13) respectively, with a common implicitness parameter $\alpha \in [0, 1]$ in both components, the mapping $G$ in (5.2) is

$$(5.4) \qquad G(\lambda\Delta) = (1 - \alpha\lambda\Delta)^{-1}(1 + (1 - \alpha)\lambda\Delta).$$

Inequality (5.3) for the $G$ in (5.4) is equivalent to the inequality of complex moduli

$$|1 + (1 - \alpha) \lambda \Delta|^2 < |1 - \alpha \lambda \Delta|^2,$$

which in turn is equivalent to

(5.5)                    $(1 - 2\alpha) \left(\lambda_1^2 + \lambda_2^2\right) \Delta^2 + 2 \lambda_1 \Delta < 0$

where $\lambda = \lambda_1 + \imath \lambda_2$. For $\frac{1}{2} \leq \alpha \leq 1$ this is satisfied by all $\lambda$ with $\lambda_1 < 0$, so the schemes are then A-stable. For $0 \leq \alpha < \frac{1}{2}$ we can write (5.5) as

$$(\lambda_1 \Delta + A)^2 + (\lambda_2 \Delta)^2 < A^2,$$

where $A = (1 - 2\alpha)^{-1}$, so the region of absolute stability is the interior of the circle with radius $A$ and centered at $-A + 0\imath$.

**Exercise 12.5.1**    *Show that the implicit order 1.5 strong Runge-Kutta scheme (3.5) is A-stable.*

**Exercise 12.5.2**    *Show that the mapping $G$ for the implicit order 2.0 strong Runge-Kutta scheme (3.8) is given by*

$$G(\lambda \Delta) = \left(1 + \frac{1}{2} \lambda \Delta\right)^{-1} \left(1 + \frac{3}{2} \lambda \Delta + \lambda^2 \Delta^2\right)$$

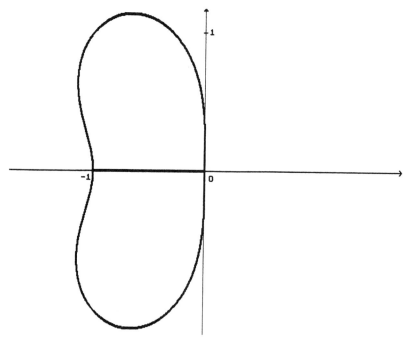

**Figure 12.5.1**    Region of absolute stability of the Runge-Kutta scheme (12.3.8).

and that the corresponding region of absolute stability satisfies the polar coordinate inequality

$$4r \cos^2 \theta + (2 + 3r^2) \cos \theta + r^3 < 0$$

with $\frac{1}{2}\pi < \theta < \frac{3}{2}\pi$, where $r = \sqrt{\lambda_1^2 + \lambda_2^2}\, \Delta$ and $\theta = \arctan(\lambda_2/\lambda_1)$.

For some higher order schemes terms such as $L^0 a$ are required. These may have to be determined from the real 2-dimensional form of (5.1), that is

$$(5.6) \qquad d \begin{pmatrix} X_t^1 \\ X_t^2 \end{pmatrix} = \begin{bmatrix} \lambda_1 & -\lambda_2 \\ \lambda_2 & \lambda_1 \end{bmatrix} \begin{pmatrix} X_t^1 \\ X_t^2 \end{pmatrix} dt + \begin{pmatrix} 1 \\ 0 \end{pmatrix} dW_t$$

where $X_t = X_t^1 + \iota X_t^2$, and then converted back into complex notation. For example, $L^0 a$ for (5.6) becomes $\lambda^2 X$.

**Exercise 12.5.3**   *Determine $L^0 a$ for*

$$a\left(X_t^1, X_t^2\right) = \begin{bmatrix} \lambda_1 & -\lambda_2 \\ \lambda_2 & \lambda_1 \end{bmatrix} \begin{pmatrix} X_t^1 \\ X_t^2 \end{pmatrix}$$

*and show that it can be written equivalently in complex notation as $\lambda^2 X$.*

**Exercise 12.5.4**   *For the families of implicit order 1.5 and 2.0 strong Taylor schemes (2.16) and (2.20), respectively, with common implicitness parameters $\alpha, \beta \in [0,1]$ in both components show that*

$$G(\lambda \Delta) = \left(1 - \alpha \lambda \Delta - \left(\frac{1}{2} - \alpha\right) \beta \lambda^2 \Delta^2\right)^{-1}$$

$$\times \left(1 + (1 - \alpha)\lambda \Delta + \left(\frac{1}{2} - \alpha\right)(1 - \beta) \lambda^2 \Delta^2\right)$$

*and that their region of absolute stability is given by the polar coordinate inequality*

$$\left\{2 + (1 - \alpha)(1 - 2\alpha)r^2\right\} \cos \theta + 2(1 - 2\alpha)r \cos^2 \theta$$

$$+ \frac{1}{4}(1 - 2\alpha)^2(1 - 2\beta)r^3 < 0$$

*with $\frac{1}{2}\pi < \theta < \frac{3}{2}\pi$. Hence deduce that these schemes are A-stable if and only if $\alpha = \frac{1}{2}, \beta \in [0,1]$ or $\alpha, \beta \in [\frac{1}{2}, 1]$.*

Determining the region of absolute stability is more complicated when the implicitness parameters have different values in the different components of the scheme. In these cases it is often more manageable to use the real 2-dimensional equation (5.6) and the corresponding real 2-dimensional scheme. The complex equation (5.2) then takes the real 2-dimensional vector form

(5.7)                    $Y_{n+1} = G(\lambda_1 \Delta, \lambda_2 \Delta) Y_n + \text{noise terms}$

where $G(\lambda_1 \Delta, \lambda_2 \Delta)$ is a real $2 \times 2$ matrix and the region of absolute stability is now determined from the real inequalities

(5.8)                    $\lambda_1 < 0 \quad \text{and} \quad |G(\lambda_1 \Delta, \lambda_2 \Delta)| < 1$

involving the matrix norm $|G| = |[g^{i,j}]| = \sqrt{\sum_{i,j=1}^2 (g^{i,j})^2}$.

For the family of implicit Milstein schemes (2.13) we have

$$G(\lambda_1 \Delta, \lambda_2 \Delta) = \begin{bmatrix} 1 - \alpha_1 \lambda_1 \Delta & \alpha_1 \lambda_2 \Delta \\ -\alpha_2 \lambda_2 \Delta & 1 - \alpha_2 \lambda_1 \Delta \end{bmatrix}^{-1}$$

$$\times \begin{bmatrix} 1 + (1-\alpha_1)\lambda_1 \Delta & -(1-\alpha_1)\lambda_2 \Delta \\ (1-\alpha_2)\lambda_2 \Delta & 1 + (1-\alpha_2)\lambda_1 \Delta \end{bmatrix}$$

and the second inequality in (5.8) is equivalent to

$$\left| \begin{bmatrix} 1 + (1-\alpha_1)\lambda_1 \Delta & -(1-\alpha_1)\lambda_2 \Delta \\ (1-\alpha_2)\lambda_2 \Delta & 1 + (1-\alpha_2)\lambda_1 \Delta \end{bmatrix} \right|^2 < \left| \begin{bmatrix} 1 - \alpha_1 \lambda_1 \Delta & \alpha_1 \lambda_2 \Delta \\ -\alpha_2 \lambda_2 \Delta & 1 - \alpha_2 \lambda_1 \Delta \end{bmatrix} \right|^2$$

which reduces to

$$(1 - \alpha_1 - \alpha_2)(\lambda_1^2 + \lambda_2^2)\Delta^2 + 2\lambda_1 \Delta < 0.$$

These implicit Milstein schemes are thus A-stable for implicitness parameters $\alpha_1, \alpha_2 \in [0, 1]$ satisfying

$$1 \leq \alpha_1 + \alpha_2 \leq 2.$$

For example, one component could be fully explicit and the other fully implicit, if this were computationally convenient. When $0 \leq \alpha_1 + \alpha_2 < 1$ the regions of absolute stability are the interior of circles of radius $A$ and centre $-A + 0\imath$, where $A = (1 - \alpha_1 - \alpha_2)^{-1}$.

**Exercise 12.5.5**   *Determine the regions of absolute stability for the families of implicit order 1.5 and 2.0 strong Taylor schemes (2.16) and (2.20), respectively, with different implicitness parameters $\alpha_1, \beta_1$ and $\alpha_2, \beta_2$ in the two components of the schemes.*

## 12.6   Convergence Proofs

The orders of strong convergence of the implicit schemes presented in this chapter follow from Theorem 11.5.1 if they are based on the Ito-Taylor expansion and from Theorem 11.5.2 if they are based on the Stratonovich-Taylor expansion. However, we must first show that the schemes can be represented as strong Ito or Stratonovich schemes as required by these theorems. In doing so we shall gain some insight into how implicit schemes can be derived in a systematic way.

We shall begin with the general strong Ito schemes which we defined in (11.5.1) by the recursive relation

$$(6.1) \qquad Y_{n+1} = Y_n + \sum_{\alpha \in A_\gamma \setminus \{v\}} g_{\alpha,n} I_\alpha + R_n,$$

where the conditions (11.5.2)–(11.5.4) must be satisfied. Applying the Ito-Taylor expansion (5.5.3) with $\mathcal{A} = A_2$ and $f(t,x) \equiv x$ at time $t + \Delta$ to the Ito process $X^{t,x}$ starting at $x$ at time $t$, we obtain the expansion

$$(6.2) \qquad X_{t+\Delta}^{t,x} = x + \sum_{\alpha' \in A_2 \setminus \{v\}} f_{\alpha'} I_{\alpha'} + \tilde{R}_2(t)$$

$$= x + a\,\Delta + \frac{1}{2} L^0 a\,\Delta^2 + M_\Delta(t)$$

with

$$M_\Delta(t) = \sum_{j=1}^m \left( b^j\,\Delta W^j + L^0 b^j I_{(0,j)} + L^j a I_{(j,0)} \right)$$

$$+ \sum_{j_1,j_2=1}^m \left[ L^{j_1} b^{j_2} I_{(j_1,j_2)} + L^0 L^{j_1} b^{j_2} I_{(0,j_1,j_2)} \right.$$

$$\left. + L^{j_1} L^0 b^{j_2} I_{(j_1,0,j_2)} + L^{j_1} L^{j_2} a I_{(j_1,j_2,0)} \right]$$

$$+ \sum_{j_1,j_2,j_3=1}^m L^{j_1} L^{j_2} b^{j_3} I_{(j_1,j_2,j_3)}$$

$$+ \sum_{j_1,j_2,j_3,j_4=1}^m L^{j_1} L^{j_2} L^{j_3} b^{j_4} I_{(j_1,j_2,j_3,j_4)} + \tilde{R}_2(t),$$

where $\tilde{R}_2(t)$ represents the remainder term. Here the coefficient functions are evaluated at $(t,x) \in \Re^+ \times \Re^d$ and the multiple stochastic integrals are over the time interval $[t, t+\Delta]$, that is $I_\alpha = I_{\alpha,t,t+\Delta}$.

The only nonrandom multiple Ito integrals in (6.2) are contained in the terms $a\,\Delta$ and $\frac{1}{2} L^0 a\,\Delta^2$. Consequently, these are the only terms that we can change in forming an implicit scheme if we are to avoid the type of problem due to inverses of Gaussian random variables in the fully implicit Euler scheme (9.8.7).

Applying the Ito-Taylor expansion (5.5.3) again, but now with $\mathcal{A} = A_{1.5}$ and $f(t,x) = a(t,x)$, and rearranging, we get

(6.3)     $a(t, x)$ $=$ $a\left(t + \Delta, X^{t,x}_{t+\Delta}\right) - \displaystyle\sum_{\alpha' \in A_{1.5}\backslash\{v\}} a_{\alpha'} I_{\alpha'} + \tilde{R}_{1.5}(t)$

$=$ $a\left(t + \Delta, X^{t,x}_{t+\Delta}\right) - L^0 a \Delta + N_\Delta(t)$

with

$N_\Delta(t)$ $=$ $-\dfrac{1}{2} L^0 L^0 a \Delta^2 - \displaystyle\sum_{j=1}^{m} \left( L^j a \Delta W^j + L^0 L^j a\, I_{(0,j)} + L^j L^0 a\, I_{(j,0)} \right)$

$- \displaystyle\sum_{j_1,j_2=1}^{m} L^{j_1} L^{j_2} a\, I_{(j_1,j_2)}$

$- \displaystyle\sum_{j_1,j_2,j_3=1}^{m} L^{j_1} L^{j_2} L^{j_3} a\, I_{(j_1,j_2,j_3)} + \tilde{R}_{1.5}(t),$

where again the coefficient functions are evaluated at $(t, x)$ and the multiple stochastic integrals are over the time interval $[t, t + \Delta]$.

Repeating this procedure once more, but with $\mathcal{A} = A_1$ and $f(t, x) = L^0 a(t, x)$, we obtain

(6.4)          $L^0 a(t, x) = L^0 a\left(t + \Delta, X^{t,x}_{t+\Delta}\right) + P_\Delta(t)$

with

$P_\Delta(t)$ $=$ $-L^0 L^0 a \Delta - \displaystyle\sum_{j=1}^{m} L^j L^0 a \Delta W^j$

$- \displaystyle\sum_{j_1,j_2=1}^{m} L^{j_1} L^{j_2} L^0 a\, I_{(j_1,j_2)} + \tilde{R}_1(t).$

Taking real numbers $\alpha, \beta \in [0, 1]$, we can insert (6.3) and (6.4) into (6.2) as follows

(6.5)  $X^{t,x}_{t+\Delta}$ $=$ $x + \alpha\, a \Delta + (1 - \alpha) a \Delta + \dfrac{1}{2} L^0 a \Delta^2 + M_\Delta(t)$

$=$ $x + \left\{\alpha \left[ a\left(t + \Delta, X^{t,x}_{t+\Delta}\right) + N_\Delta(t) \right] + (1 - \alpha) a \right\} \Delta$

$+ \left( \dfrac{1}{2} - \alpha \right) L^0 a \Delta^2 + M_\Delta(t)$

$=$ $x + \left\{\alpha \left[ a\left(t + \Delta, X^{t,x}_{t+\Delta}\right) + N_\Delta(t) \right] + (1 - \alpha) a \right\} \Delta$

$+ \left( \dfrac{1}{2} - \alpha \right) \left\{ \beta L^0 a + (1 - \beta) L^0 a \right\} \Delta^2 + M_\Delta(t)$

$=$ $x + \left\{ \alpha a\left(t + \Delta, X^{t,x}_{t+\Delta}\right) + (1 - \alpha) a \right\} \Delta$

$$+ \left( \frac{1}{2} - \alpha \right) \left\{ \beta L^0 a \left( t + \Delta, X_{t+\Delta}^{t,x} \right) + (1 - \beta) L^0 a \right\} \Delta^2$$

$$+ \alpha N_\Delta(t) \Delta + \left( \frac{1}{2} - \alpha \right) \beta P_\Delta(t) \Delta^2 + M_\Delta(t).$$

This contains implicit terms with $a \left( t + \Delta, X_{t+\Delta}^{t,x} \right)$ and $L^0 a \left( t + \Delta, X_{t+\Delta}^{t,x} \right)$. It is the starting point for the derivation of implicit strong Taylor schemes, with those terms being discarded which are not required to satisfy the conditions (11.5.2) – (11.5.4) of Theorem 11.5.1 for the resulting truncation to have a desired order of convergence. To justify this we need first to show that the approximation $Y$ satisfies an a priori estimate

$$E \left( \max_{0 \leq n \leq n_T} |Y_n|^{2q} \right) < \infty$$

for $q = 1, 2, \ldots$ and to assume that the coefficient functions $f_\alpha$ present have polynomial growth. In this way it can be shown that the implicit strong Taylor schemes and, under corresponding regularity assumptions also the other one-step schemes of this chapter, converge with the asserted strong order.

The derivation of implicit two-step schemes starts from the identity

$$(6.6) \quad X_{t+2\Delta}^{t,x} = (1 - \xi) \phi X_{t+\Delta}^{t,x} + [1 - \phi(1 - \xi)] x$$

$$+ \left( X_{t+2\Delta}^{t,x} - X_{t+\Delta}^{t,x} \right) + [1 - \phi(1 - \xi)] \left( X_{t+\Delta}^{t,x} - x \right)$$

for all $\phi, \xi \in [0, 1]$. Using (6.5) to express the increments

$$\left( X_{t+2\Delta}^{t,x} - X_{t+\Delta}^{t,x} \right) \quad \text{and} \quad \left( X_{t+\Delta}^{t,x} - x \right),$$

we end up with a two-step representation for $X_{t+2\Delta}^{t,x}$. An example of such a representation will be given in (6.7) at the end of the section. This expansion can then be used to construct two-step strong Taylor schemes of the desired strong order. The proof of their strong convergence then follows by an obvious, slight generalization of Theorem 11.5.1. There is also no difficulty in replacing the derivatives in these two-step schemes by the appropriate finite differences to obtain the corresponding derivative free schemes, as was done for the explicit two-step schemes in Section 4 of Chapter 11. The strong convergence with the asserted order of the other implicit two-step schemes in Section 4 of this chapter follows in much the same way as for the Taylor schemes.

A similar expansion to (6.5) based on the Stratonovich-Taylor formula can also be derived. It turns out to be the same as (6.5), but with $\underline{a}$, $\underline{L}^j$ and $J_\alpha$ instead of $a$, $L^j$ and $I_\alpha$.

We conclude this section by stating the general multi-dimensional, implicit two-step expansion announced above in its Stratonovich version, which describes all of the terms which are needed to derive an implicit two-step order

2.0 strong Taylor scheme. It is derived from (6.6) with $\phi = 1$ and $\gamma^k = \xi$. We shall use the abbreviations

$$X_n = X_{\tau_n}^{\tau_{n-1}, X_{n-1}}, \qquad X_{n+1} = X_{\tau_{n+1}}^{\tau_{n-1}, X_{n-1}}$$

and $f = f(\tau_n, X_n)$ for the functions $f = \underline{a}, b, \underline{L}^0 \underline{a}, \ldots$, where $X_{n-1}$ is the value of $X$ at $\tau_{n-1}$. The $k$th component is then

(6.7)
$$X_{n+1}^k = (1 - \gamma_k) X_n^k + \gamma_k X_{n-1}^k$$

$$+ \left[ \alpha_{2,k} \underline{a}^k \left( \tau_{n+1}, X_{n+1} \right) + \left( 1 - \alpha_{2,k} + \gamma_k \alpha_{1,k} \right) \underline{a}^k \right.$$

$$\left. + \gamma_k \left( 1 - \alpha_{1,k} \right) \underline{a}^k \left( \tau_{n-1}, X_{n-1} \right) \right] \Delta$$

$$+ \left[ \left( \frac{1}{2} - \alpha_{2,k} \right) \beta_{2,k} \underline{L}^0 \underline{a}^k \left( \tau_{n+1}, X_{n+1} \right) \right.$$

$$+ \left\{ \left( \frac{1}{2} - \alpha_{2,k} \right) \left( 1 - \beta_{2,k} \right) + \gamma \left( \frac{1}{2} - \alpha_{1,k} \right) \beta_{1,k} \right\} \underline{L}^0 \underline{a}^k$$

$$\left. + \gamma \left( \frac{1}{2} - \alpha_{1,k} \right) \left( 1 - \beta_{1,k} \right) \underline{L}^0 \underline{a}^k \left( \tau_{n-1}, X_{n-1} \right) \right] \Delta^2$$

$$- \sum_{j=1}^{m} \left[ \alpha_{2,k} \underline{L}^j \underline{a}^k \Delta W_n^j + \gamma \alpha_{1,k} \underline{L}^j \underline{a}^k \left( \tau_{n-1}, X_{n-1} \right) \Delta W_{n-1}^j \right] \Delta$$

$$- \sum_{j_1, j_2 = 1}^{m} \left[ \alpha_{2,k} \underline{L}^{j_1} \underline{L}^{j_2} \underline{a}^k J_{(j_1, j_2), \tau_n, \tau_{n+1}} \right.$$

$$\left. + \gamma \alpha_{1,k} \underline{L}^{j_1} \underline{L}^{j_2} \underline{a}^k \left( \tau_{n-1}, X_{n-1} \right) J_{(j_1, j_2), \tau_{n-1}, \tau_n} \right] \Delta$$

$$+ V_n^k + \gamma V_{n-1}^k$$

with

$$V_n^k = \sum_{j=1}^{m} \left[ b^{k,j} \Delta W_n^j + \underline{L}^0 b^{k,j} J_{(0,j), \tau_n, \tau_{n+1}} + \underline{L}^j \underline{a}^k J_{(j,0), \tau_n, \tau_{n+1}} \right]$$

$$+ \sum_{j_1, j_2 = 1}^{m} \left[ \underline{L}^{j_1} b^{k,j_2} J_{(j_1, j_2), \tau_n, \tau_{n+1}} + \underline{L}^0 \underline{L}^{j_1} b^{k,j_2} J_{(0, j_1, j_2), \tau_n, \tau_{n+1}} \right.$$

$$\left. + \underline{L}^{j_1} \underline{L}^0 b^{k,j_2} J_{(j_1, 0, j_2), \tau_n, \tau_{n+1}} + \underline{L}^{j_1} \underline{L}^{j_2} \underline{a}^k J_{(j_1, j_2, 0), \tau_n, \tau_{n+1}} \right]$$

$$+ \sum_{j_1, j_2, j_3 = 1}^{m} \underline{L}^{j_1} \underline{L}^{j_2} b^{k,j_3} J_{(j_1, j_2, j_3), \tau_n, \tau_{n+1}}$$

$$+ \sum_{j_1,j_2,j_3,j_4=1}^{m} \underline{L}^{j_1} \underline{L}^{j_2} \underline{L}^{j_3} b^{k,j_4} J_{(j_1,j_2,j_3,j_4),\tau_n,\tau_{n+1}}$$

+ higher order terms,

where the parameters $\alpha_{1,k}$, $\alpha_{2,k}$, $\beta_{1,k}$, $\beta_{2,k}$, $\gamma_k \in [0,1]$ for $k = 1, \ldots, d$.

Variations of the above schemes can be obtained by extrapolating instead of interpolating the constituent schemes, that is by choosing the parameters $\alpha_{1,k}$, $\alpha_{2,k}$, $\beta_{1,k}$, $\beta_{2,k}$ and $\gamma_k$ outside of the interval $[0,1]$.

**Exercise 12.6.1**    *Show under appropriate assumptions that the two-step order 1.5 strong scheme (11.4.4) converges with strong order $\gamma = 1.5$.*

**Exercise 12.6.2**    *Under sufficient conditions prove that the implicit order 2.0 strong Taylor scheme (2.19) converges with strong order $\gamma = 2.0$.*

**Exercise 12.6.3**    *the order of strong convergence of the implicit two-step order 1.5 strong scheme (4.8).*

# Chapter 13

# Selected Applications of Strong Approximations

Several applications of the strong schemes that were derived in the preceding chapters will be indicated in this chapter. These are the direct simulation of trajectories of stochastic dynamical systems, including stochastic flows, the testing of parametric estimators and Markov chain filters. In addition, some results on asymptotically efficient schemes will be presented.

## 13.1 Direct Simulation of Trajectories

An important practical application of strong approximations is the direct simulation of stochastic dynamical systems. The examples in the preceding chapters on strong approximations are typical of such direct simulations. In specific applications, like those described in Chapter 7, direct simulations of the trajectories of the stochastic differential equation can provide useful information on, or suggestions about, the qualitative behaviour of the model under investigation. In this section we shall describe some examples of direct simulations which can be interpreted as stochastic flows. Broadly speaking, a stochastic flow is a probabilistic model for the simultaneous random motion of an ensemble of particles in "space", for instance, the diffusion of aerosol particles in a gas.

### The Duffing–Van der Pol Oscillator

We shall look at a simplified version of a *Duffing–Van der Pol oscillator*

$$\ddot{x} + \dot{x} - \left(\alpha - x^2\right) x = \sigma x \xi$$

driven by multiplicative white noise $\xi$, where $\alpha$ is a real-valued parameter (see also equation (7.7.9)). The corresponding Ito stochastic differential equation is 2-dimensional, with components $X^1$ and $X^2$ representing the displacement $x$ and speed $\dot{x}$, respectively, namely

(1.1)
$$dX_t^1 = X_t^2 \, dt$$
$$dX_t^2 = \left\{ X_t^1 \left( \alpha - \left( X_t^1 \right)^2 \right) - X_t^2 \right\} dt + \sigma X_t^1 \, dW_t$$

where $W = \{W_t, t \geq 0\}$ is a 1-dimensional standard Wiener process and $\sigma \geq 0$ controls the strength of the induced multiplicative noise.

The deterministic version of (1.1) with $\sigma \equiv 0$ has the steady states

(1.2) $$X^1 = 0, \qquad X^2 = 0 \quad \text{for all} \quad \alpha$$

and

(1.3) $$X^1 = \pm\sqrt{\alpha}, \qquad X^2 = 0 \quad \text{for} \quad \alpha \geq 0,$$

the first of which is also a degenerate stationary state of the stochastic differential equation (1.1).

We shall begin by looking at the phase plane of the deterministic system.

**PC-Exercise 13.1.1**    *Use the Milstein scheme (10.3.2) with equidistant step size $\Delta = 2^{-7}$ to simulate linearly interpolated trajectories of the Duffing–Van der Pol oscillator (1.1) with $\alpha = 1.0$ and $\sigma = 0.0$ over the time interval $[0, T]$ with $T = 8$, starting at $(X_0^1, X_0^2) = (-k\epsilon, 0)$ for $k = 11, 12, \ldots, 20$, where $\epsilon = 0.2$. Plot the results on the $(X^1, X^2)$ phase plane.*

From Figure 13.1.1 we see that the typical trajectory starting with nonzero displacement and zero speed is oscillatory and is attracted to one or the other of the nontrivial steady states $(\pm 1, 0)$. We could determine the regions of attraction of these two steady states by appropriately marking each initial value on the phase plane according to the steady solution which attracts the trajectory starting there.

We shall now include noise with small strength and repeat the above calculations.

**PC-Exercise 13.1.2**    *Repeat PC-Exercise 13.1.1 with $\alpha = 1.0$ and $\sigma = 0.2$ using the same driving sample path of the Wiener process for each trajectory starting at the different initial values.*

We observe in Figure 13.1.2 that the trajectories are now random in appearance, with paths usually remaining near to each other until they come close to the origin $(0, 0)$, after which they separate and are attracted into the neighbourhood of either $(-1, 0)$ or $(1, 0)$.

Finally, we shall examine the effect of a stronger multiplicative noise in the Duffing–Van der Pol oscillator over a long period of time.

**PC-Exercise 13.1.3**    *Repeat PC-Exercise 13.1.1 with $\alpha = 1.0$ and $\sigma = 0.5$ using the same sample path of the Wiener process for each of the initial values, but now plotting the displacement component $X_t^1$ against time $t$.*

It is now quite obvious that while the noisy trajectories are initially attracted by one or the other of the points $(\pm 1, 0)$, not all of them remain indefinitely in the vicinity of the same point. Instead, after spending a period of time near one of the points, the trajectories switch over to the other point, which first happens in Figure 13.1.3 at about the time $t = 128$. This might be interpreted as a form of *tunneling*.

To convince ourselves of the reliability of the above results, we could repeat the calculations using a smaller step size or some other strong scheme. While

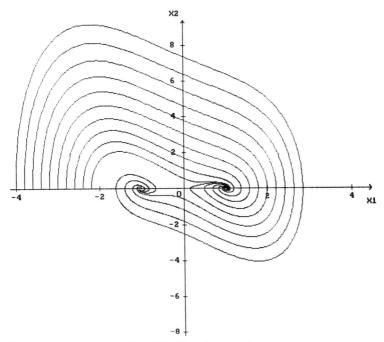

**Figure 13.1.1** Results of PC-Exercise 13.1.1
for the Duffing–Van der Pol oscillator without noise.

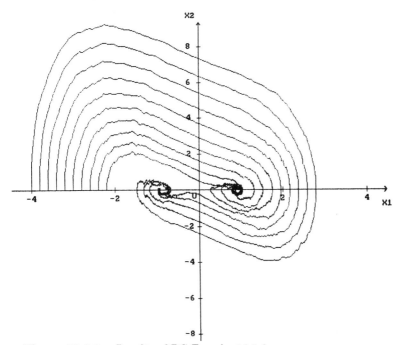

**Figure 13.1.2** Results of PC-Exercise 13.1.2
for the Duffing–Van der Pol oscillator with white noise.

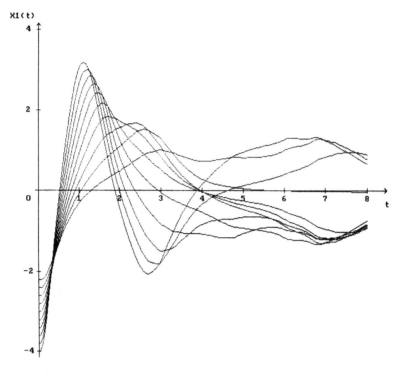

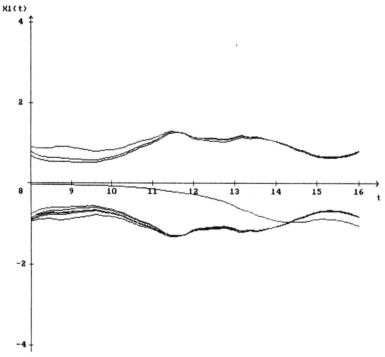

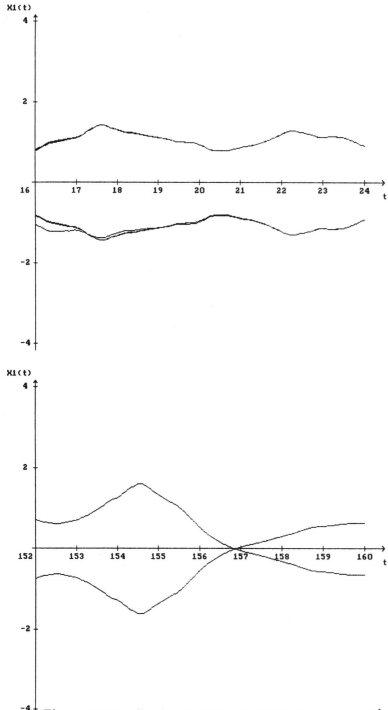

**Figure 13.1.3** Results of PC-Exercise 13.1.3: displacement $X_t^1$ versus $t$ for different particles satisfying the Duffing–Van der Pol equation with noise.

the quantitative details may then differ, the qualitative picture should be much the same. With this check we may be able to avoid results that are only an artifice of a particular numerical scheme. It turns out that higher order schemes can, for instance, ensure the preservation over a long period of time of the important diffeomorphism property that neighbouring particles remain neighbours.

### Stochastic Flow on a Circle

The above simulations of trajectories of the noisy Duffing–Van der Pol equation with the same trajectory of the driving Wiener process for different initial values is an example of a stochastic flow. Of special interest are stochastic flows on manifolds, such as the unit circle $S^1$ and the 2-dimensional torus $T^2$. These arise, for instance, in stability and bifurcation problems.

Carverhill, Chappel and Elworthy considered the *gradient stochastic flow* on $S^1$ described in terms of the Stratonovich stochastic differential equation

$$(1.4) \qquad dX_t^{0,x} = \sin\left(X_t^{0,x}\right) \circ dW_t^1 + \cos\left(X_t^{0,x}\right) \circ dW_t^2$$

with initial value $x \in [0, 2\pi)$, which is driven by two independent standard Wiener processes $W^1$ and $W^2$. Equation (1.4) has period $2\pi$ and can be interpreted modulo $2\pi$, which gives the standard embedding of $S^1$ in $\Re^1$.

Here we shall apply the Milstein scheme to (1.4) for different initial values $x \in [0, 2\pi)$, but always with the same realization of the driving Wiener process $W = (W^1, W^2)$, taking values modulo $2\pi$. Thus, we shall need to approximate the multiple Ito integral $I_{(1,2)}$. This will provide us, approximately at least, with a simulation of the corresponding stochastic flow on the unit circle $S^1$ identified as the interval $[0, 2\pi)$. We can imagine that a different particle is located at each initial point and interpret the resulting flow as the motion over time of this ensemble of particles.

**PC-Exercise 13.1.4**   *Simulate the gradient stochastic flow of $N = 10$ particles on the unit circle $S^1$ over the time interval $[0, T]$ with $T = 5$ by applying the Milstein scheme with step size $\Delta = 2^{-7}$ to the stochastic differential equation (1.4), modulo $2\pi$, for the initial values $x = 2\pi k/N$ where $k = 1, \ldots, N$. Plot the linearly interpolated trajectories on the same $x$ versus $t$ axes for $0 \leq x < 2\pi$ and $0 \leq t \leq T$.*

Figure 13.1.4 shows that most neighbouring particles move closer together and leave their initial location. Eventually, the particles appear to form a cluster which moves like a Brownian motion on the circle. Repeating the calculations with a different realization of the driving Wiener process produces another cluster, which behaves similarly.

**PC-Exercise 13.1.5**  *Repeat PC-Exercise 13.1.4 using a different seed for the random number generator.*

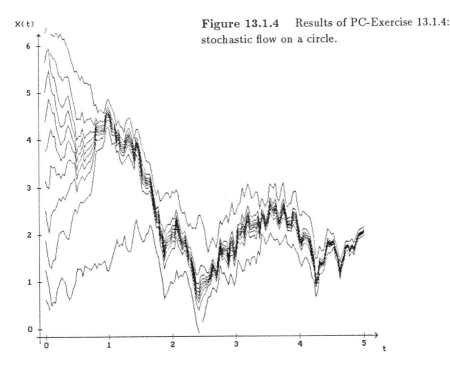

X(t)

Figure 13.1.4   Results of PC-Exercise 13.1.4: stochastic flow on a circle.

The dynamical behaviour observed here is a consequence of the flow's having a negative Lyapunov exponent.

## Stochastic Flow on a Torus

It is interesting to generalize the 1-dimensional flow on the circle $S^1$ to a 2-dimensional flow on the torus $T^2$. We recall that a torus is a surface of revolution formed by revolving a circle about a non-intersecting axis in $\Re^3$. The standard torus $T^2$ can be identified with the rectangle $[0, 2\pi)^2$ in $\Re^2$, with each point characterized by two angular coordinates.

Baxendale proposed a parametrized stochastic flow on $T^2$ in terms of a 2-dimensional Stratonovich stochastic differential equation

$$(1.5) \qquad dX_t^{0,x} = \sum_{j=1}^{4} b^j \left( X_t^{0,x} \right) \circ dW_t^j$$

with diffusion coefficients

$$(1.6) \qquad b^1(x) = b^1 \left( x^1, x^2 \right) = \begin{pmatrix} \cos \alpha \\ \sin \alpha \end{pmatrix} \sin \left( x^1 \right),$$

$$b^2(x) = b^2 \left( x^1, x^2 \right) = \begin{pmatrix} \cos \alpha \\ \sin \alpha \end{pmatrix} \cos \left( x^1 \right),$$

$$b^3(x) = b^3(x^1, x^2) = \begin{pmatrix} -\sin\alpha \\ \cos\alpha \end{pmatrix} \sin(x^2),$$

$$b^4(x) = b^4(x^1, x^2) = \begin{pmatrix} -\sin\alpha \\ \cos\alpha \end{pmatrix} \cos(x^2).$$

Here $W^1$, $W^2$, $W^3$ and $W^4$ are independent, 1-dimensional standard Wiener processes and $\alpha \in [0, \pi/2]$ is a parameter. Baxendale's stochastic flow on $T^2$ is obtained from (1.5) by interpreting the solution $X_t^{0,x}$ modulo $2\pi$ in both components. It can thus be plotted on the rectangle $[0, 2\pi)^2$ with the points $(x, 2\pi)$ and $(2\pi, y)$ being identified with $(x, 0)$ and $(0, y)$, respectively, for all $x$, $y$ in $[0, 2\pi]$. We can visualize it by applying a numerical scheme to (1.5) with the same realization of the driving Wiener process $W = (W^1, W^2, W^3, W^4)$ for a grid of initial values in $[0, 2\pi]^2$ and plotting the grid of calculated values at selected times.

**PC-Exercise 13.1.6**     *Use the Milstein scheme with step size $\Delta = 0.01$ and the same realization of the driving Wiener process $W$ to simulate 225 trajectories of (1.5) with parameter $\alpha = 0.1$ starting at the points of a uniform $15\times15$ grid in $[0, 2\pi]^2$. Plot the grid of calculated values at time $T = 0.5$, using line segments to join those corresponding to adjacent initial values. Continue the calculations for larger values of $T$.*

Figure 13.1.5 illustrates the fact that the two components of the flow are coupled for $\alpha = 0.1$, with the initial grid becoming more and more distorted as the flow evolves in time. Eventually, all of the particles will cluster into a single moving point.

**PC-Exercise 13.1.7**     *Repeat PC-Exercise 13.1.6 for parameter $\alpha = 1.0$ for the times $T = 0.5$, 1.0, 2.5 and 3.0. Plot the points at each of these times in $[0, 2\pi]^2$.*

From Figure 13.1.6 we see that after some time the particles cluster along strings and form randomly moving substructures. It can be shown that the flow here has a negative Lyapunov exponent, which accounts for the contraction to form the strings, and a positive Lyapunov exponent which accounts for the random motion on the strings. If we repeated the calculations for the uncoupled case $\alpha = 0.0$, we would obtain a single randomly moving cluster of particles since both of the Lyapunov exponents are now negative.

## 13.2   Testing Parametric Estimators

We have already considered the problem of estimating parameters in the drift coefficient of a stochastic differential equation in Section 4 of Chapter 6. Even if the statistical consistency of a proposed estimator has been established theoretically, it is still of practical interest to test the estimator by direct simulation, for instance, because the observations are often only available at discrete times. We shall do this here for two specific examples.

### The Ornstein-Uhlenbeck Process

We shall consider the Ornstein-Uhlenbeck process (see Example 4.2.1)

$$(2.1) \qquad X_t = X_0 + \alpha \int_0^t X_s \, ds + W_t,$$

supposing that the true value of $\alpha$ is $-1$ and that the process starts at $X_0 = 0$. From (6.4.8) we have the maximum likelihood estimator

$$(2.2) \qquad \hat{\alpha}(T) = \int_0^T X_s \, dX_s \bigg/ \int_0^T (X_s)^2 \, ds.$$

Usually, only time discrete observations at discretization instants are available, so only an approximation of this estimator can be evaluated, the simplest being

$$(2.3) \qquad \hat{\alpha}_1(T) = \sum_{n=0}^{n_T-1} X_{\tau_n} \left( X_{\tau_{n+1}} - X_{\tau_n} \right) \bigg/ \sum_{n=0}^{n_T-1} (X_{\tau_n})^2 \, \Delta_n$$

where

$$\Delta_n = \tau_{n+1} - \tau_n.$$

We observe that $\hat{\alpha}_1(T)$ maximizes the discrete time maximum likelihood ratio (6.4.6).

We shall also consider a variation of this approximate estimator. Using the Ito formula (3.3.6), we can write the integral in the denominator of (2.2) as

$$\int_{\tau_n}^{\tau_{n+1}} X_s \, dX_s = \frac{1}{2} \left\{ (X_{\tau_{n+1}})^2 - (X_{\tau_n})^2 - \Delta_n \right\}.$$

In addition, we can approximate the integral in the numerator of (2.2) by

$$\int_{\tau_n}^{\tau_{n+1}} (X_s)^2 \, ds \approx \frac{1}{2} \left\{ (X_{\tau_{n+1}})^2 + (X_{\tau_n})^2 \right\} \Delta_n.$$

Inserting these into (2.2), we obtain a second approximate estimator

$$(2.4) \quad \hat{\alpha}_2(T) = \left( (X_T)^2 - (X_0)^2 - T \right) \bigg/ \sum_{n=0}^{n_T-1} \left\{ (X_{\tau_{n+1}})^2 + (X_{\tau_n})^2 \right\} \Delta_n.$$

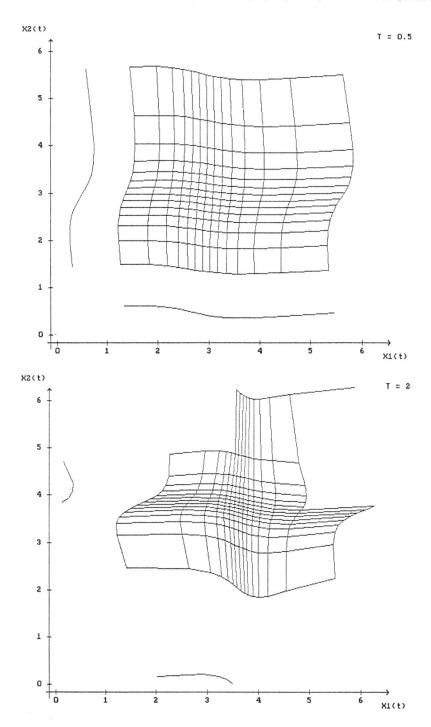

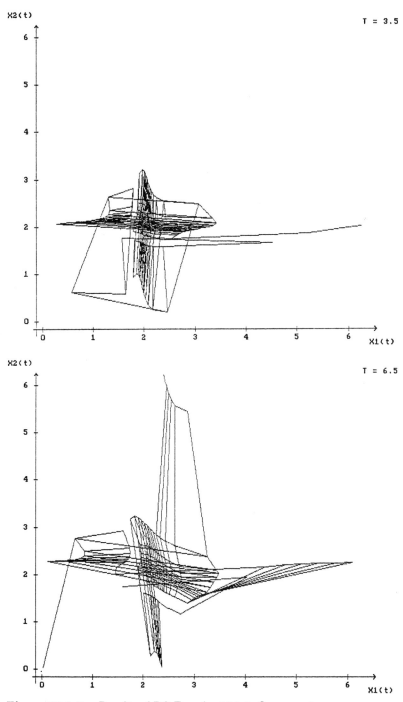

**Figure 13.1.5** Results of PC-Exercise 13.1.6: flow on a torus plotted as a moving grid with $\alpha = 0.1$.

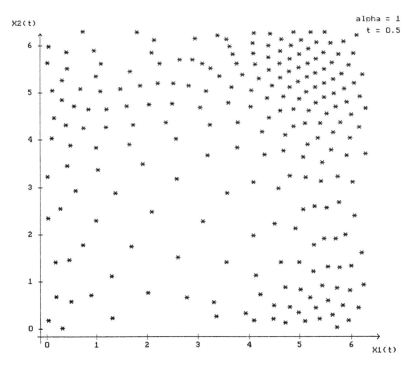

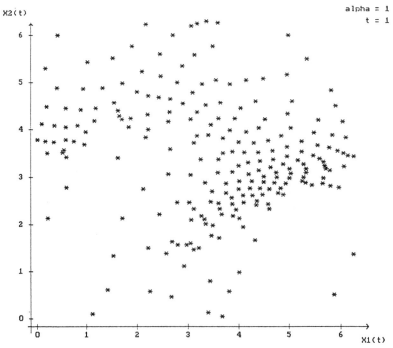

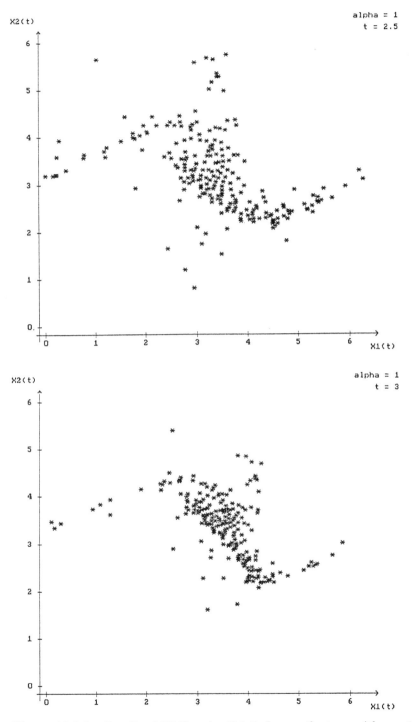

**Figure 13.1.6** Results of PC-Exercise 13.1.7: flow on the torus with $\alpha = 1.0$.

In the following PC-Exercise a simulated trajectory of the Ornstein-Uhlenbeck process (2.1) will be used to compare the two estimators $\hat{\alpha}_1(T)$ and $\hat{\alpha}_2(T)$ for sufficiently large times $T$.

**PC-Exercise 13.2.1**    *Simulate a single trajectory of the process (2.1) with $\alpha = -1$ and $X_0 = 0$ using the order 2.0 strong Taylor scheme (10.5.4) with equidistant step size $\Delta_n = \Delta = 2^{-3}$ and evaluate the estimators $\hat{\alpha}_1(T)$ and $\hat{\alpha}_2(T)$ with the same step size $\Delta$. Plot linearly interpolated values for the same step sizes of the estimators against $T$ for $T \in [0, 6400]$.*

We observe from Figure 13.2.1 that both estimators converge asymptotically to the true value $\alpha = -1$ of the parameter, with $\hat{\alpha}_2(T)$ converging more rapidly.

## A Population Growth Model

Several elementary models of population growth were introduced in Section 2 of Chapter 6 and in Section 1 of Chapter 7. Here we shall consider a similar model described by the Ito equation

$$(2.5) \qquad X_t = X_0 + \beta \int_0^t X_s \left(1 - X_s\right) ds + \int_0^t \sqrt{X_s}\, dW_s$$

with initial value $X_0 = 1.0$, where $\beta$ represents the growth rate of the population which must often be estimated in practical applications.

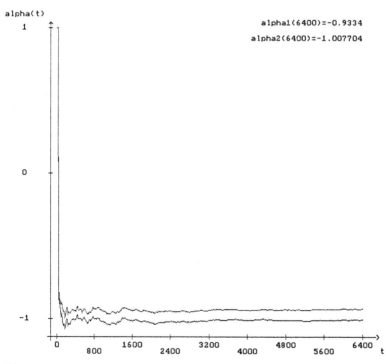

Figure 13.2.1    Results of PC-Exercise 13.2.1: the estimators $\hat{\alpha}_1(T)$ and $\hat{\alpha}_2(T)$.

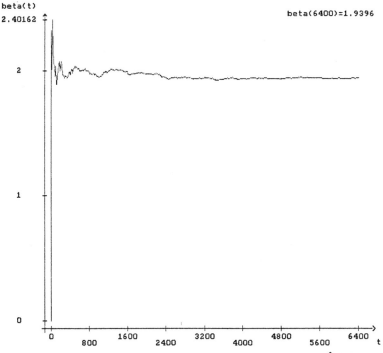

**Figure 13.2.2**   Results of PC-Exercise 13.2.2: the estimator $\hat{\beta}_1(T)$.

From Heyde, Sørenson and others an estimator for $\beta$ is

(2.6)         $$\hat{\beta}(T) = \int_0^T (1 - X_s) \, dX_s \bigg/ \int_0^T (1 - X_s)^2 \, X_s \, ds.$$

By the Ito formula (3.3.6), we can write

$$\int_{\tau_n}^{\tau_{n+1}} X_s \, dX_s = \frac{1}{2} \left\{ (X_{\tau_{n+1}})^2 - (X_{\tau_n})^2 - \int_{\tau_n}^{\tau_{n+1}} X_s \, ds \right\}.$$

Hence, with analogous approximations to those used above, we obtain the approximate estimator

$$\hat{\beta}_1(T) = \left\{ X_T - X_0 - \frac{1}{2} \left\{ (X_T)^2 - (X_0)^2 \right\} + \frac{1}{4} \sum_{n=0}^{n_T-1} (X_{\tau_n} + X_{\tau_{n+1}}) \, \Delta_n \right\}$$

(2.7)         $$\times \left\{ \frac{1}{2} \sum_{n=0}^{n_T-1} \left[ (1 - X_{\tau_{n+1}})^2 \, X_{\tau_{n+1}} + (1 - X_{\tau_n})^2 \, X_{\tau_n} \right] \Delta_n \right\}^{-1}.$$

We shall now test this estimator for the true parameter value $\beta$.

**PC-Exercise 13.2.2**    *Use the explicit order* 1.0 *strong scheme (11.1.3) with equidistant step size* $\Delta = 2^{-3}$ *to simulate an approximate trajectory of the Ito process (2.5) with* $\beta = 2.0$ *and initial value* $X_0 = 0.5$. *Evaluate the approximate estimator* $\hat{\beta}_1(T)$ *and plot linearly interpolated values of* $\hat{\beta}_1(T)$ *against* $T$ *for* $T \in [0, 6400]$.

# 13.3  Discrete Approximations for Markov Chain Filters

In Section 6 of Chapter 6 we mentioned a nonlinear filter (6.6.10) which allows the computation of the conditional probabilities for the states of a finite state Markov chain based on noisy observations. In this section we shall apply strong approximations of stochastic differential equations to approximate the optimal continuous time Markov chain filter, which is the solution of an Ito stochastic differential equation.

Let $(\Omega, \mathcal{A}, P)$ be the underlying probability space and suppose that the *state process* $\xi = \{\xi_t, t \in [0,T]\}$ is a continuous time homogeneous Markov chain (see (1.6.17)–(1.6.21)) on the finite state space $\mathcal{S} = \{a_1, a_2, \ldots, a_d\}$. Its $d$-dimensional probability vector $p(t)$, with components

$$p^i(t) = P(\xi_t = a_i)$$

for $i = 1, \ldots, d$, then satisfies the vector ordinary differential equation

$$\frac{dp}{dt} = A\,p,$$

where $A$ is the intensity matrix. In addition, suppose that the *observation process* $W = \{W_t, t \in [0,T]\}$ is the solution of the stochastic equation

$$(3.1) \qquad\qquad W_t = \int_0^t h(\xi_s)\,ds + W_t^*,$$

where $W^* = \{W_t^*, t \in [0,T]\}$ with $W_0^* = 0$ is a Wiener process with respect to the probability measure $P$ and is independent of $\xi$. Here $\mathcal{Y}_t$ denotes the $\sigma$-algebra generated by the observations $W_s$ for $0 \le s \le t \le T$. Our reason for denoting the observation process by $W$ rather than by $Y$, as we did in Chapter 6, will be made clear below. While we only consider a 1-dimensional observation process here, most of the results that we shall present also hold in a natural extension to the general multi-dimensional case.

Our task is to filter as much information about the state process $\xi$ as we can from the observation process $W$. With this aim we shall evaluate the conditional expectation

$$E\left(g(\xi_T)\,\middle|\,\mathcal{Y}_T\right)$$

for a given function $g : \mathcal{S} \to \Re$.

We note that $W$ is, generally, not a Wiener process with respect to the probability measure $P$, but is only a "drifted" Wiener process. However, by applying the Girsanov transformation (4.8.17), we can construct a probability measure $\dot{P}$, where

$$(3.2) \qquad d\dot{P} = L_T^{-1} \, dP \qquad \text{with}$$

$$(3.3) \qquad L_T = \exp\left(-\frac{1}{2} \int_0^T h^2(\xi_s) \, ds + \int_0^T h(\xi_s) \, dW_s\right),$$

such that $W$ is a Wiener process with respect to $\dot{P}$. Moreover, $\xi$ and $W$ are independent under $\dot{P}$. This set up allows a much easier analysis of the filtering problem.

We write the un-normalized conditional probability $X_t^i$ for the state $a_i \in \mathcal{S}$ at time $t$ as the conditional expectation

$$X_t^i = \dot{E}\left(I_{\{a_i\}}(\xi_t) L_t \mid \mathcal{Y}_t\right)$$

with respect to the probability measure $\dot{P}$. It follows from the Kallianpur-Striebel formula that the conditional probabilities of $\xi_t$ given $\mathcal{Y}_t$ are

$$(3.4) \qquad P(\xi_t = a_i \mid \mathcal{Y}_t) = E\left(I_{\{a_i\}}(\xi_t) \mid \mathcal{Y}_t\right) = X_t^i \Big/ \sum_{k=1}^{d} X_t^k$$

for $a_i \in \mathcal{S}$ and $t \in [0, T]$, where the $d$-dimensional process $X_t = \{X_t^1, \ldots, X_t^d\}$ of the un-normalized conditional distribution satisfies the homogeneous linear stochastic equation, known as the Zakai equation,

$$(3.5) \qquad X_t = p(0) + \int_0^t A \, X_s \, ds + \int_0^t H \, X_s \, dW_s$$

for $t \in [0, T]$. Here the second integral is an Ito integral and $H$ is the $d \times d$ diagonal matrix with $ii$th component $h(a_i)$. Suppose that we are given a function $g : \mathcal{S} \to \Re$ such that

$$E\left(|g(\xi_t)|^2\right) < \infty$$

for $t \in [0, T]$, which is no restriction since $\mathcal{S}$ is finite. Then the optimal least squares estimate for $g(\xi_t)$ with respect to the observations $W_s$ for $0 \le s \le t$, that is the $\sigma$-algebra $\mathcal{Y}_t$, is given by the conditional expectation

$$(3.6) \qquad \Pi_t(g) = E\left(g(\xi_t) \mid \mathcal{Y}_t\right)$$

$$= \sum_{k=1}^{d} g(a_k) \, P\left(\xi_t = a_k \mid \mathcal{Y}_t\right)$$

$$= \sum_{k=1}^{d} g\left(a_k\right) X_t^k \bigg/ \sum_{k=1}^{d} X_t^k,$$

which we shall call the *optimal filter*.

To compute the optimal filter (3.6) we need, in general, to know a complete sample path of the observation process $W$ on the subinterval $[0,t]$ and to have a way of solving the Ito equation (3.5). In practice, however, it is impossible to detect $W$ completely on $[0,t]$, so electronic devices are used to obtain increments of the integral observations over small time intervals, which might be interpreted in the simplest case as the increments of $W$

$$\int_{\tau_0}^{\tau_1} dW_s, \ \ldots, \ \int_{\tau_n}^{\tau_{n+1}} dW_s, \ \ldots$$

or, with more sophisticated devices, could also include multiple Ito integrals such as

$$\int_{\tau_0}^{\tau_1} \int_{\tau_0}^{s} dW_r \, ds, \ \ldots, \ \int_{\tau_n}^{\tau_{n+1}} \int_{\tau_n}^{s} dW_r \, ds, \ \ldots.$$

With such observations of multiple Ito integrals, it is possible to construct a strong time discrete approximation $Y^\delta$ with maximum step size $\delta$ of the solution $X$ of the stochastic differential equation (3.5). The corresponding approximate Markov chain filter for a given function $g$ is then

$$(3.7) \qquad\qquad \Pi_t^\delta(g) = \sum_{k=1}^{d} g\left(a_k\right) Y_t^{\delta,k} \bigg/ \sum_{k=1}^{d} Y_t^{\delta,k}$$

for $t \in [0,T]$. If we use the Euler scheme (10.2.4), for example, it can be shown (see Newton (1983)) that we have an error of the form

$$E\left(\left|\Pi_t(g) - \Pi_t^\delta(g)\right|\right) \le K \, \delta^{1/2}.$$

We saw in Chapter 10 that the achievable order of strong convergence for a time discrete approximation is restricted if it uses only the increments of the Wiener process. For instance, for the noncommutative noise system considered in Example 10.3.4 such an approximation can achieve at most the strong order $\gamma = 0.5$, which, as the above estimate shows, can be already obtained using the simple Euler scheme. For a higher order of strong convergence we need, in general, to use a time discrete approximation which involves higher order multiple Ito integrals. We have discussed such schemes extensively in Chapters 10 and 11.

Given an equidistant time discretization of the interval $[0,T]$ with step size $\delta = \Delta = T/N$ for some $N = 1, 2, \ldots$, we shall define the *partition $\sigma$-algebra* $\mathcal{P}_N^1$ as the $\sigma$-algebra generated by the increments

$$\Delta W_0 = \int_0^\Delta dW_s, \ \Delta W_1 = \int_\Delta^{2\Delta} dW_s, \ \ldots, \ \Delta W_{N-1} = \int_{(N-1)\Delta}^{N\Delta} dW_s.$$

Thus, $\mathcal{P}_N^1$ contains the information about the increments of $W$ for this time discretization. Moreover, it is all that is needed to apply the Milstein scheme to the linear stochastic differential equation (3.5), since the noise appears here diagonally and, hence, commutatively (see (10.3.11) and (10.3.14)). The Milstein scheme (10.3.12) then simplifies to

$$(3.8) \qquad Y_{n+1} = \left[ I + A\,\Delta + H\,\Delta W_n + \frac{1}{2}\left((\Delta W_n)^2 - \Delta\right) H^2 \right] Y_n$$

with initial value $Y_0 = X_0$, where $I$ is the $d \times d$ unit matrix. It follows from Theorem 10.3.5 that (3.8) converges with strong order $\gamma = 1.0$. In addition, Newton has shown that the anologue for the Milstein scheme of the approximate filter (3.7) also converges with this order, that is its error satisfies

$$E\left(\left|\Pi_t(g) - \Pi_t^\delta(g)\right|\right) \le K\,\delta.$$

We can obtain higher accuracy if we use the order 1.5 strong Taylor approximation (10.4.5), but then we would need the additional integral observations

$$\Delta Z_n = I_{(1,0),n\Delta,(n+1)\Delta} = \int_{n\Delta}^{(n+1)\Delta} \int_{n\Delta}^{s} dW_r\,ds,$$

on the intervals $[n\Delta, (n+1)\Delta]$ for $n = 0, \ldots, N-1$, to ensure strong convergence of order $\gamma = 1.5$. If these were not available, we could instead look for some "best" scheme amongst the class of order 1.0 strong schemes. This, as we shall see in the next section, yields the *order* 1.0 *asymptotically efficient scheme*

$$(3.9) \qquad Y_{n+1} = \left[ I + A\,\Delta + H\,\Delta W_n + \frac{1}{2}\left((\Delta W_n)^2 - \Delta\right) H^2 \right.$$
$$+ \frac{1}{2}A^2\,\Delta^2 + \frac{1}{2}\left(A\,H + H\,A\right)\Delta W_n\,\Delta$$
$$\left. + \frac{1}{6}H^3\left((\Delta W_n)^3 - 3\Delta W_n\,\Delta\right) \right] Y_n$$

with $Y_0 = X_0$, which is due to Newton. We can obtain (3.9) directly from the order 1.5 strong Taylor scheme (10.4.5) by replacing all of the multiple Ito integral expressions there that are not $\mathcal{P}_N^1$-measurable by their $\mathcal{P}_N^1$-conditional expectations, that is by replacing each $\Delta Z_n$ by

$$\dot{E}\left(\Delta Z_n \,|\, \mathcal{P}_N^1\right) = \dot{E}\left(I_{(1,0),n\Delta,(n+1)\Delta} \,|\, \mathcal{P}_N^1\right)$$
$$= \dot{E}\left(J_{(1,0),n\Delta,(n+1)\Delta} \,|\, \mathcal{P}_N^1\right) = \frac{1}{2}\Delta W_n\,\Delta,$$

which follows from (5.2.36) and (5.8.9), where $\dot{E}$ denotes the expectation with respect to the probability measure $\dot{P}$ under which $W$ is a Wiener process.

In the next section we shall see that one cannot obtain better accuracy by repeating this procedure for strong Taylor schemes of higher order since the leading error coefficient is already contained in the scheme (3.9). However, we can obtain better accuracy if we do the conditioning with respect to the *partition σ-algebra* $\mathcal{P}_N^{1.5}$ generated by the observations $\Delta W_0, \ldots, \Delta W_{N-1}$, $\Delta Z_0, \ldots, \Delta Z_{N-1}$ over the time interval $[0, T]$. With this information we can implement order 1.5 schemes such as the order 1.5 strong Taylor scheme. In addition, we could look for the "best" scheme among the class of order 1.5 schemes, for example by replacing the multiple stochastic integrals in the order 2.0 strong Taylor scheme (10.5.2) which are not $\mathcal{P}_N^{1.5}$-measurable by their $\mathcal{P}_N^{1.5}$-conditional expectations. In particular, we need to replace the Stratonovich integrals $J_{(0,1,1)}$, $J_{(1,0,1)}$, $J_{(1,1,0)}$ by

$$\dot{E}\left(J_{(0,1,1)}\,\big|\,\mathcal{P}_N^{1.5}\right) = \frac{5}{12}\Delta\,(\Delta W_n)^2 + \frac{1}{12}\Delta^2 - \frac{1}{2}\Delta W_n\,\Delta Z_n,$$

$$\dot{E}\left(J_{(1,0,1)}\,\big|\,\mathcal{P}_N^{1.5}\right) = -\frac{1}{12}\Delta\,(\Delta W_n)^2 - \frac{1}{12}\Delta^2 + \Delta W_n\,\Delta Z_n - \frac{(\Delta Z_n)^2}{\Delta}$$

$$\dot{E}\left(J_{(1,1,0)}\,\big|\,\mathcal{P}_N^{1.5}\right) = \frac{1}{6}\Delta\,(\Delta W_n)^2 - \frac{1}{2}\Delta W_n\,\Delta Z_n + \frac{(\Delta Z_n)^2}{\Delta},$$

respectively, where we have used the relations (5.8.9) to evaluate the expectations. For the linear stochastic differential equation (3.5) we thus obtain the *order 1.5 asymptotically efficient scheme*

$$(3.10) \quad Y_{n+1} = \left[ I + \underline{A}\Delta + H\,\Delta W_n + \frac{1}{2}(\Delta W_n)^2\,H^2 \right.$$

$$+ \frac{1}{2}\underline{A}^2\,\Delta^2 + \underline{A}\,H\,\Delta Z_n + H\,\underline{A}\,\{\Delta W_n\,\Delta - \Delta Z_n\}$$

$$+ \frac{1}{3!}H^3\,(\Delta W_n)^3 + \frac{1}{4!}H^4\,(\Delta W_n)^4$$

$$+ H^2\,\underline{A}\left\{ \frac{5}{12}\Delta\,(\Delta W_n)^2 + \frac{1}{12}\Delta^2 - \frac{1}{2}\Delta W_n\,\Delta Z_n \right\}$$

$$- H\,\underline{A}\,H\left\{ \frac{1}{12}\Delta\,(\Delta W_n)^2 + \frac{1}{12}\Delta^2 - \Delta W_n\,\Delta Z_n + \frac{(\Delta Z_n)^2}{\Delta} \right\}$$

$$+ \underline{A}\,H^2\left\{ \frac{1}{6}\Delta\,(\Delta W_n)^2 - \frac{1}{2}\Delta W_n\,\Delta Z_n + \frac{(\Delta Z_n)^2}{\Delta} \right\} \right] Y_n$$

with $Y_0 = X_0$ and

$$\underline{A} = A - \frac{1}{2}H^2.$$

**Example 13.3.1**    *Consider the random telegraphic noise process introduced in Example 1.6.8, that is the two state continuous time Markov chain $\xi$ on the state space $S = \{-1, +1\}$ with intensity matrix*

$$A = \begin{bmatrix} -0.5 & 0.5 \\ 0.5 & -0.5 \end{bmatrix}$$

*and initial probability vector $p(0) = (0.9, 0.1)$. Suppose that the observation process $W$ satisfies the stochastic equation (3.1) with*

$$h(a_k) = \begin{cases} 5 & : \quad a_k = +1 \\ 0 & : \quad a_k = -1 \end{cases}.$$

*We want to determine the actual state of the chain on the basis of these observations, that is to compute the conditional probability vector*

$$\hat{p}(t) = \left( P\left(\xi_t = -1 \,\middle|\, \mathcal{Y}_t\right), \; P\left(\xi_t = +1 \,\middle|\, \mathcal{Y}_t\right) \right).$$

*We could say that $\xi_t$ has most likely the value $+1$ if $P\left(\xi_t = +1 \,\middle|\, \mathcal{Y}_t\right) \geq 0.5$, and the value $-1$ otherwise. To do this in practice, we need to evaluate the conditional probability*

$$P\left(\xi_t = +1 \,\middle|\, \mathcal{Y}_t\right) = E\left(I_{\{+1\}}\left(\xi_t\right) \,\middle|\, \mathcal{Y}_t\right) = \Pi_t\left(I_{\{+1\}}\right),$$

*which is the optimal filter here. We can use a filter $\Pi_t^\delta\left(I_{\{+1\}}\right)$ based on a time discrete approximation to obtain a good approximation of $\Pi_t\left(I_{\{+1\}}\right)$.*

**PC-Exercise 13.3.2**    *Suppose in Example 13.3.1 that we have a realization of the Markov chain on the interval $[0, 4]$ with*

$$\xi_t = \begin{cases} +1 & : \quad 0 \leq t < \frac{1}{2} \\ -1 & : \quad \frac{1}{2} \leq t \leq 4. \end{cases}$$

*Compute the approximate filters $\Pi_t^\delta\left(I_{\{+1\}}\right)$ for the same realization of the Wiener process $W^*$ using the Euler, Milstein, order 1.5 strong Taylor and order 1.5 asymptotically efficient schemes with equidistant step size $\delta = \Delta = 2^{-7}$. Plot and compare the calculated sample paths.*

We see from Figure 13.3.1 that the Euler scheme succeeds in detecting the jump in the unobserved state at time $t = 0.5$, but also computes meaningless negative "probabilities". The printout for the Milstein scheme in Figure 13.3.2 shows better numerical stability. In Figure 13.3.4, the order 1.5 asymptotically efficient scheme appears to detect the unobserved state a little better than the order 1.5 strong Taylor scheme in Figure 13.3.3.

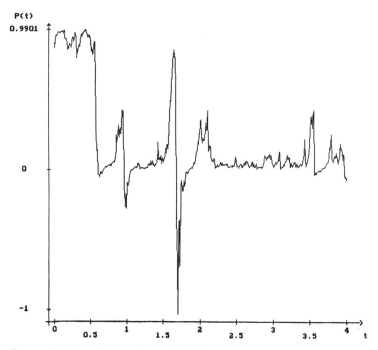

**Figure 13.3.1**    Euler scheme in PC-Exercise 13.3.2

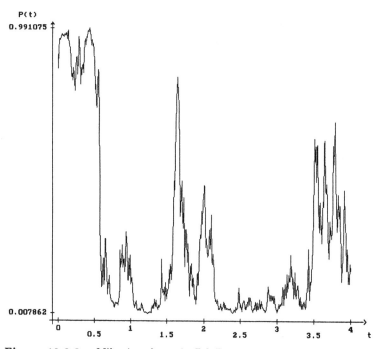

**Figure 13.3.2**    Milstein scheme in PC-Exercise 13.3.2

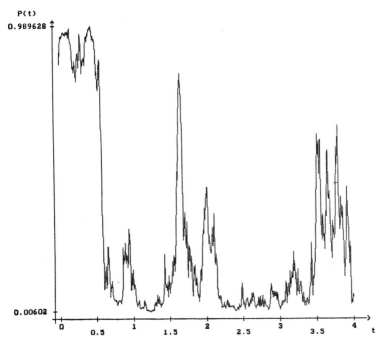

**Figure 13.3.3** 1.5 Taylor scheme in PC-Exercise 13.3.2

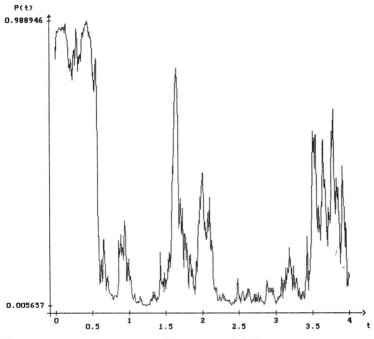

**Figure 13.3.4** 1.5 asympt. eff. scheme in PC-Ex. 13.3.2

**PC-Exercise 13.3.3** *Simulate the observation process of Example 13.3.1 using a realization of the telegraphic noise process as in PC-Exercise 13.3.2. Apply each of the four approximate Markov chain filters in PC-Exercise 13.3.2 to detect the state of the Markov chain at time $T = 1$ and estimate the frequency of failure of the filter. Take equidistant step sizes $\delta = 2^{-3}$, $2^{-4}$ and $2^{-5}$ and generate $M = 20$ batches with $N = 100$ simulations. Plot the 90%-confidence intervals against $\delta$.*

Example 13.3.1 is rather nice, but in practice the intensity matrix of some Markov chain filters yields a stochastic differential equation (3.5) which is stiff. In these cases we need to use the implicit schemes from Chapter 12 to ensure that the numerical results are reliable.

**Example 13.3.4** *Consider Example 13.3.1 with the intensity matrix replaced by*

$$A = \left[ \begin{array}{cc} -50.0 & 50.0 \\ 50.0 & -50.0 \end{array} \right]$$

*and the initial probability vector $p(0) = (0.9, 0.1)$. Now the state process usually jumps very quickly between $+1$ and $-1$.*

**PC-Exercise 13.3.5** *Repeat PC-Exercise 13.3.2 for Example 13.3.4 using the explicit and implicit Euler and Milstein schemes, using parameters $\alpha_1 = \alpha_2 = 1.0$ in the implicit versions.*

It is apparent from Figure 13.3.5 with the Euler scheme results above and the Milstein scheme results below that the implicit filters in Figure 13.3.6 detect the state of the system better in some sense than their explicit counterparts. The corresponding 90% confidence intervals should also show that the implicit filters perform better than their explicit versions for this particular system.

**PC-Exercise 13.3.6** *Repeat PC-Exercise 13.3.3 for Example 13.3.4 comparing the explicit approximate filters with the implicit Markov chain filters based on the corresponding explicit and implicit Euler and Milstein schemes.*

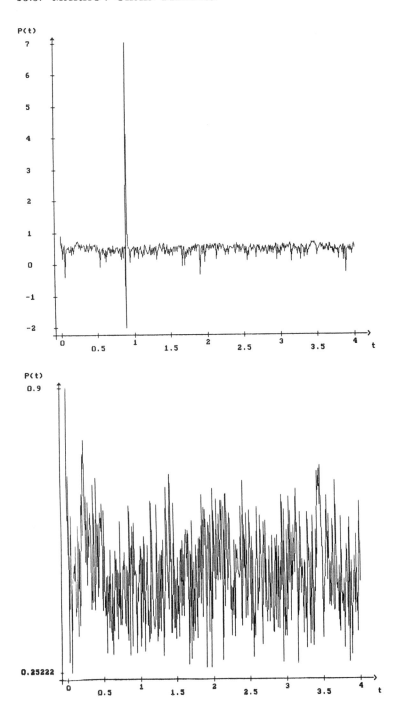

**Figure 13.3.5** Explicit Euler and Milstein schemes

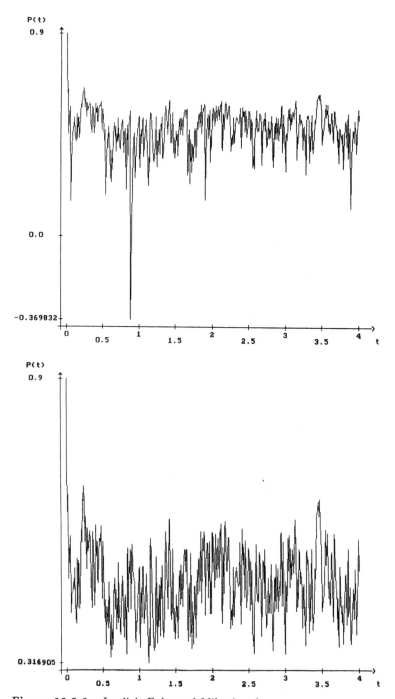

**Figure 13.3.6**   Implicit Euler and Milstein schemes

## 13.4 Strong Asymptotically Efficient Schemes

We saw in the previous section, in the context of filtering problems, that we can often achieve only a restricted strong order of convergence for time discretizations because we have access to only the simplest integral observations of the driving Wiener process. In such situations we can look for those schemes which minimise the leading coefficient in the expansion of the mean-square errors as power series in the maximum step size. In this section we shall summarize some results on such schemes, due to Newton, which we shall call *asymptotically efficient*.

Let $X = \{X_t, \ t \in [0, T]\}$ be a $d$-dimensional Ito process satisfying the stochastic differential equation

$$(4.1) \qquad X_t = X_0 + \int_0^t a\,(X_s)\ ds + \int_0^t b\,(X_s)\ dW_s$$

for $t \in [0, T]$, where $W$ is a 1-dimensional Wiener process and, for simplicity, suppose that $X_0$ is nonrandom. As before, we take an equidistant time discretization with step size $\delta = \Delta = T/N$ for some $N = 1, 2, \ldots$ and denote by $\mathcal{P}_N^1$ the partition $\sigma$-algebra generated by the increments

$$\Delta W_0 = \int_0^\Delta dW_s, \quad \Delta W_1 = \int_\Delta^{2\Delta} dW_s, \ \ldots, \ \Delta W_{N-1} = \int_{(N-1)\Delta}^{N\Delta} dW_s.$$

For any sequence $\{Z^\delta, \ \delta \in (0,1)\}$ of $\mathcal{P}_N^1$-measurable random variables we have

$$(4.2) \qquad \liminf_{\delta \to 0} \frac{E\left(\left|\delta^{-1}c^\mathsf{T}\left(X_T - Z^\delta\right)\right|^2 \big| \mathcal{P}_N^1\right) + 1}{E\left(\left|\delta^{-1}c^\mathsf{T}\left(X_T - E\left(X_T \,\big| \mathcal{P}_N^1\right)\right)\right|^2 \big| \mathcal{P}_N^1\right) + 1} \geq 1$$

for all $c \in \Re^d$, w.p.1. Thus the conditional expectation $E\left(X_T \big| \mathcal{P}_N^1\right)$ is the best $\mathcal{P}_N^1$-measurable least squares approximation of $X_T$. Compare this with (2.2.19). This suggests the following concept of efficiency for $\mathcal{P}_N^1$-measurable approximations of the solution of (4.1):

A time discrete approximation $Y^\delta$ is *order* 1.0 *asymptotically efficient* if

$$(4.3) \qquad \liminf_{\delta \to 0} \frac{E\left(\left|\delta^{-1}c^\mathsf{T}\left(X_T - Y_{n_T}^\delta\right)\right|^2 \big| \mathcal{P}_N^1\right) + 1}{E\left(\left|\delta^{-1}c^\mathsf{T}\left(X_T - E\left(X_T \,\big| \mathcal{P}_N^1\right)\right)\right|^2 \big| \mathcal{P}_N^1\right) + 1} = 1$$

for all $c \in \Re^d$, w.p.1.

We shall now list some schemes which will turn out to be order 1.0 asymptotically efficient. If we replace the $\Delta Z_n = I_{(1,0)}$ terms in the order 1.5 strong Taylor scheme (10.4.5) by the conditional expectations

$$E\left(\Delta Z_n \big| \mathcal{P}_N^1\right) = \frac{1}{2}\,\Delta W_n\,\Delta,$$

we obtain the scheme

$$(4.4) \qquad Y_{n+1} \;=\; Y_n + a\,\Delta + b\,\Delta W_n + \frac{1}{2}L^1 b\left\{(\Delta W_n)^2 - \Delta\right\}$$

$$+ \frac{1}{2}\left(L^1 a + L^0 b\right)\Delta W_n\,\Delta + \frac{1}{2}L^0 a\,\Delta^2$$

$$+ \frac{1}{2}L^1 L^1 b\left\{\frac{1}{3}(\Delta W_n)^2 - \Delta\right\}\Delta W_n$$

where, as previously,

$$L^0 = \sum_{k=1}^{d} a^k\,\frac{\partial}{\partial x^k} + \frac{1}{2}\sum_{k,l=1}^{d} b^k b^l\,\frac{\partial^2}{\partial x^k \partial x^l}$$

and

$$L^1 = \sum_{k=1}^{d} b^k\,\frac{\partial}{\partial x^k}.$$

The scheme (3.9) for the linear stochastic differential equation (3.5) is a special case of this scheme. Another example is the following Runge-Kutta type scheme:

$$(4.5) \qquad Y_{n+1} \;=\; Y_n + \frac{1}{2}\left(a_0 + a_1\right)\Delta$$

$$+ \frac{1}{40}\left(37 b_0 + 30 b_2 - 27 b_3\right)\Delta W_n$$

$$+ \frac{1}{16}\left(8 b_0 + b_1 - 9 b_2\right)(3\Delta)^{1/2}$$

where

$$a_0 = a\left(Y_n\right), \qquad b_0 = b\left(Y_n\right),$$

$$a_1 = a\left(Y_n + a_0\,\Delta + b_0\,\Delta W_n\right), \qquad b_1 = b\left(Y_n - \frac{2}{3}b_0\left\{\Delta W_n + (3\Delta)^{1/2}\right\}\right),$$

$$b_2 = b\left(Y_n + \frac{2}{9}b_0\left\{3\Delta W_n + (3\Delta)^{1/2}\right\}\right),$$

and

$$b_3 = b\left(Y_n - \frac{20}{27}a_0\,\Delta + \frac{10}{27}\left(b_1 - b_0\right)\Delta W_n - \frac{10}{27}b_1(3\Delta)^{1/2}\right).$$

Finally, using the corrected drift

$$\underline{a} = a - \frac{1}{2}L^1 b$$

of the Stratonovich stochastic differential equation corresponding to (4.1), we also have the following Stratonovich version of a Runge-Kutta type scheme:

$$(4.6) \qquad Y_{n+1} = Y_n + \frac{1}{2}\left(\underline{a}_0 + \underline{a}_1\right)\Delta + \frac{1}{6}\left(b_0 + 2b_1 + 2b_2 + b_3\right)\Delta W_n$$

where

$$\underline{a}_0 = \underline{a}\left(Y_n\right), \qquad b_0 = b\left(Y_n\right), \qquad b_1 = b\left(Y_n + \frac{1}{2}b_0\,\Delta W_n\right),$$

$$\underline{a}_1 = \underline{a}\left(Y_n + \frac{1}{2}\underline{a}\left(3\Delta - (\Delta W_n)^2\right) + b_2\,\Delta W_n\right),$$

$$b_2 = b\left(Y_n + \frac{1}{4}\underline{a}\left(3\Delta + (\Delta W_n)^2\right) + \frac{1}{2}b_1\,\Delta W_n\right)$$

and

$$b_3 = b\left(Y_n + \frac{1}{2}\underline{a}\left(3\Delta - (\Delta W_n)^2\right) + b_2\,\Delta W_n\right).$$

Newton established the following result.

**Theorem 13.4.1** *that a and b are Lipschitz continuous with all derivatives of a up to and including order 3 and all derivatives of b up to and including order 4 of polynomial growth. Then, each of the schemes (4.4), (4.5) and (4.6) is an order 1.0 asymptotically efficient scheme.*

We refer the reader to Newton (1991) for the proof, which involves stochastic Taylor expansions and a detailed analysis of the influence of the leading error coefficient. In particular, Newton showed that the $\mathcal{P}_N^1$-conditional distribution of the normalized error

$$\eta = \frac{1}{\Delta}\left(X_T - E\left(X_T \mid \mathcal{P}_N^1\right)\right)$$

converges with probability one to a normal distribution with mean zero and covariance matrix $\Psi_T$, with $\{\Psi_t,\ t \in [0,T]\}$ being the unique solution of the matrix Stratonovich equation

$$(4.7) \qquad \Psi_T = \int_0^T \left[\nabla\underline{a}\left(X_t\right)^{\mathsf{T}}\Psi_t + \Psi_t^{\mathsf{T}}\nabla\underline{a}\left(X_t\right)\right.$$

$$\left. + \frac{1}{12}L[\underline{a},b]\left(X_t\right)L[\underline{a},b]\left(X_t\right)^{\mathsf{T}}\right] dt$$

$$+ \int_0^T \left[\nabla b\left(X_t\right)^{\mathsf{T}}\Psi_t + \Psi_t^{\mathsf{T}}\nabla b\left(X_t\right)\right] \circ dW_t,$$

where $\nabla$ is the gradient operator and $L[a, b]$ is the Lie bracket operator of $\underline{a}$ and $b$ defined by

$$L[\underline{a}, b](x) = (\nabla \underline{a}(x))^{\mathsf{T}} b(x) - (\nabla b(x))^{\mathsf{T}} \underline{a}(x).$$

It can be shown that many order 1.0 strong schemes, such as the Milstein scheme, are not order 1.0 asymptotically efficient. These asymptotically efficient schemes require more computational effort than, say, the Euler or Milstein schemes, but can give better results in certain circumstances.

Simulation studies comparing the Euler, Milstein and order 1.5 strong Taylor schemes with the order 1.0 asymptotically efficient schemes (4.4), (4.5) and (4,6) provide an indication of the advantages and disadvantages of such schemes. As an example, we shall consider the computation of approximate Markov chain filters to detect from noisy observations the current state of the continuous time Markov chain $\xi$ on the state space $\mathcal{S}$ given in Example 13.3.1

**PC-Exercise 13.4.2**    *Repeat PC-Exercise 13.3.3 for Example 13.3.1 using the order 1.0 asymptotically efficient schemes (4.4), (4.5) and (4.6).*

The dynamics of the different components in the above example all have nearly the same time scales, so the explicit schemes presented in this section provide satisfactory numerical results. However, they may not be adequate or reliable for a stiff stochastic differential equation, such as the one in Example 13.3.4, and implicit schemes may have to be used.

# Chapter 14

# Weak Taylor Approximations

In this chapter we shall use truncated stochastic Taylor expansions as we did in Chapter 10, but now to derive time discrete approximations appropriate for the weak convergence criterion. We shall call the approximations so obtained weak Taylor approximations and shall investigate the corresponding weak Taylor schemes. As with strong approximations, the desired order of weak convergence also determines the truncation that must be used. However, this will be different from the truncation required for strong convergence of the same order, in general involving fewer terms. In the final section we shall state and prove a convergence theorem for general weak Taylor approximations. Throughout the chapter we shall use the notation and abbreviations introduced in Chapter 10.

## 14.1   The Euler Scheme

We have already considered the Euler scheme in varying degrees of detail in Chapters 9 and 10. The Euler approximation is the simplest weak Taylor approximation and attains the order of weak convergence $\beta = 1.0$ under suitable smoothness and growth conditions on the drift and diffusion coefficients, which was first shown by Milstein and Talay.

We recall from (10.2.5) that for the general multi-dimensional case $d$, $m = 1, 2, \ldots$ the $k$th component of the *Euler scheme* has the form

$$
(1.1) \qquad Y_{n+1}^k = Y_n^k + a^k \Delta + \sum_{j=1}^{m} b^{k,j} \Delta W^j,
$$

with initial value $Y_0 = X_0$, where

$$
\Delta = \tau_{n+1} - \tau_n \quad \text{and} \quad \Delta W^j = W_{\tau_{n+1}}^j - W_{\tau_n}^j.
$$

In what follows we shall use the Euler approximation $Y^\delta = \{ Y^\delta(t), t \in [0,T] \}$ defined in (10.2.16), that is the solution of the Ito equation formed by freezing the drift and diffusion coefficients at their values at the beginning of each subinterval of the time discretization $(\tau)_\delta$. For notational clarity we shall often omit the superscript $\delta$ from $Y^\delta$ in complicated expressions.

The Euler scheme (1.1) corresponds to the truncated Ito-Taylor expansion (5.5.3) which contains only the ordinary time integral and the simple Ito integral. We shall see from a general convergence result for weak Taylor approximations, to be stated in Theorem 14.5.1 of Section 5, that the Euler approximation has order of weak convergence $\beta = 1.0$ if, amongst other assumptions, $a$ and $b$ are four times continuously differentiable. This means that the Euler scheme (1.1) is the order 1.0 weak Taylor approximation.

**Exercise 14.1.1**    *Show that the 1-dimensional Ito process X satisfying*

$$dX_t = \frac{3}{2} X_t \, dt + \frac{1}{10} X_t \, dW_t^1 + \frac{1}{10} X_t \, dW_t^2$$

*with the initial value $X_0 = 0.1$, where $W^1$ and $W^2$ are two independent standard Wiener processes, has expectation*

$$E(X_T) = 0.1 \exp\left(\frac{3}{2} T\right).$$

**PC-Exercise 14.1.2**    *For the Ito process X in Exercise 14.1.1 simulate M = 20 batches each of N = 100 trajectories of the Euler approximation $Y^\delta$ using the Euler scheme (1.1) with equidistant time steps of step size $\delta = \Delta = 2^{-2}$. Determine the 90%-confidence interval for the mean error*

$$\mu = E\left(Y^\delta(T)\right) - E(X_T)$$

*at time $T = 1$. Repeat the calculations for step sizes $2^{-3}$, $2^{-4}$ and $2^{-5}$, and plot $\log_2 |\mu|$ versus $\log_2 \Delta$.*

For weak convergence we only need to approximate the measure induced by the Ito process $X$, so we can replace the Gaussian increments $\Delta W^j$ in (1.1) by other random variables $\Delta \hat{W}^j$ with similar moment properties. We can thus obtain a simpler scheme by choosing more easily generated noise increments. This leads to the *simplified weak Euler scheme*

$$(1.2) \qquad\qquad Y_{n+1}^k = Y_n^k + a^k \, \Delta + \sum_{j=1}^{M} b^{k,j} \, \Delta \hat{W}^j,$$

where the $\Delta \hat{W}^j$ for $j = 1, 2, \ldots, m$ must be independent $\mathcal{A}_{\tau_{n+1}}$-measurable random variables with moments satisfying the conditions

$$\left| E\left(\Delta \hat{W}^j\right) \right| + \left| E\left(\left(\Delta \hat{W}^j\right)^3\right) \right| + \left| E\left(\left(\Delta \hat{W}^j\right)^2\right) - \Delta \right| \le K \Delta^2$$

for some constant $K$; see also (5.12.7). A very simple example of such a $\Delta \hat{W}^j$ in (1.2) is a two-point distributed random variable with

$$P\left(\Delta \hat{W}^j = \pm\sqrt{\Delta}\right) = \frac{1}{2}.$$

**Exercise 14.1.3**    *Verify that the above two-point distributed random variables satisfy the moment conditions required by the simplified weak Euler scheme (1.2).*

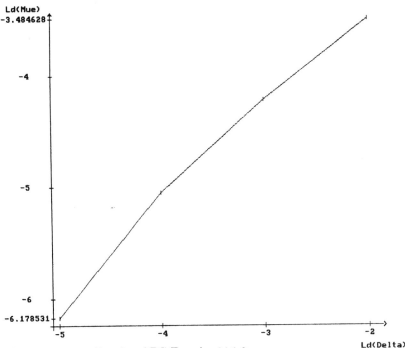

**Figure 14.1.1**   Results of PC-Exercise 14.1.2.

**PC-Exercise 14.1.4**   *Repeat PC-Exercise 14.1.2 using the simplified weak Euler scheme (1.2) with the noise increments generated by two-point distributed random variables. Compare the results with those of PC-Exercise 14.1.2.*

As mentioned above, we shall be able to conclude from Theorem 14.5.1 that the Euler scheme (1.1) converges weakly with order $\beta = 1.0$ when the drift and diffusion coefficients $a$ and $b$ are four times continuously differentiable. Many stochastic differential equations occuring in applications, however, do not possess this degree of smoothness. Nevertheless, we can show directly that the Euler scheme still converges weakly, but with order $\beta < 1.0$ when the coefficients are only Hölder continuous. For greater clarity we shall prove this here only for a $d$-dimensional Ito process with $m = 1$, that is with scalar noise, but first we shall need to introduce some additional terminology. The more practically minded reader can omit the rest of this section.

We say that a function $f \colon \Re^k \to \Re^l$ is *Hölder continuous with index* $\nu \in (0,1)$ if there exists a positive constant $K$ such that

$$|f(x) - f(y)| \le K\,|x - y|^\nu$$

for all $x$, $y \in \Re^k$. For $x = (x^1, \ldots, x^d) \in \Re^d$ and $\eta = (\eta_1, \ldots, \eta_d)$ with $\eta_j = 0$, $1, \ldots$ for $j = 1, \ldots, d$ we shall abbreviate partial differentiation by

$$\partial_x^\eta := \left(\frac{\partial}{\partial x^1}\right)^{\eta_1} \cdots \left(\frac{\partial}{\partial x^d}\right)^{\eta_d}, \qquad \text{with}$$

$$\partial_t^k := \left( \frac{\partial}{\partial t} \right)^k$$

for $k = 1, 2, \ldots$, and write

$$\partial_x^l u = \partial_x^\eta u,$$

where $l = |\eta| = \eta_1 + \ldots + \eta_d \geq 1$ with

$$\partial_t^0 u = \partial_x^0 u = u.$$

We shall denote by $\mathcal{H}_T^{(l)}$ the space of functions $u : [0, T] \times \Re^d \to \Re$ with partial derivatives $\partial_t^r \partial_x^s u$ Hölder continuous with index $l - [l]$ in $x$ and index $(l - 2r - s)/2$ in $t$ for all $2r + s \leq l$, where $l \in \mathcal{L} := (0, 1) \cup (1, 2) \cup (2, 3)$. Here $[l]$ is the integer part of $l$, that is the largest integer that does not exceed $l$. On $\mathcal{H}_T^{(l)}$ we shall use the norm

(1.3)    $\displaystyle \|u\|_T^{(l)} \;\; = \sum_{2r+s \leq [l]} \sup_{t,x} |\partial_t^r \partial_x^s u(t, x)|$

$$+ \sum_{2r+s=[l]} \sup_{t, x \neq x'} \left( \frac{|\partial_t^r \partial_x^s u(t, x) - \partial_t^r \partial_x^s u(t, x')|}{|x - x'|^{l - [l]}} \right)$$

$$+ \sum_{0 < l - 2r - s < 2} \sup_{t \neq t', x} \left( \frac{|\partial_t^r \partial_x^s u(t, x) - \partial_t^r \partial_x^s u(t', x)|}{|t - t'|^{(l - 2r - s)/2}} \right).$$

In addition, we shall denote by $\mathcal{H}^{(l)}$ the space of all functions $u \in \mathcal{H}_T^{(l)}$ which do not depend on the time parameter $t$, and by $\| \cdot \|^{(l)}$ its norm, which can be obtained from (1.3) by omitting all time dependency.

We recall that $k \wedge l$ is the minimum of $k$ and $l$, and that $(x, y)$ is the Euclidean inner product (1.4.36) on $\Re^d$. Finally, we define the $d \times d$-matrix

(1.4)                    $B(t, x) = b(t, x) \, b(t, x)^\top$

for all $(t, x) \in [0, T] \times \Re^d$, where $b$ is the vector diffusion coefficient and $b^\top$ its transpose.

**Theorem 14.1.5**    Let $X = \{X_t, \, t \in [0, T]\}$ be a d-dimensional Ito process with scalar noise, so $m = 1$, for which all moments of $X_T$ exist, with

(1.5)                    $(B(t, x)\zeta, \zeta) \geq \lambda |\zeta|^2$

for all $t \in [0, T]$ and $x, \zeta \in \Re^d$ and some positive constant $\lambda$, and

(1.6)                    $a, B \in \mathcal{H}_T^{(l)}$

for some $l \in \mathcal{L}$. In addition, suppose that $Y^\delta = \{ Y_t^\delta, \, t \in [0, T]\}$ is the Euler approximation (10.2.16) corresponding to a time discretization $(\tau)_\delta$ and that $g : \Re^d \to \Re$ satisfies

(1.7)                    $g \in \mathcal{H}^{(l+2)}.$

*Then*

(1.8) $$\left| E\left(g\left(X_T\right)\right) - E\left(g\left(Y^\delta(T)\right)\right)\right| \leq K\,\delta^{\chi(l)}$$

*with*

(1.9) $$\chi(l) = \begin{cases} l/2 & : \quad l \in (0,1) \\ 1/(3-l) & : \quad l \in (1,2) \\ 1 & : \quad l \in (2,3) \end{cases}$$

*where $K$ does not depend on $\delta$.*

We note from this theorem, which is due to Mikulevicius and Platen, that the Euler approximation converges weakly for some order $\beta > 0$ when $a$ and $B$ are Hölder continuous. For it to have order $\beta = 1.0$, we need these coefficients to be only a little more than twice continuously differentiable, rather than four times as required by Theoerm 14.5.1. We note that it is easy to prove (1.9) in the case $l = 1$ with $\chi(1) = \frac{1}{2}$ using the Lipschitz continuity of $g$ and Theorem 10.2.2 under assumptions like those in Theorem 14.1.5.

In the proof of Theorem 14.1.5 we shall use the following lemma.

**Lemma 14.1.6** *Suppose that $a$ and $B$ are bounded. Then for each $\eta \in (0,1]$ there exists a positive constant $K_\eta$ such that*

(1.10) $$\left| E\left(f(s, Y^\delta(s)) - f\left(\tau_{n_s}, Y^\delta_{n_s}\right) \left| \mathcal{A}_{\tau_{n_s}}\right.\right)\right| \leq K_\eta \|f\|_T^{(l)} \delta^{\chi(l)}$$

*for all $s \in [0,T]$ and $f \in \mathcal{H}_T^{(l)}$ with $l \in [\eta, 1) \cup (1,2) \cup (2,3)$, where $\chi(l)$ was defined in (1.9) and $n_t$ in (9.1.15).*

**Proof** Let $w \in \mathcal{C}_0^\infty(\Re^d)$ be a smooth nonnegative function with support in $\{x \in \Re^d : |x| \leq 1 \}$ and

$$\int_{\Re^d} w(x)\,dx = 1.$$

For $\epsilon$, $h \in (0,1)$ we define the mollifier function

$$w_\epsilon(x) = \frac{1}{\epsilon^d}\,w\left(\frac{x}{\epsilon}\right)$$

and the convolution

$$f^{h,\epsilon}(t,x) = \frac{1}{h}\int_t^{t+h}\int_{\Re^d} f\left(u \wedge T, y\right) w_\epsilon(x - y)\,dy\,du$$

for $(t,x) \in [0,T] \times \Re^d$. Then $f^{h,\epsilon}$ is continuously differentiable in $t$ and smooth in the variable $x$ and satisfies the inequalities

(1.11) $$\sup_{t,x}\left|f(t,x) - f^{h,\epsilon}(t,x)\right| \leq \|f\|_T^{(l)}\left(h^{(l/2)\wedge 1} + \epsilon^{l\wedge 1}\right),$$

(1.12) $$\sup_{t,x}\left|\partial_x^i f^{h,\epsilon}(t,x)\right| \leq K \|f\|_T^{(l)} \epsilon^{(l-i)\wedge 0},$$

(1.13)                    $\sup_{t,x} \left| \partial_t f^{h,\epsilon}(t,x) \right| \leq K \|f\|_T^{(l)} h^{(-1+l/2)\wedge 0}$,

for $i = 1$ and 2, where the positive constant $K$ does not depend on $f$, $l$, $\epsilon$ or $h$.

For convenience we shall now write $Y$ instead of $Y^\delta$. By the triangle inequality we have

(1.14)    $\left| E \left( f(s, Y(s)) - f(\tau_{n_s}, Y_{n_s}) \, \Big| \mathcal{A}_{\tau_{n_s}} \right) \right|$

$$\leq \quad 2 \sup_{t,x} \left| f(t,x) - f^{h,\epsilon}(t,x) \right|$$

$$+ \left| E \left( f^{h,\epsilon}(s, Y(s)) - f^{h,\epsilon}(\tau_{n_s}, Y_{n_s}) \, \Big| \mathcal{A}_{\tau_{n_s}} \right) \right|.$$

From the Ito formula (3.3.6), (1.6), (1.12) and (1.13) we obtain

(1.15)    $\left| E \left( f^{h,\epsilon}(s, Y(s)) - f^{h,\epsilon}(\tau_{n_s}, Y_{n_s}) \, \Big| \mathcal{A}_{\tau_{n_s}} \right) \right|$

$$\leq \left| E \left( \int_{\tau_{n_s}}^s \left[ \partial_t f^{h,\epsilon}(u, Y(u)) \right. \right. \right.$$

$$+ \frac{1}{2} \sum_{i,j=1}^d B^{i,j} \left( \tau_{n_s}, Y_{n_s} \right) \partial_{x^i} \partial_{x^j} f^{h,\epsilon}(u, Y(u))$$

$$\left. \left. + \sum_{i=1}^d a^i \left( \tau_{n_s}, Y_{n_s} \right) \partial_{x^i} f^{h,\epsilon}(u, Y(u)) \right] du \, \Big| \mathcal{A}_{\tau_{n_s}} \right) \right|$$

$$\leq K \|f\|_T^{(l)} \left( h^{(-1+l/2)\wedge 0} + \epsilon^{(l-2)\wedge 0} \right) \delta,$$

where the positive constant $K$ does not depend on $f$, $l$, $\epsilon$ or $h$.

Using (1.11) and (1.15) in (1.14) we thus have

$$\left| E \left( f(s, Y(s)) - f(\tau_{n_s}, Y_{n_s}) \, \Big| \mathcal{A}_{\tau_{n_s}} \right) \right|$$

$$\leq K \|f\|_T^{(l)} \left\{ \inf_{h \in (0,1)} \left( h^{(l/2)\wedge 1} + h^{(-1+l/2)\wedge 0}\delta \right) \right.$$

$$\left. + \inf_{\epsilon \in (0,1)} \left( \epsilon^{l \wedge 1} + \epsilon^{(l-2)\wedge 0}\delta \right) \right\}.$$

The asserted bound (1.10) then follows when we evaluate these infima for the different cases of the index $l$. $\square$

**Proof of Theorem 14.1.5**   We recall the operator

$$L^0 = \partial_t + \sum_{k=1}^{d} a^k \, \partial_{x^k} + \frac{1}{2} \sum_{k,l=1}^{d} B^{k,l} \partial_{x^k} \partial_{x^l}$$

introduced in (10.1.1). It follows from Theorem 5.2 on page 361 of Ladyzhen-skaya, Solonikov and Uraltseva (1968) that there exists a unique solution $v \in \mathcal{H}_T^{(l+2)}$ of the final value problem

(1.16) $$L^0 v = 0$$

(1.17) $$v(T, x) = g(x)$$

with

(1.18) $$\|v\|_T^{(l+2)} \leq K \|g\|^{(l+2)}.$$

From (1.16) and the Ito formula (3.3.6) we have

(1.19) $$E(v(0, X_0)) = E(v(T, X_T)) = E(g(X_T)).$$

Thus, from (1.16), (1.17), (1.19) and the Ito formula (3.3.6) we obtain

$$|E(g(X_T)) - E(g(Y(T)))|$$

$$= |E(v(T, Y(T))) - E(v(0, X_0))|$$

$$= |E(v(T, Y(T))) - E(v(0, Y_0))|$$

$$= \left| E\left( \int_0^T \left[ \frac{1}{2} \sum_{k,l=1}^{d} B^{k,l}(\tau_{n_s}, Y_{n_s}) \partial_{x^k} \partial_{x^l} v(s, Y(s)) \right. \right. \right.$$

$$+ \sum_{k=1}^{d} a^k(\tau_{n_s}, Y_{n_s}) \partial_{x^k} v(s, Y(s))$$

$$\left. \left. \left. + \partial_t v(s, Y(s)) - L^0 v(s, Y(s)) \right] ds \right) \right|$$

$$\leq \int_0^T \left| E\left( \sum_{k,l=1}^{d} \{ B^{k,l}(\tau_{n_s}, Y_{n_s}) - B^{k,l}(\tau_{n_s}, Y(s)) \} \partial_{x^k} \partial_{x^l} v(s, Y(s)) \right) \right| ds$$

$$+ \int_0^T \left| E\left( \sum_{k=1}^{d} [a^k(\tau_{n_s}, Y_{n_s}) - a^k(\tau_{n_s}, Y(s))] \partial_{x^k} v(s, Y(s)) \right) \right| ds$$

$$\leq \int_0^T E\left( \left\{ \sum_{k,l=1}^{d} \left| E\left( B^{k,l}(\tau_{n_s}, Y_{n_s}) \partial_{x^k} \partial_{x^l} v(\tau_{n_s}, Y_{n_s}) \right) \right. \right. \right.$$

$$-B^{k,l}\left(\tau_{n_s}, Y(s)\right) \partial_{x^k} \partial_{x^l} v\left(s, Y(s)\right) \Big| \mathcal{A}_{\tau_{n_s}}\right) \Big|$$

$$+\left|E\left(B^{k,l}\left(\tau_{n_s}, Y_{n_s}\right)\left[\partial_{x^k}\partial_{x^l} v\left(\tau_{n_s}, Y_{n_s}\right) - \partial_{x^k}\partial_{x^l} v\left(s, Y(s)\right)\right] \Big| \mathcal{A}_{\tau_{n_s}}\right)\right|\right\}\right) ds$$

$$+\int_0^T \sum_{k=1}^d E\left(\left\{\left|E\left(a^k\left(\tau_{n_s}, Y_{n_s}\right) \partial_{x^k} v\left(s, Y_{n_s}\right)\right.\right.\right.\right.$$

$$-a^k\left(\tau_{n_s}, Y(s)\right) \partial_{x^k} v\left(s, Y(s)\right) \Big| \mathcal{A}_{\tau_{n_s}}\right) \Big|$$

$$+\left|E\left(a^k\left(\tau_{n_s}, Y_{n_s}\right)\left[\partial_{x^k} v\left(\tau_{n_s}, Y_{n_s}\right) - \partial_{x^k} v\left(s, Y(s)\right)\right] \Big| \mathcal{A}_{\tau_{n_s}}\right)\right|\right\}\right) ds.$$

Now, the functions $B^{k,l}\partial_{x^k}\partial_{x^l} v$, $\partial_{x^k}\partial_{x^l} v$, $a^k\partial_{x^k} v$ and $\partial_{x^k} v$ belong to the space $\mathcal{H}_T^{(l)}$, so we can use (1.6) and Lemma 14.1.6 to obtain the desired inequality (1.8). □

## 14.2  The Order 2.0 Weak Taylor Scheme

As with strongly convergent schemes, we can derive more accurate weak Taylor schemes by including further multiple stochastic integrals from the stochastic Taylor expansion. However, now the objective is to obtain more information about the probability measure of the underlying Ito process, rather than about its sample paths. Most of the schemes in this section are due originally to either Milstein or Talay.

Here we shall consider the weak Taylor scheme, which will be of weak order $\beta = 2.0$, obtained by adding all of the double stochastic integrals from the Ito-Taylor expansion (5.5.3) to the Euler scheme. In the autonomous 1-dimensional case $d = m = 1$ we have the *order* 2.0 *weak Taylor scheme*

$$(2.1) \qquad Y_{n+1} = Y_n + a\,\Delta + b\,\Delta W + \frac{1}{2}\,bb'\left\{(\Delta W)^2 - \Delta\right\}$$

$$+a'b\,\Delta Z + \frac{1}{2}\left(aa' + \frac{1}{2}\,a''b^2\right)\Delta^2$$

$$+\left(ab' + \frac{1}{2}\,b''b^2\right)\left\{\Delta W\,\Delta - \Delta Z\right\}.$$

Once again $\Delta Z$ represents the double Ito integral $I_{(1,0)}$. We can use (10.4.3) to generate the pair of correlated Gaussian random variables $\Delta W$ and $\Delta Z$.

We have much more freedom with the weak convergence criterion than with the strong convergence criterion as to how we can generate the noise increments in a time discretization scheme. For instance, we can avoid the second random variable $\Delta Z$ in the preceding scheme (2.1) and only use a single random variable $\Delta \hat{W}$ with analogous moment properties to $\Delta W$ with $\Delta Z$ replaced by $\frac{1}{2}\Delta \hat{W}\Delta$. In the above autonomous 1-dimensional case we have the *simplified order 2.0 weak Taylor scheme*

$$(2.2) \qquad Y_{n+1} = Y_n + a\,\Delta + b\,\Delta \hat{W} + \frac{1}{2}\,bb'\left\{\left(\Delta \hat{W}\right)^2 - \Delta\right\}$$

$$+\frac{1}{2}\left(a'b + ab' + \frac{1}{2}b''b^2\right)\Delta \hat{W}\,\Delta$$

$$+\frac{1}{2}\left(aa' + \frac{1}{2}a''b^2\right)\Delta^2,$$

where $\Delta \hat{W}$ must be $\mathcal{A}_{\tau_{n+1}}$-measurable and satisfy the moment conditions

$$(2.3) \qquad \left|E\left(\Delta \hat{W}\right)\right| + \left|E\left(\left(\Delta \hat{W}\right)^3\right)\right| + \left|E\left(\left(\Delta \hat{W}\right)^5\right)\right|$$

$$+ \left|E\left(\left(\Delta \hat{W}\right)^2\right) - \Delta\right| + \left|E\left(\left(\Delta \hat{W}\right)^4\right) - 3\Delta^2\right| \le K\,\Delta^3$$

for some constant $K$. These conditions will follow from condition (5.12) in the general convergence result Theorem 14.5.2. Also see (5.12.9). We remark that the $\mathcal{A}_{\tau_{n+1}}$-measurability holds automatically if we choose the $\Delta \hat{W}$ at each step to be independent.

An $N(0;\Delta)$ Gaussian random variable certainly satisfies the moment conditions (2.3) and so does a three-point distributed random variable $\Delta \hat{W}$ with

$$(2.4) \qquad P\left(\Delta \hat{W} = \pm\sqrt{3\Delta}\right) = \frac{1}{6}, \qquad P\left(\Delta \hat{W} = 0\right) = \frac{2}{3}.$$

**Exercise 14.2.1** *Show that a three-point distributed random variable with (2.4) satisfies the moment conditions (2.3).*

**PC-Exercise 14.2.2** *Consider the Ito process $X$ from (9.4.2) satisfying the linear stochastic differential equation*

$$dX_t = a\,X_t\,dt + b\,X_t\,dW_t$$

*on the time interval $[0,T]$, where $T = 1$, with initial value $X_0 = 0.1$ and $a = 1.5$, $b = 0.01$. Generate $M = 20$ batches each of $N = 100$ trajectories of the order 2.0 weak Taylor approximation $Y^\delta$ using the order 2.0 weak Taylor scheme (2.1) with equidistant time steps $\delta = \Delta = 2^{-2}$. Determine the 90%-confidence interval for the mean error*

$$\mu = E\left(Y^\delta(T)\right) - E\left(X_T\right).$$

*Repeat the calculations for step sizes $2^{-3}$, $2^{-4}$ and $2^{-5}$, and plot $\log_2 |\mu|$ against $\log_2 \Delta$.*

Figure 14.2.1 indicates a higher than linear dependence of the mean error on the step size $\Delta$ for the order 2.0 weak Taylor scheme (2.1).

**PC-Exercise 14.2.3**    *Repeat PC-Exercise 14.2.2 for the simplified order 2.0 weak Taylor scheme (2.2) with noise increments generated by the three-point distributed random variable (2.4). Compare the results with those of PC-Exercise 14.2.2.*

In the multi-dimensional case $d = 1, 2, \ldots$ with $m = 1$, the $k$th component of the *order 2.0 weak Taylor scheme with scalar noise* is given by

$$(2.5) \qquad Y_{n+1}^k = Y_n^k + a^k \Delta + b^k \Delta W + \frac{1}{2} L^1 b^k \left\{ (\Delta W)^2 - \Delta \right\}$$

$$+ \frac{1}{2} L^0 a^k \Delta^2 + L^0 b^k \left\{ \Delta W \Delta - \Delta Z \right\}$$

$$+ L^1 a^k \Delta Z,$$

where the operators $L^0$ and $L^1$ were defined in (10.1.1) and (10.1.3); see also (5.3.1) and (5.3.2).

In the general multi-dimensional case $d$, $m = 1, 2, \ldots$ the $k$th component of the *order 2.0 weak Taylor scheme* takes the form

$$(2.6) \qquad Y_{n+1}^k = Y_n^k + a^k \Delta + \frac{1}{2} L^0 a^k \Delta^2$$

$$+ \sum_{j=1}^{m} \left\{ b^{k,j} \Delta W^j + L^0 b^{k,j} I_{(0,j)} + L^j a^k I_{(j,0)} \right\}$$

$$+ \sum_{j_1,j_2=1}^{m} L^{j_1} b^{k,j_2} I_{(j_1,j_2)}.$$

Here we have multiple Ito integrals involving different components of the Wiener process. As we saw in Chapter 10, these are generally not easy to generate. Consequently (2.6) is more of theoretical interest than of practical use. However, for weak convergence we can substitute simpler random variables for the multiple Ito integrals. In this way we obtain from (2.6) the following *simplified order 2.0 weak Taylor scheme* with $k$th component

$$(2.7) \qquad Y_{n+1}^k = Y_n^k + a^k \Delta + \frac{1}{2} L^0 a^k \Delta^2$$

$$+ \sum_{j=1}^{m} \left\{ b^{k,j} + \frac{1}{2} \Delta \left( L^0 b^{k,j} + L^j a^k \right) \right\} \Delta \hat{W}^j$$

$$+ \frac{1}{2} \sum_{j_1,j_2=1}^{m} L^{j_1} b^{k,j_2} \left( \Delta \hat{W}^{j_1} \Delta \hat{W}^{j_2} + V_{j_1,j_2} \right).$$

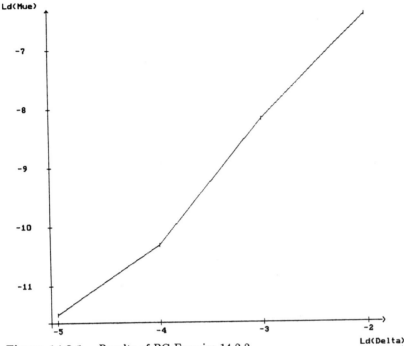

**Figure 14.2.1** Results of PC-Exercise 14.2.2.

Here the $\Delta \hat{W}^j$ for $j = 1, 2, \ldots, m$ are independent random variables satisfying the moment conditions (2.3) and the $V_{j_1,j_2}$ are independent two-point distributed random variables with

$$(2.8) \qquad\qquad P\left(V_{j_1,j_2} = \pm\Delta\right) = \frac{1}{2}$$

for $j_2 = 1, \ldots, j_1 - 1$,

$$(2.9) \qquad\qquad V_{j_1,j_1} = -\Delta$$

and

$$(2.10) \qquad\qquad V_{j_1,j_2} = -V_{j_2,j_1}$$

for $j_2 = j_1 + 1, \ldots, m$ and $j_1 = 1, \ldots, m$. We can obviously choose the $\Delta \hat{W}^j$ in (2.7) as in (2.4). See also (5.12.9) regarding the choice of random variables.

The next theorem provides conditions under which the above schemes converge weakly with order $\beta = 2.0$.

**Theorem 14.2.4** *Suppose in the autonomous case that $Y_0$ and $X_0$ have the same probability law with all moments finite. In addition, suppose that $a$ and $b$ are six times continuously differentiable with all of these derivatives uniformly bounded and that the products $b^{k,j}b^{l,j}$ for $j = 1, \ldots, m$ and $k, l = 1, \ldots, d$ have a linear growth bound. Then the schemes (2.1), (2.2), (2.5), (2.6) and (2.7) all converge weakly with order $\beta = 2.0$.*

The proof of a similar theorem was first given by Talay. The proof of Theorem 14.2.4 will also follow from Theorem 14.5.2, which we shall state and prove in Section 5. We shall see in Theorem 14.5.2 that the assumptions here can be considerably weakened. By taking the time $t$ to be the first component of the Ito process the theorem also covers the nonautonomous case.

## 14.3    The Order 3.0 Weak Taylor Scheme

We shall now consider weak Taylor schemes of order 3.0. These generally contain a large number of terms, because, as we shall see from Theorem 14.5.2, we need to include all of the multiple Ito integrals of multiplicity three from the Ito-Taylor expansion in order to ensure that the scheme converges weakly with order 3.0. We shall see later that such higher order weak schemes have a crucial theoretical importance and allow the construction of much simpler schemes with the same weak order of convergence.

In the general multi-dimensional case $d$, $m = 1$, 2, ... the $k$th component of the *order 3.0 weak Taylor scheme*, which is due to Platen, takes the form

$$
(3.1) \quad Y_{n+1}^k = Y_n^k + a^k \Delta + \sum_{j=1}^m b^{k,j} \Delta W^j + \sum_{j=0}^m L^j a^k I_{(j,0)}
$$

$$
+ \sum_{j_1=0}^m \sum_{j_2=1}^m L^{j_1} b^{k,j_2} I_{(j_1,j_2)} + \sum_{j_1,j_2=0}^m L^{j_1} L^{j_2} a^k I_{(j_1,j_2,0)}
$$

$$
+ \sum_{j_1,j_2=0}^m \sum_{j_3=1}^m L^{j_1} L^{j_2} b^{k,j_3} I_{(j_1,j_2,j_3)}.
$$

As we have already mentioned, this scheme is mainly of theoretical interest because the multiple integrals of higher multiplicity are difficult to generate and, in addition, the corresponding coefficient functions become rather complicated. To obtain more usable schemes we shall look at some special cases.

In the scalar case $d = 1$ with scalar noise $m = 1$ we propose the following *simplified order 3.0 weak Taylor scheme*

$$
(3.2) \quad Y_{n+1} = Y_n + a \Delta + b \Delta \tilde{W} + \frac{1}{2} L^1 b \left\{ \left( \Delta \tilde{W} \right)^2 - \Delta \right\}
$$

$$
+ L^1 a \, \Delta \tilde{Z} + \frac{1}{2} L^0 a \, \Delta^2 + L^0 b \left\{ \Delta \tilde{W} \, \Delta - \Delta \tilde{Z} \right\}
$$

$$
+ \frac{1}{6} \left( L^0 L^0 b + L^0 L^1 a + L^1 L^0 a \right) \Delta \tilde{W} \, \Delta^2
$$

$$
+ \frac{1}{6} \left( L^1 L^1 a + L^1 L^0 b + L^0 L^1 b \right) \left\{ \left( \Delta \tilde{W} \right)^2 - \Delta \right\} \Delta
$$

$$
+ \frac{1}{6} L^0 L^0 a \, \Delta^3 + \frac{1}{6} L^1 L^1 b \left\{ \left( \Delta \tilde{W} \right)^2 - 3\Delta \right\} \Delta \tilde{W},
$$

where $\Delta\tilde{W}$ and $\Delta\tilde{Z}$ are correlated Gaussian random variables with

$$\Delta\tilde{W} \sim N(0;\Delta), \qquad \Delta\tilde{Z} \sim N\left(0;\frac{1}{3}\Delta^3\right)$$

and covariance

$$E\left(\Delta\tilde{W}\,\Delta\tilde{Z}\right) = \frac{1}{2}\Delta^2.$$

Simpler random variables than these Gausssian ones could also be used for the noise increments, provided they satisfy the moment properties which will follow from condition (5.12) of Theorem 14.5.2. See (5.12.10) too.

**PC-Exercise 14.3.1**  *Repeat PC-Exercise 14.2.2 with $a = -5.0$ and $b = 0.1$ for the simplified order 3.0 weak Taylor scheme (3.2).*

Figure 14.3.1 suggests an approximately cubic dependence of the mean error on the step size for the scheme (3.2).

Another special case of practical importance involves additive noise (10.2.7). For the general multi-dimensional case $d$, $m = 1, 2, \ldots$ an order 3.0 *weak Taylor scheme for additive noise* has componentwise form

$$(3.3) \qquad Y_{n+1}^k = Y_n^k + a^k\,\Delta + \sum_{j=1}^{m} b^{k,j}\,\Delta\tilde{W}^j + \frac{1}{2}L^0 a^k\,\Delta^2 + \frac{1}{6}L^0 L^0 a^k\,\Delta^3$$

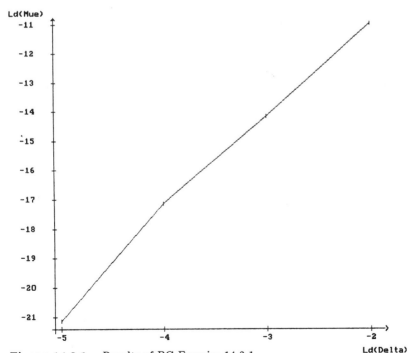

**Figure 14.3.1**  Results of PC-Exercise 14.3.1.

$$+ \sum_{j=1}^{m} \left[ L^j a^k \, \Delta \tilde{Z}^j + L^0 b^{k,j} \left\{ \Delta \tilde{W}^j \Delta - \Delta \tilde{Z}^j \right\} \right.$$

$$\left. + \frac{1}{6} \left( L^0 L^0 b^{k,j} + L^0 L^j a^k + L^j L^0 a^k \right) \Delta \tilde{W}^j \Delta^2 \right]$$

$$+ \frac{1}{6} \sum_{j_1, j_2 = 1}^{m} L^{j_1} L^{j_2} a^k \left\{ \Delta \tilde{W}^{j_1} \Delta \tilde{W}^{j_2} - I_{\{j_1 = j_2\}} \Delta \right\} \Delta.$$

Here the $\Delta \tilde{W}^j$ and $\Delta \tilde{Z}^j$ for each $j = 1, \ldots, m$ are correlated Gaussian random variables as in the previous scheme (3.2), but are independent for different $j$.

Conditions under which the above schemes converge weakly with order $\beta = 3.0$ will be provided by Theorem 14.5.2.

## 14.4   The Order 4.0 Weak Taylor Scheme

To construct the order 4.0 weak Taylor scheme we need also to include all of the fourth order multiple Ito integrals from the Ito-Taylor expansion (5.5.3). Since many terms are involved, we shall write $a^k$ as $b^{k,0}$ in order to simplify the resulting formulae. Then, in the general multi-dimensional case $d, m = 1$, $2, \ldots$ we can write the $k$th component of the *order 4.0 weak Taylor scheme* as

(4.1)
$$Y_{n+1}^k = Y_n^k + \sum_{l=1}^{4} \sum_{j_1, \ldots, j_l = 0}^{m} L^{j_1} \cdots L^{j_{l-1}} b^{k, j_l} I_{(j_1, \ldots, j_l)}.$$

In practice the order 4.0 weak Taylor scheme, which is also due to Platen, seems to be useful only in special cases, for example for scalar stochastic differential equations with additive noise, that is with $b(t, x) \equiv b(t)$ for all $(t, x)$.

For the 1-dimensional case $d = m = 1$ with additive noise, we obtain from (4.1) the following *simplified order 4.0 weak Taylor scheme for additive noise*:

(4.2)
$$Y_{n+1} = Y_n + a \Delta + b \Delta \tilde{W} + \frac{1}{2} L^0 a \Delta^2 + L^1 a \Delta \tilde{Z}$$

$$+ L^0 b \left\{ \Delta \tilde{W} \Delta - \Delta \tilde{z} \right\}$$

$$+ \frac{1}{3!} \left\{ L^0 L^0 b + L^0 L^1 a \right\} \Delta \tilde{W} \Delta^2$$

$$+ L^1 L^1 a \left\{ 2 \Delta \tilde{W} \Delta \tilde{Z} - \frac{5}{6} \left( \Delta \tilde{W} \right)^2 \Delta - \frac{1}{6} \Delta^2 \right\}$$

$$+ \frac{1}{3!} L^0 L^0 a \Delta^3 + \frac{1}{4!} L^0 L^0 L^0 a \Delta^4$$

$$+\frac{1}{4!}\left\{L^1L^0L^0a+L^0L^1L^0a+L^0L^0L^1a+L^0L^0L^0b\right\}\Delta\tilde{W}\,\Delta^3$$

$$+\frac{1}{4!}\left\{L^1L^1L^0a+L^0L^1L^1a+L^1L^0L^1a\right\}\left\{\left(\Delta\tilde{W}\right)^2-\Delta\right\}\Delta^2$$

$$+\frac{1}{4!}L^1L^1L^1a\,\Delta\tilde{W}\left\{\left(\Delta\tilde{W}\right)^2-3\Delta\right\}\Delta.$$

Here $\Delta\tilde{W}$ and $\Delta\tilde{Z}$ are correlated Gaussian random variables with $\Delta\tilde{W}\sim N(0;\Delta)$, $\Delta\tilde{Z}\sim N(0;\frac{1}{3}\Delta^3)$ and $E(\Delta\tilde{W}\,\Delta\tilde{Z})=\frac{1}{2}\Delta^2$, which we have already used several times. See (10.4.3) for the generation of such random variables.

The weak convergence with order $\beta=4.0$ of the above scheme will follow under suitable conditions from Theorem 14.5.2.

**PC-Exercise 14.4.1**   *Consider the Ito process $X$ satisfying the linear stochastic differential equation with additive noise*

$$dX_t = a\,X_t\,dt + b\,dW_t$$

*on the time interval $[0, T]$, where $T = 1.0$, with initial value $X_0 = 0.1$, $a = 2.0$ and $b = 0.01$. Simulate $M = 20$ batches each with $N = 100$ trajectories of the order 4.0 weak Taylor scheme (4.2) for equidistant step size $\delta = \Delta = 2^0$ and evaluate the 90%-confidence interval for the mean error $\mu$ at time $T$. Repeat the calculations for the step sizes $2^{-1}$ and $2^{-2}$ and plot $\log_2|\mu|$ versus $\log_2\Delta$.*

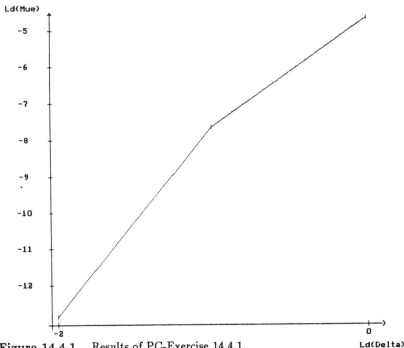

**Figure 14.4.1**    Results of PC-Exercise 14.4.1.

## 14.5    General Weak Taylor Approximations

In the preceding sections we have considered weak Taylor approximations of up to order $\beta = 4.0$. Higher order schemes can be obtained by adding in an appropriate way more terms from the Ito-Taylor expansion. We shall show here that a weak Taylor scheme of order $\beta = 1.0, 2.0, \ldots$ needs all of the multiple Ito integrals from the Ito-Taylor expansion of up to and including order $\beta$. This rule leads to weak Taylor schemes differing from the strong Taylor schemes (10.6.3) of the corresponding order. These will involve an index set $\Gamma_\beta$, to be defined in (5.3), rather than the set $\mathcal{A}_\gamma$ defined by (10.6.2). This underlines the clear difference between weak and strong convergences in stochastic numerical analysis, as well as in other branches of stochastics.

In what follows we shall use the notation from Chapter 5 where we formulated the Ito-Taylor expansion. In particular, for the function $f(t, x) \equiv x$ we have the Ito coefficient functions (5.3.3)

$$(5.1) \qquad\qquad f_\alpha(t, x) = L^{j_1} \cdots L^{j_{l-1}} b^{j_l}(t, x)$$

for all $(t, x) \in \Re \times \Re^d$ and all multi-indices $\alpha = (j_1, \ldots, j_l) \in \mathcal{M}$, where we have used $b^0 = a$ and the differential operators $L^j$ for $j = 0, 1, \ldots, m$ are given by (5.3.1) and (5.3.2). We note that the $f_\alpha$ do not depend on $t$ in the autonomous case. In addition, for a multi-index $\alpha = (j_1, \ldots, j_l) \in \mathcal{M}$ we define the multiple Ito integral

$$(5.2) \qquad I_{\alpha, \tau_n, \tau_{n+1}} = \int_{\tau_n}^{\tau_{n+1}} \int_{\tau_n}^{s_l} \cdots \int_{\tau_n}^{s_2} dW_{s_1}^{j_1} \ldots dW_{s_{l-1}}^{j_{l-1}} dW_{s_l}^{j_l},$$

where we set $dW_s^0 = ds$.

We shall see that a weak Taylor scheme of order $\beta = 1.0, 2.0, 3.0, \ldots$ is associated with the set of multi-indices

$$(5.3) \qquad\qquad \Gamma_\beta = \{\alpha \in \mathcal{M} : l(\alpha) \le \beta\},$$

where $l(\alpha)$ denotes the number of components of the multi-index $\alpha$. Obviously, $\Gamma_\beta$ contains all multi-indices with $\beta$ or fewer components. We recall from Exercise 5.4.3 that the sets $\Gamma_\beta$ are hierarchical sets for all $\beta = 1.0, 2.0, 3.0, \ldots$. We can thus use them to form Ito-Taylor expansions.

Let $(\tau)_\delta$ be a time discretization as defined in (9.5.3). In the general multidimensional case $d, m = 1, 2, \ldots$ for $\beta = 1.0, 2.0, 3.0, \ldots$ we define the *weak Taylor scheme of order* $\beta$ by the vector equation

$$(5.4) \qquad\qquad \begin{aligned} Y_{n+1} &= Y_n + \sum_{\alpha \in \Gamma_\beta \backslash \{v\}} f_\alpha(\tau_n, Y_n) \, I_{\alpha, \tau_n, \tau_{n+1}} \\[2mm] &= \sum_{\alpha \in \Gamma_\beta} f_\alpha(\tau_n, Y_n) \, I_{\alpha, \tau_n, \tau_{n+1}} \end{aligned}$$

with coefficient functions $f_\alpha$ corresponding to $f(t,x) \equiv x$. Here $Y_0 = X_0$ is assumed to be a nonrandom constant. It is easy to see that for $\beta = 1.0$, $2.0$, $3.0$ and $4.0$ the schemes given by (5.4) coincide with the weak Taylor schemes presented earlier in the chapter.

We recall that $C_P^l(\Re^d, \Re)$ denotes the space of $l$ times continuously differentiable functions $g : \Re^d \to \Re$ for which $g$ and all of its partial derivatives of orders up to and including $l$ have polynomial growth; see Theorem 4.8.6 and the paragraph preceding it. We state the following theorem in terms of an autonomous Ito diffusion process $X$ with drift $a = a(x)$ and diffusion coefficient $b = b(x)$. It goes back to Talay in the case $\beta = 2.0$ and was proved by Platen in the general case $\beta \geq 1.0$.

**Theorem 14.5.1**    *In the autonomous case let $Y^\delta$ be a weak Taylor approximation of order $\beta$ for some $\beta = 1.0$, $2.0$, $\ldots$ corresponding to a time discretization $(\tau)_\delta$. Suppose that $a$ and $b$ are Lipschitz continuous with components $a^k$, $b^{k,j} \in C_P^{2(\beta+1)}(\Re^d, \Re)$ for all $k = 1, \ldots, d$ and $j = 0, 1, \ldots, m$, and that the $f_\alpha$ corresponding to $f(t,x) \equiv x$ satisfy the linear growth bound*

$$(5.5) \qquad |f_\alpha(t, x)| \leq K \, (1 + |x|),$$

*where $K < \infty$, for all $x \in \Re^d$, $t \in [0, T]$ and $\alpha \in \Gamma_\beta$. Then for each $g \in C_P^{2(\beta+1)}(\Re^d, \Re)$ there exists a constant $C_g$, which does not depend on $\delta$, such that*

$$(5.6) \qquad \left| E \left( g \left( X_T \right) - g \left( Y_T^\delta \right) \right) \right| \leq C_g \, \delta^\beta,$$

*that is $Y^\delta$ converges weakly with order $\beta$ to $X$ at time $T$ as $\delta \to 0$.*

The assertion of Theorem 14.5.1 will follow from the more general result in Theorem 14.5.2 and from Exercise 14.5.3, where it is to be shown that the above weak Taylor approximations satisfy the conditional moment and regularity assumptions of this theorem.

We note that Theorem 14.5.1 also applies to the nonautonomous case if we take the time $t$ as the first component of the Ito process $X$. However, it imposes stronger than necessary conditions on the derivatives of the coefficients $a$ and $b$ with respect to the time variable $t$, since the same result can be proved directly using regularity conditions of lower order on the time derivatives than on state variable derivatives.

If we interpolate the weak Taylor approximations in the same way as we did for the strong Taylor approximations in (10.6.4), then we can show that such an interpolated weak Taylor approximation of order $\beta$ converges weakly on the whole interval $[0, T]$ with order $\beta$ as $\delta \to 0$. This means that there is a constant $C_g$ independent of $\delta$ such that (5.6) holds not only for $t = T$, but for all $t \in [0, T]$.

We emphasize the important fact that the order of weak convergence does *not* depend on the specific choice of the test function $g$. All we need is that $g$ is sufficiently smooth and of polynomial growth. The convergence of the first, second and higher order moments of the approximations $Y_T^\delta$ to those of $X_T$ is automatically covered with the obvious choices of $g$.

Now we shall state a more general result on weak convergence for a wide class of time discrete approximations, which includes time discrete Markov chains on a discrete state space as well as weak Taylor approximations. For simplicity we shall also do this for the autonomous case.

**Theorem 14.5.2**    *Let $Y^\delta$ be a time discrete approximation of an autonomous Ito process $X$ corresponding to a time discretization $(\tau)_\delta$, such that all moments of the initial value $X_0$ exist, that is*

$$(5.7) \qquad\qquad\qquad E\left(|X_0|^i\right) < \infty$$

*for $i = 1, 2, \ldots$, and such that $Y_0^\delta$ converges weakly with order $\beta$ to $X_0$ as $\delta \to 0$ for some fixed $\beta = 1.0, 2.0, \ldots$. Suppose that the drift and diffusion coefficients are Lipschitz continuous with components*

$$(5.8) \qquad\qquad\qquad a^k, \, b^{k,j} \in C_P^{2(\beta+1)}\left(\Re^d, \Re\right)$$

*for all $k = 1, \ldots, d$ and $j = 1, \ldots, m$ and satisfy the linear growth bound*

$$(5.9) \qquad\qquad\qquad |a(x)| + |b(x)| \le K\left(1 + |x|\right),$$

*where $K < \infty$, for all $x \in \Re^d$. In addition, suppose that for each $p = 1, 2, \ldots$ there exist constants $K < \infty$ and $r \in \{1, 2, \ldots\}$, which do not depend on $\delta$, such that for each $q \in \{1, \ldots, p\}$*

$$(5.10) \qquad\qquad E\left(\max_{0\le n\le n_T}\left|Y_n^\delta\right|^{2q}\,\Big|\mathcal{A}_0\right) \le K\left(1 + \left|Y_0^\delta\right|^{2r}\right)$$

*and*

$$(5.11) \quad E\left(\left|Y_{n+1}^\delta - Y_n^\delta\right|^{2q}\,\Big|\mathcal{A}_{\tau_n}\right) \le K\left(1 + \max_{0\le k\le n}\left|Y_k^\delta\right|^{2r}\right)(\tau_{n+1} - \tau_n)^q$$

*for $n = 0, 1, \ldots, n_T - 1$, and such that*

$$(5.12) \qquad \left|E\left(\prod_{h=1}^l\left(Y_{n+1}^{\delta,p_h} - Y_n^{\delta,p_h}\right) - \prod_{h=1}^l\left(\sum_{\alpha\in\Gamma_\beta\backslash\{v\}} f_\alpha^{p_h}\left(\tau_n, Y_n^\delta\right) I_{\alpha,\tau_n,\tau_{n+1}}\right)\Big|\mathcal{A}_{\tau_n}\right)\right|$$

$$\le K\left(1 + \max_{0\le k\le n_T}\left|Y_k^\delta\right|^{2r}\right)\delta^\beta\left(\tau_{n+1} - \tau_n\right)$$

*for all $n = 0, 1, \ldots, n_T - 1$ and $(p_1, \ldots, p_l) \in \{1, \ldots, d\}^l$, where $l = 1, \ldots, 2\beta + 1$ and $Y^{\delta,p_h}$ denotes the $p_h$th component of $Y^\delta$. Then the time discrete approximation $Y^\delta$ converges weakly with order $\beta$ as $\delta \to 0$ to the Ito process $X$ at time $T$.*

We shall prove Theorem 14.5.2 at the end of the section. The most important assumption of the theorem is condition (5.12) which provides a rule on how to construct a time discrete approximation $Y^\delta$ of weak order $\beta$. In it the conditional moments of the increments of $Y^\delta$ are compared with those of the corresponding weak Taylor approximation. We shall see in the proof that they behave approximately like those of the Ito process itself. The initial moment condition (5.7) together with the Lipschitz continuity of the coefficients $a$ and $b$ ensure the existence of moments of all orders of the Ito process $X$. From condition (5.8) we see that more smoothness of $a$ and $b$ is needed for higher order convergence. Condition (5.9) ensures the regularity of the corresponding weak Taylor approximation. It holds if, for instance, the coefficients $a$ and $b$ and all of their derivatives are bounded. Conditions (5.10) and (5.11) require the regularity of the time discretization itself. In particular, (5.10) is necessary to make the weak convergence criterion meaningful.

Summarizing the above, we can obtain a higher order weak approximation $Y^\delta$ if we construct it in such a way that the conditional moments of its increments approximate those of the Ito process $X$ sufficiently closely and if we have sufficient smoothness of the drift and diffusion coefficients, together with some regularity of the approximation itself.

**Exercise 14.5.3**    *Show that the assumptions of Theorem 14.5.1 imply those of Theorem 14.5.2.*

**Exercise 14.5.4**    *Check whether or not condition (5.12) is satisfied by the simplified weak Taylor schemes given in (2.2), (2.7), (3.2) and (3.3); see (5.12.8) – (5.12.10).*

In the special case of a nonrandom time discretization $(\tau)_\delta$ it can be easily seen from the proof of Theorem 14.5.2 to follow that for each $g \in C_P^{2(\beta+1)}(\Re^d, \Re)$ there exists a finite constant $C_g$ not depending on $\delta$ such that

$$(5.13) \qquad \max_{0 \le n \le n_T} \left| E\left(g\left(X_{\tau_n}\right)\right) - E\left(g\left(Y_n^\delta\right)\right) \right| \le C_g \, \delta^\beta.$$

In this sense, as we mentioned already in the first theorem, we have uniform weak convergence of order $\beta$ on the interval $[0, T]$.

We now list some useful results in preparation for the proof of Theorem 14.5.2. We shall make extensive use of the diffusion process

$$(5.14) \qquad X_t^{s,y} = y + \int_s^t a\left(X_u^{s,y}\right) du + \int_s^t b\left(X_u^{s,y}\right) dW_u,$$

for $s \le t \le T$, which starts at $y \in \Re^d$ at time $s \in [0, T]$ and has the same drift and diffusion coefficients as the Ito process $X = \{X_t; 0 \le t \le T\}$ which we are approximating. As in (4.8.11), for a given function $g \in C_P^{2(\beta+1)}(\Re^d, \Re)$ we define the functional

$$(5.15) \qquad u(s, y) = E\left(g\left(X_T^{s,y}\right)\right)$$

for $(s, y) \in [0, T] \times \Re^d$. Hence we have

$$(5.16) \qquad u(0, X_0) = E\left(g\left(X_T^{0,X_0}\right)\right) = E\left(g\left(X_T\right)\right).$$

From Theorem 4.8.6 we know that $u$ is the unique solution of the Kolmogorov backward equation

$$(5.17) \qquad L^0 u(s, y) = 0$$

for $(s, y) \in (0, T) \times \Re^d$ with the final time condition

$$u(T, y) = g(y)$$

for all $y \in \Re^d$. Moreover, we have

$$(5.18) \qquad u(s, \cdot) \in C_P^{2(\beta+1)}(\Re^d, \Re)$$

for each $y \in \Re^d$. By the Ito formula (3.3.6) it then follows that

$$(5.19) \qquad E\left(u\left(\tau_n, X_{\tau_n}^{\tau_{n-1}, y}\right) - u\left(\tau_{n-1}, y\right) \middle| \mathcal{A}_{\tau_{n-1}}\right) = 0$$

for all $n = 1, \ldots, n_T$ and $y \in \Re^d$. As with the estimate (4.5.13) it can be shown that for each $p = 1, 2, \ldots$ there exists a finite constant $K$ such that

$$(5.20) \qquad E\left(\left|X_{\tau_n}^{\tau_{n-1}, y} - y\right|^{2q} \middle| \mathcal{A}_{\tau_{n-1}}\right) \leq K\left(1 + |y|^{2q}\right)\left(\tau_n - \tau_{n-1}\right)^q$$

for all $q = 1, \ldots, p$, all $n = 1, \ldots, n_T$ and all $y \in \Re^d$.

We shall write

$$(5.21) \qquad \eta_{n,\beta}^y = \sum_{\alpha \in \Gamma_\beta} I_\alpha \left[f_\alpha\left(\tau_{n-1}, y\right)\right]_{\tau_{n-1}, \tau_n},$$

for each $y \in \Re^d$ and $n = 1, \ldots, n_T$, and, as in Section 11 of Chapter 5,

$$(5.22) \qquad F_{\vec{p}}(y) = \prod_{h=1}^{l} y^{p_h}$$

for all $y = (y^1, \ldots, y^d) \in \Re^d$ and $\vec{p} = (p_1, \ldots, p_l) \in P_l$, where

$$P_l = \{1, \ldots, d\}^l$$

for $l = 1, 2, \ldots$. Also, in what follows the $K$ and $r \in \{1, 2, \ldots\}$ will denote constants, which will generally take different values in the different places that they appear.

According to Lemma 5.11.7 there then exist finite constants $K$ and $r \in \{1, 2, \ldots\}$ such that

$$(5.23) \quad \left| E\left(F_{\vec{p}}\left(\eta_{n,\beta}^{Y_{n-1}^\delta} - Y_{n-1}^\delta\right) - F_{\vec{p}}\left(X_{\tau_n}^{\tau_{n-1}, Y_{n-1}^\delta} - Y_{n-1}^\delta\right) \middle| \mathcal{A}_{\tau_{n-1}}\right)\right|$$

$$\leq K\left(1 + \left|Y_{n-1}^\delta\right|^r\right) \delta^\beta\left(\tau_n - \tau_{n-1}\right)$$

for each $n = 1, \ldots, n_T$ and $\vec{p} \in P_l$ for $l = 1, \ldots, 2(\beta + 1)$. In addition, by Lemma 5.11.4 for each $p = 1, 2, \ldots$ there exist finite constants $K$ and $r \in \{1, 2, \ldots \}$ such that

$$(5.24) \quad E\left( \left| F_{\vec{p}}\left( \eta^y_{n,\beta} - y \right) \right|^{2q} \big| \mathcal{A}_{\tau_{n-1}} \right) \leq K \left( 1 + |y|^{2r} \right) \delta^\beta \left( \tau_n - \tau_{n-1} \right)^{ql}$$

for each $q = 1, \ldots, p$, $n = 1, \ldots, n_T$ and $\vec{p} \in P_l$ for $l = 1, \ldots, 2(\beta + 1)$.

**Proof of Theorem 14.5.2**  We need to estimate the difference

$$(5.25) \quad H := \left| E\left( g\left( Y^\delta_{n_T} \right) \right) - E\left( g\left( X_T \right) \right) \right|.$$

In view of (5.16) and the final time condition for the Kolmogorov backward equation (5.17) we have

$$H = \left| E\left( u\left( T, Y^\delta_{n_T} \right) - u\left( 0, X_0 \right) \right) \right|.$$

Since $Y^\delta_0$ converges weakly to $X_0$ with order $\beta$ as $\delta \to 0$ we obtain

$$H \leq \left| E\left( \sum_{n=1}^{n_T} \left( u\left( \tau_n, Y^\delta_n \right) - u\left( \tau_{n-1}, Y^\delta_{n-1} \right) \right) \right) \right| + K\,\delta^\beta.$$

We shall write henceforth $Y$ for $Y^\delta$ except where we wish to emphasize the role of $\delta$. From the relation (5.19) we have

$$(5.26) \quad H \leq \left| E\left( \sum_{n=1}^{n_T} \left[ \{ u\left( \tau_n, Y_n \right) - u\left( \tau_{n-1}, Y_{n-1} \right) \} \right. \right. \right.$$

$$\left. \left. \left. - \{ u\left( \tau_n, X^{\tau_{n-1},Y_{n-1}}_{\tau_n} \right) - u\left( \tau_{n-1}, Y_{n-1} \right) \} \right] \right) \right| + K\,\delta^\beta$$

$$\leq H_1 + H_2 + K\,\delta^\beta$$

where

$$(5.27) \quad H_1 = \left| E\left( \sum_{n=1}^{n_T} \left( \{ u\left( \tau_n, Y_n \right) - u\left( \tau_{n-1}, Y_{n-1} \right) \} \right. \right. \right.$$

$$\left. \left. \left. - \left\{ u\left( \tau_n, \eta^{Y_{n-1}}_{n,\beta} \right) - u\left( \tau_{n-1}, Y_{n-1} \right) \right\} \right) \right) \right|$$

and

$$(5.28) \quad H_2 = \left| E\left( \sum_{n=1}^{n_T} \left( \left\{ u\left( \tau_n, \eta^{Y_{n-1}}_{n,\beta} \right) - u\left( \tau_n, Y_{n-1} \right) \right\} \right. \right. \right.$$

$$\left. \left. \left. - \{ u\left( \tau_n, X^{\tau_{n-1},Y_{n-1}}_{\tau_n} \right) - u\left( \tau_n, Y_{n-1} \right) \} \right) \right) \right|.$$

Using the smoothness of $u$ expressed in (5.18) we can expand the increments of $u$ in $H_1$ by the deterministic Taylor formula, obtaining

$$(5.29) \quad H_1 \;=\; \left| E\left( \sum_{n=1}^{n_T} \left\{ \left[ \sum_{l=1}^{2\beta+1} \frac{1}{l!} \sum_{\vec{p}\in P_l} \partial_y^{\vec{p}} u\left(\tau_n, Y_{n-1}\right) F_{\vec{p}}\left(Y_n - Y_{n-1}\right) \right.\right.\right.$$

$$\left.\left.\left. + R_{n,\beta}\left(Y_n\right) \right]\right.\right.$$

$$\left.\left.- \left[ \sum_{l=1}^{2\beta+1} \frac{1}{l!} \sum_{\vec{p}\in P_l} \partial_y^{\vec{p}} u\left(\tau_n, Y_{n-1}\right) F_{\vec{p}}\left(\eta_{n,\beta}^{Y_n} - Y_{n-1}\right) + R_{n,\beta}\left(\eta_{n,\beta}^{Y_{n-1}}\right) \right] \right\}\right) \right|.$$

Here the remainder terms have the form

$$R_{n,\beta}(Z) \;=\; \frac{1}{(2\beta+2)!} \sum_{\vec{p}\in P_{2(\beta+1)}} \partial_y^{\vec{p}} u\left(\tau_n, Y_{n-1} + \theta_{\vec{p},n}(Z)\left(Z - Y_{n-1}\right)\right)$$

$$(5.30) \qquad\qquad\qquad\qquad \times F_{\vec{p}}\left(Z - Y_{n-1}\right)$$

for $Z = Y_n$ and $\eta_{n,\beta}^{Y_{n-1}}$, respectively, where $\theta_{\vec{p},n}(Z)$ is a $d \times d$ diagonal matrix with

$$(5.31) \qquad\qquad\qquad \theta_{\vec{p},n}^{k,k}(Z) \in (0,1)$$

for $k = 1, \ldots, d$.

From (5.29) we have

$$H_1 \;\leq\; E\left( \sum_{n=1}^{n_T} \left\{ \sum_{l=1}^{2\beta+1} \frac{1}{l!} \sum_{\vec{p}\in P_l} \left| \partial_y^{\vec{p}} u\left(\tau_n, Y_{n-1}\right) \right| \right.\right.$$

$$(5.32) \qquad \times \left| E\left( F_{\vec{p}}\left(Y_n - Y_{n-1}\right) - F_{\vec{p}}\left(\eta_{n,\beta}^{Y_{n-1}} - Y_{n-1}\right) \middle| \mathcal{A}_{\tau_{n-1}} \right) \right|$$

$$\left.\left. + E\left( \left|R_{n,\beta}\left(Y_n\right)\right| \middle| \mathcal{A}_{\tau_{n-1}} \right) + E\left( \left|R_{n,\beta}\left(\eta_{n,\beta}^{Y_{n-1}}\right)\right| \middle| \mathcal{A}_{\tau_{n-1}} \right) \right\}\right)$$

and from (5.30), (5.18), (5.31) and (5.11)

$$(5.33) \quad E\left( \left|R_{n,\beta}\left(Y_n\right)\right| \middle| \mathcal{A}_{\tau_{n-1}} \right)$$

$$\leq\; K \sum_{\vec{p}\in P_{2(\beta+1)}} \left( E\left( \left| \partial_y^{\vec{p}} u\left(\tau_n, Y_{n-1} + \theta_{\vec{p},n}(Y_n)\left(Y_n - Y_{n-1}\right)\right) \right|^2 \middle| \mathcal{A}_{\tau_{n-1}} \right) \right)^{1/2}$$

$$\times \left( E\left( \left|F_{\vec{p}}\left(Y_n - Y_{n-1}\right)\right|^2 \middle| \mathcal{A}_{\tau_{n-1}} \right) \right)^{1/2}$$

$$\leq\; K \left( E\left( 1 + \left|Y_n\right|^{2r} + \left|Y_n - Y_{n-1}\right|^{2r} \middle| \mathcal{A}_{\tau_{n-1}} \right) \right)^{1/2}$$

$$\times \left( E\left( |Y_n - Y_{n-1}|^{4(\beta+1)} \big| \mathcal{A}_{\tau_{n-1}} \right) \right)^{1/2}$$

$$\leq K \left( 1 + \max_{0 \leq k \leq n-1} |Y_k|^{2r} \right) \delta^\beta (\tau_n - \tau_{n-1})$$

for $n = 1, \ldots, n_T$. In a similar way from (5.30), (5.21), (5.8) and the moment properties of multiple Ito integrals (see Lemma 5.7.5), we find that

$$(5.34) \quad E\left( \left| R_{n,\beta}\left( \eta_{n,\beta}^{Y_{n-1}} \right) \right| \big| \mathcal{A}_{\tau_{n-1}} \right) \leq K \left( 1 + |Y_{n-1}|^{2r} \right) \delta^\beta (\tau_n - \tau_{n-1})$$

for $n = 1, \ldots, n_T$. Hence, applying (5.18), (5.23), (5.21), (5.12), (5.33), (5.34), (5.11) and (5.10) to (5.32), we obtain the estimate

$$
\begin{aligned}
(5.35) \quad H_1 &\leq E\left( \sum_{n=1}^{n_T} K \left( 1 + \max_{0 \leq k \leq n-1} |Y_k|^{2r} \right) \right) \delta^\beta (\tau_n - \tau_{n-1}) \\
&\leq K \delta^\beta \left( 1 + E\left( \max_{0 \leq k \leq n_T} |Y_k|^{2r} \right) \right) \\
&\leq K \delta^\beta \left( 1 + E\left( |Y_0|^{2r} \right) \right) \\
&\leq K \delta^\beta.
\end{aligned}
$$

Here the constant $K$ differs from line to line, as indicated earlier.

We can derive an estimate for $H_2$ in an analogous way, but now using the inequality (5.23) instead of (5.12). To begin we have

$$
\begin{aligned}
(5.36) \quad H_2 &\leq E\Big( \sum_{n=1}^{n_T} \Big\{ \sum_{l=1}^{2\beta+1} \frac{1}{l!} \sum_{\vec{p} \in P_l} |\partial_y^{\vec{p}} u(\tau_n, Y_{n-1})| \\
&\quad \times \left| E\left( F_{\vec{p}}\left( \eta_{n,\beta}^{Y_{n-1}} - Y_{n-1} \right) - F_{\vec{p}}\left( X_{\tau_n}^{\tau_{n-1}, Y_{n-1}} - Y_{n-1} \right) \big| \mathcal{A}_{\tau_{n-1}} \right) \right| \\
&\quad + E\left( \left| R_{n,\beta}\left( \eta_{n,\beta}^{Y_{n-1}} \right) \right| \big| \mathcal{A}_{\tau_{n-1}} \right) + E\left( \left| R_{n,\beta}\left( X_{\tau_n}^{\tau_{n-1}, Y_{n-1}} \right) \right| \big| \mathcal{A}_{\tau_{n-1}} \right) \Big\} \Big) \\
&\leq E\Big( \sum_{n=1}^{n_T} \Big\{ K \left( 1 + |Y_{n-1}|^{2r} \right) \delta^\beta (\tau_n - \tau_{n-1}) \\
&\quad + E\left( \left| R_{n,\beta}\left( X_{\tau_n}^{\tau_{n-1}, Y_{n-1}} \right) \right| \big| \mathcal{A}_{\tau_{n-1}} \right) \Big\} \Big)
\end{aligned}
$$

for $n = 1, \ldots, n_T$. We can estimate the remainder in (5.36) as in (5.34), but now using (5.20). Hence we obtain

$$H_2 \leq E\left( \sum_{n=1}^{n_T} K \left( 1 + |Y_k|^{2r} \right) \right) \delta^\beta (\tau_n - \tau_{n-1})$$

$$\leq \ K \, \delta^\beta \left( 1 + E \left( \max_{0 \leq k \leq n_T} |Y_k|^{2r} \right) \right).$$

Then, from (5.10) and the fact that $Y_0^\delta$ has finite moments we have

(5.37)                                   $H_2 \leq K \, \delta^\beta.$

Combining (5.25), (5.26), (5.35) and (5.37) we have the desired result that

$$H = \left| E \left( g \left( Y_{n_T}^\delta \right) \right) - E \left( g \left( X_T \right) \right) \right| \leq C_g \, \delta^\beta,$$

where the constant $C_g$ depends on the particular function $g$ that we have been using. This completes the proof of Theorem 14.5.2. $\square$


## 14.6   Leading Error Coefficients

In Theorem 14.5.2 we described conditions under which time discrete approximations converge weakly with a given order. For theoretical, and sometimes practical, purposes it is helpful to know at least the structure of the leading coefficients of the error expansion with respect to powers of the step size $\delta$. Using the same notation as in the previous section we shall formulate a theorem here which provides a characterization of these leading error coefficients. This theorem was first proved by Talay and Tubaro for the cases $\beta = 1.0$ and $2.0$.

We shall fix $\beta \in \{1.0, 2.0, \ldots\}$ and choose an order $\beta$ weak approximation $Y^\delta$ of an autonomous Ito processs $X = \{X_t; \ t \in [0, T]\}$, with $Y_0^\delta = X_0$, corresponding to an equidistant time discretization $(\tau)_\delta$ of $[0, T]$ with step size $\delta$. In addition, we shall suppose that the following are satisfied: the conditions (5.7) for the moments of $X_0$; the linear growth bound (5.9) for the Lipschitz continuous drift and diffusion coefficients; conditions (5.10) and (5.11) concerning the regularity of $Y^\delta$; and

(6.1)                          $g, a^k, b^{k,j} \in \mathcal{C}_P^\infty \left( \Re^d, \Re \right)$

for each $k = 1, \ldots, d$ and $j = 1, \ldots, m$. Finally, we shall assume for all $n = 0$, $1, \ldots, n_T - 1$, $l = 1, \ldots, 4\beta + 1$ and $\vec{p} = (p_1, \ldots, p_l) \in P_l := \{1, \ldots, d\}^l$ that the partial derivative of $u$ (see (14.5.15))

$$\frac{\partial^l}{\partial y^{p_1} \cdots \partial y^{p_l}} \, u(\cdot, y) = \partial_y^{\vec{p}} u(\cdot, y) \in \mathcal{C}_P^\infty ([0, T], \Re)$$

for all $y \in \Re^d$ and that

$$(6.2) \left| E \left( \prod_{h=1}^l \left( Y_{n+1}^{\delta, p_h} - Y_n^{\delta, p_h} \right) \right) - \prod_{h=1}^l \left( \sum_{\alpha \in \Gamma_{2\beta+1} \backslash \{v\}} f_\alpha^{p_h} \left( \tau_n, Y_n^\delta \right) I_{\alpha, \tau_n, \tau_{n+1}} \right) \right.$$

$$\left| \mathcal{A}_{T_n} \right) - \sum_{\gamma=\beta}^{2\beta-1} c_{\gamma,\vec{p}} \left( Y_n^\delta \right) \delta^{\gamma+1} \right|$$

$$\leq K \left( 1 + \max_{0 \leq k \leq n} \left| Y_k^\delta \right|^{2r} \right) \delta^{2\beta+1},$$

for certain functions $c_{\gamma,\vec{p}} \in C_P^\infty \left( \Re^d, \Re \right)$, $\gamma = \beta, \ldots, 2\beta-1$, where $Y^{\delta, p_h}$ denotes the $p_h$th component of $Y^\delta$.

**Theorem 14.6.1** *Under the above assumptions, for an order $\beta$ weak approximation $Y^\delta$ the error expansion of the functional satisfies*

$$(6.3) \qquad \left| E \left( g \left( Y^\delta(T) \right) \right) - E \left( g \left( X_T \right) \right) - \sum_{\gamma=\beta}^{2\beta-1} \psi_{g,\gamma}(T) \delta^\gamma \right| \leq C_g \delta^{2\beta},$$

*where the leading error coefficients $\psi_{g,\gamma}(T)$ for $\gamma = \beta, \ldots, 2\beta - 1$ are well defined real numbers which, like the constant $C_g$, do not depend on $\delta$.*

The main assertion of the theorem is the fact that there exist leading error coefficients at least up to the power $2\beta - 1$ of $\delta$ which do not depend on $\delta$. This will be used in Section 3 of Chapter 15 to construct extrapolation methods.

In the 1-dimensional case $d = m = 1$ where $Y^\delta$ is the Euler approximation (14.1.1) the leading error coefficient is

$$(6.4) \qquad \psi_{g,1}(T) = \sum_{l=1}^{5} \frac{1}{l!} \sum_{\vec{p} \in P_l} \int_0^T c_{1,\vec{p}}(X_s) \, \partial_y^{\vec{p}} u(s, X_s) \, ds$$

where

$$(6.5) \qquad c_{1,(1)} = -\frac{1}{2} \left\{ aa' + \frac{1}{2} b^2 a'' \right\}$$

$$c_{1,(1,1)} = -\frac{1}{2} \left\{ 2abb' + b^2 \left( 2a' + (b')^2 + bb'' \right) \right\}$$

$$c_{1,(1,1,1)} = 3b^2 bb'$$

$$c_{1,(1,1,1,1)} = 0, \qquad c_{1,(1,1,1,1,1)} = 0.$$

Fortunately, the explicit form of the leading error coefficients is not important in many applications.

For the proof of Theorem 14.6.1 we shall need the following lemma.

**Lemma 14.6.2** *Suppose that the above asumptions hold and let $\beta \in \{1, 2, \ldots\}$, $\delta \in (0, 1)$ and $n = 1, \ldots, n_T$ be given. Then, for any function $w : [0, T] \times \Re^d \to \Re$ with $w(t, \cdot) \in C_P^\infty \left( \Re^d, \Re \right)$ for all $t \in [0, T]$ and $w(\cdot, x) \in C_P^\infty \left( [0, T], \Re \right)$ for all $x \in \Re^d$, there exist a finite constant $K$ and functions $\phi_0, \ldots, \phi_{2\beta-1} : [0, T] \times \Re^d \to \Re$ depending on $w$ such that*

(6.6) $\left| \delta\, E\left(w\left(\tau_n, X_{\tau_{n-1}}\right)\right) - \sum_{r=0}^{2\beta-1} \delta^r \int_{\tau_{n-1}}^{\tau_n} E\left(\phi_r\left(s, X_s\right)\right)\, ds \right| \le K\, \delta^{2\beta+1}.$

**Proof**   Applying the Ito-Taylor expansion (5.5.3) with hierarchical set $\Gamma_{2\beta-1}$ and using the properties of multiple Ito integrals, we obtain in a straight forward manner an expansion for $E(\delta w(\tau_n, X_{\tau_{n-1}}))$ and the estimate

$$\left| E\left( \int_{\tau_{n-1}}^{\tau_n} w\left(s, X_s\right)\, ds \right) - E\left(\delta w\left(\tau_{n-1}, X_{\tau_{n-1}}\right)\right) \right.$$

$$\left. - E\left( \sum_{r_1=1}^{2\beta-1} \delta\left(L^0\right)^{r_1} w\left(\tau_{n-1}, X_{\tau_{n-1}}\right) \frac{\delta^{r_1}}{(r_1+1)!} \right) \right| \le K\, \delta^{2\beta+1}.$$

Similarly expanding $E\left(\delta(L^0)^{r_1} w(\tau_{n-1}, X_{\tau_{n-1}})\right)$ we obtain

$$\left| E\left( \int_{\tau_{n-1}}^{\tau_n} w\left(s, X_s\right)\, ds \right) - E\left(\delta w\left(\tau_{n-1}, X_{\tau_{n-1}}\right)\right) \right.$$

$$- E\left( \sum_{r_1=1}^{2\beta-1} \left[ \int_{\tau_{n-1}}^{\tau_n} \left(L^0\right)^{r_1} w\left(s, X_s\right)\, ds\, \frac{\delta^{r_1}}{(r_1+1)!} \right.\right.$$

$$\left.\left.\left. - \sum_{r_2=1}^{2\beta-2} \delta\left(L^0\right)^{r_2}\left(L^0\right)^{r_1} w\left(\tau_{n-1}, X_{\tau_{n-1}}\right) \frac{\delta^{r_1+r_2}}{(r_1+1)!\,(r_2+1)!} \right] \right) \right|$$

$$\le K\, \delta^{2\beta+1}.$$

Continuing in the same way we get

(6.7)   $\left| E\left(\delta w\left(\tau_{n-1}, X_{\tau_{n-1}}\right)\right) - \sum_{r=0}^{2\beta-1} \delta^r \zeta_r \int_{\tau_{n-1}}^{\tau_n} E\left(\left(L^0\right)^r w\left(s, X_s\right)\right)\, ds \right|$

$$\le K\, \delta^{2\beta+1},$$

where $\zeta_0, \ldots, \zeta_{2\beta-1}$ are fixed real numbers.

On the other hand from the deterministic Taylor expansion we have

(6.8)   $\left| \delta E\left(w\left(\tau_n, X_{\tau_{n-1}}\right)\right) - \delta E\left(w\left(\tau_{n-1}, X_{\tau_{n-1}}\right)\right) \right.$

$$\left. - \delta \sum_{l=1}^{2\beta-1} E\left( \left(\frac{\partial}{\partial t}\right)^l w\left(\tau_{n-1}, X_{\tau_{n-1}}\right) \right) \frac{\delta^l}{l!} \right| \le K\, \delta^{2\beta+1}.$$

Then, applying (6.7) but analogously to $\delta\, E((\frac{\partial}{\partial t})^l w(\tau_{n-1}, X_{\tau_{n-1}}))$ in the above formula, we obtain the desired estimate (6.6). $\square$

**Proof of Theorem 14.6.1** The proof is similar to that of Theorem 14.5.2. We shall use the notation introduced there and analogous arguments without specifying further details.

As in (5.25)–(5.28) we can write

$$(6.9) \qquad \bar{H} = E\left(g\left(Y_{n_T}^\delta\right)\right) - E\left(g\left(X_T\right)\right) = \bar{H}_1 + \bar{H}_2$$

with

$$(6.10) \quad \bar{H}_1 = E\left(\sum_{n=1}^{n_T}\left[\left\{u\left(\tau_n, Y_n^\delta\right) - u\left(\tau_n, Y_{n-1}^\delta\right)\right\}\right.\right.$$
$$\left.\left. - \left\{u\left(\tau_n, \eta_{n,2\beta}^{Y_{n-1}^\delta}\right) - u\left(\tau_n, Y_{n-1}^\delta\right)\right\}\right]\right)$$

and

$$(6.11) \quad \bar{H}_2 = E\left(\sum_{n=1}^{n_T}\left[\left\{u\left(\tau_n, \eta_{n,2\beta}^{Y_{n-1}^\delta}\right) - u\left(\tau_n, Y_{n-1}^\delta\right)\right\}\right.\right.$$
$$\left.\left. - \left\{u\left(\tau_n, X_{\tau_n}^{\tau_{n-1}, Y_{n-1}^\delta}\right) - u\left(\tau_n, Y_{n-1}^\delta\right)\right\}\right]\right).$$

Using (6.10) and the smoothness of $u$, we can expand the increments of $u$ in $\bar{H}_1$ by the usual deterministic Taylor formula to obtain

$$(6.12) \quad \bar{H}_1 = E\left(\sum_{n=1}^{n_T}\left\{\left[\sum_{l=1}^{4\beta+1}\frac{1}{l!}\sum_{\vec{p}\in P_l}\left(\partial_y^{\vec{p}}u\left(\tau_n, Y_{n-1}^\delta\right)\right)\right.\right.\right.$$
$$\left.\times F_{\vec{p}}\left(Y_n^\delta - Y_{n-1}^\delta\right) + R_{n,2\beta}\left(Y_n^\delta\right)\right]$$
$$- \left[\sum_{l=1}^{4\beta+1}\frac{1}{l!}\sum_{\vec{p}\in P_l}\left(\partial_y^{\vec{p}}u\left(\tau_n, Y_{n-1}^\delta\right)\right)\right.$$
$$\left.\left.\times F_{\vec{p}}\left(\eta_{n,2\beta}^{Y_{n-1}^\delta} - Y_{n-1}^\delta\right) + R_{n,2\beta}\left(\eta_{n,2\beta}^{Y_{n-1}^\delta}\right)\right]\right\}\right)$$

$$= E\left(\sum_{n=1}^{n_T}\left\{\sum_{l=1}^{4\beta+1}\frac{1}{l!}\sum_{\vec{p}\in P_l}\left(\partial_y^{\vec{p}}u\left(\tau_n, Y_{n-1}^\delta\right)\right)\right.\right.$$
$$\times E\left(F_{\vec{p}}\left(Y_n^\delta - Y_{n-1}^\delta\right) - F_{\vec{p}}\left(\eta_{n,2\beta}^{Y_{n-1}^\delta} - Y_{n-1}^\delta\right)\Big|\mathcal{A}_{\tau_{n-1}}\right)$$
$$\left.\left. + R_{n,2\beta}\left(Y_n^\delta\right) - R_{n,2\beta}\left(\eta_{n,2\beta}^{Y_{n-1}^\delta}\right)\right\}\right).$$

Applying (6.2) we have from (6.12)

$$\left| \bar{H}_1 - \sum_{l=1}^{4\beta+1} \frac{1}{l!} \sum_{\vec{p} \in P_l} \sum_{\gamma=\beta}^{2\beta-1} \sum_{n=1}^{n_T} E\left(\left(\partial_y^{\vec{p}} u\left(\tau_n, Y_{n-1}^{\delta}\right)\right) c_{\gamma,\vec{p}}\left(Y_{n-1}^{\delta}\right)\right) \delta^{\gamma+1} \right| \leq K \, \delta^{2\beta}.$$

Now, from (5.13) we can conclude that

$$(6.13) \quad \left| \bar{H}_1 - \sum_{l=1}^{4\beta+1} \frac{1}{l!} \sum_{\vec{p} \in P_l} \sum_{\gamma=\beta}^{2\beta-1} \delta^{\gamma} \sum_{n=1}^{n_T} \delta \, E\left(\left(\partial_y^{\vec{p}} u\left(\tau_n, X_{\tau_{n-1}}\right)\right) c_{\gamma,\vec{p}}\left(X_{\tau_{n-1}}\right)\right) \right|$$

$$\leq K \, \delta^{2\beta}.$$

Hence, with the help of (6.6), it follows from (6.13) that

$$(6.14) \quad \left| \bar{H}_1 - \sum_{l=1}^{4\beta+1} \frac{1}{l!} \sum_{\vec{p} \in P_l} \sum_{\gamma=\beta}^{2\beta-1} \delta^{\gamma} \int_0^T E\left(\left(\partial_y^{\vec{p}} u\left(s, X_s\right)\right) c_{\gamma,\vec{p}}\left(X_s\right)\right) ds \right|$$

$$= \left| \bar{H}_1 - \sum_{\gamma=\beta}^{2\beta-1} \left\{ \sum_{l=1}^{4\beta+1} \frac{1}{l!} \sum_{\vec{p} \in P_l} \int_0^T E\left(\left(\partial_y^{\vec{p}} u\left(s, X_s\right)\right) c_{\gamma,\vec{p}}\left(X_s\right)\right) ds \right\} \delta^{\gamma} \right|$$

$$\leq K \, \delta^{2\beta}.$$

Similarly as (5.37), it follows from (6.11) that

$$(6.15) \qquad\qquad\qquad \left| \bar{H}_2 \right| \leq K \, \delta^{2\beta}.$$

Thus, combining (6.9), (6.14) and (6.15), we obtain the desired estimate

$$(6.16) \qquad\qquad \left| \bar{H} - \sum_{\gamma=\beta}^{2\beta-1} \psi_{g,\gamma}(T) \, \delta^{\gamma} \right| \leq K \, \delta^{2\beta},$$

where for $\gamma = \beta, \ldots, 2\beta - 1$ the $\psi_{g,\gamma}(T)$ are real numbers which do not depend on $\delta$. $\quad\square$

# Chapter 15

# Explicit and Implicit Weak Approximations

We saw in the previous chapter that higher order weak Taylor schemes require the determination and evaluation of derivatives of various orders of the drift and diffusion coefficients. As with strong schemes, we can also derive Runge-Kutta like weak approximations which avoid the use of such derivatives. Here too, these will not be simply heuristic generalizations of deterministic Runge-Kutta schemes. We shall also introduce extrapolation methods, implicit schemes and predictor-corrector methods in this chapter.

## 15.1 Explicit Order 2.0 Weak Schemes

In the autonomous case $d = 1, 2, \ldots$ with scalar noise $m = 1$ Platen proposed the following *explicit order* 2.0 *weak scheme:*

$$
\begin{aligned}
(1.1) \quad Y_{n+1} &= Y_n + \frac{1}{2} \left( a\left(\bar{\Upsilon}\right) + a \right) \Delta \\
&\quad + \frac{1}{4} \left( b\left(\bar{\Upsilon}^+\right) + b\left(\bar{\Upsilon}^-\right) + 2b \right) \Delta \hat{W} \\
&\quad + \frac{1}{4} \left( b\left(\bar{\Upsilon}^+\right) - b\left(\bar{\Upsilon}^-\right) \right) \left\{ \left(\Delta \hat{W}\right)^2 - \Delta \right\} \Delta^{-1/2}
\end{aligned}
$$

with supporting values
$$
\bar{\Upsilon} = Y_n + a\,\Delta + b\,\Delta \hat{W},
$$
and
$$
\bar{\Upsilon}^{\pm} = Y_n + a\,\Delta \pm b\sqrt{\Delta}.
$$

Here $\Delta \hat{W}$ must be $\mathcal{A}_{\tau_{n+1}}$-measurable and satisfy the moment conditions (14.2.3). For instance, $\Delta \hat{W}$ could be Gaussian or it could be three-point distributed with

$$
(1.2) \qquad P\left(\Delta \hat{W} = \pm\sqrt{3\Delta}\right) = \frac{1}{6}, \qquad P\left(\Delta \hat{W} = 0\right) = \frac{2}{3}.
$$

Obviously, for each iteration we need to evaluate the drift $a$ at two points and the diffusion coefficient $b$ at three points. In addition, we need to generate one random variable. Comparing (1.1) with the corresponding simplified weak Taylor scheme (14.2.2), we see that (1.1) avoids the derivatives in (14.2.2) by using additional supporting values.

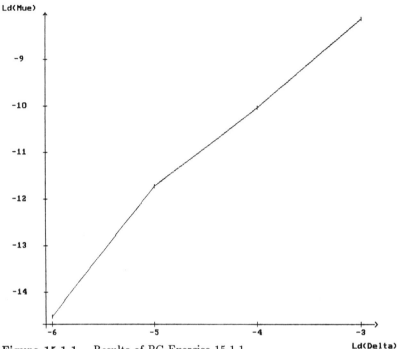

**Figure 15.1.1**    Results of PC-Exercise 15.1.1.

**PC-Exercise 15.1.1**    *Repeat PC-Exercise 14.2.2 with $a = 1.5$ and $b = 0.01$ using the explicit order 2.0 weak scheme (1.1) with $\delta = \Delta = 2^{-3}, \ldots, 2^{-6}$ and compare the results with those of PC-Exercise 14.2.2.*

There is a multi-dimensional counterpart of (1.1). For the autonomous case with $d$, $m = 1, 2, \ldots$ the *multi-dimensional explicit order 2.0 weak scheme* has the vector form

$$(1.3) \qquad Y_{n+1} = Y_n + \frac{1}{2} \left( a \left( \bar{\Upsilon} \right) + a \right) \Delta$$

$$+ \frac{1}{4} \sum_{j=1}^{m} \left[ \left( b^j \left( \bar{R}_+^j \right) + b^j \left( \bar{R}_-^j \right) + 2 b^j \right) \Delta \hat{W}^j \right.$$

$$+ \sum_{\substack{r=1 \\ r \neq j}}^{m} \left( b^j \left( \bar{U}_+^r \right) + b^j \left( \bar{U}_-^r \right) - 2 b^j \right) \Delta \hat{W}^j \, \Delta^{-1/2} \Bigg]$$

$$+ \frac{1}{4} \sum_{j=1}^{m} \left[ \left( b^j \left( \bar{R}_+^j \right) - b^j \left( \bar{R}_-^j \right) \right) \left\{ \left( \Delta \hat{W}^j \right)^2 - \Delta \right\} \right.$$

$$+ \sum_{\substack{r=1 \\ r \neq j}}^{m} \left( b^j \left( \bar{U}_+^r \right) - b^j \left( \bar{U}_-^r \right) \right) \left\{ \Delta \hat{W}^j \Delta \hat{W}^r + V_{r,j} \right\} \Bigg] \Delta^{-1/2}$$

with supporting values

$$\bar{\Upsilon} = Y_n + a\,\Delta + \sum_{j=1}^{m} b^j\,\Delta\hat{W}^j, \qquad \bar{R}_{\pm}^j = Y_n + a\,\Delta \pm b^j\,\sqrt{\Delta}$$

and

$$\bar{U}_{\pm}^j = Y_n \pm b^j\,\sqrt{\Delta},$$

where the random variables $\Delta\hat{W}^j$ and $V_{r,j}$ are defined in (14.2.8)–(14.2.10). In Section 5 of Chapter 14 we stated conditions on $a$ and $b$ under which the scheme (1.3) converges with weak order $\beta = 2.0$; see Theorem 14.5.2.

For *additive noise* the explicit weak scheme (1.3) reduces to

$$(1.4) \qquad Y_{n+1} \;=\; Y_n + \frac{1}{2}\left\{ a\left(Y_n + a\,\Delta + \sum_{j=1}^{m} b^j\,\Delta\hat{W}^j\right) + a \right\}\Delta$$

$$+ \sum_{j=1}^{m} b^j\,\Delta\hat{W}_j.$$

In the deterministic case, that is with $b \equiv 0$, this is just the Heun method (8.1.12).

Several other schemes, which are not completely derivative free, have also appeared in the literature. For example, for the nonautonomous 1-dimensional case $d = m = 1$ Milstein proposed the scheme

$$(1.5) \qquad Y_{n+1} \;=\; Y_n + \frac{1}{2}b\,\Delta\hat{W} + \frac{1}{2}\left(a - bb'\right)\Delta + \frac{1}{2}bb'\left(\Delta\hat{W}\right)^2$$

$$+ \frac{1}{2}a\left(\tau_{n+1}, Y_n + a\,\Delta + b\,\Delta\hat{W}\right)\Delta$$

$$+ \frac{1}{4}b\left(\tau_{n+1}, Y_n + a\,\Delta + \frac{1}{\sqrt{3}}b\,\Delta\hat{W}\right)\Delta\hat{W}$$

$$+ \frac{1}{4}b\left(\tau_{n+1}, Y_n + a\,\Delta - \frac{1}{\sqrt{3}}b\,\Delta\hat{W}\right)\Delta\hat{W},$$

where the random variable $\Delta\hat{W}$ is as for the scheme (1.1). At each step here we have to evaluate the drift $a$ at two points, the diffusion coefficient $b$ at four points and its derivative $b'$ at one point, as well as generating a single random variable.

Another scheme involving the derivative $b'$ was proposed by Talay. In the autonomous 1-dimensional case $d = m = 1$ it has the form

$$(1.6) \qquad Y_{n+1} \;=\; Y_n + \sqrt{2}\,b\,(\bar{\Upsilon})\,\Delta\tilde{W} + \frac{1}{\sqrt{2}}b\left\{\Delta\hat{W} - \Delta\tilde{W}\right\}$$

$$+ \left( a\left( \bar{\Upsilon} \right) - \frac{1}{2} b\left( \bar{\Upsilon} \right) b'\left( \bar{\Upsilon} \right) \right) \Delta$$

$$+ \frac{1}{2} bb' \left( \Delta \hat{W} \right)^2 + \frac{1}{2} b\left( \bar{\Upsilon} \right) b'\left( \bar{\Upsilon} \right) \left( \Delta \tilde{W} \right)^2$$

$$- \frac{1}{2} bb' \left( \Delta \hat{W} + \Delta \tilde{W} \right)^2$$

with the supporting value

$$\bar{\Upsilon} = Y_n + \frac{1}{\sqrt{2}} b\,\Delta \hat{W} + \frac{1}{2} \left( a - \frac{1}{2} bb' \right) \Delta + \frac{1}{4} bb' \left( \Delta \hat{W} \right)^2,$$

where $\Delta \hat{W}$ and $\Delta \tilde{W}$ are independent random variables satisfying the moment conditions (14.2.3), that is they can be chosen like $\Delta \hat{W}$ in the scheme (1.1). Each step here requires $a$, $b$ and $b'$ to be evaluated at two points, together with the generation of two random variables.

**Exercise 15.1.2**    *Show that the scheme (1.1) satisfies condition (14.5.12) of Theorem 14.5.2 for weak convergence with order $\beta = 2.0$.*

**PC-Exercise 15.1.3**    *Repeat PC-Exercise 14.2.2 for the schemes (1.5) and (1.6) with Gaussian $\Delta \hat{W}$ and $\Delta \tilde{W}$. Compare the results with those of PC-Exercise 15.1.1.*

## 15.2   Explicit Order 3.0 Weak Schemes

Here we shall present some order 3.0 weak schemes which do not involve derivatives of the drift and diffusion coefficients. These are naturally more complicated than the order 2.0 weak schemes proposed in the previous section, so we shall first consider the simpler setting of additive noise.

In the autonomous case $d = 1, 2, \ldots$ with $m = 1$ we have in vector form the *explicit order* 3.0 *weak scheme for scalar additive noise*

$$(2.1) \qquad Y_{n+1} \;=\; Y_n + a\,\Delta + b\,\Delta \hat{W}$$

$$+ \frac{1}{2} \left( a_\zeta^+ + a_\zeta^- - \frac{3}{2} a - \frac{1}{4} \left( \tilde{a}_\zeta^+ + \tilde{a}_\zeta^- \right) \right) \Delta$$

$$+ \sqrt{\frac{2}{\Delta}} \left( \frac{1}{\sqrt{2}} \left( a_\zeta^+ - a_\zeta^- \right) - \frac{1}{4} \left( \tilde{a}_\zeta^+ - \tilde{a}_\zeta^- \right) \right) \zeta\,\Delta \hat{Z}$$

$$+ \frac{1}{6} \left[ a\left( Y_n + \left( a + a_\zeta^+ \right) \Delta + (\zeta + \rho)\, b\,\sqrt{\Delta} \right) - a_\zeta^+ - a_\rho^+ + a \right]$$

$$\times \left[ (\zeta + \rho)\, \Delta \hat{W}\, \sqrt{\Delta} + \Delta + \zeta\,\rho \left\{ \left( \Delta \hat{W} \right)^2 - \Delta \right\} \right]$$

with

$$a_\phi^\pm = a\left(Y_n + a\Delta \pm b\sqrt{\Delta}\,\phi\right)$$

and

$$\tilde{a}_\phi^\pm = a\left(Y_n + 2a\Delta \pm b\sqrt{2\Delta}\,\phi\right),$$

where $\phi$ is either $\zeta$ or $\rho$. Here we use two correlated Gaussian random variables $\Delta\hat{W} \sim N(0;\Delta)$ and $\Delta\hat{Z} \sim N(0;\frac{1}{3}\Delta^3)$ with $E(\Delta\hat{W}\Delta\hat{Z}) = \frac{1}{2}\Delta^2$, together with two independent two-point distributed random variables $\zeta$ and $\rho$ with

$$P(\zeta = \pm 1) = P(\rho = \pm 1) = \frac{1}{2}.$$

In this scheme, which is due to Platen, we have to evaluate the drift coefficient $a$ at six points for each time step.

**PC-Exercise 15.2.1**  *Consider the scalar linear stochastic differential equation with additive noise (4.4.1)*

$$dX_t = a\,X_t\,dt + b\,dW_t$$

*on the time interval $[0,T]$ with $T = 1$, $X_0 = 0.1$ at time $t_0 = 0$, $a = -4.0$ and $b = 0.1$. Simulate $M = 20$ batches each of $N = 100$ values $Y^\delta(T)$ of the explicit order 3.0 weak scheme (2.1) with equidistant step size $\delta = \Delta = 2^0$ and determine the 90%-confidence interval for the mean error*

$$\mu = E\left(Y^\delta(T)\right) - E\left(X_T\right),$$

*using the known fact that $E(X_1) = 0$. Repeat the calculations for step sizes $\Delta = 2^{-1}$ and $2^{-2}$ and plot $\log_2 |\mu|$ against $\log_2 \Delta$.*

In the autonomous case $d = 1, 2, \ldots$ with general scalar noise $m = 1$ we have the following generalization of the scheme (2.1) which we shall call the *explicit order 3.0 weak scheme for scalar noise:*

$$
\begin{aligned}
(2.2) \quad Y_{n+1} &= Y_n + a\Delta + b\Delta\hat{W} + \frac{1}{2}H_a\Delta + \frac{1}{\Delta}H_b\Delta\hat{Z} \\
&\quad + \sqrt{\frac{2}{\Delta}}G_a\zeta\Delta\hat{Z} + \frac{1}{\sqrt{2\Delta}}G_b\zeta\left\{\left(\Delta\hat{W}\right)^2 - \Delta\right\} \\
&\quad + \frac{1}{6}F_a^{++}\left(\Delta + (\zeta+\rho)\sqrt{\Delta}\,\Delta\hat{W} + \zeta\rho\left\{\left(\Delta\hat{W}\right)^2 - \Delta\right\}\right) \\
&\quad + \frac{1}{24}\left(F_b^{++} + F_b^{-+} + F_b^{+-} + F_b^{--}\right)\Delta\hat{W} \\
&\quad + \frac{1}{24\sqrt{\Delta}}\left(F_b^{++} - F_b^{-+} + F_b^{+-} - F_b^{--}\right)\left\{\left(\Delta\hat{W}\right)^2 - \Delta\right\}\zeta \\
&\quad + \frac{1}{24\Delta}\left(F_b^{++} + F_b^{--} - F_b^{-+} - F_b^{+-}\right)\left\{\left(\Delta\hat{W}\right)^2 - 3\right\}\Delta\hat{W}\,\zeta\rho
\end{aligned}
$$

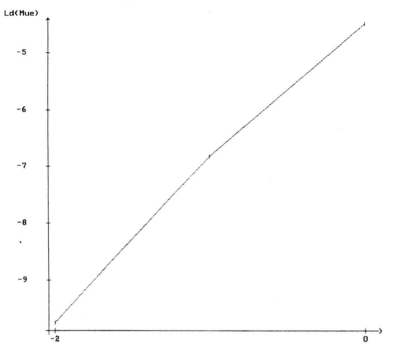

**Figure 15.2.1**   Results of PC-Exercise 15.2.1.

$$+\frac{1}{24\sqrt{\Delta}}\left(F_b^{++}+F_b^{-+}-F_b^{+-}-F_b^{--}\right)\left\{\left(\Delta\hat{W}\right)^2-\delta\right\}\rho$$

with

$$H_g = g^+ + g^- - \frac{3}{2}g - \frac{1}{4}\left(\tilde{g}^+ + \tilde{g}^-\right),$$

$$G_g = \frac{1}{\sqrt{2}}\left(g^+ - g^-\right) - \frac{1}{4}\left(\tilde{g}^+ - \tilde{g}^-\right),$$

$$F_g^{+\pm} = g\left(Y_n + (a+a^+)\,\Delta + b\zeta\,\sqrt{\Delta} \pm b^+\,\rho\sqrt{\Delta}\right) - g^+$$
$$-g\left(Y_n + a\,\Delta \pm b\,\rho\sqrt{\Delta}\right) + g,$$

$$F_g^{-\pm} = g\left(Y_n + (a+a^-)\,\Delta - b\zeta\,\sqrt{\Delta} \pm b^-\,\rho\sqrt{\Delta}\right) - g^-$$
$$-g\left(Y_n + a\,\Delta \pm b\,\rho\sqrt{\Delta}\right) + g$$

where

$$g^\pm = g\left(Y_n + a\,\Delta \pm b\zeta\,\sqrt{\Delta}\right)$$

and

$$\tilde{g}^\pm = g\left(Y_n + 2a\,\Delta \pm \sqrt{2}b\zeta\,\sqrt{\Delta}\right),$$

with $g$ being equal to either $a$ or $b$. The random variables $\Delta \hat{W}$, $\Delta \hat{Z}$, $\zeta$ and $\rho$ here are as specified earlier for the scheme (2.1).

It remains an open and challenging task to derive simpler derivative free order 3.0 weak schemes, at least for important classes of stochastic differential equations.

**PC-Exercise 15.2.2**   *Repeat PC-Exercise 14.2.2 using the explicit order 3.0 weak scheme (2.2) and compare the results with those of PC-Exercise 15.1.3.*

# 15.3   Extrapolation Methods

In Chapter 8 we mentioned the Richardson extrapolation method (8.1.17) which improves the order $\beta = 1.0$ of a deterministic Euler approximation by extrapolating the calculations for a given step size and twice the step size to give a result of order $\beta = 2.0$ accuracy. Here we shall use a similar approach for the approximate evaluation of the functional

$$E\left(g\left(X_T\right)\right)$$

of an autonomous Ito process $X = \{X_t, t \in [0, T]\}$ at time $T$, where $g$ is a given smooth function. It will turn out that extrapolation provides an efficient, yet simple way of obtaining a higher order weak approximation.

Only equidistant time discretizations of the time interval $[0, T]$ with $\tau_{n_T} = T$ will be used in what follows. As before, we shall denote the time discrete approximation under consideration with step size $\delta > 0$ by $Y^\delta$, with value $Y^\delta(\tau_n)$ $= Y_n^\delta$ at the discretization times $\tau_n$, and the correponding approximation with twice this step size by $Y^{2\delta}$, and so on.

As an introduction, suppose that we have simulated the functional

$$E\left(g\left(Y^\delta(T)\right)\right)$$

for an order 1.0 weak approximation using, say, the Euler scheme (14.1.1) or the simplified Euler scheme (14.1.2) with step size $\delta$. Suppose that we repeat the calculations with double the step size $2\delta$ to simulate the functional

$$E\left(g\left(Y^{2\delta}(T)\right)\right).$$

We can then combine these two functionals to obtain the *order* 2.0 *weak extrapolation*

$$(3.1) \qquad V_{g,2}^\delta(T) = 2E\left(g\left(Y^\delta(T)\right)\right) - E\left(g\left(Y^{2\delta}(T)\right)\right),$$

which was proposed by Talay and Tubaro.

**PC-Exercise 15.3.1**   *Consider the Ito process $X$ satisfying the linear stochastic differential equation*

$$dX_t = a\,X_t\,dt + b\,X_t\,dW_t$$

*with $X_0 = 0.1$, $a = 1.5$ and $b = 0.01$ on the time interval $[0, T]$ where $T = 1$. Use the Euler scheme (14.1.1) to simulate the order 2.0 weak extrapolation $V_{g,2}^\delta(T)$ for $g(x) = x$ and $\delta = 2^{-3}$. Generate $M = 20$ batches of $N = 100$ trajectories each and determine the 90% confidence interval for*

$$\mu_2 = V_{g,2}^\delta(T) - E\left(g\left(X_T\right)\right).$$

*Repeat the calculations for step sizes $\delta = 2^{-4}$, $2^{-5}$ and $2^{-6}$ and plot the results on $\log_2 |\mu_2|$ versus $\log_2 \delta$ axes.*

We can use an order 2.0 weak approximation $Y^\delta$, such as the order 2.0 weak Taylor scheme (14.2.1), the simplified order 2.0 weak Taylor scheme (14.2.2) or the explicit order 2.0 weak scheme (15.1.1), and extrapolate to obtain a fourth order approximation of the functional. The *order 4.0 weak extrapolation* has the form

$$(3.2) \qquad V_{g,4}^\delta(T) \;\; = \;\; \frac{1}{21}\left[32 E\left(g\left(Y^\delta(T)\right)\right) - 12 E\left(g\left(Y^{2\delta}(T)\right)\right)\right.$$

$$\left. + E\left(g\left(Y^{4\delta}(T)\right)\right)\right].$$

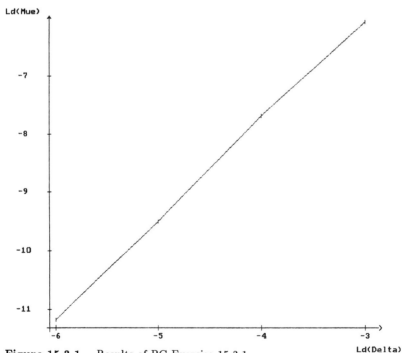

**Figure 15.3.1**     Results of PC-Exercise 15.3.1.

**PC-Exercise 15.3.2** *Repeat PC-Exercise 15.3.1 with $a = -5.0$ and $b = 2.0$ using the order 4.0 weak extrapolation (3.2) and the order 2.0 weak Taylor scheme (14.2.1) for $\delta = 2^{-2}$, $2^{-3}$ and $2^{-4}$, with*

$$\mu_4 = V_{g,4}^\delta(T) - E\left(g\left(X_T\right)\right)$$

*instead of $\mu_2$.*

Similarly, we can extrapolate an order 3.0 weak approximation $Y^\delta$, such as the order 3.0 weak Taylor scheme (14.3.1), the simplified order 3.0 weak Taylor scheme (14.3.2) or the order 3.0 weak schemes (15.2.1) and (15.2.2), to obtain a sixth order approximation of the functional. The *order 6.0 weak extrapolation* is defined by

$$(3.3) \quad V_{g,6}^\delta(T) \;=\; \frac{1}{2905}\left[4032E\left(g\left(Y^\delta(T)\right)\right) - 1512E\left(g\left(Y^{2\delta}(T)\right)\right)\right.$$

$$\left. +448E\left(g\left(Y^{3\delta}(T)\right)\right) - 63E\left(g\left(Y^{4\delta}(T)\right)\right)\right].$$

In general, for any $\beta = 1.0, 2.0, \ldots$ it turns out that we can obtain an order $2\beta$ weak approximation by extrapolating an order $\beta$ weak approximation. Let $Y^\delta$ be an order $\beta$ weak approximation of an autonomous Ito diffusion $X$ on an interval $[0, T]$. Then for a given sequence of step sizes

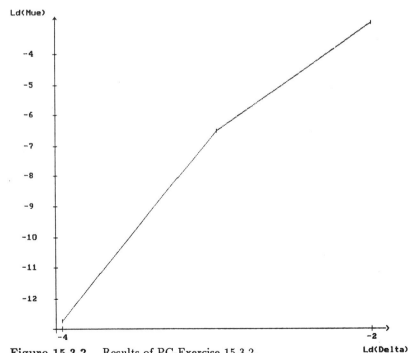

**Figure 15.3.2**    Results of PC-Exercise 15.3.2.

(3.4) $$\delta_l = d_l\, \delta$$

for $l = 1, \ldots, \beta + 1$ with

(3.5) $$0 < d_1 < \cdots < d_{\beta+1} < \infty$$

an *order $2\beta$ weak extrapolation* is given by the expression

(3.6) $$V_{g,2\beta}^{\delta}(T) = \sum_{l=1}^{\beta+1} a_l\, E\left(g\left(Y^{\delta_l}(T)\right)\right),$$

where

(3.7) $$\sum_{l=1}^{\beta+1} a_l = 1$$

and

(3.8) $$\sum_{l=1}^{\beta+1} a_l\, (d_l)^{\gamma} = 0$$

for each $\gamma = \beta, \ldots, 2\beta - 1$.

We note that $\beta + 1$ simulations of functionals are necessary for the evaluation of an order $2\beta$ weak extrapolation. It is clear that the extrapolations (3.1), (3.2) and (3.3) are covered by the above definition. Moreover, we can see from (3.5) that sequences of step sizes other than the ones proposed are also possible.

**Exercise 15.3.3**   *Show that (3.2) and (3.3) satisfy conditions (3.6)–(3.8) for $\beta = 2.0$ and $3.0$, respectively.*

The following theorem shows that an order $2\beta$ weak extrapolation does indeed converge weakly with order $2\beta$.

**Theorem 15.3.4**   *Under the assumptions of Theorem 14.6.1 for any given $\beta = 1.0, 2.0, \ldots$ the order $2\beta$ weak extrapolation (3.6) satisfies the estimate*

(3.9) $$\left|V_{g,2\beta}^{\delta}(T) - E\left(g\left(X_T\right)\right)\right| \le C_g\, \delta^{2\beta},$$

*where the constant $C_g$ does not depend on $\delta$.*

**Proof**   The proof is a straight forward application of the leading error expansion given in (14.6.3). From (3.6), (14.6.3) and (3.4) we easily obtain

(3.10) $$\left|V_{g,2\beta}^{\delta}(T) - E\left(g\left(X_T\right)\right)\right|$$

$$= \left|\sum_{l=1}^{\beta+1} a_l\, E\left(g\left(Y^{\delta_l}(T)\right)\right) - E\left(g\left(X_T\right)\right)\right|$$

$$\le \left|\sum_{l=1}^{\beta+1} a_l \left\{E\left(g\left(X_T\right)\right) + \sum_{\gamma=\beta}^{2\beta-1} \psi_{g,\gamma}(T)\,(\delta_l)^{\gamma}\right\} - E\left(g\left(X_T\right)\right)\right| + K\,\delta^{2\beta},$$

where the $\psi_{g,\gamma}(T)$ are the coefficients in the leading error expansion. Applying (3.7), (3.4) and (3.8), it follows from (3.10) that

$$\left| V_{g,2\beta}^{\delta}(T) - E\left(g\left(X_T\right)\right)\right| \le K\,\delta^{2\beta},$$

which completes the proof. $\square$

**Exercise 15.3.5**   *Show that*

$$(3.11) \qquad V_{g,4}^{\delta}(T) \;=\; \frac{1}{11}\left[18E\left(g\left(Y^{\delta}(T)\right)\right) - 9E\left(g\left(Y^{2\delta}(T)\right)\right)\right.$$

$$\left. +2E\left(g\left(Y^{3\delta}(T)\right)\right)\right].$$

*is an order* 4.0 *weak extrapolation if* $Y^{\delta}$ *is the order* 2.0 *weak Taylor scheme.*

# 15.4   Implicit Weak Approximations

In Section 8 of Chapter 9 we became aware of the necessity of implicit schemes for the numerical integration of stiff stochastic differential equations. We noticed that the use of discrete bounded random variables in weak schemes allows the construction of fully implicit schemes, that is with the diffusion coefficient implicit as well as the drift coefficient.

In this section we shall describe a few implicit weak schemes which are suitable for stiff stochastic differential equations. In general, however, such schemes also require an algebraic equation to be solved at each time step, thus imposing an additional computational burden.

## The Implicit Euler Scheme

The simplest implicit weak scheme is the *implicit Euler scheme,* which in the general multi-dimensional case $d$, $m = 1, 2, \ldots$ has the form

$$(4.1) \qquad Y_{n+1} = Y_n + a\left(\tau_{n+1}, Y_{n+1}\right)\Delta + \sum_{j=1}^{m} b^j\left(\tau_n, Y_n\right)\Delta\hat{W}^j,$$

where the $\Delta\hat{W}^j$ for $j = 1, \ldots, m$ and $n = 1, 2, \ldots$ are independent two-point distributed random variables with

$$(4.2) \qquad P\left(\Delta\hat{W}^j = \pm\sqrt{\Delta}\right) = \frac{1}{2}.$$

We can also form a *family of implicit Euler schemes*

$$(4.3) \qquad Y_{n+1} = Y_n + \{(1 - \alpha)a\,(\tau_n, Y_n) + \alpha a\,(\tau_{n+1}, Y_{n+1})\}\,\Delta$$

$$+ \sum_{j=1}^{m} b^j\,(\tau_n, Y_n)\,\Delta \hat{W}^j$$

with the $\Delta \hat{W}^j$ as in (4.1). The parameter $\alpha$ here can be interpreted as the
degree of implicitness. With $\alpha = 0.0$ the scheme (4.3) reduces to the simplified Euler scheme (14.1.2), whereas with $\alpha = 0.5$ it represents a stochastic
generalization of the trapezoidal method (8.1.11). Under assumptions of suffi-
cent regularity, it follows from Theorem 14.5.2 that the schemes (4.3) converge
weakly with order $\beta = 1.0$. According to (12.5.4)-(12.5.5) the implicit Euler
scheme (4.3) is A-stable for $\alpha \in [0.5, 1]$, whereas for $\alpha \in [0, 0.5)$ its region of
absolute stability is the interior of the circle of radius $r = (1 - 2\alpha)^{-1}$ centered
at $-r + 0\imath$.

**PC-Exercise 15.4.1**     *Consider the 2-dimensional Ito process $X$ in Example
12.2.1 satisfying the stochastic differential equation (12.2.4) with coefficient
matrices given by (12.2.6) with $a = 5.0$ and $b = 0.001$ and initial value $X_0 =
1.0$ on the interval $[0, T]$ where $T = 1$. Compute $M = 20$ batches each of $N =
100$ values $Y^\delta(T)$ of the implicit Euler scheme (4.3) with $\alpha = 0.0$, $0.5$ and $1.0$
for step size $\delta = \Delta = 2^{-3}$. Evaluate the 90% confidence intervals for the mean
error*

$$\mu = E\left(Y^\delta(T)\right) - E\left(X_T\right)$$

*at time $T = 1$. Repeat the calculations for step sizes $\delta = 2^{-4}$, $2^{-5}$ and $2^{-6}$ and
plot the results on separate $\mu$ versus $\delta$ axes for the three cases $\alpha = 0.0$, $0.5$ and
$1.0$. Finally, replot the results on $\log_2 |\mu|$ versus $\log_2 \delta$ axes.*

## The Fully Implicit Euler Scheme

In view of the definitions of Ito and Stratonovich stochastic integrals, we cannot
construct a meaningful implicit Euler scheme simply by making implicit the
diffusion coefficient in the implicit Euler scheme in an analogous way to the
drift coefficient. As we saw in Section 9 of Chapter 4 the solution of such a
scheme would not, in general, converge to that of the given Ito equation. To
be meaningful an implicit Euler scheme should be at least weakly consistent
(see (9.7.5)-(9.7.7)).

**Exercise 15.4.2**     *Show that the scheme (9.8.8) is in general not weakly consistent.*

To obtain a weakly consistent implicit approximation we need to appropriately
modify the drift term.

In the 1-dimensional autonomous case $d = m = 1$ the *fully implicit Euler
scheme* has the form

$$(4.4) \qquad Y_{n+1} = Y_n + \bar{a}(Y_{n+1})\,\Delta + b(Y_{n+1})\,\Delta\hat{W},$$

where $\Delta\hat{W}$ is as in (4.2) and $\bar{a}$ is the corrected drift coefficient defined by

$$(4.5) \qquad \bar{a} = a - bb'.$$

Note how the correction term here differs by the absence of the factor $\frac{1}{2}$ from that in the corrected drift $\underline{a}$ of the corresponding Stratonovich stochastic differential equation (4.9.11).

In the general multi-dimensional case $d, m = 1, 2, \ldots$ we have, in vector notation, a *family of implicit Euler schemes*

$$(4.6) \quad Y_{n+1} = Y_n + \{\alpha\bar{a}_\eta(\tau_{n+1}, Y_{n+1}) + (1-\alpha)\bar{a}_\eta(\tau_n, Y_n)\}\,\Delta$$

$$+ \sum_{j=1}^{m} \{\eta b^j(\tau_{n+1}, Y_{n+1}) + (1-\eta)b^j(\tau_n, Y_n)\}\,\Delta\hat{W}^j,$$

where the $\Delta\hat{W}^j$ are as in (4.1) and the corrected drift coefficient $\bar{a}_\eta$ is defined by

$$(4.7) \qquad \bar{a}_\eta = a - \eta \sum_{j_1, j_2 = 1}^{m} \sum_{k=1}^{d} b^{k, j_1} \frac{\partial b^{j_2}}{\partial x^k}.$$

The choice $\alpha = \eta = 1$ in (4.6) gives us the fully implicit Euler scheme. For $\eta = 0.5$ the corrected drift $\bar{a}_\eta = \underline{a}$ is the corrected drift of the corresponding Stratonovich equation, and for $\alpha = 0.5$ the scheme (4.6) yields further stochastic generalizations of the deterministic trapezoidal method (8.1.11).

Once again Theorem 14.5.2 provides conditions which ensure that the implicit Euler schemes (4.6) converge weakly with order $\beta = 1.0$. They are similar to those of the explicit Euler scheme (14.1.1).

**Exercise 15.4.3** *Show that the scheme (4.6) for $\alpha = \eta = 0.5$ and $d = m = 1$ is weakly consistent.*

**Exercise 15.4.4** *Using the definition in Section 8 of Chapter 9 check whether the scheme (4.6) with $\alpha = \eta = 1$ is A-stable.*

**PC-Exercise 15.4.5** *Repeat PC-Exercise 15.4.1 for the implicit Euler scheme (4.6) with $\alpha = \eta = 1$.*

## The Implicit Order 2.0 Weak Taylor Scheme

We shall now adapt the order 2.0 weak Taylor scheme to obtain implicit schemes.

In the autonomous 1-dimensional case $d = m = 1$ the *implicit order 2.0 weak Taylor scheme* has the form

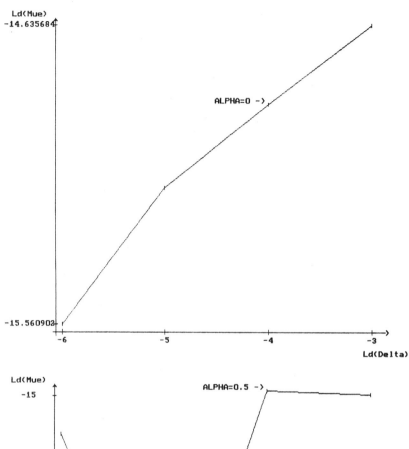

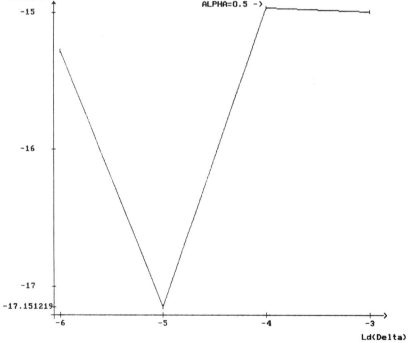

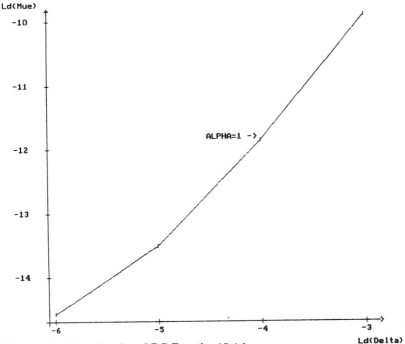

**Figure 15.4.1** Results of PC-Exercise 15.4.1.

$$(4.8) \quad Y_{n+1} = Y_n + a\left(Y_{n+1}\right)\Delta + b\,\Delta\hat{W}$$

$$-\frac{1}{2}\left\{a\left(Y_{n+1}\right)a'\left(Y_{n+1}\right) + \frac{1}{2}b^2\left(Y_{n+1}\right)a''\left(Y_{n+1}\right)\right\}\Delta^2$$

$$+\frac{1}{2}bb'\left\{\left(\Delta\hat{W}\right)^2 - \Delta\right\}$$

$$+\frac{1}{2}\left\{-a'b + ab' + \frac{1}{2}b''b^2\right\}\Delta\hat{W}\,\Delta$$

where $\Delta\hat{W}$ is $N(0;\Delta)$-Gaussian or three-point distributed with

$$(4.9) \qquad P\left(\Delta\hat{W} = \pm\sqrt{3\Delta}\right) = \frac{1}{6}, \qquad P\left(\Delta\hat{W} = 0\right) = \frac{2}{3}.$$

For the general multi-dimensional case $d$, $m = 1, 2, \ldots$ Milstein proposed the following *family of implicit order* 2.0 *weak Taylor schemes:*

$$(4.10) \qquad Y_{n+1} = Y_n + \left\{\alpha a\left(\tau_{n+1}, Y_{n+1}\right) + (1-\alpha)a\right\}\Delta$$

$$+\frac{1}{2}\sum_{j_1,j_2=1}^{m} L^{j_1}b^{j_2}\left(\Delta\hat{W}^{j_1}\Delta\hat{W}^{j_2} + V_{j_1,j_2}\right)$$

$$+ \sum_{j=1}^{m} \left\{ b^j + \frac{1}{2} \left( L^0 b^j + (1 - 2\alpha) L^j a \right) \Delta \right\} \Delta \hat{W}^j$$

$$+ \frac{1}{2} (1 - 2\alpha) \left\{ \beta L^0 a + (1 - \beta) L^0 a \left( \tau_{n+1}, Y_{n+1} \right) \right\} \Delta^2,$$

where $\alpha, \beta \in [0, 1]$ and the random variables $\Delta \hat{W}^j$ and $V_{j_1, j_2}$ can be chosen as in (14.2.4) and (14.2.8)–(14.2.10).

When $\alpha = 0.5$ the scheme (4.10) simplifies to

$$(4.11) \qquad Y_{n+1} \;=\; Y_n + \frac{1}{2} \left\{ a \left( \tau_{n+1}, Y_{n+1} \right) + a \right\} \Delta$$

$$+ \sum_{j=1}^{m} b^j \, \Delta \hat{W}^j + \frac{1}{2} \sum_{j=1}^{m} L^0 b^j \, \Delta \hat{W}^j \, \Delta$$

$$+ \frac{1}{2} \sum_{j_1, j_2 = 1}^{m} L^{j_1} b^{j_2} \left( \Delta \hat{W}^{j_1} \Delta \hat{W}^{j_2} + V_{j_1, j_2} \right).$$

We note that the last two terms in (4.11) vanish for additive noise with $b \equiv$ *const*. It is easy to check that the scheme (4.11) is A-stable.

**PC-Exercise 15.4.6**    *Repeat PC-Exercise 15.4.1 with $a = 5.0$, $b = 10^{-6}$ and $\delta = \Delta = 2^{-5}, \ldots, 2^{-8}$ for the implicit order 2.0 weak Taylor scheme (4.11).*

## The Implicit Order 2.0 Weak Scheme

We can also avoid derivatives in the above implicit schemes by using similar approximations as in the explicit order 2.0 weak scheme.

In the autonomous case $d = 1, 2, \ldots$ with scalar noise $m = 1$ Platen proposed the following *implicit order 2.0 weak scheme*

$$(4.12) \qquad Y_{n+1} \;=\; Y_n + \frac{1}{2} \left( a + a \left( Y_{n+1} \right) \right) \Delta$$

$$+ \frac{1}{4} \left( b \left( \bar{\Upsilon}^+ \right) + b \left( \bar{\Upsilon}^- \right) + 2b \right) \Delta \hat{W}$$

$$+ \frac{1}{4} \left( b \left( \bar{\Upsilon}^+ \right) - b \left( \bar{\Upsilon}^- \right) \right) \left\{ \left( \Delta \hat{W} \right)^2 - \Delta \right\} \Delta^{-1/2}$$

with supporting values

$$\bar{\Upsilon}^{\pm} = Y_n + a \Delta \pm b \sqrt{\Delta},$$

where $\Delta \hat{W}$ must be as in (4.8).

There is a multi-dimensional counterpart of (4.12). For the autonomous case with $d$, $m = 1$, $2$, ... the *implicit order 2.0 weak scheme* has the vector form

(4.13)
$$Y_{n+1} = Y_n + \frac{1}{2}\left(a + a\left(Y_{n+1}\right)\right)\Delta$$

$$+\frac{1}{4}\sum_{j=1}^{m}\left[b^j\left(\bar{R}_+^j\right) + b^j\left(\bar{R}_-^j\right) + 2b^j\right.$$

$$+\sum_{\substack{r=1\\r\neq j}}^{m}\left(b^j\left(\bar{U}_+^r\right) + b^j\left(\bar{U}_-^r\right) - 2b^j\right)\Delta^{-1/2}\right]\Delta\hat{W}^j$$

$$+\frac{1}{4}\sum_{j=1}^{m}\left[\left(b^j\left(\bar{R}_+^j\right) - b^j\left(\bar{R}_-^j\right)\right)\left\{\left(\Delta\hat{W}^j\right)^2 - \Delta\right\}\right.$$

$$+\sum_{\substack{r=1\\r\neq j}}^{m}\left(b^j\left(\bar{U}_+^r\right) - b^j\left(\bar{U}_-^r\right)\right)\left\{\Delta\hat{W}^j\Delta\hat{W}^r + V_{r,j}\right\}\right]\Delta^{-1/2}$$

with supporting values

$$\bar{R}_\pm^j = Y_n + a\,\Delta \pm b^j\,\sqrt{\Delta}$$

and

$$\bar{U}_\pm^j = Y_n \pm b^j\,\sqrt{\Delta},$$

where the random variables $\Delta\hat{W}^j$ and $V_{r,j}$ are as in (4.10). We note that the scheme (4.13) is A-stable.

In conclusion, we remark that specific assumptions for the implicit schemes (4.10) and (4.13) to converge with weak order 2.0 follow again from Theorem 14.5.2. An expansion of the Ito process, which is useful for deriving such implicit schemes, will be given at the end of Section 6 of this chapter.

**PC-Exercise 15.4.7**   *Repeat PC-Exercise 15.4.6 for the implicit order 2.0 weak scheme (4.13).*

## 15.5   Predictor-Corrector Methods

We mentioned in Chapter 8 that deterministic predictor-corrector methods are used mainly because of their numerical stability, which they inherit from the implicit counterparts of their corrector schemes. In addition, the difference between the predicted and the corrected values at each time step provides an indication of the local error. In principle, these advantages carry over to the stochastic case. Here we shall describe a few predictor-corrector methods, due to Platen, for stochastic differential equations. They use as their predictors weak Taylor or explicit weak schemes and as their correctors the corresponding implicit schemes made explicit by using the predicted value $\bar{Y}_{n+1}$ instead of $Y_{n+1}$ on the right hand side of the implicit scheme. The asserted weak order of convergence of these schemes follows from Theorem 14.5.2 and expansions that will be given in Section 6 at the end of the chapter.

## An order 1.0 Predictor-Corrector Method

In the autonomous 1-dimensional scalar noise case, $d = m = 1$ we have the *modified trapezoidal method of weak order* $\beta = 1.0$ with corrector

$$(5.1) \qquad Y_{n+1} = Y_n + \frac{1}{2} \left\{ a \left( \bar{Y}_{n+1} \right) + a \right\} \Delta + b \, \Delta \hat{W}$$

and predictor, the weak Euler scheme,

$$(5.2) \qquad \bar{Y}_{n+1} = Y_n + a \, \Delta + b \, \Delta \hat{W}.$$

Here the $\Delta \hat{W}$ can be chosen as Gaussian $N(0; \Delta)$ distributed random variables or as two-point distributed random variables with

$$(5.3) \qquad P \left( \Delta \hat{W} = \pm \sqrt{\Delta} \right) = \frac{1}{2}.$$

Using the family of implicit Euler schemes (4.6) as corrector, in the general multi-dimensional case $d, m = 1, 2, \ldots$ we can form the following *family of order* 1.0 *weak predictor-corrector methods* with corrector

$$(5.4) \quad Y_{n+1} \;\; = \;\; Y_n + \left\{ \alpha \bar{a}_\eta \left( \tau_{n+1}, \bar{Y}_{n+1} \right) + (1 - \alpha) \bar{a}_\eta \left( \tau_n, Y_n \right) \right\} \Delta$$

$$+ \sum_{j=1}^{m} \left\{ \eta b^j \left( \tau_{n+1}, \bar{Y}_{n+1} \right) + (1 - \eta) b^j \left( \tau_n, Y_n \right) \right\} \Delta \hat{W}^j$$

for $\alpha, \eta \in [0, 1]$, where

$$(5.5) \qquad \bar{a}_\eta = a - \eta \sum_{j_1, j_2 = 1}^{m} \sum_{k=1}^{d} b^{k, j_1} \frac{\partial b^{j_2}}{\partial x^k},$$

and with predictor

$$(5.6) \qquad \bar{Y}_{n+1} = Y_n + a \, \Delta + \sum_{j=1}^{m} b^j \, \Delta \hat{W}^j,$$

where the $\Delta \hat{W}^j$ are as in (5.3).

We note that the corrector (5.4) with $\eta > 0$ allows us to include some degree of implicitness in the the diffusion term too.

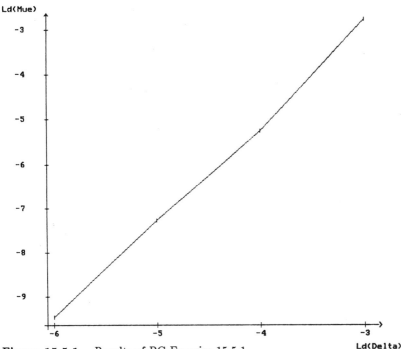

**Figure 15.5.1**    Results of PC-Exercise 15.5.1.

**PC-Exercise 15.5.1**    *Repeat PC-Exercise 15.4.1 with $a = 5.0$ and $b = 0.1$ for the predictor-corrector method (5.4)–(5.6) with $\alpha = 0.5$ and $\eta = 0$.*

From Figure 15.5.1 we can see that the predictor-corrector method (5.4)–(5.6) with $\alpha = 0.5$ and $\eta = 0$ is numerically stable and, in fact, of weak order 2.0 for the example under consideration. The higher than expected order here is a consequence of the special structure of both the scheme and the particular example.

## Order 2.0 Weak Predictor-Corrector Methods

We shall now combine the order 2.0 weak Taylor scheme and its implicit counterpart to form an order 2.0 predictor-corrector method.

In the autonomous 1-dimensional scalar noise case, $d = m = 1$, a possible *order 2.0 weak predictor-corrector method* has corrector

$$(5.7) \qquad Y_{n+1} = Y_n + \frac{1}{2} \left\{ a \left( \bar{Y}_{n+1} \right) + a \right\} \Delta + \Psi_n$$

with

$$\Psi_n = b \, \Delta \hat{W} + \frac{1}{2} bb' \left\{ \left( \Delta \hat{W} \right)^2 - \Delta \right\} + \frac{1}{2} \left( ab' + \frac{1}{2} b^2 b'' \right) \Delta \hat{W} \, \Delta$$

and predictor

(5.8)        $\bar{Y}_{n+1} = Y_n + a\Delta + \Psi_n$

$$+\frac{1}{2}a'b\,\Delta\hat{W}\,\Delta + \frac{1}{2}\left(aa' + \frac{1}{2}a''b^2\right)\Delta^2,$$

where the $\Delta\hat{W}$ are $N(0;\Delta)$ Gaussian or three-point distributed with

(5.9)        $P\left(\Delta\hat{W} = \pm\sqrt{3\Delta}\right) = \frac{1}{6}, \qquad P\left(\Delta\hat{W} = 0\right) = \frac{2}{3}.$

For the general multi-dimensional case $d,\ m = 1,\ 2,\ \ldots$ this generalizes to a method with corrector

(5.10)        $Y_{n+1} = Y_n + \frac{1}{2}\left\{a\left(\tau_{n+1}, \bar{Y}_{n+1}\right) + a\right\}\Delta + \Psi_n,$

where

$$\Psi_n = \sum_{j=1}^{m}\left\{b^j + \frac{1}{2}L^0 b^j\Delta\right\}\Delta\hat{W}^j + \frac{1}{2}\sum_{j_1, j_2=1}^{m}L^{j_1}b^{j_2}\left(\Delta\hat{W}^{j_1}\,\Delta\hat{W}^{j_2} + V_{j_1, j_2}\right),$$

and predictor corresponding to the order 2.0 weak Taylor scheme

(5.11)        $\bar{Y}_{n+1} = Y_n + a\Delta + \Psi_n + \frac{1}{2}L^0 a\,\Delta^2 + \frac{1}{2}\sum_{j=1}^{m}L^j a\,\Delta\hat{W}^j\,\Delta,$

where the independent random variables $\Delta\hat{W}^j$ and $V_{j_1, j_2}$ can be chosen as in (14.2.7).

**PC-Exercise 15.5.2**    *Repeat PC-Exercise 15.4.1 with $a = 5.0$ and $b = 0.8$ for the predictor-corrector method (5.10) – (5.11).*

Figure 15.5.2 shows for this particular example that the predictor-corrector method (5.10)–(5.11) is numerically stable and of weak order $\beta \geq 3.0$.

We can also formulate a weak order 2.0 predictor-corrector method which avoids the need to determine and evaluate the derivatives of $a$ and $b$.

In the autonomous case $d = 1,\ 2,\ \ldots$ with scalar noise $m = 1$ a *derivative free order 2.0 weak predictor-corrector method* has corrector

(5.12)        $Y_{n+1} = Y_n + \frac{1}{2}\left\{a\left(\bar{Y}_{n+1}\right) + a\right\}\Delta + \phi_n,$

where

$$\phi_n = \frac{1}{4}\left(b\left(\bar{\Upsilon}^+\right) + b\left(\bar{\Upsilon}^-\right) + 2b\right)\Delta\hat{W}$$

$$+\frac{1}{4}\left(b\left(\bar{\Upsilon}^+\right) - b\left(\bar{\Upsilon}^-\right)\right)\left\{\left(\Delta\hat{W}\right)^2 - \Delta\right\}\Delta^{-1/2}$$

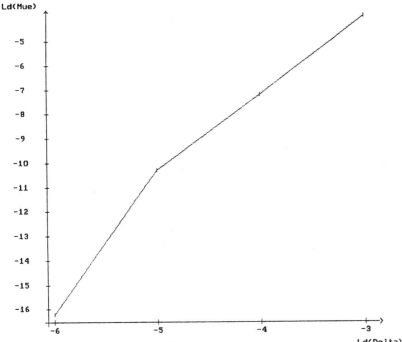

**Figure 15.5.2** Results of PC-Exercise 15.5.2.

with supporting values

$$\bar{\Upsilon}^{\pm} = Y_n + a\,\Delta \pm b\sqrt{\Delta},$$

and with predictor

(5.13)  $\qquad \bar{Y}_{n+1} = Y_n + \dfrac{1}{2}\left\{a\left(\bar{\Upsilon}\right) + a\right\}\Delta + \phi_n$

with the supporting value

$$\bar{\Upsilon} = Y_n + a\,\Delta + b\,\Delta\hat{W}.$$

Here the $\Delta\hat{W}$ can be chosen as in (5.8). Essentially, $\bar{\Upsilon}$ is used as an initial predictor which is corrected in (5.13), with the output $\bar{Y}_{n+1}$ being itself corrected in (5.12).

We also have a multi-dimensional counterpart of (5.12)–(5.13). For the autonomous case with $d$, $m = 1, 2, \ldots$ the method has corrector

(5.14)  $\qquad Y_{n+1} = Y_n + \dfrac{1}{2}\left\{a\left(\bar{Y}_{n+1}\right) + a\right\}\Delta + \phi_n,$

where

$$\phi_n \;=\; \frac{1}{4}\sum_{j=1}^{m}\left[b^j\left(\bar{R}^j_+\right) + b\left(\bar{R}^j_-\right) + 2b^j\right.$$

$$+ \sum_{\substack{r=1 \\ r \neq j}}^{m} \left( b^j \left( \bar{U}_+^r \right) + b \left( \bar{U}_-^r \right) - 2b^j \right) \Delta^{-1/2} \right] \Delta \hat{W}^j$$

$$+ \frac{1}{4} \sum_{j=1}^{m} \left[ \left( b^j \left( \bar{R}_+^j \right) - b \left( \bar{R}_-^j \right) \right) \left\{ \left( \Delta \hat{W} \right)^2 - \Delta \right\}$$

$$+ \sum_{\substack{r=1 \\ r \neq j}}^{m} \left( b^j \left( \bar{U}_+^r \right) - b \left( \bar{U}_-^r \right) \right) \left( \Delta \hat{W}^j \, \Delta \hat{W}^r + V_{r,j} \right) \right] \Delta^{-1/2}$$

with supporting values

$$\bar{R}_\pm^j = Y_n + a\,\Delta \pm b^j \sqrt{\Delta} \quad \text{and} \quad \bar{U}_\pm^j = Y_n \pm b^j \sqrt{\Delta},$$

and with predictor

(5.15)                     $$\bar{Y}_{n+1} = Y_n + \frac{1}{2} \left\{ a \left( \tilde{\Upsilon} \right) + a \right\} \Delta + \phi_n$$

with the supporting value

$$\tilde{\Upsilon} = Y_n + a\,\Delta + \sum_{j=1}^{m} b^j \, \Delta \hat{W}^j.$$

Here the independent random variables $\Delta \hat{W}^j$ and $V_{r,j}$ can be chosen as in (14.2.7).

**PC-Exercise 15.5.3**     *Repeat PC-Exercise 15.4.1 with $a = 5.0$ and $b = 0.08$ using the predictor-corrector method (5.12)–(5.13).*

At each step in the above schemes we first compute the predicted approximate value $\bar{Y}_{n+1}$ and then the corrected value $Y_{n+1}$. Their difference

$$Z_{n+1} = \bar{Y}_{n+1} - Y_{n+1}$$

provides us with information about the local error at each step, which we could use on-line to improve the simulations. For instance, if the mean of $Z_{n+1}$ is rather large, we should change to a finer time step and repeat the calculation.

## 15.6   Convergence of Weak Schemes

We shall now indicate the key steps required by Theorem 14.5.2 to show the weak convergence, with the asserted order, of various schemes introduced in this chapter. Before looking at specific schemes, we shall write down for the simple 1-dimensional case $d = m = 1$ the conditional moments of the increments of weak Taylor schemes, which satisfy the crucial condition (14.5.12). This will make it easier to verify this condition for the specific schemes.

## Conditional Moments of Increments

Following the usual convention, we shall use the error abbreviation $O(\Delta^k)$ to indicate that the remaining terms satisfy

$$\lim_{\Delta \to 0} O\left(\Delta^k\right) / \Delta^k < \infty.$$

We shall also omit explicit dependence on $Y_n$, absorbing it into the $O(\Delta^k)$ term.

In the case $\beta = 1.0$, it is easily shown that the Euler scheme (14.1.1) satisfies the conditions

$$(6.1) \qquad E\left(Y_{n+1} - Y_n \,|\, A_{\tau_n}\right) \;=\; a\,\Delta + O\left(\Delta^2\right),$$

$$E\left(|Y_{n+1} - Y_n|^2 \,|\, A_{\tau_n}\right) \;=\; b^2\,\Delta + O\left(\Delta^2\right),$$

$$E\left((Y_{n+1} - Y_n)^3 \,|\, A_{\tau_n}\right) \;=\; O\left(\Delta^2\right).$$

For $\beta = 2.0$, the order 2.0 weak Taylor scheme (14.2.1) satisfies the conditions

$$(6.2) \qquad E\left(Y_{n+1} - Y_n \,|\, A_{\tau_n}\right) = a\,\Delta + \frac{1}{2}\left(aa' + \frac{1}{2}b^2 a''\right)\Delta^2 + O\left(\Delta^3\right),$$

$$E\left((Y_{n+1} - Y_n)^2 \,|\, A_{\tau_n}\right) \;=\; b^2\,\Delta + \frac{1}{2}\Big[2a\,(a + bb') + b^2\,(2a' + (b')^2$$

$$+\, bb'')\Big]\,\Delta^2 + O\left(\Delta^3\right),$$

$$E\left((Y_{n+1} - Y_n)^3 \,|\, A_{\tau_n}\right) = 3b^2\,(a + bb')\,\Delta^2 + O\left(\Delta^3\right),$$

$$E\left((Y_{n+1} - Y_n)^4 \,|\, A_{\tau_n}\right) = 3b^4\,\Delta^2 + O\left(\Delta^3\right),$$

$$E\left((Y_{n+1} - Y_n)^5 \,|\, A_{\tau_n}\right) = O\left(\Delta^3\right).$$

Finally, in the case $\beta = 3.0$, we find that the order 3.0 weak Taylor scheme (14.3.1) satisfies

$$(6.3) \quad E\left(Y_{n+1} - Y_n \,|\, A_{\tau_n}\right) = a\,\Delta + \frac{1}{2}\left(aa' + \frac{1}{2}b^2 a''\right)\Delta^2 + A\,\Delta^3 + O\left(\Delta^4\right)$$

with

$$A = \frac{1}{6}\left[a\left((a')^2 + a''(a+bb')\right) + \frac{1}{2}b^2\left\{2aa''' + 3a'a'' + 2a'''bb'\right.\right.$$

$$\left.\left. + a''\left(bb'' + (b')^2\right) + \frac{1}{2}a^{(iv)}b^2\right\}\right];$$

$$E\left((Y_{n+1} - Y_n)^2 \mid \mathcal{A}_{\tau_n}\right) = b^2\Delta + \frac{1}{2}\left[2a(a+bb') + b^2(2a' + (b')^2\right.$$

$$\left. + bb'')\right]\Delta^2 + B\,\Delta^3 + O\left(\Delta^4\right)$$

with

$$B = \frac{1}{6}\left[2a\left\{a\left(3a' + (b')^2 + bb''\right) + bb'\left(3a' + (b')^2\right)\right\}\right.$$

$$+ ab^2\left(7a'' + 8b'b'' + 2bb'''\right) + b^2\left\{4(a')^2 + 7bb'a''\right.$$

$$+ (b')^2\left(4a' + (b')^2\right) + 8b(b')^2 b'' + 4bb''a'$$

$$\left.\left. + \frac{1}{2}b^2\left(4a''' + 5(b'')^2 + 8b'b''' + bb^{(iv)}\right)\right\}\right];$$

$$E\left((Y_{n+1} - Y_n)^3 \mid \mathcal{A}_{\tau_n}\right) = 3b^2(a+bb')\,\Delta^2 + C\,\Delta^3 + O\left(\Delta^4\right)$$

with

$$C = a^2(a + 3bb') + \frac{1}{2}ab^2\left(9a' + 11(b')^2 + 7bb''\right)$$

$$+ \frac{1}{2}b^2\left\{bb'\left(10a' + 8(b')^2\right) + \frac{1}{2}b^2(7a'' + 28b'b'' + 4bb'')\right\};$$

$$E\left((Y_{n+1} - Y_n)^4 \mid \mathcal{A}_{\tau_n}\right) = 3b^4\,\Delta^2 + D\,\Delta^3 + O\left(\Delta^4\right)$$

with

$$D = 2b^2\left[a(3a + 9bb') + \frac{1}{2}b^2\left(6a' + 19(b')^2 + 7bb''\right)\right];$$

$$E\left((Y_{n+1} - Y_n)^5 \mid \mathcal{A}_{\tau_n}\right) = 15b^4(a + 2bb')\,\Delta^3 + O\left(\Delta^4\right);$$

$$E\left((Y_{n+1} - Y_n)^6 \mid \mathcal{A}_{\tau_n}\right) = 15b^6\,\Delta^3 + O\left(\Delta^4\right);$$

and

$$E\left((Y_{n+1} - Y_n)^7 \mid \mathcal{A}_{\tau_n}\right) = O\left(\Delta^4\right).$$

Having these expressions for the conditional moments of the increments of the weak Taylor approximations, it is now easy to check whether or not the condition (14.5.12) is satisfied, provided we can show that the a priori estimates (14.5.10) and (14.5.11) hold. Such a priori estimates are readily obtained by standard methods using the Gronwall inequality in Lemma 4.5.17 under the assumption, for instance, of boundedness of all of the derivatives of $a$ and $b$.

The mixed moments of multiple Ito integrals described in (5.12.7) will be useful in determining the multi-dimensional generalizations of (6.1)–(6.3).

## Expansion for Implicit Schemes

Here we shall describe expansions of Ito processes which will enable us to derive second order weak implicit and predictor-corrector schemes. For this we shall slightly modify our notation for multiple stochastic integrals and coefficient functions in Chapter 5, now writing

$$f_{(j),n} = f_{(j)}\left(\tau_n, X_{\tau_n}\right), \qquad f_{(j_1,j_2),n} = f_{(j_1,j_2)}\left(\tau_n, X_{\tau_n}\right),$$

$$I_{(j),n} = I_{(j)}\left[1\right]_{\tau_n, \tau_{n+1}}, \qquad I_{(j_1,j_2),n} = I_{(j_1,j_2)}\left[1\right]_{\tau_n, \tau_{n+1}}, \qquad X_n = X_{\tau_n}$$

for $n = 0, 1, 2, \ldots$ and $j, j_1, j_2 = 1, \ldots, m$.

Thus, for a sufficiently smooth function $f$ we obtain

$$(6.4) \quad f\left(X_{n+1}\right) \approx f\left(X_n\right) + \sum_{j=0}^{m} f_{(j),n} I_{(j),n} + \sum_{j_1,j_2=0}^{m} f_{(j_1,j_2),n} I_{(j_1,j_2),n}$$

for $n = 0, 1, \ldots$ by retaining only the terms of local weak order less than three from the Ito-Taylor expansion (5.5.3) of $f$. See Section 11 of Chapter 5 for more details. In particular, with the choice $f(x) \equiv x$ in (6.4), we obtain the following approximation for the increment of the Ito process:

$$(6.5) \quad X_{n+1} - X_n \approx \sum_{j=0}^{m} f_{(j),n} I_{(j),n} + \sum_{j_1,j_2=0}^{m} f_{(j_1,j_2),n} I_{(j_1,j_2),n}$$

for $n = 0, 1, \ldots$. Here we have omitted all those terms that are not relevant for a second order weak approximation, and shall continue to do so in the expressions that follow.

From (6.5) we have

$$(6.6) \quad X_{n+1} - X_n \approx \left[l_0 f_{(0),n} + (1 - l_0) f_{(0),n}\right] \Delta + \sum_{j=1}^{m} f_{(j),n} I_{(j),n}$$

$$+ \sum_{j_1,j_2=0}^{m} f_{(j_1,j_2),n} I_{(j_1,j_2),n}$$

for any $l_0 \in \Re^1$. Inserting expansion (6.4) applied to the coefficient function $f_{(0),n+1}$ after rearranging into (6.6), we then obtain

$$(6.7) \quad X_{n+1} - X_n \approx l_0 \left\{ f_{(0),n+1} - \sum_{j=0}^{m} f_{(j,0),n} \, I_{(j),n} \right\} \Delta$$

$$+ \sum_{j=1}^{m} f_{(j),n} \, I_{(j),n} + \sum_{j_1,j_2=0}^{m} f_{(j_1,j_2),n} \, I_{(j_1,j_2),n}$$

$$+ (1 - l_0) \, f_{(0),n} \, \Delta.$$

As in the order 2.0 simplified weak Taylor scheme, we can substitute $\Delta \hat{W}^j$ for $I_{(j),n}$, $\hat{I}_{(0,j),n} = \hat{I}_{(j,0),n} = \frac{1}{2}\Delta \hat{W}^j \Delta$ for $I_{(0,j),n}$ and $I_{(j,0),n}$, and

$$\hat{I}_{(j_1,j_2),n} = \frac{1}{2} \left\{ \Delta \hat{W}^{j_1} \, \Delta \hat{W}^{j_2} + V_{j_1,j_2} \right\}$$

for $I_{(j_1,j_2),n}$ with $j_1, j_2 \in \{1, \ldots, m\}$ and $j_1 \neq j_2$. This gives us

$$(6.8) \quad X_{n+1} - X_n \approx \left\{ l_0 f_{(0),n+1} + (1 - l_0) \, f_{(0),n} \right\} \Delta$$

$$+ \sum_{j=1}^{m} f_{(j),n} \, \Delta \hat{W}^j + \sum_{j_1=0}^{m} \sum_{j_2=1}^{m} f_{(j_1,j_2),n} \, \hat{I}_{(j_1,j_2),n}$$

$$+ (1 - 2l_0) \sum_{j=0}^{m} f_{(j,0),n} \, \hat{I}_{(j,0),n}.$$

For any $h_0 \in \Re^1$ we can insert the remaining terms of the expansion (6.4) applied to the coefficient $f_{(0),n+1}$ into (6.8) to get

$$(6.9) \quad X_{n+1} - X_n \approx \left\{ l_0 f_{(0),n+1} + (1 - l_0) \, f_{(0),n} \right\} \Delta$$

$$+ \frac{1}{2} \left\{ h_0 \, (1 - 2l_0) \, f_{(0,0),n+1} + (1 - h_0) \, f_{(0,0),n} \right\} \Delta^2$$

$$+ \sum_{j=1}^{m} \left[ f_{(j),n} + \frac{1}{2} \, f_{(0,j),n} \, \Delta + \frac{1}{2} \, (1 - 2l_0) \, f_{(j,0),n} \, \Delta \right] \Delta \hat{W}^j$$

$$+ \sum_{j_1,j_2=1}^{m} f_{(j_1,j_2),n} \, \hat{I}_{(j_1,j_2),n}.$$

From expansion (6.9), via Theorem 14.5.2, we can now derive the implicit weak schemes (4.8), (4.10) and (4.11), and with an application of the deterministic Taylor expansion the schemes (4.12) and (4.13). The corrector expressions (5.7), (5.10), (5.12) and (5.14) also follow from (6.9).

# Chapter 16

# Variance Reduction Methods

In this chapter we shall describe several methods which allow a reduction in the variance of functionals of weak approximations of Ito diffusions. One method changes the underlying probability measure by means of a Girsanov transformation, another uses general principles of Monte-Carlo integration. Unbiased estimators are also constructed.

## 16.1   Introduction

In the preceding chapters we used weak approximations of an Ito process $X$ satisfying the stochastic equation

$$(1.1) \qquad X_t = X_0 + \int_0^t a\left(s, X_s\right) ds + \sum_{j=1}^m \int_0^t b^j\left(s, X_s\right) dW_s^j.$$

to evaluate functionals of the form

$$(1.2) \qquad E\left(g\left(X_T\right)\right),$$

where $g$ is a given function. Until now we have only considered *direct* time discrete approximations $Y$ of the Ito process $X$, in the sense that the solution of the stochastic equation (1.1) is discretized directly. In this chapter we shall introduce additional classes of approximations which allow a reduction in the variance of an estimator of the functional (1.2).

When applying a direct, time discrete weak approximation $Y$ previously, we evaluated (1.2) with the functional

$$(1.3) \qquad E\left(g\left(Y_{n_T}\right)\right)$$

using the estimator

$$(1.4) \qquad \eta_N = \frac{1}{N} \sum_{r=1}^N g\left(Y_{n_T}\left(\omega_r\right)\right),$$

which is just the arithmetic mean of $N$ independent simulations of the random variable $g\left(Y_{n_T}\right)$. Here $Y_{n_T}\left(\omega_r\right)$ denotes the $r$th simulation of $Y$ at time $\tau_{n_T} = T$.

We have already seen in Section 4 of Chapter 9 that we can represent the difference of the estimator (1.4) and the desired functional (1.2)

$$(1.5) \qquad \hat{\mu} = \eta_N - E\left(g\left(X_T\right)\right)$$

by the sum of the systematic error

$$(1.6) \qquad \mu_{sys} = E\left(\hat{\mu}\right) = E\left(g\left(Y_{n_T}\right)\right) - E\left(g\left(X_T\right)\right)$$

and the statistical error

$$\mu_{stat} = \hat{\mu} - \mu_{sys},$$

which, for $N \to \infty$, is asymptotically Gaussian distributed with mean zero and variance

$$(1.7) \qquad \mathrm{Var}\left(\mu_{stat}\right) = \mathrm{Var}\left(\hat{\mu}\right) = \frac{1}{N}\,\mathrm{Var}\left(g\left(Y_{n_T}\right)\right).$$

In particular, we noticed that the length of confidence intervals decreased only with order $N^{-1/2}$ as $N \to \infty$. Hence, in order to obtain sufficiently small confidence intervals it is important to begin with a small variance in the random variable $g\left(Y_{n_T}\right)$. With a direct simulation method one tries to fix the variance of $g\left(Y_{n_T}\right)$ to a value which is close to that of the variance of $g\left(X_T\right)$. However, this variance, which depends completely on $g$ and the given stochastic differential equation, may sometimes be extremely large. This problem leads to the question of whether it is possible to construct other estimators which have nearly the same expectation, but smaller variance.

One way to construct such an estimator will be described in Section 2. It is based on a Girsanov transformation of the underlying probability measure and involves a modified Ito process $\tilde{X}$ and a correcting process $\Theta$, which are chosen so that the values of their functionals are related by

$$(1.8) \qquad E\left(g\left(\tilde{X}_T\right)\Theta\right) = E\left(g\left(X_T\right)\right),$$

that is so that $g\left(\tilde{X}_T\right)\Theta$ and $g\left(X_T\right)$ have the same expectations. We shall see that we can control the variance of $g\left(\tilde{X}_T\right)\Theta$, and reduce it considerably, with an appropriate choice of $\tilde{X}_T$.

In Section 3 we shall consider other variance reducing estimators which are derived using the general principles of Monte-Carlo integration. Finally, in Section 4 we shall describe unbiased estimators $\eta$ which avoid any systematic weak errors, that is with

$$(1.9) \qquad E\left(\eta\right) = E\left(g\left(X_T\right)\right).$$

With such estimators we can focus our attention completely on the reduction of the variance of $\eta$.

The aim of the following sections is to illustrate these variance reduction methods in the context of special classes of problems and examples. However, in general, it may not be easy to construct an adapted and efficient variance reducing or unbiased estimator for a given, specific problem.

## 16.2 The Measure Transformation Method

Suppose we are given a $d$-dimensional Ito process $X^{s,x} = \{X_t^{s,x}, s \leq t \leq T\}$, starting at $x \in \Re^d$ at time $s \in [0, T)$, in terms of the stochastic equation

$$(2.1) \qquad X_t^{s,x} = x + \int_s^t a\left(z, X_z^{s,x}\right) dz + \sum_{j=1}^m \int_s^t b^j\left(z, X_z^{s,x}\right) dW_z^j.$$

Our aim is to approximate the functional

$$(2.2) \qquad u(s, x) = E\left(g\left(X_T^{s,x}\right)\right)$$

for a given real-valued function $g$ and time $s = 0$.

If we assume that the function $g$ and the drift and diffusion coefficients $a$ and $b$ are sufficiently smooth (see Theorem 4.8.6), then the function $u$ defined by (2.2) satisfies the Kolmogorov backward equation (2.4.7). That is,

$$(2.3) \qquad L^0 u(s, x) = 0$$

for $(s, x) \in (0, T) \times \Re^d$ with

$$(2.4) \qquad u(T, y) = g(y)$$

for all $y \in \Re^d$, where $L^0$ is the operator

$$L^0 = \frac{\partial}{\partial s} + \sum_{k=1}^d a^k \frac{\partial}{\partial x^k} + \frac{1}{2} \sum_{k,l=1}^d \sum_{j=1}^m b^{k,j} b^{l,j} \frac{\partial^2}{\partial x^k \partial x^l}.$$

Milstein proposed the use of the Girsanov transformation (4.8.16)–(4.8.17) to transform the underlying probability measure $P$ so that the process $\tilde{W}$ defined by

$$(2.5) \qquad \tilde{W}_t^j = W_t^j - \int_0^t d^j\left(z, \tilde{X}_z^{0,x}\right) dz$$

is a Wiener process with respect to the transformed probability measure $\tilde{P}$ with Radon-Nikodym derivative

$$(2.6) \qquad \frac{d\tilde{P}}{dP} = \Theta_t / \Theta_0.$$

Here the Ito process $\tilde{X}^{0,x}$ satisfies the stochastic equation

$$(2.7) \qquad \tilde{X}_t^{0,x} = x + \int_0^t a\left(z, \tilde{X}_z^{0,x}\right) dz + \sum_{j=1}^m \int_0^t b^j\left(z, \tilde{X}_z^{0,x}\right) d\tilde{W}_z^j$$

$$= x + \int_0^t \left[ a\left(z, \tilde{X}_z^{0,x}\right) - \sum_{j=1}^m b^j\left(z, \tilde{X}_z^{0,x}\right) d^j\left(z, \tilde{X}_z^{0,x}\right) \right] dz$$

$$+ \sum_{j=1}^m \int_0^t b^j\left(z, \tilde{X}_z^{0,x}\right) dW_z^j$$

and the correction process $\Theta$ satisfies the equation

(2.8) $$\Theta_t = \Theta_0 + \sum_{j=1}^m \int_0^t \Theta_z d^j\left(z, \tilde{X}_z^{0,x}\right) dW_z^j$$

with $\Theta_0 \neq 0$, where the $d^j$ denote given real-valued functions for $j = 1, \ldots, m$. We note that $\tilde{X}^{0,x}$ is $d$-dimensional, whereas $\Theta$ is only 1-dimensional.

Obviously, the process $\tilde{X}^{0,x}$ in (2.7) is an Ito process with respect to $\tilde{P}$ with the same drift and diffusion coefficients as the Ito process $X^{s,x}$ in (2.1). From this fact and (2.6), it follows that

(2.9) $$E\left(g\left(X_T^{0,x}\right)\right) = \int g\left(X_T^{0,x}\right) dP$$

$$= \int g\left(\tilde{X}_T^{0,x}\right) d\tilde{P}$$

$$= \int g\left(\tilde{X}_T^{0,x}\right) \Theta_T/\Theta_0 \, dP = E\left(g\left(\tilde{X}_T^{0,x}\right) \Theta_T/\Theta_0\right).$$

Hence, we can estimate the expectation of the random variable

(2.10) $$g\left(\tilde{X}_T^{0,x}\right) \Theta_T/\Theta_0$$

to evaluate the functional (2.2). This result does not depend on the choice of the functions $d^j$, $j = 1, \ldots, m$, so we can use them as parameters to reduce the variance of the random variable (2.10).

The following situation is interesting from a theoretical viewpoint. Suppose that $u(t, x) > 0$ everywhere and that the corresponding solutions of (2.7) and (2.8) exist. In addition, we choose the parameter functions $d^j$ as

(2.11) $$d^j(t, x) = -\frac{1}{u(t, x)} \sum_{k=1}^d b^{k,j}(t, x) \frac{\partial u}{\partial x^k}(t, x)$$

for all $(t, x) \in [0, T] \times \Re^d$ and $j = 1, \ldots, m$. Then, from the Ito formula (3.4.6) with the aid of (2.7), (2.8), (2.11) and (2.3), it follows that

(2.12) $$u\left(t, \tilde{X}_t^{0,x}\right) \Theta_t = u(0, x)\Theta_0$$

for all $t \in [0, T]$.

**Exercise 16.2.1**    *Show (2.12).*

Combining (2.4) and (2.12), we can conclude that

$$(2.13) \qquad u(0, x) = g\left(\tilde{X}_T^{0,x}\right) \Theta_T/\Theta_0.$$

Hence, with the choice of parameter functions (2.11) the variable

$$(2.14) \qquad g\left(\tilde{X}_T^{0,x}\right) \Theta_T/\Theta_0$$

is nonrandom and the variance is reduced to zero.

If we could simulate the expression (2.14) approximately using our proposed weak approximations for the stochastic equations (2.7) and (2.8), we would expect, in general, to obtain a small variance for the estimator (see PC-Exercise 16.2.3 below). Unfortunately, for the construction of the parameter functions in (2.11) we need to know the solution $u$ of the Kolmogorov backward equation, but this is exactly what we are trying to determine by means of the simulation. Nevertheless, the above discussion shows that it is possible to obtain a substantial reduction in the variance of the estimator by an application of a measure transformation.

A practical way of implementing such a method is to, somehow, find or guess a function $\bar{u}$ which is similar to the solution $u$ of the Kolmogorov backward equation (2.3)–(2.4). We can then use $\bar{u}$ instead of $u$ to define parameter functions of the form

$$(2.15) \qquad d^j(t, x) = -\frac{1}{\bar{u}(t, x)} \sum_{k=1}^{d} b^{k,j}(t, x) \frac{\partial \bar{u}}{\partial x^k}(t, x)$$

for all $(t, x) \in [0, T] \times \Re^d$ and $j = 1, \ldots, m$. Then

$$g\left(\tilde{X}_T^{0,x}\right) \Theta_T/\Theta_0$$

will still be a random variable, but with small variance if $\bar{u}$ is chosen sufficiently close to $u$.

The weak schemes, especially the higher order ones, that we studied in the preceding chapters can be readily applied to provide weak approximations of the process $(\tilde{X}^{0,x}, \Theta)$ to estimate the functional

$$E\left(g\left(\tilde{X}_T^{0,x}\right) \Theta_T/\Theta_0\right) = E\left(g\left(X_T^{0,x}\right)\right).$$

We assumed above that $u$ is strictly positive, which follows if $g$ is strictly positive, but this is often not the case. When $g > -c$ for some $c > 0$ the function $\bar{g} + 2c$ is strictly positive, whereas if $g$ is unbounded we can write $g$ as the difference of its positive and negative parts

$$g = g^+ - g^- = (g^+ + c) - (g^- + c)$$

for $c > 0$. We can then apply the above method with functionals involving the positive functions $\bar{g}$, $g^+ + c$ and $g^- + c$ rather than with $g$ itself.

The above method has been found to be worthwhile, both in terms of providing smaller confidence intervals and requiring less computational effort, in cases where the variance of the arithmetic averaging estimator (1.4) is large.

**Exercise 16.2.2** *Check that* $g\left(\tilde{X}_T^{0,x}\right) \Theta_T/\Theta_0$ *is nonrandom and equal to*

$$E\left(g\left(X_T^{0,x}\right)\right)$$

*for the linear stochastic equation*

$$(2.16) \qquad X_t = x + \int_0^t \alpha\, X_z\, dz + \int_0^t \beta\, X_z\, dW_z$$

*with* $d = m = 1$ *when* $g(y) \equiv y^2$ *and* $d^1$ *is defined by (2.11)*.

**PC-Exercise 16.2.3** *Evaluate* $E\left((X_T)^2\right)$ *using the measure transformation method for the linear problem in Exercise 16.2.2 with* $\alpha = 1.5$, $\beta = 1.0$, $x = 0.1$ *and* $T = 1$. *Apply the Euler scheme with equidistant step size* $\delta = \Delta = 2^{-4}$ *to generate* $M = 20$ *batches each with* $N = 15$ *trajectories. Determine the 90% confidence intervals for the mean error* $\mu$. *Repeat the calculations for step sizes* $\Delta = 2^{-5}$, $2^{-6}$ *and* $2^{-7}$. *Plot the confidence intervals against* $\Delta$.

## 16.3 Variance Reduced Estimators

In this section we shall describe another way of constructing variance reduced random variables for the simulation of functionals of Ito processes. These results are due to Wagner and are based on the theory of Monte-Carlo integration.

We begin with a $d$-dimensional Ito process

$$(3.1) \qquad X_t = X_0 + \int_0^t a\left(s, X_s\right)\, ds + \int_0^t b\left(s, X_s\right)\, dW_s$$

with scalar noise, $m = 1$, and wish to evaluate the functional

$$(3.2) \qquad E\left(g\left(X_T\right)\right) < \infty.$$

Suppose that we are given a time discretization

$$0 = \tau_0 < \tau_1 < \cdots < \tau_N = T$$

of the interval $[0, T]$. As in Section 7 of Chapter 1, we shall denote by $p(s, y; t, x)$ the transition density of the Ito process (3.1) for the transition from the point $y$ at time $s$ to the point $x$ at time $t$, where $t \geq s$. In addition, we shall denote the probability measure of the initial value $X_0$ by $\rho(dx)$. Using the Markov property and the Chapman-Kolmogorov equation (1.7.5), we can represent the functional (3.2) in the form

$$
E\left(g\left(X_T\right)\right) \;=\; \int_{\Re^d} \cdots \int_{\Re^d} \int_{\Re^d} \prod_{i=1}^{N} p\left(\tau_{i-1}, x_{i-1}; \tau_i, x_i\right) g\left(x_N\right)
$$

(3.3)
$$
\times \rho\left(dx_0\right) dx_1 \cdots dx_N.
$$

If we set

(3.4)
$$
d\mu(\zeta) = \rho\left(dx_0\right) dx_1 \cdots dx_N
$$

and

(3.5)
$$
F(\zeta) = \prod_{i=1}^{N} p\left(\tau_{i-1}, x_{i-1}; \tau_i, x_i\right) g\left(x_N\right),
$$

where $\zeta = \left(x_0, \ldots, x_N\right) \in \Gamma := \left(\Re^d\right)^{N+1}$, then we can write (3.3) as the finite-dimensional integral

(3.6)
$$
E\left(g\left(X_T\right)\right) = \int_{\Gamma} F(\zeta)\, d\mu(\zeta).
$$

A simple Monte-Carlo estimator for this integral is the one-point estimator

(3.7)
$$
\eta_1 = \frac{F(\zeta)}{D(\zeta)},
$$

where $\zeta$ is a random variable with density $D$ with respect to the measure $d\mu$. If we assume that

(3.8)
$$
D\left(\left(x_0, \ldots, x_N\right)\right) > 0
$$

for all $\left(x_0, \ldots, x_N\right) = \zeta \in \Gamma$ with $g(x_N) \neq 0$, then it follows from (3.6) and (3.7) that

(3.9)
$$
\begin{aligned}
E\left(\eta_1\right) &= E\left(\frac{F(\zeta)}{D(\zeta)}\right) \\[2mm]
&= \int_{\Gamma} \frac{F(\zeta)}{D(\zeta)} D(\zeta)\, d\mu(\zeta) \\[2mm]
&= \int_{\Gamma} F(\zeta)\, d\mu(\zeta) = E\left(g\left(X_T\right)\right).
\end{aligned}
$$

The estimator (3.7) is thus unbiased. However, in general, the function $F$ here is not known explicitly.

## The Variance Reducing Euler Estimator

The estimator (3.7) is nevertheless still useful if we can somehow approximate or estimate the finite-dimensional density of the Ito process

$$(3.10) \qquad Q\left((x_0, \ldots, x_N)\right) = \prod_{i=1}^{N} p\left(\tau_{i-1}, x_{i-1}; \tau_i, x_i\right)$$

by a similar, but known density. For instance, we could use the density of a weak time discrete approximation of the Ito process, the simplest being the Euler approximation.

In what follows we shall represent the density of the Euler approximation in the form

$$(3.11) \qquad \tilde{Q}\left(\zeta\right) = \prod_{i=1}^{N} \tilde{p}_{a,\sigma}\left(\tau_{i-1}, x_{i-1}; \tau_i, x_i\right)$$

for all $\zeta = (x_0, \ldots, x_N) \in \Gamma$ with the density for Gaussian increments

$$(3.12) \qquad \tilde{p}_{a,\sigma}(s, y; t, x) = \frac{1}{\sqrt{det\, \sigma(s, y) \left(2\pi(t - s)\right)^d}}$$

$$\times \exp\left(-\frac{1}{2(t - s)}\left(\sigma^{-1}(s, y)A, A\right)\right),$$

where

$$A = x - y - a(s, y)(t - s).$$

Here we are assuming that the symmetric matrix

$$(3.13) \qquad \sigma(s, y) = b(s, y)b(s, y)^{\mathsf{T}}$$

is strictly positive definite for all $(s, y) \in [0, T] \times \Re^d$ and use the standard notation for the determinant $det$, the scalar product $(\cdot, \cdot)$, the matrix inverse $\sigma^{-1}$ and the vector and matrix transpose $b^{\mathsf{T}}$.

We define the *variance reducing Euler estimator* as

$$(3.14) \qquad \tilde{\eta}_E = \frac{\tilde{Q}(\zeta)}{D(\zeta)}\, g\left(x_N\right),$$

where $\zeta = (x_0, \ldots, x_N) \in \Gamma$ is a random variable with density $D$ with respect to $d\mu$. We note that the systematic error of (3.14) does not depend on $D$ and thus coincides with that of the *direct Euler estimator* obtained from (3.14) by replacing $D$ with $\tilde{Q}$ defined in (3.11), which, in turn, is equivalent to the systematic error of the weak Euler scheme discussed in Section 1 of Chapter 14. An appropriate choice of the free parameter $D$ in (3.14) then provides a means for reducing the variance of the estimator.

## The Optimal Density

In order to determine the optimal version $D_{opt}$ of $D$ we need to assume that we know the transition density $p(s, y; t, x)$ of the Ito process. For the one-point estimator (3.7) it is easy to show that the random variable

(3.15)
$$D_{opt}(\zeta) = |F(\zeta)| \Big/ \int_{\Gamma} |F(z)| \, d\mu(z)$$

$$= Q(\zeta) |g(x_N)| \Big/ E(|g(X_T)|),$$

where $\zeta = (x_0, \ldots, x_N) \in \Gamma$, has variance

(3.16)
$$\text{Var}\left(\frac{F(\zeta)}{D_{opt}(\zeta)}\right) = \left(\int_{\Gamma} |F(\zeta)| \, d\mu(\zeta)\right)^2 - \left(\int_{\Gamma} F(\zeta) \, d\mu(\zeta)\right)^2.$$

**Exercise 16.3.1** *Show (3.16).*

On the other hand, from the Lyapunov inequality (1.4.12), for any given density $D$ we obtain the inequality

(3.17)
$$\text{Var}\left(\frac{F(\zeta)}{D(\zeta)}\right) = E\left(\left|\frac{F(\zeta)}{D(\zeta)}\right|^2\right) - \left(E\left(\frac{F(\zeta)}{D(\zeta)}\right)\right)^2$$

$$\geq \left(E\left(\frac{|F(\zeta)|}{D(\zeta)}\right)\right)^2 - \left(\int_{\Gamma} F(\zeta) \, d\mu(\zeta)\right)^2$$

$$\geq \left(\int_{\Gamma} |F(\zeta)| \, d\mu(\zeta)\right)^2 - \left(\int_{\Gamma} F(\zeta) \, d\mu(\zeta)\right)^2.$$

This shows that the minimum possible variance, which is zero when $g$ is only positive or only negative, is attained with the $D_{opt}$ defined in (3.15), which justifies its being designated the *optimal density*.

Usually the transition density $p(s, y; t, x)$ of the Ito process, and hence $D_{opt}$, is not known. The method of choosing $D$ close to $D_{opt}$ in the one-point estimator (3.7) is known as *importance sampling* because the density is high in those regions making the most important contributions to the functional.

When $g$ changes sign, we can write it as the sum of its positive and negative parts and apply the above procedure to each part separately. The near optimal choice of the corresponding densities $D^+$ and $D^-$ is then called *stratified sampling*, but we shall not say anymore about it here.

## The Direct Euler Estimator

The optimal density $D_{opt}$ obviously depends on the shape of the given function $g$. Here we shall consider the simple case that $g$ takes only two distinct values. Such a function can be easily transformed to another, which we shall also denote by $g$ for simplicity, satisfying

(3.18)
$$|g(y)| \equiv const$$

for all $y \in \Re^d$. From (3.15) we then obtain

(3.19)                                $D_{opt}(\zeta) = Q(\zeta)$

for all $\zeta \in \Gamma$, which is just the finite-dimensional density (3.10) of the original Ito process (3.1). Thus, we can approximate $D_{opt}$ in (3.19) simply by the density of the Euler approximation $\tilde{Q}$ defined in (3.11) or by other explicitly given densities of weak schemes. If we use the Euler density $\tilde{Q}$ in place of $D$ in the variance reducing Euler estimator (3.14), we obtain the *direct Euler estimator*

(3.20)                                $\tilde{\eta}_E = g(x_N)$

for all $\zeta = (x_0, \dots, x_N) \in \Gamma$. Consequently, the direct Euler estimator (3.20) is a good variance reducing estimator for functions $g$ satisfying (3.18).

## Ito Processes with Constant Coefficients

Another useful special case is when the drift and diffusion coefficients of the Ito process are constants, since the results obtained can also be used to provide an approximation of the optimal density when these coefficients differ only slightly from constants.

We shall assume here that

$$a(s, y) \equiv a \qquad \text{and} \qquad b(s, y) b(s, y)^\top = \sigma(s, y) \equiv \sigma$$

for all $(s, y) \in [0, T] \times \Re^d$, where $a$ and $\sigma > 0$ are constants. It is obvious from (3.10) and (3.11) that the density of the Ito process then coincides with that of the Euler scheme, that is

(3.21)                                $Q(\zeta) = \tilde{Q}(\zeta)$

for all $\zeta \in \Gamma$. For this case it then follows from (3.15) that the optimal density is

(3.22)                        $\tilde{D}_{a,\sigma}(\zeta) = d_{x_0}(x_N) \prod_{i=1}^{N} D_{i, x_{i-1}, x_N}(x_i),$

where

(3.23)        $d_{x_0}(x_N) = \tilde{p}_{a,\sigma}(0, x_0; T, x_N) |g(x_n)| \Big/ E(|g(X_T)|)$

and

(3.24)        $D_{i, x_{i-1}, x_N}(x_i) = \dfrac{\tilde{p}_{a,\sigma}(\tau_{i-1}, x_{i-1}; \tau_i, x_i) \, \tilde{p}_{a,\sigma}(\tau_i, x_i; T, x_N)}{\tilde{p}_{a,\sigma}(\tau_{i-1}, x_{i-1}; T, x_N)}$

for all $\zeta = (x_0, \dots, x_N) \in \Gamma$, with $\tilde{p}_{a,\sigma}$ defined in (3.12). It is not difficult to show that $D_{i, x_{i-1}, x_N}(x_i)$ is the density of a Gaussian random variable $x_i$ with vector mean

(3.25)                        $x_{i-1} \dfrac{T - \tau_i}{T - \tau_{i-1}} + x_N \dfrac{\tau_i - \tau_{i-1}}{T - \tau_{i-1}}$

and covariance matrix

(3.26)
$$\sigma \frac{(T - \tau_i)(\tau_i - \tau_{i-1})}{T - \tau_{i-1}}$$

for given $i$, $x_{i-1}$, $x_N$.

**Exercise 16.3.2**    *Show (3.25) and (3.26) for $d = 1$, $a = const$, $\sigma = const$.*

Before we can begin simulating the conditional Gaussian random variables $x_1, \ldots, x_{N-1}$ step by step, we need to generate the starting value $x_0$ according to $p(dx_0)$ and then $x_N$ using the density $d_{x_0}(x_N)$. From (3.23) and (3.24) we note that only the second of these densities depends on the particular function $g$ under consideration. It is useful to find approximations for $d_{x_0}(x_N)$ for special classes of functions $g$, for instance, the indicator functions.

## Functionals involving Indicator Functions

In Chapters 14 and 15 we restricted attention to functions $g$ which were sufficiently smooth. We shall now investigate what happens in the 1-dimensional case $d = m = 1$ if we use the indicator function

(3.27)
$$g(y) = I_{[\underline{c}, \bar{c}]}(y) = \begin{cases} 1 & : \quad y \in [\underline{c}, \bar{c}] \\ 0 & : \quad \text{otherwise} \end{cases}$$

for all $y \in \Re^1$ where $-\infty < \underline{c} < \bar{c} < \infty$. Obviously,

(3.28)
$$E(g(X_T)) = P(X_T \in [\underline{c}, \bar{c}]),$$

so with this functional we are estimating the probability that $X_T$ lies in the interval $[\underline{c}, \bar{c}]$. A possible approximation for $d_{x_0}(x_N)$ in (3.22)–(3.23) is the uniform density

(3.29)
$$d(x_N) = \frac{1}{\bar{c} - \underline{c}} I_{[\underline{c}, \bar{c}]}(x_N),$$

for a uniformly distributed random variable on $[\underline{c}, \bar{c}]$.

We shall use the following example in PC-Exercises to illustrate the possibilities of the variance reduction technique.

**Example 16.3.3**    *Let $X$ be the 1-dimensional Ito process with initial value $X_0 = 0$, driven by additive noise with drift and diffusion coefficients*

$$a(s, y) = \frac{1}{2} - \frac{1}{2}\sin y \quad and \quad b(s, y) \equiv 1.$$

*Consider the functional (3.2) with the indicator function*

$$g(y) = I_{[0.3, 0.4]}(y)$$

*for $y \in \Re^1$ and final time $T = 1$.*

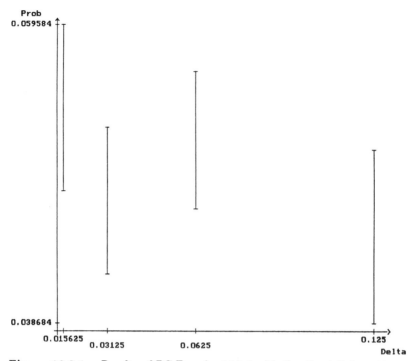

**Figure 16.3.1**     Results of PC-Exercise 16.3.4 with the direct Euler estimator.

**PC-Exercise 16.3.4**     *For Example 16.3.3 use the direct Euler estimator (3.20) to simulate $M = 20$ batches each of $N = 200$ trajectories and estimate the probability $P(X_T \in [0.3, 0.4])$ for equidistant step sizes $\Delta = \tau_{i+1} - \tau_i = 2^{-3}$, ..., $2^{-6}$. Plot the corresponding 90% confidence intervals against $\Delta$.*

**PC-Exercise 16.3.5**     *Repeat PC-Exercise 16.3.4 using the variance reducing Euler estimator (3.14), taking as the density $D$ the expression $\tilde{D}_{a,\sigma}$ in (3.22) with $a = 0.5$, $\sigma = 1.0$ and approximating $d_{x_0}(x_N)$ in (3.22) by the $d(x_N)$ in (3.29) with $\underline{c} = 0.3$ and $\bar{c} = 0.4$.*

A comparison of Figures 16.3.1 and 16.3.2 shows that a considerable reduction in the variance of the estimation of the desired functional can be achieved with the variance reducing Euler estimator. We remark that it would require an immense amount of computer time to obtain a comparable variance with the direct Euler estimator.

# 16.4   Unbiased Estimators

The weak approximations that we have considered so far all produce some systematic error. In this section we shall see how ideas from the theory of Monte-Carlo integration can be used to construct estimators which have no bias. The

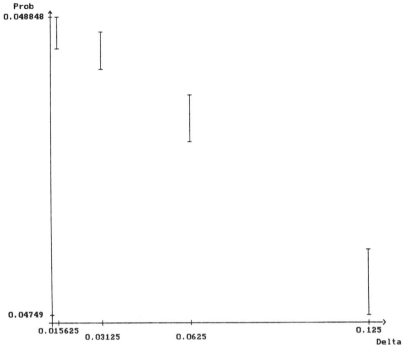

**Figure 16.3.2**   Results of PC-Exercise 16.3.5
with the variance reduced Euler estimator.

following results are due to Wagner, who obtained unbiased estimators by replacing the transition densities $p(\tau_{i-1}, x_{i-1}; \tau_i, x_i)$ of the Ito process in the
estimator (3.5)–(3.7) by the corresponding unbiased Monte-Carlo estimators
$q(\tau_{i-1}, x_{i-1}; \tau_i, x_i)$. These unbiased estimators result from the application of
a powerful Monte-Carlo method due to von Neumann and Ulam to an integral
equation which is related to the Kolmogorov backward equation. They allow
us to introduce the *variance reducing unbiased estimator*

$$(4.1) \qquad \eta_u = \frac{\prod\limits_{i=1}^{N} q\left(\tau_{i-1}, x_{i-1}; \tau_i, x_i\right)}{D\left((x_0, \ldots, x_N)\right)}\, g\left(x_N\right),$$

where $\zeta = (x_0, \ldots, x_N) \in \Gamma$ is a random variable with density $D$ with respect
to the measure $d\mu$ (see (3.4)) and $q(\tau_{i-1}, x_{i-1}; \tau_i, x_i)$ for given $i$, $\tau_{i-1}$, $x_{i-1}$,
$\tau_i$, $x_i$ denotes a random variable with the property

$$(4.2) \qquad E\left(q\left(\tau_{i-1}, x_{i-1}; \tau_i, x_i\right)\right) = p\left(\tau_{i-1}, x_{i-1}; \tau_i, x_i\right).$$

If we choose the density $D$ in the estimator (4.1) to be the Euler density $\tilde{Q}$
given in (3.11), then we call $\eta_u$ the *direct unbiased Euler density*. However, here
we can also reduce the variance by chosing $D$ close to the optimal density $D_{opt}$,

in which case we shall call $\eta_u$ the *unbiased variance reducing Euler estimator*.

We shall describe the unbiased estimator $q$ for the transition density $p$ in the simple case of additive noise of the kind

$$(4.3) \qquad\qquad b(s,y)b(s,y)^\top \equiv \sigma I,$$

where $\sigma > 0$ is a constant and $I$ is the $d \times d$ identity matrix. In addition, we shall assume that the drift $a(s, y)$ is Hölder continuous and bounded.

To simulate the unbiased estimator $q$, for each $i \in \{1, \ldots, N\}$, we need to generate a number $l_i$ of auxiliary random time instants $\tau_{i,j}$ in the interval $[\tau_{i-1}, \tau_i]$ for $j = l_i, l_i - 1, \ldots, 1$ with

$$(4.4) \qquad\qquad \tau_{i-1} = \tau_{i,l_i} < \tau_{i,l_i-1} < \cdots < \tau_{i,1} < \tau_{i,0} = \tau_i.$$

The number $l_i$, which is random, is obtained automatically by the following procedure: starting with $\tau_{i,0} = \tau_i$, for $j = 1, 2, \ldots$ choose

$$\tau_{i,j} = \tau_{i-1}$$

with probability

$$(4.5) \qquad\qquad P\left(\tau_{i,j} = \tau_{i-1}\right) = \exp\left(-\left(\tau_{i,j-1} - \tau_{i-1}\right)\right),$$

otherwise choose $\tau_{i,j}$ as a random variable taking values in the interval $[\tau_{i-1}, \tau_{i,j-1}]$ with probability density

$$(4.6) \qquad\qquad e_{\tau_{i-1},\tau_{i,j-1}}\left(\tau_{i,j}\right) = \frac{\exp(\tau_{i,j} - \tau_{i-1})}{\exp(\tau_{i,j-1} - \tau_{i-1}) - 1};$$

repeat this procedure until $\tau_{i-1}$ is chosen as the value for $\tau_{i,j}$ and take this $j$ as the number $l_i$.

Then, at the auxiliary times $\tau_{i,j}$ we need to generate auxiliary points $x_{i,j} \in \Re^d$ of the trajectory for each $j = 1, \ldots, l_i - 1$ and $i = 1, 2, \ldots, N$. For convenience we shall also write

$$(4.7) \qquad\qquad x_{i,0} = x_{i-0} \qquad \text{and} \qquad x_{i,l_i} = x_i.$$

Analogously with (3.25)–(3.26), we generate the auxiliary points $x_{i,j}$ for $j = 1, \ldots, l_i - 1$ and $i = 1, 2, \ldots, N$ as Gaussian random variables with vector mean

$$(4.8) \qquad\qquad x_{i,j-1} \frac{\tau_i - \tau_{i,l_i-j}}{\tau_i - \tau_{i,l_i-j+1}} + x_i \frac{\tau_{i,l_i-j} - \tau_{i,l_i-j+1}}{\tau_i - \tau_{i,l_i-j+1}}$$

and covariance matrix

$$(4.9) \qquad\qquad \sigma I \frac{\left(\tau_i - \tau_{i,l_i-j}\right)\left(\tau_{i,l_i-j} - \tau_{i,l_i-j+1}\right)}{\tau_i - \tau_{i,l_i-j+1}}.$$

The unbiased estimator $q$ for the transition density $p$ is then given by

$$(4.10) \qquad q\left(\tau_{i-1}, x_{i-1}; \tau_i, x_i\right) = \Phi\left(\tau_{i-1}, x_{i-1}; \tau_i, x_i\right)$$

$$+\tilde{p}_{0,\sigma I}\left(\tau_{i-1}, x_{i-1}; \tau_i, x_i\right)\exp\left(\tau_i - \tau_{i-1}\right)$$

$$\times \sum_{k=1}^{l_i-1}\left[\prod_{j=1}^{k}\frac{K\left(\tau_{i,l_i-j}, x_{i,j}; \tau_{i,l_i-j+1}, x_{i,j-1}\right)}{\tilde{p}_{0,\sigma I}\left(\tau_{i,j}, x_{i,l_i-j}; \tau_{i,j-1}, x_{i,l_i-j+1}\right)}\right]$$

$$\times\frac{\Phi\left(\tau_{i-1}, x_{i-1}; \tau_{i,k}, x_{i,l_i-k}\right)}{\tilde{p}_{0,\sigma I}\left(\tau_{i-1}, x_{i-1}; \tau_{i,k}, x_{i,l_i-k}\right)\exp\left(\tau_{i,k} - \tau_{i-1}\right)},$$

where

$$\Phi(s, y; t, x) = \frac{1}{\sqrt{(2\pi\sigma(t-s))^d}}\,\exp\left(-\frac{|x - y - a(t,x)(t-s)|^2}{2\sigma(t-s)}\right)$$

and

$$K(s, y : t, x) = \Phi(s, y; t, x)\frac{(a(s,y) - a(t,x), x - y - a(t,x)(t-s))}{\sigma(t-s)}.$$

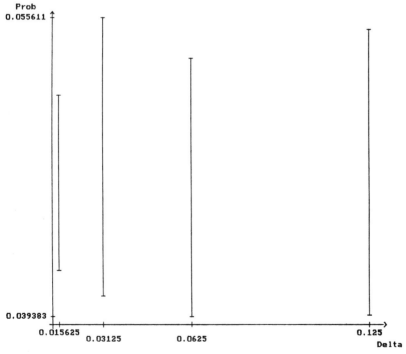

**Figure 16.4.1** Results of PC-Exercise 16.4.2 with the direct unbiased Euler estimator.

Wagner proved the following theorem, the proof of which we shall omit, using general results from the theory of Monte-Carlo integration.

**Theorem 16.4.1**     *Suppose for the variance reducing estimator (4.1) that the above assumptions hold, that D satisfies condition (3.8) and that ρ and g are such that*

(4.11) $$\int_{\Re^d} E\left(|g\left(y + W_t\right)|\right) \rho\left(dy\right) < \infty$$

*for all $t > 0$, where $W$ is a d-dimensional Wiener process. Then the variance reducing unbiased estimator (4.1) satisfies*

(4.12) $$E\left(\eta_u\right) = E\left(g\left(X_T\right)\right).$$

We shall use Example 16.3.3 from the last section to illustrate how unbiased estimators work. Our direct unbiased Euler estimator has the form

(4.13) $$\eta_N = \frac{\displaystyle\prod_{i=1}^{N} q\left(\tau_{i-1}, x_{i-1}; \tau_i, x_i\right)}{\tilde{Q}\left(\left(x_0, \ldots, x_N\right)\right)} g\left(x_N\right),$$

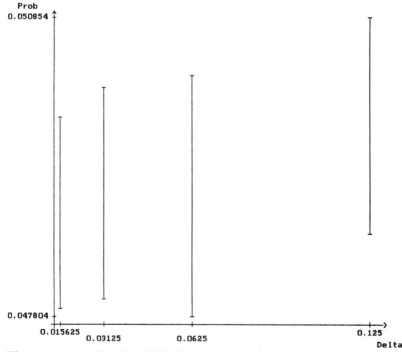

**Figure 16.4.2**     Results of PC-Exercise 16.4.3 with the variance reducing unbiased Euler estimator.

where the $x_0$, ..., $x_N$ are determined from a realization of the Euler scheme and the $q(\tau_{i-1}, x_{i-1}; \tau_i, x_i)$ are generated by (4.10) using the auxiliary times $\tau_{i,j}$ and points $x_{i,j}$.

**PC-Exercise 16.4.2** *Repeat PC-Exercise 16.3.4 with the direct unbiased Euler estimator (4.13).*

To construct a variance reducing unbiased Euler estimator

$$
(4.14) \qquad \tilde{\eta}_N = \frac{\displaystyle\prod_{i=1}^{N} q\left(\tau_{i-1}, x_{i-1}; \tau_i, x_i\right)}{\tilde{D}_{a,\sigma}\left((x_0, \ldots, x_N)\right)} \, g\left(x_N\right)
$$

we can use the same $\tilde{D}_{a,\sigma}$ as in the last section, which was given by (3.21) and (3.25).

**PC-Exercise 16.4.3** *Repeat PC-Exercise 16.3.4 with the variance reducing unbiased Euler estimator (4.14).*

A comparison of Figures 16.4.1 and 16.4.2 shows that a considerable reduction in variance is also possible for unbiased estimators.

# Chapter 17

# Selected Applications
# of Weak Approximations

In this final chapter we indicate several examples of applications of weak approximations. We begin with the evaluation of Wiener function space integrals, which generalize stochastic quadrature formulae, and then use weak schemes to approximate invariant measures. Finally, we compute the top Lyapunov exponents for linear stochastic differential equations. We believe that the techniques outlined here bear much potential for the development of effective numerical methods for higher dimensional partial differential equations, in particular nonlinear ones.

## 17.1  Evaluation of Functional Integrals

In a number of problems in mathematical physics and other fields Wiener function space integrals, which are also called functional integrals or Wiener integrals, play an important role. For instance, they arise in the analysis of wave scattering in random media (see Blankenship and Baras (1981)) and representations of the Schrödinger equation, which is just the time reversed Kolmogorov backward equation (4.8.15). The corresponding partial differential equation has the form

$$(1.1) \qquad \frac{\partial}{\partial t} u(t, x) + \frac{1}{2} \sum_{k=1}^{m} \frac{\partial^2}{\partial x^k \partial x^k} u(t, x) + V(t, x) u(t, x) = 0$$

for $0 \leq t \leq T$ and $x \in \Re^m$ with final time condition

$$u(T, x) = f(T, x),$$

where $V$ and $f$ are given. From the Feynman-Kac formula (4.8.14) we have

$$(1.2) \qquad u(t, x) = E\left( f(T, W_T) \exp\left( \int_t^T V(s, W_s) \, ds \right) \Big| W_t = x \right),$$

where $W = \{W_s, \ s \geq t\}$ is an $m$-dimensional Wiener process starting at $x = (x^1, \ldots, x^m) \in \Re^m$ at time $t$. The expectation here is with respect to the probability measure of the Wiener process $W$.

The evaluation of the functional integral (1.2) is a typical problem for the weak approximations that we considered in Chapters 14–16. On the other hand, evaluating $u(t, x)$ in (1.2) for all possible $(t, x)$ is the same as solving the parabolic equation (1.1). Weak aproximations for stochastic differential equations thus provide a tool for solving second order partial differential equations, which can be particular efficient in higher dimensions if parallel computers are available.

For given, sufficiently smooth functions $f$, $\phi$ and $a$ we introduce the functional

$$(1.3) \qquad F = E\left( f(T, W_T) \phi\left(\int_0^T a(s, W_s)\, ds\right)\right),$$

where $W$ is an $m$-dimensional Wiener process starting at $x = (x^1, \ldots, x^m) \in \Re^m$ at time $t = 0$. Using the $(m+1)$-dimensional Ito process $X = \{X_t = (X_t^1, \ldots, X_t^{m+1}), t \geq 0\}$ which satisfies the stochastic equation

$$(1.4) \qquad X_t^k = x^k + \int_0^t dW_s^k, \qquad \text{for} \quad k = 1, \ldots, m,$$

$$X_t^{m+1} = \int_0^t a\left(s, \tilde{X}_s\right)\, ds,$$

where $\tilde{X}_s = (X_s^1, \ldots, X_s^m)$, we can rewrite the functional (1.3) in the form

$$(1.5) \qquad F = E\left( f\left(T, \tilde{X}_T\right) \phi\left(\int_0^T a\left(s, \tilde{X}_s\right)\, ds\right)\right)$$

$$= E\left( f\left(T, \tilde{X}_T\right) \phi\left(X_T^{m+1}\right)\right)$$

$$= E\left(g\left(X_T\right)\right),$$

where

$$g(y) = f\left(T, (y^1, \ldots, y^m)\right) \phi\left(y^{m+1}\right).$$

The expression (1.5) is now a functional of an multi-dimensional Ito process, so we can approximate it using a weak scheme. We note that the structure of the stochastic equation (1.4) is quite simple, so there are good propects of finding efficient weak schemes of high order with which to evaluate the functional (1.5).

**Example 17.1.1**     *For $m = T = 1$ in (1.4)–(1.5) with*

$$f(t, y^1) \equiv 1, \qquad \phi\left(y^2\right) = \exp\left(y^2\right), \qquad a\left(t, y^1\right) = -\frac{1}{2}\left(y^1\right)^2$$

*and $\tilde{X}_0 = 0$ we obtain the functional*

$$F = E\left(\exp\left(-\frac{1}{2}\int_0^T (W_s)^2\, ds\right)\right).$$

**Exercise 17.1.2**  *Show for Example 17.1.1 that the function*

$$u(t,x) = \exp\left(\frac{1}{2}(T-t) + \frac{1}{2}\left(\frac{1 - e^{2(T-t)}}{1 + e^{2(T-t)}}\right)x^2 + \frac{1}{2}\ln\left(\frac{2}{1 + e^{2(T-t)}}\right)\right)$$

*for $t \in [0,T]$ and $x \in \Re^1$ satisfies the equation (1.1) with $f \equiv 1$ and $V(t,x) = \frac{1}{2}x^2$, and that the functional (1.5) takes the value*

$$F = u(0,0) = \sqrt{\frac{2e^T}{1 + e^{2T}}}.$$

We shall now apply some of the previously discussed weak schemes with equidistant step size $\Delta$ to the problem (1.4)–(1.5).

The Euler scheme (14.1.1) for (1.4) is

$$(1.6) \qquad Y_{n+1}^k = Y_n^k + \Delta \hat{W}_n^k, \qquad \text{for } k = 1, \ldots, m,$$

$$Y_{n+1}^{m+1} = Y_n^{m+1} + a\left(\tau_n, \tilde{Y}_n\right)\Delta$$

with $\tilde{Y}_0 = x$ and $Y_0^{m+1} = 0$, where $\tilde{Y}_n = (Y_n^1, \ldots, Y_n^m)$. Here the $\Delta \hat{W}_n^k$ are independent $N(0; \Delta)$ Gaussian random variables or two-point distributed random variables with

$$P\left(\Delta \hat{W}_n^k = \pm\sqrt{\Delta}\right) = \frac{1}{2}$$

for $k = 1, \ldots, m$ and $n = 0, 1, \ldots$. Using (1.6) with a step size $\delta = \Delta$, we define the random variable

$$(1.7) \qquad \tilde{F}_1^\delta = f\left(T, \tilde{Y}_{n_T}\right)\phi\left(\Delta \sum_{n=0}^{n_T-1} a\left(\tau_n, \tilde{Y}_n\right)\right)$$

with $n_T = T/\Delta$. Its expectation provides an approximation for the functional

$$(1.8) \qquad F = E\left(f\left(T, \tilde{X}_T\right)\phi\left(\int_0^T a\left(s, \tilde{X}_x\right)ds\right)\right)$$

with weak order $\beta = 1.0$, that is

$$(1.9) \qquad \left|F - E\left(\tilde{F}_1^\delta\right)\right| \le K\delta.$$

**PC-Exercise 17.1.3**  *Apply the Euler scheme with step size $\delta = \Delta = 2^{-3}$ for Example 17.1.1 (see Exercise 17.1.2) to simulate $M = 20$ batches each of $N = 100$ realizations of the random variable $\tilde{F}_1^\delta$ and evaluate the 90% confidence interval for the mean error*

$$\mu = F - E(\tilde{F}_1^\delta).$$

*Repeat the calculations for step sizes $\delta = \Delta = 2^{-4}$, $2^{-5}$ and $2^{-6}$. Plot the confidence intervals on $\mu$ versus $\delta$ axes and also $\log_2|\mu|$ versus $\log_2 \delta$ axes.*

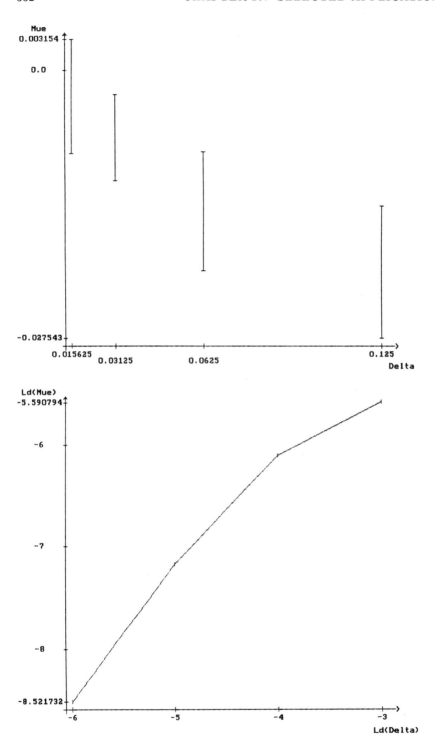

**Figure 17.1.1**    Results of PC–Exercise 17.1.3 with the Euler scheme.

If we apply the order 2.0 weak extrapolation method (15.3.1) to the above Euler scheme with step sizes $\Delta = \delta$ and $\Delta = 2\delta$, where $T/2\delta$ is an integer, we obtain the random variable

$$(1.10) \qquad \tilde{F}_2^\delta = 2\tilde{F}_1^\delta - \tilde{F}_1^{2\delta}.$$

Under the assumptions of Theorem 15.3.4, the expectation $E(\tilde{F}_2^\delta)$ approximates $F$ with weak order $\beta = 2.0$, that is

$$(1.11) \qquad \left| F - E\left( \tilde{F}_2^\delta \right) \right| \leq K \delta^2.$$

When $f(T, y) \equiv 1$ and the $\Delta \hat{W}_n^k$ are Gaussian, the expression (1.10) reduces to the stochastic quadrature formula in Chorin (1973).

The implicit weak scheme (15.4.13) for (1.4) takes the form

$$(1.12) \qquad Y_{n+1}^k = Y_n^k + \Delta \hat{W}_n^k, \qquad \text{for} \quad k = 1, \ldots, m,$$

$$Y_{n+1}^{m+1} = Y_n^{m+1} + \frac{1}{2} \left\{ a\left( \tau_{n+1}, \tilde{Y}_{n+1} \right) + a\left( \tau_n, \tilde{Y}_n \right) \right\} \Delta$$

with $\tilde{Y}_0 = x$ and $Y_0^{m+1} = 0$, where $\tilde{Y}_n = \left( Y_n^1, \ldots, Y_n^m \right)$, and is in fact explicit because the $\tilde{Y}_n$ do not include the $Y^{m+1}$ component. Here the $\Delta \hat{W}_n^k$ can be chosen as independent $N(0; \Delta)$ Gaussian random variables or three-point distributed random variables with

$$P\left( \Delta \hat{W}_n^k = \pm\sqrt{3\Delta} \right) = \frac{1}{6}, \qquad P\left( \Delta \hat{W}_n^k = 0 \right) = \frac{2}{3}$$

for $k = 1, \ldots, m$ and $n = 0, 1, \ldots$. We define the random variable

$$(1.13) \qquad \tilde{F}_3^\delta = f\left( T, \tilde{Y}_{n_T} \right) \phi \left( \Delta \sum_{n=0}^{n_T} a\left( \tau_n, \tilde{Y}_n \right) \right.$$

$$\left. - \frac{1}{2}\Delta \left\{ a\left( \tau_0, x \right) + a\left( \tau_{n_T}, \tilde{Y}_{n_T} \right) \right\} \right)$$

using (1.12) with step size $\delta = \Delta$ and $n_T = T/\Delta$. Its expectation $E(\tilde{F}_3^\delta)$ then approximates $F$ with weak order $\beta = 2.0$. We can interpret (1.13) as a stochastic generalization of the trapezoidal formula for Riemann integrals.

**PC-Exercise 17.1.4**   *Repeat PC-Exercise 17.1.3 with (1.13), $\Delta = 2^{-1}, \ldots,$* *$2^{-4}$ and the implicit order 2.0 weak scheme (1.12).*

We can apply the order 4.0 extrapolation method (15.3.11) to the implicit order 2.0 weak scheme (1.12) with step sizes $\Delta = \delta$, $2\delta$ and $3\delta$, where $T/6\delta$ is an integer, to obtain the random variable

$$(1.14) \qquad \tilde{F}_4^\delta = \frac{1}{11} \left[ 18\tilde{F}_3^\delta - 9\tilde{F}_3^{2\delta} + 2\tilde{F}_3^{3\delta} \right].$$

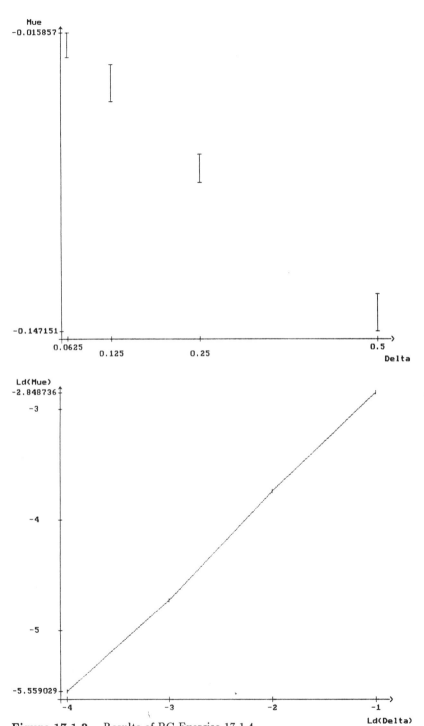

**Figure 17.1.2**    Results of PC-Exercise 17.1.4
with the implicit order 2.0 weak scheme.

Then, under the assumptions of Theorem 15.3.4, the expectation $E(\tilde{F}_4^\delta)$ approximates the functional $F$ with weak order 4.0, that is

$$\left| F - E\left(\tilde{F}_4^\delta\right) \right| \leq K\,\delta^4.$$

Another order 4.0 weak extrapolation is given in (15.3.2). It uses step sizes $\Delta = \delta$, $2\delta$ and $4\delta$ with $T/4\delta$ an integer and the random variable

(1.15) $$\tilde{F}_5^\delta = \frac{1}{21}\left[32\tilde{F}_3^\delta - 12\tilde{F}_3^{2\delta} + \tilde{F}_3^{4\delta}\right].$$

The expectation $E(\tilde{F}_5^\delta)$ also approximates $F$ with weak order 4.0, that is

$$\left| F - E\left(\tilde{F}_5^\delta\right) \right| \leq K\,\delta^4$$

under the assumptions of Theorem 15.3.4.

**PC-Exercise 17.1.5** *Repeat PC-Exercise 17.1.4 with (1.15) and the implicit order 2.0 weak scheme (1.12).*

Similarly, we could use an order 3.0 weak scheme for additive noise, such as the order 3.0 weak Taylor scheme (14.3.3), to approximate $F$ with weak order 3.0. Extrapolating as in (15.3.3) with step sizes $\Delta = \delta$, $2\delta$, $3\delta$ and $4\delta$ would then provide us with an order 6.0 weak approximation for the functional $F$.

**Exercise 17.1.6** *Derive the order 3.0 weak Taylor scheme (14.3.3) for (1.4).*

**Exercise 17.1.7** *Write down an order 6.0 weak approximation method for the functional $F$.*

We shall now consider the special case of an exponential functional with $f \equiv 1$ and $\phi(y^{m+1}) = \exp(y^{m+1})$ in (1.3), which then has the form

(1.16) $$\hat{F} = E\left(\exp\left(\int_0^T a\left(s, W_s\right)\,ds\right)\right),$$

where $W = \{W_t,\ t \geq 0\}$ is an $m$-dimensional Wiener process starting at $x = (x^1, \ldots, x^m) \in \Re^m$. It is easy to see that we have

(1.17) $$\hat{F} = E\left(X_T^{m+1}\right)$$

for the Ito process $X = \{X_t = (X_t^1, \ldots, X_t^{m+1}),\ t \geq 0\}$ satisfying the stochastic equation

(1.18) $$X_t^k = x^k + \int_0^t dW_s^k, \qquad \text{for}\quad k = 1, \ldots, m,$$

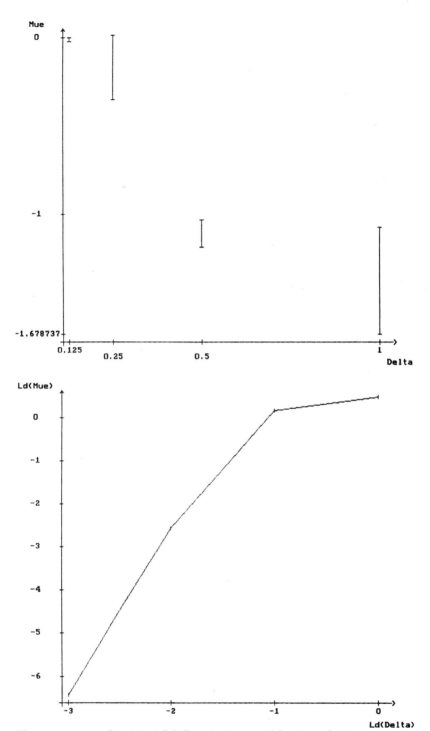

**Figure 17.1.3**    Results of PC-Exercise 17.1.5 with extrapolation.

$$X_t^{m+1} = 1 + \int_0^t a\left(s, \tilde{X}_s\right) X_s^{m+1} \, ds,$$

where once again $\tilde{X}_s = (X_s^1, \ldots, X_s^m)$.

Gladyshev and Milstein proposed an order 4.0 weak scheme specifically for the stochastic equation (1.18). This is a one-step scheme in the first $m$ components, which are uncoupled, and a two-step scheme in the last component, namely

(1.19)
$$Y_{n+1}^k = Y_n^k + \Delta \hat{W}_n^k, \qquad \text{for} \quad k = 1, \ldots, m,$$

$$Y_{n+2}^{m+1} = Y_n^{m+1} + \frac{1}{3} \{K_1 + 2K_2 + 2K_3 + K_4\}$$

with

$$K_1 = a\left(\tau_n, \tilde{Y}_n\right) Y_n^{m+1} \Delta$$

$$K_2 = a\left(\tau_{n+1}, \tilde{Y}_{n+1}\right) \left\{Y_n^{m+1} + \frac{1}{2}K_1\right\} \Delta$$

$$K_3 = a\left(\tau_{n+1}, \tilde{Y}_{n+1}\right) \left\{Y_n^{m+1} + \frac{1}{2}K_3\right\} \Delta$$

$$K_4 = a\left(\tau_{n+2}, \tilde{Y}_{n+2}\right) \left\{Y_n^{m+1} + K_3\right\} \Delta$$

with $\tilde{Y}_0 = x$, $Y_0^{m+1} = 1$, where $\tilde{Y}_n = (Y_n^1, \ldots, Y_n^m)$. Here $T/2\Delta$ is an integer and the $\Delta \hat{W}_n^k$ are independent $N(0; \Delta)$ Gaussian random variables for $k = 1$, ..., $m$ and $n = 0, 1, \ldots$. The proof that this scheme converges with weak order 4.0 is similar to that of Theorem 15.7.1. Note the similarity of the last component of (1.19) with the deterministic Runge-Kutta scheme (8.2.8). Although it is a two-step scheme, we do not need a starting routine here as we can just use an even indexed term $Y_{n_T}^{m+1}$ and these depend only on their even indexed predecessors.

Under sufficient smoothness of $a$, it is easy to show that the expectation of the random variable

(1.20)
$$\tilde{F}_6^\delta = Y_{n_T}^{m+1}$$

for step size $\delta = \Delta$ approximates the functional $\hat{F}$ in (1.16) with weak order $\beta = 4.0$, that is

$$\left|\hat{F} - E\left(\tilde{F}_6^\delta\right)\right| \leq K \delta^4.$$

We note that the above schemes may result in estimators with rather large variances. This problem can be handled with the variance reduction techniques discussed in Chapter 16, although the nice, simple structure of the stochastic equations (1.4) and (1.18) may then be lost. Other weak approximations from Chapters 14 and 15 are appropriate for the corresponding stochastic equations that arise when such variance reduction techniques are used.

To conclude this section we shall apply the measure transformation method described in Section 2 of Chapter 16 with the parameter function

$$(1.21) \qquad d^1(t, y^1) = \frac{T-t}{1+T-t}\, y^1$$

to Example 17.1.1. From (16.2.7) and (16.2.8) we then obtain the system of stochastic equations

$$(1.22) \qquad Z_t^1 = x + \int_0^t dW_s - \int_0^t d\left(s, Z_s^1\right)\, ds$$

$$Z_t^2 = 1 - \frac{1}{2}\int_0^t \left(Z_s^1\right)^2 Z_s^2\, ds$$

$$Z_t^3 = 1 + \int_0^t d\left(s, Z_s^1\right) Z_s^3\, dW_s.$$

According to (16.2.9) we have

$$(1.23) \qquad \hat{F} = u(t, y) = E\left(Z_T^2\, Z_T^3\right).$$

We can use a weak approximation $Y$ to approximate the solution of the stochastic system (1.22) in (1.23) and expect the functional

$$(1.24) \qquad \tilde{F}_7^\delta = Y_{n_T}^2\, Y_{n_T}^3$$

to have a relatively small variance.

**PC-Exercise 17.1.8**     *Repeat PC-Exercise 17.1.4 with variance reduction using (1.22)– (1.24) and the implicit order 2.0 weak scheme (15.4.12).*

A comparison of Figures 17.1.1–17.1.4 shows that the variance reduction technique yields an enormous improvement in the result relative to the necessary computation time.

In this section we have seen how the solutions of the Schrödinger equation can be approximated numerically by solving an associated stochastic differential equation. We remark that it is possible to extend the described method to more general parabolic partial differential equations, in particular higher dimensional ones. This is still, however, very much an open area of research.

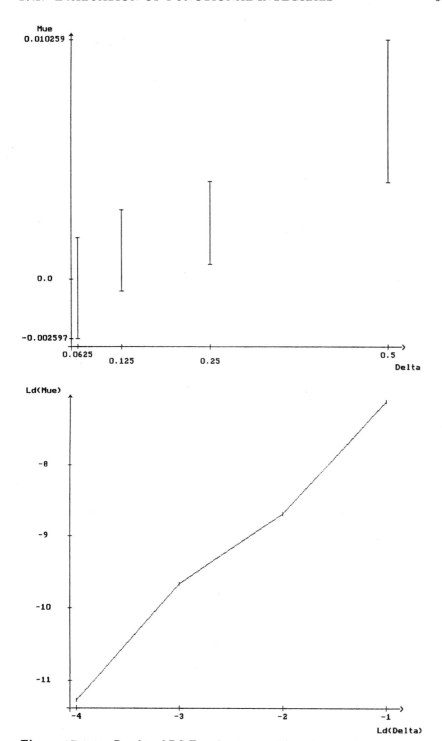

**Figure 17.1.4** Results of PC-Exercise 17.1.8 with variance reduction.

## 17.2   Approximation of Invariant Measures

In this section we shall consider an autonomous $d$-dimensional Ito process

$$(2.1) \qquad X_t = X_0 + \int_0^t a(X_s)\, ds + \sum_{j=1}^m \int_0^t b^j(X_s)\, dW_s^j$$

for $t \geq 0$, which is driven by an $m$-dimensional Wiener process $W = (W^1, \ldots, W^m)$. We shall assume that $X = \{X_t, t \geq 0\}$ is ergodic. Thus, by (4.8.18), it has a unique invariant probability law $\mu$ such that

$$(2.2) \qquad \lim_{t \to \infty} \frac{1}{t} \int_0^t f(X_s)\, ds = \int_{\Re^d} f(y)\, d\mu(y),$$

w.p.1, for all $\mu$–integrable functions $f : \Re^d \to \Re$ and any admissible initial value $X_0$. We recall that Theorem 4.8.8 provides a sufficient condition for an Ito process $X$ to be ergodic.

In many applications we know on theoretical grounds that an Ito process $X$ has a unique invariant probability law $\mu$ and are interested in evaluating a functional of the form

$$(2.3) \qquad F = \int_{\Re^d} f(y)\, d\mu(y)$$

for a given function $f : \Re^d \to \Re$, but we do not know $\mu$ explicitly. For instance, when $f(y) \equiv y$ the functional (2.3) is simply the asymptotic first moment $\lim_{t\to\infty} E(X_t)$. In the next section we shall consider another example where it is the upper Lyapunov exponent of a linear stochastic system (2.1).

An example of an ergodic Ito process is the Ornstein–Uhlenbeck process

$$(2.4) \qquad X_t = X_0 - \int_0^t X_s\, ds + \int_0^t \sqrt{2}\, dW_s,$$

which has as its invariant measure $\mu$ the standard Gaussian law $N(0; 1)$ with density (see (1.2.12))

$$(2.5) \qquad p(y) = \frac{1}{\sqrt{2\pi}} \exp\left(-\frac{1}{2} y^2\right).$$

Using (2.2) with $f(y) = y^2$, we see that the asymptotic second moment of the Ornstein–Uhlenbeck process satisfies

$$(2.6) \quad \lim_{t \to \infty} \frac{1}{t} \int_0^t (X_s)^2\, ds = \int_{-\infty}^{\infty} y^2 p(y)\, dy = \lim_{t \to \infty} E\left((X_t)^2\right) = F = 1.$$

Thus, if we did not already know the answer, we could use the expression

$$(2.7) \qquad F_T = \frac{1}{T} \int_0^T (X_s)^2\, ds$$

for large $T$ as an estimate for the asymptotic second moment of $X$. We note that $F_T$ here involves just one sample path of the Ito process $X$, though taken over a rather long time interval $[0, T]$. We can determine it approximately by evaluating the sum

$$(2.8) \qquad F_T^\delta = \frac{1}{n_T} \sum_{n=0}^{n_T - 1} \left( Y_n^\delta \right)^2$$

for a single trajectory of a time discrete approximation $Y^\delta$ with equidistant step size $\delta = \Delta = T/n_T$. Later we shall see that $F_T^\delta$ converges under appropriate assumptions to the desired functional $F$ given in (2.3) as $T \to \infty$ and $\delta \to 0$.

This method, proposed by Talay, of using a single simulated trajectory to approximate a functional of the form (2.3) can also be used for other functions $f$ and more general ergodic Ito processes $X$. We shall specify some additional weak schemes below which are useful for this purpose.

When the invariant law $\mu$ is absolutely continuous it has a density, which we shall denote by $p$. Under appropriate conditions this invariant probability density satisfies the stationary Kolmogorov backward equation (2.4.7)

$$(2.9) \qquad \mathcal{L}p = \sum_{k=1}^{d} a^k \frac{\partial p}{\partial y^k} + \frac{1}{2} \sum_{i,j=1}^{d} b^i b^j \frac{\partial^2 p}{\partial y^i \partial y^j} = 0,$$

as well as the stationary Fokker–Planck equation (2.4.5)

$$(2.10) \qquad \mathcal{L}^* p = 0,$$

where $\mathcal{L}$ is the elliptic operator defined in (2.4.6) and $\mathcal{L}^*$ is its adjoint. Experience has shown that standard difference methods are usually inadequate for such partial differential equations in higher dimensions. The simulation method discussed above thus offers an alternative means for calculating the density $p$ of an invariant law $\mu$ or, equivalently, the stationary solution of a Fokker–Planck equation, and may sometimes be the only effective method available.

Talay has proposed a criterion for assessing time discrete approximations $Y^\delta$ which are used to calculate limits of the form

$$(2.11) \qquad F^\delta = \lim_{T \to \infty} \frac{1}{n_T} \sum_{n=0}^{n_T - 1} f\left( Y_n^\delta \right),$$

where $\delta = T/n_T$ is kept fixed. A time discrete approximation $Y^\delta$ is said to *converge with respect to the ergodic criterion with order* $\beta > 0$ to an ergodic Ito process $X$ as $\delta \to 0$ if for each $f \in C_P^\infty(\Re^d, \Re)$ there exist a positive constant $C_f$, which does not depend on $\delta$, and a $\delta_0 > 0$ such that

$$(2.12) \qquad \left| F^\delta - F \right| \le C_f \, \delta^\beta$$

for $\delta \in (0, \delta_0)$. Here $F$ and $F^\delta$ are defined in (2.3) and (2.11), respectively, and $C_P^\infty(\Re^d, \Re)$ denotes the space of smooth functions which, together with their derivatives of all orders, have at most polynomial growth.

In view of the ergodicity relationship (2.2) we can interpret $F$ as $E(f(X_\infty))$ and $F^\delta$ as $E(f(Y_\infty^\delta))$. Thus, the ergodic convergence criterion (2.12) can be considered to be an extension of the weak convergence criterion (9.7.4) to the infinite time horizon $T = \infty$.

Under appropriate assumptions it follows (see Talay (1990)) that the multi-dimensional Euler scheme

$$(2.13) \qquad Y_{n+1} = Y_n + a\,\Delta + \sum_{j=1}^{m} b^j\,\Delta\hat{W}_n^j,$$

where the $\Delta\hat{W}_n^j$ are independent $N(0;1)$ Gaussian random variables, converges with respect to the ergodic criterion (2.12) with order $\beta = 1.0$.

The *simplified order* 2.0 *weak Taylor scheme* (14.2.7) has the form

$$(2.14) \qquad Y_{n+1} \;=\; Y_n + a\,\Delta + \frac{1}{2}L^0 a\,\Delta^2$$

$$+ \sum_{j=1}^{m} \left\{ b^j + \frac{1}{2}\left(L^0 b^j + L^j a\right)\Delta \right\}\Delta\hat{W}_n^j$$

$$+ \frac{1}{2}\sum_{j_1,j_2=1}^{m} L^{j_1} b^{j_2}\left(\Delta\hat{W}_n^{j_1}\,\Delta\hat{W}_n^{j_2} + V_{j_1,j_2}\right)$$

where the $\Delta\hat{W}_n^j$ are independent $N(0;1)$ Gaussian random variables and the $V_{j_1,j_2}$ are as in (14.2.8)–(14.2.10). We shall see from Theorem 17.2.1 to be stated below that this scheme converges with respect to the ergodic criterion with order $\beta = 2.0$. The same is true for the multi-dimensional *explicit order* 2.0 *weak scheme* (15.1.3)

$$(2.15) \qquad Y_{n+1} = Y_n + \frac{1}{2}\left(a\left(\bar{\Upsilon}\right) + a\right)\Delta$$

$$+ \frac{1}{4}\sum_{j=1}^{m}\left[ b^j\left(\bar{R}_+^j\right) + b^j\left(\bar{R}_-^j\right) + 2b^j \right.$$

$$+ \sum_{\substack{r=1 \\ r\neq j}}^{m}\left( b^j\left(\bar{U}_+^r\right) + b^j\left(\bar{U}_-^r\right) - 2b^j \right)\Bigg]\Delta\hat{W}_n^j\,\Delta^{-1/2}$$

$$+ \frac{1}{4}\sum_{j=1}^{m}\left[ \left( b^j\left(\bar{R}_+^j\right) - b^j\left(\bar{R}_-^j\right) \right)\left\{ \left(\Delta\hat{W}_n^j\right)^2 - \Delta \right\} \right.$$

$$+ \sum_{\substack{r=1 \\ r\neq j}}^{m}\left( b^j\left(\bar{U}_+^r\right) - b^j\left(\bar{U}_-^r\right) \right)\left\{\Delta\hat{W}_n^j\Delta\hat{W}_n^r + V_{r,j}\right\}\Bigg]\Delta^{-1/2}$$

with supporting values

$$\bar{\Upsilon} = Y_n + a\,\Delta + \sum_{j=1}^{m} b^j\,\Delta\hat{W}_n^j, \qquad \bar{R}_{\pm}^j = Y_n + a\,\Delta \pm b^j\,\sqrt{\Delta} \qquad \text{and}$$

$$\bar{U}_\pm^j = Y_n \pm b^j \sqrt{\Delta},$$

where the random variables $\Delta\hat{W}_n^j$ and $V_{j_1,j_2}$ are as in the scheme (2.13).

A theorem of the following type is proved in Talay (1990).

**Theorem 17.2.1**  *Let $m = 1$ and suppose that the drift and diffusion coefficients $a^k$, $b^k$ for $k = 1, \ldots, d$ have bounded derivatives of all orders, with the $b^k$ also bounded; that there is a constant $C_1 > 0$ such that*

$$(2.16) \qquad \sum_{k,l=1}^{d} b^k(x)b^l(x)\zeta^k\zeta^l \geq C_1 |\zeta|^2$$

*for all $x$, $\zeta = (\zeta^1, \ldots, \zeta^d) \in \Re^d$; and that there exists a compact set $K$ and a constant $C_2 > 0$ such that*

$$(2.17) \qquad x^\top a(x) \leq -C_2 |x|^2$$

*for all $x \in \Re^d \setminus K$.*
*Then, the schemes (2.14) and (2.15) with Gaussian $\Delta\hat{W}_n$ converge with respect to the ergodic criterion with order $\beta = 2.0$ as $\delta \to 0$.*

We anticipate that most of the weak schemes of a given weak order $\beta$ introduced in Chapters 14 and 15 also converge with respect to the ergodic criterion with the same order $\beta$ under appropriate assumptions.

**PC-Exercise 17.2.2**  *For the Ornstein–Uhlenbeck process (2.4) with $X_0 = 0$, use the Euler scheme (2.13) with step size $\delta = \Delta = 2^{-3}$ to simulate the sum $F^\delta$ in (2.8) with $T \in [0, 25600]$ to estimate the second moment of the invariant law. Plot the linearly interpolated values of $F^\delta$ against $T$.*

**PC-Exercise 17.2.3**  *Repeat PC-Exercise 17.2.2 using the scheme (2.15).*

**PC-Exercise 17.2.4**  *Consider PC-Exercise 17.2.2, but now estimating the probabilities*

$$\mu([r\epsilon, (r+1)\epsilon]) = \int_{r\epsilon}^{(r+1)\epsilon} d\mu(y)$$

*for $r = -10, -9, \ldots, 8, 9$ and $\epsilon = 0.3$. Plot the results in a histogram.*

Such simple 1-dimensional equations are not typically encountered in applications. However, we note that good numerical results are also obtained in the multi-dimensional case, with the required computational effort increasing only polynomially with the dimension of the state.

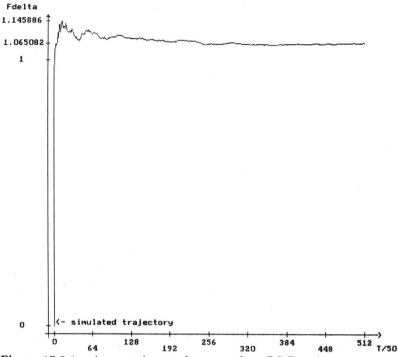

**Figure 17.2.1**    Asymptotic second moment from PC-Exercise 17.2.2.

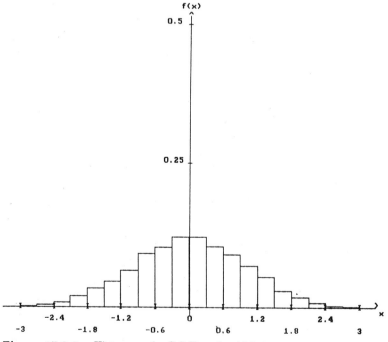

**Figure 17.2.2**    Histogram for PC-Exercise 17.2.4.

## 17.3 Approximation of Lyapunov Exponents

Investigations of the stability of a stochastic dynamical system are of crucial importance in engineering and other fields of applications. An indication can be seen in the examples of helicopter rotor blade stability and satellite orbital stability outlined in Section 5 of Chapter 7.

The stability of such systems is characterized by the negativeness of their upper Lyapunov exponents, which we defined in Section 3 of Chapter 6. In general, it is not possible to determine Lyapunov exponents explicitly, so a numerical approach is needed. Talay (1989b) has proposed such a method, to be described here, which uses simulations of approximate trajectories of a system to evaluate an approximation of its upper Lyapunov exponents. Essentially, it is an application of the results of the last section on the approximation of functionals of invariant measures.

We shall start from a $d$-dimensional linear Stratonovich stochastic differential equation

$$(3.1) \qquad dZ_t = AZ_t\,dt + \sum_{k=1}^{m} B^k Z_t \circ dW_t^k,$$

where $d \geq 2$ and $W$ is an $m$-dimensional Wiener process. As we have seen in Section 3 of Chapter 6, we obtain equations like (3.1) when we linearize nonlinear stochastic differential equations about stationary solutions.

The projection

$$S_t = \frac{Z_t}{|Z_t|}$$

of the solution $Z_t$ of (3.1) onto the unit sphere $S^{d-1}$ then satisfies the Stratonovich stochastic differential equation (see (6.3.23))

$$(3.2) \qquad dS_t = h\left(S_t, A\right)dt + \sum_{k=1}^{m} h\left(S_t, B^k\right) \circ dW_t^k,$$

where for any $d \times d$ matric $C$

$$(3.3) \qquad h(s, C) = \left(C - \left(s^\mathsf{T} C s\right) I\right) s.$$

Under appropriate assumptions on the matrices $A$, $B^1$, ..., $B^m$, from (6.3.24) we see that the upper Lyapunov exponent $\lambda_1$ of the system (3.1) is given by

$$(3.4) \qquad \lambda_1 = \int_{S^{d-1}} q(s)\,d\bar{\mu}(s)$$

where

$$(3.5) \qquad q(s) = s^\mathsf{T} A s + \sum_{k=1}^{m} \left(\frac{1}{2} s^\mathsf{T} \left(B^k + \left(B^k\right)^\mathsf{T}\right) s - \left(s^\mathsf{T} B^k s\right)^2\right)$$

and $\bar{\mu}$ is the invariant probability measure of the process $S = \{S_t,\ t \geq 0\}$ on $S^{d-1}$.

The problem of approximating Lyapunov exponents is thus one of approximating a functional of an invariant probability law of a diffusion process, so we can use the methods proposed in the last section. This, it turns out, is not an easy task on account of the sensitivity of the results to the numerical stability of the numerical scheme that is used. To illustrate this we shall look at an example due to Baxendale for which an explicit formula for the upper Lyapunov exponent is known. Here $d = 2$, $m = 1$ and equation (3.1) is

$$(3.6) \qquad dZ_t = AZ_t \, dt + BZ_t \circ dW_t$$

with coefficient matrices

$$(3.7) \qquad A = \begin{bmatrix} a & 0 \\ 0 & b \end{bmatrix}, \qquad B = \begin{bmatrix} 0 & -\sigma \\ \sigma & 0 \end{bmatrix}$$

where $a$, $b$ and $\sigma$ are real-valued parameters. Baxendale has shown that the top Lyapunov exponent $\lambda_1$ here is given by

$$(3.8) \qquad \lambda_1 = \frac{1}{2}(a + b) + \frac{1}{2}(a - b) \, \frac{\displaystyle\int_0^{2\pi} \cos(2\theta) \exp\left(\frac{a-b}{2\sigma^2} \cos(2\theta)\right) d\theta}{\displaystyle\int_0^{2\pi} \exp\left(\frac{a-b}{2\sigma^2} \cos(2\theta)\right) d\theta}.$$

The functions (3.3) and ( 3.5) for this example are

$$(3.9) \qquad h(s, A) = \begin{pmatrix} s^1 \left(a - \left[a\left(s^1\right)^2 + b\left(s^2\right)^2\right]\right) \\ s^2 \left(b - \left[a\left(s^1\right)^2 + b\left(s^2\right)^2\right]\right) \end{pmatrix},$$

$$(3.10) \qquad h(s, B) = \begin{pmatrix} -\sigma s^2 \\ \sigma s^1 \end{pmatrix}$$

and

$$(3.11) \qquad q(s) = a\left(s^1\right)^2 + b\left(s^2\right)^2.$$

The projected process $S_t$ here lives on the unit circle $S^1$, so we can represent it in terms of polar coordinates, specifically, the polar angle $\phi$. The resulting process

$$(3.12) \qquad \phi_t = \arctan\left(\frac{S_t^2}{S_t^1}\right)$$

satisfies the stochastic equation

$$(3.13) \qquad \phi_t = \phi_0 + \frac{1}{2}(b - a) \int_0^t \sin(2\phi_u) \, du + \sigma \int_0^t dW_u,$$

which we interpret modulo $2\pi$. Moreover, it follows from (3.11) that

$$(3.14) \qquad q(S_t) = \tilde{q}(\phi_t) = a\left(\cos \phi_t\right)^2 + b\left(\sin \phi_t\right)^2.$$

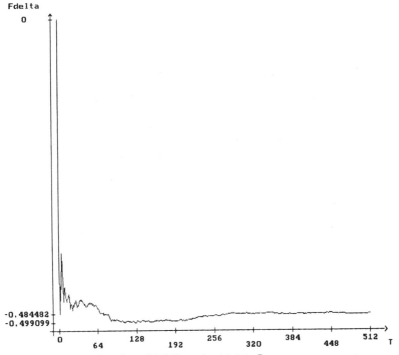

**Figure 17.3.1** Results of PC-Exercise 17.3.2: Lyapunov exponent approximated with the explicit order 2.0 weak scheme.

**Exercise 17.3.1** *Show that $\phi_t$ satisfies (3.13).*

We shall consider the case where the upper Lyapunov exponent is negative, so the system is stable and many potential numerical schemes are asymptotically numerically stable. For $a = 1.0$, $b = -2.0$ and $\sigma = 10$ formula (3.8) for the upper Lyapunov exponent gives $\lambda_1 = -0.489\ldots$. We shall investigate the behaviour of the approximate functional

$$(3.15) \qquad F_T^\delta = \frac{1}{n_T} \sum_{n=0}^{n_T-1} \tilde{q}\left(Y_n^\delta\right)$$

with these parameter values, using different time discrete approximations $Y^\delta$ of $\phi$.

**PC-Exercise 17.3.2** *Simulate the Lyapunov exponent $\lambda_1$ of (3.6) for the parameters $a = 1.0$, $b = -2.0$ and $\sigma = 10$ by (3.15) using the explicit order 2.0 weak scheme (2.15) with $\delta = \Delta = 2^{-9}$, $T = 512$ and $Y_0 = \phi_0 = 0.0$. Plot the linearly interpolated values of $F_t^\delta$ against $t$ for $0 \le t \le T$.*

We see from Figure 17.3.1 that $F_t^\delta$ tends to value of $\lambda_1$ as $t \to \infty$.

In contrast, however, when the Stratonovich stochastic differential equation (3.2) is stiff, particularly for dimensions $d \ge 3$, we may encounter numerical difficulties with many schemes. It will be necessary in such cases to use a weak implicit scheme to simulate the approximate trajectories for use in (3.15). As with many of the theoretical results and practical applications described in this book this is also an area open to more extensive research.

# Solutions of Exercises

## Solutions of Exercises for Chapter 1

**Exercise 1.4.1** An anti-derivative of $xp(x) = x/(1 + x^2)$ is $\ln \sqrt{1 + x^2}$, so the one-sided improper integrals $\int_{-\infty}^{0} xp(x)\, dx$ and $\int_{0}^{\infty} xp(x)\, dx$ both diverge. Hence the two-sided improper integral $\int_{-\infty}^{\infty} xp(x)\, dx$ diverges.

**Exercise 1.4.2** $E((X - \mu)^2) = E(X^2 - 2X\mu + \mu^2) = E(X^2) - 2\mu E(X) + \mu^2 = E(X^2) - \mu^2$ since $\mu = E(X)$.

**Exercise 1.4.3**

a). Let $X$ be Poisson distributed with parameter $\lambda > 0$. Then $p_i = \frac{\lambda^i}{i!} e^{-\lambda}$ for $i = 0$, $1, \ldots$, so

$$E(X) = \sum_{i=0}^{\infty} i\, p_i = \lambda e^{-\lambda} \sum_{i=1}^{\infty} \frac{\lambda^{i-1}}{(i-1)!} = \lambda e^{-\lambda} e^{\lambda} = \lambda$$

and

$$
\begin{aligned}
E(X^2) &= \sum_{i=0}^{\infty} i^2\, p_i = \lambda e^{-\lambda} \sum_{i=1}^{\infty} i\, \frac{\lambda^{i-1}}{(i-1)!} \\
&= \lambda e^{-\lambda} \left( \sum_{i=1}^{\infty} \frac{\lambda^{i-1}}{(i-1)!} + \sum_{i=1}^{\infty} (i-1) \frac{\lambda^{i-1}}{(i-1)!} \right) \\
&= \lambda e^{-\lambda} \left( \sum_{i=1}^{\infty} \frac{\lambda^{i-1}}{(i-1)!} + \lambda \sum_{i=2}^{\infty} \frac{\lambda^{i-2}}{(i-2)!} \right) = \lambda e^{-\lambda} \left( e^{\lambda} + \lambda e^{\lambda} \right) = \lambda + \lambda^2.
\end{aligned}
$$

Hence, by Exercise 1.4.2, $\mathrm{Var}(X) = E(X^2) - (E(X))^2 = \lambda + \lambda^2 - \lambda^2 = \lambda$.

b). Let $X$ be $U(a, b)$ uniformly distributed. Then

$$E(X) = \int_{a}^{b} \frac{x}{b - a}\, dx = \frac{1}{2(b - a)} (b^2 - a^2) = \frac{1}{2}(b + a)$$

and

$$E(X^2) = \int_{a}^{b} \frac{x^2}{b - a}\, dx = \frac{1}{3(b - a)} (b^3 - a^3) = \frac{1}{3}(b^2 + ab + a^2).$$

Hence $\mathrm{Var}(X) = \frac{1}{3}(b^2 + ab + a^2) - \frac{1}{4}(b + a)^2 = \frac{1}{12}(b - a)^2$.

c). Let $X$ be exponentially distributed with parameter $\lambda > 0$. Then

$$E(X) = \int_{0}^{\infty} x\lambda e^{-\lambda x}\, dx = \lim_{x \to \infty} \frac{1}{\lambda} \left( 1 - (\lambda x + 1)e^{-\lambda x} \right) = \frac{1}{\lambda} \qquad \text{and}$$

$$E(X^2) = \int_0^\infty x^2 \lambda e^{-\lambda x}\, dx + \lim_{x\to\infty} \frac{1}{\lambda^2}\left(2 - (\lambda^2 x^2 2\lambda x + 2)e^{-\lambda x}\right) = \frac{1}{\lambda^2},$$

so $\mathrm{Var}(X) = 2\lambda^{-2} - (\lambda^{-1})^2 = \lambda^{-2}$.

d). For a standard Gaussian random variable $X$ the expectation $E(X) = 0$ because $x\,p(x) = \frac{1}{\sqrt{2\pi}} x e^{-\frac12 x^2}$ has an anti-derivative $-\frac{1}{\sqrt{2\pi}} e^{-\frac12 x^2}$ so the one-sided improper integrals exist with

$$\int_{-\infty}^0 x p(x)\, dx = \lim_{x\to-\infty}\left(e^{-\frac12 x^2} - 1\right) = -1, \quad \int_0^\infty x p(x)\, dx = \lim_{x\to\infty}\left(1 - e^{-\frac12 x^2}\right) = 1,$$

from which it follows that $E(X) = \int_{-\infty}^\infty x\,p(x)\, dx$ exists and has the value $-1 + 1 = 0$. Integrating by parts

$$\frac{1}{\sqrt{2\pi}}\int_0^\infty x^2 e^{-\frac12 x^2}\, dx \;=\; \frac{1}{\sqrt{2\pi}}\lim_{x\to\infty}\left(-x e^{-\frac12 x^2}\right) + \frac{1}{\sqrt{2\pi}}\int_0^\infty e^{-\frac12 x^2}\, dx$$

$$=\; 0 + \frac{1}{\sqrt{2\pi}}\frac{\sqrt{2\pi}}{2} = \frac12$$

and similarly $\int_{-\infty}^0 x^2 p(x)\, dx = \frac12$. Hence the two-sided improper integral exists, giving

$$\mathrm{Var}(X) = E(X^2) - 0^2 = \frac{1}{\sqrt{2\pi}}\int_{-\infty}^\infty x^2 e^{-\frac12 x^2}\, dx = \frac12 + \frac12 = 1.$$

**Exercise 1.4.7**  Let $X \sim N(\mu; \sigma^2)$ and let $k = 0, 1, 2, \ldots$. Using the substitutions $z = (x - \mu)/\sigma$ and $t = z^2/2$

$$E((X - \mu)^{2k}) \;=\; \frac{1}{\sqrt{2\pi\sigma^2}}\int_{-\infty}^\infty (x - \mu)^{2k} e^{-\frac{(x-\mu)^2}{2\sigma^2}}\, dx = \sigma^{2k}\sqrt{\frac{2}{\pi}}\int_0^\infty z^{2k} e^{-\frac12 z^2}\, dz$$

$$=\; \sigma^{2k} 2^{(2k-1)/2}\sqrt{\frac{2}{\pi}}\int_0^\infty t^{(2k-1)/2} e^{-t}\, dt$$

$$=\; \sigma^{2k} 2^{(2k-1)/2}\sqrt{\frac{2}{\pi}}\,\Gamma\left(k + \frac12\right)$$

$$=\; 1\cdot 3\cdot 5\cdots(2k-1)\sigma^2 = (2k-1)!!\sigma^2,$$

where $\Gamma$ is the Gamma function.
With similar substitutions and integration by parts for $k \geq 1$

$$\int_0^\infty (x - \mu)^{2k+1} e^{-\frac{(x-\mu)^2}{2\sigma^2}}\, dx = \sigma^{2k+1}\int_0^\infty z^{2k+1} e^{-\frac12 z^2}\, dz$$

$$= \sigma^{2k+1}\left(\lim_{z\to\infty}\left(-z^{2k} e^{-\frac12 z^2}\right) + 2k\int_0^\infty z^{2k-1} e^{-\frac12 z^2}\, dz\right),$$

which gives a reduction formula $I(2k+1) = 2k\, I(2k-1)$. As in the solution of Exercise 1.4.3(d) $I(1) = \sqrt{2\pi}/2$. A similar reduction formula holds for the integral from $-\infty$

to 0, but with $I(1)$ having the opposite sign (the integrand is an odd function). Hence the two-sided improper integrals all exist and vanish, giving

$$E((X - \mu)^{2k+1}) = \frac{1}{\sqrt{2\pi\sigma^2}} \int_{-\infty}^{\infty} (x - \mu)^{2k+1} e^{-\frac{(x-\mu)^2}{2\sigma^2}} \, dx = 0$$

for each $k = 0, 1, 2, \ldots$.

Finally $E(|X|^{-1}) = \infty$ follows from $|x|^{-1} e^{-\frac{1}{2}x^2} \geq |x|^{-1}$ and the divergence of the improper integral $\int_{-\infty}^{\infty} |x|^{-1} \, dx$.

**Exercise 1.4.8** Let $X$ be a random variable with probability density $p$ and let $g : [0, \infty) \to [0, \infty)$ be a nondecreasing function such that $E(g(|X|)) < \infty$. Since $p(x) \geq 0$

$$E(g(|X|)) = \int_{-\infty}^{\infty} g(|x|)p(x) \, dx \geq \int_{|x| \geq a} g(|x|)p(x) \, dx$$

$$\geq \int_{|x| \geq a} g(a)p(x) \, dx = g(a)P(|X| \geq a)$$

for all $a > 0$

$$P(|X| \geq a) \leq E(g(|X|))/g(a).$$

Using $g(x) = x$ for a non-negative random variable $X \geq 0$ gives the Markov inequality and $g(x) = x^2$ the Chebyshev inequality.

**Exercise 1.4.9** Let $X$ be exponentially distributed with parameter $\lambda > 0$ and let $a > 0$. Then

$$E(X \mid X \geq a) = \int_a^{\infty} xp(x) \, dx \Big/ \int_a^{\infty} p(x) \, dx = \int_a^{\infty} x\lambda e^{-\lambda x} \, dx \Big/ \int_a^{\infty} \lambda e^{-\lambda x} \, dx$$

$$= \lim_{x \to \infty} \left\{ (a + \lambda^{-1})e^{-\lambda a} - (x + \lambda^{-1})e^{-\lambda x} \right\} \Big/ \lim_{x \to \infty} \left\{ e^{-\lambda a} - e^{-\lambda x} \right\}$$

$$= (a + \lambda^{-1})e^{-\lambda a}/e^{-\lambda a} = a + \lambda^{-1} = a + E(X).$$

**Exercise 1.4.11** Let $X_1$, $X_2$ be uniformly $U(0,1)$ distributed with joint density $p(x_1, x_2)$, let $Y_1$, $Y_2$ be the Box-Muller random variables

$$Y_1 = \sqrt{-2 \ln X_1} \, \cos(2\pi X_2), \quad Y_2 = \sqrt{-2 \ln X_1} \, \sin(2\pi X_2)$$

and let

$$x_1 = x_1(y_1, y_2) = e^{-\frac{1}{2}(y_1^2 + y_2^2)}, \quad x_2 = x_2(y_1, y_2) = \frac{1}{2\pi} \arctan\left(\frac{y_2}{y_1}\right).$$

Then the joint density $q(y_1, y_2)$ of $(Y_1, Y_2)$ is given by

$$q(y_1, y_2) = p(x_1(y_1, y_2), x_2(y_1, y_2)) \left| \det\left[ \frac{\partial x_i}{\partial y_j} \right] \right| = 1 \cdot \frac{1}{2\pi} e^{-\frac{1}{2}(y_1^2 + y_2^2)}$$

$$= \frac{1}{\sqrt{2\pi}} e^{-\frac{1}{2}y_1^2} \cdot \frac{1}{\sqrt{2\pi}} e^{-\frac{1}{2}y_2^2},$$

which is the product of the densities of two independent standard Gaussian random variables. Thus the Box-Muller variables $Y_1$, $Y_2$ are independent $N(0;1)$ random variables.

For the Polar-Marsaglia random variables apply the above procedure to the pairs of independent uniformly distributed random variables $W$, $V_1/\sqrt{W}$ and $W$, $V_2/\sqrt{W}$, respectively.

**Exercise 1.4.14** A theorem states that Gaussian random variables $X_1$, $X_2$ are independent if and only if they are uncorrelated, that is $E(X_1 X_2) = 0$. Thus $X_1 \sim N(0; h)$ and $X_2 \sim N(0; h^3/3)$ with $E(X_1 X_2) = h^2/2$ for $h > 0$ are dependent. The proof follows by showing that the joint density $p(x_1, x_2) \neq p_1(x_1)p_2(x_2)$, the product of the individual densities. Here the covariance matrix $C^{-1} = [E((X_i - \mu_i)(X_j - \mu_j))]$, so

$$C^{-1} = \begin{bmatrix} h & h^2/2 \\ h^2/2 & h^3/3 \end{bmatrix}, \quad C = \begin{bmatrix} 4/h & -6/h^2 \\ -6/h^2 & 12/h^3 \end{bmatrix}$$

with $\det C = 12/h^4$. Thus

$$p(x_1, x_2) = \frac{\sqrt{\det C}}{2\pi} \exp\left(-\frac{1}{2} \sum_{i,j=1}^{2} c^{i,j}(x_i - \mu_i)(x_j - \mu_j)\right)$$

$$= \frac{\sqrt{12}}{2\pi h^2} \exp\left(-\frac{1}{2}\left(4h^{-1}x_1^2 - 12x_1 x_2 h^{-2} + 12 x_2^2 h^{-3}\right)\right)$$

$$\neq \frac{1}{\sqrt{2\pi h}} \exp\left(-\frac{1}{2}\frac{x_1^2}{h}\right) \cdot \frac{1}{\sqrt{2\pi h^3/3}} \exp\left(-\frac{1}{2}\frac{x_1^2}{h^3/3}\right) = p_1(x_1)p_2(x_2).$$

**Exercise 1.4.15** Inequality (4.41) follows from the inequality $(a + b)^r \leq c_r(a^r + b^r)$ for $r > 0$ and $a$, $b > 0$ where $c_r = 1$ if $r < 1$ and $c_r = 2^{r-1}$ if $r \geq 1$. Inequalities (4.42) and (4.43) are just the Minkowski and Hölder inequaliti.es, respectively. Proofs in the context of $L_p$ spaces can be found, for example, in Kolmogorov and Fomin (1975) or Royden (1968).

**Exercise 1.5.1** Let $X_n = \sqrt{n}\, I_{A_n}$ where $A_n = \{\omega \in 0, 1 : 0 \leq \omega \leq 1/n\}$. Then

$$P(X_n \geq \epsilon) = P\left(\sqrt{n}\, I_{A_n} \geq \epsilon\right) = P(A_n) = 1/n$$

for $0 < \epsilon \leq \sqrt{n}$. Thus

$$\lim_{n \to \infty} P(|X_n - 0| \geq \epsilon) = \lim_{n \to \infty} P(X_n \geq \epsilon) = 0$$

for all $\epsilon > 0$, so $X_n$ converges in probability to $X = 0$. However

$$E\left(X_n^2\right) = E\left(n I_{A_n}^2\right) = E\left(n I_{A_n}\right) = n\, E\left(I_{A_n}\right) = n\, P(A_n) = n \cdot (1/n) = 1.$$

Hence $\lim_{n \to \infty} E(|X_n - X|^2) = 1$, so $X_n$ does not converge to $X$ in the mean-square sense.

**Exercise 1.5.5** The Law of Large Numbers says only that that the random variables $A_n$ converge in probability to the mean $\mu$, a deterministic number. The Central Limit Theorem says that the random variables

$$Z_n = (S_n - n\mu)/\sigma\sqrt{n} = \sqrt{n}(A_n - \mu)/\sigma$$

converge in distribution to an standard Gausian random variable. It thus also provides information about how the errors $A_n - \mu$ are distributed about the limiting value 0.

**Exercise 1.6.3** Let $W_t$ be a standard Wiener process and let $0 \le s < t$. Then

$$
\begin{aligned}
C(s,t) &= E\left((W_t - E(W_t))(W_s - E(W_s))\right) = E\left(W_t\, W_s\right) \\
&= E\left((W_t - W_s + W_s)\, W_s\right) = E\left((W_t - W_s)\, W_s\right) + E\left(W_s^2\right) \\
&= E\left(W_t - W_s\right) E\left(W_s\right) + E\left(W_s^2\right) = 0 \cdot 0 + s = s
\end{aligned}
$$

since $W_s$ and $W_t - W_s$ are independent for $s < t$. Analogously, $C(s,t) = t$ for $t < s$. Hence

$$
C(s,t) = \min\{s,t\} = \frac{1}{2}\left(|s+t| - |s-t|\right).
$$

This is not a function of $(t - s)$ only, so the Wiener process is not stationary in the wide-sense.

**Exercise 1.6.6** For any $\lambda \in [0,1]$

$$
\begin{aligned}
\bar{p}_\lambda P &= (\lambda/2, \lambda/2, 1 - \lambda) \begin{bmatrix} 0.5 & 0.5 & 0 \\ 0.5 & 0.5 & 0 \\ 0 & 0 & 1 \end{bmatrix} \\
&= \left(\frac{1}{2}(\lambda/2 + \lambda/2), \frac{1}{2}(\lambda/2 + \lambda/2), 1 - \lambda\right) \\
&= (\lambda/2, \lambda/2, 1 - \lambda) = \bar{p}_\lambda.
\end{aligned}
$$

**Exercise 1.6.10** For any initial probability vector $p(0) = (p, 1 - p)$ with $p \in [0,1]$

$$
\begin{aligned}
p(t) &= p(0)\, P(t) = (p, 1 - p) \begin{bmatrix} (1 + e^{-t})/2 & (1 - e^{-t})/2 \\ (1 - e^{-t})/2 & (1 + e^{-t})/2 \end{bmatrix} \\
&= \left(\frac{1}{2}(1 - e^{-t}) + pe^{-t}, \frac{1}{2}(1 + e^{-t}) - pe^{-t}\right) \\
&\to \left(\frac{1}{2}, \frac{1}{2}\right) = \bar{p} \quad \text{as} \quad t \to \infty.
\end{aligned}
$$

**Exercise 1.6.11** Let $0 \le t_1 \le t_2 \le \ldots \le t_n$ and integers $0 \le i_1 \le i_2 \ldots \le i_n$. In view of the independent increments of the Poisson process $N_t$

$$
\begin{aligned}
P\left(N_{t_1} = i_1, N_{t_2} = i_2\right) &= P\left(N_{t_1} = i_1, N_{t_2} - N_{t_1} = i_2 - i_1\right) \\
&= P\left(N_{t_1} = i_1\right) P\left(N_{t_2} - N_{t_1} = i_2 - i_1\right) \\
&= \frac{t_1^{i_1}}{i_1!} e^{-t_1} \frac{(t_2 - t_1)^{i_2 - i_1}}{(i_2 - i_1)!} e^{-(t_2 - t_1)}.
\end{aligned}
$$

Hence

$$
P\left(N_{t_1} = i_1, N_{t_2} = i_2, \ldots, N_{t_n} = i_n\right)
$$

$$
\begin{aligned}
&= P\left(N_{t_1} = i_1\right) P\left(N_{t_2} - N_{t_1} = i_2 - i_1\right) \cdots P\left(N_{t_n} - N_{t_{n-1}} = i_n - i_{n-1}\right) \\
&= \frac{t_1^{i_1}(t_2 - t_1)^{i_2 - i_1} \cdots (t_n - t_{n-1})^{i_n - i_{n-1}}}{i_1!(i_2 - i_1)! \cdots (i_n - i_{n-1})!} e^{-t_n} \qquad \text{so}
\end{aligned}
$$

$$P\left(N_{t_n} = i_n \Big| N_{t_{n-1}} = i_{n-1}\right) = \frac{P\left(N_{t_{n-1}} = i_{n-1}, N_{t_n} = i_n\right)}{P\left(N_{t_{n-1}} = i_{n-1}\right)}$$

$$= \frac{(t_n - t_{n-1})^{i_n - i_{n-1}}}{(i_n - i_{n-1})!} e^{-(t_n - t_{n-1})}$$

$$= \frac{P\left(N_{t_1} = i_1, N_{t_2} = i_2, \dots, N_{t_{n-1}} = i_{n-1}, N_{t_n} = i_n\right)}{P\left(N_{t_1} = i_1, N_{t_2} = i_2, \dots, N_{t_{n-1}} = i_{n-1}\right)}$$

$$= P\left(N_{t_n} = i_n \Big| N_{t_1} = i_1, N_{t_2} = i_2, \dots, N_{t_{n-1}} = i_{n-1}\right).$$

Thus the Poisson process is a Markov process. It is a homogeneous Markov process because

$$P(N_{t+h} - N_t = i) = \frac{h^i}{i!} e^{-h} = P(N_{s+h} - N_s = i)$$

for all $s, t \geq 0$ and integers $i \geq 0$, and similarly for all of the finite-dimensional distributions. In fact, it is a homogeneous Markov chain on $\{0, 1, 2, \dots\}$ with transition probabilities $p^{i,j}(t) = 0$ if $j < i$ and $p^{i,j}(t) = t^{j-i} e^{-t}/(j-i)!$ if $i \leq j$.

**Exercise 1.7.1**   For the transition densities of a standard Wiener process

$$\int_{-\infty}^{\infty} p(s, x; \tau, z) p(\tau, z; t, y) \, dz$$

$$= \int_{-\infty}^{\infty} \frac{1}{2\pi \sqrt{(\tau - s)(t - \tau)}} \exp\left(-\frac{1}{2}\left(\frac{(z - x)^2}{\tau - s} + \frac{(y - z)^2}{t - \tau}\right)\right) dz$$

$$= \frac{1}{\sqrt{2\pi(t - s)}} \exp\left(-\frac{(y - x)^2}{2(t - s)}\right) \int_{-\infty}^{\infty} \frac{1}{\sqrt{2\pi}} \exp\left(-\frac{1}{2} u^2\right) du$$

$$= p(s, x; t, y) \cdot 1 = p(s, x; t, y)$$

with the substitution

$$u = u(z) = \left(z - \frac{x(t - \tau) + y(\tau - s)}{t - s}\right) \sqrt{\frac{t - s}{(\tau - s)(t - \tau)}}.$$

This is the Chapman-Kolmogorov equation in terms of transition probabilities. The parameter $\tau$ cancels out here.

**Exercise 1.7.2**   For a standard Wiener process $a(s, x) \equiv 0$ and $b(s, x) \equiv 1$ because

$$a(s, x) = \lim_{t \downarrow s} \frac{1}{t - s} E\left(W_t - W_s \Big| W_s = x\right)$$

$$= \lim_{t \downarrow s} \frac{1}{t - s} E(W_t - W_s) = \lim_{t \downarrow s} \frac{1}{t - s} \cdot 0 = 0$$

and

$$b^2(s, x) = \lim_{t \downarrow s} \frac{1}{t - s} E\left((W_t - W_s)^2 \Big| W_s = x\right)$$

$$= \lim_{t \downarrow s} \frac{1}{t - s} E\left((W_t - W_s)^2\right) = \lim_{t \downarrow s} \frac{1}{t - s} \cdot (t - s) = 1.$$

For the Ornstein-Uhlenbeck process with paramter $\gamma = 1$ the substitution

$$u = u(y) = \left(y - xe^{-(t-s)}\right)/\sqrt{1 - e^{-2(t-s)}}$$

and $\epsilon \to \infty$ gives

$$
\begin{aligned}
a(s, x) &= \lim_{t \downarrow s} \frac{1}{t-s} \int_{-\infty}^{\infty} (y - x) p(s, x; t, y) \, dy \\
&= \lim_{t \downarrow s} \frac{1}{t-s} \int_{-\infty}^{\infty} \frac{y - x}{\sqrt{2\pi(1 - e^{-2(t-s)})}} \exp\left(-\frac{(y - xe^{-(t-s)})^2}{2(1 - e^{-2(t-s)})}\right) dy \\
&= \lim_{t \downarrow s} \frac{1}{t-s} \int_{-\infty}^{\infty} \frac{1}{\sqrt{2\pi}} \left(u\sqrt{1 - e^{-2(t-s)}} - x\left(1 - e^{-(t-s)}\right)\right) \exp\left(-\frac{1}{2}u^2\right) du \\
&= -x \cdot \lim_{t \downarrow s} \frac{1 - e^{(t-s)}}{t-s} = -x \cdot 1 = -x
\end{aligned}
$$

and

$$
\begin{aligned}
b^2(s, x) &= \lim_{t \downarrow s} \frac{1}{t-s} \int_{-\infty}^{\infty} (y - x)^2 p(s, x; t, y) \, dy \\
&= \lim_{t \downarrow s} \frac{1}{t-s} \int_{-\infty}^{\infty} \frac{(y - x)^2}{\sqrt{2\pi(1 - e^{-2(t-s)})}} \exp\left(-\frac{(y - xe^{-(t-s)})^2}{2(1 - e^{-2(t-s)})}\right) dy \\
&= \lim_{t \downarrow s} \left(\frac{1 - e^{-2(t-s)}}{t-s} + \frac{x^2(1 - e^{-(t-s)})^2}{t-s}\right) \\
&= (2 + x^2 \cdot 0) \cdot 1 = 2,
\end{aligned}
$$

using

$$\frac{1}{\sqrt{2\pi}} \int_{-\infty}^{\infty} u^2 \exp\left(-\frac{1}{2}u^2\right) du = \frac{1}{\sqrt{2\pi}} \int_{-\infty}^{\infty} \exp\left(-\frac{1}{2}u^2\right) du = 1$$

and

$$\frac{1}{\sqrt{2\pi}} \int_{-\infty}^{\infty} u \exp\left(-\frac{1}{2}u^2\right) du = 0.$$

Hence $a(s, x) = -x$ and $b(s, x) = \sqrt{2}$ for the Ornstein-Uhlenbeck process with parameter $\gamma = 1$.

**Exercise 1.7.3** Since $a(s, x) = -x$ and $b^2(s, x) = 2$ for the Ornstein-Uhlenbeck process with parameter $\gamma = 1$, the Kolmogorov forwards and backwards equations are

$$\frac{\partial p}{\partial t} + \frac{\partial}{\partial y}(yp) + \frac{\partial^2 p}{\partial y^2} = 0 \quad \text{and} \quad \frac{\partial p}{\partial s} - x\frac{\partial}{\partial x}(p) + \frac{\partial^2 p}{\partial x^2} = 0,$$

respectively. It can then be verified directly that (7.4) satisfies the backwards equation.

**Exercise 1.7.4** Integrate the differential equation

$$\frac{d}{dy}\left((\nu y - y^2)\bar{p}\right) - \frac{1}{2}\frac{d^2}{dy^2}(2y\bar{p}) = 0$$

over the half-line $y \geq 0$ and use the assumed asymptotic behaviour of $\bar{p}$ to obtain the first order separable differential equation

$$(\nu y - y^2)\bar{p}) - \frac{d}{dy}(y\bar{p}) = 0, \quad \text{ie} \quad \frac{d}{dy}\bar{p} = \left(\frac{\nu - 2}{y} - 1\right)\bar{p}.$$

This has the general solution $\bar{p} = N y^{\nu-2} e^{-y}$, which has a singularity at $y = 0$ when $\nu < 2$. The constant of integration $N$ is so that $\int_0^\infty \bar{p}(y)\,dy = 1$. For $\nu < 1$ the integral diverges and the only stationary probability density is the delta function $\delta(y)$ centered on 0, whereas for $1 \leq \nu < 2$ the singularity is integrable. For all $\nu \geq 1$ the improper integral converges and a suitable normalizing constant $N$ can be determined.

**Exercise 1.7.5**   For a standard Wiener process the increments $W_t - W_s \sim N(0; |t - s|)$, so from Exercise 1.4.7 with $k = 2$ and $\sigma^2 = |t - s|$ it follows that

$$E\left(|W_t - W_s|^4\right) = 1 \cdot 3\sigma^4 = 3|t - s|^2.$$

**Exercise 1.8.1**   $\text{Var}\,(S_N(t) - S_N(s)) = 0$ if $t_k^N \leq s \leq t < t_{k+1}^N$. Let $t_j^N \leq s \leq t_{j+1}^N < \cdots < t_k^N \leq t < t_{k+1}^N$. Then

$$\text{Var}\,(S_N(t) - S_N(s)) = \text{Var}\left(\sum_{i=j+1}^k X_i \sqrt{\Delta t}\right) = \text{Var}\left(\sum_{i=j+1}^k X_i\right)\Delta t$$

$$= \sum_{i=j+1}^k \text{Var}\,(X_i)\,\Delta t = (k - j)\,\Delta t = \left[\frac{t - s}{\Delta t}\right]\Delta t$$

$$\to \quad t - s \quad \text{as} \quad \Delta t = \frac{1}{N} \to 0.$$

Here $[x]$ denotes the integer part of $x$.

**Exercise 1.8.3**   For $t_n^N \leq t < t_{n+1}^N$ define

$$S_N(t) = S_N\left(t_n^N\right) + \left(S_N\left(t_{n+1}^N\right) - S_N\left(t_n^N\right)\right)\frac{t - t_n^N}{t_{n+1}^N - t_n^N}$$

$$= S_N\left(t_n^N\right) + \sqrt{N}\left(t - t_n^N\right)X_{n+1},$$

so

$$E\,(S_N(t)) = E\left(S_N\left(t_n^N\right)\right) + \sqrt{N}\left(t - t_n^N\right)E\,(X_{n+1}) = 0 + \sqrt{N}\left(t - t_n^N\right)0 = 0$$

and, in view of the independence of $S_N\left(t_n^N\right)$ and $X_{n+1}$,

$$\text{Var}\,(S_N(t)) = E\left((S_N(t))^2\right) = E\left(\left(S_N(t_n^N)\right)^2\right) + N\left(t - t_n^N\right)^2 E\left(X_{n+1}^2\right)$$

$$= n + N\left(t - t_n^N\right)^2 = \left[\frac{t_n^N}{\Delta t}\right]\Delta t + \frac{\left(t - t_n^N\right)^2}{\Delta t}$$

$$\to \quad t + 0 = t \quad \text{as} \quad N \to \infty \quad \text{since} \quad \left|t - t_n^N\right| < \Delta t = 1/N.$$

Here $[x]$ denotes the integer part of $x$. However for $t_n^N \leq s_1 < s_2 < s_3 < t_{n+1}^N$

$$S_N(s_2) - S_N(s_1) = \sqrt{N}\,(s_2 - s_1)\,X_{n+1}, \quad S_N(s_3) - S_N(s_2) = \sqrt{N}\,(s_3 - s_2)\,X_{n+1},$$

so $S_N$ does not have independent increments.

**Exercise 1.8.4**  For a standard Wiener process

$$\lim_{t \to \infty} E\left(\frac{W(t)}{t}\right) = \lim_{t \to \infty} \frac{1}{t} E\left(W(t)\right) = \lim_{t \to \infty} \frac{1}{t} \cdot 0 = 0$$

and

$$\lim_{t \to \infty} E\left(\frac{W^2(t)}{t^2}\right) = \lim_{t \to \infty} \frac{1}{t^2} E\left(W^2(t)\right) = \lim_{t \to \infty} \frac{t}{t^2} = \lim_{t \to \infty} \frac{1}{t} = 0,$$

from which it follows that $\lim_{t \to \infty} W(t)/t \to 0$ w.p.1.

**Exercise 1.8.5**  Recall that a linear combination of Gaussian variables is also Gaussian. Let $W$ be a standard Wiener process and define $X(t) = W(t+s) - W(s)$ for all $t \ge 0$ and some fixed $s \ge 0$. Then $X(t) \sim N(0;t)$ because

$$E(X(t)) = E(W(t+s)) - E(W(s)) = 0 - 0 = 0$$

and

$$
\begin{aligned}
E\left(X(t)^2\right) &= E(W(t+s)^2) - 2E((W(t+s) - W(s))W(s)) - E(W(s)^2) \\
&= t + s - 0 \cdot 0 - s = t.
\end{aligned}
$$

Moreover $X(t+h) - X(t) = W(t+s+h) - W(t+s)$, so $X$ also has independent increments and is thus a standard Wiener process.

Now define $Y(t) = t\,W(1/t)$ for all $t > 0$. From Exercise 1.8.4 $t\,W(1/t) = W(s)/s \to 0$ as $s = 1/t \to 0$, so define $Y(0) = 0$. Then $Y(t) \sim N(0;t)$ because

$$E(Y(t)) = E(t\,W(1/t)) = t\,E(W(1/t)) = 0$$

and

$$E\left(Y(t)^2\right) = t^2 E\left(W(t)^2\right) = t^2 \cdot (1/t) = t.$$

That $Y$ has independent increments follows from

$$Y(t) - Y(s) = t\,W(1/t) - s\,W(1/s) = t\,(W(1/t) - W(1/s)) + (t - s)\,W(1/s)$$

and the independence of the terms in the final expression. Hence $Y$ is also a standard Wiener process.

**Exercise 1.8.8**  The Brownian bridge process $B_{0,x}^{T,y}$ is Gaussian since it is a linear transformation of a Gaussian process with

$$E\left(B_{0,x}^{T,y}(t)\right) = \frac{1}{T}\left(x(T - t) + ty + (T - t)E(W(t))\right) = \frac{1}{T}\left(x(T - t) + ty\right) = \mu(t)$$

and

$$E\left((B_{0,x}^{T,y}(t))^2\right) = \mu(t)^2 + \frac{(T - t)^2}{T^2} E(W^2(t)) = \mu(t)^2 + \frac{(T - t)^2}{T^2} t,$$

so

$$\mathrm{Var}\left(B_{0,x}^{T,y}(t)\right) = \frac{t}{T^2}(T - t)^2.$$

The expectation and variance remain the same under the substitution $(x, y, t) \to (y, x, T - t)$, so $B_{0,y}^{T,x}(T - t)$ has the same distribution as $B_{0,x}^{T,y}(t)$. Finally, with $0 \le s \le t \le T$

$$
\begin{aligned}
C(s, t) &= E\left(\left(B_{0,x}^{T,y}(s) - \mu(s)\right)\left(B_{0,x}^{T,y}(t) - \mu(t)\right)\right) \\
&= E\left(\frac{T - s}{T} W(s) \frac{T - t}{T} W(t)\right) = \frac{(T - s)(T - t)}{T^2} E(W(s)W(t)) \\
&= \frac{(T - s)(T - t)}{T^2} s = s\left(1 - \frac{s}{T}\right)\left(1 - \frac{t}{T}\right).
\end{aligned}
$$

**Exercise 1.8.10**  $E\left(X^h(t)\right) = 0$ for all $h > 0$, so for $s + h \le t$

$$
\begin{aligned}
E\left(X^h(s) X^h(t)\right) &= \frac{1}{h^2} E\left((W(t + h) - W(t))(W(s + h) - W(s))\right) \\
&= \frac{1}{h^2} E\left(W(t + h) - W(t)\right) E\left(W(s + h) - W(s)\right) = 0 \cdot 0 = 0
\end{aligned}
$$

and for $s \le t \le s + h$

$$
\begin{aligned}
E\left(X^h(s) X^h(t)\right) &= \frac{1}{h^2} E\left((W(t + h) - W(t))(W(s + h) - W(s))\right) \\
&= \frac{1}{h^2} E\left((W(s + h) - W(t))^2\right) = \frac{1}{h^2}(s + h - t) \\
&= \frac{1}{h}\left(1 - \frac{t - s}{h}\right).
\end{aligned}
$$

Similar expressions hold on reversing the roles of $s$ and $t$, so

$$
C(s, t) = E\left(X^h(s) X^h(t)\right) = \frac{1}{h} \max\left\{0, 1 - \frac{|t - s|}{h}\right\} = c(t - s).
$$

Hence

$$
\begin{aligned}
S_h(\nu) &= \frac{1}{h} \int_{\infty}^{\infty} c(s) \cos(-2\pi\nu s)\, ds = \frac{1}{h} \int_{-h}^{h}\left(1 - \frac{|s|}{h}\right)\cos(-2\pi\nu s)\, ds \\
&= \frac{2}{h} \int_{0}^{h}\left(1 - \frac{s}{h}\right)\cos(-2\pi\nu s)\, ds = \frac{2}{h}\frac{1 - \cos(2\pi\nu h)}{4\pi^2\nu^2 h} \\
&= \left(\frac{\sin(\pi\nu h)}{\pi\nu h}\right)^2.
\end{aligned}
$$

**Exercise 1.8.11**  For the Ornstein-Uhlenbeck process with parameter $\gamma > 0$

$$
\begin{aligned}
S(\nu) &= \int_{\infty}^{\infty} e^{-\gamma|s|} e^{-2\pi\imath\nu s}\, ds = 2 \int_{0}^{\infty} c(s) e^{(\gamma - 2\pi\imath\nu)s}\, ds \\
&= \frac{1}{\gamma + 2\pi\imath\nu} + \frac{1}{\gamma - 2\pi\imath\nu} = \frac{2\gamma}{\gamma^2 + 4\pi^2\nu^2}.
\end{aligned}
$$

Obviously $S(\nu) \to 0$ as $\gamma \to \infty$, so the covariance $c(t)$ tends to a delta function centered on 0, the covariance function of Gaussian white noise. Thus the Ornstein-Uhlenbeck behaves more and more like Gaussian white noise as its parameter $\gamma$ increases without bound.

**Exercise 1.9.1** Let the $X_i$ be a sequence of i.i.d. $U(0,1)$ random variables and let $A_n = \frac{1}{n}\sum_{i=0}^{n} X_i$. Then

$$\mu_n = E\left(A_n\right) = \frac{1}{2} \quad \text{and} \quad \text{Var}\left(A_n\right) = \frac{1}{12n}.$$

By the Chebyshev inequality

$$P\left(|A_n - \mu_n| \geq a\right) \leq \frac{1}{a^2}\,\text{Var}\left(A_n\right) = \frac{1}{12na^2},$$

which is no larger the $1 - \alpha$ if

$$n \geq \bar{n}(a,\alpha)) = \frac{1}{12a^2\alpha}.$$

By the Central Limit Theorem with $Z_n = (A_n - \mu_n)\sqrt{12/n}$ and $b = a\sqrt{12n}$

$$P\left(|A_n - \mu_n| < a\right) = P\left(|Z_n| < b\right) \approx 2\Phi(b),$$

which is no larger the $1 - \alpha$ if

$$n \geq \tilde{n}(a,\alpha)) = \frac{1}{12}\left(\frac{\Phi^{-1}(1 - \alpha/2)}{a\alpha}\right)^2.$$

For example with $a = 0.1$ and $\alpha = 0.05$

$$\bar{n}(0.1, 0.05) = \frac{1000}{6} \approx 167, \quad \tilde{n}(0.1, 0.05) = \left(\frac{\Phi^{-1}(1 - 0.025)}{0.1\sqrt{12}}\right)^2 \approx 32.$$

## Solutions of Exercises for Chapter 2

**Exercise 2.2.1** Let $x_i^{(n)} = i/n$ for $i = 0, 1, \ldots, n$. Since $f(x) = 2x$ is increasing on $[0, 1]$ the lower rectangular sums

$$
\begin{aligned}
L_n &= \sum_{i=0}^{n-1} f(x_i^{(n)})\left(x_{i+1}^{(n)} - x_i^{(n)}\right) = \sum_{i=0}^{n-1} \frac{2i}{n}\left(\frac{i+1}{n} - \frac{i}{n}\right) \\
&= \frac{2}{n^2}\sum_{i=0}^{n-1} i = \frac{2}{n^2}\frac{(n-1)n}{2} = 1 - \frac{1}{n} \to 1 \quad \text{as} \quad n \to \infty
\end{aligned}
$$

and the upper rectangular sums

$$
\begin{aligned}
U_n &= \sum_{i=o}^{n-1} f(x_{i+1}^{(n)})\left(x_{i+1}^{(n)} - x_i^{(n)}\right) = \sum_{i=0}^{n-1} \frac{2(i+1)}{n}\left(\frac{i+1}{n} - \frac{i}{n}\right) \\
&= \frac{2}{n^2}\sum_{k=1}^{n} k = \frac{2}{n^2}\frac{n(n+1)}{2} = 1 + \frac{1}{n} \to 1 \quad \text{as} \quad n \to \infty.
\end{aligned}
$$

Hence the Riemann inegral $\int_0^1 2x\,dx$ exists and equals 1.

**Exercise 2.2.2**   Here $0 \leq f(x) \leq 2$ and the subsets $E_i^n = E_i([0,1];n) = \{x \in [0,1] : 2i/n \leq f(x) \leq 2(i+1)/n\} = [i/n,(i+1)/n]$. For the simple function $\phi_n(x) = 2i/n$ for $x \in E_i^n$ the Lesbegue integral

$$
\begin{aligned}
\int_{[0,1]} \phi_n\,d\mu &= \sum_{i=0}^{n-1} \frac{2i}{n}\,\mu\left(E_i^n\right) = \sum_{i=0}^{n-1} \frac{2i}{n}\frac{1}{n} \\
&= \frac{2}{n^2}\sum_{i=0}^{n-1} i = \frac{2}{n^2}\frac{(n-1)n}{2} = 1 - \frac{1}{n} \to 1 \quad \text{as} \quad n \to \infty,
\end{aligned}
$$

so the Lesbegue integral $\int_{[0,1]} f\,d\mu$ exists and equals 1.

**Exercise 2.2.4**   Here $S = \{\emptyset, A, A^c, \Omega\}$ and $T = \{\emptyset, A, A^c, B, B^c, A \cup B, (A \cup B)^c, A \cap B, (A \cap B)^c, \Omega\}$, so

$$
E(X|S)(\omega) = \frac{1}{P(C)}\int_C X\,dP \quad \text{for} \quad \omega \in C \quad \text{where} \quad \emptyset \neq C \in T
$$

and

$$
E(X|T)(\omega) = \frac{1}{P(C)}\int_C X\,dP \quad \text{for} \quad \omega \in C \quad \text{where} \quad \emptyset \neq C \in T.
$$

Hence for $\omega \in C = A$, $A^c$ or $\Omega$

$$
\begin{aligned}
E(E(X|T)|S)(\omega) &= \frac{1}{P(C)}\int_C E(X|T)\,dP \\
&= \frac{1}{P(C)}\int_C \left(\frac{1}{P(C)}\int_C X\,dP\right)dP \\
&= \frac{1}{P(C)^2}\int_C X\,dP \int_C dP = \frac{1}{P(C)^2}\int_C X\,dP \cdot P(C) \\
&= \frac{1}{P(C)}\int_C X\,dP = E(X|S)(\omega).
\end{aligned}
$$

**Exercise 2.2.5**   Let $f^*(Y) = E(X|Y)$ so $E(f^*(Y)|Y) = E(X|Y)$. Since

$$
E((X - f^*(Y))(f^*(Y) - f(Y))) = E(E(X - f^*(Y)|Y)(f^*(Y) - f(Y)))
$$
$$
= E(E(X - X|Y)(f^*(Y) - f(Y))) = 0
$$

it follows that

$$
\begin{aligned}
E\left((X - f(Y))^2\right) &= E\left((X - f^*(Y) + f^*(Y) - f(Y))^2\right) \\
&= E\left((X - f^*(Y))^2\right) + E\left((f^*(Y) - f(Y))^2\right) \\
&\quad + 2E((X - f^*(Y))(f^*(Y) - f(Y))) \\
&\geq E\left((X - f^*(Y))^2\right).
\end{aligned}
$$

**Exercise 2.2.6**   Let $F = F_1 - F_2$ where $F_1$ and $F_2$ are bounded and monotonic on $[a,b]$. Then $F_1$ and $F_2$ have bounded variation on $[a,b]$. Since

$$
V_a^b(F) = V_a^b(F_1 - F_2) \leq V_a^b(F_1) + V_a^b(F_2)
$$

it follows that $F$ has bounded variation on $[a,b]$. Now let $F$ have bounded variation on $[a,b]$ and define $F_1$ and $F_2$ by

$$F(x) = V_a^x(F) - (V_a^x(F) - F(x)) = F_1(x) - F_2(x)$$

for each $x \in [a,b]$. The functions $F_1$ and $F_2$ are bounded and monotonic on $[a,b]$.

**Exercise 2.3.2**  Since

$$E\left((W_t^2 - t)^2\right) = E\left(W_t^4\right) - 2tE\left(W_t^2\right) + t^2 = 3t^2 - 2t \cdot t + t^2 = 2t^2$$

the maximal martingale inequality with $p = 2$ gives

$$P\left(\sup_{0 \leq s \leq t} \left|W_t^2 - t\right| \geq a\right) \leq \frac{1}{a^2} E\left((W_t^2 - t)^2\right) = \frac{1}{a^2} 2t^2$$

and the Doob inequality with $p = 2$

$$E\left(\sup_{0 \leq s \leq t} \left|W_t^2 - t\right|^2\right) \leq 4 E\left((W_t^2 - t)^2\right) = 8t^2.$$

**Exercise 2.4.1**  Integrating by parts once for the first order partial derivatives and twice for the second order gives

$$\int_{-\infty}^{\infty} f a^i \frac{\partial g}{\partial x_i} \, dx_i = \lim_{x_i \to \pm\infty} a^i f g - \int_{-\infty}^{\infty} g \frac{\partial}{\partial x_i}(a^i f) \, dx_i = - \int_{-\infty}^{\infty} g \frac{\partial}{\partial x_i}(a^i f) \, dx_i$$

and

$$\begin{aligned}
\int_{\Re^2} f d^{i,j} \frac{\partial^2 g}{\partial x_i \partial x_j} \, dx_i dx_j &= \int_{\Re} \lim_{x_i \to \pm\infty} d^{i,j} f \frac{\partial^2 g}{\partial x_j} \, dx_j - \int_{\Re^2} \frac{\partial}{\partial x_i}(d^{i,j} f) \frac{\partial g}{\partial x_j} \, dx_i dx_j \\
&= \lim_{x_i \to \pm\infty} g \frac{\partial}{\partial x_i}(d^{i,j} f) + \int_{\Re^2} g \frac{\partial}{\partial x_j}\left(\frac{\partial}{\partial x_i}(d^{i,j} f)\right) \, dx_i dx_j
\end{aligned}$$

since the boundary limits vanish by assumption, so

$$\begin{aligned}
\int_{\Re^d} f \mathcal{L} g \, dx &= \int_{\Re^d} f \left(\sum_{i=1}^{d} a^i \frac{\partial g}{\partial x_i} + \frac{1}{2} \sum_{i,j=1}^{d} d^{i,j} \frac{\partial^2 g}{\partial x_i \partial x_j}\right) dx \\
&= \int_{\Re^d} g \left(-\sum_{i=1}^{d} \frac{\partial}{\partial x_i}(a^i f) + \frac{1}{2} \sum_{i,j=1}^{d} \frac{\partial}{\partial x_i}\left(\frac{\partial}{\partial x_j}(d^{i,j} f)\right)\right) dx \\
&= \int_{\Re^d} g \mathcal{L}^* f \, dx.
\end{aligned}$$

**Exercise 2.4.3**  Let $C_i = 2^{m/2}$ for $i = 2^m + k$ where $k = 1, 2, \ldots, 2^m$ and let $[t^j, t_*^j]$ be the interval on which $H_j(x)$ is nonzero. For $j < i$ the interval $[t^i, t_*^i]$ is either disjoint from $[t^j, t_*^j]$ or contained in it, so

$$\int_0^1 H_i(x) H_j(x) \, dx = C_j \int_{t^j}^{t_*^j} H_i(x) \, dx = 0$$

because $H_i$ is an odd function on $[t^j, t^j_*]$. In addition for each $j \geq 1$

$$\int_{t^j}^{t^j_*} H_j(x) H_j(x)\, dx = \left(2^{m/2}\right)^2 2^{-m-1} + \left(2^{m/2}\right)^2 2^{-m-1} = \frac{1}{2} + \frac{1}{2} = 1$$

for each $j \geq 1$. Thus the Haar functions form an orthonormal system of functions. The Karhunen-Loève expansion in terms of the Haar functions $W_t(\omega) = \sum_{n=1}^{\infty} Z_n(\omega) H_n(t)$ has coefficients

$$Z_n = \int_0^1 W_t H_n(t)\, dt = \int_{t^n}^{t^n_*} W_t H_n(t)\, dt = C_n \left(W_{t^n_{\#}} - W_{t^n}\right) - C_n \left(W_{t^n_*} - W_{t^n_{\#}}\right),$$

where $t^n_{\#}$ is the midpoint of the interval $[t^n, t^n_*]$, which are thus $N(0; 1)$ distributed random variables.

## Solutions of Exercises for Chapter 3

**Exercise 3.2.7**   Let $f$ be a nonrandom function and define step functions $f^{(n)}$ by $f^{(n)}(t) = f(t^n_j)$ for $t \in [t^n_j, t^n_{j+1})$ where $j = 0\ 1, 2, \ldots n-1$. Then

$$\int_0^T f(t)\, dW_t = \lim_{n \to \infty} Z_T^{(n)} \quad \text{mean square convergence}$$

where

$$\begin{aligned} Z_T^{(n)} &= \int_0^T f^{(n)}(t)\, dW_t = \sum_{j=1}^{n-1} f^{(n)}(t^n_j) \left(W_{t^n_{j+1}} - W_{t^n_j}\right) \\ &= f^{(n)}(T) W_T - f^{(n)}(0) W_0 - \sum_{j=0}^{n-1} W_{t^n_{j+1}} \left(f^{(n)}(t^n_{j+1}) - f^{(n)}(t^n_j)\right) \\ &\to f(T) W_T - \int_0^T W_t f'(t)\, dt \quad \text{in mean square,} \end{aligned}$$

giving

$$\int_0^T f(t)\, dW_t = f(T) W_T - \int_0^T W_t f'(t)\, dt.$$

This will also hold for continuously differentiable random functions which are $\mathcal{A}_t$-measurable and have independent increments.

**Exercise 3.2.8**   Here $\tilde{t} = \tilde{t}(t) = \int_0^t (2s)^2\, ds = \frac{4}{3} t^3$ has inverse $t = t(\tilde{t}) = \left(\frac{3}{4}\tilde{t}\right)^{1/3}$, so

$$\tilde{Z}_{\tilde{t}} = Z_{t(\tilde{t}}} = \int_0^{t(\tilde{t})} 2s\, dW_s$$

is a standard Wiener process with respect to the $\sigma$-algebras $\tilde{\mathcal{A}}_{\tilde{t}}$.

**Exercise 3.2.9**   Let $a = \{\omega \in \Omega : \sup_{t_0 \leq s \leq t} |Z_s(\omega)| > N\}$ and $B = \{\omega \in \Omega : \int_{t_0}^t f(s,\omega)^2\, ds \leq M\}$. Then

$$P(A) = P(A \cap (B^c \cup B)) = P(A \cap B^c) + P(A \cap B)$$

$$\leq \; P(B^c) + P\left(I_B \int_{t_0}^t f(s,\omega)^2 \, ds \geq N\right) \leq P(B^c) + \frac{1}{N^2} E\left((I_B Z_t)^2\right)$$

$$\leq \; P(B^c) + \frac{1}{N^2} E\left(I_B \int_0^t E\left(f(s)^2\right) ds\right) \leq P(B^c) + \frac{M}{N^2}.$$

**Exercise 3.3.4**  Let $Y_t = U(t, X_t)$ where $dX_t = f_t \, dW_t$ and $U(t, x) = x^{2n}$, so $U_t(t, x) = 0$, $U_x(t, x) = 2nx^{2n-1}$ and $U_{xx}(t, x) = 2n(2n-1)x^{2n-2}$. Thus by the Ito formula

$$d\left(X_t^{2n}\right) \;=\; dY_t = \frac{1}{2} f_t^2 U_{xx} \, dt + f_t U_x \, dW_t$$

$$= \; n(2n-1) f_t^2 X_t^{2n-2} \, dt + 2n f_t X_t^{2n-1} \, dW_t.$$

For $f_t \equiv 1$ and hence $X_t = W_t$ this reduces to

$$d\left(W_t^{2n}\right) = n(2n-1) W_t^{2n-2} \, dt + 2n W_t^{2n-1} \, dW_t.$$

**Exercise 3.4.2**  Use Example 3.4.1. With $e_t^i = 0$ and $f_t^i = 1$, $X_t^i = W_t^i$ for $i = 1$, 2. If $W^1$ and $W^2$ are independent (4.10) applies giving

$$d\left(W_t^1 W_t^2\right) = W_t^2 \, dW_t^1 + W_t^1 \, dW_t^2,$$

whereas if $W^1 = W^2 = W$ then (4.10) applies giving

$$d\left((W_t)^2\right) = 1 \, dt + 2 W_t \, dW_t.$$

**Exercise 3.4.3**  Apply the Ito formula to $Y_t = U(X_t)$ with $U(x) = \cos x$, $\sin x$ and $X_t = W_t$, that is with $e_t = 0$ and $f_t = 1$, to obtain

$$d\left(\cos W_t\right) = \frac{1}{2} \frac{d^2}{dx^2}\left(\cos W_t\right) dt + \frac{d}{dx}\left(\cos W_t\right) dW_t = -\frac{1}{2} \cos W_t \, dt - \sin W_t \, dW_t$$

and

$$d\left(\sin W_t\right) = \frac{1}{2} \frac{d^2}{dx^2}\left(\sin W_t\right) dt + \frac{d}{dx}\left(\sin W_t\right) dW_t = -\frac{1}{2} \sin W_t \, dt + \cos W_t \, dW_t.$$

**Exercise 3.4.4**  Apply the Ito formula $U(x) = e^x$, $xe^x$ and $X_t = W_t$, that is with $e_t = 0$ and $f_t = 1$, to obtain

$$dY_t^1 \;=\; d\left(e^{W_t}\right) = \frac{1}{2} \frac{d^2}{dx^2}\left(e^x\right)\Big|_{x=W_t} dt + \frac{d}{dx}\left(e^x\right)\Big|_{x=W_t} dW_t$$

$$= \; \frac{1}{2} e^{W_t} \, dt + e^{W_t} \, dW_t = \frac{1}{2} Y_t^1 \, dt + Y_t^1 \, dW_t$$

and

$$dY_t^2 \;=\; d\left(W_t e^{W_t}\right) = \frac{1}{2} \frac{d^2}{dx^2}\left(xe^x\right)\Big|_{x=W_t} dt + \frac{d}{dx}\left(xe^x\right)\Big|_{x=W_t} dW_t$$

$$= \; \frac{1}{2}\left(W_t e^{W_t} + 2e^{W_t}\right) dt + \left(W_t e^{W_t} + e^{W_t}\right) dW_t$$

$$= \; \frac{1}{2}\left(Y_t^2 + 2Y_t^1\right) dt + \left(Y_t^2 + Y_t^1\right) dW_t.$$

**Exercise 3.5.2**    Using ordinary calculus for the separable differential equation $dy = e^{-y}dw$ gives

$$w = \int_0^w dw = \int_0^y e^y \, dy = e^y - 1$$

so $y = \ln(1+w)$. Thus the Stratonovich SDE has solution $Y_t = \ln(1 + W_t)$. Applying the Ito formula to $U(x) = \ln(1 + x)$ and $dX_t = dW_t$ then gives

$$
\begin{aligned}
dY_t &= \frac{1}{2}\frac{d^2}{dx^2}\left(\ln(1+x)\right)\Big|_{x=W_t} dt + \frac{d}{dx}\left(\ln(1+x)\right)\Big|_{x=W_t} dW_t \\
&= \frac{-1}{2(1+W_t)^2} dt + \frac{1}{1+W_t} dW_t = -\frac{1}{2}e^{-2Y_t} dt + e^{-Y_t} dW_t.
\end{aligned}
$$

# Solutions of Exercises for Chapter 4

**Exercise 4.2.2**    $X_t$ is Gaussian as the linear combination of Gaussian random variables because $\int_{t_0}^t f(t)dW_t$ is Gaussian for a deterministic $f$ and $X_{t_0}$ is assumed to be deterministic or Gaussian.

**Exercise 4.2.3**    Applying the Ito formula to $U(x) = x^2$

$$d\left(X_t^2\right) = \left((a_1 X_t + a_2)\cdot 2X_t + \frac{1}{2}(b_1 X_t + b_2)^2\cdot 2\right) dt + (b_1 X_t + b_2)\cdot 2X_t \, dW_t,$$

so

$$P'(t)\,dt = \frac{d}{dt}E\left(X_t^2\right)\,dt = E\left(d\left(X_t^2\right)\right)$$

$$= E\left[\left(2a_1 X_t^2 + 2a_2 X_t + b_1^2 X_t^2 + 2b_1 b_2 X_t + b_2^2\right)\,dt\right] + E\left[\left(2b_1 X_t^2 + 2b_2 X_t\right)\,dW_t\right]$$

$$= \left(2a_1 E\left(X_t^2\right) + 2a_2 E\left(X_t\right) + b_1^2 E\left(X_t^2\right) + 2b_1 b_2 E\left(X_t\right) + b_2^2\right)\,dt + 0$$

since the expectation of the Ito integral (differential form here) vanishes. Hence

$$P' = \left(2a_1 + b_1^2\right) P + 2\left(a_2 + b_1 b_2\right) m + b_2^2$$

where $m = E\left(X_t\right)$.

**Exercise 4.2.4**    For the Langevin equation (1.7)

$$m' = -am, \qquad P' = -2aP + b^2.$$

These are first order linear ordinary differential equations with solutions

$$m(t) = m(0)e^{-at}, \qquad P(t) = P(0)e^{-2at} + \frac{b^2}{2a}\left(1 - e^{-2at}\right).$$

For the bilinear SDE (1.7) $m' = am$ and $P' = \left(2a + b^2\right)P$ with solutions

$$m(t) = m(0)e^{at}, \qquad P(t) = P(0)e^{(2a+b^2)t}.$$

**Exercise 4.3.2**    Since $(h^{-1})'(w) = 1/h'(x) = g(x)$ and $(h^{-1})''(w) = g(x)g'(x)$ the Ito formula applied to $X_t = h^{-1}(W_t + h(X_0))$ gives

$$
\begin{aligned}
dX_t &= \frac{1}{2}\left(h^{-1}\right)''\left(W_t + h(X_0)\right) dt + \left(h^{-1}\right)'\left(W_t + h(X_0)\right) dW_t \\
&= \frac{1}{2}g\left(X_t\right)g'\left(X_t\right) dt + g\left(X_t\right) dW_t.
\end{aligned}
$$

**Exercise 4.3.3**   Applying the Ito formula to $X_t = h^{-1}(\alpha t + W_t + h(X_0))$ gives

$$
\begin{aligned}
dX_t &= \left\{ \alpha \left(h^{-1}\right)'(\alpha t + W_t + h(X_0)) + \frac{1}{2}\left(h^{-1}\right)''(\alpha t + W_t + h(X_0)) \right\} dt \\
&\quad + \left(h^{-1}\right)'(\alpha t + W_t + h(X_0))\, dW_t \\
&= \left\{ \alpha g(X_t) + \frac{1}{2} g(X_t)\, g'(X_t) \right\} dt + g(X_t)\, dW_t.
\end{aligned}
$$

**Exercise 4.3.4**   $h' = 1/g$ and $h'' = -g'/g^2$. Applying the Ito formula to $Y_t = h(X_t)$ for

$$
dX_t = a\, dt + b\, dW_t = \left(\beta g h + \frac{1}{2} g g'\right) dt + g\, dW_t
$$

gives

$$
dY_t = \left(ah' + \frac{1}{2} b^2 h''\right) dt + bh'\, dW_t = \beta h(X_t)\, dt + dW_t = \beta Y_t + dW_t.
$$

This Langevin equation has solution $Y_t = h(X_t) = e^{\beta t} Y_0 + e^{\beta t} \int_0^t e^{-\beta s}\, dW_s$, so the original SDE has solution

$$
X_t = h^{-1}\left(e^{\beta t} h(X_0) + e^{\beta t} \int_0^t e^{-\beta s}\, dW_s, \right).
$$

**Exercise 4.3.5**   The equivalent Stratonovich SDE is $dX_t = g(X_t) o dW_t$. By ordinary calculus the separable ODE $dh = dx/g(x) = dw$ has solution $h(x) = w + const.$. Hence the Stratonovich SDE has solution $h(X_t) = W_t + h(X_0)$, that is $X_t = h^{-1}(W_t + h(X_0))$.

**Exercise 4.5.8**   The $X_t^{(\nu)}$ converge to the solution $X_t^{(0)} = x(t; 0, x_0)$ of the ordinary differential equation $\dot{x} = a(t, x)$ with the same initial value $x(0) = x_0$. This follows from Theorem 4.5.6 with $a^{(\nu)}(t, x) \equiv a(t, x)$ and $b^{(\nu)}(t, x) \equiv \nu$ for all $\nu \geq 0$.

**Exercise 4.6.2**   $X_t = e^{-(t-s)} X_s + e^{-(t-s)} \int_s^t e^{-(\tau-s)}\, dW_\tau$ is a diffusion process because

$$
\begin{aligned}
\frac{1}{t-s} E(X_t - x) &= \frac{1}{t-s}\left(1 - e^{-(t-s)}\right) x + \frac{1}{t-s} E\left(e^{-(t-s)} \int_s^t e^{-(\tau-s)}\, dW_\tau\right) \\
&= \frac{1}{t-s}\left(1 - e^{-(t-s)}\right) x + 0 \to -x \quad \text{as } t \to s+
\end{aligned}
$$

and $\frac{1}{t-s} E\left(|X_t - x|^2\right)$

$$
\begin{aligned}
&= \frac{1}{t-s}\left(1 - e^{-(t-s)}\right)^2 x^2 + \frac{2}{t-s}\left(1 - e^{-(t-s)}\right) x E\left(e^{-(t-s)} \int_s^t e^{-(\tau-s)}\, dW_\tau\right) \\
&\quad + \frac{1}{t-s} E\left(\left|e^{-(t-s)} \int_s^t e^{-(\tau-s)}\, dW_\tau\right|^2\right) \\
&= \frac{1}{t-s}\left(1 - e^{-(t-s)}\right)^2 x^2 + 0 + \frac{1}{t-s} e^{-2(t-s)} E\left(\int_s^t e^{-2(\tau-s)}\, d\tau\right) \\
&\to 0 + 0 + 2 = 2 \quad \text{as } t \to s+,
\end{aligned}
$$

using l'Hôpital's rule.

**Exercise 4.6.3**   From (1.7.10) with $\epsilon \to \infty$ and $t \to s+$

$$\frac{1}{t-s} E\left(|X_t - x|^2\right) = \frac{1}{t-s} \int_\Re (y-x)^2 P(s,y;t,x)\, dx \to b^2(s,x).$$

**Exercise 4.7.3**   Let $Y_t = f(X_t)$ where $dX_t = -X_t\, dt + \sqrt{2}\, dW_t$. By the Ito formula

$$\begin{aligned}
dY_t &= \left(-X_t f'(X_t) + \frac{1}{2}(\sqrt{2})^2 f''(X_t)\right) dt + \sqrt{2} f'(X_t)\, dW_t \\
&= \mathcal{L}f(X_t)\, dt + \sqrt{2} f'(X_t)\, dW_t.
\end{aligned}$$

Integrating and writing $M_t$ for the martingale $\int_0^t \sqrt{2} f'(X_s)\, dW_s$

$$f(X_t) = f(X_0) + \int_0^t \mathcal{L}f(X_t)\, dt + M_t.$$

**Exercise 4.8.1**   For $A$ the matrix in the drift coefficient, $A^{2k} = (-1)^k I$ and $A^{2k+1} = (-1)^k A$. By (8.8)

$$\begin{aligned}
\Phi_t &= \exp(At) = \sum_{l=0}^\infty \frac{1}{l!} A^l t^l = \sum_{k=0}^\infty \frac{1}{(2k)!} A^{2k} t^{2k} + \sum_{k=0}^\infty \frac{1}{(2k+1)!} A^{2k+1} t^{2k+1} \\
&= \sum_{k=0}^\infty \frac{1}{(2k)!} (-1)^k t^{2k} I + \sum_{k=0}^\infty \frac{1}{(2k+1)!} (-1)^k t^{2k+1} A = \cos t\, I + \sin t\, A.
\end{aligned}$$

Now $\Phi_t^{-1} = \Phi_t^\top$ here, so by (8.5)

$$X_t = \Phi_t X_0 + \Phi_t \int_0^t \Phi_s^{-1} \begin{pmatrix} 0 \\ \sigma^2 \end{pmatrix} dW_s = \Phi_t X_0 + \sigma^2 \Phi_t \int_0^t \begin{pmatrix} \sin s \\ \cos s \end{pmatrix} dW_s.$$

**Exercise 4.8.2**   Equation (8.6) reduces to the deterministic matrix differential equation $d\Phi_{t,t_0} = A\Phi_{t,t_0}\, dt$ which has solution given by (8.8).

**Exercise 4.8.3**   This is a multi-dimensional version of Exercise 4.2.3. See Arnold (1974), pages 131- 132, for details.

**Exercise 4.8.4**   Let $dX_t = AX_t\, dt + BX_t\, dW_t$. Here $B^2 = -I$ and $A = -2I$ commutes with $B$, so by (8.5) and (8.8) the solution $X_t = \Phi_t X_0$ where

$$\Phi_t = \exp\left(\left(A - \frac{1}{2} B^2\right) t + BW_t\right) = \exp\left(-\frac{3}{2} It + BW_t\right).$$

**Exercise 4.8.5**   Here the SDE (8.4) has $1 \times 1$ matrices $A = -\frac{1}{2}$ and $B^1 = B^2 = 1$, which all commute, so the solution is

$$X_t = X_0 \Phi_t = X_0 \exp\left(-\frac{3}{2} t + W_t^1 + W_t^2\right).$$

**Exercise 4.8.7**   For $dX_t = a(t, X_t)\, dt + b(t, X_t)\, dW_t$ and $Y_t = u(t, X_t)$ the Ito formula gives

$$dY_t = \left( \frac{\partial}{\partial t} + \mathcal{L} \right) u(t, X_t)\, dt + b(t, X_t) \frac{\partial}{\partial x} u(t, X_t)\, dW_t.$$

Integrating and taking conditional expectations

$$E\left( u(T, X_T) | X_s = x \right) = u(s, x) + E\left( \int_s^T \left( \frac{\partial}{\partial t} + \mathcal{L} \right) u(t, X_t)\, dt \Big| X_s = x \right) + 0,$$

since the expectation of the Ito integral vanishes. Thus

$$E\left( f(X_t) | X_s = x \right) = u(s, x)$$

if

$$u(T, X_T) = f(X_T) \quad \text{and} \quad \left( \frac{\partial}{\partial t} + \mathcal{L} \right) u = 0.$$

**Exercise 4.8.9**   Theorem 4.8.8 can be applied with

$$x \cdot a(x) = x(1 - x) = x - x^2 \le \frac{1}{2} x^2 - x^2 = -\frac{1}{2} x^2,$$

since $x \le \frac{1}{2} x^2$ for all $x$ with $|x| \ge 2$.

**Exercise 4.9.1**   With subscripts for partial derivatives of $U$, which are all evaluated at $(t, X_t)$, the Ito formula for $Y_t = U(t, X_t)$ and $dX_t = a\, dt + b\, dW_t$ is

$$
\begin{aligned}
dY_t &= \left( U_t + aU_x + \frac{1}{2} b^2 U_{xx} \right) dt + bU_x\, dW_t \\
&= (U_t + \underline{a} U_x)\, dt + \left( \frac{1}{2}(bU_x)(bU_x)_x + bU_x\, dW_t \right) = (U_t + \underline{a} U_x)\, dt + bU_x \circ dW_t \\
&= U_t\, dt + U_x\, (\underline{a}\, dt + b \circ dW_t) = U_t\, dt + U_x \circ dX_t
\end{aligned}
$$

since $\underline{a} = a - \frac{1}{2} bb_x$.

**Exercise 4.9.2**   Here $\underline{a} = a - \frac{1}{2} bb_x = 1 - \frac{1}{2} 2x^{1/2} 2\frac{1}{2} x^{-1/2} = 1 - 1 = 0$, so

$$dX_t = 1\, dt + 2\sqrt{X_t}\, dW_t = 2\sqrt{X_t} \circ dW_t.$$

Hence $x^{-1/2} dx = 2 \circ dW_t$, so $2x^{1/2} = 2W_t + const$ and the desired solution is $X_t = (W_t + \sqrt{X_0})^2$.

**Exercise 4.9.3**   The Stratonovich SDE is $dX_t = BX_t \circ dW_t$ since $B^\mathsf{T} B = -I$, so

$$\underline{a}(x) = -\frac{1}{2} x - \frac{1}{2} B^\mathsf{T} Bx = -\frac{1}{2} x + \frac{1}{2} Ix = 0.$$

## Solutions of Exercises for Chapter 5

**Exercise 5.2.1**  $\alpha = (0,0,0)$: $l(\alpha) = 3$, $n(\alpha) = 3$, $-\alpha = (0,0)$, $\alpha- = (0,0)$;
$\alpha = (2,0,1)$: $l(\alpha) = 3$, $n(\alpha) = 1$, $-\alpha = (0,1)$, $\alpha- = (2,0)$;
$\alpha = (0,1,0,0,2)$: $l(\alpha) = 5$, $n(\alpha) = 3$, $-\alpha = (1,0,0,2)$, $\alpha- = (0,1,0,0)$.

**Exercise 5.2.2**  $I_{(0,0,0)}[1]_{0,T} = \int_0^T \int_0^{s_3} \int_0^{s_2} ds_1\, ds_2\, ds_3 = \frac{1}{3!}T^3$,

$I_{(2,0,1)}[1]_{0,T} = \int_0^T \int_0^{s_3} \int_0^{s_2} dW_{s_1}^2\, ds_2\, dW_{s_3}^1$,    $I_{(1,2)}[1]_{0,T} = \int_0^T \int_0^{s_2} dW_{s_1}^1\, dW_{s_2}^2$.

**Exercise 5.2.5**  $I_{(0,1)} = I_{(0)}I_{(1)} - I_{(1,0)}$,    $I_{(1,1)} = \frac{1}{2}\left(\left(I_{(1)}\right)^2 - I_{(0)}\right)$,

$I_{(1,1,1)} = \frac{1}{3!}\left(\left(I_{(1)}\right)^3 - 3I_{(1)}I_{(0)}\right)$.

**Exercise     5.2.6**  Exercise     5.2.5     with     $I_{(1,1,1,1)}$  =
$\frac{1}{4!}\left(\left(I_{(1)}\right)^4 - 6\left(I_{(1)}\right)^2 I_{(0)} + 3\left(I_{(0)}\right)^2\right)$.

**Exercise 5.2.7**  $E\left(I_{(1,0)}\right) = \int E\left(W_s\right)\, ds = 0$,

$E\left(\left(I_{(0,1)}\right)^2\right) = E\left(\int s\, dW_s \int s\, dW_s\right) = \int s^2\, ds = \frac{1}{3}\Delta^3$,

$E\left(I_{(1)}I_{(0,1)}\right) = E\left(\int dW_s \int s\, dW_s\right) = \int s\, ds = \frac{1}{2}\Delta^2$,

$E\left(\left(I_{(1,0)}\right)^2\right) = E([\Delta I_{(1)} - I_{(0,1)}]^2) = \Delta^2 E\left(\left(I_{(1)}\right)^2\right) - 2\Delta E\left(I_{(1)}I_{(0,1)}\right)$
$+ E\left(\left(I_{(0,1)}\right)^2\right) = \Delta^3\left(1 - 1 + \frac{1}{3}\right) = \frac{1}{3}\Delta^3$.

**Exercise 5.2.9**  Apply (2.3.4).

**Exercise 5.2.10**  $J_{(0,1)} = J_{(0)}J_{(1)} - J_{(1,0)}$,    $J_{(1,1)} = \frac{1}{2}\left(J_{(1)}\right)^2$,    $J_{(1,1,1)} = \frac{1}{3!}\left(J_{(1)}\right)^3$.

**Exercise 5.3.1**  $f_{(1,0)} = ba'$,    $f_{(1,1,1)} = b\left((b')^2 + bb''\right)$.

**Exercise 5.3.2**  Apply (5.3.3).    $f_{(j_1,j_2,j_3,j_4)} = b^{j_1}\left[b^{j_2\prime}\left(b^{j_3\prime}b^{j_4\prime} + b^{j_3}b^{j_4\prime\prime}\right) + b^{j_2}\left(b^{j_3\prime\prime}b^{j_4\prime} + 2b^{j_3\prime}b^{j_4\prime\prime} + b^{j_3}b^{j_4\prime\prime\prime}\right)\right]$.

**Exercise 5.3.3**  $\underline{f}_{(1,0)} = b\underline{a}'$,    $\underline{f}_{(1,1,1)} = b\left[bb'' + (b')^2\right]$.

**Exercise 5.3.4**  Use Exercise 5.3.2 as $\underline{f}_{(j_1,j_2,j_3,j_4)} = f_{(j_1,j_2,j_3,j_4)}$.

**Exercise 5.4.1**  $\{v,(1)\}$,    $\{v,(0),(1),(0,1)\}$.

**Exercise 5.4.2**  $B\left(\{v,(1)\}\right) = \{(0),(0,1),(1,1)\}$,    $B\left(\{v,(0),(1),(0,1)\}\right) = \{(1,1),$

$(1, 0)$, $(0, 0)$, $(1, 0, 1)$, $(0, 0, 1)\}$.

**Exercise 5.4.3**  Yes.

**Exercise 5.4.4**  $B(\Gamma_r) = \{\alpha \in \mathcal{M} : l(\alpha) = r + 1\}$ for $r = 1, 2, \ldots$.

**Exercise 5.4.5**  Yes.

**Exercise 5.5.2**

$$
\begin{aligned}
X_t &= X_0 + a(0, X_0) \int_0^t ds + b(0, X_0) \int_0^t dW_s \\
&\quad + b(0, X_0) \frac{\partial}{\partial x} b(0, X_0) \int_0^t \int_0^{s_2} dW_{s_1} \, dW_{s_2}.
\end{aligned}
$$

**Exercise 5.5.3**  For $\alpha_r = (1, 0, 0, \ldots, 0)$ with $l_r(\alpha_r) = r$

$$
X_t = X_0 + \sum_{r=1}^{l} \left[ (-1)^r \frac{1}{r!} t^r X_0 + (-1)^{r-1} I_{\alpha_r} \right].
$$

**Exercise 5.6.2**

$$
\begin{aligned}
X_t &= X_0 + \underline{a}(0, X_0) \int_0^t ds + b(0, X_0) \int_0^t dW_s \\
&\quad + b(0, X_0) \frac{\partial}{\partial x} b(0, X_0) \int_0^t \int_0^{s_2} dW_{s_1} \circ dW_{s_2}.
\end{aligned}
$$

**Exercise 5.6.3**  The same result as for Exercise 5.5.3 as $a = \underline{a}$, $I_{(0,\ldots,0)} = J_{(0,\ldots,0)}$ and $I_{(1,0,\ldots,0)} = J_{(1,0,\ldots,0)}$

**Exercise 5.6.5**  Analogous to the proofs of Lemma 5.5.5 and Theorem 5.5.1, respectively.

**Exercise 5.8.1**  Integrate $J_{(j,0),t}$ from (5.8.6) with repect to time: $J_{(j,0,0),t} = \int_0^t J_{(j,0),s} \, ds$ to obtain the desired formula.

**Exercise 5.8.2**  Evaluate the integrals in (5.8.5)–(5.8.7) with $\Delta = t$.

**Exercise 5.8.3**  Verify that the random variables in (5.8.10) are precisely those needed in (5.8.11) to give the formulae (5.8.9).

**Exercise 5.10.3**  Analogous to the proofs of Propositions 5.9.1 and 5.9.2, respectively.

**Exercise 5.12.2**  The proof of Corollary 5.12.1 is a slight modification of the proof of Proposition 5.11.1.

**Exercise 5.12.4**  See the remarks after Lemma 5.12.3.

## Solution of Exercises for Chapter 6

**Exercise 6.2.2** Let $\zeta_t = \pm 1$ with $E(\zeta_t) = 0$ and $E(\zeta_t^2) = 1$ and let

$$\Delta X_k^{(N)} = X_{k+1}^{(N)} - X_k^{(N)} = a\,\frac{1}{N} + b\,\frac{1}{\sqrt{N}}\,\zeta_t.$$

Then

$$N\,E\left(\Delta X_k^{(N)}\right) = a + b\sqrt{N}\,E\left(\zeta_t\right) = a + 0 \to a$$

and

$$N\,E\left(\left(\Delta X_k^{(N)}\right)^2\right) = \frac{1}{N}a^2 + 2ab\,\frac{1}{\sqrt{N}}\,E\left(\zeta_t\right) + b^2\,E\left(\zeta_t^2\right) = \frac{1}{N}a^2 + 0 + b^2 \to b^2$$

as $N \to \infty$.

**Exercise 6.3.3** For $V(x) = x^2$ and $a < -\frac{1}{2}b^2 < 0$

$$LV(x) = ax(1 + x^2)\cdot 2x + \frac{1}{2}(bx)^2 \cdot 2 = \left(2a + b^2\right)x^2 + 2ax^4 \le 0.$$

**Exercise 6.3.4** Here the linearization of the Ito SDE about $(x_1, x_2) = (0, 0)$ has the same coefficients in Ito and Stratonovich forms

$$dZ_t = AZ_t\,dt + BZ_t\,dW_t = AZ_t\,dt + BZ_t \circ dW_t$$

where

$$A = \left[\frac{\partial a^i}{\partial x_j}\Big|_{x_1 = x_2 = 0}\right] = \begin{bmatrix} 0 & 1 \\ -1 - 2c & -b \end{bmatrix}$$

and

$$B = \left[\frac{\partial b^i}{\partial x_j}\Big|_{x_1 = x_2 = 0}\right] = \begin{bmatrix} 0 & 0 \\ 1 & 0 \end{bmatrix},$$

since the correction term vanishes. The Stratonovich version of the original nonlinear SDE has drift coefficients

$$\underline{a}^1 = a^1, \quad \underline{a}^2 = a^2 - (a)^2(x_2)^3 + ax_2 \sin x_1,$$

which has the same linearization $AZ$ about $(0, 0)$. Hence the linearization of the nonlinear Stratonvich SDE coincides with that of the nonlinear Ito SDE in our case.

**Exercise 6.3.5** The solution is $X_t = X_0 \exp(at + bW_t)$ so

$$\frac{1}{t}\ln|X_t| = \frac{1}{t}\ln|X_0| + a + b\frac{1}{t}W_t \to 0 + a + 0 = a = \lambda \quad \text{as} \quad t \to \infty.$$

Thus the Lyapunov exponent $\lambda$ is negative when $a$ is negative, whatever the value of $b$.

**Exercise 6.3.6** For the matrices $A$ and $B$ in the solution to Exercise 6.3.4

$$s^\top A s = -2c s_1 s_2 - b(s_2)^2, \quad q^1(s) = s^\top B s = s_1 s_2, \quad s^\top\left(B + B^\top\right)s = 2s_1 s_2.$$

Hence

$$\begin{aligned} q(s) &= s^\top A s + \frac{1}{2}s^\top\left(B + B^\top\right)s - \left(s^\top B s\right)^2 \\ &= -2c s_1 s_2 - b\left(s_2\right)^2 + s_1 s_2 - \left(s_1 s_2\right)^2 = (1 - 2c)s_1 s_2 - b\left(s_2\right)^2 - \left(s_1 s_2\right)^2, \end{aligned}$$

$$h(s, A) = \left(A - s^\top A s I\right) s = \begin{pmatrix} 2c(s_1)^2 s_2 + bs_1(s_2)^2 + s_2 \\ -(1 + 2c)s_1 - bs_2 + 2cs_1(s_2)^2 + b(s_2)^3 \end{pmatrix}.$$

and

$$h(s, B) = \left(B - s^\top B s I\right) s = \begin{pmatrix} -(s_1)^2 s_2 \\ s_1 - s_1(s_2)^2 \end{pmatrix}.$$

The equations (3.22)–(3.23) are then

$$dR_t = R_t q(S_t)\, dt + R_t q^1(S_t) \circ dW_t, \quad dS_t = h(S_t, A)\, dt + h(S_t, B) \circ dW_t.$$

**Exercise 6.3.7**  Let $P = [p^{i,j}]$, where $p^{1,2} = p^{2,1}$, and $p = (p_1, p_2, p_3) = (p^{1,1}, p^{1,2}, p^{2,2})$. Then with $\gamma = 1 + 2c$

$$P' = AP + PA^\top + BPB^\top = \begin{bmatrix} 2p^{1,2} & -\gamma p^{1,1} - bp^{1,2} + p^{2,2} \\ -\gamma p^{1,1} - bp^{1,2} + p^{2,2} & p^{1,1} - 2\gamma p^{1,2} - 2bp^{2,2} \end{bmatrix},$$

so $p' = \mathcal{A}p$ where

$$\mathcal{A} = \begin{bmatrix} 0 & 2 & 0 \\ -\gamma & -b & 1 \\ 1 & -2\gamma & -2b \end{bmatrix}.$$

Its eigenvalues are the roots of the cubic

$$\lambda^3 + 3b\lambda^2 + 2\left(2\gamma + b^2\right)\lambda - 2 + 4b\gamma = 0$$

and the zero solution is mean-square asymptotically stable when the real parts of these eigenvalues are all negative.

**Exercise 6.4.1**  Here

$$\hat{\alpha}(T) = \int_0^T X_t\, dX_t \Big/ \int_0^T X_t^2\, dt \quad \text{and} \quad \sigma^2 = \left(\frac{1}{\sqrt{\pi}} \int_{\Re} x^2 e^{-x^2}\, dx\right)^{-1} = 2.$$

**Exercise 6.5.1**  Let subscripts denote partial derivatives of $H$. The HJB equation

$$H_s + Cx^2 + Ax H_x + \frac{1}{2}\sigma^2 H_{xx} + \min_{u \in \Re}\left\{Gu^2 + Mu H_x\right\} = 0$$

achieves its minimum where $\frac{d}{du}\left\{Gu^2 + Mu H_x\right\} = 2Gu + M H_x = 0$, that is at $u^* = -\frac{1}{2}MG^{-1}H_x$. The HJB equation is then

$$H_s + Cx^2 + Ax H_x + \frac{1}{2}\sigma^2 H_{xx} - \frac{1}{4}M^2 G^{-1}\left(H_x\right)^2 = 0.$$

Assuming $H(s, x) = S(s)x^2 + a(s)$ with $a' = -S(s)\sigma^2$ and cancelling $x^2$ this gives

$$S' + C - 2AS + M^2 G^{-1} S^2 = 0.$$

**Exercise 6.5.2**  Here $A = -4$, $C = 8$, $G = R = 1$, $M = 2$ and $\sigma = 4$. The optimal control $u^* = -2x$ so the optimal path satisfies the SDE $dX_t^* = -6X_t^*\, dt + dW_t$ with solution $X_t^* = X_0 e^{-6t + W_t}$. $S$ satisfies the Riccati equation $S' + -4\left(s^2 - 2s + 2\right)$ with $S(T) = 1$. This is a separable ODE with solution $S(s) = 1 + \tan(4T - 4s)$. Thus

$a' = -16S$ with $a(T) = 0$ has solution $a(s) = 16(T - s) - 4\ln(\cos(4t - 4s))$. This gives the optimal cost

$$H(s.x) = S(s)x^2 + a(s) = x^2 + x^2\tan(4T - 4s) + 16(T - s) - 4\ln(\cos(4T - 4s)).$$

**Exercise 6.6.1**   Let $C = H^2/\Gamma^2 > 0$. Then the Riccati equation (6.6) takes the form

$$S' = -CS^2 + 2AS + B^2 = -C\left(S - S^-\right)\left(S - S^+\right),$$

with real-valued $S^\pm = (A \pm \sqrt{A^2 + B^2C})/C$. This is a separable differential equation solved by integrating

$$-C\left(S^+ - S^-\right)dt = \frac{S^+ - S^-}{(S - S^-)(S - S^+)}\,dS = \left(\frac{1}{(S - S^+)} - \frac{1}{(S - S^-)}\right)dS$$

to obtain

$$S(t) = \left(S^+ - S^-C_1e^{-\alpha t}\right)\Big/\left(1 - C_1e^{-\alpha t}\right)$$

where $\alpha = 2\sqrt{A^2 + B^2C}$ and $C_1$ is the integration constant.

**Exercise 6.6.2**   Here $A = B = 0$ and $H = \Gamma = 1$, so $S' = -S^2$ with $S(0) = \sigma^2$ has solution $S(t) = \sigma^2/(1 + \sigma^2 t)$ and

$$d\hat{X}_t = -S(t)\hat{X}_t\,dt + S(t)\,dY_t.$$

Let $I(t) = \exp\left(\int_0^t S(s)\,ds\right)$. Then $d(\hat{X}_t I(t)) = S(t)I(t)\,dY_t$ so integrating with $\hat{X}_0 = Y_0 = 0$ gives

$$\hat{X}_t I(t) = S(t)I(t)Y_t - \int_0^t Y_s(S(t)I(t))'\,dt = S(t)I(t)Y_t$$

as $(SI)' = S'I + SI' = (S' - S^2)I = 0$. Hence $\hat{X}_t = S(t)Y_t$.

**Exercise 6.6.3**   Here $k = 1$ and $h_1(i) = (-1)^{i+1}$ for $i = 1, 2$ so

$$dQ_t = AQ_t\,dt + H_1Q_t\,dW_t = \begin{bmatrix} -0.5 & 0.5 \\ 0.5 & -0.5 \end{bmatrix}Q_t\,dt + \begin{bmatrix} 1 & 0 \\ 0 & -1 \end{bmatrix}Q_t\,dW_t$$

where $Q^\top = (Q(1), Q(2))$.

## Solutions of Exercises for Chapter 8

**Exercise 8.1.4**   The modified Euler or Heun method is the second order Runge-Kutta method with parameters given in equation (8.2.7). Its local discretization error is derived in Section 2 of Chapter 8 in the context of these Runge-Kutta methods. For the trapezoidal method let $y(t)$ be the solution with $y(t_k) = y_n$ and let $A(t) = a(t, y(t))$. Using the Taylor expansion of $A$ at $t_n$ with second order remainder term

$$\begin{aligned}
y(t_{n+1}) - y_n &= \int_{t_n}^{t_{n+1}} A(s)\,ds \\
&= \int_{t_n}^{t_{n+1}}\left(A(t_n) + A'(t_n)(s - t_n) + \frac{1}{2!}A''(\theta_n(s))(s - t_n)^2\right)ds \\
&= a(t_n, y_n)\Delta_n + \frac{1}{2}A'(t_n)\Delta_n^2 + \frac{1}{2!}\int_{t_n}^{t_{kn+1}} A''(\theta_n(s))(s - t_n)^2\,ds
\end{aligned}$$

$$= \frac{1}{2} a(t_n, y_n) \Delta_n + \frac{1}{2} \left( a(t_n, y_n) + A'(t_n) \Delta_n \right) \Delta_n + \frac{1}{2!} \int_{t_n}^{t_{n+1}} A''(\theta_n(s))(s - t_n)^2 \, ds$$

so

$$y(t_{n+1}) - y_n - \frac{1}{2} \left( a(t_n, y_n) + a(t_{n+1}, y(t_{n+1})) \right) \Delta_n$$

$$= -\frac{1}{2^2} A(\theta_n) \Delta_n^3 + \frac{1}{2!} \int_{t_n}^{t_{n+1}} A''(\theta_n(s))(s - t_n)^2 \, ds$$

Here $\theta_n(s) \in [t_n, s]$ and $\theta_n \in [t_n, t_{n+1}]$. The local discretization error is thus of order three.

**Exercise 8.1.6** Let $P_{n+1}$ be the interpolating quadratic for the points $(t_{n-2}, a_{n-2})$, $(t_{n-1}, a_{n-1})$ and $(t_n, a_n)$ where $a_n = a(t_n, y(t_n))$. It is known that there is a continuous function $E(t)$ such that

$$a(t, y(t)) - P_{n+1}(t) = E(t) \prod_{i=n-2}^{n} (t - t_i).$$

The Adams-Bashford method (1.14) is derived from

$$y(t_{k+1}) - y(t_k) = \int_{t_k}^{t_{k+1}} P_{n+1}(t) \, dt,$$

so its local discretization error is given by

$$l_{n+1} = \int_{t_k}^{t_{k+1}} \left( a(t, y(t)) - P_{n+1}(t) \right) dt = \int_{t_k}^{t_{k+1}} E(t) \prod_{i=n-2}^{n} (t - t_i) \, dt$$

which is fourth order.

**Exercise 8.3.3** It needs only be shown that the second term in $\Psi$ satisfies a Lipschitz condition.

$$\left| a \left( t' + \Delta', x' + a(t', x') \Delta' \right) - a \left( t + \Delta, x + a(t, x) \Delta \right) \right|$$

$$\leq K \left( |t' + \Delta' - t - \Delta| + |x' + a(t', x') \Delta' - x - a(t, x) \Delta| \right)$$

$$\leq K \left( |t' - t| + |\Delta' - \Delta| + |x' - x| + |a(t', x') - a(t, x)||\Delta| + |a(t, x)||\Delta' - \Delta| \right)$$

$$\leq K \left( |t' - t| + |\Delta' - \Delta| + |x' - x| + K \left( |t' - t| + |x' - x| \right) |\Delta| + L |\Delta' - \Delta| \right).$$

This is global in $(t, x)$ and local in $\Delta$. However $\Delta$ can be restricted without loss of generality to $\Delta \leq 1$, with the values of $\Psi$ being frozen at $\Psi(t, x, 1)$ for larger $\Delta$. The modified $\Psi$ is the globally Lipschitz in all three variables.

**Exercise 8.3.5** Let $E_n = y_n - \bar{y}_n$. Then

$$\begin{aligned} |E_{n+1}| &= |E_n + \Delta \left( \Psi(t_n, y_n, \Delta) - \Psi(t_n, \bar{y}_n, \Delta) \right)| \\ &\leq |E_n| + \Delta |\Psi(t_n, y_n, \Delta) - \Psi(t_n, \bar{y}_n, \Delta)| \\ &\leq |E_k| + \Delta K |y_n - \bar{y}_n| \leq (1 + K\Delta) |E_n|. \end{aligned}$$

Hence with $M = (1 + K\Delta)^{nT}$

$$|E_n| \leq (1 + K\Delta)^n |E_0| \leq (1 + K\Delta)^{nT} |E_0| \leq M |E_0|$$

for $n = 0, 1, \ldots, n_T$.

**Exercise 8.3.6**  Solving $y_{n+1} = y_n + \frac{1}{2}\lambda\Delta y_n + \frac{1}{2}\lambda\Delta y_{n+1}$ gives

$$\left(1 - \frac{1}{2}\lambda\Delta\right) y_{n+1} = \left(1 + \frac{1}{2}\lambda\Delta\right) y_n \quad \text{or} \quad y_{n+1} = \frac{1 + \frac{1}{2}\lambda\Delta}{1 - \frac{1}{2}\lambda\Delta} y_n.$$

Absolute stability holds for

$$\left|\frac{1 + \frac{1}{2}\lambda\Delta}{1 - \frac{1}{2}\lambda\Delta}\right| < 1 \quad \text{i.e. for} \quad \left|1 + \frac{1}{2}\lambda\Delta\right| < \left|1 - \frac{1}{2}\lambda\Delta\right|,$$

i.e. for the real part of $\lambda\Delta$ being negative. The region of absolute stability is thus the left half of the complex plane.

**Exercise 8.3.7**  The trapezoidal method is A-stable because its region of absolute stability is the left half of the complex plane.

**Exercise 8.3.8**  For $a(t, x) = \lambda x$ the Adams-Bashford method (1.14) is

$$y_{n+1} = y_n + \frac{1}{12}\Delta\left(23\lambda y_n - 16\lambda y_{n-1} + 5\lambda y_{n-2}\right)$$

and the Adams-Moulton method (2.12) is

$$y_{n+1} = y_n + \frac{1}{12}\Delta\left(5\lambda y_{n+1} - 8\lambda y_n + \lambda y_{n-1}\right).$$

Hence the required polynomials are, respectively,

$$\zeta^3 - \left(1 + \frac{23}{12}\lambda\Delta\right)\zeta^2 + \frac{4}{3}\lambda\Delta\zeta - \frac{5}{12}\lambda\Delta = 0,$$

and

$$\left(1 - \frac{5}{12}\lambda\Delta\right)\zeta^2 - \left(1 + \frac{2}{3}\lambda\Delta\right)\zeta + \frac{1}{12}\lambda\Delta = 0.$$

## Solutions of Exercises for Chapter 9

**Exercise 9.1.1**  $Y_{n_t} = X_0 + a\tau_{n_t} + bW_{\tau_{n_t}} \sim N\left(X_0 + a\tau_{n_t}; b^2\tau_{n_t}\right).$

**Exercise 9.4.4**

$$
\begin{aligned}
\mathrm{Var}\left(\mu_{stat}\right) &= \mathrm{Var}\left(\hat{\mu} - \mu_{sys}\right) \\
&= \mathrm{Var}\left(\frac{1}{MN}\sum_{j=1}^{M}\sum_{k=1}^{N} Y_{T,k,j} - E\left(X_T\right) - \left(E\left(Y(T)\right) - E\left(X_T\right)\right)\right) \\
&= E\left(\frac{1}{MN}\sum_{j=1}^{M}\sum_{k=1}^{N}\left(Y_{T,k,j} - E\left(Y(T)\right)\right)^2\right) \\
&= \frac{1}{(MN)^2}\sum_{j=1}^{M}\sum_{k=1}^{N} E\left(\left(Y_{T,k,j} - E\left(Y(T)\right)\right)^2\right) = \frac{1}{MN}\mathrm{Var}(Y(T)).
\end{aligned}
$$

**Exercise 9.6.1** Yes. The Euler scheme approximation of an Ito diffusion with constant drift and diffusion coefficients coincides with the Ito diffusion itself so we have $\gamma = \infty$.

**Exercise 9.6.3**

$$E\left(\left|E\left(\frac{Y_{n+1}^\delta - Y_n^\delta}{\Delta_n}\,\Big|\,\mathcal{A}_{\tau_n}\right) - a\left(\tau_n, Y_n^\delta\right)\right|^2\right) = 0,$$

$$E\left(\frac{1}{\Delta_n}\left|Y_{n+1}^\delta - Y_n^\delta - E\left(Y_{n+1}^\delta - Y_n^\delta\,\big|\,\mathcal{A}_{\tau_n}\right) - b\left(\tau_n, Y_n^\delta\right)\Delta W_n\right|^2\right) = 0.$$

**Exercise 9.6.4**

$$E\left(\left|E\left(\frac{Y_{n+1} - Y_n}{\Delta_n}\,\Big|\,\mathcal{A}_{\tau_n}\right) - a\left(Y_n\right)\right|^2\right) = \left(\frac{1}{2}b'\left(Y_n\right)b\left(Y_n\right)\right)^2 + O\left(\Delta_n\right),$$

$$E\left(\frac{1}{\Delta_n}\left|Y_{n+1} - Y_n - E\left(Y_{n+1} - Y_n\,\big|\,\mathcal{A}_{\tau_n}\right) - b\left(Y_n\right)\Delta W_n\right|^2\right) = O\left(\Delta_n\right).$$

For $b(x) \equiv const.$ the scheme is strongly consistent.

**Exercise 9.6.5** No. The trajectories do not converge pathwise to each other.

**Exercise 9.7.1** Yes. The probability measures of both Wiener processes are identical.

**Exercise 9.7.2**

$$E\left(\left|E\left(\frac{Y_{n+1} - Y_n}{\Delta_n}\,\Big|\,\mathcal{A}_{\tau_n}\right) - a\left(Y_n\right)\right|^2\right) = 0,$$

$$E\left(\left|E\left(\frac{1}{\Delta_n}\left(Y_{n+1} - Y_n\right)^2\,\big|\,\mathcal{A}_{\tau_n}\right) - b^2\left(Y_n\right)\right|^2\right) = O\left(\Delta_n^2\right).$$

**Exercise 9.7.3** We obtain the same relations as in Exercise 9.7.2.

**Exercise 9.7.5** In the absence of noise it follows from (9.7.6) that

$$\lim_{\Delta \to 0}\left|\frac{Y_{n_t+1} - Y_{n_t}}{\Delta} - a\left(Y_{n_t}\right)\right| = \lim_{\Delta \to 0}\left|\psi\left(\tau_{n_t}, Y_{n_t}, \Delta\right) - a\left(Y_{n_t}\right)\right| = 0$$

and so similarly to (8.3.3) we obtain

$$\lim_{\Delta \to 0}\psi(t, y, \Delta) = a(t, y),$$

whenever $Y_n$ is fixed at the value $y$.

**Exercise 9.7.6** Using conditional expectations we have

$$E\left(X_t X_{t+h}\right) = E\left(E\left(X_{t+h}\,\big|\,\mathcal{A}_t\right)X_t\right).$$

Under appropriate assumptions it follows from Theorem 4.8.6 that the function

$$\tilde{\mu}_{t,h}(x) = E\left(X_{t+h}\,\big|\,X_t = x\right)$$

is sufficiently smooth and of polynomial growth together with its derivatives. Obviously, the same also holds for the function $g(x) = \tilde{\mu}_{t,h}(x)\,x$, so we can write

$$E(X_t X_{t+h}) = E(\tilde{\mu}_{t,h}(X_t)\,X_t) = E(g(X_t))$$

which is a functional in the usual form.

**Exercise 9.8.1**  As in the proof of Theorem 4.5.3 from the Lipschitz continuity of $a$ and $b$ we obtain

$$Z_t = \sup_{0 \le s \le t} E\left(\left|Y^\delta_{\tau_s} - \bar{Y}^\delta_{\tau_s}\right|^2\right) \le \left|Y^\delta_0 - \bar{Y}^\delta_0\right|^2 + K \int_0^t Z_s\, ds.$$

The Gronwall inequality and the Chebyshev inequality then yield the desired limit (9.8.1).

**Exercise 9.8.4**  We have

$$Y^\delta_{n+1} - \bar{Y}^\delta_{n+1} = Y^\delta_n - \bar{Y}^\delta_n - 16\left(Y^\delta_n - \bar{Y}^\delta_n\right)\delta = (1 - 16\delta)^{n+1}\left(Y^\delta_0 - \bar{Y}^\delta_0\right).$$

For $\delta < \Delta_0 = 1/8$ and $\epsilon > 0$ by the Chebyshev inequality we obtain

$$\lim_{|Y^\delta_0 - \bar{Y}^\delta_0| \to 0} \lim_{T \to \infty} P\left(\sup_{t_0 \le t \le T} \left|Y^\delta_{\tau_{n_t}} - \bar{Y}^\delta_{\tau_{n_t}}\right| \ge \epsilon\right) \le \lim_{|Y^\delta_0 - \bar{Y}^\delta_0| \to 0} P\left(\left|Y^\delta_0 - \bar{Y}^\delta_0\right| \ge \epsilon\right) = 0.$$

**Exercise 9.8.5**  With $X_t = X^1_t + \imath X^2_t$ and $\lambda = \lambda_1 + \imath \lambda_2$ we have

$$dX^1_t = \left(\lambda_1 X^1_t - \lambda_2 X^2_t\right)dt + dW_t$$
$$dX^2_t = \left(\lambda_2 X^1_t + \lambda_1 X^2_t\right)dt.$$

## Solutions of Exercises for Chapter 10

**Exercise 10.3.1**

$$E\left(\left|E\left(\frac{Y_{n+1} - Y_n}{\Delta}\,\bigg|\,\mathcal{A}_{\tau_n}\right) - a(Y_n)\right|^2\right) = 0,$$

$$E\left(\frac{1}{\Delta}\left|Y_{n+1} - Y_n - E\left(Y_{n+1} - Y_n \,\big|\, \mathcal{A}_{\tau_n}\right) - b(Y_n)\Delta W_n\right|^2\right) = \frac{1}{2}E\left(\left|bb'\right|^2\right)\Delta.$$

**Exercise 10.3.6**  It follows from (5.5.3) that

$$X^2_s = X^2_{\tau_{n_s}} + \int_{\tau_{n_s}}^s W^1_{\tau_{n_s}}\,dW^2_s + \int_{\tau_{n_s}}^s \left(W^1_s - W^1_{\tau_{n_s}}\right)dW^2_s$$

for all $s \in [0, T]$, so we have

$$E\left(\left|Y^2_{\tau_{n_T}} - X^2_T\right|^2\right) = E\left(\left|\int_0^T \left(\frac{1}{2}\Delta W^1_{n_s} - \left(W^1_s - W^1_{\tau_{n_s}}\right)\right)dW^2_s\right|^2\right) = \frac{1}{4}\Delta T.$$

**Exercise 10.4.1**  For the terms (10.4.1) look at (5.5.4). From (5.2.22) we have $I_{(0,1)} = \Delta W_n \Delta - \Delta Z_n$. For strong consistency it suffices that

$$E\left(\left|E\left(\frac{Y_{n+1} - Y_n}{\Delta}\,\bigg|\,\mathcal{A}_{\tau_n}\right) - a(Y_n)\right|^2\right) = \frac{1}{2}\left(aa' + \frac{1}{2}b^2 a''\right)\Delta^2,$$

$$E\left(\frac{1}{\Delta}\left|Y_{n+1} - Y_n - E\left(Y_{n+1} - Y_n \mid \mathcal{A}_{\tau_n}\right) - b\left(Y_n\right)\Delta W_n\right|^2\right) \le K\left\{\frac{1}{2}E\left(\left|bb'\right|^2\right)\Delta\right.$$

$$\left. +\frac{1}{3}E\left(\left|a'b\right|^2\right)\Delta^3 + \frac{1}{3}E\left(\left|ab' + \frac{1}{2}b^2b''\right|^2\right)\Delta^3 + \frac{1}{3!}E\left(\left|b\left(bb'' + (b')^2\right)\right|^2\right)\Delta^3\right\}.$$

**Exercise 10.4.3**   It follows from (5.2.16) and (5.2.22) for $j_1 \ne j_2$, $j_1 \ne j_3$, $j_2 \ne j_3$
that

$$I_{(j_1,j_2,j_3)} + I_{(j_2,j_1,j_3)} + I_{(j_2,j_3,j_1)} = I_{(j_1)}I_{(j_2,j_3)},$$

$$I_{(j_3,j_2,j_1)} + I_{(j_3,j_1,j_2)} + I_{(j_1,j_3,j_2)} = I_{(j_1)}I_{(j_3,j_2)},$$

$$I_{(j_1)}\left(I_{(j_2,j_3)} + I_{(j_3,j_2)}\right) = I_{(j_1)}I_{(j_2)}I_{(j_3)},$$

$$2\left(I_{(j_1,j_2,j_2)} + I_{(j_2,j_1,j_2)} + I_{(j_2,j_2,j_1)}\right) = 2I_{(j_1)}I_{(j_2,j_2)} = I_{(j_1)}\left(\left(I_{(j_2)}\right)^2 - \Delta\right),$$

$$6I_{(j_1,j_1,j_1)} = I_{(j_1)}\left(\left(I_{(j_1)}\right)^2 - 3\Delta\right).$$

**Exercise 10.5.1**   Use (5.8.10) and (5.8.11) to see that the multiple Stratonovich
integrals are just as described.

**Exercise 10.5.3**   Take the hierarchical set

$$A_{2.0} = \{\alpha \in \mathcal{M} : l(\alpha) + n(\alpha) \le 4\}$$

and use the truncated Stratonovich expansion (5.6.3) with $f(t, x) \equiv x$ to obtain the
representation for the time increments of the scheme (10.5.3).
In the same way as in Exercise 10.4.1 note that the relations (9.6.5) and (9.6.6) hold
true, but now including still more higher order terms on their right hand sides than
in Exercise 10.4.1.

**Exercise 10.5.4**   Use (5.3.9) to see which coefficient functions become zero in
(10.5.3).

**Exercise 10.6.1**   for $\gamma = 0.5$, $1.0$, $1.5$, ... and $\alpha \in A_\gamma$ we have

$$l(\alpha) + n(\alpha) \le 2\gamma \quad \text{or} \quad l(\alpha) = n(\alpha) = \gamma + \frac{1}{2}.$$

Hence we also have $l(-\alpha) + n(-\alpha) \le 2\gamma$ for each $\alpha \in A_\gamma \setminus \{v\}$, which means that $-\alpha$
$\in A_\gamma$. But this gives condition (5.4.3) in the definition of an hierarchical set.

**Exercise 10.6.2**   $A_0 = \{v\}$,    $A_{0.5} = A_0 \cup \{(0), (1)\}$,    $A_{1.0} = A_{0.5} \cup \{(1, 1)\}$,

$$A_{1.5} = A_{1.0} \cup \{(0, 1), (1, 0), (0, 0), (1, 1, 1)\},$$

$$A_{2.0} = A_{1.5} \cup \{(0, 1, 1), (1, 0, 1), (1, 1, 0), (1, 1, 1, 1)\}.$$

# Solutions of Exercises for Chapter 11

**Exercise 11.1.2**

$$E\left(\left|E\left(\frac{Y_{n+1} - Y_n}{\Delta} \mid \mathcal{A}_{\tau_n}\right) - a\left(Y_n\right)\right|^2\right) \le \left(\frac{1}{2}aa'\right)^2\Delta^2 + \left(\frac{1}{4}a''b^2\right)^2\Delta^2 + O\left(\Delta^3\right),$$

$$E\left(\frac{1}{\Delta}\left|Y_{n+1}-Y_n-E\left(Y_{n+1}-Y_n\,|\,\mathcal{A}_{\tau_n}\right)-b\left(Y_n\right)\Delta W_n\right|^2\right)\le\left(\frac{1}{2}a''ab\right)^2\Delta^2+O\left(\Delta^3\right).$$

**Exercise 11.1.4** The commutativity condition (10.3.13) is satisfied and by (10.1.4) we have $\underline{a}=-\frac{3}{2}x$. Hence it follows from (11.1.11) and (11.1.12) that

$$Y_{n+1}=Y_n\left(1-\frac{3}{2}\Delta+\frac{1}{2}\left(2-\frac{3}{2}\Delta+\Delta W^1+\Delta W^2\right)\left(\Delta W^1+\Delta W^2\right)\right).$$

**Exercise 11.3.1** Use the Ito formula (3.3.6) for the function

$$X_t=f\left(t,W_t\right)=(1+t)^2\left(1+t+W_t\right)$$

to obtain the desired stochastic equation.

**Exercise 11.5.3** Using the deterministic Taylor formula it follows from (11.2.1) that

$$
\begin{aligned}
Y_{n+1} &= Y_n+a\,\Delta+b\,\Delta W+\frac{1}{\sqrt{\Delta}}\left\{a'b\sqrt{\Delta}+a''ab\Delta^{3/2}+\cdots\right\}I_{(1,0)}\\
&\quad+\left\{a'a+\frac{1}{2}a''\left(b^2+a^2\Delta\right)+\cdots\right\}I_{(0,0)}\\
&\quad+\left\{b'b+b''ba\Delta+\cdots\right\}I_{(1,1)}\\
&\quad+\left\{b'a+\frac{1}{2}b''\left(a^2\Delta+b^2\right)+\cdots\right\}I_{(0,1)}\\
&\quad+\frac{G}{2\Delta}I_{(1,1,1)}+\cdots
\end{aligned}
$$

with

$$
\begin{aligned}
G &= \left\{b\left(\bar{\Phi}_+\right)-b\left(\bar{\Phi}_-\right)\right\}-\left\{b\left(\bar{\Upsilon}_+\right)-b\left(\bar{\Upsilon}_-\right)\right\}\\
&= \left\{\left[b\left(\bar{\Phi}_+\right)-b\left(\bar{\Upsilon}_+\right)\right]-\left[b\left(\bar{\Phi}_-\right)-b\left(\bar{\Upsilon}_+\right)\right]\right\}\\
&\quad-\left\{\left[b\left(\bar{\Upsilon}_+\right)-b\right]-\left[b\left(\bar{\Upsilon}_-\right)-b\right]\right\}\\
&= \left\{2b'\left(\bar{\Upsilon}_+\right)b\left(\bar{\Upsilon}_+\right)\sqrt{\Delta}\right\}-\left\{2b'b\sqrt{\Delta}\right\}+\cdots\\
&= 2\Delta\left(bb'\right)'\left(b+a\sqrt{\Delta}\right)+\cdots.
\end{aligned}
$$

Hence from the coefficient functions of the order 1.5 strong Taylor scheme (10.4.1) it can be seen that the conditions (11.5.1)–(11.5.4) of Theorem 11.5.1 are satisfied under appropriate asumptions on $a$ and $b$.

**Exercise 11.5.4** From (11.3.2) we have

$$
\begin{aligned}
Y_{n+1} &= Y_n+\frac{1}{2}\left\{\left(\underline{a}\left(\bar{\Upsilon}_+\right)-\underline{a}\right)-\left(\underline{a}\left(\bar{\Upsilon}_-\right)-\underline{a}\right)\right\}\Delta+a\,\Delta+b\,\Delta W\\
&= Y_n+\underline{a}\,\Delta+b\,\Delta W+\frac{1}{2}\underline{a}'\underline{a}\,\Delta^2+\underline{a}'b\,\Delta Z\\
&\quad+\frac{1}{2}\underline{a}''\Delta\left(\frac{1}{4}\underline{a}^2\,\Delta^2+\underline{a}b\,\Delta Z+b^2\left(\Delta Z\right)^2\Delta^{-1}+\left(2J_{(1,1,0)}\Delta-b^2\left(\Delta Z\right)^2\Delta^{-2}\right)\right)\\
&= Y_n+\underline{a}\,\Delta+b\,\Delta W+\frac{1}{2}\underline{a}'\underline{a}\,\Delta^2+\underline{a}'b\,\Delta Z+\underline{a}''b^2J_{(1,1,0)}\\
&\quad+\frac{1}{8}\underline{a}''\underline{a}^2\,\Delta^3+\frac{1}{2}\underline{a}''\underline{a}b\,\Delta Z\,\Delta+\cdots.
\end{aligned}
$$

Hence using Theorem 11.5.2 and comparing the above with the order 2.0 strong Taylor scheme (10.5.4) with additive noise we see that the scheme (11.3.2) is of strong order $\gamma = 2.0$.

## Solutions of Exercises for Chapter 12

**Exercise 12.2.2**    For $B = bI$ we have $h(s, B) \equiv 0$, so equation (6.3.23) reduces to the ordinary differential equation $\dot{s} = h(s, \underline{A})s$ where $\underline{A} = A - \frac{1}{2}B^2$. Assuming that the matrix $\underline{A}$ is diagonalizable with eigenvalues $\lambda_1(\underline{A})$ and $\lambda_1(\underline{A})$, so $\underline{A}P = P\Lambda$ where $P^{-1} = P^{\mathsf{T}}$ and

$$\Lambda = \left[ \begin{array}{cc} \lambda_1(\underline{A}) & 0 \\ 0 & \lambda_2(\underline{A}) \end{array} \right],$$

the differential equation (6.2.23) transforms to $\underline{\dot{s}} = h(\underline{s}, \Lambda)\underline{s}$, that is

$$\underline{\dot{s}}^1 = \left( \left( 1 - \left( \underline{s}^1 \right)^2 \right) \lambda_1(\underline{A}) - \left( \underline{s}^2 \right)^2 \lambda_2(\underline{A}) \right) \underline{s}^1$$

$$\underline{\dot{s}}^2 = \left( \left( 1 - \left( \underline{s}^2 \right)^2 \right) \lambda_2(\underline{A}) - \left( \underline{s}^1 \right)^2 \lambda_1(\underline{A}) \right) \underline{s}^2$$

where $\underline{s} = (\underline{s}^1, \underline{s}^2)^{\mathsf{T}} = P^{-1}s$ ( in the nondiagonalizable case we can use the Jordan canonical form). The lim sup in formula (6.3.24) thus attains the value equal to the larger of the real parts of the eigenvalues of $\underline{A}$ and gives the top Lyapunov exponent. Since the sum of the Lyapunov exponents and the sum of the real parts of the eigenvalues both equal the trace of the matrix $\underline{A} = A - \frac{1}{2}B^2$, we see that the second Lyapunov exponent is equal to the real part of the other eigenvalue.

**Exercise 12.3.4**    From (12.3.8) we get

$$Y_{n+1} = Y_n + b\,\Delta W + \underline{a}\,\Delta + Q$$

with

$$
\begin{aligned}
Q &= \left\{ [\underline{a}\,(\bar{Y}_+) - \underline{a}] + [\underline{a}\,(\bar{Y}_-) - \underline{a}] - \frac{1}{2}[\underline{a}\,(Y_{n+1}) - \underline{a}] \right\} \Delta \\
&= 2\left\{ \underline{a}' \left( \frac{1}{2}\underline{a}\,\Delta + b\,\Delta Z \Delta^{-1} \right) + \frac{1}{2}\underline{a}''b^2 \left( (\Delta Z)^2 + \left( \bar{\zeta} \right)^2 \right) \Delta^{-2} \right\} \Delta \\
&\quad - \frac{1}{2}\left\{ \underline{a}' \left( \frac{1}{2}\underline{a}\,\Delta + b\,\Delta W \right) + \frac{1}{2}\underline{a}''b^2 \,(\Delta W)^2 \right\} \Delta + \cdots \\
&= \frac{1}{2}\underline{a}'\underline{a}\Delta^2 + \underline{a}''b \left( 2\left[ \frac{1}{2}\Delta Z + \frac{1}{4}\Delta W\,\Delta \right] - \frac{1}{2}\Delta W\,\Delta \right) \\
&\quad + \underline{a}''b^2 \left\{ \frac{1}{4}\left[ (\Delta Z)^2 + \Delta Z\,\Delta W\,\Delta + \frac{1}{4}(\Delta W\,\Delta)^2 \right] \Delta^{-1} \right. \\
&\quad + J_{(1,1,0)} - \frac{1}{2}(\Delta Z)^2\,\Delta^{-1} + \frac{1}{8}\Delta\,(\Delta W)^2 - \frac{1}{4}\Delta\,(\Delta W)^2 \\
&\quad \left. + \frac{1}{16}\left( 2\Delta Z\,\Delta^{-1} - \Delta W \right)^2 \Delta \right\} + \cdots \\
&= \frac{1}{2}\underline{a}'\underline{a}\,\Delta + \underline{a}'b\,\Delta Z + \underline{a}''b^2\,J_{(1,1,0)} + \cdots .
\end{aligned}
$$

Comparing this with the scheme (10.5.4) and using Theorem 11.5.2 we can conclude that (12.3.8) has strong order $\gamma = 2.0$.

**Exercise 12.5.2**  For the test equation (12.5.2) the implicit order 2.0 strong Runge-Kutta scheme gives

$$\left(1 + \frac{1}{2}\lambda\Delta\right) Y_{n+1} = \left(1 + \frac{3}{2}\lambda\Delta + \lambda^2\Delta^2\right) Y_n + \Delta W_n + 2\lambda b\bar{\eta}_n,$$

so

$$G(\lambda\Delta) = \left(1 + \frac{1}{2}\lambda\Delta\right)^{-1} \left(1 + \frac{3}{2}\lambda\Delta + \lambda^2\Delta^2\right).$$

Hence the complex modulus inequality $|G(\lambda\Delta)| < 1$ takes the form

$$\left|1 + \frac{3}{2}\lambda\Delta + \lambda^2\Delta^2\right|^2 < \left|1 + \frac{1}{2}\lambda\Delta\right|^2.$$

Writing $\lambda = \lambda_1 = \imath\lambda_2$ and using polar coordinates with $\lambda_1 = r\cos\theta$ and $\lambda_2 = r\sin\theta$, this becomes

$$4r^2\cos^2\theta + 3r\left(1 + r^2\right)\cos\theta + 1 + \frac{1}{4}r^2 + r^4 < r\cos\theta + 1 + \frac{1}{4}r^2.$$

As $r > 0$ this simplifies to

$$4r\cos^2\theta + \left(2 + 3r^2\right)\cos\theta + 1 + r^3 < 0.$$

**Exercise 12.5.3**  The time and second order spatial partial derivatives of $a^1 = \lambda_1 x^1 - \lambda_2 x^2$ and $a^2 = \lambda_2 x^1 + \lambda_1 x^2$, so $L^0 a^k = a^1 \frac{\partial a^k}{\partial x^1} + a^2 \frac{\partial a^k}{\partial x^2}$ for $k = 1$ and 2. Hence

$$L^0 a^1 = \left((\lambda_1)^2 - (\lambda_2)^2\right) x^1 - 2\lambda_1\lambda_2 x^2, \quad L^0 a^2 = 2\lambda_1\lambda_2 x^1 + \left((\lambda_1)^2 - (\lambda_2)^2\right) x^2.$$

Moreover, in complex notation we have

$$\lambda^2 x = (\lambda_1 + \imath\lambda_2)(x^1 + \imath x^2) = ((\lambda_1)^2 - (\lambda_2)^2 + 2\imath\lambda_1\lambda_2)(x^1 + \imath x^2) = L^0 a^1 + \imath L^0 a^2.$$

**Exercise 12.5.4**  Using complex notation with $a = \lambda x$ and $b = 1 + \imath 0$ we have $L^0 a = \lambda^2 x$ and $L^j b \equiv 0$. Thus with $\alpha_k \equiv \alpha$ and $\beta_k \equiv \beta$ the implicit order 1.5 strong Taylor scheme (12.2.16) gives

$$\begin{aligned}
Y_{n+1} &= Y_n + \{\alpha\lambda Y_{n+1} + (1 - \alpha)\lambda Y_n\} \Delta \\
&\quad + \left(\frac{1}{2} - \alpha\right) \{\beta\lambda^2 Y_{n+1} + (1 - \beta)\lambda^2 Y_n\} \Delta^2 + \text{noise terms,}
\end{aligned}$$

which we can write as $Y_{n+1} = G(\lambda\Delta)Y_n + noise$ where $G(\lambda\Delta)$ equals

$$\left(1 - \alpha\lambda\Delta - \left(\frac{1}{2} - \alpha\right)\beta\lambda^2\Delta^2\right)^{-1} \left(1 + (1 - \alpha)\lambda\Delta + \left(\frac{1}{2} - \alpha\right)(1 - \beta)\lambda^2\Delta^2\right).$$

The implicit order 2.0 strong Taylor scheme (12.2.20) gives the same expression except that the noise terms differ. The inequality $|G(\lambda\Delta)| < 1$ is equivalent to

$$\left|1 + (1 - \alpha)\lambda\Delta + \left(\frac{1}{2} - \alpha\right)(1 - \beta)\lambda^2\Delta^2\right|^2 < \left|1 - \alpha\lambda\Delta - \left(\frac{1}{2} - \alpha\right)\beta\lambda^2\Delta^2\right|^2$$

Using $\lambda = r\cos\theta + \imath r\sin\theta$, so $\lambda^2 = r^2\cos 2\theta + \imath r^2 \sin 2\theta$, and simplifying we obtain the stated polar coordinate. For $\theta = \pm\pi/2$ we have $\cos\theta = 0$ and the left hand side of the inequality reduces to $\frac{1}{4}(1 - 2\alpha)^2(1 - 2\beta)r^3$ which is either negative or identically zero for the stated parameter values. Also with $\cos\pi = -1$ the left hand side of the inequality is the cubic

$$-2 + 2(1 - 2\alpha)r - (1 - \alpha - \beta)(1 - 2\alpha)r^2 + \frac{1}{4}(1 - 2\alpha)^2(1 - 2\beta)r^3,$$

which has no positive real root if and only if the parameters are as stated. From this we see that the region of absolute stability contains the entire left hand side of the complex plane if and only if the parameters are as stated.

**Exercise 12.5.5**  We use real coordinates with $a^k$ and $L^0 a^k$ from Exercise 12.5.3, writing both schemes in the form $Y_{n+1} = A^{-1}BY_n + noise$ where the $2\times 2$ matrices $A$ and $B$ have components

$$
\begin{aligned}
a^{1,1} &= 1 - \alpha_1\lambda_1\Delta - \frac{1}{2}(1 - 2\alpha_1)\beta_1((\lambda_1)^2 - (\lambda_2)^2)\Delta^2 \\
a^{1,2} &= \alpha_1\lambda_2\Delta + \frac{1}{2}(1 - 2\alpha_1)\beta_1 2\lambda_1\lambda_2\Delta^2 \\
a^{2,1} &= -\alpha_2\lambda_2\Delta - \frac{1}{2}(1 - 2\alpha_2)\beta_2 2\lambda_1\lambda_2\Delta^2 \\
a^{2,2} &= 1 - \alpha_2\lambda_1\Delta - \frac{1}{2}(1 - 2\alpha_2)\beta_2((\lambda_1)^2 - (\lambda_2)^2)\Delta^2 \\
b^{1,1} &= 1 + (1 - \alpha_1)\lambda_1\Delta + \frac{1}{2}(1 - 2\alpha_1)(1 - \beta_1)((\lambda_1)^2 - (\lambda_2)^2)\Delta^2 \\
b^{1,2} &= -(1 - \alpha_1)\lambda_2\Delta - \frac{1}{2}(1 - 2\alpha_1)(1 - \beta_1)2\lambda_1\lambda_2\Delta^2 \\
b^{2,1} &= (1 - \alpha_2)\lambda_2\Delta - \frac{1}{2}(1 - 2\alpha_2)(1 - \beta_2)2\lambda_1\lambda_2\Delta^2 \\
b^{2,2} &= 1 + (1 - \alpha_2)\lambda_1\Delta + \frac{1}{2}(1 - 2\alpha_1)(1 - \beta_2)((\lambda_1)^2 - (\lambda_2)^2)\Delta^2.
\end{aligned}
$$

Then $G = A^{-1}B$ and $|G| < 1$ is equivalent to $|B|^2 < |A|^2$, that is $sum^2_{i,j=1}(b^{i,j})^2 < \sum^2_{i,j=1}(A^{i,j})^2$. With polar coordinates $r\cos\theta = \lambda_1\Delta$ and $r\sin\theta = \lambda_2\delta$, so $2\lambda_1\lambda_2\Delta^2 = r^2\sin 2\theta$ and $((\lambda_1)^2 - (\lambda_2)^2)\Delta^2 = r^2\cos 2\theta$, we obtain the equivalent inequality

$$\frac{1}{8}\left[(1 - 2\alpha_1)^2(1 - 2\beta_1) + (1 - 2\alpha_2)^2(1 - 2\beta_2)\right]r^3$$

$$\left[2 + \frac{1}{2}\{(1 - 2\alpha_1)(1 - \alpha_1 - \beta_1) + (1 - 2\alpha_2)(1 - \alpha_2 - \beta_2)\}r^2\right]\cos\theta$$

$$2(1 - \alpha_1 - \alpha_2)r\cos^2\theta < 0.$$

This decribses the region of absolute stability and reduces to the inequality in Exercise 12.5.4 when $\alpha_k = \alpha$ and $\beta_k = \beta$.

**Exercise 12.6.1**  Use representation (12.6.7) with $\gamma = 1.0$, $\alpha_{1,k} = \beta_{1,k} = 1.0$, $\alpha_{2,k} = \beta_{2,k} = 0$ and observe that the scheme (11.4.4) coincides with it up to terms which do not disturb the order $\gamma = 1.5$ of strong convergence. Under appropriate conditions order $\gamma = 1.5$ strong convergence can then be shown as in the proof of Theorem 11.5.1.

**Exercise 12.6.2**   Similarly to the previous exercise, using $\gamma_k = 0$ and $\alpha_{2,k} = \frac{1}{2}$.

**Exercise 12.6.3**   Use (12.6.7) with $\gamma_k = 1.0$, $\alpha_{1,k} = \alpha_{2,k} = \frac{1}{2}$ and compare with the scheme (12.4.8), neglecting those terms which are not necessary for strong convergence with order $\gamma = 1.5$

# Solutions of Exercises for Chapter 14

**Exercise 14.1.1**   $E(X_t) = E(X_0) + \frac{3}{2} \int_0^t E(X_s)\, ds = E(X_0) \exp\left(\frac{3}{2} t\right)$.

**Exercise 14.1.3**   $E(\Delta \hat{W}^j) = E((\Delta \hat{W}^j)^3) = 0$,    $E((\Delta \hat{W}^j)^2) = \frac{1}{2}\Delta + \frac{1}{2}\Delta = \Delta$.

**Exercise 14.2.1**   $E(\Delta \hat{W}) = E((\Delta \hat{W})^3) = E((\Delta \hat{W})^5) = 0$,

$$E((\Delta \hat{W})^2) = \frac{1}{6} 3\Delta + \frac{1}{6} 3\Delta = \Delta, \quad E((\Delta \hat{W})^4) = \frac{1}{6} 9\Delta^2 + \frac{1}{6} 9\Delta^2 = 3\Delta^2.$$

**Exercise 14.5.3**   According to the definition of a weak Taylor scheme (14.5.4) $Y_0 = X_0$ is a nonrandom constant, so it follows that (14.5.7) holds. Also (14.5.8) and (14.5.9) follow directly from the assumptions of Theorem 14.5.1. With Lemma 5.7.5 it is easy to see from (14.5.4) that (14.5.11) holds. The relation (14.5.12) is trivial for a weak Taylor scheme. It remains to show that the moments of $Y^\delta$ are bounded. For simplicity we shall consider only the case $d = m = 1$ and denote all constants by $K$. Using the notation of Chapter 5 we can write

$$Y_t = X_0 + \int_0^t a_s\, ds + \int_0^t b_s\, dW_s$$

with

$$a_s = \sum_{\substack{\alpha \in \Gamma_\beta \\ j_l(\alpha) = 0}} f_\alpha\left(\tau_{n_s}, Y_{n_s}\right) I_{\alpha-,\tau_{n_s},s}, \qquad b_s = \sum_{\substack{\alpha \in \Gamma_\beta \\ j_l(\alpha) = 1}} f_\alpha\left(\tau_{n_s}, Y_{n_s}\right) I_{\alpha-,\tau_{n_s},s}$$

Applying the Doob inequality (2.3.7) we obtain

$$Z_t = \left( E\left( \sup_{0 \le s \le t} |Y_s|^{2q} \right) \right)^{1/q} \le K \left\{ \left( E\left( |X_0|^{2q} \right) \right)^{1/q} \right.$$

$$+ \left( E\left( \sup_{0 \le s \le t} \left| \int_0^s a_z\, dz \right|^{2q} \right) \right)^{1/q} + \left( E\left( \sup_{0 \le s \le t} \left| \int_0^s b_z\, dW_z \right|^{2q} \right) \right)^{1/q} \right\}$$

$$\le K \left\{ 1 + \left( E\left( \left| \int_0^t |a_z|\, dz \right|^{2q} \right) \right)^{1/q} + \left( \frac{2q}{2q-1} \right)^2 \left( E\left( \left| \int_0^t |b_z|\, dW_z \right|^{2q} \right) \right)^{1/q} \right\}.$$

As in the proof of Lemma 5.7.5 we get

$$Z_t \le K \left\{ 1 + \int_0^t \left( E\left( |a_z|^{2q} \right) \right)^{1/q} dz + \int_0^t \left( E\left( |b_z|^{2q} \right) \right)^{1/q} dz \right\}$$

and hence by an application of Lemma 5.7.5 and the growth condition (14.5.5)

$$Z_t \le K \left\{ 1 + \int_0^t Z_s \, ds \right\}.$$

Finally we use the Gronwall inequality (Lemma 4.5.1) to conclude that $Z_t$ is bounded for all $t \in [0, T]$.

**Exercise 14.5.4**  We shall fix $\beta = 1.0, 2.0, \ldots$ and denote by $\hat{I}_{\alpha,t_0,t}$ the weak approximation of a multiple Ito integral $I_{\alpha,t_0,t}$ for any $\alpha \in \Gamma_\beta \setminus \{v\}$. If we can show for all choices of multi-indices $\alpha_k \in \Gamma_\beta \setminus \{v\}$ with $k = 1, 2, \ldots, l$ and $l = 1, \ldots, 2\beta + 1$ that

$$\left| E \left( \prod_{k=1}^l I_{\alpha_k,t_0,t} - \prod_{k=1}^l \hat{I}_{\alpha_k,t_0,t} \,\middle|\, \mathcal{A}_{t_0} \right) \right| \le K \, (t - t_0)^{\beta+1},$$

see (5.12.2) too, then it is obvious that condition (14.5.12) also holds for the simplified schemes under consideration.  On the other hand the relations (5.12.8)–(5.12.10) provide examples of weak approximations of multiple Ito integrals which satisfy the above condition and are used in the schemes mentioned.

## Solutions of Exercises for Chapter 15

**Exercise 15.1.2**  Omitting higher order terms with respect to condition (14.5.12) we have from (15.1.1)

$$
\begin{aligned}
Y_{n+1} - Y_n \;=\;& a\Delta + \frac{1}{2} \left( a\left(\bar{Y}\right) - a \right) \Delta + b\Delta\hat{W} \\
&+ \frac{1}{4} \left\{ \left( b\left(\bar{Y}^+\right) - b \right) + \left( b\left(\bar{Y}^-\right) - b \right) \right\} \Delta\hat{W} \\
&+ \frac{1}{4} \left\{ \left( b\left(\bar{Y}^+\right) - b \right) + \left( b\left(\bar{Y}^-\right) - b \right) \right\} \left\{ \left(\Delta\hat{W}\right)^2 - \Delta \right\} \Delta^{-1/2} \\
&+ \cdots \\
\;=\;& a\Delta + b\Delta\hat{W} + \frac{1}{2} aa'\Delta^2 + \frac{1}{2} a'b\Delta\hat{W}\,\Delta + \frac{1}{2} a''b^2 \left(\Delta\hat{W}\right)^2 \Delta \\
&+ \frac{1}{2} ab'\,\Delta\hat{W}\,\Delta + \frac{1}{4} b''b^2\,\Delta\hat{W}\,\Delta + \frac{1}{2} b'b \left\{ \left(\Delta\hat{W}\right)^2 - \Delta \right\} \\
&+ \cdots
\end{aligned}
$$

Examining the simplified order 2.0 weak Taylor scheme (14.2.2) then makes it easy to verify condition (14.5.12) for the scheme (15.1.1).

**Exercise 15.3.5**  From Theorem 14.6.1 we have for $\delta \in (0, 1)$

$$E \left( g \left( Y^\delta(T) \right) \right) - E \left( g \left( X_T \right) \right) = \sum_{\gamma=2}^3 \psi_{g,\gamma}(T) \, \delta^\gamma + O \left( \delta^4 \right).$$

Thus from (15.3.11) we obtain

$$
\begin{aligned}
V_{g,4}^\delta(T) - E \left( g \left( X_T \right) \right) \;=\;& \frac{1}{11} \left[ \psi_{g,2}(T) \left\{ 18 - 9 \cdot 4 + 2 \cdot 9 \right\} \delta^2 \right. \\
&\left. + \psi_{g,3}(T) \left\{ 18 - 9 \cdot 8 + 2 \cdot 27 \right\} \delta^3 \right] + O \left( \delta^4 \right) \\
\;=\;& O \left( \delta^4 \right).
\end{aligned}
$$

**Exercise 15.6.3** We have to show (9.7.6)–(9.7.7). With (15.6.6) we obtain

$$
E\left(\left|E\left(\frac{1}{\Delta}\left(Y_{n+1}-Y_n\right)\mid A_{\tau_n}\right)-a\left(\tau_n,Y_n\right)\right|^2\right)
$$

$$
= E\left(\left|E\left(\alpha\left(\bar{a}_\eta\left(\tau_{n+1},Y_{n+1}\right)-\bar{a}_\eta\left(\tau_n,Y_n\right)\right)+\left(\bar{a}_\eta\left(\tau_n,Y_n\right)-a\right)\right.\right.\right.
$$

$$
\left.\left.\left.+\eta\left(b\left(\tau_{n+1},Y_{n+1}\right)-b\right)\Delta\hat{W}_n^1\,\Delta^{-1}\mid A_{\tau_n}\right)\right|^2\right)\le K\,\delta
$$

and

$$
E\left(\left|E\left(\frac{1}{\Delta}\left(Y_{n+1}-Y_n\right)^2-b^2\mid A_{\tau_n}\right)\right|^2\right)
$$

$$
= E\left(\left|E\left(\left(b\left(\tau_n,Y_n\right)\Delta\hat{W}_n^1\right)^2\,\Delta^{-1}-b^2\mid A_{\tau_n}\right)\right|^2\right)+O(\delta)\le K\,\delta.
$$

## Solutions of Exercises for Chapter 16

**Exercise 16.2.1** We apply the Ito formula (3.4.6) to the function

$$
f\left(t,\tilde{X}_t,\Theta_t\right)=u\left(t,\tilde{X}_t\right)\Theta_t
$$

with $\tilde{X}_t$ from (16.2.7) and $\Theta_t$ from (16.2.8). With (16.2.11) and (16.2.3) this yields

$$
u\left(t,\tilde{X}_t\right)\Theta_t=u\left(0,x\right)\Theta_0
$$

$$
+\int_0^t\left\{\Theta_z\frac{\partial}{\partial t}u\left(z,\tilde{X}_z\right)\sum_{k=1}^d\left[a^k\left(z,\tilde{X}_z\right)-\sum_{j=1}^m b^{k,j}\left(z,\tilde{X}_z\right)d^j\left(z,\tilde{X}_z\right)\right]\Theta_z\frac{\partial}{\partial x^k}u\left(z,\tilde{X}_z\right)\right.
$$

$$
+\frac{1}{2}\sum_{i,k=1}^d b^{k,j}\left(z,\tilde{X}_z\right)b^{i,j}\left(z,\tilde{X}_z\right)\Theta_z\frac{\partial^2}{\partial x^k\partial x^i}u\left(z,\tilde{X}_z\right)
$$

$$
\left.+\sum_{k=1}^d b^{k,j}\left(z,\tilde{X}_z\right)d^j\left(z,\tilde{X}_z\right)\Theta_z\frac{\partial}{\partial x^k}u\left(z,\tilde{X}_z\right)\right\}dz
$$

$$
+\sum_{j=1}^m\int_0^t\left\{\sum_{k=1}^d b^{k,j}\left(z,\tilde{X}_z\right)\Theta_z\frac{\partial}{\partial x^k}u\left(z,\tilde{X}_z\right)+\Theta_z u\left(z,\tilde{X}_z\right)d^j\left(z,\tilde{X}_z\right)\right\}dW_z^j
$$

$$
=u(0,x)\Theta_0.
$$

**Exercise 16.2.2** It follows from (4.4.6) that

$$
u(s,x)=E\left(\left(X_T^{s,x}\right)^2\right)=x^2\exp\left(\left(2a+b^2\right)(t-s)\right).
$$

Also from (16.2.11) we have $d^1(t,y)=-2b$, so from (4.4.6) again we get

$$
\tilde{X}_t^2=x^2\exp\left(\left(2a+3b^2\right)(t-s)+2b\left(W_t-W_s\right)\right)
$$

and

$$
\Theta_t=\Theta_0\exp\left(-2b^2(t-s)-2b\left(W_t-W_s\right)\right).
$$

This means

$$g\left(\tilde{X}_T^{0,x}\right)\Theta_T/\Theta_0 = \left(\tilde{X}_T^{0,x}\right)^2\Theta_T/\Theta_0$$
$$= x^2\exp\left(\left(2a+b^2\right)T\right) = u(0,x) = E\left(g\left(X^{0,x}\right)\right).$$

**Exercise 16.3.1** From (16.3.15) it follows that

$$\begin{aligned}
\mathrm{Var}\left(\frac{F(\xi)}{D_{opt}(\xi)}\right) &= E\left(\left(\frac{F(\xi)}{D_{opt}(\xi)}\right)^2\right) - \left(E\left(\frac{F(\xi)}{D_{opt}(\xi)}\right)\right)^2 \\
&= \int_\Gamma\frac{(F(\xi))^2}{D_{opt}(\xi)}\,d\mu(\xi) - \left(\int_\Gamma\frac{F(\xi)}{D_{opt}(\xi)}\,D_{opt}(\xi)\,d\mu(\xi)\right)^2 \\
&= \left(\int_\Gamma|F(\xi)|\,d\mu(\xi)\right)^2 - \left(\int_\Gamma F(\xi)\,d\mu(\xi)\right)^2.
\end{aligned}$$

**Exercise 16.3.2** From (16.3.24) and (16.3.12) we obtain

$$D_{i,x_{i-1},x_N}(x_i) = \sqrt{2\pi\,\frac{(\tau_i-\tau_{i-1})(T-\tau_i)}{T-\tau_{i-1}}}\,\sigma\,\exp(V)$$

with

$$V = \frac{V^*}{2(T-\tau_{i-1})(\tau_i-\tau_{i-1})(T-\tau_i)}\,\sigma^{-1}$$

where

$$\begin{aligned}
V^* &= -\left[(T-\tau_i)+(\tau_i-\tau_{i-1})\right](T-\tau_i)(x_i-x_{i-1}-a(\tau_i-\tau_{i-1}))^2 \\
&\quad -\left[(T-\tau_i)+(\tau_i-\tau_{i-1})\right](\tau_i-\tau_{i-1})(x_N-x_i-a(T-\tau_i))^2 \\
&\quad +(\tau_i-\tau_{i-1})(T-\tau_i)\left(\{x_N-x_i-a(T-\tau_i)\}\right. \\
&\qquad\qquad \left.+\{x_i-x_{i-1}-a(\tau_i-\tau_{i-1})\}\right) \\
&= \left[(T-\tau_{i-1})x_i-(T-\tau_i)x_{i-1}-(\tau_i-\tau_{i-1})x_N\right]^2
\end{aligned}$$

so

$$V = \left[x_i-x_{i-1}\frac{T-\tau_i}{T-\tau_{i-1}}-x_N\frac{\tau_i-\tau_{i-1}}{T-\tau_{i-1}}\right]^2\left[2\sigma\frac{(T-\tau_i)(\tau_i-\tau_{i-1})}{T-\tau_{i-1}}\right]^{-1}$$

Thus we have a Gaussian density with respect to $x_i$ with the asserted mean (16.3.25) and variance (16.3.26).

# Solutions of Exercises for Chapter 17

**Exercise 17.1.2** The function $u$ satisfies $u(T,x) = 1 = f(x)$,

$$\frac{\partial u(t,x)}{\partial t} = u(t,x)\left\{2x^2\frac{a}{(1+a)^2}-\frac{1}{2}+\frac{a}{1+a}\right\}$$

and

$$\frac{\partial^2 u(t,x)}{\partial x^2} = u(t,x)\left\{x^2\frac{(1-a)^2}{(1+a)^2}+\frac{1-a}{1+a}\right\}$$

where $a = \exp(2(T - t))$, from which it follows that

$$\frac{\partial u}{\partial t} + \frac{1}{2}\frac{\partial^2 u}{\partial x^2} - \frac{1}{2}x^2 = 0.$$

**Exercise 17.1.6**   With $\tilde{Y}_n = (Y_n^1, \dots, Y_n^m)$ and $L^0 = \frac{\partial}{\partial t} + \frac{1}{2}\sum_{k=1}^m \frac{\partial^2}{\partial x^k \partial x^k}$ we have

$$
\begin{aligned}
Y_{n+1}^k &= Y_n^k + \Delta \tilde{W}_n^k \quad \text{for} \quad k = 1, \dots, m \\
Y_{n+1}^{m+1} &= Y_n^{m+1} + a\left(\tau_n, \tilde{Y}_n\right) \Delta + \frac{1}{2}L^0 a\left(\tau_n, \tilde{Y}_n\right) \Delta^2 + \frac{1}{6}L^0 L^0 a\left(\tau_n, \tilde{Y}_n\right) \Delta^3 \\
&\quad + \sum_{j=1}^m \left[ \frac{\partial}{\partial x^j} a\left(\tau_n, \tilde{Y}_n\right) \Delta \tilde{Z}^j + \frac{1}{6}\left(L^0 \frac{\partial}{\partial x^j} + \frac{\partial}{\partial x^j}L^0\right) a\left(\tau_n, \tilde{Y}_n\right) \Delta \tilde{W}^j \Delta^2 \right] \\
&\quad + \frac{1}{6}\sum_{j_1, j_2 = 1}^m \frac{\partial}{\partial x^{j_1}}\frac{\partial}{\partial x^{j_2}} a\left(\tau_n, \tilde{Y}_n\right) \left\{ \Delta \tilde{W}^{j_1} \Delta \tilde{W}^{j_2} - I_{\{j_1 = j_2\}}\Delta \right\} \Delta.
\end{aligned}
$$

**Exercise 17.1.7**   Extrapolate the scheme from Exercise 17.1.6 with the order 6.0 weak extrapolation method (15.3.3).

**Exercise 17.3.1**   We note that $\left(s_t^1\right)^2 + \left(s_t^2\right)^2 = 1$ for $t \geq 0$. Then it follows from the Ito formula that

$$
\begin{aligned}
\phi_t &= \phi_0 + \int_0^t \left[ -s_z^1 s_z^2 \left( a - \left( a\left(s_z^1\right)^2 + b\left(s_z^2\right)^2 \right) \right) \right. \\
&\qquad\qquad\qquad \left. + s_z^1 s_z^2 \left( b - \left( a\left(s_z^1\right)^2 + b\left(s_z^2\right)^2 \right) \right) \right] dz \\
&\quad + \int_0^t \left[ \left(-\sigma s_z^2\right)\left(-s_z^2\right) + \sigma s_z^1 s_z^1 \right] dW_z \\
&= \phi_0 + \int_0^t (b - a)s_z^1 s_z^2 \, dz + \int_0^t \sigma \, dW_z \\
&= \phi_0 + \frac{1}{2}(b - a)\int_0^t \sin\left(2\phi_z\right) \, dz + \sigma \int_0^t dW_z.
\end{aligned}
$$

# Bibliographical Notes

**Chapter 1: Probability and Statistics**

Ash (1970), Chung (1975), Karlin & Taylor (1970, 1981), Papoulis (1966), Parzen (1962) and Shiryayev (1984) provide comprehensive introductions to probability and stochastic processes. Gardiner (1983) is a useful handbook on stochastic methods in physics and other sciences. See Groeneveld (1979) for a computational approach to elementary probability and basic statistical tests.

**1.3**  Ripley (1983a) gives a tutorial introduction to pseudo-random number generation. Books including this subject are Ermakov (1975), Morgan (1984), Ripley (1983b) and Rubinstein (1981) and Yakowitz (1977). See also Brent (1974), Box & Muller (1958) and Marsaglia & Bray (1964). Random number generation on supercomputers is considered in Petersen (1988) and Anderson (1990). Some new nonlinear congruential pseudo-random number generators are proposed in Eichenauer & Lehn (1986), Niederreiter (1988) and Eichenauer-Hermann (1991).

**1.4 & 1.5**  See Ash (1970), Ito (1984), Jacod & Shiryaev (1987), Shiryayev (1984).

**1.6**  Markov chains are treated in Chung (1975) and Karlin & Taylor (1970). See also Shiryayev (1984) or Cinlar (1975).

**1.7 & 1.8**  See Jazwinski (1970), Papoulis (1966), Parzen (1962), Skorokhod (1982), van Kampen (1981b), Wong & Hajek (1985) and Stroock & Varadhan (1982).

**1.9**  See Groeneveld (1979), Kleijnen (1974), (1975) and Liptser & Shiryayev (1977) in addition to the books listed in 1.3.

**Chapter 2: Probability Theory and Stochastic Processes**

**2.1 & 2.2**  Axiomatically developed probability theory originated with Kolmogorov (1933). Introductory texts are Ash (1970) and Ito (1984), with Doob (1953), Loeve (1977) and Feller (1957), (1972) providing more advanced treatments. See Dynkin (1965) and Wong (1971) for conditional probabilities. Background material on measure theory and integration can be found in Ash (1972), Kolmogorov & Fomin (1975) and Royden (1968).

**2.3**  See Doob (1953), Dynkin (1965) and Gikhman & Skorohod (1979). Martingales are treated in Métivier (1982), Métivier & Pellaumail (1980) and Protter (1990).

**2.4**  See Hida (1980), Ito & McKean (1974), Karatzas & Shreve (1988), Risken (1984) and Stroock & Varadhan (1982). Limit theorems for random walks are given in Skorohod & Slobodenjuk (1970).

### Chapter 3: Stochastic Calculus

The books by Elliot (1980), Gikhman & Skorohod (1979), Karatzas & Shreve (1988), McKean (1969), McShane (1974), Rao (1979) and Wong & Hajek (1985) are standard texts on stochastic calculus. Introductory treatments can also be found in texts on stochastic differential equations such as Arnold (1974), Gard (1988), Gikhman & Skorokhod (1972a, b), Øksendal (1985) and Pugachev & Sinitsyn (1987), as well as on other applications of stochastic processes. Ikeda & Watanabe (1981), (1989) and Stroock & Varadhan (1982) provide more advanced treatments. The papers of Ito (1951a, b) are historically significant. See Emery (1989) for stochastic calculus on manifolds.

**3.2**     Pugachev & Sinitsyn (1987) also give a list of explicit stochastic integrals.

**3.5**     See Stratonovich (1966, 1968), McShane (1974), and Wong & Zakai (1965a, 1965b, 1969). Protter (1990) develops stochastic calculus using semimartingales. See also Bichteler (1981), Jacod (1979), Métivier & Pellaumail (1980) and Meyer (1976).

### Chapter 4: Stochastic Differential Equations

Gikhman & Skorokhod (1972 a,b) and Ikeda & Watanabe (1981), (1989) are standard treatises on stochastic differential equations. More introductory texts are Arnold (1974), Gard (1988), Øksendal (1985) and Pugachev & Sinitsyn (1987). Honerkamp (1990), Horsthemke & Lefever (1984), Schuss (1980), Sobczyk (1990) and van Kampen (1981b) emphasize applications. See also Syski (1967), Friedman (1975), Stroock (1979), Stroock & Varadhan (1982), Wong (1971) and Wong & Hajek (1985).

**4.1**     Random differential equations are studied in Barucha-Reid (1979), Bunke (1972) and Soong (1973). Doss (1977) and Sussmann (1978) examine the relationship between ordinary and stochastic differential equations. See also McShane (1974) and van Kampen (1981a).

**4.2**     See Arnold (1974), Richardson (1964), McKenna & Morrison (1970), (1971).

**4.3**   See Arnold (1974), Gard (1988) and Gikhman & Skorokhod (1972a).

**4.4**     Several examples are from Arnold (1974), Gard (1988), Horsthemke & Lefever (1984) and Pugachev & Sinitsyn (1987). The complex-valued stochastic differential equations are taken from Klauder & Petersen (1985a, b).

**4.5**     Existence and uniqueness proofs are given in many books. Weaker assumptions are considered in Gikhman & Skorokhod (1972a). Non-Lipschitzian examples are considered by Balakrishnan (1985). Properties of strong solutions are discussed by Karandikar (1981). Existence and uniqueness theorems for more general types of stochastic differential equations are considered by Gikhman & Skorokhod (1972a), Ikeda & Watanabe (1981), (1989), McShane (1974) and Protter (1990). Also see Zhank & Padgett (1984).

**4.6**   See Ito & McKean (1974) and Stroock & Varadhan (1982). See Ikeda & Watanabe (1981), (1989), Krylov (1980) and Stroock & Varadhan (1982).

**4.8**     See Mikulevicius (1983) and Stroock & Varadhan (1982). The ergodic result is from Hasminski (1980).

**4.9** Stratonovich (1963), (1966), (1968). Stratonovich equations are included in most books on SDEs, in particular, see Ikeda & Watanabe (1981) or Arnold (1974).

## Chapter 5: Stochastic Taylor Expansions

**5.1** The Ito-Taylor formula was first derived and used in Wagner & Platen (1978) and Platen & Wagner (1982). A generalization for Ito process with jump component is described in Platen (1982b) and for semi-martingale equations in Platen (1981b), (1982a). Azencott (1982) gave a generalization. Sussmann (1988) describes product expansions of exponential Lie series. In Kloeden & Platen (1990) the Stratonovich-Taylor formula is derived. Yen (1988) found a stochastic Taylor formula for two-parameter stochastic differential equations.

**5.2** In Liske, Platen & Wagner (1982) relations between multiple stochastic integrals are discussed. A more detailed investigation about relations between multiple Ito and Stratonovich integrals can be found in Kloeden & Platen (1991a).

**5.3** See Platen & Wagner (1982) and Platen (1984).

**5.4** See Platen (1984).

**5.5** See Wagner & Platen (1978), Platen & Wagner (1982) and Platen (1984).

**5.6** See Kloeden & Platen (1990).

**5.7** See Platen & Wagner (1982), Platen (1981a), (1984). First attempts to approximate multiple stochastic integrals are made in Liske, Platen & Wagner (1982) and Liske (1982). The section is based on the results in Kloeden, Platen & Wright (1991). See also Milstein (1988).

**5.9** See Platen & Wagner (1982).

**5.10** The result is similar to that in Platen & Wagner (1982).

## Chapter 6: Modelling with Stochastic Differential Equations

**6.1** See Horsthemke & Lefever (1984), McShane (1971), (1974), (1976), Turelli (1977) and van Kampen (1981a). The Wong-Zakai Theorem is stated and proved in Wong & Zakai (1965a, 1965b, 1969). A more general convergence result is due to Papanicolaou & Kohler (1974).

**6.2** Jacod & Shiryaev (1987), Kushner (1974), (1984), Platen & Rebolledo (1985) investigate the convergence of Markov chains to diffusions. Applications to genetics can be found in Kimura & Ohta (1971).

**6.3** Early literature on stochastic stability is surveyed by Kozin (1969). Lyapunov function techniques are discussed by Hasminski (1980) and Kushner (1967). See also Arnold (1974), Mortensen (1969), and Gard (1988). The role of Lyapunov exponents in stochastic stability is reviewed in the conference proceedings by Arnold & Wihstutz (1986). For explicit examples of Lyapunov exponents see Arnold & Kloeden (1989) and Auslender & Milstein (1982).

**6.4** This section follows Kozin (1983). Also see Le Breton (1976), Liptser & Shiryayev (1977), Basawa & Prakasa Rao (1980), Bellach (1983), Florence-Zmirou (1989), Dacunha-Castelle & Florence-Zmirou (1986), Donhal (1987), Sørensen (1989) and Heyde (1989).

**6.5**  Aström (1970), Davis (1984), Krylov (1980) and Kushner (1971), (1977), (1984).

**6.6**  Books on filtering include Kallianpur (1980), Davis (1984), Jazwinski (1970) and Pugachev & Sinitsyn (1987). See Wonham (1965), Fujisaki, Kallianpur & Kunita (1972), Zakai (1969), Clark (1978), Di Masi & Runggaldier (1981), Newton (1984), (1986a,b), Picard (1984), (1986a, b, c) and Hernandez (1988) for developments in nonlinear filtering including numerical methods.

## Chapter 7: Applications of Stochastic Differential Equations

Bellman (1964), Chandrasekhar (1943 ), Horsthemke & Lefever (1984), Ricciardi (1977), Schuss (1980), van Kampen (1981b), Gardiner (1983) and Sobczyk (1991) contain many examples of applications of stochastic differential equations.

**7.1**  Population dynamics: Gard (1988), Gard & Kannan (1976), Schoener (1973), Turelli (1977). Protein Kinetics: Arnold, Horsthemke & Lefever (1978). See Ehrhardt (1983) for the noisy Brusselator. Genetics: Kimura & Ohta (1971), Levkinson (1977), Shiga (1985). See Benson (1980) for chemical kinetics.

**7.2**  Experimental Psychology: Schöner, Haken & Kelso (1986). Neuronal Activity: Kallianpur (1987), Giorno, Lansky, Nobile & Ricciardi (1988), Lansky & Lanska (1987).

**7.3**  Finance and Option Pricing : Merton (1971), (1973), Black & Scholes (1973), Barrett & Wright (1974), Fahrmeier & Beeck (1976), Karatzas & Shreve (1988), Karatzas (1989), Föllmer & Sondermann (1986) and Föllmer (1991). Dothan (1990) is a book on prices in financial markets. Johnson & Shanno (1987) determined option prices with simulations using the Euler scheme.

**7.4**  Turbulent Diffusion: Bywater & Chung (1973), Drummond, Duane, Horgan (1984), Durbin (1983), Obukov (1959), Pope (1985), Yaglom (1980). Radio-Astronomy: Chandrasekhar (1954), Le Gland (1981).

**7.5**  Heliocopter Rotor Stability: Pardoux & Pignol (1984), Pardoux & Talay (1988), Talay (1988a). Satellite Orbital Stability: Sagirow (1970). Satellite Altitude Dynamics: Balakrishnan (1985).

**7.6**  Biological Waste Treatment: Harris (1976), (1977), (1979), Miweal, Ognean & Straja (1987). Hydrology: Bodo, Thomson & Unny (1987), Finney, Bowles & Windham (1983), Unny (1984), Unny & Karmeshu (1983). Air Quality: Haghighat, Chandrashekar & Unny (1987), Haghighat, Fazio & Unny (1988).

**7.7**  Seismology: Bolotin (1960), Fischer & Engelke (1983), Kozin (1977), Shinozuka (1972), Shinozuka & Sato (1967). Structural Mechanics: Friedrich, Lange & Lindner (1987), Hennig & Grunwald (1984), Shinozuka & Wen (1972), Wedig (1988), Shinozuka (1971), (1972), Shinozuka & Jan (1972), Shinozuka & Sato (1967).

**7.8**  Fatigue Cracking: Sobczyk (1986). Optical Bistability: Gragg (1981), Horsthemke & Lefever (1984), Smith & Gardiner (1988). Nemantic Liquid

Crystals: Horsthemke & Lefever (1984). Ivanov & Shvec (1979) simulated collisions in plasmas.

**7.9** Blood clotting: Fogelson (1984). Celluar Energetics: Veuthey & Stucki (1987).

**7.10** Josephson Tunneling Junctions: Horsthemke & Lefever (1984). Communications: Horsthemke & Lefever (1984). Stochastic Annealing: Geman & Hwang (1986), Goldstein (1988).

## Chapter 8: Time Discrete Approximation of Deterministic Stochastic Differential Equations

Blum (1972), Bulirsch & Stoer (1980), Butcher (1987), Dahlquist & Bjorck (1974), Gear (1971), Grigorieff (1972), Henrici (1962), Wilkes (1966) and Lambert (1973) are a selection of textbooks on numerical methods for ordinary differential equations.

**8.2** See Butcher (1987) for Runge-Kutta methods.

**8.3** Dahlquist & Bjorck (1974), Dahlquist (1963). See Lambert (1973) for stiff equations.

**8.4** Henrici (1962) gives a statistical analysis of roundoff errors.

## Chapter 9: Introduction to Stochastic Time Discrete Approximations

Different approaches have been suggested for the numerical solution of SDEs. A very general method is mentioned by Boyce (1978), Kohler & Boyce (1974) and allows, in principle, the study of general random systems by Monte-Carlo simulations not using the special structure of SDEs. Kushner (1977) proposed as approximating processes time discrete finite state Markov chains which seem to be efficient mainly in low dimensional problems with bounded domains (see also Section 17.4 ). Dashevski & Liptser (1966) and Fahrmeier (1976) used analog computers to handle SDEs numerically.

A widely applicable and efficient approach is the simulation of approximate sample paths on digital computers. These methods are well particularly suited to parallel computers. See Petersen (1987), (1990).

Clements & Anderson (1973), Clark & Cameron (1980), Fahrmeier (1976), Rümelin (1982), Wright (1974) and others show that not all heuristic time discrete approximations converge in a useful sense. Consequently a careful and systematic investigation of different methods is needed. Surveys or more systematic treatments can be found in Gard (1988), Milstein (1988a), Pardoux & Talay (1985), Hernandez (1989), Kloeden & Platen (1989) and Talay (1990b).

**9.1** Maruyama (1955) was one of the first to use and investigate the mean-square convergence of the Euler method. A corresponding result for the Ito process with jump component can be found in Gikhman & Skorokhod(1979); see also Atalla (1986). Yamada (1976) and Allain (1974) considered strong approximations. Results concerning the weak convergence of the Euler approximation are contained for instance in Billingsley (1968), Grigelionis & Mikulevicius (1981), Jacod & Shiryaev (1987), Platen & Rebolledo (1985), Mikulevicius & Platen (1986). See also Franklin (1965). Gorostiza (1980) and

Newton (1990) investigated approximations with random time steps jumping only from threshold to threshold. See Tudor & Tudor (1983) and Tudor (1989) for extensions to multi-parameter and delay SDEs.

**9.2**  Simulation studies for special examples of SDEs can be found for example in Pardoux & Talay (1985), Liske & Platen (1987), Newton (1991), Klauder & Petersen (1985a) or Kloeden, Platen & Schurz (1992). Leblond & Talay (1986) developed an expert system which writes FORTRAN computer programs for numerical schemes for specified drift and diffusion coefficients. See also Talay (1989a).

**9.3**  Higher order time discrete strong approximations of Ito diffusions have been proposed and investigated for instance by Milstein (1974), McShane (1974), Wright (1974), Rao, Borwankar & Ramakrishna (1974), Glorennec (1977), Kloeden & Pearson (1977), Wagner & Platen (1978), Clark (1978), Nikitin & Razevig (1978), Razevig (1980), Platen (1980a, b), Clark & Cameron (1980), Talay (1982a), (1982b), Jannsen (1982a), (1984a, b), Rümelin (1982), Shimizu & Kawachi (1984), Tetzlaff & Zschiesche (1984), Chang (1987), Milstein (1988a, b), Golec & Ladde (1989), Drummond, Duane & Horgan (1983), Guo (1982), (1984), Casasus (1982), Drummond, Hoch & Horgan (1986), Kozlov & Petryakov (1986), Greiner, Strittmatter & Honerkamp (1987). Strong approximations for Ito processes with jumps can be found in Dsagnidse & Tschitaschvili (1975), Wright (1980) and Platen (1982b). Bally (1989a, b), Protter (1985) and Marcus (1978), (1981) considered strong approximations for semi-martingale equations.

**9.4**  Weak approximations of higher order are proposed and investigated for instance in Milstein (1978), Helfand (1979), Greenside & Helfand (1989), Talay (1984), (1986), (1987a), Platen (1984), (1987), Pardoux & Talay (1985), Milstein (1985), Mikulevicius & Platen (1986). Artemev (1984), (1985), Kanagawa (1985), (1989), Klauder & Petersen (1985a, b), Averina & Artemev (1986), Haworth & Pope (1986), Römisch (1983), Römisch & Wakolbinger (1987), Petersen (1987), (1990), Chang (1987), Milstein (1988a), Wagner (1989a, b), Gelbrich (1989). Mikulevicius & Platen (1988) considered the case of a SDE with jump component. Gerardi, Marchetti & Rosa (1984) simulated diffusions with boundary conditions. Platen (1983), (1985) approximated first exit times.

**9.5**  For the case of discontinuous coefficients see Jannsen (1984a). The definitions follow Platen (1984) and Mikulevicius & Platen (1988).

**9.6**  See 9.3.

**9.7**  See 9.4.

**9.8**  Definitions on numerical stability are contained in Pardoux & Talay (1985), Klauder & Petersen (1985a), Petersen (1987), Milstein (1988a), Drummond & Mortimer (1990), Hernandez & Spigler (1990) or Kloeden & Platen (1991c).

## Chapter 10: Strong Taylor Approximations

**10.1**  See 5.2 and 5.3

**10.2**  See 9.1. The proof of Theorem 10.2.2 follows Platen (1981a).

**10.3**  The scheme originates from Milstein (1974). Clark & Cameron (1980) showed that one needs in the non-commutative case the double Ito integrals to obtain first strong order as described in the example. Doss (1977), Sussmann (1978), Yamato (1979) and Talay (1983a, b) studied a similar question.

**10.4 & 10.5** See Wagner & Platen (1978), Platen (1981a), Platen (1984), Milstein (1988a) and Kloeden & Platen (1991c).

**10.6** The proof of the theorem can be found in Platen (1984) and (1981a).

**10.7** The result is new.

**10.8** The lemma is contained in Platen (1981a).

## Chapter 11: Explicit Strong Approximations

**11.1** Clements & Anderson (1973) and Wright (1974) showed by computational experiments that not all stochastic generalizations of well established numerical methods as Runga-Kutta schemes converge to the desired solution. Rümelin (1982) investigated systematically Runge-Kutta type schemes for stochastic differential equations with strong order 1.0. The schemes (1.3), (1.5), (1.7), (1.9) and (1.11) are contained in Platen (1984). A slightly more complicated version of the scheme (1.11) can be also found in Gard (1988).

**11.2** The proposed schemes (2.1), (2.7), (2.10), (2.13), (2.16) and (2.19) are mentioned in Platen (1984) or are new. The method (2.20) is due to Chang (1987).

**11.3** The scheme (3.2) was also proposed by Chang (1987). Its generalization (3.3) seems to be new.

**11.4** (4.8) is due to Lépingle & Ribémond (1989). A 1.5 strong order two step method for the case of additive noise can be found in Milstein (1988a). The schemes (4.4), (4.5), (4.6) and (4.7) seem to be new or can be found in Kloeden & Platen (1991c).

**11.5** The proof of Theorem 11.5.1 was given in Platen (1984).

## Chapter 12: Implicit Strong Approximations

**12.1** Implicit schemes were proposed or investigated by Talay (1982b), Klauder & Petersen (1985), Milstein (1988a), Smith & Gardiner (1988), McNeil & Craig (1988), Drummond & Mortimer (1990), Hernandez & Spigler (1989), Petersen (1990) and Kloeden & Platen (1991c). The described implicit Euler and Milstein schemes are straight forward implicit counterparts of their explicit versions and for were mentioned in Talay (1982b) or Milstein (1988a). In Milstein (1988a) an implicit 1.5 strong order scheme for additive noise can be found. The other implicit order 1.5 strong schemes are described in Kloeden & Platen (1991c) or new.

**12.3** All the schemes are mainly new or proposed in Kloeden & Platen (1991c). An implicit 1.5 strong method of Runge-Kutta type for additive noise is also given in Milstein (1988a).

**12.4** The implicit two-step schemes seem to be new.

**12.5** A-stability was discussed, for instance, by Milstein (1988a), Hernandez & Spigler (1989), Petersen (1990) or Kloeden & Platen (1991c). The last paper also defines stiff stochastic differential equations.

## Chapter 13: Selected Applications of Strong Approximations

**13.1** The influence of periodic excitations and noise on Duffing-Van der Pol oscillators is investigated by several authors, e.g. Ebeling, Herzel, Richert

& Schimansky-Geier (1986). The stochastic flow on the circle was considered by Carverhill, Chappel & Elworthy (1986) and Baxendale (1986). The last author also studied the given example of families of stochastic flows on the torus. Further references including isotropic flows can be found in Baxendale & Harris (1986) or Darling (1990) and Kloeden, Platen & Schurz (1991). The basic theory on stochastic flows and diffeomorphisms is described in Ikeda & Watanabe (1989) or Kunita (1984), (1990). See Emery (1989) for stochastic calculus on manifolds

**13.2** There exists an extensive literature on parametric estimation for stochastic differential equations, e.g. Sørensen (1989), Heyde (1989), Küchler & Sørenson (1989), Brown & Hewitt (1975), Novikov (1972), Liptser & Shiryayev (1978), Kozin (1983), Bellach (1983), Lanska (1979), Taraskin (1974), Linkov (1984), Balakrishnan (1976). A few authors such as Kazimierczyk (1989) and Kloeden, Platen, Schurz & Sørensen (1991) have studied the behaviour of estimators numerically.

**13.3** Discrete Approximations for Markov chain filters were extensively considered in Clark (1978), (1982), Newton (1984), (1986b) (1991). The section follows the results of Newton (1984) and Kloeden & Platen(1991d). The proposed order 1.5 asymptotically efficient scheme is new.

**13.4** The proposed 1.0 asymptotically efficient schemes are due to Newton (1991).

## Chapter 14: Weak Taylor Approximations

**14.1** The Euler scheme appears in Milstein (1978) as a weak scheme. Theorem 14.1.5 is proved in Mikulevicius & Platen (1986).

**14.2** The order 2.0 weak Taylor schemes mentioned here were proposed by Milstein (1978) and Talay (1984). Talay (1984) proved the second order weak convergence for a whole class of schemes.

**14.3** Order 3.0 weak Taylor schemes are given in Platen (1984) and for additive noise in Milstein (1988a).

**14.4** The order 4.0 weak Taylor scheme for additive noise is new.

**14.5** In Talay (1984) the assertion of Theorem 14.5.1 was proved for weak order 2.0. In Platen (1984) this result was generalized to the general case. Theorem 14.5.2 is proved in Platen (1984).

**14.6** The first result on extrapolation methods for stochastic differential equations is that of Talay and Tubaro (1990). They derived an expansion of the leading error coefficients for the Euler scheme. In Kloeden & Platen (1991b) the general representation of the leading error coefficients described in Theorem 14.6.1 was proved.

## Chapter 15: Explicit and Implicit Weak Approximations

**15.1** The class of order 2.0 weak schemes was characterized in Talay (1984). The completely derivative free explicit order 2.0 weak schemes (1.1) and (1.3) were proposed in Platen (1984). Milstein (1985), (1988a) describes the scheme (1.5) and Talay (1984) proposed the method (1.6) where he used also random variables as defined in (14.2.8)–(14.2.10).

**15.2** The presented schemes are described in Platen (1984).

**15.3** By the use of the Euler approximation Talay & Tubaro (1990) proved the order 2.0 weak extrapolation method (3.1). In Kloeden & Platen (1991b) the order 4.0 and 6.0 weak extrapolations are described together with the proof of Theorem 15.3.4.

**15.4** Some results on implicit weak schemes can be found in Milstein (1985), (1988a) for the case of additive noise. The schemes (4.12) and (4.13) are new.

**15.5** Results on predictor-corrector methods can be found in Platen (1991).

**15.6** The remarks on the convergence proofs for explicit and implicit weak schemes relate to those in Platen (1984).

### Chapter 16: Variance Reduced Approximations

**16.1** Variance reduction is a basic technique in Monte-Carlo integration. Monographs on Monte-Carlo theory are Hammersly & Handscomb (1964), Ermakov (1975), Mikhailov (1974), Rubinstein (1981), Ermakov & Mikhailov (1982), Ermakov, Nekrution & Sipin (1984), Kalos & Whitlock (1986).

**16.2** The results presented here are from Milstein (1988a).

**16.3 & 16.4** In these sections we follow Wagner (1988a, b) and (1989a,b). Variance reduction using Hermite polynomials was proposed by Chang (1987) applying Chorin's Monte-Carlo estimator for Gaussian random variables, see Chorin (1971), (1973a, b) and Altz & Hitl (1979).

### Chapter 17: Selected Applications of Weak Approximations

**17.1** In Kac (1949) and Feynman & Hibbs (1965) representations of the solutions of the Schrödinger equation can be found. Chow (1972) describes applications for function space integrals to problems in wave propagation in random media. Donsker & Kac (1950) already studied functional integrals numerically as early as 1950. Fortet (1952), Gelfand & Chentsov (1956), Gelfand, Frolov & Chentsov (1958) continued this direction. Stochastic quadrature formulas are derived in Cameron (1951), Vladimirov (1960), Konheim & Miranker (1967), Tatarski (1976), Sabelfeld (1979), Yanovich (1976) and Egorov, Sobolevski & Yanovich (1985). Second order approximation formulas were given in Fosdick (1965), Fosdick & Jordan (1968). Chorin (1973a) found a surprisingly simple method which Blankenship & Baras (1981) and Hald (1985) generalized to wide classes of functionals. Gladyschev & Milstein (1984) and Milstein (1988a) proposed the scheme (1.19). The approximations (1.14) and (1.15) seem to be new. Example 17.1.1 is due to Milstein (1988a). Dyadkin & Zhukova (1968) and Wagner (1987a, b) avoided any bias in the computation of functional integrals. Variance reduction is considered in Chorin (1973a), Maltz & Hitzl (1979), Ventzel, Gladyschev & Milstein (1985), Fosdick & Jordan (1968), Kalos (1984), De Raedt & Lagendijk (1985), Binder (1985), Kalos & Whitlock (1986), Wagner (1988b). Clark (1984), (1985) considered strong approximations for stochastic linear integrals.

**17.2** The results of this section are due to Talay (1990a).

**17.3** The section presents results on numerical approximations of Lyapunov exponents of stochastic differential systems given in Talay (1989b) and Grorud

& Talay (1990). Important references on Lyapunov exponents for linear and nonlinear stochastic systems are Arnold (1987), Arnold & San Martin (1986) and Arnold & Wihstutz (1986).

# References

Allain, M.F.
  1974    Sur quelques types d' approximation des solutions d' équations
          différentielles stochastiques. Thèse 3ème cycle, Univ. Rennes.

Altz, F.H.M., and Hitl, D.L.
  1979    Variance reduction in Monte-Carlo computations using multi-dimensional
          Hermite polynomials. J. Comput. Phys. **32**, 345-376.

Anderson, S.L.
  1990    Random number generators on vector supercomputers and other advanced
          structures. SIAM Review **32**, 221-251.

Arnold, L.
  1974    Stochastic Differential Equations. Wiley, New York.

  1987    Lyapunov exponents of nonlinear stochastic systems. In Nonlinear Stochas-
          tic Dynamic Engineering Systems, Proc. IUTAM Sympos. Innsbruck, 1987
          ( G.I. Schueller & F. Ziegler, editors), Springer

Arnold, L., Horsthemke, W. , and Lefever, R.
  1978    White and coloured external noise and transition phenomena in nonlinear
          systems. Z. Phys. **B29**, 867.

Arnold, L. , and Kloeden, P.E.
  1989    Explicit formulae for the Lyapunov exponents and rotation number of two-
          dimensional systems with telegraphic noise. SIAM J. Applied Math. **49**,
          1242-1274.

Arnold, L., and San Martin, L.
  1986    A control problem related to the Lyapunov spectrum of stochastic flows.
          Matematica Applicada e Computational **5** (1), 31-64 .

Arnold, L., and Wihstutz, V. (editors)
  1986    Lyapunov Exponents. Springer Lecture Notes in Mathematics Vol. 1186.

Artemev, S.S
  1984    Comparison of certain methods of numerical solution of stochastic differ-
          ential equations. Preprint 84-474, Akad. Nauk SSSR, Vychisl. Tsentr.
          Novosibirsk, 25pp. (In Russian)

  1985    A variable step algorithm for the numerical solution of stochastic differ-
          ential equations. Methody Mekhaniki Sploshnoi Sredy **16** (2), 11-23. (In
          Russian).

Ash, R.B.
  1970    Basic Probability Theory. Wiley and Sons, New York.

1972    Real Analysis and Probability. Academic Press, New York and London.

Aström, K.L.
1970    Introduction to Stochastic Control Theory. AcademicPress, New York .

Atalla, M.A.
1986    Finite-difference approximations for stochastic differential equations. In Probabilistic Methods for the Investigation of Systems with an Infinite Number of Degrees of Freedom. Collection of Scientific Works. Kiev, pp. 11-16. (In Russian).

Auslender, E.I., and Milstein, G.N.
1982    Asymptotic Expansion of the Lyapunov index for linear stochastic systems with small noise. Prikl. Matem. Mekhan. **46**, 358-365. (In Russian).

Averina, T.A., and Artemev, S.S.
1986    A new family of numerical methods for solving stochastic differential equations. Dokl. Akad. Nauk SSSR **288** (4), 777-780. (In Russian).

Azencott, R.
1982    Formule de Taylor stochastique et développement asymptotique d' intégrales de Feynman. Springer Lecture Notes in Mathematics Vol. 921, pp. 237-285.

Balakrishnan, A.V.
1976    Likelihood ratios for time continuous data models; white noise approach. Springer Lecture Notes in Control and Inform. Sc. Vol. 2.

1985    On a class of stochastic differential equations which do not satisfy a Lipschitz condition. Springer Lecture Notes in Control and Inform. Sc. Vol. 78, pp. 27-35.

Bally, V.
1989a   Approximation for the solutions of stochastic differential equations, I: $L^p$-convergence. Stochastics Stoch. Rep. **28**, 209-246.

1989b   Approximation for the solutions of stochastic differential equations, II: Strong convergence. Stochastics Stoch. Rep. **28**, 357-385.

Barrett, J.F., and Wright, D.J.
1974    The random nature of stockmarket prices. Operations Research **22**, 175-177.

Barucha-Reid, A.T. (editor)
1979    Approximate Solution of Random Equations. North Holland, New York .

Basawa, I. V., and Prakasa Rao, B.L.S.
1980    Statistical Inference for Stochastic Processes. Academic Press, London.

Baxendale, P.
1986    Asymptotic behaviour of stochastic flows of diffeomorphisms. Springer Lecture Notes in Mathematics Vol. 1203.

Baxendale, P., and Harris, T.
1986    Isotropic stochastic flows. Annals Probab. **14**, 1155-1179.

Black, F., and Scholes, M.
1973    The pricing of options and corporate liabilities. J. Pol. Econ. **81**, 637-659.

Bellach, B.
1983a   Parameter estimators in linear stochastic differential equations and their asymptotic properties. Math. Operationsforschung Statist., Ser. Statist. **14** (1), 141-191.

1983b   Parametric estimators in linear stochastic differential equations. In Random Vibrations and Reliability ( K. Hennig, editor). Akademie-Verlag, Berlin, pp. 137-144.

Bellman, R. (editor)
1964    Stochastic Processes in Mathematical Physics and Engineering, Proc. Sympos. Appl. Math., Vol. 16, Amer. Math. Soc., Providence RI.

Benson, S.W.
1980    The Foundations of Chemical Kinetics. McGraw-Hill, New York.

Bichteler, K.
1981    Stochastic integration and $L^p$-theory of semi-martingales. Ann. Prob. **9**, 49-89.

Billingsley, P.
1968    Convergence of Probability Measures. Wiley, New York

Binder, K.
1985    The Monte-Carlo method for the study of phase transitions: a review of recent progress. J. Comput.. Phys. **59**, 1-55.

Blankenship, G.L., and Baras, J. S.
1981    Accurate evaluation of stochastic Wiener integrals with applications to scattering in random media and nonlinear filtering. SIAM J. Appl. Math. **41**, 518-552.

Blum, E.K.
1972    Numerical Analysis and Computation Theory and Practice. Addison-Wesley, Reading MA.

Bodo, B.A., Thompson, M.E., and Unny, T.E.
1987    A review on stochastic differential equations for applications in hydrology. J. Stoch. Hydrol. Hydraulics **1**, 81-100.

Bolotin, V. V.
1960    Statistical theory of the seismic design of structures. Proc. 2nd WEEE Japan, p 1365.

Bouleau, N.
1990    On effective computation of expectations in large or infinite dimension. J. Comput. Appl. Math. **31**, 23-34.

Box, G., and Muller, M.
1958    A note on the generation of random normal variables. Ann. Math. Stat. **29**, 610-611.

Boyce, W.E.
  1978   Approximate solution of random ordinary differential equations. Adv.
         Appl. Prob. **10**, 172-184.

Brent, R. P.
  1974   A Gaussian pseudo number generator. Commun. Assoc. Comput. Mach.
         **17**, 704-706.

Brown, B. M., and Hewitt, J. T.
  1972   Asymptotic likelihood theorem for diffusion processes. J. Appl. Prob. **12**,
         228-238.

Bunke, H.
  1972   Gewöhnliche Differentialgleichungen mit zufälligen Parameter. Akademie-
         Verlag, Berlin.

Butcher, J.C.
  1987   The Numerical Analysis of Ordinary Differential Equations. Runge- Kutta
         and General Linear Methods. Wiley, Chirchester.

Bywater, R.J., and Chung, P.M.
  1973   Turbulent flow fields with two dynamically significant scales. AIAA papers
         73-646, 991-10.

Cameron, R. H.
  1951   A "Simpson's rule" for the numerical evaluation of Wiener's integrals in
         function space. Duke Math. J. **18**, 111-130.

Caverhill, A., Chappel, M., and Elworthy, K.D.
  1986   Characteristic exponents for stochastic flows. Springer Lecture Notes in
         Mathematics Vol. 1158.

Casasus, L.
  1982   Sobre la resolution numerica de ecuaciones differentiales estocastias.
         Acta IX Jornadas Matematicas Hispano-Lusas, Vol. II, Universidad de
         Salamancha.

Chandrasekhar, S.
  1943   Stochastic Problems in Physics and Astronomy. Rev. Mod. Physics **15**,
         2-89.

Chang, C.C.
  1987   Numerical solution of stochastic differential equations with constant diffu-
         sion coefficients. Math. Comp.ut **49**, 523-542.

  1988   Random vortex methods for the Navier-Stokes equations. J. Comput Phys.
         **76**, 281-300.

Chorin, A. J.
  1971   Hermite expansions in Monte-Carlo computation. J. Comput. Phys. **8**,
         472-482.

  1973a  Accurate evaluation of Wiener integrals. Math. Comput. **27**, 1-15.

  1973b  Numerical study of slightly viscous flow. J. Fluid Mech. **57**, 785-786.

Chow, P.L.
  1972  Applications of function space integrals to problems in wave propagation in random media. J. Math. Phys. **13**, 1224-1236.

Chung, K.L.
  1975  Elementary Probability with Stochastic Processes. Springer.

Cinlar, E.
  1975  Introduction to Stochastic Processes. Prentice-Hall, Englewood Cliffs, N.J.

Clark, J.M.C.
  1978  The design of robust approximations to the stochastic differential equations of nonlinear filtering. In Communication Systems and Random Process Theory (J.K. Skwirzynski, editor). Sijthoff & Noordhoff, Alphen naan den Rijn, pp. 721-734.

  1982  An efficient approximation scheme for a class of stochastic differential equations. Springer Lecture Notes in Control and Inform. Sc., Vol. 42, pp. 69-78.

  1984  Asymptotically optimal ratic formulae for stochastic integrals. Proc. 23rd Conf. Decisions and Control, Las Vegas. IEEE, pp. 712-715.

  1985  A nice discretization for stochastic line integrals. Springer Lecture Notes in Control and Inform. Sci. Vol. 63, 131-142.

  1989a  The discretization of stochastic differential equations: a primer. In Road-Vehicle Systems and Related Mathematics: Proc. 2nd DMV-GAMM Workshop, Torino, 1987 (H. Neunzert, editor). Teubner, Stuttgart, and Kluwer Academic Publishers, Amsterdam, pp. 163-179.

  1989b  The simulation of pinned diffusions. Proc. 29th Conf. Decision and Control, Honolulu, pp. 1418-1420.

Clark, J.M.C., and Cameron, R.J.
  1980  The maximum rate of convergence of discrete approximations for stochastic differential equations. Springer Lecture Notes in Control and Inform. Sc. Vol. 25, pp. 162-171.

Clements, D.J., and Anderson, B.D.O.
  1973  Well behaved Ito equations with simulations that always misbehave. IEEE Trans. Autom. Control **AC-18**, 676-677.

Dacunha-Castelle, D., and Florence-Zmirou, D.
  1986  Estimators of the coefficients of a diffusion from discrete observations. Stochastics **19**, 263-284.

Darling, R.W.R.
  1990  Isotropic stochastic flows: a survey. Report ICM 90-011. Intitute for Constructive Mathematics, Univ. South Florida, Tampa. 21 pages .

Dahlquist, G.
  1963  A special stability problem for linear multistep methods. BIT **3**, 27-43.

Dahlquist, G., and Bjorck, A.
  1974  Numerical Methods. Prentice-Hall, New York.

Dashevski, M.L., and Liptser, R.S.
1966    Simulation of stochastic differential equations connected with the disorder problem by means of analog computer. Autom. Remote Control **27**, 665-673. (In Russian)

Davis, M.H.A.
1984    Lectures on Stochastic Control and Nonlinear Filtering. Tata Institute of Fundamental Research, Bombay. Springer.

De Raedt, H., and Lagendijk, A.
1985    Monte-Carlo simulation of quantum statistical lattice models. Physics Reports **127**, 233-307.

Di Masi, G., and Rungaldier, W.
1981    An approximation to optimal nonlinear filtering with discontinuous observations. In Stochastic Systems: The Mathematics of Filtering and Identification and Applications, Proc. NATO Study Institute, pp. 583-590.

Dohnal, G.
1987    On estimating the diffusion coefficient. J. Appl. Prob. **24**, 105-114.

Donsker, M.D., and Kac, M.
1950    A sampling method for determining the lowest eigenvalue and principal eigenfuctions of Schrödinger's equations. J. Res. Nat. Bur. Standards **44**, 551-557.

Doob, J.L.
1953    Stochastic Processes. Wiley and Sons, New York.

Doss, H.
1977    Liens entre équations différentielles stochastiques et ordinaires. Ann. Inst. Henri Poincaré **13**, 99-125.

Dothan, M.U.
1990    Prices in Financial Markets. Oxford Univ. Press, Oxford.

Drummond, I.T., Duane, S., and Horgan, R.R.
1983    The stochastic method for numerical simulations: higher order corrections. Nuc. Phys. **B220 FS8**, 119-136.

1984    Scalar diffusion in simulated helical turbulence with molecular diffusivity. J. Fluid Mech. **138**, 75-91.

Drummond, I.T., Hoch, A., and Horgan, R.R.
1986    Numerical integration of stochastic differential equations with variable diffusivity. J. Phys. A: Math. Gen. **19**, 3871-3881.

Drummond, P. D., and Mortimer, I. K.
1990    Computer simulation of multiplicative stochastic differential equations. J. Comput. Phys. (to appear)

Dsagnidse, A.A., and Tschitashvili, R.J.
1975    Approximate integration of stochastic differential equations. Tbilisi State University, Inst. Appl. Math. "Trudy IV", Tibilisi, pp. 267-279. (In Russian)

Durbin, P. A.
1983   Stochastic differential equations and turbulent dispersion. NASA RP-1103.

Dyadkin, I.G., and Zhykova, S.A.
1968   About a Monte-Carlo algorithm for the solution of the Schrödinger equation. Zh. Vychisl. Matem. i Matem. Fiz. **8**, 222-229.

Dynkin, E. B.
1965   Markov Processes, Vol. I and Vol. II. Springer.

Ebeling, W.; Herzel, H.; Richert, W., and Schimansky-Geier, L.
1986   Influence of noise on Duffing-Van der Pol oscillators. Z. Angew. Mech. **6** (3), 141-146.

Egorov, A.D., Sobolevski, P.I., and Yanovich, L. A.
1985   Approximation methods for the computation of Wiener integrals. Nauka i Technika, Minsk. (In Russian).

Ehrhardt, M.
1983   Invariant probabilities for systems in a random environment — with applications to the Brusselator. Bull. Math. Biol. **45**, 579-590.

Eichenauer, J., and Lehn, J.
1986   A non-linear congruential pseudo random number generator. Statist. Paper **27**, 315-326.

Eichenauer-Hermann, J.
1991   Inverse congruential pseudo random number generators avoid the planes. Math. Comput. (to appear).

Elliot, R. J.
1980   Stochastic Calculus and Applications. Springer.

Emery, M.
1989   Stochastic Calculus in Manifolds. Springer.

Ermakov, S. M.
1975   The Monte-Carlo Method and Related Questions. Nauka, Moscow. (In Russian)

Ermakov, S.M., Nekrutin, V.V., and Sipin, A.S.
1984   Stochastic Processes for the Solution of Classical Equations in Mathematical Physics. Nauka, Moscow (in Russian).

Fahrmeier, L.
1974   Schwache Konvergenz gegen Diffusionsprozesse. Z. angew. Math. Mech. **54**, 245.

1976   Approximation von stochastischen Differenzialgleichungen auf Digital- und Hybridrechnern. Computing **16**, 359-371.

Fahrmeier, L., and Beeck, H.
1976   Zur Simulation stetiger stochastischer Wirtschaftsmodelle. 7th Prague Conference on Information Theory, Statistics, Decision Functions and Random Processes.

Faure, O.
1991   Numerical pathwise approximations of stochastic differential equations. (to appear)

Feller, W.
1957   An Introduction to Probability Theory and its Applications, Vol. 1, 2nd edition. Wiley, New York.

1972   An Introduction to Probability Theory and its Applications, Vol. 2. Wiley, New York.

Feynman, R.P., and Hibbs, A.R.
1965   Quantum Mechanics and Path-Integrals. McGraw-Hill, New York.

Finney, B. A., Bowles, D. S., and Windham, M. P.
1983   Random differential equations in river quality modelling. Water Resources Res. **18**, 122- 134.

Fischer, U., and Engelke, M.
1983   Polar cranes for nuclear power stations in earthquake areas. In Random Vibrations and Reliability (K. Hennig, editor). Akademie-Verlag, Berlin, pp. 45-54.

Florens-Zmirou, D.
1984   Théorème de limite centrale pour une diffusion et sa discrétisée. Comptes Rendus Acad. Sci. Paris, Série 1, **299**, 19.

1989   Approximate discrete time schemes for statistics of diffusion processes. Statistics **20**, 547 -557.

Fogelson, A. L.
1984   A mathematical model and numerical method for studying platelet adhesion and aggregation during blood clotting. J. Comput. Phys. **50**, 111-134.

Föllmer, H.
1991   Probabilistic aspects of options. R. Nevanlinna Institute Research Report. (to appear)

Föllmer, H., and Sondermann, D.
1986   Hedging of non-redundant contingent claims. In Contributions to Mathematical Economics (W. Hildebrandt & A. Mas-Colell, editors), pp. 205-223.

Fortet, R.
1952   On the estimation of an eigenvalue by an additive functional of a stochastic process, with special reference to Kac-Donsker method. J. Res. Nat. Bur. Standards **48**, 68-75

Fosdick, L.D.
1965   Approximation of a class of Wiener integrals. Math. Comput. **19**, 225-233.

Fosdick, L.D., and Jordan, H.F.
1968   Approximation of a conditional Wiener integral. J. Comput. Phys. **3**, 1-16.

Franklin, J. N.
  1965  Difference methods for stochastic ordinary differential equations. Math. Comput. **19**, 552- 561.

Friedman, A.
  1975  Stochastic Differential Equations and Applications, Vol. I and Vol. II. Academic Press, New York.

Friedrich, H., Lange, C., and Lindner, E.
  1987  Simulation of Stochastic Processes, Akad. der Wiss. der DDR, Institut für Mechanik, FMC Series no. 35, Karl-Marx-Stadt (Chemnitz), 154 pp.

Fujisaki, M., Kallianpur, G., and Kunita, H.
  1972  Stochastic differential equations for the nonlinear filtering problem. Osaka J. Math. **9**, 19-40.

Gard, T.C.
  1988  Introduction to Stochastic Differential Equations. Marcel Dekker, New York.

Gard, T.C., and Kannan, D.
  1976  On a stochastic differential equations modeling of predator–prey evolution. J. Appl. Prob. **13**, 429-443.

Gardiner, C. W.
  1983  Handbook of Stochastic Methods for Physics, Chemistry and Natural Sciences. Springer.

Gear, C.W.
  1971  Numerical Initial Value Problems in Ordinary Differential Equations. Prentice-Hall, Englewood Cliffs, N.J.

Gelfand, I.M., Frolov, A.S., and Chentsov, N.N.
  1958  Computation of Wiener integrals by the Monte-Carlo method. Izv. Vysch. Uchehn. Zaved. Mat. **5**, 32-45. (In Russian).

Geman, S., and Hwang, C.
  1986  Diffusions for global optimization. SIAM J. Control Optim. **24**, 1031-1043.

Gelbrich, M.
  1989  $L^p$-Wasserstein-Metriken und Approximation stochastischer Differentialgleichungen, Dissertation A, Humboldt-Universität Berlin

Gerardi, A., Marchetti, F., and Rosa, A.M.
  1984  Simulation of diffusions with boundary conditions. Systems and Control Letters **4**, 253.

Gikhman, I. I., and Skorokhod, A. V.
  1972a Stochastic Differential Equations, Springer.

  1972b Stochastic Differential Equations and their Applications. Naukova Dumka, Kiev. (In Russian).

  1979  The Theory of Stochastic Processes, Vol. I-III. Springer.

Giorno, V.; Lansky, P.; Nobile, A. G., and Ricciardi, L. M.
1988    Diffusive approximation and first passage time for a model neuron III. A
        birth and death process. Biol. Cybern. 58, 387-404.

Gladyshev, C. A., and Milstein, G. N.
1985    Runge-Kutta method for the computation of Wiener integrals of exponen-
        tial type. Zh. Vychisl. Mat. i Mat. Fiz. 24, 1136-1149. (In Russian).

Glorennec, P.Y.
1977    Estimation a priori des erreurs dans la résolution numérique d' équations
        dif'ferentielles stochastiques. Séminaire de Probabilités, Univ. Rennes,
        Facsicule 1, 57-93.

Goldstein, L.
1988    Mean-square rates of convergence in the continuous time simulation an-
        nealing algorithm in $\Re^d$. Adv. Appl. Math. 9, 35-39.

Golec, J., and Ladde, G.
1989    Euler-type approximation for systems of stochastic differential equations.
        J. Appl. Math. Simul. 28, 357-385.

Gorostiza, L.G.
1980    Rate of convergence of an approximate solution of stochastic differential
        equations. Stochastics 3, 267-276. Erratum Stochastics 4 (1981), 85.

Gragg, R. F.
1981    Stochastic Switching in Absorptive Optical Bistability. Ph.D. Thesis, Univ.
        Texas at Austin.

Greenside, H.S., and Helfand, E.
1981    Numerical integration of stochastic differential equations II. Bell System
        Tech. J. 60, 1927-1940.

Greiner, A., Strittmatter, W., and Honerkamp, J.
1987    Numerical integration of stochastic differential equations.   J. Statist.
        Physics 51, 95-108.

Grigelonius, B., and Mikulevicius, R.
1981    On weak convergence of semimartingales. Lietuvos Matem. Rink. 21 (3),
        9-25.

Grigorieff, R. D.
1972    Numerik gewöhnlicher Differentialgleichungen. Teubner, Stuttgart.

Groeneveld, R.A.
1979    An Introduction to Probability and Statistics using BASIC. Marcel-Dekker,
        New York.

Grorud, A., and Talay, D.
1990    Approximation of Lyapunov exponents of nonlinear stochastic differential
        systems. INRIA Report no. 1341.

Guo, S. J.
1982    On the mollifier approximation for solutions of stochastic differential equa-
        tions. J. Math. Kyoto Univ. 22, 243-254.

1984     Approximation theorems based on random partitions for stochastic differential equations and applications. Chinese Ann. Math. **5**, 169-183.

Haghighat, F., Chandrashekar, M., and Unny, T. E.
1987     Thermal behaviour in buildings under random conditions. Appl. Math. Modelling **11**, 349- 356.

Haghighat, F., Fazio, P., and Unny, T. E.
1988     A predictive stochastic model for indoor air quality. Building and Environment **23**, 195- 201.

Hald, O.H.
1985     Approximation of Wiener integrals. Report PAM-303, Univ. California, Berkeley.

Hammersley, J.M., and Handscomb, D.C.
1964     Monte-Carlo Methods. Methuen, London.

Harris, C.J.
1976     Simulation of nonlinear stochastic equations with applications in modelling water pollution. In Mathematical Models for Environmental Problems, 9C.A. Brebbi, editor). Pentech Press, London, pp. 269-282.

1977     Modelling, simulation and control of stochastic systems. Inter. J. Systems Sc. **8**, 393-411.

1979     Simulation of multivariate nonlinear stochastic systems. Inter. J. Numer. Meth. Eng. **14**, 37-50.

Harrison, J. M., and Pliska, S.R.
1981     Martingales and stochastic integrals in the theory of continuous trading. Stoch. Proc. Appl. **11**, 215-260.

Hasminski, R. Z.
1980     Stochastic Stability of Differential Equations. Sijthoff & Noordhoff, Alphen naan den Rijn.

Haworth, D.C., and Pope, S.B.
1986     A second-order Monte-Carlo method for the solution of the Ito stochastic differential equation. Stoch. Anal. Appl. **4**, 151-186.

Helfand, E.
1979     Numerical integration of stochastic differential equations. Bell System Tech. J. **58**, 2239-2299.

Hennig, K., and Grunwald, G.
1984     Treatment of flow induced pendulum oscillations. Kernenergie **27**, 286-292.

Henrici, P.
1962     Discrete Variable Methods in Ordinary Differential Equations. Wiley, New York.

Hernandez, D.B.
1988     Computational studies in continuous time nonlinear filtering. In Analysis and Control of Nonlinear Systems (C.I. Byrnes, C.E. Martin and R.E. Sacks, editors), North-Holland, Amsterdam, pp. 239-246.

1989     Sistemi dinamici stocastici. Pitagore Editrice, Bologna.

Hernandez, D.B., and Spigler, R.
1990     Numerical stability of implicit Runge-Kutta methods for stochastic differential equations. Preprint, University of Padua.

Heyde, C.C.
1989     Quasi-likelihood and optimality for estimating functions: some current unifying themes. Bull. Int. Statist. Inst. **53**, Book 1, 19-29.

Hida, T.
1980     Brownian Motion. Springer.

Honerkamp, J.
1990     Stochastische Dynamische Systeme. VCH-Verlag, Weinheim.

Horsthemke, W., and Lefever, R.
1984     Noise Induced Transitions. Springer.

Ikeda, N., and Watanabe, S.
1981     Stochastic Differential Equations and Diffusion Processes. North-Holland, Amsterdam.

1989     Stochastic Differential Equations and Diffusion Processes. 2nd edition. North-Holland, Amsterdam.

Ito, K.
1951a    On Stochastic Differential Equations. Memoirs Amer. Math. Soc. **4**.

1951b    Multiple Wiener integrals. J. Math. Soc. Japan **3**, 157-169.

1984     Introduction to Probability Theory. Cambridge Univ. Press, Cambridge.

Ito, K., and McKean Jr, H. P.
1974     Diffusion Processes and their Sample Paths. Springer.

Ivanov, M. F., and Shvec, V. F.
1979     Numerical solution of stochastic differential equations for modelling collisions in a plasma. Chisl. Metody Mekh. Sploshn. Sredy, Gas Dinavnika, **10**, 64-70. (In Russian)

Jacod, J.
1979     Calcul stochastique et problèmes de martingales. Springer Lecture Notes in Mathematics Vol. 714.

Jacod, J., and Shiryaev, A.N.
1987     Limit Theorems for Stochastic Processes. Springer.

Janssen, R.
1982     Diskretisierung stochastischer Differentialgleichungen. Preprint no. 51, FB Mathematik, Universität Kaiserslautern.

1984a    Difference-methods for stochastic differential equations with discontinuous coefficients. Stochastics **13**, 199-212.

1984b    Discretization of the Wiener process in difference methods for stochastic differential equations. Stoch. Processes Appl. **18**, 361-369.

Jazwinski, A.
1970    Stochastic Processes and Filtering Theory. Academic Press, London.

Johnson, H., and Shanno, D.
1987    Option pricing when the variance is changing. J. Finan. Quant. Anal. **22**, 143-151.

Kac, M.
1949    On distributions of certain Wiener functionals. Trans. Amer. Math. Soc. **65**, 1-13.

Kalos, M.H. (editor).
1984    Monte-Carlo Methods in Quantum Problems (NATO ASI, Series C, Vol. 125). Reidel, Boston.

Kalos, M.H., and Whitlock, P.A.
1986    Monte-Carlo Methods. Vol. 1, Basics. Wiley, New York.

Kanagawa, S.
1985    The rate of convergence in Maruyama's invariance principle. Proc. 4th Inter. Vilnius Confer. Prob. Theory Math. Statist., Vilnius 1985, 142-144.

1989    The rate of convergence for approximate solutions of stochastic differential equations. Tokyo J. Math. **12**, 33-48.

Kallianpur, G.
1980    Stochastic Filtering Theory. Springer.

1987    Weak convergence of stochastic neuronal models. Springer Lecture Notes in Biomathematics Vol. 70, 116-145.

Karandikar, R. L.
1981    Pathwise solutions of stochastic differential equations. Sankhya **43A**, 121-132.

Karatzas, I.
1989    Optimization problems in the theory of continuous trading. SIAM J. Control Optim. **27**, 1227-1259.

Karatzas, I., and Shreve, S.E.
1988    Brownian Motion and Stochastic Calculus. Springer.

Karlin, S., and Taylor, H. M.
1970    A First Course in Stochastic Processes, 2nd edition. Academic Press, New York

1981    A Second Course in Stochastic Processes. Academic Press, New York

Kazimierczyk, P.
1989    Consistent ML estimator for drift parameters of both ergodic and nonergodic diffusions. Springer Lecture Notes in Control and Inform. Sci. Vol. 136, 318-327.

Kimura, M., and Ohta, T.
1971    Theoretical Aspects of Population Genetics. Princeton Univ. Press, Princeton.

Klauder, J.R., and Petersen, W.P.

1985a    Numerical integration of multiplicative-noise stochastic differential equations. SIAM J. Numer. Anal. **22**, 1153-1166.

1985b    Spectrum of certain non-self-adjoint operators and solutions of Langevin equations with complex drift. J. Statist. Physics **39**, 53-72.

Kleijnen, J.P.C.

1974    Statistical Techniques in Simulation, Part I. Marcel-Dekker, New York.

1975    Statistical Techniques in Simulation, Part II. Marcel-Dekker, New York.

Kloeden, P.E., and Pearson, R.A.

1977    The numerical solution of stochastic differential equations. J. Austral. Math. Soc., Series B, **20**, 8-12.

Kloeden, P.E., and Platen, E.

1989    A survey of numerical methods for stochastic differential equations. J. Stoch. Hydrol. Hydraul. **3**, 155-178.

1990    The Stratonovich- and Ito-Taylor expansions. Math. Nachr. (to appear).

1991a    Relations between multiple Ito and Stratonovich integrals. Stoch. Anal. Appl. (to appear)

1991b    Extrapolation methods for the weak approximation of Ito diffusions. Preprint P-Math- 16/90, Berlin.

1991c    Higher-order implicit strong numerical schemes for stochastic differential equations. J. Statist. Physics (to appear).

1991d    Higher order approximate Markov chain filters (to appear).

Kloeden, P.E.; Platen, E.; Hofmann, N., and Schurz, H.

1992    The stochastic Taylor formula and higher order numerical schemes for stochastic differential equations. (to appear)

Kloeden, P.E., Platen, E., and Schurz, H.

1991    The numerical solution of nonlinear stochastic dynamical systems: a brief introduction. J. Bifur. Chaos **1**, (to appear)

1992    The Numerical Solution of Stochastic Differential Equations through Computer Experiments. Springer. (to appear)

Kloeden, P.E.; Platen, E.; Schurz, H., and Sørenson, M.

1991    On the effects of discretization on estimators of diffusions. (to appear)

Kloeden, P.E., Platen, E., and Wright, I.

1991    The approximation of multiple stochastic integrals. J. Stoch. Anal. Appl. (to appear).

Konheim, A.G., and Miranker, W.L.

1967    Numerical evaluation of path-integrals. Math. Comput. **21**, 49-65.

Knuth, D.E.
1979   The Art of Computer Programming, Vol. 1: Fundamental Algorithms. Addison-Wesley, Reading MA.

1981   The Art of Computer Programming, Vol. 2: Seminumerical Algorithms. Addison-Wesley, Reading MA.

Kohler, W. E., and Boyce, W. E.
1974   A numerical analysis of some first order stochastic initial value problems. SIAM J. Appl. Math. **27**, 167-179.

Kolmogorov, A.N.
1933   Grundbegriffe der Wahrscheinlichkeitsrechnung. Springer.

Kolmogorov, A.N., and Fomin, S.V.
1975   Introductory Real Analysis. Dover, New York.

Kozin, F.
1969   A survey of stability of stochastic systems. Automatica **5**, 95-112.

1977   An approach to characterizing, modelling and analyzing earthquake excitation records. CISM Lecture Notes Vol. 225, pp. 77-109. Springer.

1983   Estimation of parameters for systems driven by white noise. In Random Vibrations and Reliability ( K. Hennig, editor). Akademie-Verlag, Berlin, pp. 163-172.

Kozlov, R. I., and Petryakov, M. G.
1986   The construction of comparison systems for stochastic differential equations and numerical methods. Nauka Sibirsk Otdel. Novosibirsk, pp. 45-52. (In Russian).

Krylov, N. V.
1980   Controlled Diffusion Processes. Springer.

Küchler, U., and Sørensen, M.
1989   Exponential families of stochastic processes: A unifying semimartingale approach. Inter. Statist. Rev. **57**, 123-144.

Kunita, H.
1984   Stochastic differential equations and stochastic flows of diffeomorphisms. Springer Lecture Notes in Mathematics Vol. 1097, 143-303.

Kurth, P., and Zschiesche, U.
1987   Numerische Approximation von Markowprozessen. Wiss. Zeitschrift TU Magdeburg **31** (5), 26-29.

Kushner, H.J.
1967    Stochastic Stability and Control. Academic Press, New York.

1971    Introduction to Stochastic Control. Holt, New York

1974    On the weak convergence of interpolated Markov chains to a diffusion.
        Ann. Prob. **2**, 40-50.

1977    Probability Methods for Approximations in Stochastic Control and for El-
        liptic Equations. Academic Press, New York.

1984    Approximation and Weak Convergence Methods for Random Processes
        with Applications to Stochastic Systems Theory. MIT Press, Cambridge
        MA.

Kushner, H.J., and Di Masi, G.
1978    Approximation for functionals and optimal control on jump diffusion pro-
        cesses. J. Math. Anal. Appl. **63**, 772-800.

Ladyzhenskaya, O., Solonnikov, V. A., and Uraltseva, N. N.
1968    Linear and quasilinear equations of the parabolic type. Amer. Math. Soc.
        Providence RI.

Lambert, J.D.
1973    Computational Methods in Ordinary Differential Equations. Wiley, New
        York.

Lanska, V.
1979    Miminum contrast estimation in diffusion processes. J. Appl. Prob. **16**,
        65-75.

Lansky, P., and Lanska, V.
1987    Diffusion approximation of the neuronal model with synaptic reversal po-
        tentials. Biol. Cybern. **56**, 19-26.

Leblond, J., and Talay, D.
1986    Simulation of diffusion processes with PRESTO building systems like
        PRESTO with ADAGIO. Proc. French-Belgian Statisticians' Congress,
        November 1986, Rouen. (J.P. Raoult, editor).

Le Breton, A.
1976    On continuous and discrete sampling for parameter estimation in diffusion
        type processes. Math. Prog. Stud. **5**, 124-144.

Le Gland, F.
1981    Estimation de paramètres dans les processus stochastiques en observation
        incomplète: Application à un problème de radio-astronomie. Dr. Ing.
        Thesis, Univ. Paris IX (Dauphine).

Lépingle, D., and Ribémont, B.
1990    Un schéma multipas d' approximation de l' équation de Langevin. Stoch.
        Processes Appl. (to appear).

Levkinson, B.
1977    Diffusion approximations in population genetics — how good are they?
        Proc. Conf. Stochastic Differential Equations and Applications, Park City
        Utah 1076. Academic Press, New York.

## REFERENCES

Linkov, J.N.
19784 On the asymptotic behaviour of likelihood functions for certain problems for semi-martingales. Teor. Sluch. Proc., **12**(1), 141-191. (In Russian)

Liptser, R. S., and Shiryayev, A. N.
1977 Statistics of Random Processes, Vol. I and Vol. II. Springer.

Liske, H.
1982 On the distribution of some functional of theWiener process. Theory of Random Processes, Naukova Dumka Kiev, **10**, 50-54. (In Russian)

Liske, H., and Platen, E.
1987 Simulation studies on time discrete diffusion approximations. Math. Comp. Simul. **29**, 253-260.

Liske, H., Platen, E., and Wagner, W.
1982 About mixed multiple Wiener integrals. Preprint P-Math-23/82, IMath, Akad. der Wiss. der DDR, Berlin.

Loève, M.
1977 Probability Theory I, 4th edition. Springer.

Malz, F.H., and Hitzl, D.L.
1979 Variance reduction in Monte-Carlo computations using multi-dimensional Hermite polynomials J. Comput. Phys. **32**, 345-376.

Marcus, S. I.
1978 Modelling and analysis of stochastic differential equations driven by point processes. IEEE Trans. Inform. Theory **24**, 164-172.

1981 Modelling and approximations of stochastic differential equations driven by semimartingales. Stochastics 4, 223-245.

Marsaglia, G., and Bray, T. A.
1964 A convenient method for generating normal variables. SIAM Review **6**, 260-264.

Maruyama, G.
1955 Continuous Markov processes and stochastic equations. Rend. Circolo Math. Palermo 4, 48-90.

McKean Jr, H. P.
1969 Stochastic Calculus. Academic Press, New York.

McKenna, J., and Morrison, J. A.
1970 Moments and correlation functions of a stochastic differential equation. J. Math. Phys. **11**, 2348-2360.

1971 Moments of solutions of a class of stochastic differential equations. J. Math. Phys. **12**, 2126-2136.

McNeil, K.J., and Craig, I.J.D.
1988 An unconditional stable numerical scheme for nonlinear quantum optical systems— quantised limit cycle behaviour in second harmonic generation. Univ. Waikato, Research Report no. 166.

McShane, E.J.
1971    Stochastic differential equations and models of random processes. Proc. Sixth Berkeley Sympos. Prob. Math. Stat., pp. 263-294.

1974    Stochastic Calculus and Stochastic Models. Academic Press, New York.

1976    The choice of a stochastic model for a noisy system. Math. Prog. Study 6, 79-92.

Merton, R.C.
1971    Optimum consumption and portfolio rules in a continuous-time model. J. Econ. Theory 3, 373-413.

1973    Theory of rational option pricing. Bell J. Econ. Manag. Sc. 4, 141-183.

Métivier, M.
1982    Semimartingales: A Course on Stochastic Processes. De Gruyter, New York.

Métivier, M., and Pellaumail, J.
1980    Stochastic Integration. Academic Press, New York.

Meyer, P. A.
1976    Théorie des Intégrales Stochastiques. In Séminaire de Probabilités X, Springer Lecture Notes in Mathematics Vol. 511, pp. 245-400.

Michailov, G.A.
1974    Some problems with the Monte-Carlo theory. Nauka. Novosirbirsk. (In Russian).

Mikulevicius, R.
1983    On some properties of solutions of stochastic differential equations. Lietuvos Matem. Rink. 4, 18-31.

Mikulevicius R., and Platen, E.
1986    Rate of convergence of the Euler approximation for diffusion processes. Preprint P-Math-38/86, IMath, Akad. der Wiss. der DDR, Berlin. Math. Nachr. (to appear)

1988    Time discrete Taylor approximations for Ito processes with jump component. Math. Nachr. 138, 93-104.

Milstein, G.N.
1974    Approximate integration of stochastic differential equations. Theor. Prob. Appl. 19, 557-562.

1978    A method of second-order accuracy integration of stochastic differential equations. Theor. Prob. Appl. 23, 396-401.

1985    Weak approximation of solutions of systems of stochastic differential equations. Theor. Prob. Appl. 30, 750-766.

1988a   The Numerical Integration of Stochastic Differential Equations. Urals Univ. Press, Sverdlovsk. 225 pp. (In Russian).

1988b A theorem on the order of convergence of mean-square approximations of solutions of systems of stochastic differential equations. Theor. Prob. Appl. **32**, 738-741.

Miwail, R., Ognean, T., and Straja, S.
1987 Stochastic modelling of a biochemical reactor. Hungar. J. Indus. Chem. **15**, 55-62.

Morgan, B.J.J.
1984 Elements of Simulation. Chapman & Hall, London.

Mortenson, R. E.
1969 Mathematical problems of modelling stochastic nonlinear dynamical systems. J. Stat. Phys. **1**, 271-296.

Newton, N.J.
1984 Discrete Approximations for Markov Chains. Ph.D. Thesis, Univ. London.

1986a An asymptotically efficient difference formula for solving stochastic differential equations. Stochastics **19**, 175-206.

1986b Asymptotically optimal discrete approximations for stochastic differential equations. In Theory and Applications of Nonlinear Control systems (C. Byrnes and A. Lindquist, editors) . North-Holland, Amsterdam.

1990 An efficient approximation for stochastic differential equations on the partition of symmetrical first passage times. Stochastics Stoch. Rep. **29**, 227-258.

1991 Asymptotically efficient Runge-Kutta methods for a class of Ito and Stratonovich equations. SIAM J. Appl. Math. 51, 542-567.

Niederreiter, H.
1988 Remarks on nonlinear pseudo random numbers. Metrika **35**, 321-328.

Nikitin, N.N., and Razevig, V.D.
1978 Methods of computer simulation of stochastic differential equations. Zh. Vichisl. Mat. i Mat. Fiz. **18**, 106-117. (In Russian)

Novikov, A. A.
1972 Sequential estimation of the parameters of diffusion type processes. Math. Notes **12**, 812-818.

Obukhov, A. M.
1959 Description of turbulence in terms of Lagrangian variables. Adv. Geophys. **6**, 113-115.

Øksendal, B.
1985 Stochastic Differential Equations. Springer.

Papanicolaou, G.C., and Kohler, W. E.
1974 Asymptotic theory of mixing stochastic differential equations. Comm. Pure Appl.Math. **27**, 641-668.

Papoulis, A.
  1965    Probability, Random Variables and Stochastic Processes. McGraw-Hill, New York.

Pardoux, E., and Pignol, M.
  1984    Etude de la stabilité de la solution d'une EDS bilinéaire à coefficients périodiques: Application au mouvement d'une pale hélicoptère. Springer Lecture Notes in Control and Inform. Sc. Vol. 63, pp. 92-103.

Pardoux, E., and Talay, D.
  1985    Discretization and simulation of stochastic differential equations. Acta Appl. Math. 3, 23-47.

  1988    Stability of nonlinear differential systems with parametric excitation. In Nonlinear Stochastic Dynamic Engineering Systems. (G.I. Schueller and F. Ziegler, editors). Proc. IUTAM Sympos., Innsbruck 1987. Springer.

Parzen, E.
  1962    Stochastic Processes. Holden-Day, San Francisco.

Petersen, W.P.
  1987    Numerical simulation of Ito stochastic differential equations on supercomputers. Springer IMA Series Vol. 7, 215-228.

  1988    Some vectorized random number generators for uniform, normal and Poisson distributions for CRAY X-MP. J. Supercomputing 1, 318-335.

  1990    Stability and accuracy of simulations for stochastic differential equations. IPS Research Report No. 90-02, ETH Zürich.

Picard, J.
  1984    Approximation of nonlinear filtering problems and order of convergence. Springer Lecture Notes in Control and Inform. Sc. Vol. 61, pp. 219-236.

  1986a   Nonlinear filtering of one-dimensional diffusions in the case of a high signal-to-noise-ratio. SIAM J. Applied Math. 46, 1098-1125.

  1986b   Filtrage de diffusions vectorielles faiblement bruitées. Springer Lecture Notes in Control and Inform. Sc. Vol. 83.

  1986c   An estimate of the error in time discretization of nonlinear filtering problems. Theory and Applications of Nonlinear Control Systems (C.I. Byrnes and A. Lindquist, editors). North-Holland, Amsterdam, pp. 401-412.

Platen, E.
  1980a   Weak convergence of approximation of Ito integral equations. Z. angew. Math. Mech. 60, 609-614.

  1980b   Approximation of Ito integral equations. Springer Lecture Notes in Control and Inform. Sc. Vol. 25, pp. 172-176.

  1981a   An approximation method for a class of Ito processes. Lietuvos Matem. Rink. 21, 121-133.

  1981b   A Taylor formula for semimartingales solving a stochastic differential equation. Springer Lecture Notes in Control and Inform. Sc. Vol. 36, pp. 163-172.

1982a  A generalized Taylor formula for solutions of stochastic differential equations. Sankhya **44A** , 163-172.

1982b  An approximation method for a class of Ito processes with jump component. Lietuvos Matem. Rink. **22**, 121-133.

1983  Approximation of first exit times of diffusions and approximate solution of parabolic equations. Math. Nachr. **111**, 127-146.

1984  Zur zeitdiskreten Approximation von Itoprozessen. Diss. B., IMath, Akad. der Wiss. der DDR, Berlin.

1985  On first exit times for diffusions. Springer Lecture Notes in Control and Inform. Sc. Vol. 69, pp. 192-195.

1987  Derivative free numerical methods for stochastic differential equations. Springer Lecture Notes in Control and Inform. Sc. Vol. 96, pp. 187-193.

1990  Higher order weak approximation to Ito diffusions by Markov chains. Preprint P-Math-02/90, 25 pages.

1991  Weak predictor-corrector methods for stochastic differential equations. (to appear).

Platen, E., and Rebolledo, R.
1985  Weak convergence of semimartingales and discretization methods. Stoch. Processes. Appl. **20**, 41-58.

Platen, E., and Wagner, W.
1982  On a Taylor formula for a class of Ito processes. Prob. Math. Statist. **3**, 37-51.

Pope, S. B.
1985  Pdf methods in turbulent reactive flows. Prog. Energy Comb. Sc. **11**, 119-192.

Protter, P.
1985  Approximations of solutions of stochastic differential equations driven by semimartingales. Ann. Prob. **13**, 716-743.

1990  Stochastic Integration and Differential Equations. Springer.

Pugachev, V.S., and Sinitsyn, I.N.
1987  Stochastic Differential Systems: Analysis and Filtering. Wiley, New York.

Rao, M. M.
1979  Stochastic Processes and Integration, Sijthoff & Noordhoff, Alphen naan den Rijn.

Rao, N.J., Borwankar, J.D., and Ramakrishna, D.
1974  Numerical solution of Ito integral equations. SIAM J. Control **12**, 124-139.

Razevig, V. D.
1980  Digital modelling of multi-dimensional dynamic systems under random perturbations. Autom. Remote Control **4**, 177-186. (In Russian)

Ricciardi, L. M.
1977  Diffusion Processes and Related Topics in Biology. Springer Lecture Notes in Biomathematics Vol. 14.

Richardson, J. M.
1964  The application of truncated hierarchy techniques in the solution of a stochastic linear differential equation. In Stochastic Processes in Mathematical Physics and Engineering. Proc. Sympos. Appl. Math. Vol. 16 (R. Bellman, editor). Amer. Math. Soc., Providence RI, pp. 290-302.

Ripley, B. D.
1983a  Computer generation of random variables: a tutorial. Inter. Statist. Rev. **45**, 301-319.

1983b  Stochastic Simulation. Wiley, New York.

Risken, H.
1984  The Fokker-Planck Equation: Methods of Solution and Applications. Springer.

Römisch, W.
1983  On an approximate method for random differential equations. In Problems of Stochastic Analysis in Applications. Wiss. Beiträge der Ingenieurhochschule Zwickau, Sonderheft (J. vom Scheidt, editor), pp. 327-337.

Römisch, W., and Wakolbinger, A.
1987  On the convergence rates of approximate solutions of stochastic equations. Springer Lecture Notes in Control and Inform. Sc. Vol. 96, pp. 204-212.

Royden, H.L.
1968  Real Analysis. 2nd Edition. Macmillan, London.

Rubinstein, R.Y.
1981  Simulation and the Monte-Carlo Method. Wiley, New York.

Rümelin, W.
1982  Numerical treatment of stochastic differential equations. SIAM J. Numer. Anal. **19**, 604-613.

Sabelfeld, K. K.
1979  On the approximate computation of Wiener integrals by the Monte-Carlo method. Zh. Vichisl. Mat. i Mat. Fiz. **19**, 29-43. (In Russian).

Sagirow, P.
1970  Stochastic Methods in The Dynamics of Satellites. ICMS Lecture Notes Vol. 57. Springer.

Saito, Y., and Mitsui, T.
1991  Discrete approximations for stochastic differential equations. (to appear).

Schöner, G., Haken, H., and Kelso, J.A.S.
1986  A stochastic theory of phase transitions in human hand movement. Biol. Cybern. **53**, 247-257.

Schoener, T. W.
  1973    Population growth regulated by intraspecific competition for energy or
          time: some simple representations. Theor. Pop. Biol. **4**, 56-84.

Shurko, I. O.
  1987    Numerical solution of linear systems of stochastic differential equations.
          In: Numerical Methods for Statistics and Modeling. Collected Scientific
          Works. Novosibirsk, pp. 101-109. (In Russian)

Schuss, Z.
  1980    Theory and Applications of Stochastic Differential Equations. Wiley, New
          York.

Shiga, T.
  1985    Mathematical results on the stepping stone model of population genetices.
          In Population Genetics and Molecular Evolution. ( T. Ohta and K. Aoki,
          editors). Springer, pp. 267-279.

Shimizu, A., and Kawachi, T.
  1984    Approximate solutions of stochastic differential equations. Bull. Nagoya
          Inst. Tech. **36**, 105-108.

Shinozuka, M.
  1971    Simulation of multivariate and multidimensional random differential pro-
          cesses. J. Acoust. Soc. Amer. **49**, 357-367.

  1972    Monte-Carlo solution of structural dynamics. J. Comp. Structures 2, 855-
          874.

Shinozuka, M., and Jan, C. M.
  1972    Digital simulation of random processes and its applications. J. Sound and
          Vibrat. **25**, 111-128.

Shinozuka, M., and Sato,Y.
  1967    Simulation of nonstationary random processes. J. Eng. Mech. Div. ASCE
          **93 EM1**, 11.

Shinozuka, M., and Wen, Y. K.
  1972    Monte-Carlo solution of nonlinear vibrations. AIAA J. **10**, 37-40.

Shiryayev, A. N.
  1984    Probability Theory. Springer.

Skorokhod, A. V.
  1982    Studies in the Theory of Stochastic Processes. Dover, New York.

Skorokhod, A. V., and Slobodenjuk, N. P.
  1970    Limit Theorems for Random Walks. Naukova Dumka, Kiev. (In Russian).

Smith, A.M., and Gardiner, C.W.
  1988    Simulation of nonlinear quantum damping using the positive representa-
          tion. Univ. Waikato Research Report.

Sobczyk, K.
  1986    Modelling of random fatigue crack growth. Eng. Fracture Mech. **24**,
          609-623.

1991    Stochastic Differential Equations. Kluwer, Dordrecht.

Soong, T. T.
1973    Random Differential Equations in Science and Engineering. Academic
        Press, New York.

Sørensen, M.
1989    On quasi-likelihood for semi-martingales. Stoch. Processes Appl. (to
        appear).

Stoer, J., and Bulirsch, R.
1980    Introduction to Numerical Analysis. Springer.

Stratonovich, R. L.
1963    Topics in the Theory of Random Noise, Vol. I and Vol. II. Gordon &
        Breach, New York.

1966    A new representation for stochastic integrals and equations. SIAM J. Con-
        trol 4, 362-371.

1968    Conditional Markov Processes and their Application to the Theory of Op-
        timal Control. American Elsevier, New York.

Stroock, D. W.
1979    Lectures on Topics in Stochastic Differential Equations. Tata Institute of
        Fundamental Research, Bombay. Springer.

Stroock, D. W., and Varadhan, S. R. S.
1982    Multidimensional Diffusion Processes. Springer.

Sussmann, H.
1978    On the gap between deterministic and stochastic differential equations.
        Ann. Prob. 6, 19-41.

1988    Product expansions of exponential Lie series and the discretization of
        stochastic differential equations. Springer IMA Series Vol. 10, pp. 563-582.

Syski, R.
1967    Stochastic Differential Equations. Chapter in Modern Nonlinear Equations
        by T. J. Sastry. McGraw-Hill, New York.

Talay, D.
1982a   Convergence pour chaque trajectoire d'un schéma d' approximation des
        EDS. Comptes Rendus Acad. Sc. Paris, Série I, 295, 249-252.

1982b   Analyse Numérique des Equations Différentielles Stochastiques. Thèse
        3ème cycle, Univ. Provence.

1983a   How to discretize a stochastic differential equation. Springer Lecture Notes
        in Mathematics Vol. 972, pp. 276-292

1983b   Résolution trajectorielle et analyse numérique des équations différentielles
        stochastiques. Stochastics 9, 275-306.

1984    Efficient numerical schemes for the approximation of expectations of func-
        tionals of the solution of an SDE and applications. Springer Lecture Notes
        in Control and Inform. Sc. Vol. 61, pp. 294-313.

1986    Discrétisation d' une EDS et calcul approché d' ésperances de fonctionnelles de la solution. Math. Mod. Numer. Anal. **20**, 141-179.

1987    Classification of discretization schemes of diffusions according to an ergodic criterium. Springer Lecture Notes in Control and Inform. Sc. Vol. 91, pp. 207-218.

1988a   Calcul numérique des exposants de Lyapounov d' une pale d' hélicoptère. Publ. de la R.C.P. de Mécanique Aléatoire.

1988b   Simulation and numerical analysis of stochastic differential systems. In Effective Stochastic Analysis (P. Krée and W. Wedig, editors). Springer.

1989a   PRESTO: Mode d' emploi. INRIA Report no. 106.

1989b   Approximation of upper Lyapunov exponents of bilinear stochastic differential equations. INRIA Report no. 965. (to appear in SIAM J. Numer. Anal.)

1990    Second order discretization schemes of stochastic differential systems for the computation of the invariant law. Stochastics Stoch. Rep. **29**, 13-36.

Talay, D., and Tubaro, L.
1991    Expansions of the global error for numerical schemes solving stochastic differential equations. Stoch. Processes Appl. (to appear)

Taraskin, A. F.
1974    On the asymptotic normality of vector-valued stochastic integrals and estimates of a multi-dimensional diffusion process. Theory Prob. Math. Statist. **2**, 209-224.

Tatarskij, V.I.
1976    On the numerical approximation of conditioned Wiener integrals and some Feynman functional integrals. Metody Monte-Karlo v vychislitelnoj matematike i matematichesko fizike, Novosibirsk, pp. 75-90 (In Russian).

Tetzlaff, U., and Zschiesche, H.
1984    Approximate solutions for Ito differential equations using the Taylor expansion for semigroups of operators. Wiss. Z. TH Leuna-Merseburg **26**, 332-339.

Tudor, C.
1989    Approximation of delay stochastic differential equations with constant retardation by usual Ito equations. Rev. Roum. Math. Pures Appl. **34**, 55-64.

Tudor, C., and Tudor, M.
1983    On the approximation in ratic mean for the solution of two parameter stochastic differential equations in Hilbert spaces. Anal. Univ. Bucuresti **32**, 73-88.

Turelli, M.
1977    Random environments and stochastic calculus. Theor. Pop. Biol. **12**, 140-178.

Unny, T.E.
1984    Numerical integration of stochastic differential equations in catchment modelling. Water Resources Res. **20**, 360-368.

Unny, T.E., and Karmeshu
1983    Stochastic nature of outputs from conceptual reservoir model cascades. J. Hydrol. **68**, 161-180.

van Kampen, N.G.
1981a   Ito versus Stratonovich. J. Stat. Phys. **24**, 175-187.

1981b   Stochastic Processes in Physics and Chemistry. North-Holland, Amsterdam.

Ventzel, A. D., Gladyshev, C. A., and Milstein, G. N.
1985    Piecewise constant approximation for the Monte-Carlo calculation of Wiener integrals. Theory Prob. Appl. **24**, 745-752.

Veuthey, A.-L., and Stucki, J.
1987    The adenylate kinase reaction acts as a frequency filter towards fluctuations of ATP utilization in the cell. Biophys. Chem. **26**, 19-28.

Vladimirov, V. S.
1960    On the numerical approximation of Wiener integrals. Usp. Mat. Nauk **15**, 129-135 (In Russian).

Wagner, W.
1987a   Unbiased Monte-Carlo evaluation of certain functional integrals. J. Comput. Phys. **71**, 21-33.

1987b   Unbiased Monte-Carlo evaluation of functionals of solutions of stochastic differential equations variance reduction and numerical examples. Preprint P-Math-30/87 Inst. Math. Akad. der Wiss. der DDR.

1988a   Unbiased multi-step estimators for the Monte-Carlo evaluation of certain functionals. J. Comput. Physics **79**, 336-352.

1988b   Monte-Carlo evaluation of functionals of solutions of stochastic differential equations. Variance reduction and numerical examples. Stoch. Anal. Appl. **6**, 447-468.

1989a   Unbiased Monte-Carlo estimators for functionals of weak solutions of stochastic differential equations. Stochastics Stoch. Rep. **28**, 1-20.

1989b   Stochastische numerische Vefahren zur Berechnung von Funktionalintegralen. Report R-Math 02/89 Inst. Math. Akad. d. Wiss. DDR. 149 pp.

Wagner, W., and Platen, E.
1978    Approximation of Ito integral equations. Preprint ZIMM, Akad. der Wiss. der DDR, Berlin.

Wedig, W.
1988    Pitchfork and Hopf bifurcations in stochastic systems — effective methods to calculate Lyapunov exponents. In Effective Stochastic Analysis (P. Krée and W. Wedig, editors). Springer.

Wilkes, M. V.
  1966    A Short Introduction to Numerical Analysis. Cambridge Univ. Press, Cambridge UK.

Wong, E.
  1971    Stochastic Processes in Information and Dynamical Systems, McGraw-Hill, New York.

Wong, E., and Hajek, B.
  1985    Stochastic Processes in Engineering Systems. Springer.

Wong, E., and Zakai, M.
  1965a   On the relation between ordinary and stochastic differential equations . J. Eng. Sc. **3**, 213-229.

  1965b   On the convergence of ordinary integrals to stochastic integrals. Ann. Math. Statist. **36**, 1560-1564.

  1969    Riemann-Stieltjes approximation of stochastic integrals. Z. Wahrsch. verw. Gebiete **12**, 87-97.

Wonham, W.M.
  1965    Some applications of stochastic differential equations to optimal nonlinear filtering. SIAM J. Control **2**, 347-369.

Wright, D.J.
  1974    The digital simulation of stochastic differential equations. IEEE Trans. Autom. Control **AC-19**, 75-76.

  1980    Digital simulation of Poisson stochastic differential equations. Int. J. Systems. Sc. **11**, 781-785.

Yaglom, A.M.
  1980    Application of stochastic differential equations to the description of turbulent equations. Springer Lecture Notes in Control and Inform. Sc. Vol. 25, pp. 1- 13.

Yakowitz, S. J.
  1977    Computational Probability and Simulation. Addison-Wesley, Reading MA.

Yamada, T.
  1976    Sur l' approximation des solutions d' équations différentielles stochastiques. Z. Wahrsch. verw. Gebiete **36**, 153-164.

Yamato, A.M.
  1979    Stochastic differential equations and nilpotent Lie algebras. Z. Wahrsch. verw. Gebiete **47**, 213-239.

Yanovich, L. A.
  1976    Approximate calculation of continual integrals with respect to Gaussian measures. Nauka i Tekhnika. Minsk. (In Russian).

Yen, V.V.
  1988    A stochastic Taylor formula for functionals of two-parameter semimartingales. Acta Vietnamica, **13** (2), 45-54.

Zakai, M.
  1969    On the optimal filtering of diffusion processes. Z. Wahrsch. verw. Gebiete
          **11**, 230-343.

Zhank, B. G., and Padgett, W. J.
  1984    The existence and uniqueness of solutions to stochastic differential-
          difference equations. Stoch. Anal. Appl. **2**, 335-345.

# Index